[redacted]

# Area-Wide Control of Fruit Flies and Other Insect Pests

*JOINT PROCEEDINGS of the*
INTERNATIONAL CONFERENCE ON AREA-WIDE CONTROL OF INSECT PESTS, May 28–June 2, 1998
*and the*
FIFTH INTERNATIONAL SYMPOSIUM ON FRUIT FLIES OF ECONOMIC IMPORTANCE, June 1–5, 1998
*Penang, Malaysia*

Edited by Keng-Hong Tan

Penerbit Universiti Sains Malaysia
Pulau Pinang
2000

*Perpustakaan Negara Malaysia* *Cataloguing-in-Publication Data*

Area-wide control of fruit flies and other insect pests : joint proceedings of the International Conference in Area-Wide Control of Insect Pests, May 28–June 2, 1998, and the Fifth International Symposium on Fruit Flies of Economic Importance, June 1–5, 1998, Penang, Malaysia / edited by Keng-Hong Tan.

ISBN 983-861-195-6

1. Insect pests—Control—Congresses. 2. Fruit-flies—Control—Congresses. 3. Fruit-flies—Economicaspects—Congresses. I. Tan, Keng Hong. II. International Conference on Area-Wide Control of Insect Pests (1998 : Pulau Pinang). III. International Symposium on Fruit Flies of Economic Importance (5th : 1998 : Pulau Pinang).
628.9657 SB945.F8

Cover design: Mohd Murad Shahiran
Printed by: Sinaran Bros. Sdn. Bhd.

# DEDICATION

This book is dedicated in memory of the life and work of the late Dr. Edward F. Knipling (1909-2000), who pioneered the area-wide approach to insect pest management. He also is known as the father of the sterile insect technique and promoter of other biological approaches to insect pest control. Dr. Knipling was not only a scientist, but also a philanthropist and a man committed to the search of solutions to problems faced by poor farmers in the developing world.

# TABLE OF CONTENTS

## G) BIOCHEMISTRY AND PHYSIOLOGY OF FRUIT FLIES



## H) SEMIOCHEMICALS AND FIELD MONITORING OF FRUIT FLIES



## I) BEHAVIOUR OF FRUIT FLIES



## J) BIOSYSTEMATICS AND BIODIVERSITY OF FRUIT FLIES



## K) NATURAL ENEMIES AND BIOCONTROL OF FRUIT FLIES

## *CONTRIBUTING AUTHORS' NAMES AND ADDRESSES*

***H. Ahmad***
Institut Haiwan, P.O. Box 520, 86009, Kluang, Johor, MALAYSIA

***H. Ako***
University of Hawaii at Manoa, Department of Entomology, 3050 Maile Way, Gilmore 310, Honolulu, HI 96822 USA

***M.A.J. Al-Izzi***
Entomology Department, IAEC, P.O. Box 765, Baghdad, IRAQ

***A. A. Al-Taweel***
Entomology Department, IAEC, P.O. Box 765, Baghdad, IRAQ

***M. Alleck***
Entomology Division, Ministry of Agriculture, Fisheries and Cooperatives, Reduit, MAURITIUS

***R. Allsopp***
Tsetse Control Division, Department of Animal Health and Production, Ministry of Agriculture, P.O. Box 14 Maun, BOTSWANA

***A. Allwood***
Project on Regional Management of Fruit Flies in the Pacific, Secretariat of the Pacific Community, Private Mail, Suva, FIJI

***A. Alviña***
Instituto de Entomología, Universidad Metropolitana de Ciencias de la Educación, AV.J.P. Alessandri 774, Casilla 147, Santiago, CHILE

***I. Amouroux***
FNGDS, 149 rue de Bercy, 75012, Paris, FRANCE

***M.H. Andreasen***
London School of Hygiene and Tropical Medicine, London WC1E 7HT, UNITED KINGDOM

***G. Aranda***
Del. Prov. Agric. Dep. Sanidad Vegetal, Malaga, SPAIN

***K.F. Armstrong***
Ecology and Entomology Group, Lincoln University, Canterbury, NEW ZEALAND

***P.W. Atkinson***
Department of Entomology, University of California, Riverside, CA, 92521, USA

*A. Barbosa*
Programa Madeira-Med, Direcção Regional de Agricultura, Madeira, PORTUGAL

*S. Barbosa*
Secção Autónoma de Protecção Integrada, Instituto Superior de Agronomia (UTL), Lisboa, PORTUGAL

*B.N. Barnes*
ARC-Fruit, Vine and Wine Research Institute (INFRUITEC Center), Stellenbosch, SOUTH AFRICA

*A. Bartolucci*
ISCAMen. Boulogne Sur Mer 3050 (5500) Mendoza, ARGENTINA

*R.C. Bautista*
USDA-ARS, Tropical Fruit, Vegetable and Ornamental Crop Research Laboratory, 2727 Woodlawn Drive, Honolulu, Hawaii 96822, USA

*U.R. Bernier*
USDA, Agricultural Research Service, Center for Medical Agricultural and Veterinary Entomology, P.O. Box 14565, Gainesville, FL 32604, USA

*H.G. Bignayan*
BPI-NMRDC, Jordan, Guimaras, PHILIPPINES

*K.A. Bloem*
OK-CM SIR Program, Box 1080, Osoyoos, BC, V0H 1V0, CANADA; Current address: USDA, APHIS, PPQ, Central Biological Control at Florida, A&M University, Tallahassee, Florida, USA

*S. Bloem*
USDA-ARS YARL, 5230 Konnowac Pass Rd., Wapato, WA 98951, USA. Current address: 429 Meadow Ridge Drive, Tallahassee, Florida 31312-1577, USA

*C. Bonvicini*
Department of Animal Biology, Laboratory of Zoology, University of Pavia, Piazza Botta 9, I-27100 Pavia, ITALY

*S. Broughton*
Queensland Department of Primary Industries, P.O. Box 652, Cairns, Queensland 4870, AUSTRALIA

*C. Cáceres*
USDA, El Pino Modular Mass Rearing Facility, Medfly Program, GUATEMALA. Current address: c/o IAEA Laboratories, Seibersdorf, Entomology Unit, A-2444 Seirsdorf, AUSTRIA

*J. Caldeira*
Programa Madeira-Med, Direcção Regional de Agricultura, Madeira, PORTUGAL

*C.O. Calkins*
USDA, Agricultural Research Service, Yakima Agricultural Research Laboratory, Wapato, Washington, WA 98951, USA

*M. Calvitti*
Enea Casaccia - Division of Agriculture and biotechnologies via Anguillarese 301, S.M. di Galeria (Rome) ITALY

*C.M. Cameron*
Ecology and Entomology Group, Lincoln University, Canterbury, NEW ZEALAND

*P. Caplong*
Federation Departementale de Groupements de Defense des Cultures de la Guyane, Cayenne, FRENCH GUAYANA

*J.R. Carey*
Department of Entomology, University of California, Davis, USA

*D.A. Carlson*
USDA, Agricultural Research Service, Center for Medical Agricultural and Veterinary Entomology, P.O. Box 14565, Gainesville, FL 32604, USA

*J.E. Carpenter*
Insect Biology and Population Management Research Laboratory, Agricultural Research Service, United States Department of Agriculture, Tifton, Georgia, USA

*J.R. Carpenter*
College of Tropical Agriculture and Human Resources, University of Hawaii-Manoa, Honolulu, Hawaii 96822, USA

*J. Carrillo*
Programa Moscamed, 6a Calle 1-36, Zona 10, Edificio Valsari, Office 203, GUATEMALA C.A.

*E.D. Casagrande*
AgriSenseBCS Ltd., Treforest Industrial Estate, Pontypridd, S. Wales, CF37 5SU, UNITED KINGDOM

*J.C. Casin*
Centro Nacional de Sanidad Vegetal (CNSV), Ciudad de La Habana, CUBA

*J.P. Cayol*
Entomology Unit, FAO/IAEA Agriculture and Biotechnology Laboratory, A-2444 Seibersdorf, AUSTRIA. Current address: CIRAD-FLHOR Guyane, Campus Agronomique, BP 701 F-97387 Kourou Cedex, GUYANE FRANCAISE

*H.T. Chan*
Tropical Fruit, Vegetable and Ornamental Crops Research Laboratory, USDA-ARS, Hilo, Hawaii, USA

*L.D. Chandler*
USDA-ARS Northern Grain Insects Research Laboratory, Brookings, South Dakota, USA

*T. Cheikh*
Institut National Agronomique de Tunisie, 43 Ave. Charles Nicolle cité Mahrajène, 1082 Tunis Mahrajene, TUNISIA

*M. Cherni*
Institut National Agronomique de Tunisie, 43 Ave. Charles Nicolle cité Mahrajène, 1082 Tunis Mahrajene, TUNISIA

*U. Cirio*
Enea Casaccia - Division of Agriculture and biotechnologies via Anguillarese 301, S.M. di Galeria (Rome) ITALY

*A. Colombo*
ISCAMen. Boulogne Sur Mer 3050 (5500) Mendoza, ARGENTINA

*J.R. Coppedge*
USDA-ARS College Station, Texas, USA

*S.A. Covacha*
BPI-NMRDC, Jordan, Guimaras, PHILIPPINES

*J.M. Crampton*
Liverpool School of Tropical Medicine, Liverpool L3 5QA, UNITED KINGDOM; The University of Liverpool, Liverpool, L69 7ZB, UNITED KINGDOM

*G.L. Cunningham*
USDA-APHIS, 4700 River Road, Riverdale, Maryland 20737, USA

*C.F. Curtis*
London School of Hygiene and Tropical Medicine, London WC1E 7HT, UNITED KINGDOM

*L. Dantas*
Programa Madeira-Med, Direcção Regional de Agricultura, Madeira, PORTUGAL

*O. De Longo*
ISCAMen. Boulogne Sur Mer 3050 (5500) Mendoza, ARGENTINA

*G. De Martino*
Dipartimento di Genetica, Biologia Generale e Molecolare, Via Mezzocannone 8, 80134, Università "Federico II" di Napoli, Naples, ITALY

*M.H. Dhouibi*
Institut National Agronomique de Tunisie, 43 Avenue Charles Nicolle cité Mahrajène, 1082 Tunis Mahrajene, TUNISIA

*F. Di Paola*
Dipartimento di Genetica, Biologia Generale e Molecolare, Via Mezzocannone 8, 80134, Università "Federico II" di Napoli, Naples, ITALY

*D. Didelot*
Ministère de l'Agriculture et la Pêche, Jardin Botanique, BP 5002, Cayenne, GUYANA FRANCAISE

*R. Dokmaihom*
Plant Protection Service Division, Department of Agriculture Extension, Jatuchak, Bangkok 10900, THAILAND

*R.V. Dowell*
California Department of Food and Agriculture, 1220 N Street, Sacramento, CA 95814, USA

*R.A.I. Drew*
Australian School of Environmental Studies, Griffith University, Nathan Campus, Queensland 4111, AUSTRALIA

*V.A. Dyck*
IAEA P.O. Box 100, A-1400, Vienna, AUSTRIA

*C.R. Edwards*
Purdue University, W. Lafayette, Indiana, USA

*I. Escobar*
Caja Rural de Granada, Granada, SPAIN

*D.K. Eyles*
ARC-Fruit, Vine and Wine Research Institute (INFRUITEC Center), Stellenbosch, SOUTH AFRICA

*U. Feldmann*
Insect and Pest Control Section, Joint FAO/IAEA Division, P.O. Box 100, A-1400, Vienna, AUSTRIA

*K. Fisher*
FAO/IAEA Agricultural and Biotechnology Laboratory, Seibersdorf, A-2444 AUSTRIA. Current address: Entomologist, California Department of Food & Agriculture, Hawaii Fruit Fly Rearing Facility, 41-650 Ahiki Street, Waimanalo, Hawaii 96795 USA

*E.R. Frampton*
Ministry of Agriculture and Forestry, Gerald Street, Lincoln, NEW ZEALAND

*G. Franz*
Entomology Unit, FAO/IAEA Agriculture and Biotechnology Laboratory, Agency's Laboratories, A-2444 Seibersdorf, AUSTRIA

*D. Frías*
Instituto de Entomología, Universidad Metropolitana de Ciencias de la Educación, AV.J.P. Alessandri 774, Casilla 147, Santiago, CHILE

*M. Frommer*
Fruit Fly Research Centre, School of Biological Sciences, A12, University of Sydney, NSW 2006, AUSTRALIA

*E.G. Gaitan*
BPI-NMRDC, Jordan, Guimaras, PHILIPPINES

*A. Gariou-Papalexiou*
Department of Biology, Division of Genetics, Cell and Developmental Biology, University of Patras, GREECE

*G. Gasperi*
Department of Animal Biology, Laboratory of Zoology, University of Pavia, Piazza Botta 9, I-27100 Pavia, ITALY

*E.F. Gersabeck*
USDA-APHIS, 4700 River Road, Riverdale, Maryland 20737, USA

*P. Gleeson*
Queensland Department of Primary Industries, P.O. Box 652, Cairns, Queensland 4870, AUSTRALIA

*P.J. Gomes*
Joint FAO/IAEA Division, International Atomic Energy Agency, P. O. Box 100, A-1400 Vienna, AUSTRIA. Current address: USDA-APHIS-IS, c/o AmEmbassy Guatemala City, Unit 3319, APO AA 34024-3319 USA

*P. Gomez-Riera*
ISCAMen. Boulogne Sur Mer 3050 (5500) Mendoza, ARGENTINA

*G.H. Grant*
Veterinary Services, P.O. Box 309, Hope Gardens, Kingston, JAMAICA

*P.D. Greany*
Center for Medical, Agricultural and Veterinary Entomology, ARS-USDA, Gainesville, FL, USA

*W.J. Grefenstette*
USDA-APHIS, 4700 River Road, Riverdale, Maryland 20737, USA

*R.L. Griffin*
Coordinator, Secretariat of the International Plant Protection Convention, FAO-AGP, Room B-761, Viale delle Terme di Caracalla, Rome, ITALY

*P. Guerin*
Institute of Zoology, University of Neuchâtel, Neuchâtel, SWITZERLAND

***B. Gungah***
Entomology Division, Ministry of Agriculture, Fisheries and Cooperatives, Reduit, MAURITIUS

***D.L. Hancock***
Queensland Department of Primary Industries, P.O. Box 652, Cairns, Queensland 4870, AUSTRALIA

***J.W. Hansen***
FAO-AGA, Animal Health Service, Viale delle Terme di Caracalla, I-00100, Rome, ITALY

***E.J. Harris***
USDA-ARS, Tropical Fruit, Vegetable and Ornamental Crop Research Laboratory, 2727 Woodlawn Drive, Honolulu, Hawaii 96822, USA

***N. Hawlitsky***
INRA Versaille FRANCE

***A.K.W. Hee***
School of Biological Sciences, Universiti Sains Malaysia, 11800 Penang, MALAYSIA

***P. Hernández M.***
Comisión Nacional de Sanidad Agropecuaria, MEXICO

***J.-M. Hougard***
Institut Francais de Recherche Scientifique pour le Developpement et Cooperation, Vector Control Unit, Onchocerciasis Control Programme, P.O. Box 549, Ouagadougou, BURKINA FASO

***P.E. Howse***
School of Biological Sciences, The University, Southampton SO16 7PX, UNITED KINGDOM

***A. Imyim***
Plant Protection Service Division, Department of Agriculture Extension, Jatuchak, Bangkok 10900, THAILAND

***O. Iwahashi***
College of Agriculture, University of the Ryukyus, Nishihara, Okinawa 903-0213, JAPAN

***E.B. Jang***
USDA-ARS, Tropical Fruit, Vegetable and Ornamental Crop Research Laboratory, P.O. Box 4459, Hilo, Hawaii 96720, USA

***F.A. Jassim***
Faculty of Veterinary Services, Ministry of Agriculture, Baghdad, IRAQ

***O.T. Jones***
AgriSenseBCS Ltd., Treforest Industrial Estate, Pontypridd, S. Wales, CF37 5SU, UNITED KINGDOM

***K.G. Juma***
Commission of Agriculture and Livestock, P.O. Box 159, Zanzibar, TANZANIA

***J.P. Kabayo***
Department of Biochemistry, Makerere University, Kampala, UGANDA. Current address: IAEA, Wagramer strasse 5, P.O. Box 100, A-1400 Vienna AUSTRIA

***H. Kakinohana***
Okinawa Prefectural Agricultural Experiment Station, 4-222, Sakoyama-cho, Naha, Okinawa 903-0814, JAPAN. Current address: Agricultural Management & Promotion Division, Department of Agriculture, Forestry Fisheries, Okinawa Prefectural Government, 1-2-2 Izumizaki, Naha, 900-0021 JAPAN

***K.Y. Kaneshiro***
Hawaiian Evolutionary Biology Program, University of Hawaii at Manoa, Department of Entomology, 3050 Maile Way, Gilmore 310, Honolulu, HI 96822, USA

***M Karras***
Imperial College of Science, Technology and Medicine, London SW7 2BB, UNITED KINGDOM

***B.I. Katsoyannos***
Laboratory of Applied Zoology and Parasitology, Department of Agriculture, University of Thessaloniki, GREECE

***P. Keawchoung***
Atomic Energy for Peace of Thailand, 16 Vibhavadi – Rangsit Rd. Bangkok 10900, THAILAND

***V. Kellman***
IICA Office in Guyana, 18 Brickdam, Stabroek, Georgetown, GUYANA FRANCAISE

***N. Kiwia***
Tsetse and Trypanosomiasis Research Institute, P.O. Box 1026, Tanga, TANZANIA

***W. Klassen***
Tropical Research and Education Center, IFAS, University of Florida, Homestead, FL 33031, USA

***A.L. Knight***
USDA, Agricultural Research Service, Yakima Agricultural Research Laboratory, Wapato, Washington, USA

***T. Kocacitak***
Department of Biology, Emory University, Atlanta, Georgia, USA

*A. Kovaleski*
EMBRAPA/Centro Nacional de Pesquisa de Uva e Vinho. CP 177, CEP 95200-000, Vacaria, RS, BRAZIL

*T. Krisaane*
Groupement Interprofessionnel de Dattes de Toseur, TUNISIA

*D. Lance*
USDA-APHIS-PPQ, P.O. Box 1064, Waimanalo, Hawaii 96795, USA. Current address: USDA-APHIS-PPQ,Otis ANGB, MA 02542, USA

*P.A. Langley*
Insect Investigations Ltd., School of Pure & Applied Biology, University of Wales, Cardiff, UNITED KINGDOM

*P.O. Lawrence*
Department of Entomology and Nematology, University of Florida, Gainesville, FL. 32611-0620, USA

*R.A. Leopold*
US Department of Agriculture, Agricultural Research Service, Biosciences Research Laboratory, Fargo, North Dakota 58105, USA

*P. Liedo*
El Colegio de la Frontera Sur, Apartado postal 36, 30700 Tapachula, Chiapas, MEXICO

*V. Limohpasmanee*
Atomic Energy for Peace of Thailand, 16 Vibhavadi – Rangsit Rd. Bangkok 10900, THAILAND

*D.A. Lindquist*
Friedlg. 25/2, 1190 Vienna, AUSTRIA

*C. Lira*
Programa Moscamed, 6a Calle 1-36, Zona 10, Edificio Valsari, Office 203, GUATEMALA, C.A.

*J.A. Lockwood*
Entomology Section, Department of Renewable Resources, University of Wyoming, Laramie, Wyoming, USA

*M. Loosjes*
De Groene Vlieg, Duivenwaardsedijk 1, Nieuwe Tonge, NL-3244LG, THE NETHERLANDS

*C. Louis*
Dept. Of Biology, University of Crete and Instituete of Molecular Biology and Biotechnology, FORTH, Heraklion, Crete, GREECE

***J.C. Lucchesi***
Department of Biology, Emory University, Atlanta, Georgia, USA

***D. Luger***
Entomology Unit, FAO/IAEA Agriculture and Biotechnology Laboratory, IAEA Laboratories, A-2444 Seibersdorf, AUSTRIA

***F. Luyimbazi***
Ministry of Agriculture, Animal Resources and Fisheries, P. O. Box 102, Kampala, UGANDA

***R.J. Mahon***
CSIRO, Institut Haiwan, P.O. Box 520, 86009, Kluang, Johor, MALAYSIA. Current address: CSIRO Entomology, GPO Box 1700, Canberra, ACT 2601, AUSTRALIA

***A.R. Malacrida***
Department of Animal Biology, Laboratory of Zoology, University of Pavia, Piazza Botta 9, I-27100 Pavia, ITALY

***A. Malavasi***
IICA Office in Suriname, Letitia Vriesdelaan 11, Paramaribo, SURINAME. Current address: Depto. Biologia, IBUSP. CP 11.461, CEP 05422-970, Sao Paulo, SP, BRAZIL

***I.I. Malele***
Tsetse and Trypanosomiasis Research Institute, P.O. Box 1026, Tanga, TANZANIA

***R.L. Mangan***
USDA, ARS, Kika de la Garza Subtropical Agricultural Research Center, Crop Quality and Fruit Insects Research Unit, 2301 South International Boulevard, Weslaco, TX 78596, USA

***E.C. Manoto (deceased)***
formerly DOST, PNRI, Diliman, Quezon City, PHILIPPINES

***A.G. Manoukas***
Institute of Biology, NCSR "Demokritos", Agia Paraskevi 15130, Athens, GREECE

***R.P. Marañon***
BPI-NMRDC, Jordan, Guimaras, PHILIPPINES

***S.D. McCombs***
University of Hawaii at Manoa, Department of Entomology, 3050 Maile Way, Gilmore 310, Honolulu, HI 96822, USA

***D. McInnis***
Tropical Fruit, Vegetable and Ornamental Crop Research Laboratory, USDA-ARS Honolulu, Hawaii 96823, USA

*B.A. McPheron*
Pennsylvania State University, University Park, Pennsylvania, USA

*G.T. McQuate*
USDA-ARS, Tropical Fruit, Vegetable and Ornamental Crop Research Laboratory Hilo, Hawaii, USA

*S. Meecheepsom*
Plant Protection Service Division, Department of Agriculture Extension, Jatuchak, Bangkok 10900, THAILAND

*R.H. Messing*
University of Hawaii, Department of Entomology, 7370 Kuamo'o Rd., Kapa'a, HI 96746, USA

*A. Mexia*
Secção Autónoma de Protecção Integrada, Instituto Superior de Agronomia (UTL), Lisboa, PORTUGAL

*F. Meyer*
California Department of Food and Agriculture, 1220 N Street, Sacramento, CA 95814, USA

*D. Midgarden*
FAO, Letitia Vriesdelaan 11, Paramaribo, SURINAME

*P. Montoya*
Programa Moscamed, Comisión Nacional de Sanidad Agropecuaria, SAGAR, Apartado postal 368, 30700 Tapachula, Chiapas, MEXICO

*D.S. Moreno*
USDA ARS, Kika de la Garza Subtropical Agricultural Research Center, Crop Quality and Fruit Insects Research Unit, 2301 South International Boulevard, Weslaco, TX 78596, USA

*I.B. Moussa*
Institut National Agronomique de Tunisie, 43 Avenue Charles Nicolle cité Mahrajène, 1082 Tunis Mahrajene, TUNISIA

*F. Mramba*
Tsetse and Trypanosomiasis Research Institute, P.O. Box 1026, Tanga, TANZANIA

*A.R. Msangi*
Tsetse and Trypanosomiasis Research Institute, P.O. Box 1026, Tanga, TANZANIA

*J.D. Mumford*
Imperial College of Science, Technology and Medicine, Silwood Park, Ascot, Berkshire SL5 7PY, UNITED KINGDOM

***W.A. Mussa***
Commission of Agriculture and Livestock, P.O. Box 159, Zanzibar, TANZANIA

***A. Navarro***
Centro Nacional de Sanidad Vegetal (CNSV), Ciudad de La Habana, CUBA

***R.Y. Niino-Duponte***
College of Tropical Agriculture and Human Resources, University of Hawaii-Manoa, Honolulu, Hawaii, USA

***R. Nishida***
Laboratory of Chemical Ecology, Faculty of Agriculture, Kyoto University, Sakyo-ku, Kyoto 606-011, JAPAN

***G.B. Obra***
DOST, PNRI, Diliman, Quezon City, PHILIPPINES

***D.A. O'Brochta***
Center for Agricultural Biotechnology, University of Maryland Biotechnology Institute, College Park, MD, 20742, USA

***L.M. Ogwal***
COCTU, Entebbe, UGANDA

***F.P. Oloo***
Ministry of Agriculture, Veterinary Department, P. O. Kabete, Nairobi, KENYA

***W. Olson***
Butte County Cooperative Extension, University of California, Oroville, CA, USA

***E. Opiyo***
Entomology Unit, FAO/IAEA Agriculture and Biotechnology Laboratory, IAEA Laboratories, A-2444 Seibersdorf, AUSTRIA

***S.B. Opp***
Department of Biological Sciences, California State University, Hayward, CA, USA

***R. Osborne***
Queensland Department of Primary Industries, P.O. Box 652, Cairns, Queensland 4870, AUSTRALIA

***J. Pacheco***
Programa Madeira-Med, Direcção Regional de Agricultura, Madeira, PORTUGAL

***H. Pan***
IAEA P.O. Box 100, A-1400, Vienna, AUSTRIA

***A. Pane***
Dipartimento di Genetica, Biologia Generale e Molecolare, Via Mezzocannone 8, 80134, Università "Federico II" di Napoli, Naples, ITALY

***A. Pannuti***
Department of Biology, Emory University, Atlanta, Georgia, USA

***N.T. Papadopoulos***
Laboratory of Applied Zoology and Parasitology, Department of Agriculture, University of Thessaloniki, GREECE

***A.G. Parker***
IAEA, P.O. Box 100, A-1400, Vienna, AUSTRIA

***R. Parkes***
Liverpool School of Tropical Medicine, Liverpool L3 5QA, UNITED KINGDOM

***S.L. Peck***
USDA-ARS, Tropical Fruit, Vegetable and Ornamental Crop Research Laboratory Hilo, Hawaii, USA

***R.G. Pegram***
Caribbean *Amblyomma* Programme, Food and Agriculture Organisation, BARBADOS

***R. Pereira***
Programa Madeira-Med, Direcção Regional de Agricultura, Madeira, PORTUGAL

***I. Pérez***
Centro Nacional de Sanidad Vegetal (CNSV), Ciudad de La Habana, CUBA

***S. Permalloo***
Entomology Division, Ministry of Agriculture, Fisheries and Cooperatives, Reduit, MAURITIUS

***T.K. Phillemon-Motsu***
Tsetse Control Division, Department of Animal Health and Production, Ministry of Agriculture, P.O. Box 14 Maun, BOTSWANA

***C. Pickel***
Sutter/Yuba County Cooperative Extension, University of California, Yuba City, CA, USA

***L.C. Polito***
Dipartimento di Genetica, Biologia Generale e Molecolare, Via Mezzocannone 8, 80134, Università "Federico II" di Napoli, Naples, ITALY

***M.M. Ramadan***
Hawaii Dept. of Agriculture, Plant Pest Control Branch, 1428 S. King St., Honolulu, HI 96814, USA

***P.C. Remotti***
Enea Casaccia - Division of Agriculture and biotechnologies via Anguillarese 301, S.M. di Galeria (Rome) ITALY

*P. Rendón*
USDA-APHIS-PPQ, Methods Development Station, c/o AmEmbassy – Guatemala City, APO AA 34024-3319, USA

*S.S. Resilva*
DOST, PNRI, Diliman, Quezon City, PHILIPPINES

*J. Reyes F.*
Comisión Nacional de Sanidad Agropecuaria (CONASA), Guillermo Perez Valenzuela No. 127, Colonia del Carmen, Coyoacan, D.F. 041000, MEXICO

*M.R. Reyes*
DOST, PNRI, Diliman, Quezon City, PHILIPPINES

*K.M. Reynolds*
USDA-ARS, Western Regional Research Center, Albany, CA, USA

*O. Ribeiro*
Ministério da Agricultura, D.D.I.V. Brasília DF 70.235-070, BRAZIL

*G. Richardson*
USDA, ARS, Yakima Agricultural Research Laboratory, Oroville, Washington, USA

*R.C. Roberson*
California Department of Food and Agriculture, Sacramento, CA, USA

*A.S. Robinson*
Entomology Unit, FAO/IAEA Agriculture and Biotechnology Laboratory, A-2444 Seibersdorf, AUSTRIA

*M. Robson*
Fruit Fly Research Centre, School of Biological Sciences, A12, University of Sydney, NSW 2006, AUSTRALIA

*M.C. Romig*
Australian School of Environmental Studies, Griffith University, Nathan Campus, Queensland 4111, AUSTRALIA

*J.P. Ros*
Inst. Nal. Invest. Agrarias. José abascal 56, Madrid, SPAIN

*G. Saccone*
Dipartimento di Genetica, Biologia Generale e Molecolare, Via Mezzocannone 8, 80134, Università "Federico II" di Napoli, Naples, ITALY

*M.P. Salema*
Joint FAO/IAEA Division, International Atomic Energy Agency, P. O. Box 100, A-1400 Vienna, Austria

*K.M. Saleh*
Commission of Agriculture and Livestock, P.O. Box 159, Zanzibar, TANZANIA

*G. Santiago M.*
Comisión Nacional de Sanidad Agropecuaria (CONASA), Guillermo Perez Valenzuela No. 127, Colonia del Carmen, Coyoacan, D.F. 041000, MEXICO

*M. Santoro*
Dipartimento di Genetica, Biologia Generale e Molecolare, Via Mezzocannone 8, 80134, Università "Federico II" di Napoli, Naples, ITALY

*A.van Sauers-Muller*
Ministry of Agriculture, Animal Husbandry and Fisheries, Letitia Vriesdelaan 10, Paramaribo, SURINAME

*C.J. Schofield*
ECLAT Coordinator, LSHTM, London WC1 E7HT, UNITED KINGDOM

*S.I. Seewooruthun*
Entomology Division, Ministry of Agriculture, Fisheries and Cooperatives, Reduit, MAURITIUS

*T.E. Shelly*
Hawaiian Evolutionary Biology Program, University of Hawaii at Manoa, Department of Entomology, 3050 Maile Way, Gilmore 310, Honolulu, HI 96822, USA
Current address: USDA-APHIS, Box 1040, Waimanalo, HI 96795, USA

*I. A. Siddiqui*
Special Assistant to the Secretary of Agriculture for Trade, U.S. Department of Agriculture, 14th & Independence Ave., SW, Washington, DC 20250, USA

*N. Silva*
Programa Madeira-Med, Direcção Regional de Agricultura, Madeira, PORTUGAL

*R.E. Sinden*
Imperial College of Science, Technology and Medicine, London SW7 2BB, UNITED KINGDOM

*K.D. Slagell*
K & K Aircraft Inc., Bridgewater, VA, USA

*J.W. Snow*
International Atomic Energy Agency, P.O. Box 100, A-1400 Vienna, AUSTRIA

*A.R. Soonnoo*
Entomology Division, Ministry of Agriculture, Fisheries and Cooperatives, Reduit, MAURITIUS

*E.L. Spaugy*
Los Angeles County Agricultural Commissioner, 12300 Lower Azusa Rd., Arcadia, Los Angeles, California, USA

*J.P. Spencer*
USDA-ARS, Tropical Fruit, Vegetable and Ornamental Crop Research Laboratory, 2727 Woodlawn Drive Honolulu, Hawaii 96822, USA. Current address: USDA-APHIS-IS, AmEmbassy- Guatemala City, Unit 3319, APO AA 34024-3319, USA

*R.T. Staten*
USDA, APHIS, PPQ, Phoenix Plant Protection Center, Phoenix, AZ, USA

*J. Stewart*
Regional Medfly Program, Mexico-Guatemala-United States, USDA-APHIS-IS, GUATEMALA

*S. Stowell*
Liverpool School of Tropical Medicine, Liverpool, L3 5QA, UNITED KINGDOM

*R.L. Sugayama*
Depto. Biologia, IBUSP. CP 11.461, CEP 05422-970, Sao Paulo, SP, BRAZIL

*B.D. Sutton*
Department of Agriculture and Consumer Affairs, State of Florida, Gainesville, FL 32611, USA

*J. Sved*
Fruit Fly Research Centre, School of Biological Sciences, A12, University of Sydney, NSW 2006, AUSTRALIA

*K.H. Tan*
School of Biological Sciences, Universiti Sains Malaysia, 11800 Penang, MALAYSIA

*F.J. Garcia Tapia*
Caja Rural de Granada, Granada, SPAIN

*G. Testa*
Dipartimento di Genetica, Biologia Generale e Molecolare, Via Mezzocannone 8, 80134, Università "Federico II" di Napoli, Naples, ITALY

*J.J. Tollefson*
Iowa St. University, Ames, Iowa, USA

*M. Tóth*
Plant Protection Institute, Hungarian Academy of Sciences, Budapest, HUNGARY

*N. Tweddle*
Emergency Disease Strategies Section, National Office of Animal and Plant Health, GPO Box 858, Canberra, ACT 2601, AUSTRALIA

*I. Ujváry*
Plant Protection Institute, Hungarian Academy of Sciences, Budapest, HUNGARY

*K.L. Underwood*
School of Biological Sciences, The University, Southampton SO16 7PX, UNITED KINGDOM

*M. Vargas T.*
FAO, Division Ministry of Agriculture, c/o P.O. Box 309, Hope Gardens, Kingston, JAMAICA

*R.I. Vargas*
USDA-ARS, Tropical Fruit Vegetable and Ornamental Crop Research Laboratory, P.O. Box 4459, Hilo, HI 96720, USA

*L.L. Vázquez*
Instituto de Investigaciones de Sanidad Vegetal (INISAV), Calle 110# 514 e/5ta B y 5ta F. Playa. Ciudad de La Habana, CUBA

*A. Villaseñor*
Medfly Commision Mexico-Guatemala 6 Calle 1-36, Zona 10 Edif, Valsari Of. 303, Guatemala City, GUATEMALA

*T.T. Vo*
Policy Analysis and Development, Policy and Program Development, Animal and Plant Health Inspection Service, U.S. Department of Agriculture, Riverdale, Maryland, USA

*M.J.B. Vreysen*
IAEA P.O. Box 100, A-1400, Vienna, AUSTRIA

*M.L. Walters*
USDA, APHIS, PPQ, Phoenix Plant Protection Center, Phoenix, AZ, USA

*S.L. Wee*
School of Biological Sciences, Universiti Sains Malaysia, 11800 Penang, MALAYSIA

*G.E. Wilde*
Kansas St. University, Manhattan, Kansas, USA

*D.D. Wilson*
United States Department of Agriculture, Animal and Plant Health Inspection Service, 4700 River Road, Riverdale, Maryland 20737 USA

*J.H. Wyss*
USDA, APHIS, International Services, Panama City, PANAMA

*M. Yamagishi*
Okinawa Prefectural Fruit Fly Eradication Project Office, Maji 123, Naha 902-0061 JAPAN

***G. Yannopoulos***
Department of Biology, Division of Genetics, Cell and Developmental Biology, University of Patras, GREECE

***H. Yu***
Fruit Fly Research Centre, School of Biological Sciences, A12, University of Sydney, NSW 2006, AUSTRALIA

***H. Zaaraoui***
Groupement Interprofessionnel de Dattes de Toseur, TUNISIA

***A. Zacharopoulou***
Department of Biology, Division of Genetics, Cell and Developmental Biology, University of Patras, GREECE

***N.F. Zamora***
BPI-NMRDC, Jordan, Guimaras, PHILIPPINES

***J. Zavala***
Regional Medfly Program, Mexico-Guatemala-United States, Comisión Nacional de Sanidad Agropecuaria, MEXICO

***Z-R. Zhu***
IAEA, P.O. Box 100, A-1400, Vienna, AUSTRIA

***E.N. Zografou***
Institute of Biology, NCSR "Demokritos", Agia Paraskevi 15130, Athens, GREECE

***R.A. Zucchi***
Departamento de Entomologia, Escola Superior de Agricultura Luiz de Queiroz, Universidade de São Paulo, BRAZIL

# FOREWORD

The organisation of these proceedings on the Area-wide Approach to the Control of Insect Pests is appropriate and timely. There is increasing interest in the holistic approach to dealing with major insect pest problems . This interest has been prompted by the steady progress scientists have made in the development of the sterile insect technique for eliminating the screw-worm from North America, the melon fly from Okinawa, the elimination and containment of the medfly in various countries and the progress that scientists have made in eradicating tsetse fly populations from isolated areas. Increased interest has also been shown by agriculturalists because of the realisation that the farm-to-farm reactive method of insect control is only a temporary solution to problems and that pests continue to be about as numerous as ever from year-to-year. In the meantime, there is increasing public concern over the environmental hazards created by the use of broad-spectrum insecticides to deal with insect pest problems.

While there has been progress in the area-wide approach to insect control it has not advanced to the extent that it should have. There are many other important insects that would be good candidates for area-wide management. Our agricultural leaderships in both the public and private sectors and many pest management scientists do not fully appreciate the large economic and environmental benefits that can be realised by directing control efforts against total pest populations in a fully organised manner. The sterile insect technique provides a feasible way to manage total insect pest populations. However, other techniques and strategies appropriately integrated into management programmes can increase the effectiveness and efficiency of area-wide management programmes. These include the augmentation of mass-produced biological organisms and the use of semiochemicals such as the insect sex pheromones.

These proceedings will give pest management scientists from many countries the opportunity to exchange information on the area-wide approach to insect pest management – an approach that if fully developed can be highly effective, low in cost and at the same time make a major contribution to alleviating the environmental concerns associated with primary reliance on broad-spectrum insecticides for controlling insect pests.

***E.F. Knipling***
USDA Collaborator,
2623 N. Military Road,
Arlington, Virginia 22207, USA

December, 1998

# PREFACE

With the world population attaining the six billion mark, the urgency of increasing quality food production and reducing the spread of diseases transmitted by insects, without affecting our fragile environment, will be of paramount importance. Losses currently experienced in agricultural production, due to insect pests and through diseases transmitted by insect vectors, are very high especially in developing and poor countries. Many insect pests and vectors are of economic importance, and several such as fruit flies, mosquitoes and tsetse flies have attracted international concerns. Most pests are traditionally controlled through heavy reliance on pesticides which can cause environmental pollution, pesticide resistance, and pest resurgence. The control, management or eradication of insect pests and vectors with minimal adverse impact on our food quality, environment, health and well-being should be of great concern to many agriculturists, biological and physical scientists as well as to national and international agencies responsible for pest control. Steps taken by the various concerned agencies to improve and implement the area-wide control will hopefully lead us into the next millennium free from major insect pests and vectors while at the same time protect our precarious global environment.

This volume is the culmination of proceedings conducted in two recent international meetings, FAO/IAEA International Conference on Area-Wide Control of Insect Pests, 28 May – 2 June 1998, and the Fifth International Symposium on Fruit Flies of Economic Importance, 1 - 5 June 1998, held in Penang, Malaysia. Over three hundred papers (both oral contributions and posters) were presented at the two meetings. The manuscripts submitted by authors are divided according to broad topics into eighteen sections originally defined by the organisers as corresponding to the sessions of the meetings. The organisers identified one to several individuals in each of the sessions to deliver an oral presentation of general and/or a specific interest, the subject matter of which is related to the respective sessions in the meetings.

This book is organised into parts that follow the sequence of the two meetings. Due to space and financial constraints, an international panel selected ninety-one papers for inclusion into this book. It contains an opening session and three parts, each with one to several sections consisting of invited contribution(s) and selected poster(s). Overview or review chapters form the major part of these joint proceedings. At the end of this volume are appended a) the final programmes of the two meetings reflecting further the diverse topics presented but not included in this book, and b) taxonomic, author and subject indices.

Moderation is the editing philosophy adopted. An attempt to improve readability and to standardise format and style was made without affecting the authors' inferences, interpretations and conclusions. The kind cooperation provided by the authors in checking edited proofs is greatly appreciated.

I am indebted to the Food and Agriculture Organisation of the United Nations, International Atomic Energy Agency, International Fruit Fly Steering Committee, Universiti Sains Malaysia, and Working Group on Malaysian Fruit Flies for their administrative and financial support; and to all the authors for their respective contributions to this book. The editorial assistance provided by Josephine Choo, S.T.

Tan, T.C. Baldwin and L.C. Sim is gratefully acknowledged. My special thanks go to J. Hendrichs, P.J. Gomes, and A.S. Robinson for their support and selection of papers for inclusion in this volume. The computer and typing skills of J. K. Chan and S.L. Wee plus the assistance of Penerbit Universiti Sains Malaysia (USM Press) in designing the cover and publishing this book are much appreciated.

Tan Keng-Hong
Universiti Sains Malaysia,
11800 Penang, Malaysia.

# OPENING SESSION

# Opening Address

*M. P. Salema*

Deputy Director, Joint FAO/IAEA Division, International Atomic Energy Agency,
P. O. Box 100, A-1400 Vienna, Austria

Honourable Chief Minister of Penang,
Vice-Chancellor of the Universiti Sains Malaysia,
Local Conference Organiser,
Representative of the FAO Regional Office,
Distinguished Conference Participants,
Ladies and Gentleman,

On behalf of the Directors General of the International Atomic Energy Agency, in Vienna, and the Food and Agriculture Organisation of the United Nations, in Rome, and on my own behalf, I wish to extend a warm welcome to all participants of this International Conference on "Area-Wide Control of Insect Pests: Integrating the Sterile Insect and Related Nuclear and Other Techniques" at this beautiful location of Penang.

I would like to thank, most heartily, the Government of Malaysia for having accepted to host this conference and for facilitating its execution through the Universiti Sains Malaysia. I would like to place on record the appreciation by the International Atomic Energy Agency and the Food and Agriculture Organisation of the United Nations to the Government of Malaysia for being a reliable partner in promoting the peaceful use of nuclear technology in food and agriculture. I would also like to thank the Local Organising Committee for its cooperation and the excellent local arrangements made.

The International Atomic Energy Agency, IAEA, serves the United Nations family as the global forum for scientific and technical cooperation in the peaceful uses of atomic energy. Its aim is to accelerate and enlarge the contribution of atomic energy to peace, health and prosperity. Within this context, it assists research on and the development of, practical application of atomic energy for peaceful uses.

The Food and Agriculture Organisation, FAO, is the leading agency in the United Nations for food and agriculture, and rural development. Its aim is to eliminate world hunger and rural poverty by assisting countries to increase agricultural production and by promoting the institutional and policy reforms required for sustainable development.

In 1964, the International Atomic Energy Agency, and the Food and Agriculture Organisation of the United Nations, in recognition of the commonality in their objectives and the complementarity of their roles, joined forces and established the

Joint FAO/IAEA Division of Nuclear Techniques in Food and Agriculture, and later the associated FAO/IAEA Agriculture and Biotechnology Laboratory. The mission of the Division is to strengthen capacity for using nuclear methods in various fields in food and agriculture, and to disseminate these through international cooperation in research, training and other outreach activities in Member States of IAEA and FAO.

In the field of insect pest control, the Joint Division has the mandate to help reduce insect-caused pre-and post-harvest losses as well as pesticide use and its negative impact on the environment. It does this through the development and implementation of environment-friendly nuclear and related technologies.

Following the important success achieved by Dr. E. F. Knipling and his group in the USA in the 1950s in the development and application of the Sterile Insect Technique against the New World Screw-worm fly, Member States requested the Joint Division to establish a programme to promote the development and application of the Sterile Insect Technique against key insect pests. As a result, since the early 1960s, the Joint Division has supported national, trans-boundary and sub-regional insect pest control programmes against major pests, and has played an important role in promoting the Sterile Insect Technique and related technologies.

The Sterile Insect Technique is the most environment-friendly method of pest management. It has proven highly effective against several key insect pests in controlling or even eradicating them across provincial, national, or regional boundaries. Unlike pesticides, which generally have a broad-spectrum lethal activity, the Sterile Insect Technique is species-specific, that is, it affects exclusively the reproduction of the target pest and not other useful insects. Sterile insects, mass-reared in large factories, are released by air on a sustained and regular basis into the target areas, substantially reducing fertile matings. As a consequence, the Sterile Insect Technique has no impact on the food or environment, does not affect beneficial species, and is compatible with most other types of control. Cost-benefit analyses have shown that the Sterile Insect Technique is economically competitive or even superior to conventional methods when considered over longer time frames. For developing countries, the use of the Sterile Insect Technique is also attractive because it reduces the need of expending hard currency for pesticide imports and creates labour opportunities in mass rearing factories. Finally, being area-wide in nature, the Sterile Insect Technique provides economic and environment benefits equally for large commercial farmers and small subsistence farmers, and the population at large.

For some key insect pests, considerable progress has been made over the last 35 years in the development and applicability of the Sterile Insect Technique. This includes large reductions in costs of mass rearing insects, as well as advances in technologies to process and distribute sterile insects and to monitor them in the field. For these pests, the use of the Sterile Insect Technique is no longer considered experimental - important successes for purposes of control, exclusion or even eradication have been achieved. Major successes include the eradication of the screw-worm from North Africa, North America and most of Central America, and of various fruit flies from Mexico, Chile, Japan, Australia and the USA. We are particularly proud that recently the Sterile Insect Technique was successfully used to eradicate the tsetse fly from the Zanzibar Island of Tanzania, indicating a very promising future for employing the technique in tsetse control or eradication programmes on the African continent.

In spite of these successes, most major insect pests still reduce global food output by about 25 to 35%. They transmit diseases that affect humans, livestock and crops, and continue to represent important barriers to free international trade in agricultural products. These high losses due to insects attacking crops, transmitting disease and inhibiting trade occur in spite of pesticide applications in agriculture continuing to increase world-wide. In 1995, world pesticide consumption reached 2.6 million metric tons of active ingredients, costing about US$32 billion annually. In developing countries, according to the most recent World Bank report on the environment, pesticide sales are on a strong upswing, and many highly toxic insecticides remain popular. In spite of Integrated Pest Management successes in reducing pesticide use in some crops such as rice, a significant overall increase in pesticide use is likely over the next decade at least, with pesticide sales growing at a rate of about 5% per year in India, China or Brazil. Pesticide use is particularly intense where such export crops as cotton, coffee, fruits, vegetables and flowers predominate, and for which the Integrated Pest Management approach has been less successful.

The present heavy reliance on chemical pesticides to control insect pests cannot be sustained in the future. Increasingly, there is pesticide resistance in target pests, as well as the emergence of new pests as a result of the elimination of their natural enemies by insecticides. Furthermore, public concerns have grown considerably as a result of widespread environmental pollution, contamination of groundwater, and the presence of pesticide residues in food and fibre. In view of this unsustainable reliance on chemical pesticides and public concerns, new and more environment-friendly techniques, approaches and strategies in the war against insect pests are called for.

Integrated Pest Management has been the dominant paradigm of pest control during the last 30 years with the objective of reducing pesticide use. However, traditional Integrated Pest Management is a localised strategy, with localised short-term objectives which are achieved by remedial interventions, generally based on insecticides to maintain local pest populations below economic injury levels. As pest insects from neighbouring populations keep moving into the treated areas, the effectiveness of these uncoordinated and reactive farm-by-farm, orchard-by orchard, or herd-by herd controls is therefore at best temporary, resulting in the frequent need to re-apply and eventual over-reliance and overuse of pesticides. Furthermore, for some crops such as vegetables and fruit, the economic damage thresholds are so low that the presence of even a few individuals of the pest often trigger the initiation of insecticide applications.

The concept of Area-wide Integrated Pest Management, which is the main focus of this International Conference, implies more than just extending local strategies to large areas. Area-wide management has longer-term objectives, and intervention strategies are planned and implemented on a regional scale. Area-wide Integrated Pest Management requires coordination between farmers and addresses the management of the total population of a pest in an area or region. This involves a strategy very different from traditional Integrated Pest Management in that the large-scale spatial distribution of the pest species has to be considered, both in cultivated as well as non-cultivated and urban areas. It also involves considering the temporal distribution of the pest to determine the periods when the pest is most susceptible to preventive, rather than remedial, interventions. When producers of a given area or region organise themselves to take area-wide integrated action and target all individuals of the pest population, much less inputs are required and the control is usually more effective and sustainable. The area-wide Integrated Pest Management approach is central to the effective integration of the Sterile Insect Technique with other pest control technologies. At the same time, as will be described by several speakers in this Conference, it is also very effective when other pest management tools are integrated on an area-wide basis.

In summary, present trends show that we can look forward with confidence towards a more rational approach to manage major insect pests. Mounting public opposition against the risk of pesticide residues in food and the negative effects of the permanent insecticide treatments will continue favouring non-pesticide technologies. At the same time, the economic feasibility of using area-wide approaches will become increasingly apparent, firstly, with a more realistic accounting of the negative environmental effects of synthetic pesticide applications, and secondly, with further improvement in the cost-effectiveness of these methods as a result of continuing investment in applied Research and Development. Finally, world trade in agricultural commodities will continue to increase rapidly, together with the need for countries to overcome non-tariff trade barriers based on the International Plant Protection Convention. One can predict that these global trends, that are forcing growers towards further rationalising and integrating of activities into large production and trading associations, will strongly encourage an area-wide approach to pest control that integrates, for each pest and situation, the most appropriate pest management techniques.

However, as pointed out by Dr. Knipling in the Preface to the Book of Abstracts of this Conference, agricultural leadership in academia as well as public and private sectors, generally does not fully appreciate the large economic and environmental benefits that can be realised by directing control efforts against total pest populations in a systematic manner. A number of questions come to mind, which I hope will be debated and discussed by scientists at this Conference. Why is the area-wide approach not more widely applied in view of its obvious effectiveness, favourable economics and better sustainability? What are the major factors limiting its acceptance and implementation? Does the area-wide approach have to involve Government intervention? How can the area-wide approach be partially or totally commercialised? How can ownership and participation by academia in area-wide control be strengthened? What other major insect pests are viable targets for area-wide control? What other insect pest control approaches can be components of Area-wide Integrated Pest Management?

I am confident that the over two hundred scientists from over seventy countries and five international organisations represented at this Conference will critically review these and related issues and come up with recommendations that will move this field forward, and guide us over, the next 5-10 years in providing better support to our Member Countries in the field of insect pest control. The International Atomic Energy Agency and the Food and Agriculture Organisation of the United Nations look forward to receiving your recommendations.

Once again, I wish to thank the Government of Malaysia for accepting to host this important Conference, and the Universiti Sains Malaysia for the excellent local organisation.

I wish you successful deliberations and a pleasant stay in this beautiful country.

Thank you.

# Welcome Remarks

*P. J. Gomes*

Chairperson, International Fruit Fly Steering Committee

Welcome everyone to Penang, Malaysia for the 5th International Symposium on Fruit Flies of Economic Importance. The fact that this is the 5th Symposium says a lot about how well we address the needs of the fruit fly community at large. The theme of this Symposium is: Fruit Flies: Current Global Scenario. As we come to the close of the second millennium and begin the next, it is indeed quite fitting to take a closer look at where we presently are with regard to the tools and approaches available to control these destructive pests and to consider where we want to be in the future.

When we talk about fruit flies, we must think and act globally to address the many concerns that these pests present both to agricultural producers and consumers. In order to meet the increasing demand for food security and safety, our agricultural systems must produce greater quantities of fresh fruits and vegetables in a more efficient and environmentally sound manner. Not only must we concern ourselves with the quality of goods produced, but we also must address new international standards for quality, nutrition, and freedom from pests and harmful residues. This Symposium aims to satisfy these needs.

Despite many improvements that have occurred since the last international fruit fly Symposium in Clearwater, Florida, fruit flies still predominate as major phytosanitary concerns among countries involved in the trade of fresh fruits and vegetables. Although world trade is expanding at a rapid rate, the mere presence of certain fruit fly species prevents the movement of fresh commodities from one country to another. Recent assessments on the economic and social impacts of certain fruit flies reveal that losses, direct and indirect, easily total in the millions of dollars each year. More importantly, potential benefits of their control can easily total 10-20 times those projected losses. This is why we now gather here in Penang, so that we, as an international community of fruit fly workers and researchers, can present and discuss new options and strategies for bringing about better exclusion, detection, and control of fruit fly pests.

Reflecting back four years to the last Symposium, I am pleased to report significant progress in many areas, both new and old, for dealing with these pests. This progress will be highlighted by the 33 speakers, 160 posters presentations and 14 discussion sessions involving 180 of our colleagues representing more than 40 countries and several international organisations. In addition, we consider ourselves very fortunate that the Food and Agricultural Organisation and the International Atomic Energy Agency approached us with their intention to convene its International Conference on Area-wide Control of Insect pests Integrating the Sterile insect and Related Nuclear and other techniques at the same venue. This provides for even greater opportunities for interaction and discussion with some of the world's best and most knowledgeable pest control specialists and scientists. This adds to the global dimension of both events, and their participation is most welcome. A joint proceedings of the Conference and Symposium will be published very soon.

On behalf of the International Steering Committee and all the Symposium participants, I would like to thank the Universiti Sains Malaysia, the Working Group on Malaysian Fruit Flies, and members of the Secretariat and Local Organising Committee for all their hard work arranging this Symposium. I especially wish to thank Professor Tan Keng Hong and his devoted staff for their many hours of preparation for a successful Symposium. I wish you all a productive and enjoyable stay here in beautiful Penang.

# PART I.

## AREA-WIDE APPROACH: CONCEPTS AND ECONOMICS

# Pest Management Strategies: Area-wide and Conventional

*D.A. Lindquist*

Friedlg. 25/2, 1190 Vienna, Austria

## INTRODUCTION

Man has fought a war with insects for centuries. With few exceptions, man has been on the defensive.

Neither side has won the war, but both have won battles. Man has managed to contain the damage caused by insects that transmit diseases and limit food and fibre production well enough that the human population has increased significantly over the centuries. The insect control successes have been primarily a result of utilising toxic chemicals against pest insects. Alternatives to insecticides to reduce the damage caused by insects are now receiving more attention because of the perceived environmental problems with the continuous use of insecticides in a defensive battle against insects.

One of the curious and discouraging things about insect control is the lack of high technology methods available for the producer or home owner. The insect has no such problem.

Insects use an outstanding array of high technology methods of fighting the war. These include genetics to develop resistance to insecticides and other methods of control, genetic selection of the strongest individuals, genetic adaptability that prevents natural or man-made disasters from eliminating the species. The insect sensory mechanisms are outstanding. A few molecules are all that are needed to locate mates or food.

Producers and homeowners continue to use the same low technology method that has been used for many years: **a sprayer and an insecticide, the universal tools to control insects.**

### Conventional Insect Control

The usual approach to insect control is to treat the commodity only after a damaging population of insects has developed. In other words, the producer, home owner or casual gardener fights a defensive battle. He reacts to an insect attack. When he sees the enemy or the damage caused by the enemy, he loads up his sprayer with an insecticide and mounts a counter attack.

Most insect control procedures are applied by an individual producer on his own relatively small production area. This conventional insect control approach encourages the producer to make his own decisions about whether or not any insect control is to be used, which insect control method or product to use, when to use it, how to use it, who applies it, etc.

Advice to producers on insect control is usually available from government extension personnel, private insect control consultants or representatives of companies that sell insecticides or other insect control materials or methods.

The conventional procedure results in great variability in the efficacy of insect control because each producer makes his own decisions.

**The objective of conventional insect control is to protect the commodity. This is usually accomplished by treating the commodity, be it cows or corn or rice in a warehouse.**

Conventional insect control requires virtually no planning to achieve results. The "programme" is short-term, frequently measured in days until the next insecticide application is needed. It is reactive (defensive).

## Area-wide Insect Control

Area-wide insect control is applied against an important insect pest over a relatively large area involving many individual producers of the same or similar crops. The "area" is a combination of geography and the range of hosts of the target insect pest. The term "area" in "area-wide" refers to the area where the target insect population survives. The area is not limited to production of the major crop(s) to be protected.

It is very likely that a large part of the cost of an area-wide programme will be fighting the target pest away from the commercial production – before the commercial crops are susceptible – on wild or alternate hosts or abandoned orchards, untreated host plants in homeowners' gardens, etc.

In most cases, area-wide insect control will be the responsibility of a separate organisation hired by the producers.

A separate organisation can plan an aggressive offense against the target pest population over the entire area. High technology systems can be effectively utilised to plan the population management programme. Included will be satellite imagery to detect alternate hosts, sensitive methods to detect movement of the pest populations, computer programmes to predict changes in the pest insect population based on biological parameters, a systems approach to utilise natural enemies on an area-wide basis, genetic analysis to detect the development of resistance and utilisation of systems to delay the development of resistance over the total area.

Further, area-wide programmes encourage the use of specialised methods of insect control that are not effective or are not used on a farm by farm basis. These include the sterile insect technique (SIT), male annihilation, inundative releases of parasites, mating inhibitors, large-scale trap cropping with very attractive plants, treatment of alternate hosts on public lands and hosts in private gardens, etc.

**The objective of area-wide control is to reduce the pest population within the target area to a non-economic level. This is accomplished by attacking the entire insect pest population in the target area.**

Conventional insect control attempts to protect the plant or animal, is carried out by individual producers over a small area with little planning, is short-term, low technology and is a reactive (defense) approach to insect control.

Area-wide insect control attempts to reduce the pest population to a non-economic level over a large area involving many individual producers, is conducted by a special organisation which will carry out a thoroughly planned long-term proactive (offense) approach to insect control. High technology systems that reduce costs and environmental problems and increase efficacy will be used.

## SOME EXAMPLES

### Insects

**There are more examples of area-wide insect control than most plant and animal protection specialists believe. Included are:**

- all SIT programmes,
- all insect eradication programmes,
- most forest insect control programmes,
- in certain cases, host plant resistance are applied on an area-wide basis,
- mosquito control programmes,
- the Black Fly programme in West Africa, and
- fumigation of silos or buildings.

Most of us are familiar with cockroaches. Many of us live in apartment houses. Cockroach control in apartment buildings offer a simple example of area-wide control. Treating an individual apartment will control the cockroaches in that apartment for a few days or perhaps a few weeks. But the pests will return rapidly, moving into the treated apartment from adjacent untreated apartments. Thus the individual apartment owner has great difficulty controlling cockroaches unless his neighbours also treat against cockroaches.

If the people in the apartment building agree that every apartment will be treated during one week, the roach population is greatly reduced, except in the garage, utility rooms and the garbage room. These are common areas and not owned or occupied by any one individual. The apartment dwellers may say, "Why should I pay for treating the garbage room. I do not live there and I do not care if there are cockroaches there". (This is the same attitude that many producers have when they are asked to pay for insect control on land that is not theirs.)

Thus to achieve long lasting roach control in an apartment building, the entire structure must be effectively treated.

The target of this approach is the total roach population in the building and not the cockroaches in any one apartment.

When correctly done, cockroaches will have been essentially eradicated from the building. They will of course re-invade the building from neighboring structures or by being carried into the building with food stuff. However, long term control has been achieved and the cost of roach control for the individual apartment dweller has been significantly reduced.

### Non-insects

Other agricultural pests are also susceptible to the area-wide concept. Examples are weeds, plant diseases, nematodes, rats, birds, etc.

The management of human diseases is becoming more area-wide oriented. The eradication of smallpox is the best example. The mass inoculation of school children against various diseases is a type of area-wide disease management. The treatment is to the individual child but the objective is more than to protect that single child. It is to prevent the spread of the disease from child to child with a possibility of a large epidemic.

## Producer Support

The producers (all of them) should be members of a producer organisation or a special organisation established for the area-wide programme. If some producers are opposed to the programme, difficulties can be expected unless there are means to "encourage" them to join the programme.

The producers will be the primary source of funds to operate the programme. Detailed agreement must be reached with the producer organisation for several years' funding.

Producer support is mandatory for successful area-wide programmes.

## Legal

A well planned legal agreement between the operational organisation and the producers' organisation must be negotiated. It must identify the degree of insect control, penalties for poor results, responsibilities of both the producers and the operational programme, insurance coverage, including personal liability insurance, etc. Possible environmental problems must be covered. The legal document must also very specifically identify funding sources, accounting procedures and possibly insurance against fraud.

The legal agreement must be negotiated professionally and not between friends.

## Operational Organisation

A professional organisation must be established to operate the programme. Personnel must be trained to their assignments. All must understand clearly their duties and responsibilities, accountability, rewards for excellent performance and penalties for poor performance. All must have well prepared job descriptions.

Professionalism at all levels and in all contacts is essential.

## Technology

Technical plans for the area-wide campaign must be prepared in detail. The technology will change over time as the operational staff become more familiar with the problem and new technology is developed by research scientists. Automation should be used wherever possible as it reduces human error, increases efficiency, and reduces the relatively high and ever increasing cost of labour.

# CONCLUSIONS

Area-wide insect control:

- will provide improved efficacy, cost saving and environment,
- is a long-term approach,
- offers excellent opportunity to use high technology methods,
- is an offensive strategy not a defensive one,
- usually requires a separate operational organisation, and

- will initially target a single species but eventually will expand to crop protection.

The requirements for success in area-wide insect control programmes include:

- use only proven tested control methods,
- research is not conducted within the programme,
- prevent political control of the programme, and
- establish an independent organisation to operate the programme.

**An area-wide insect control programme is a long-term planned campaign against a pest insect population in a relatively large predefined area with the objective of reducing the insect population to a non-economic status.**

# Area-Wide Approaches to Insect Pest Management: History and Lessons

*Waldemar Klassen*

Tropical Research and Education Center, IFAS, University of Florida, Homestead, FL 33031, USA

## INTRODUCTION

World agriculture is now entering a very trying era because currently our numbers (Figure 1) are expanding by more than 90 million additional people per year. Demographers project that our growth will not drop below 90 million people per year until about 2020 (United Nations 1993, Nygaard 1998).

The challenge is to increase food production every three or four years sufficiently to feed an additional population equivalent to that of Western Europe or North America.

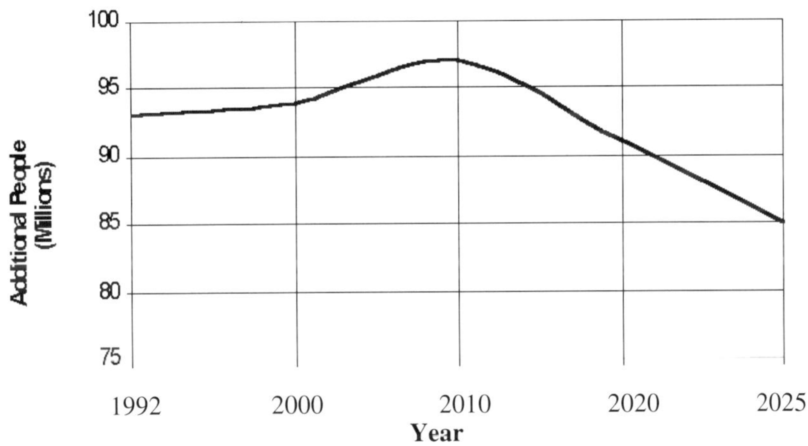


Figure 1. The number of additional people added to the world population each year.

The land available for agriculture on a per capita basis (Figure 2) is becoming progressively more limited so than in 2010, on average, 1 hectare in developing countries will have to feed 5 people, and in South Asia, 1 hectare will have to feed 8 people (Alexandratos 1995, Klassen 1995).

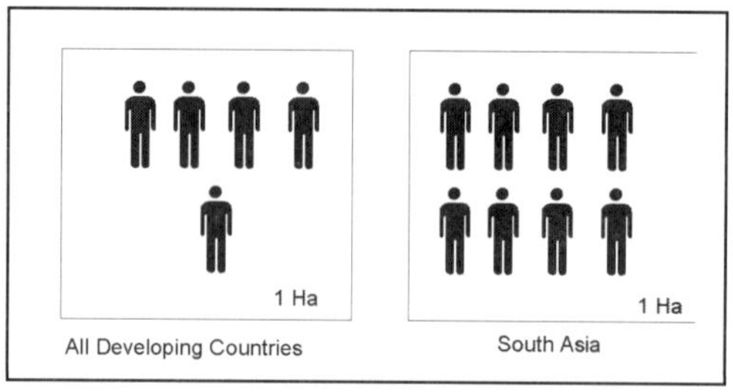


Figure 2. Number of people whose food will have to be produced on one hectare of land in 2010.

On an average, 66 percent of the additional food must come from increased yields (Figure 3), and in South Asia, fully 80 percent must come from increased yields. The balance will come from expanding the area cultivated and use of intensified cropping systems. However, this is not a simple matter since pest populations tend to be favoured by yield-boosting measures.

Since population growth rates recede as people overcome poverty, and since increasing food production is the principal means of overcoming poverty in many countries, it is imperative that in the decades immediately ahead major improvements be made in reducing losses to pests and in other yield enhancing measures (UN Department of International Economic and Social Affairs 1993, Alexandratos 1995, Klassen 1995).

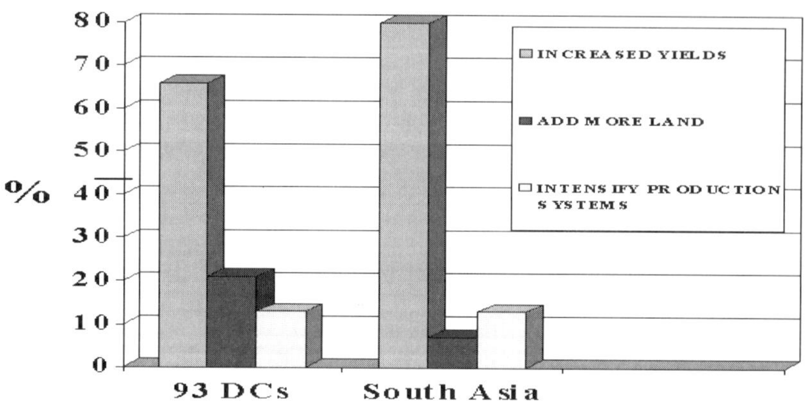


Figure 3. Sources of yield increases in 93 developing countries and in South Asia.

## NEAR INVINCIBILITY OF INSECTS

Insects continue to be the major competitor of man. Although we have been able to increase food production greatly, we have not made headway in reducing overall losses caused by insects. Our insect competitors have proved much more resourceful than we could imagine, and we are forced to rethink our approaches of dealing with them.

The great antiquity of insects gives them a tremendous advantage (Metcalf 1991). Insects first appear in the fossil record of the Lower Devonian Period about 390 million years before the present (Myr BP). The Class Insecta has diversified into 27 orders, about 945 families and several million species. During these 400 million years, insects have surmounted very diverse challenges, and the winning gene combinations have been retained in their germlines. Within their gene pools, insects have the genetic information to meet almost any imaginable challenge.

Insects have a tremendous numerical advantage. For every human on earth, there are 200 million insects (Metcalf 1991). The desert locust, *Schistocerca gregaria* Forskal may form swarms of as many 13 x $10^9$ individuals per swarm – twice more individuals than the number of people on the whole earth (Uvarov 1977). Metcalf (1991) stated, "Such swarms eat their own weight of vegetation daily, an estimated 1 x $10^4$ metric tons, and can cause almost total destruction of agriculture wherever they alight." An outbreak of the desert locust is very dangerous since it may spread in 60 countries over 20 percent of the earth's land surface and threaten 10 percent of the world's human population with starvation (FAO 1998).

## ORIGIN OF AREA-WIDE APPROACHES

Probably, migratory locusts were one of the plagues that caused prehistoric man to attempt forms of group or area-wide control. Since migratory locust swarms can be seen approaching from a distance and descending onto crops, it seems likely that people banded together and used whatever means at hand to stamp out as many as possible. No doubt invasions of armyworms, leafcutter ants and other insects caused people to cooperate in combating them (Howard 1930).

In China since 707 B.C., more than 800 outbreaks of *Locusta migratoria manilensis* L. have been recorded along the floodplains (Figure 4) of the Hwang, Huai and Chang Jiang rivers. In 1929, an outbreak devastated 4.5 million hectares of cropland. Consequently, about 120 million people were mobilised to modify the floodplains by damming, terracing and reforestation. Over almost 30 centuries, the Chinese slowly developed an area-wide pest management programme that now folds together knowledge of biology, ecology, forecasting cultural practices and water management (Metcalf and Kelman 1980, Metcalf 1991).

During the late 1920s, catastrophic locust plagues were widespread in Africa and Southwest Asia. Boris Uvarov and Zena Waloff of the British Ministry of Overseas Development, responded by establishing the International Unit of Locust Research. This unit became the Antilocust Research Centre and it provided the focal point for international cooperation in coping with plagues of the desert locust, the red locust and the African migratory locust. The Centre created databases and a sustained regular flow of information on the status of locust populations throughout their ranges. The Centre developed a system of monthly forecasting. Uvarov was able to interest FAO in creating the International Desert Locust Information Service to coordinate forecasting and the planning of campaigns. Leadership in these vitally important functions has been assigned to FAO's Locust Group (Waloff and Popov 1990).

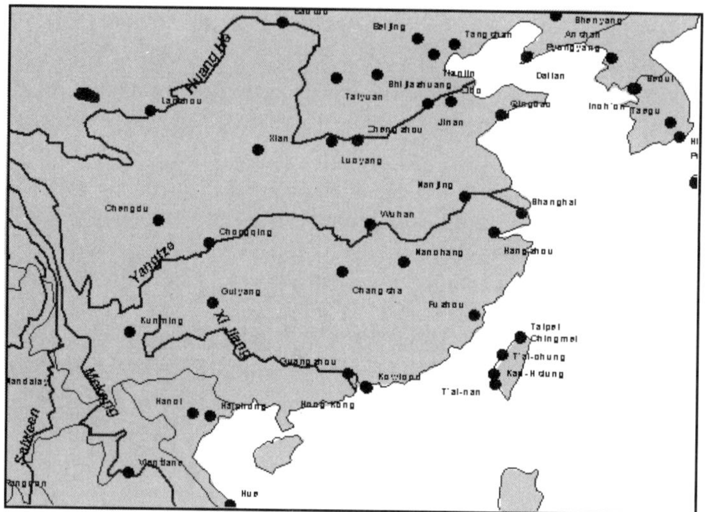


Figure 4. River basins in China afflicted by locust plagues.

Doubtless, the scientific pioneers of area-wide approaches were influenced strongly by concepts from the field of public health and hygiene. About 2,500 years ago, the Greek spirit and the Roman capacity for organisation had produced a highly developed system of hygiene in what is now southern Europe. The Romans procured safe supplies of water by means of aqueducts, practised daily bathing and removed garbage from cities. Why did they do this? Varro (116 BC-27 BC), who served Pompey and Julius Caesar, asserted that malaria is caused by minute living creatures. He wrote, "In damp places there grow tiny creatures, too small for us to see, which make their way into our bodies....and give rise to grave illness" (Sigerist 1958). However, with the collapse of the Roman Empire and the storms of folk migrations, classical hygiene eroded. Nevertheless, raging outbreaks of malaria, typhoid, typhus and bubonic plague during the later Middle Ages re-awakened concepts of hygiene and public health. Doctors and public authorities joined forces to erect walls against these plagues. Dr. Johann Peter Frank (1745-1821) had considerable success in persuading the rulers of Europe during the late 1700s and early 1800s to establish public hygiene policies and to enforce them vigorously. While only a 21-year old student at the University of Strasbourg, Frank called for "systematic action by the authorities" to intervene in the lives of all citizens in order to forestall or halt epidemics (Sigerist 1958). The discoveries of Pasteur, Koch and others on the nature of diseases were foundation stones for rational policies of public health.

Thus through collective action within communities, even without an overall national plan and central coordination, malaria in southern Europe and North America largely disappeared in consequence of education, the universal adoption of window screens, destruction of habitats of Anopheles larvae and the treatment of all cases with quinine (Collins and Pasketwitz 1995).

Investigations conducted in the late 1800s and in early years of this century on the transmission by mosquitoes of deadly diseases led to the widespread use of area-wide programmes. Yellow fever, dengue, filariasis and malaria were shown to be transmitted by various species of mosquito. In 1892, Howard and in 1900, Ross began to recommend that the habitat of mosquito larvae over extensive areas be either treated with kerosene or drained. These practices were first implemented in West Africa to combat malaria (Stage 1952), and soon adopted by communities in many countries. By 1912, the New Jersey Mosquito Extermination Association was founded, and it provided the model for the organisation and operation of area-wide mosquito abatement districts of which there are about 260 in the United States and a thousand or more world-wide.

## POWER OF SOCIAL CONCERNS

Some of the programmes conducted on an area-wide basis – especially those aimed at eradication – have aroused opposition. The strategy of eradication emerged just over one century ago as the brainchild of Charles Henry Fernald of the University of Massachusetts. Under Fernald's leadership, Massachusetts attempted to eradicate an introduced pest, the gypsy moth, *Lymantria dispar* L., in an 11-year campaign from 1890 to 1901. Initially, the primary eradicant was Paris green spray (Forbush and Fernald 1896). The use of Paris green, which suffered from modest efficacy and phytotoxicity, had to be abandoned because of adverse public reaction.

Forbush and Fernald (1896) noted, "Considerable opposition to the use of Paris green for spraying was manifested by many people living in the infested towns. A mass meeting of opponents of the spraying was held in Medford. One citizen, who attempted to cut the hose attached to one of the spraying tanks, and threatened with violence the employees of the Board who had entered upon his land, was arrested and fined. Others neutralised the effects of the spraying by turning the garden hose upon trees and shrubs that had been sprayed, and washing off the solution. The opposition to the spraying affected the results of the work unfavourably to a considerable extent." Clearly, apathy by many members of the public had turned into outrage.

Thus the social dimension of pest management was revealed in a new way with emphasis on stakeholders who are not primarily concerned with the economic dimension of the pest problem. Repeatedly, applied entomologists have learned the hard way that the ecological, environmental, social and human health criteria are more important to some stakeholders than the economic criterion (Rabb 1972, Dreistadt 1983, Scribner 1983, Myers, Savoie and van Randen 1998).

One hundred years passed before social scientists such as Sandman (1987) and Starr (1985) taught us to analyse programmes for outrage factors, and to be proactive in dealing with issues in order to avert outrage. Indeed, our inadequate expertise in coping with social and political aspects has resulted in many setbacks and difficulties (Perkins 1982).

In 1955, the World Health Assembly urged WHO to lead and organise world-wide eradication of malaria (Wright, Fritz and Hanworth 1972). By 1959, almost 65 percent of the people at risk were protected and this percentage rose to 74 percent by 1970. Malaria eradication was claimed in 37 countries, and the incidence of malaria had dropped dramatically in many countries. In Sri Lanka, the number of cases dropped from more than 2.8 million per year to just 17 in 1963 (Metcalf 1991), but then the effort floundered as WHO was unable to deal with the widespread hue and cry for local control. In 1969, the Global Programme disintegrated. Soon, DDT was banned in the United States and WHO's resources for malaria were re-allocated. Malaria resurged to more than half a million cases per year in Sri Lanka and to more than 100 million cases world-wide (Collins and Pasketwitz 1995).

In 1969 and 1970, the Indian Council for Medical Research and the WHO initiated several projects relevant to area-wide control. Unfortunately in 1974, these projects became the target of a press campaign by writers who feared that these projects were actually a USA-funded effort to develop methods of biological warfare. Since the government of India was unable to restore confidence, the projects were terminated (World Health Organisation 1976).

## LEGAL AUTHORITY

Area-wide control owes much to the love of wine by Europeans. In about 1860, the grape phylloxera, *Phylloxera vitifoliae* Fitch was transported from the United States to France. Within 25 years of its arrival, this insect had utterly destroyed 1 million hectares of vineyards or fully one-third of the productive capacity of France. In order to protect the German wine industry, the government of Germany in 1873 passed the first law that provides for quarantines and regulatory control of agricultural pests (National Academy of Sciences 1969). Other governments quickly followed the example set by Germany. Without such legislation, it would be impossible to prevent the entry or establishment of exotic pests or to limit or retard their rate of spread. Meanwhile, the French sought the assistance of the charismatic American entomologist, Charles Valentine Riley. In 1871, Riley set the French on an area-wide programme of grafting all European grapes on the phylloxera-resistant American rootstocks (Summers 1925).

The discovery in 1889 by Theobald Smith and colleagues that cattle fever is caused by a tick-transmitted parasite of red blood cells led to the initiation in 1906 of a county-by-county effort to eliminate the two *Boophilus* tick vectors from the United States (Figure 5). Many pastures were rendered tick-free by excluding all host animals until all ticks had starved to death. Livestock were dipped in an arsenical solution at two-week intervals. Quarantines were used to prevent the movement of infested cattle into areas that had been cleared. By 1943 – after 37 years of gruelling effort – the ticks had been eliminated entirely from the United States at a total cost of about US$40 million or the equivalent of the annual losses suffered before the programme was initiated (Cole and MacKellar 1956, MacKellar 1942). Quarantines have been effective in preventing these ticks from becoming re-established from their populations in Mexico. A broadly shared vision sustained this programme in spite of war and the great economic depression.

## CONTRIBUTIONS OF NATURAL ENEMIES

I believe that area-wide pest management was practised more widely during the nineteenth than during the twentieth century. During the eighteenth and nineteenth centuries, people began to understand the roles of natural enemies in preventing insect outbreaks. Further, the powerful synthetic insecticides were not available to allow smallholders independently to protect their crops and livestock. The beneficial work of coccinellids and other predators had been common knowledge for centuries, and they were collected and distributed for insect control (DeBach 1964). However, insect parasitism was discovered only around 1700 by Leuwenhoeck in the Netherlands and in 1706 by Vallisnieri in Italy. Significantly, Emperor Francis 1 of Austria ordered Vincent Kollar to publish his work on the role of natural enemies in suppressing pests.

Kollar's great work appeared in 1837 and the English translation appeared in London's Gardner's Magazine in 1840 (Debach 1964).

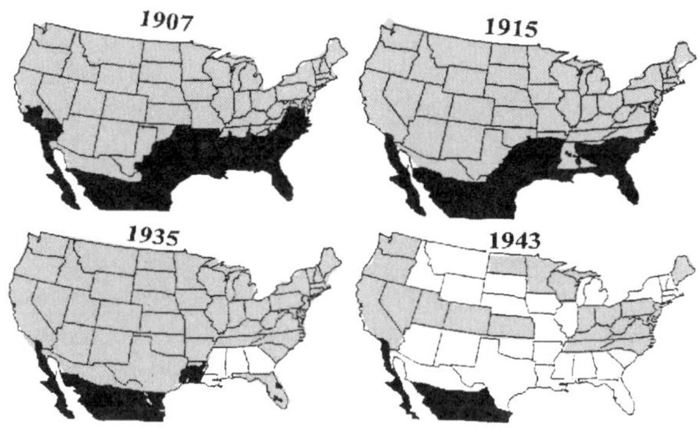


Figure 5. Progress in eradicating *Boophilus* ticks in the United States 1906-1943.

That classical biological control can provide area-wide solutions was dramatically illustrated against an exotic pest in California in 1888 and 1889. At that time an introduced pest, the cottony cushion scale, *Icerya purchasi* Maskell, was killing hundreds of thousands of citrus trees. However, Albert Koebele was able to introduce a scale predator, the vedalia beetle, *Rodolia cardinalis* Mulsant from Australia and New Zealand. Less than 11,000 vedalias were distributed, but they spread throughout the entire citrus growing area of southern California and saved the industry (see DeBach 1964, Klassen 1991). The vedalia beetle continues to effectively protect citrus in California, and nothing needs to be done other than to avoid the use of certain insecticides, which would decimate this invaluable natural enemy.

Almost exactly one hundred years after the great vedalia success, a team led by Dr. Hans Herren of the International Institute for Tropical Agriculture (IITA) successfully implemented the largest classical biological control programme in history (Figure 6). In 1973, cassava near Brazzaville and Kinshasa was found to be attacked by the cassava mealybug, *Phenacoccus manihoti* Matile-Ferrero. In a few short years, immature crawlers were dispersed by wind throughout sub-Saharan Africa.

The cassava mealybug created starvation and hardship for many of the 200 million people for whom cassava had become a staple crop. In 1981 an excellent parasitoid, *Epidinocarsis lopezi* DeSantis – found in Paraguay by A. C. Bellotti – proved capable of bringing the cassava mealybug under control. The parasite was mass reared and released by aircraft over 38 countries of sub-Saharan Africa (an area much larger than the combined area of the United States, Mexico and India) with excellent results (Herren and Neuenschwander 1991). This singular accomplishment required strong and imaginative leadership and action by IITA, generous funding by donors, and brilliant scientific and technical work by Herren's team and their cooperators in Africa, Europe and the Americas.

Uses of natural enemies and habitat modification have proved to be the only durable insect management techniques (Klassen 1991).

In many cases, natural enemies are effective only if most smallholders in an area work to conserve them. Since both pests and natural enemies are mobile, their populations distribute themselves throughout the region in which their food sources are available. Even smallholders who do not participate in the conservation programme receive some of its benefits. They get a free ride. This is a positive externality. On the other hand, the movement of natural enemies off the property of the participating farmer to that of the free rider is a negative externality (Reichelderfer et al. 1984). The brown planthopper, *Nilaparvata lugens* Stal, has been the scourge of rice production in Southeast Asia for many years. However, during the 1980s, Indonesia, with technical assistance from FAO and GTZ, simultaneously achieved substantial increases in rice production and major reductions in insecticide use. Generally, brown planthoppers are effectively controlled by indigenous spiders and other predators (Heinrichs 1991, Oka 1991). Moreover, since insecticides have a greater impact on the predators than on the pest, the brown planthopper populations are able to resurge after being sprayed. In the past, farmers induced resurgence by beginning to spray 40 days after transplanting the rice. However, cage studies showed that the smallholder who delays spraying until 65 days after transplanting saves two insecticide applications and realises a yield increase for a total benefit of US$588 per hectare.

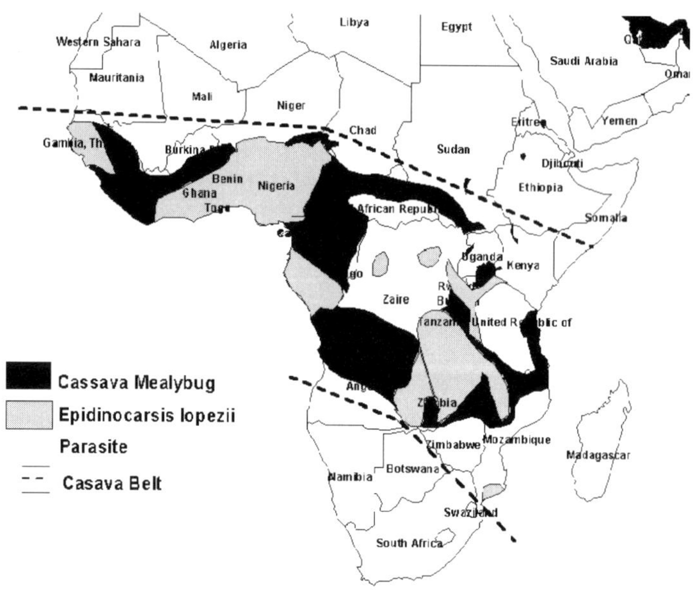


Figure 6. Biological control of the cassava mealybug in sub-Saharan Africa through the introduction of a parasite from Paraguay.

## THEORETICAL BASIS OF AREA-WIDE APPROACHES

Reichelderfer et al. (1984) modelled what happens when some smallholders delay spraying to conserve natural enemies but others do not (Figure 7). If about ten percent of smallholders conserve natural enemies, they gain only one-fifth of the potential benefit. If 30 percent of smallholders conserve natural enemies, they gain only one-quarter of the potential benefit and the free riders gain about 7 percent. When 50 percent of smallholders conserve natural enemies, they gain one-third of the potential benefit and the free riders gain about 18 percent.

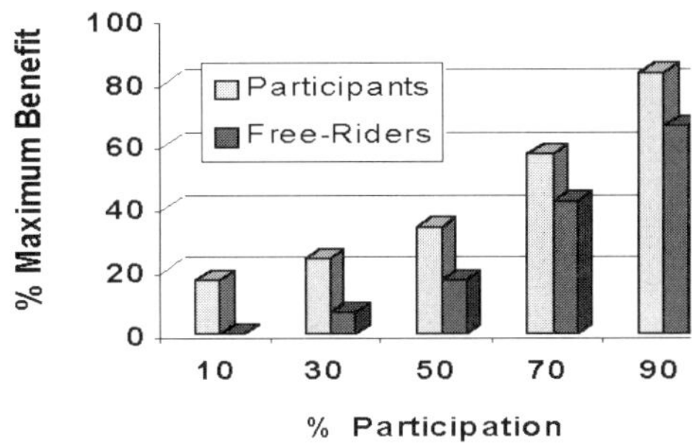


Figure 7. Benefits of increasing participation in biocontrol programme.

When 70 percent of smallholders participate, they gain almost 60 percent and the free riders gain about 40 percent, and when 90 percent participate they gain about 83 percent of the potential benefit, while the free riders gain 66 percent. Clearly a conservation programme is almost futile until about one-half of the smallholders participate, and the programme becomes progressively more beneficial as the percent participation increases toward 100.

An attempt to explain the theoretical basis of area-wide pest management was made by Knipling (1992). Knipling noted that apple growers achieve almost 100 percent control of codling moths *Cydia pomonella* L., in major apple growing areas, but that in these communities, there is often an abandoned orchard or some untreated non-commercial trees. Because of these tiny reservoirs – normally less than 5 percent of the total host resources of the pest – the codling moth is able to build to economic numbers throughout the entire apple growing region. Thus, Knipling used a simple population model to compare the overall trends of a total population in an ecosystem subjected to uniform suppression versus one in which a small part of the population escapes treatment each generation. The model showed that, in five generations, a field in which 10 percent of the population escaped treatment would produce 1000 times more insects than in a field in which the total population is suppressed uniformly in each generation. Thus, Knipling formulated the principle, *"Uniform suppressive pressure applied against the total population of the pest over a period of generations will achieve greater suppression than a higher level of suppression on most, but not all, of the population each generation."*

## COPING WITH PEST MOBILITY AND EXTERNALITIES

In some instances, high pest mobility and externalities can be met only by employing the sterile insect technique, pheromones or by the release of natural enemies on an area-wide basis. During the past five decades, theoretical models and practical experience have shown that, whenever possible, highly selective control measures should be chosen. If the pest must be strongly suppressed or eradicated, then a method that is effective against high density populations and a method that is effective against low density populations should be integrated in such a way that the former potentiates the latter, especially if the goal is eradication (Knipling 1966, 1979, Klassen 1989).

Control measures, whose effectiveness is greatest against low-density populations, include sterile insects and sex attractants. Without such techniques, it would have been impossible to eradicate the screw-worm or tropical fruit flies from southern Japan (Yamagishi et al. 1993), Australia (Fisher 1994) and Mexico (Orozco et al. 1994). Systems including the area-wide use of sex pheromones are now emerging to manage the pink bollworm, Oriental fruit moth, tomato pinworm and other lepidopterous pests (Carde and Minks 1995). Knipling (1992) promulgated his view that many parasitoids have phenomenal abilities to find their hosts and that such parasitoids are highly efficient even when the host density is low. This hypothesis needs to be evaluated in an area-wide context. For total population management of pests with fairly high economic thresholds, the integrated use of two or more methods effective against high-density populations tends to be effective and durable.

## CONTINGENCIES OFTEN DICTATE STRATEGY

Terms of reference and contingencies have a tremendous influence on the selection of strategies of area-wide pest management. Peter Geier (1970) wrote a brilliant paper entitled "Temporary Suppression, Population Management or Eradication: How and When to Choose". He stated that the title "implies that we face a true choice, and that proper attention to the facts of the matter should lead us to the "right" decision. But what is "proper attention", what are "relevant facts" in this context, and, finally, by what criteria can we judge a practical decision to be "right", or to be "wrong"?" Geier notes, "I have become convinced that the honest answer ... is simply this: as specialists, as experts, as economic entomologists, we hardly ever choose the course of our actions deliberately. That course is determined for us, not by the guidelines of our trade, but by the terms of reference, largely subjective and contingent, that govern our operations. Those are the real decision-makers. As such, they might deserve more notice than they usually receive in our specialist circles."

Geier's stricture applies to the epic struggle to remove the New World screw-worm, *Cochliomyia hominivorax* Coquerel, from North America. To accomplish this goal, different area-wide strategies had to be selected to meet different situations. As known, eradication is the destruction of every individual of a species in an area surrounded by natural or man-made barriers sufficiently effective to prevent re-invasion except through the intervention of man (Newsom 1978).

The process began in 1954 on Curacao (Figure 8) where the strategy of eradication was executed successfully, and then again in Florida. In Florida, one-quarter of the cost of the programme was borne by the cattle industry, one-quarter by the state and one-half by the federal government. Subsequently, a cost-shared programme was implemented in the states bordering on Mexico. The parasite was eliminated from Texas and New Mexico, Arizona and California. However, the parasite population in the adjacent area of Northern Mexico was managed to minimise the number that would enter the United States. Eradication could not be the immediate strategy because eradication was contingent on having an agreement between the governments of Mexico and USA to remove the parasite from Northern Mexico. It took six years to reach an agreement, and only then could the eradication strategy be resumed.

Figure 8. Campaigns against the screw-worm and changes in strategy dictated by circumstances.

Meanwhile during 1966, no screw-worms could be found in the United States for several months. Thus, the cattle industry and the state governments were able to persuade the US Secretary of Agriculture to declare that the parasite had been eradicated from the United States. This declaration caused the screw-worm to be considered a foreign pest, and thus the federal government became solely responsible for any costs that would be incurred when screw-worms reappeared in the United States. Therefore the border states and the cattle industry were able to avoid cost-sharing the programme after 1966. However, numerous screw-worm cases (Figure 9) occurred in USA until 1982.

In 1977, the late Dr. L.D. Newsom of Louisiana State University voiced a concern felt in many universities. Newsom noted that the screw-worm had not been eradicated in fact but by fiat – namely by an authoritative pronouncement of the USDA. Newsom noted that this lack of candour "has caused grave problems for our profession by generating far too much optimism for its general applicability as the ultimate in strategy of pest management." Newsom stated, "Keeping in mind that the objective of the screw-worm programme is to eradicate the pest from the United States, we can only conclude that there has been abysmal failure to achieve that objective. Reaction to such failures to achieve eradication has become characteristic and highly predictable. The typical procedure is to increase the size of the operation without bothering to do the necessary research required to explain the failure encountered at the previous level of operation. This does not seem to be a reasonable approach. Nevertheless, this is precisely what has happened with the screw-worm programme."

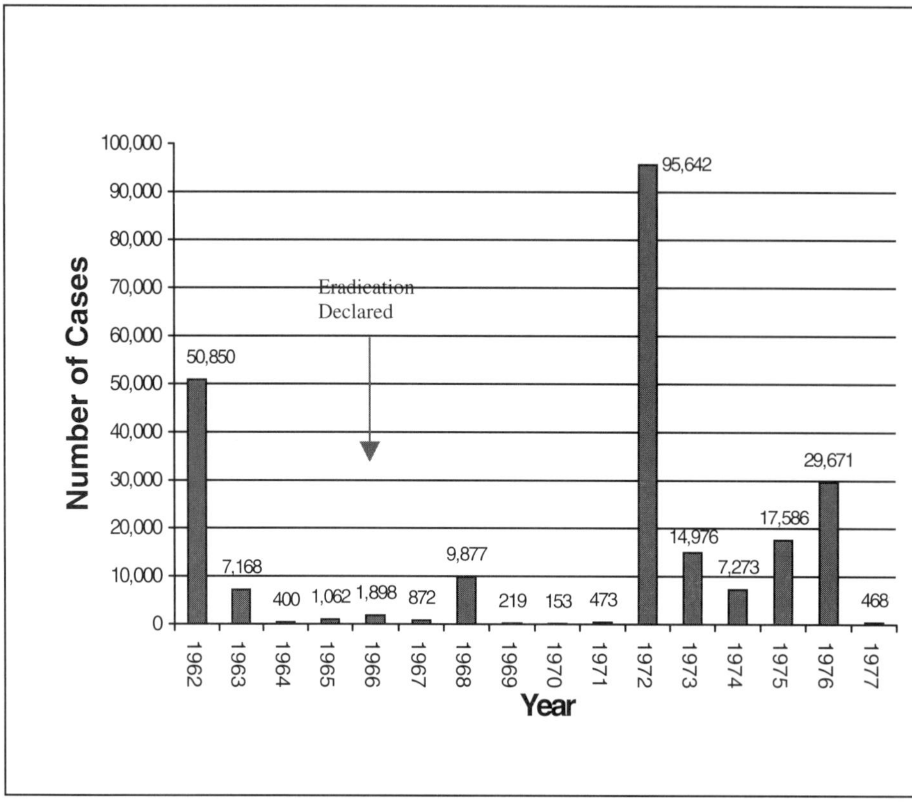


Figure 9. Number of screw-worm cases reported in the US since the inception of the southwestern programme in 1962. (Data from Lincoln and Eden, 1975, assembled from USDA, APHIS, Veterinary Services). Source: LaChance 1979.

Around this time, a movement to discredit the strategy of eradication arose in the United States, and this movement hampered work against the boll weevil, *Anthonomus grandis* Boheman. Tremendous controversy developed within the scientific community during the early 1970s when the United States Department of Agriculture (USDA) mounted a large-scale experiment to assess the feasibility of eradicating the boll weevil from the United States. The losses caused by the boll weevil were far greater in the eastern and southern parts of the pest's range than in the northern and western zones. Thus, in the event that the boll weevil was removed from the Cotton Belt, cotton acreage would shift from the west to the east – to the detriment of Texas and to the benefit of the eastern states.

This large pilot field experiment was conducted in 1971-1973 in southern Mississippi, Alabama, and Louisiana (Figure 10) to assess the feasibility of eradication. The test area included the primary eradication area (Zone 1) surrounded by three buffer zones. The eradication area had a radius of about 40 kilometres and the entire experimental area had a radius of about 120 kilometres. Very intensive suppression was implemented in the two inner zones. However, the two outer zones received only normal in-season control by growers and insecticide applications applied on a diapause control schedule. The boll weevil was suppressed below detectable levels in 203 of 236 fields in the eradication zone. All of the 33 lightly infested fields were located in the northern one-third of the eradication zone and less than 40 kilometres from substantial populations farther north in the buffer. In the southern two-thirds of the eradication zone, no reproduction could be detected in any of the 170 fields (Committee on Appropriations 1975).

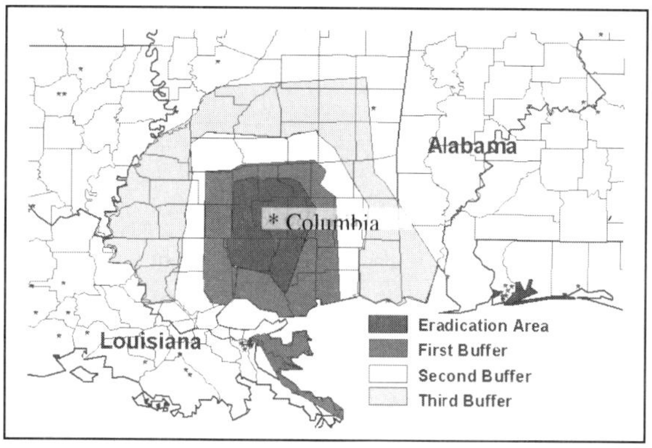


Figure 10. Pilot boll weevil eradication experiment in Louisana, Mississippi and Alabama.

The USDA leaders concluded that the experiment indicated that the technology was sufficiently effective to achieve eradication. Their experience with the screw-worm had convinced them that the achievement of total elimination of all individuals from the target area following the first application of the suppressive system is not necessary.

Eradication is accomplished iteratively. The first application of the suppressive system clears the pest from most of the target zone. Subsequently, surviving populations are delimited and suppressive measures are applied to them. In this iterative fashion, the aggregate range occupied by the pest is progressively reduced toward zero (Klassen 1989).

However, neither the experiment's Technical Guidance Committee (Knipling 1976), a committee of the Entomological Society of America nor a committee appointed by the National Academy of Sciences, recommended initiation of an eradication programme (Entomological Society of America Review Committee 1973, National Research Council 1981, Perkins 1983). Nevertheless, the cotton industry insisted that a trial eradication programme be undertaken. The programme started in Virginia and North Carolina (Carlson and Wetzstein 1993). By phases, this programme has removed the boll weevil from about 4.5 million acres in seven states. The weevil has been removed from one-third of the Cotton Belt. Cotton acreage has shifted strongly to the states that no longer have the boll weevil.

## SOME AREA-WIDE PROGRAMMES MAY DAMAGE ENVIRONMENT

The boll weevil programme relies heavily on applications of malathion to all cotton in an eradication zone. Under dry windy conditions, these applications tend to decimate the natural enemies of the beet armyworm. Thus in the Rio Grande Valley in 1995, a catastrophic outbreak of the beet armyworm was induced. Consequently, the growers terminated the eradication programme. Ecologically selective technology exists, but it costs more than the use of malathion. Unfortunately, the US Congress reduced the share of federal funds to less than 15 percent, and thus short-term costs are decisive in designing the boll weevil pest management system.

## FORESTALLING INSECTICIDE RESISTANCE

Genes for susceptibility to insecticides are beginning to come under area-wide management by the US Environmental Protection Agency (EPA) (Carlson et al. 1998). EPA controls the registration of pesticides and the release of transgenic crop cultivars. The gene for the *Bacillus thuringiensis* toxin has been incorporated into corn, cotton and other crops. Since these lepidoptera-resistant resistant corn and cotton cultivars are planted on millions of acres, there is concern that resistance to Bt will develop rapidly. To avert this, EPA has mandated a resistance management programme whereby in 1996, 3.85 percent of each cotton field had to be planted with non-Bt cotton and left untreated with insecticides. Alternatively, if the grower wished to use insecticides other than Bt, then 20 percent of the field had to be planted with non-Bt cotton. EPA also placed limits on total sales of Bt-corn in cotton producing counties in order to prevent the corn earworm from developing resistance to *Bacillus thuringiensis* (Carlson et al. 1998).

## CONTRIBUTION OF HOST RESISTANCE

The benefits of area-wide use of pest-resistant crop cultivars can scarcely be overstated. Even a difference of 50 percent between resistant and susceptible varieties, which is cumulative each generation, greatly reduces the pest's density (National Academy of Sciences 1969). Fields planted with resistant cultivars serve as traps for pests. Moreover, resistant crop cultivars greatly facilitate the effectiveness of natural enemies, cultural methods and insecticides but when used alone, insects commonly overcome host resistance in three to ten years (Klassen 1991).

## SUPPORT OF ENVIRONMENTALISTS NEEDED

Rachel Carson's book, Silent Spring, which appeared in 1962, gave very strong support to the development of ecologically selective tools for pest management (Carson 1962). However, we have failed to effectively enlist the support of the environmental community for area-wide programmes. Worse, the United States Environmental Protection Agency has a history of opposing area-wide programmes which are highly beneficial to the environment. There is a need everywhere to engage the environmental movement and to lead it to an understanding of the great benefits to the environment that can be derived from well-conceived area-wide programmes.

## URGENT NEED TO ADVANCE AREA-WIDE APPROACHES

History has produced a harmonious body of theory, principles and practices to guide and support successful area-wide pest management programmes. This conference is only a tiny part of what needs to be done to transmit this body of knowledge to the next generation of students and workers. It is very urgent that we extensively replace field-by-field management with area-wide strategies. Otherwise, we will not fulfil our obligation to future generations. In order to provide for future generations, we must make major progress in the next two decades. Let us not underestimate our splendid competitors who have had 400 million years to assemble genes for the battle. We can imagine their confident rallying cry: "vincero, vincero, vincero!".

## REFERENCES


Alexandratos, N. (ed.) 1995. World agriculture: towards 2010, an FAO study. J. Wiley & Sons, Chichester, UK, and FAO, Rome, Italy. 488 pp.

Bushland, R.C. 1985. Eradication programme in the southwestern United States, pp.12-15. *In* Symposium on eradication of the screw-worm from the United States and Mexico. 0. H. Graham [ed.]. Miscellaneous Publication 62. Entomol. Soc. Am. College Park, MD, USA.

Carde, R.T. and A.K. Minks. 1995. Control of moth pests by mating disruption: successes and constraints. Ann. Rev. Entomol. 40: 559–585.

Carlson, G.A., M. Marra and Bryan Hubbel. 1997. Transgenic technology for crop protection: the new "super seeds". Choices, 3rd Quarter, 31-36.

Carlson, G.A. and M.E. Wetzstein. 1993. Pesticides and pest management, pp. 268-318. *In* Agriculture and Environmental Resources Economics. G.A. Carlson, D. Zilberman and J. A. Miranowski (eds.). Oxford University Press, Oxford, UK.

Carson, R. 1962. Silent Spring. Houghton Mifflin, Boston, MA, USA. 368 pp.

Cole, T.W. and W.M. MacKellar. 1956. Cattle tick fever, pp. 310-313. *In* Animal diseases, USDA Yearbook of Agriculture, 1956.

Collins, F.H. and S.M. Paskewitz. 1995. Malaria: current and future prospects for control. Ann. Rev. Entomol. 40:195–219.

Committee on Appropriations, U.S. House of Representatives. 1975. The pilot boll weevil eradication experiment, pp. 775-785. *In* Hearings: Agriculture, Environmental and Consumer Protection Appropriations for 1975. Government Printing Office, Washington, DC, USA.

DeBach, P. 1964 (ed.). Biological control of insect pests and weeds. Reinhold Publishing Corporation, New York, USA. 844 pp.

Dreistadt, S. 1983. Sociopolitical impact: environmental concerns, pp. 36-44. *In* Medfly and the aftermath. Symposium conducted at the Entomol. Soc. Am. Annual Meeting, December 1982. USDA, APHIS 81-60.

Entomological Society of America Review Committee. 1973. The pilot boll weevil eradication experiment. Bull. Entomol. Soc. Am. 19: 218-221.

[FAO] Food and Agriculture Organisation. 1998. Frequently Asked Questions (FAQS) about Desert Locusts. http://www.fao.org/NEWS/GLOBAL/LOCUSTS/LOCFAQ.htm

Fisher, K. 1994. The eradication of the Queensland fruit fly, *Bactrocera tryoni,* from Western Australia, pp. 237–246. *In* Fruit Flies and the Sterile Insect Technique. C.O. Calkins, W. Klassen and P. Liedo (eds). CRC Press, Boca Raton, FL, USA. 272 pp.

Forbush, E.H. and C.H. Fernald. 1896. The gypsy moth. *Porthetria dispar (Lion.).* Wright & Potter. Boston, MA, USA.

Geier, P.W. 1970. Temporary suppression, population management, or eradication: how and when to choose, pp. 170-190. *In* Concepts of Pest Management. R.L. Rabb and F.E. Guthrie (eds.). Proc. of a Conference at North Carolina State University, Raleigh, NC, USA, 25-27 March 1970. 242 pp.

Heinrichs, E.A. 1991. Entomology in the Developing World, pp. 29-42. *In* Progress and Perspectives for the 21st Century. J. J. Menn and A.L. Steinhauer (eds.). Centennial National Symposium, Entomological Society of America, Lanham, MD, USA. 170 pp.

Herren, H.R. and P. Neuenschwander. 1991. Biological control of cassava pests in Africa. Ann. Rev. Entomol. 36:257-83.

Howard, L.0. 1930. A history of applied entomology. Smithsonian Miscellaneous Collections. Vol. 84.

Klassen, W. 1989. Eradication of introduced arthropod pests: theory and historical practice. Misc. Publ. Entomol. Soc. Amer.: 73: 1-29.

Klassen, W. 1991. Insects and agriculture, pp. 23-68. *In* Progress and Perspectives for the 21st Century. J.J. Menn and A.L. Steinhauer (eds.). Centennial National Symposium, Entomological Society of America, Lanham, MD, USA. 170 pp.

Klassen, W. 1995. World food security at 2000 and Beyond and the global pesticide situation, pp. 1-41. *In* Options 2000, invited papers at the IUPAC 8th International Congress of Pesticide Chemistry, 4-9 July 1994, Washington, DC, Am. Chem. Soc., Conf. Proc. Series, Washington, DC, 450 pp.

Knipling, E.F. 1966. Some basic principles in insect population suppression. Bull. Entomol. Soc. Am. 12(1): 7-15.

Knipling, E.F. 1972. Entomology and the management of man's environment. J. Austral. Entomol. Soc. 11: 153-167.

Knipling, E.F. 1976. Report of the Technical Guidance Committee for the pilot boll weevil eradication experiment, pp. 122-125. *In* Boll Weevil Suppression, Management, and Elimination Technology. Proceedings of a Conference, Feb. 13-15, 1974. Memphis, TN, ARS-71.USDA, Washington, D.C, USA.

Knipling, E.F. 1979. The basic principles of insect population suppression and management. USDA Agriculture Handbook. 512 pp.

Knipling, E.F. 1992. Principles of insect parasitism analysed from new perspectives. Practical implications for regulating insect populations by biological means. Agriculture handbook Number 693, ARS/USDA, Washington, DC, USA. 337 pp.

Lachance, L.E. 1979. Genetic strategies affecting the success and economy of the sterile insect release method, pp. 8–18. *In* Genetics in Relation to Insect Management. M.A. Hoy and J. J. McKelvey, Jr. (eds.). The Rockefeller Foundation, New York, USA.

MacKellar, W.M. 1942. Cattle tick fever, pp. 572-578. *In* Keeping Livestock Healthy. USDA Yearbook of Agriculture, Washington, DC, USA.

Metcalf, R.L. 1991. Insects – man's chief competitors, pp. 19-28. *In* Progress and Perspectives for the 21st Century. *J.* J. Menn and A.L. Steinhauer (eds.).

Centennial National Symposium, Entomological Society of America, Lanham, MD, USA. 170 pp.

Metcalf, R.L. and A. Kelman. 1980. Plant protection, pp. 313-344. *In* Science in Contemporary China. L. A. Orleans (ed.). Stanford University Press, CA, USA.

Myers, J.H., A. Savoie and E. van Randen. 1998. Eradication and pest management. Ann. Rev. Entomol. 43: 471 – 491.

National Academy of Sciences. 1969. Insect pest management and control. Principles of plant and animal pest control, Volume 3. National Academy of Sciences Publication 1695, Washington, DC, USA. 508 pp.

National Research Council. 1981. Cotton boll weevil: an evaluation of USDA programmes. Committee on Agriculture and Renewable Resources, Commission on Natural Resources, National Academy of Sciences Press, Washington, DC.

Newsom, L.D. 1978. Eradication of plant pests-con. Bull. Entomol. Soc. Am. 24(1): 35-40.

Nygaard, D. 1998. World population projections, 2020: 2020 IFPRI Brief 5, October1994. http//:www.cgiar.org/ifpri/2020/BRIEFS/number05.htm

Oka, N. 1991. Success and challenges of the Indonesia National Integrated Pest Management Programme in the rice-based cropping season. Crop Protection 10(3): 163 –165.

Orozco, D., W.R. Enkerlin and J. Reyes. 1994. The Moscamed program: practical achievements and contributions to science, pp. 209 –222. *In* Fruit Flies and the Sterile Insect Technique. C.O. Calkins, W. Klassen and P. Liedo (eds.). CRC Press, Boca Raton, FL, USA. 272 pp.

Perkins, J.H. 1982. Insects, experts and the insecticide crisis. The quest for new pest management strategies. Plenum. New York, USA.

Perkins, J.H. 1983. The boll weevil in North America: scientific conflicts over management of natural resources. Agric. Ecosyst. Environ. 10: 217-245.

Rabb, R.L. 1972. Principles and concepts of pest management, pp. 6-29. *In* Implementing Practical Pest Management Strategies. Purdue University Press, West Lafayette, IN, USA.

Reichelderfer, K.H., G.A. Carlson and G.A. Norton. 1984. Economic guidelines for crop pest control. FAO Plant Production and Protection Paper 58. FAO, Rome. 93 pp.

Sandman, P.M, 1987. Apathy versus hysteria: public perceptions of risk, pp. 219-231. *In* Public Perceptions of Biotechnology. L. R. Batra and W. Klassen, (eds). Agricultural Research Institute, Bethesda, MD, USA.

Scribner, J. 1983. A review of the California action program, pp.1-12. *In* Medfly and the aftermath. Symposium presented at the Entomological Society of America's Annual Meeting. December 1982. APHIS 81-60.

Sigerist, H.E. 1958. The great doctors, A biographical history of medicine. Doubleday & Co., Inc. Garden City, NY, USA. 422 pp.

Stage, H.H. 1952. Mosquitoes, pp. 476–486. *In* Insects, The Yearbook of Agriculture 1952. U.S. Government Printing Office, Washington, DC, USA. 780 pp. and LXII plates.

Starr, C. 1985. Risk management, assessment and acceptability. Risk Analysis 5(2): 97-102.

Summers, F.G. 1925. Charles Valentine Riley, Benefactor of Agriculture. The Missouri Historical Review. 19(4):611-621.

United Nations. 1993. World population prospects: the 1992 Revision. New York. 667 pp.

United Nations Department of International Economic and Social Affairs, 1991. World population prospects 1987. Fertility behaviour in the context of development. Population Studies No. 100. New York. 387 pp.

Uvarov, R.P. 1977. Grasshoppers and locusts, 2nd ed. Center for Overseas Research, London, UK.

Walloff, N. and G.B. Popov. 1990. Sir Boris Uvarov (1889-1970): The father of acridology. Ann. Rev. Entomol. 35: 1–24.

[WHO] World Health Organization. 1976. WHO-supported collaborative research projects in India: the facts. WHO Chronicle, 30: 131-139 (1976).

Wright, J.W., R.F. Fritz and J. Hanworth. 1972. Changing concepts of vector control in malaria eradication. Ann. Rev. Entomol. 17: 75-102.

Yamagishi, M., H. Kakinohana, H. Kuba, T. Kohama, Y. Nakamoto, Y. Sokei and K. Kinjo. 1993. Eradication of the melon fly from Okinawa by means of the sterile insect technique, pp. 49–60. *In* International Atomic Energy Agency, Management of Insect Pests: Nuclear and Related Molecular and Genetic Techniques, Proceedings of a Symposium, Vienna 19-23 October 1992 jointly organised by IAEA and FAO. Vienna, Austria. 669 pp.

# Economics of Area-Wide Pest Control

*J.D. Mumford*

Imperial College of Science, Technology and Medicine, Silwood Park, Ascot, Berkshire SL5 7PY, United Kingdom

## INTRODUCTION

Area-wide pest management is commonly practised throughout the world, probably much more so than is generally recognised (Lindquist 2000, Klassen 2000). Apart from highly publicised area-wide schemes such as the sterile insect technique (SIT) for fruit flies, pheromone disruption for cotton bollworms and classical biological control, there are many examples of actions such as concerted host plant eradication, enforced closed crop seasons, organised pesticide rotation for resistance management, coordination of resistant crop genotypes, etc., some going back several centuries, which should also be considered as area-wide practices. Each of these is faced with many of the economic issues generally associated with area-wide management which will be discussed below.

In general, there are to be four major questions to answer in devising an area-wide pest management programme:

- Should a particular pest be controlled locally or area-wide?
- What is an appropriate area over which management should be attempted?
- Within that area what form of control is most efficient?
- What level of organisation should be used to get the job done?

It should be noted that apart from clearly objective measures such as technical effectiveness (say, mortality) or cost efficiency (mortality per dollar), there are many subjective measures that come into the evaluation of area-wide control due to the element of risk (for example, in quarantine and eradication), the boundaries of externalities (for example, variable probabilities of pesticide drift under different conditions or target organism sensitivities) and time preferences for returns on capital investments (such as insect rearing facilities or research to develop pheromone technologies). As a result of these subjective components, it may sometimes be difficult to reach clearly agreed decisions based on objective economic analyses, even with a consensus on the data used.

## PRINCIPLES

There are three general classes of economic problems in comparing pest management options:

- annual comparisons (for example, pheromone disruption),



**Area-Wide Control of Fruit Flies and Other Insect Pests,**
ed. K.H. Tan. Penerbit Universiti Sains Malaysia, Penang, 2000.

- initial capital costs with delayed projected benefits (SIT eradication), and
- initial costs with delayed and uncertain future benefits (quarantine, pesticide resistance management).

In any of these cases, the two basic principles governing the selection of area-wide management over individual local pest management are effectiveness and efficiency.

Quarantine and eradication can only be achieved on an area-wide basis; they are unlikely to be effective otherwise. In both cases, total effectiveness is required and this must be the paramount indicator. Questions of efficiency subsequently arise in deciding which area-wide techniques to employ, not in comparing them to uncoordinated local control. However, economic efficiency should be considered at an early stage to decide whether total control, as epitomised by quarantine or eradication, is a viable option.

In suppression campaigns it is quite possible to choose a less effective control if it is much less expensive or more practical to use. Suppression may or may not be more effective by area-wide management, but the main issue is one of cost efficiency. Is area-wide suppression more cost efficient than the sum of local management? This is in fact quite a challenge because, in many areas of local pest management, private returns of 4:1 are commonly obtained, or at least perceived (Pimentel and Lehman 1993).

A basic conflict can arise between technical analyses of pest management and economic analyses because of their respective emphases on effectiveness and efficiency. Technical leaders of programmes naturally strive for the ultimate effect, such as eradication, while economists may point out that a less ambitious goal with lower technology and costs may give a greater return and pose less risk of financial loss. To this, the technologist can argue that technological development requires risk taking and a greater good could be found by extending the boundaries of space and time to include a wider potential. Unlike the technologist – who can look towards an objective perfect indicator of success – economists, like lawyers, must refer to a subjective norm, in effect the risk attitude and willingness to pay of a "reasonable" person, often as represented by their political agents in government. The latter, because they are merely agents, are often even more averse to risk than the sum of society itself (Stonehouse and Mumford 1994). Politicians in fact do not merely reflect public attitudes; they may amplify them. We must accept that there may be irreconcilable differences in the attitudes of the people running programmes and those paying for them.

## PROBLEMS

The fundamental problem of economic analysis for area-wide pest management must be in deciding (and then convincing others) that the results expected and the measures used are indeed "reasonable". Let us consider some of the component problems.

### Scale

For some projects, the scale of operations is predetermined by geographical or political boundaries, such as the eradication of *Bactrocera dorsalis* (Hendel) in Mauritius or *Ceratitis capitata* (Wied.) in Florida. Either the entire area infested or the entire area within one political jurisdiction must be included, and the costs and benefits are determined for that entire area. But where management is being considered for only part of an infested area (for example, fruit flies in South Africa, or tsetse flies in East Africa) the extent and characteristics of the area chosen can have very substantial effects on the benefit:cost calculations, due to the relative values of production, costs of controls, area of non-crop hosts, presence of related pest species, etc., in different parts of the total infested region.

For the eradication of fruit flies in South Africa, Mumford, Tween and Barnes (1997, 1998, unpublished) calculated the returns on a SIT eradication for each of 19 fruit growing districts in the Western Cape province in early 1998. In the most favourable district eradication (to be achieved in one year and lasting three years) is predicted to have a five year net present value gaining approximately US$1.5 million, while the least favourable district would have a value losing approximately US$2.0 million. By including some areas and excluding others, an optimal economic return can be obtained; fruit fly control throughout the Western Cape would not be economic over a five-year period, even under the most optimistic assumptions made in the technical plans (under current export market standards, more stringent residue restrictions in the EU would make eradication or SIT suppression more attractive). However, if only high return areas are managed, political questions arise: if public money is used, should uncontrolled areas be compensated, or should the management be privately funded by those who will benefit?

This question becomes even more complex in border areas, such as in the Middle East, where fruit flies in citrus may be controlled in Jordan, Israel and Palestine, within a few kilometres of each other (Enkerlin and Mumford 1997). The prices and market conditions for citrus are different and the level of control currently used also differ considerably because of economic circumstances in the orchards in each territory. The benefits in Israel, where fruit flies are effectively, if not efficiently, controlled by insecticides from area-wide fruit fly SIT are largely in reduced pesticide inputs with lower residues and less environmental damage and continued access to discerning high value markets. In Jordan and Palestine, however, the main advantages would be in reduced crop damage. Due to political boundaries, it may in many cases, be necessary to manage different aspects of the programme at different scales, for instance regional/international surveillance, but national or district level control, etc.

The economic advantages of being able to concentrate on high return areas are clear and this may pose additional technical challenges to area-wide management to produce sharp and effective boundaries that would allow, in effect, pest management gerrymandering.

## Time and Discount

A frequent claim for eradication programmes is that they are "forever" and that, therefore, future benefits should be calculated in perpetuity. As a result, even small annual benefits would justify large initial eradication costs. However, it is unreasonable to include infinite streams of benefits, not least because we know from experience that quarantine can fail and the process may need to be repeated. Furthermore, we must discount future benefits to balance them against present (or near future) expenditure. This is done using a discount rate that measures the value of having money now compared to some time in the future. The most common calculation of the discount rate is to subtract the national inflation rate from the bank interest rate for savings, which will generally produce a figure around 4% or 5% in developed countries. This represents the "reasonable" person's discount on the future; they put their money in the bank because they are paid this premium, otherwise they would spend it now. So, the benefit of eradication next year is worth 5% less if it is brought back to the present. Benefits in 20 years are only worth 37% of their face value when brought back to the present. In riskier economic environments discounts rates will be much greater. At 15% discount, a benefit 20 years in the future is only worth 6% of face value at present.

Two other options for calculating discount rates should also be mentioned. It could be argued that there are greater returns available than saving money in a bank, and if it was possible to gain 25% per year from some other form of investment such as stock markets or currency speculation, then the opportunity cost of foregoing this investment should also be added to the discount. At a discount rate of around 25% a benefit 20 years in the future is worth only 1% of its face value in the present. At the other extreme, it is sometimes argued that social benefit should be calculated using a social discount rate. An individual values money today more than in the future because he may die, or otherwise lose the opportunity to spend it, before the future arrives. Society, on the other hand, is relatively timeless and could place equal values on future and present good, effectively producing a discount rate of 0%.

So the total present value derived from an economic analysis must reflect a reasonable time period and discount rate. Conservative analysis would use shorter time periods and higher discount rates and may present a range of options to indicate sensitivity. Time is more arbitrary than the discount rate, since the latter can be reasonably objectively calculated. Time horizons in the range of 10-20 years are commonly used (Kehlenbeck 1996, Cox and Forrester 1992, Mumford et al. 1995, 1996), unless a project is expected to wind up sooner (Table 1).

Time horizons are mainly of concern in cases in which projects have substantial initial capital expenditure preceding a long benefit stream. Mumford et al. (1996) indicated annual input savings from pheromone control of pink bollworm in Egypt in the range of US$5 to US$9 million for 1995 (depending on the inclusion of environmental costs), but also projected that this annual benefit was likely to decline steadily over 15 years due to decentralised cotton management and a reduced cotton area (Table 2). It cannot, and should not, be assumed that input savings pertain to a static production or protection system.

Table 1. Summary of benefit indices for SIT eradication of codling moth in the Suweida and Damascus regions of Syria, without including environmental benefits (Mumford and Knight 1996).

| **Economic Benefit-Cost Indicators** | | | |
|---|---|---|---|
| Pay-back period in years to break-even (discounted net) | 12 | | |
| | **10 years** | **12 years** | **15 years** |
| Total cost US$,000 (no discount) | 11011 | 11739 | 12913 |
| Total direct benefit US$,000 (no discount) | 13300 | 16940 | 22400 |
| Net present value US$,000 | -623 | 351 | 1484 |
| Net benefit/cost (no discount) | 0.208 | 0.443 | 0.735 |
| Internal rate of return (no discount) | 0.069 | 0.114 | 0.146 |
| N/K (return on equity, discounted) | 0.845 | 1.087 | 1.369 |
| Discount rate | 0.10 | Interest on capital | 0.04 |

Table 2. Projection of input and environmental savings (E£'000) from reducing 2.5 sprays per year for pink bollworm on Egyptian cotton (Mumford et al. 1996); pheromone use - phased in from 1992.

| Year | Value of 2.5 sprays saved | Environmental value of 2.5 sprays saved | Annual benefits from adoption of pheromones from 1992 |
|---|---|---|---|
| | *(phased in from 1992-5 and out from 2000)* | | |
| 1992 | 431 | 2,831 | 2,831 |
| 1993 | 2,680 | 7,482 | 7,482 |
| 1994 | 9,496 | 18,823 | 18,823 |
| 1995 | 32,050 | 42,293 | 42,293 |
| 1996 | 28,325 | 37,644 | 37,644 |
| 1997 | 28,325 | 37,644 | 37,644 |
| 1998 | 28,325 | 37,644 | 37,644 |
| 1999 | 28,325 | 37,644 | 37,644 |
| 2000 | 25,492 | 33,879 | 33,879 |
| 2001 | 22,660 | 30,115 | 30,115 |
| 2002 | 19,827 | 26,351 | 26,351 |
| 2003 | 16,995 | 22,586 | 22,586 |
| 2004 | 14,162 | 18,822 | 18,822 |
| 2005 | 11,330 | 15,057 | 15,057 |
| 2006 | 8,497 | 11,293 | 11,293 |
| 2007 | 5,665 | 7,529 | 7,529 |
| 2008 | 2,832 | 3,764 | 3,764 |
| 2009 | 0 | 0 | 0 |

## Valuation

Two forms of net benefit calculation can be considered, input comparisons and economic surpluses, as used by Cox and Forrester (1992) to analyse the benefits of insecticide resistance management. Input comparisons are attractive because they are easily valued, since pesticides and other inputs have known prices. Such an analysis is particularly appropriate for annual comparisons of net benefits in cases where both forms of control give very high levels of control, for instance cotton bollworm control (Knipling and Stadelbacher 1983, Mumford et al. 1996). An environmental estimate may be added to the input comparison (Mumford et al. 1996).

Economic surpluses are intended to show the overall impact on the economy due to a practice. This takes into account not only the changes in pest management inputs, but also the possible stimulus to investment in other inputs such as fertiliser, seed and land that improved pest management might encourage. Very great improvements in pest management may even affect crop prices by changing production. Many of these changes are very difficult to predict, although there are examples of studies that give indications of farmer responses to improved pest management in some areas (Napit et al. 1988, Kazmierczak 1996).

Not all benefits from pest management can be calculated directly from practically observable data. Reduced pest damage and savings in present control costs are the most immediate benefits that can be estimated. Other benefits must be converted into monetary units, often by subjective means. Two examples are presented, environmental and health benefits, and aesthetic benefits.

Environmental and health benefits arising from improved (area-wide) pest management can be estimated based on reduced pesticide load. Pimentel et al. (1993) estimated that the environmental cost of pesticide use was approximately US$19.00/kg/year in the USA, over and above purchase price. This included impact on humans and other organisms based on a combination of likely exposure, value of life, cost of treatment and loss of work in the case of humans and an arbitrary value for individual animals poisoned. Mumford et al. (1996) adapted the technique to determine environmental savings from area-wide cotton pest management in Egypt using the ratio of per capita income in Egypt and USA (less than 3%) to adjust this figure for Egypt, which resulted in an estimate of an environmental cost of US$0.52/kg/year. This was felt to be a considerable underestimate, given factors such as the greater risk of exposure to insecticide in Egypt (through direct and indirect contact), the greater proximity of households to agricultural areas and the importance of the agricultural sector in Egypt. Subjective weightings were put on these factors and a modified value of US$5.00/kg/year was assumed (just over 25% of the US value). While still subjective, this figure gives at least some measure of the importance of environmental savings from reduced pesticide use on Egyptian cotton.

In addition to input and environmental savings, aesthetic benefits may need to be estimated in some area-wide pest management. Miller and Lindsay (1993) estimated the benefits of area-wide gypsy moth control in New Hampshire, USA, where the principal benefits are better visual impact of undefoliated trees for residents and possibly greater tourism revenue from visitors. The benefits were estimated using the contingent valuation method to determine the public's willingness to pay for control. Mean values were US$13-US$28 million/year (~US$38/ha/yr) and median values were US$8-US$17 million/year (~US$25/ ha/yr) (the range depending on the interpretation of non-respondent views in the surveys).

The adoption of area-wide pest management eventually becomes a political process. Kazmierczak (1996) describes the process for adopting boll weevil eradication in southeastern USA, where decisions are generally based on public referenda. These generally require at least 67% of growers in an area to agree, then all growers are required to participate, and pay the assessed charges. Because of the use of referenda and at least partial payment and management by growers, these eradication efforts have a high level of grower "ownership". Votes are influenced by the likely charge for the area-wide programme. He quotes potential reductions in "yes" votes of 5% for each US$2.50/ha assessed, so the estimation and perception of value from area-wide control can be critical to their establishment. In the case of the gypsy moth valuation noted above, the mean value may give a good estimation of the actual worth of area-wide management, while the median gives a better picture of the perceived worth to the typical voter, the "reasonable" person economists and politicians aim at.

A particularly important factor affecting adoption in Kazmierczak's review was knowledge of the success of previous area-wide pest management programmes.

## Risk

Quarantine is an aspect of pest management that involves analyses of probability and risk even more than most others (Bartlett 1996). Two models of economic analyses for quarantine can be considered. Kehlenbeck (1996) has suggested that quarantine is viewed as merely delaying the introduction, spread and damage of a pest and that the value of quarantine can be estimated by comparing the present value of the stream of losses expected with and without quarantine, up to the point, where in both cases, the damage has reached its full potential. This is a rather pessimistic view of quarantine and does not really consider risk at all, since it relies on a discreet change in the expected rate of spread, rather than a probability change.

An alternative view of quarantine is that its effect is to impose a continual shift in the probability of introduction, establishment and damage, which may be more or less depending on how rigorously it is conducted, and how vigorously it is challenged by pest invasion or transport. The pest challenge and quarantine effort probabilities may be independent (but not necessarily so, since some of the pest challenge is related to human transport, which may be deterred by quarantine). These are clearly quite difficult to assess in practice, given a lack of actuarial data, but in theory, the annual value of quarantine could be estimated by:

Annual quarantine value = annual expected cost (with Q) – annual expected cost (no Q)

in which each of the annual expected costs is the probability weighted sum of the combined costs of quarantine operations (if any) and losses from pests that evade quarantine. A very simple example of this could involve a "no Q" case in which there are two states, loss of US$0 (p=0.75) and loss of US$10 million (p=0.25). A "with Q" case could have two different annual states (due to quarantine costs and efforts) with costs of US$1 million (p=0.9) and loss+cost of US$8 million (p=0.1). In this case, both the quarantine effort and the pest challenge have changed, and the overall annual value comes to:

Annual net value = [ (-1*0.9)+(-8*0.1)] – [(0*0.75)+(-10*0.25)] = US$0.8 million

In reality, continuous probability distributions would be estimated, either directly from data on recorded introductions and damage and/or by panels of expert opinion.

## Capturing Returns

It is increasingly important in public pest management that some of the benefits are used to repay the costs of the action, as has always been the case with private pest control. Authorities inevitably wish to recoup some of the costs through charges on participants but some forms of area-wide control are more amenable to this than others. Closed seasons and insecticide resistance management are examples of strategies that are difficult to charge for directly on a per use basis, although much of the cost is borne by farmers forgoing options to plant out of season or use particular pesticides. In some cases, charging for some forms of area-wide control, such as quarantine inspection or phytosanitary certificates, is potentially a disincentive to participation and could undermine the effectiveness of the programmes.

Pheromones and SIT, however, have good potential for capturing returns through commercial sales or privatised management. This is already common in the case of pheromones, and is likely to be increasingly common for SIT where it is used for suppression, rather than eradication. In general, in eradication programmes, the period in which sterile insects are used is too short to offer a reasonable opportunity to recover the cost of private investment in a rearing facility. However, in suppression programmes, a reasonable profit over a much longer period can repay private investment in facilities and avoid the need for publicly initiated or coordinated control programmes to add to public debt (Enkerlin and Mumford 1997). The relatively easy and efficient transport of some sterile insects by air may allow commercial competition on price and quality and opportunities for specialisation and economies of scale in production facilities.

## CONCLUSIONS

There is little doubt that many forms of area-wide pest control offer effective and efficient pest management, even to the extent of pest eradication. The larger scale of operations in many cases offers opportunities for a wider range of techniques, generally with less harmful environmental impact, to be used. The professional coordination required to manage pests on an area-wide basis involves technicians who can ensure that practices retain their specified design and standard. At least in theory, area-wide control offers the opportunity to integrate economic and environmental externalities within the programme so that decisions more appropriately reflect the true costs and benefits.

The main area in which problems lie with area-wide pest control appears to be in the mechanisms for public participation. Reports over many years cite technical, economic and environmental success with the concept but there is still indifference, reluctance and antagonism. There are several reasons for this related to attitudes and perceptions (Norton and Mumford 1993). There is a natural tendency in many people to want to manage their own affairs, either out of sheer independence or a belief that they can do it better than their neighbours, despite a lack of empirical evidence. Ignorance or disbelief in the effectiveness or efficiency of area-wide programmes is a factor which can to some extent be overcome by good publicity. And finally, these attitudes may be the result of a lack of opportunity for involvement and ownership of programmes, which may be seen as being imposed from above/outside, managed by technocrats or otherwise not arising from or meeting the needs of the people directly concerned. None of these issues should be insurmountable, but it is worth noting that area-wide pest management is an activity in which social participation and attention to the "reasonable person" is as important as technical proficiency.

## REFERENCES

Bartlett, P.W. 1996. Tephritid fruit flies – A quandary for international plant quarantine. BCPC Brighton Crop Protection Conference – Pests and Diseases – 1996, pp. 1153-1160.

Cox, P.G. and N.W. Forrester. 1992. Economics of insecticide resistance management in *Heliothis armigera* (Lepidoptera: Noctuidae) in Australia. J. Econ. Entomol. 85: 1539-1550.

Enkerlin, W. and J.D. Mumford. 1997. Economic evaluation of three alternative control methods of the Mediterranean fruit fly (Diptera: Tephritidae) in Israel, Palestine and Jordan. J. Econ. Entomol. 90: 1066-1072.

Kazmierczak, R.F. 1996. Differences between landowner and producer support for the Louisiana boll weevil eradication program. J. Agric. Entomol. 13: 375-391.

Kehlenbeck, H. 1996. Model cases for evaluating the impact of pest introduction and phytosanitary measures. EPPO Bulletin, 26: 537-543.

Klassen, W. 2000. Area-wide approaches to insect pest interventions: history and lessons. pp. 21-38. *In* Area-Wide Control of Fruit Flies and Other Insect Pests. K.H. Tan (ed.). Penerbit Universiti Sains Malaysia, Penang, Malaysia.

Knipling, E.F. and E.A. Stadelbacher. 1983. The rationale for areawide management of *Heliothis* (Lepidoptera: Noctuidae) populations. Bull. Ent. Soc. Am. 29: 29-37.

Lindquist, D.A. 2000. Pest management strategies: area-wide and conventional. pp. 13-19. *In* Area-Wide Control of Fruit Flies and Other Insect Pests. K.H. Tan (ed.). Penerbit Universiti Sains Malaysia, Penang, Malaysia.

Miller, J.D. and B.E. Lindsay. 1993. Willingness to pay for a state gypsy moth control program in New Hampshire: a contingent valuation case study. J. Econ. Entomol 86: 828-837.

Mumford, J.D., A. Driouchi, Enkerlin, W., Carlson, G.A. and E.J. Buycx. 1995. Economic evaluation of damage caused by, and methods of control, the Mediterranean fruit fly in the Maghreb. International Atomic Energy Agency, Vienna, Austria. 65 pp. IAEA-TECDOC-830 (ISSN 1011-4289).

Mumford, J.D., S.J. Henderson, K. Harrap, and M.H. Birley. 1996. Pheromones. Evaluation Report EV578, Evaluation Department, Overseas Development Administration, London, 132 pp.

Mumford, J.D. and J.D. Knight. 1996. Economic analysis of alternatives of codling moth control. SYR/5/011/03, International Atomic Energy Agency, Vienna, 34 pp.

Napit, K.B., G.W. Norton, R.F. Kazmierczak, and E.G. Rajotte. 1988. Economic impact of extension integrated pest management programs in several states. J. Econ. Entomol. 81: 251-256.

Norton, G.A. and J.D. Mumford (eds.). 1993. Decision Tools for Pest Management, CAB International, Wallingford, United Kingdom.

Pimentel, D., H. Acquay, M. Biltonen, P. Rice, M. Siva, J. Nelson, V. Lipner, S. Giordano, A. Horowitz, and M. D'Amore. 1993. Assessment of environmental and economic impact of pesticide use. pp. 47-84. *In* The Pesticide Question – Environment, Economics and Ethics. D. Pimentel and H. Lehman (eds.). Chapman and Hall, London.

Pimentel, D. and H. Lehman (eds.). 1993. The pesticide question – environment, economics and ethics. Chapman and Hall, London.

Stonehouse, J.M. and J.D. Mumford. 1994. Science, risk analysis and environmental policy decisions. Environment and Trade, UNEP, Geneva. 79 pp.

# Trade Issues and Area-Wide Pest Management

*Robert L. Griffin*

Coordinator, Secretariat of the International Plant Protection Convention, FAO-AGP, Room B-761, Viale delle Treme di Caracalla, Rome, Italy

Food security and economic security are unarguably desirable objectives for all nations – indeed for the world. Equally important is the sustainability of designs that achieve these objectives without disadvantaging others or damaging the environment. Considering area-wide pest management in the context of these interrelated global policy forces is essential to fully understand its role in both the protection of plant resources and in facilitation of trade.

The case for food security begins with the realisation that there are currently about 800 million people in the world who are suffering from malnutrition due to lack of food. The World Food Summit, convened in November 1996, urgently called for coordinated world-wide action to ensure "food for all". A key strategy for realising this goal is reducing losses due to plant pests.

In this light, area-wide pest management can be viewed as a valuable addition to the toolbox of pest management strategies. It can also be one of the most sustainable and cost-effective options to consider for pest management. However, just as the problem of world hunger is not solved by a single farmer, area-wide pest management cannot be successful at the individual level. It requires commitment and cooperation to make it feasible – the same type of commitment and cooperation that was expressed at the World Food Summit.

Where economic security is concerned, one need not look far to see a world of growing economic integration and widening circles of development. As the World Trade Organisation celebrates the 50$^{th}$ anniversary of the rules-based trading system which began with the General Agreement on Tariffs and Trade (GATT) after World War II, it is clear that globalisation and the liberalisation of trade have become permanent fixtures in international policy formulation and are integral to the economic security of all nations. Now, more than ever before, the world's prosperity rests on maintaining an open international economy based on commonly agreed rules.

The significance of agriculture in this international economy is quickly evident, as we see enormous quantities of fresh and processed agricultural commodities racing across borders and seas to markets in the far corners of the globe where the availability of such products had been inconceivable only decades before. For every country in the world, the import and export of agricultural products are essential to the health of the economy as well as the population. Non-industrialised countries, in particular, rely upon agriculture as a cornerstone for commerce. Industrial countries rely upon trade in agricultural products to provide the quantity, quality and variety of goods demanded by modern consumers.

However, more trade, faster trade, and the opening of new markets for agricultural products also offer greater opportunities for the movement of pests that can have deleterious consequences. This raises significant concerns in policy formulation, especially as measures for protection can affect the free movement of commodities in trade. On the one hand, countries need to be able to import, to meet their needs and market demands, and they have the reciprocal need to have their exports accepted by others. On the other hand, countries must exercise a certain amount of care to ensure that they do not unduly jeopardise their own resources by introducing harmful new pests. This must be considered against the corresponding need to ensure that they do not ship harmful pests to other countries.

As a result, there emerges a strong need for a balanced, dynamic, multi-disciplinary approach to policies concerned with pest management – for both domestic and foreign pest management. These policies are increasingly based on international cooperation, sophisticated technologies, and the marriage of economic and biological analyses. We are currently experiencing this transition in practice based on a more holistic approach, as evidenced in part by the expanding interest in applications for area-wide pest management. The evidence indicates that this shift will bring significant benefits in increased production and trade, as well as offering more sustainable and environmentally acceptable pest management options.

There is also the concern that as tariffs and other barriers are removed, countries may impose measures under the guise of protection in order to secure market or other unfair advantages. It is the nature of quarantine to follow the old adage, "an ounce of prevention is better than a pound of cure", i.e., it is comparatively cheaper and easier to prevent the entry of pests with strongly restrictive measures than it is to deal with the result of pest introduction. But while a certain degree of care is clearly justified, unreasonably conservative policies are seen to unnecessarily restrict trade. The question revolves around the issue of justification.

As globalisation and the liberalisation of trade have matured, and international trade in agricultural products has grown in importance, it has become necessary for "free trade" and "fair trade" to evolve still further to embrace the concept of "safe trade". That is to say, disciplines are necessary to ensure that protective measures are used to the extent justified by legitimate concerns, but not as unjustified barriers to trade.

This brings the discussion to the early part of the 1990s and the last round of multilateral trade negotiations under the GATT – the Uruguay Round, and the agreements therein related to agriculture. Emerging from these negotiations was the Agreement on the Application of Sanitary and Phytosanitary Measures (SPS Agreement) which dealt specifically with the issue of measures to protect plant, animal and human health and life. The SPS Agreement provides discipline to the use of protective measures in order to prevent such measures from being used as unjustified trade barriers.

The SPS Agreement is structured around several key principles, beginning with the sovereign right of a country to put protective measures in place, but balancing this with the obligation to ensure that such measures are justified. The SPS Agreement further describes such justification as being based on scientific principles and evidence, which are considered in the framework of a systematic evaluation process known as risk assessment. A number of other very important principles and definitions are found in the SPS Agreement which are not critical to the discussions here except to say that the sum of these represents a blueprint for establishing proper measures as well as for evaluating the measures of others.

A key point to note is that the SPS Agreement gives countries the option to base their measures on risk assessment or on international standards. It then goes further by identifying the sources for such standards in the three disciplines: plant health (phytosanitary), and animal and human health (sanitary). The Codex Alimentarius is responsible for food safety, the Office International des Epizooties (OIE) addresses animal health, and the International Plant Protection Convention (IPPC) is the organisation named in the SPS Agreement as the source of international standards for phytosanitary measures, including measures for pest management where these affect trade.

Prior to the SPS Agreement, the IPPC had led a relatively quiet existence as an international treaty deposited with the Director-General of FAO. Since 1951 when it was adopted, the Convention's most distinguishing feature had been the Phytosanitary Certificate, a harmonised format to be used by countries to certify the phytosanitary status of shipments for export. It is on this certificate that pest management measures are noted, especially as they pertain to requirements which may be specified by the importing country.

The SPS Agreement suddenly placed new expectations on the IPPC, resulting in a series of consultations involving members, regional plant protection organisations and FAO with the objective of making necessary adjustments in the Convention and its activities. Included in these initiatives was the formation of a Secretariat, the launching of an ambitious programme of standard setting, and the negotiation of the New Revised Text of the IPPC to better reflect contemporary practices and the new role of the Convention in standard setting.

As the IPPC advanced into standard setting, members quickly defined priority subjects for standards. This soon led to the adoption of several important documents, including: *Principles of plant quarantine as related to international trade*; *Guidelines for pest risk analysis*; *Glossary of Phytosanitary Terms*; and the *Code of Conduct for the Import and Release of Biological Control Agents*.

Each of the above listed documents contains important elements of pest management, providing valuable guidance for the development of appropriate harmonised systems and also for evaluating the systems of others. More recent standards, such as *Guidelines for surveillance* (recently approved) and *Requirements for the establishments of pest free places of production* (near completion), provide additional information and another level of detail related specifically to pest management or components of pest management strategies.

As additional standards are added, and still greater detail is agreed upon, the standards will become increasingly more valuable. However, this framework of standards already offers significant utility to national plant protection organisations, particularly where pest management systems have an important role in trade. By using standards to the extent possible for designing and implementing pest management systems, countries reduce the level of analytical resources needed to design systems which can be expected to withstand the scrutiny of trading partners and also meet the obligations of governments under the IPPC and the SPS Agreement. The standards serve not only as models for developing measures, but also as reference points for evaluating or challenging the measures of others. They offer both conceptual and technical guidance.

Some highlights from the Convention and current standards that are relevant to area-wide pest management include:

- obligations for cooperation, information sharing, surveillance, pest listing, pest risk analysis,
- definitions for critical terms such as pest, area, and establishment,
- setting out the principles of area freedom and areas of low pest prevalence – recognising that pest presence is associated with areas rather than political boundaries, and that pest absence or low prevalence may be used as the basis for phytosanitary certification,

- describing the essential components of pest free areas – systems to establish freedom, measures to maintain freedom, and checks to verify that freedom has been maintained,
- establishing the principle of equivalence – accepting measures that are not identical but have the same result,
- elaboration of the principle of managed risk – accepting that zero-pest risk is not feasible and that the strength of measures used to manage risk should be based on the level of risk and the appropriate level of protection,
- defining and describing risk analysis – basing measures on a systematic assessment of the risk using scientific principles and evidence, and both biological and economic data,
- identifying key elements of surveillance – distinguishing between general and specific pest surveillance and outlining components of each, and
- identifying situations that enhance success – natural barriers, the isolation of growing areas, and targeting commodities that are marginal or poor hosts.

Although there is as yet no specific standard for area-wide pest management, the fundamentals for the development of such guidelines are in place. In the meantime, the existing principles and standards present no obstacles to the acceptance of area-wide pest management as a viable phytosanitary measure, and in fact reinforce its acceptance and application. By continuing to elaborate additional guidance in standards, the complexity of the concept can be reduced. Through the increased recognition and reliance on standards in policy making, its acceptance in practice can be further expanded.

There are also other factors which are not directly linked to international phytosanitary standards, but which argue for further development and more applications of area-wide pest management. A particular concern is the growing number of problems associated with traditional pesticide treatments. Field and postharvest treatments with pesticides are rapidly falling from favour, and those that can be used, often adversely affect product quality. Through the skilful management of pests at an area level, pesticides can be eliminated or dramatically reduced, thereby mitigating the inherent problems associated with safety, health, pest resistance and environmental damage. In situations where treatments are unavailable or unsatisfactory, this can be a very strong motivation for area-wide pest management.

In addition, it should be noted that the use of area-wide pest management as a phytosanitary measure toward the objective of "safe trade" is only a limited application of the concept. There are additional opportunities to address non-commercial concerns as well. For example, there are very useful applications for protecting human health, but this does not mean that phytosanitary standards cannot also be used to provide guidance. There are elements of good surveillance practice which are largely generic, and the feasibility of an area-wide pest management programme will always involve roughly similar analytical inputs such as cost/benefit analyses to support decision making. Thus, the emphasis in phytosanitary standards on technical soundness and analytical methods, allows the material to have substantial carry-over value that can often prove useful in a variety of situations where area-wide pest management may be an option.

In summary, it may be noted that the contemporary maxim, "think global and act local", has special significance for area-wide pest management because it represents exactly the type of thinking that brings the concept into focus. Although not a panacea, it is clearly a powerful option when used in the proper circumstances and with the full commitment and cooperation of implementing parties. As governments and growers become more aware of the economic and social benefits accruing from eradication or sustainable pest management over an extended period and area, the concept of area-wide pest management will become increasingly apparent and more attractive. Both the IPPC and the SPS Agreement are structured to accept and encourage area-wide pest management as a tool for promoting safe trade and contributing as much as possible to the complementary goals of food security and economic security for all countries.

## REFERENCES

FAO. 1992. International Plant Protection Convention.
FAO. 1995. ISPM: Principles of plant quarantine as related to international trade; Publication No. 1.
FAO. 1996. ISPM: Guidelines for Pest Risk Analysis; Publication No. 2.
FAO. 1996. ISPM: Requirements for the establishment of pest free areas; Publication No. 4.
FAO. 1997. New Revised Text of the International Plant Protection Convention.
FAO. 1997. Glossary of Phytosanitary Terms; Publication No. 5.
FAO. 1997. Guidelines for surveillance; Publication No. 6.
WTO. 1994. Agreement on the Application of Sanitary and Phytosanitary Measures.

# PART II.

## APPLICATION OF AREA-WIDE CONCEPT TO NON-FRUIT FLIES

### A) MEDICAL AND VETERINARY PESTS

# Success in Zanzibar: Eradication of Tsetse

*A.R. Msangi*[1]*, K.M. Saleh*[2]*, N. Kiwia*[1]*, I.I. Malele*[1]*, W.A. Mussa*[2]*, F. Mramba*[1]*, K.G. Juma*[2]*, V.A. Dyck*[3]*, M.J.B. Vreysen*[3]*, A.G. Parker*[3]*, U. Feldmann*[3]*, Z-R Zhu*[3] *and H. Pan*[3]

[1] Tsetse and Trypanosomiasis Research Institute, P.O. Box 1026, Tanga, Tanzania
[2] Commission of Agriculture and Livestock, P.O. Box 159, Zanzibar, Tanzania
[3] IAEA, P.O. Box 100, A-1400, Vienna, Austria

## INTRODUCTION

There are about 22 species of tsetse flies found nowhere else in the world except in 36 countries of sub-Saharan Africa. Tsetse flies transmit a debilitating and often fatal disease, trypanosomiasis, which causes tremendous losses of livestock, and severely limits agricultural production (it reduces output of milk and meat, causes mortality, infertility and abortion in livestock, deprives the rural population of draught power and manure to improve and increase crop production). Tsetse flies also transmit human trypanosomiasis, commonly known as "sleeping sickness". It is estimated that over 55 million people living in rural sub-Saharan Africa are at risk from this fatal disease.

Tanzania's Zanzibar Island is situated 35 km off the eastern coast and comprises two main islands, Unguja and Pemba. Previous surveys revealed that out of the seven tsetse species found on mainland Tanzania, only *Glossina austeni* Newstead infested Unguja Island. No tsetse fly was found on the island of Pemba. The fly is responsible for the cyclical transmission of trypanosomiasis in livestock, the causative agents being mainly *Trypanosoma congolense* and, to a lesser extent, *T. vivax*. It is estimated that in Zanzibar, the disease causes annual losses of US$2 million. Since fly suppression by conventional techniques has often resulted in short-term success, Tanzania has always appreciated that the long-term solution to the trypanosomiasis problem is the eradication of tsetse flies in the country.

In 1994, the International Atomic Energy Agency (IAEA) and the United Republic of Tanzania embarked on a project with the objective of eradicating tsetse flies from Zanzibar Island by applying the sterile insect technique (SIT) (Dyck et al. 1995, in press). Previous tsetse eradication efforts in Tanzania using SIT, enabled the establishment of a modest capacity on tsetse mass rearing in Tanga (Williamson et al. 1983). The Zanzibar tsetse project was successfully completed in 1997. The estimated cost was US$7,941,000 but only US$5,788,097 was actually spent. Financial assistance came from international donors such as the governments of Belgium, Canada, China, Sweden, the United Kingdom and the USA. The results and achievements of the Zanzibar tsetse fly eradication project are presented in this paper.

## MATERIALS AND METHODS

### Fly Production

The Tsetse and Trypanosomiasis Research Institute (TTRI), Tanga, Tanzania, provided the sterile male flies for release. Tsetse mass rearing procedures used were those previously described (Nash et al. 1968, Tarimo et al. 1984 and Msangi et al. 1995, in press). Flies were held in three insectaries (insectary I holding 21 trolleys, and insectaries II and III holding 30 trolleys each). About 40 male and 200 female flies were put into a round PVC production cage (20 cm diameter). The cages with flies were put onto trolleys each capable of holding 98 cages. This gave the facility a holding capacity of 1,587,600 female flies. The fly holding room was maintained at 22-24°C and 70-80% relative humidity. Deposited pupae were collected daily, machine sorted and incubated in open trays. Emerging flies were automatically marked in the ptilinum by a Day Glo fluorescent dye for identification. Flies were fed daily on defibrinated bovine blood through a silicon membrane. In the handling room, flies were immobilised at 4°C for sex separation. Female and some male flies (20%) were retained to form colony flies.

Male flies meant for release (80%) were fed twice before release, with trypanocide (samorin) added to their diet to prevent risk of disease transmission (Moloo and Kamunya 1987). The flies were chilled and placed in 0.5 litre thermos flasks, sterilised with 120 Gy, and put in small biodegradable release boxes at a rate of either 50 or 100 per box.

### Improvements in Rearing Procedure Beneficial to Fly Production

#### *No sex separation after mating*

Female and male flies were no longer separated after mating. Instead, they were caged together immediately after emergence (day zero mating). This eliminated chilling, and minimised handling which produced better quality insects. In the method used previously, flies were chilled at 4°C twice for sex separation (i.e., before and after mating).

#### *Mating ratio of 1 male : 5 females*

Upon emergence, flies were transferred to production cages at a ratio of 1 male: 5 females (only 20% of all male flies produced were retained). This increased the number of male flies for release, saved space and diet. In the old method, flies were mated at a ratio of 1 male: 2 females (50% of all male flies being retained).

#### *Bulk irradiation of males*

About 3,000 - 7,000 chilled flies were put into chilled thermos flasks and irradiated at 120 Gy in a gamma irradiator- GC 220 of Co-60. This procedure saved a lot of time and labour. In the old method, 120-160 chilled flies were put into 4 small round cages and irradiated at 120 Gy in the old gamma irradiator - IBL 337 of Cs-137.

### *Reduced membrane sterilisation time*

Membranes were heat sterilised in ovens at 120°C for six hours instead of overnight (more than 15 hours) sterilisation as done in the past. This procedure saved electricity consumption by 50%.

### *Other important improvements*

These included training of staff in the most recent methods of rearing tsetse flies and refurbishment of the three insectaries. Equipment such as a large stand-by electricity generator for continuous power, pumps which drew water from a bored hole into a newly constructed water tank with a capacity of 15,000 litres for continuous water supply, two large walk-in freezer rooms for blood storage each with capacity of 10,000 litres, a clean air station (UV Hood) for checking blood contamination, computers for data processing, communication equipment, heating mats with thermistor thermostats to provide a stable temperature, timers to synchronise the operation of heating mats and ovens, electrical protection equipment, digital thermometers for monitoring chiller temperatures, were purchased.

For every shipment, a sample of the flies for release was taken in Tanga for quality assessment prior to plane take-off. The parameters assessed were the flying response, fly mortality, sexing error, induced sterility, fly marking, feeding status, etc.

## Fly Release

The Commission of Agriculture and Livestock in Zanzibar, Tanzania, was responsible for the field operations involving fly release and monitoring. Fly suppression was done through an FAO project from 1988 to 1993, using topical application insecticides in areas with higher cattle density and insecticide impregnated screens in areas where cattle were scarce, such as in the large forested areas in the southern part of the island (Horeth-Bonntgen 1992). Releases of the sterile male flies started in 1991 with the release of adult flies on the ground in the Jozani forest as a pilot study (Vreysen et al. 1992) and terminated in July 1994. The aerial releases started in August 1994 in the southern half of the island and expanded to the whole island in mid-1996 when enough flies were produced. Flies were collected at the Tanga facility twice a week, using a light aircraft, and dispersed along the specific flight paths 1-2 km apart (Figure 1) at an altitude of 700-900 feet at a speed of 100-130 miles per hour.

## Fly Monitoring

*G. austeni* population monitoring was performed with sticky blue-white leg panels (Vreysen et al. 1996). There were about 55 fixed monitoring sites (Figure 2) with over 500 sticky panels at any one time. Each site had at least 5 sticky leg panels to trap flies. The traps were inspected daily and replaced every week. All trapped flies were checked for the presence of fluorescent dye in their head capsules using UV microscopy to distinguish released, from indigenous flies. Wild female flies were dissected for assessment of their reproductive status.

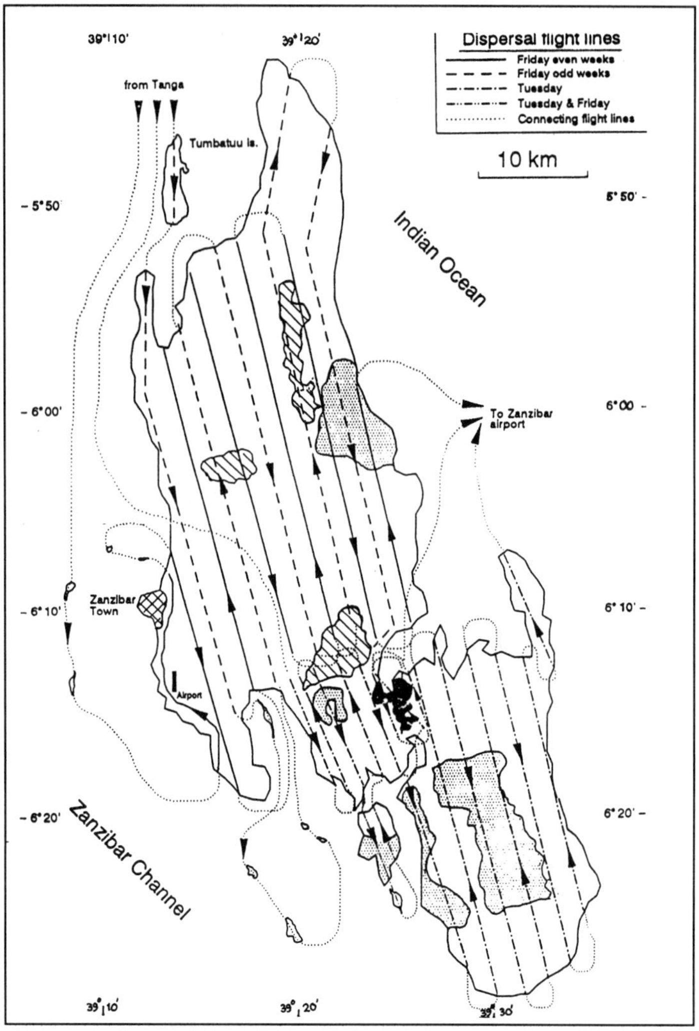


Figure 1. Flight lines for the dispersal of sterile males.

## Disease Monitoring

Disease monitoring was performed using parasitological techniques (PCV, MHCT and BCT). The entire Unguja Island was divided into 38 blocks (Figure 3) with over 1,000 sentinel cattle (about 2% of cattle population). Each block had a sentinel herd of 30-40 cattle to monitor the disease and were treated with Berenil at each collection (Pan et al. 1995, in press). Blood samples were taken every 2-5 months and examined for presence of the parasite. Animals found positive received an immediate treatment of Berenil.

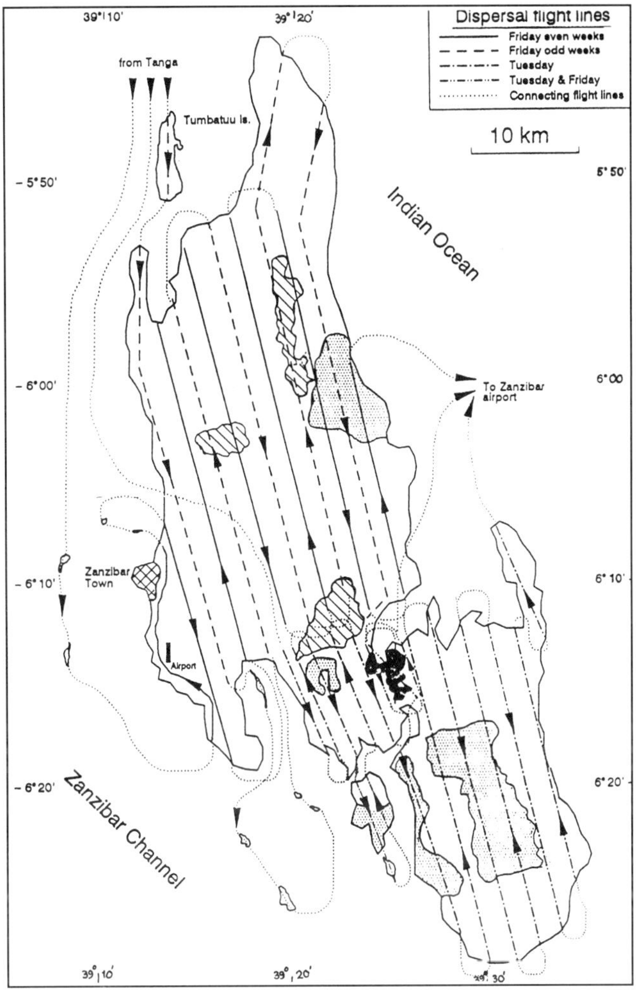


Figure 2. Fifty-five fixed monitoring sites each with at least 5 panels.

## RESULTS

### Fly Production

In 1994, the *G. austeni* colony at TTRI remained below 50,000 female flies. It then rapidly increased to 400,000 at the end of 1995. By the end of 1996, the colony had reached close to 1 million female flies (Figure 4). The FAO/IAEA laboratory at Seibersdorf, Austria, shipped a total of 3,861,242 pupae with the objective of boosting the fly colony at TTRI. This was stopped in August, 1996 when the TTRI fly colony was able to sustain itself

As the colony expanded, more flies were available for release. In 1994, slightly less than 20,000 sterile male flies were produced per week at TTRI. The production increased to 60,000 in 1995 and peaked at more than 100,000 in 1996 (Figure 5). During the project's 4 years of operation, TTRI produced 24,115,332 pupae and from these a total of 8,967,585 adult male flies (about 50,000 flies per week) were produced, sterilised and shipped to Zanzibar.

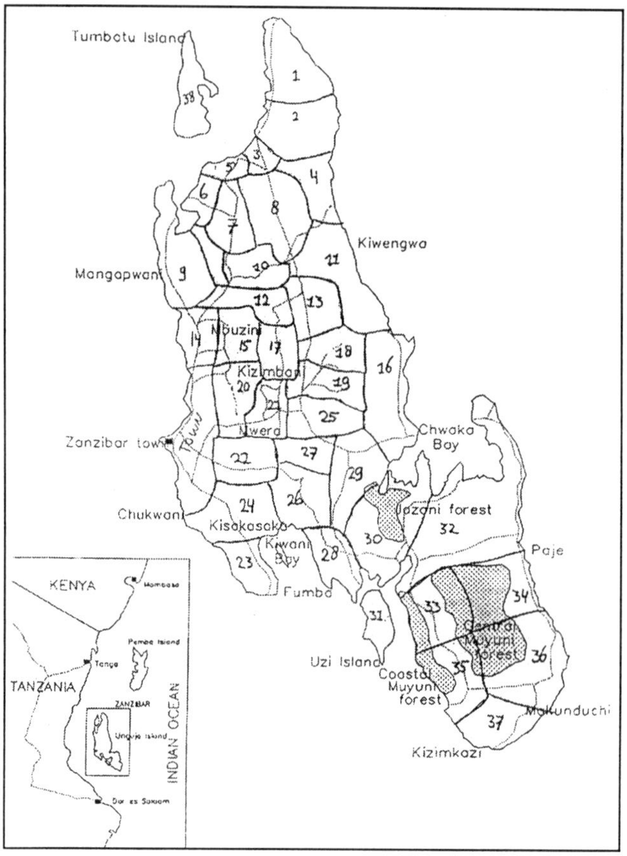


Figure 3. Thirty-eight blocks each with 30-40 sentinel cattle.

Results from quality assessment showed an increase in quality. On average, more than 93% of all flies were able to fly, only 2.6% died and sexing error was 0.66%. Induced sterility and marking was 100% (Kitwika et al. 1997).

## Fly Monitoring

For every flight, a sample of the flies for release was taken in Zanzibar to assess its quality. The data were compared with those taken at Tanga just before the shipment and indicated that the release operation did not seriously affect the quality of flies. Fly mortality increased by only 2% and proportion of non-fliers increased to 1.6% (Saleh et al. 1997).

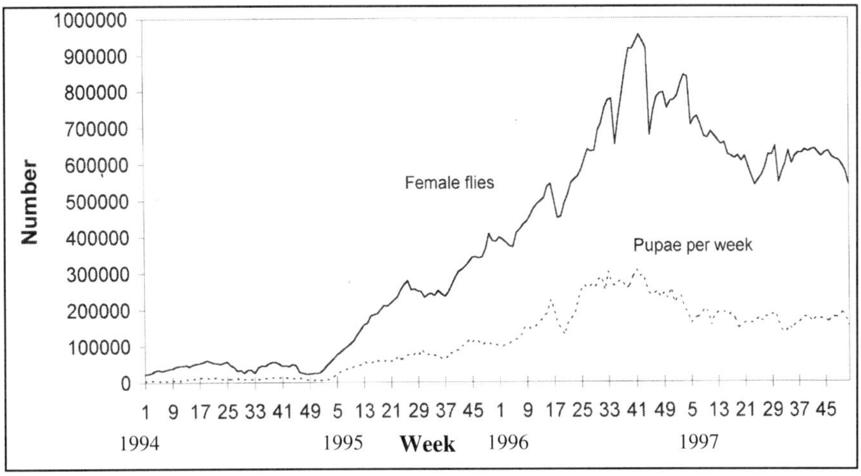


Figure 4. Number of female *G. austeni* and pupae produced at TTRI, Tanga.

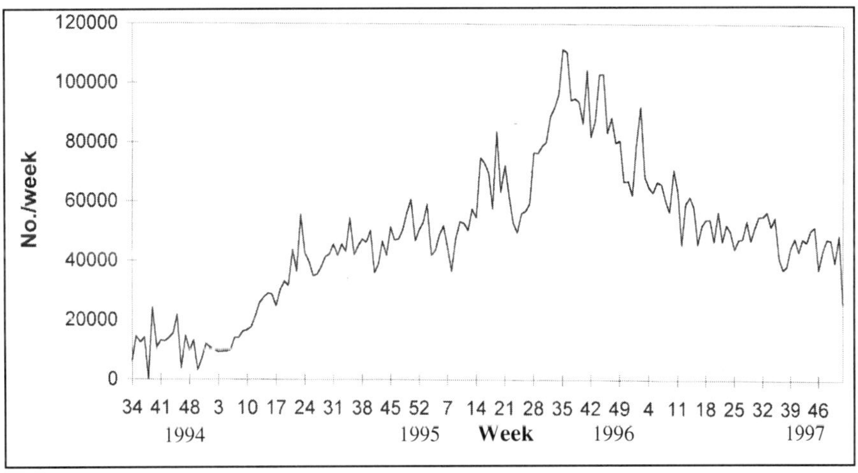


Figure 5. Weekly numbers of sterile *G. austeni* males released over Unguja by air.

Before suppression, the apparent fly density in the heavily forested areas (Jozani forest) was about 3.0 flies/panel/day. Suppression efforts in the island reduced the wild fly density by over 95%. Ground release caused little further reduction. The apparent density of the wild tsetse population on Unguja Island since the initiation of the aerial release programme is shown in Figure 6. In 1994, the wild fly apparent density was just below 0.1 flies/panel/day and was reduced further to below 0.01 flies/panel/day at the end of 1995. The population crashed at the beginning of 1996 with apparent densities of less than 0.001 flies/panel/day and the last wild fly was caught in September 1996.

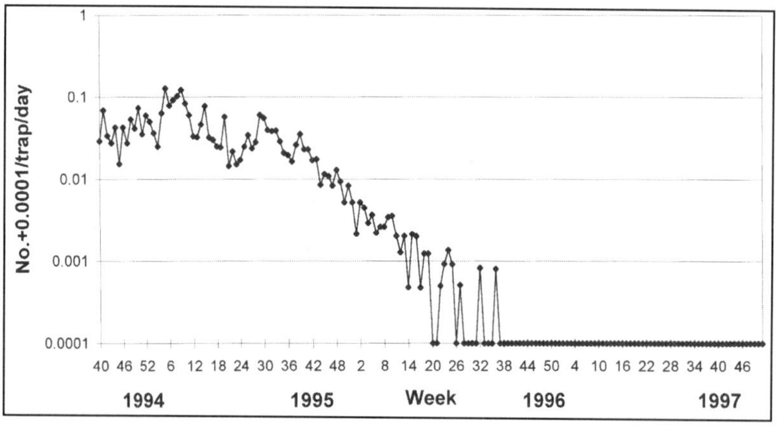


Figure 6. Apparent density of wild *G. austeni* on Unguja Island.

The ratio of sterile:wild male flies trapped increased initially due to the increasing number of released sterile male flies per week and later due to a decline in the number of wild flies. For the year 1994 up to the first quarter of 1995, the ratio was below 10:1 (sterile/wild) and this ratio induced sterility in 26% of young females. A ratio of 20:1 caused further reduction in the wild fly population. During the last quarter of 1995, the ratio of 50:1 was reached and induced sterility in 72% of young females. A ratio of 100:1 ensured 100% induced sterility in all newly emerged females (Saleh et al. 1997, in press). One year after the ratio of more than 50:1 was obtained, the last wild fly was trapped on Zanzibar Island.

## Disease Monitoring

In the northern half of Unguja, the disease monthly incidence was less than 0.1% for *T. congolense* and less than 0.4% for *T. vivax* when the project started in 1994. In 1997, the disease incidence was reduced to undetectable levels of *T. congolense* and occasional infections of *T. vivax*. In the southern half of Unguja, the monthly incidence was less than 1% for *T. congolense* and 2-4% for *T. vivax* when the project started in 1994. In 1997, the disease incidence was reduced to undetectable levels of *T. congolense* and the occasional infection of *T. vivax*. The disease incidence at the end of the project was, therefore, less than 0.1% and this was no longer a threat to cattle improvement.

## DISCUSSIONS

During the early stages of the project, the fly colony at TTRI, was faced with several limiting factors such as, poor conditions of the insectaries, lack of continuous water supply for cleaning equipment, unstable rearing conditions due to irregular power supply and an inadequate quality blood diet. The improvements made were crucial to the production of sufficient sexually sterile male tsetse flies for the project. Efforts are ongoing, through the FAO/IAEA Coordinated Research Programme (CRP), to further improve tsetse mass rearing procedures in order to increase the production capacity and quality of the flies.

The aerial release operation was efficient in dispersing flies (5% flies re-captured) especially to habitats that were difficult to reach and had very small effect on the quality of released flies. The ratio of more than 50 sterile:1 wild male was sufficient for the rapid reduction of the wild fly population. At such a high ratio, *G. austeni* was eradicated from Zanzibar Island within one year.

The decline in disease incidence suggested that released sterile male flies were not transmitting trypanosomiasis. It also suggested that released sterile male flies caused the decline and extinction of the *G. austeni* fly population on Zanzibar. Monitoring of both flies and disease in the absence of tsetse flies is currently in progress and the outcome will confirm the eradication of tsetse flies in the island.

## ACHIEVEMENTS

The Zanzibar project strengthened TTRI's capacity, both human and physical, for tsetse fly mass production and the use of SIT. The institute will play a vital role in the eradication of tsetse using SIT, where feasible, on mainland Tanzania. It also has the potential to assist other affected countries in the region.

The successful eradication of tsetse flies on Zanzibar Island created new opportunities for the Zanzibar farmers to enhance the mixed farming system which is considered the best way of achieving sustainable agricultural development and food security. Zanzibar farmers will be able to keep healthier animals as well as introduce more productive breeds. Zanzibar will no longer import trypanocidal drugs and insecticides, as well as animal products.

The success of the Zanzibar project has demonstrated SIT's strong potential for African countries. It has given a new impetus to the fight against the tsetse fly not only in Tanzania but also in other affected countries.

## REFERENCES

Dyck, V. A., M.J.B. Vreysen, F. Mramba, A.G. Parker, P.A.A. Mkonyi, I.A. Shambwana, A. Msangi and U. Feldmann. 1995. Eradication of *Glossina austeni* Newstead on Unguja Island (Zanzibar) by the sterile insect technique. Development and strategy of the project Tsetse Fly Eradication on Zanzibar. *In* Animal Trypanosomiasis Vector and disease control using nuclear technique. Second FAO/IAEA Seminar for Africa, Zanzibar, Tanzania, 27 November - 1 December 1995.

Horeth-Bonntgen, D.W. 1992. Control of *Glossina austeni* and cattle trypanosomiasis in Unguja Island by Deltamethrin pour-on application to livestock and with stationary targets in cattle free zones. pp. 213-218. *In* Tsetse Control, Diagnosis and Chemotherapy using Nuclear Techniques, Proceedings of an IAEA/FAO Seminar, Muguga, Kenya, 11-15 February 1991, IAEA-TECDOC, 634,

Kitwika, W.A., A.R. Msangi, N.E Kiwia, F. Mramba, I.I. Malele, G. Kasilagila, M.B. Byamungu, A.G. Parker and V.A. Dyck (1997). Eradication of *Glossina austeni* Newstead on Unguja Island (Zanzibar) by the sterile insect technique (SIT). Quality assessment of sterile males. Third Tanzania Entomological Association, Annual scientificconference, Zanzibar, 9-11 September 1997.

Moloo, S.K. and G.W. Kamunya. 1987. Suppressive action of Samorin on the cyclical development of pathogenic trypanosomes in *Glossina morsitans* centralis. Med. Vet. Entomol. 1: 285 - 287.

Msangi, A., N. Kiwia, F. Mramba, W. Kitwika, I. Malele, M. Byamungu, G. Ksilagila, V.A. Dyck and A. Parker. 1995. Eradication of *Glossina austeni* Newstead on Unguja Island (Zanzibar) by the sterile insect technique. 2. Mass production of *G. austeni* for SIT application in Zanzibar. Second FAO/IAEA Seminar for Africa, Zanzibar, Tanzania, 27 November – 1 December 1995.

Nash, T.A.M., A.M. Jordan and J.A. Boyle. 1968. The large scale rearing of *Glossina austeni* (Newst.) in the lab. IV. The final techq. Ann. Trop. Med. Parasit. 62: 336-341.

Pan, H.J., S.S. Kassim, F.W. Suleiman and I.A. Shambwana. 1995. Eradication of *Glossina austeni* Newstead on Unguja Island (Zanzibar) by the sterile insect technique. 5. Monitoring of transmission of *Trypanosoma* species by MHCT and Ag-ELISA. Second FAO/IAEA Seminar for Africa, Zanzibar, Tanzania, 27 November – 1 December 1995.

Saleh, K.M., M.J.B. Vreysen, S.S. Kassim, F.W. Suleiman, K.G. Juma, Z.R. Zhu, H.J. Pan, V.A. Dyck, and U. Feldmann. 1997. The successful application of the sterile insect technique (SIT) for the eradication of *Glossina austeni* (Diptera: Glossinidae) from Unguja Island (Zanzibar). ISCTRC, Mozambique.

Tarimo, C.S., N.E. Kiwia, U. Feldmann, F. Makishe and S.R. Mbise. 1984. The sterile insect technique against *G. austeni* Newstead in Tanzania. Development feeding technique in Tanga. FAO/IAEA Seminar on the sterile insect technique for tsetse fly eradication/control for developing countries in Africa. Zambia, 25-29th June 1984.

Vreysen, M.J.B., I.S. Khamis and A.M.V. Van der Vloedt. 1996. Evaluation of sticky panel to monitor population of *Glossina austeni* on Unguja Island of Zanzibar. Bull. Entomol. Res. 86: 289.

Vreysen, M.J.B., F. Mramba and I.S. Khamis. 1992. Laboratory and field observation in relation to the release of sterile *Glossina austeni* on Unguja (Zanzibar) Island. *In* Tsetse Control, Diagnosis and Chemotherapy using Nuclear Techniques, Proceedings of IAEA/FAO Sem., Muguga, Kenya, 11-15 February 1991, IAEA-TECDOC, 219-230.

Williamson, D. L., H.H. Baumgartner, A.G. Mtuya, S.A. Warner, S.A. Tarimo and D.A. Dame. 1983. Integration of insect sterility and insecticides for control of *Glossina morsitans morsitans* Westwood (Diptera: Glossinidae) in Tanzania. 1. Production of tsetse flies for release. Bull. Entomol. Res. 73: 259 - 265.

# Current Tsetse Control Operations in Botswana and Prospects for the Future

*R. Allsopp and T.K. Phillemon-Motsu*

Tsetse Control Division, Department of Animal Health and Production, Ministry of Agriculture, P.O. Box 14 Maun, Botswana

## INTRODUCTION

Several tsetse control methods have been used in Botswana over the past 70 years, ranging from bush clearing and selective game elimination, ground spraying with residual insecticides, aerial spraying with non-residual insecticides to the present odour-bait approach.

Sequential aerial spraying, initially with endosulfan or mixtures of endosulfan with one of the synthetic pyrethroids, effectively controlled tsetse but did not achieve eradication. Aerial spraying was suspended in 1991 when the tsetse population had been reduced to a relatively small population along the Linyanti and Kwando rivers bordering Namibia and to an area of 4,000-5,000 km$^2$ in the Okavango Delta.

After the 1991 aerial spraying operation the Tsetse Contol Division switched to the use of odour-baited, insecticide-impregnated targets as well as localised, ground based thermal fogging.

There has been no incidence of human and bovine trypanosomiasis since the mid-1980s and no significant insect population recovery since targets were introduced.

## THE CURRENT TSETSE SITUATION IN BOTSWANA (FIGURE 1)

Relatively little aerial spraying was carried out along the Linyanti and Kwando rivers, bordering the Namibian Caprivi Strip, between 1972 and 1991. Surveys carried out in May 1992, i.e., after the aerial spraying campaign, revealed a fairly widespread infestation and traps caught an average of 80 tsetse/day. The Division deployed and periodically re-treated a limited number of targets to provide some protection for government officers and armed forces operating in this extremely remote area. In 1995, 21 monitoring traps were deployed in the predominantly linear riverine habitat. The highest trap catch was 16 tsetse/trap/day and 16 of the traps caught no flies at all.

Patchy recovery has subsequently been reported by tour operators in the area but selective target deployment has continued and has kept this to a minimum.

Tsetse distribution in the Okavango Delta where the tsetse habitat is much more widespread, varied and fragmented has been more problematic. However, despite logistical problems, the situation remains under control.

Routine surveys increased over the past three years and tsetse have been confirmed throughout an area of about 5,000 km$^2$ in the northwest and central delta.

Until the late 1970s, the southwestern delta was a tsetse stronghold but over the past 20 years, the annual flood cycle has failed to reach the area and it has progressively dried out. Host animals such as buffaloes and elephants and many of the antelope species are now scarce in the area and tsetse, removed by aerial spraying, remain absent.

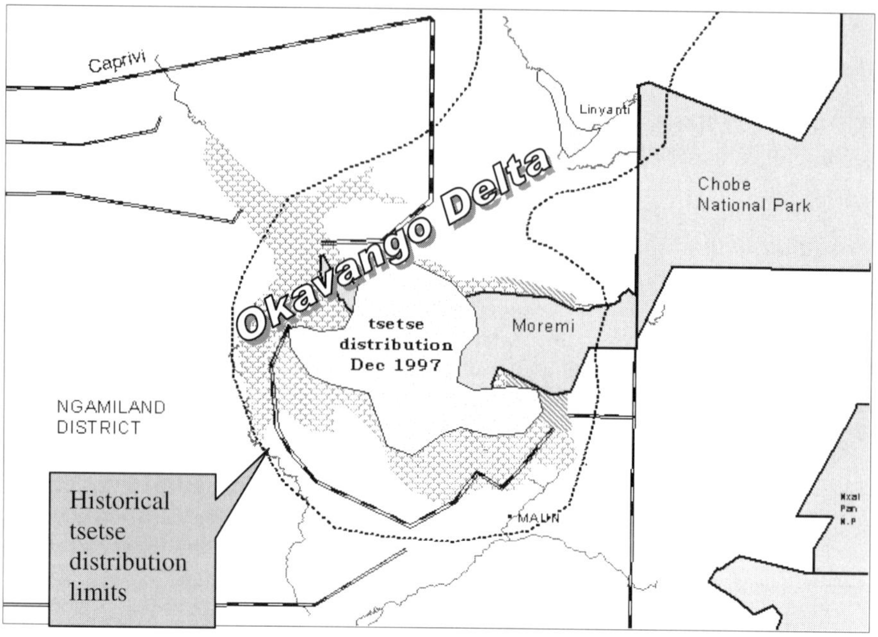


Figure 1. Distribution of tsetse flies in the Ngamiland District of northern Botswana 1997.

The annual floods have continued to regularly irrigate the eastern delta and wildlife remains spectacularly abundant. To date, surveys have not revealed the presence of tsetse in this area either.

The northern central delta was never completely cleared of tsetse by aerial spraying and a patchy population still remains. Some areas with historically high tsetse densities such as Lopis Island, where survey teams would regularly catch over 200 tsetse per day in the late 1980s, were cleared by the aerial spraying and remain fly free. Other areas such as Mombo and Guai Island, which is only about 20 km away from Lopis, continued to catch large numbers of tsetse. Mombo became so bad in 1994 that tourists were complaining, and the Division considered the re-introduction of aerial spraying. One trap deployed by helicopter on the remote and virtually inaccessible Guai Island caught 2,500 flies after being in place for 20 days.

Assignment of a full time operational field team to the Mombo/Guai area in 1996 gradually expanded the deployment of targets and ensured reasonably frequent servicing intervals. The tsetse situation was subsequently brought under control (Figure 2).

Three other specific areas have been allocated to operational field units and these too have been kept under control.

Unfortunately the Division's resources, particularly transport, have not allowed any further concerted control efforts beyond these four assigned areas. Small resurgent populations around the periphery of the "known" tsetse distribution (Figure 3) have periodically been discovered and selective target deployment by a mobile support unit has dealt effectively with these. Recently, the Division has been assisted by private sector tour operators who have accepted responsibility for maintaining targets in the vicinity of their safari camps (Figure 4).

To date, tsetse have not been eradicated from any area since 1992, i.e., since the introduction of targets.

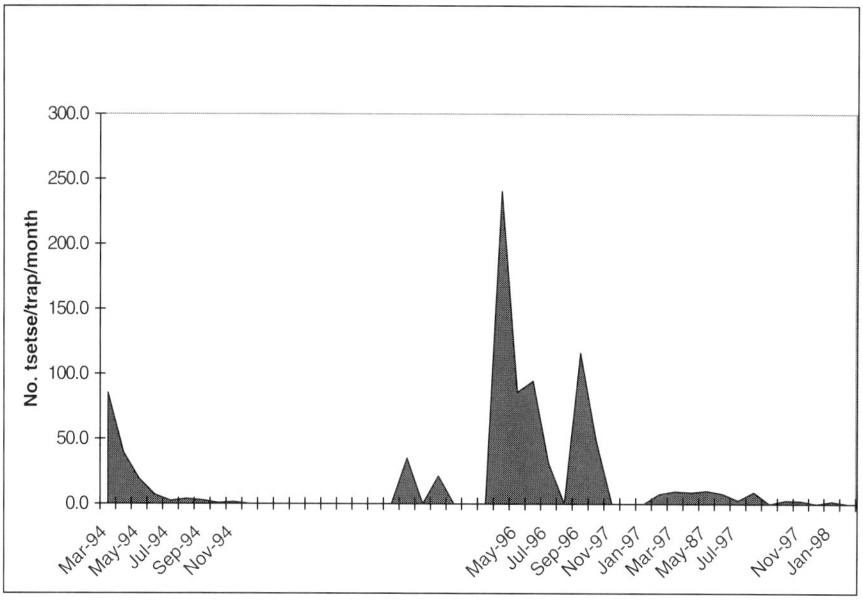


Figure 2. Mombo tsetse survey results.

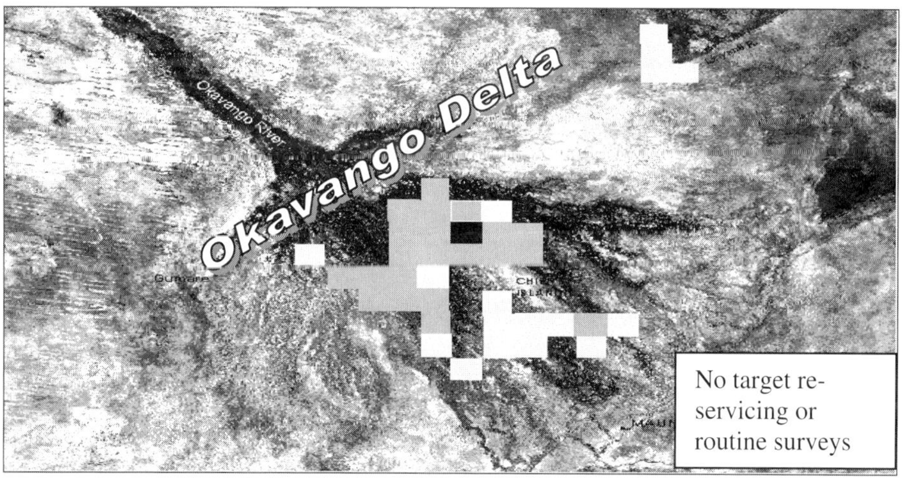


Figure 3. Patchy tsetse distribution in the Okavango Delta (total area about 20,000 km$^2$ and infested area about 5,000km$^2$).
Dark rectangle - high fly density limited to a very small area of a few hundred km$^2$; grey rectangle - medium fly density; clear rectangle - low fly density. A low density of tsetse occurs on the northern border with the Caprivi.

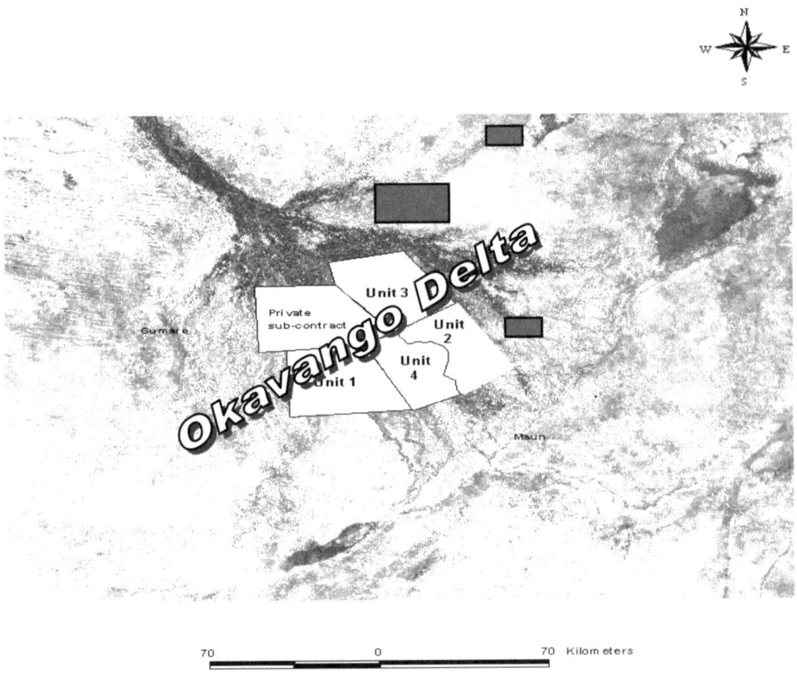


The areas allocated to Units 1-4 and to a private contractor are shown above. They amount to 6000 km$^2$ in total. In addition, small peripheral areas are infested by fragmented pockets of tsetse and these, shown in dark rectangles, are controlled jointly by the Tsetse Control Division and private tour operators.

Figure 4. Operational Control Areas.

## TARGET DEPLOYMENT IN NORTHERN BOTSWANA

The Division deployed or re-serviced 20,000 targets in 1997 (Figure 5). During the same period, 10,000 target covers had to be replaced because they were damaged by wind, wildlife or fire. Elephant damage is particularly widespread and in 1997, accounted for 60% of the total damage. Target re-servicing is an ongoing process and the aim is to re-service targets at six monthly intervals. Lack of serviceable transport seldom enables the Division to achieve this aim thus some targets may be left for over a year. On an average, targets are re-treated about every 8 months and it is estimated that there are about 12,000-15,000 fully effective targets in the delta at any one time.

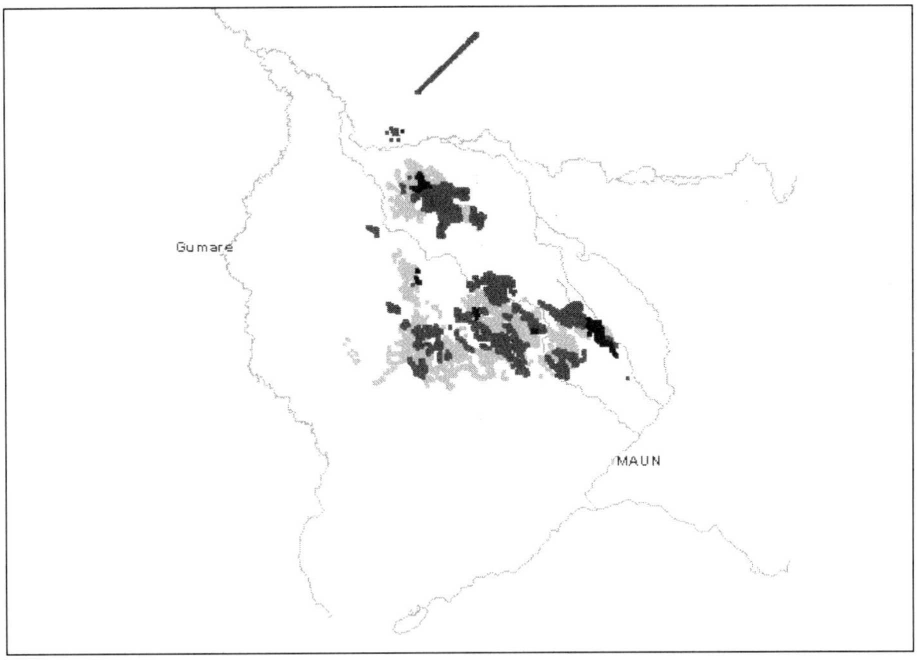


Figure 5. Distribution of targets deployed in the Okavango Delta.
Dark grey - in January 1998; grey - May-December, 1997; light grey - before May, 1997.
(The exact location of each target deployed was determined by using the geographical positioning systems (GPS). The locations were archived in a geographical information system (GIS) Arc View.)

## FUTURE PLANS

The aim of the Tsetse Control Division is to eradicate tsetse, using cost-effective, environmentally sensitive measures; thus the past few years have seen a progression from ground spraying to non-residual aerial spraying and finally to the odour-bait technique. Resources, particularly transport, are a problem and the government has recently approved collaborative ventures with the private sector to overcome these problems.

Two studies were recently completed with Department for International Development, United Kingdom (UK-DFID) support to assist the Division to develop a long-term control strategy to achieve the above aims:

- A socio-economic analysis of the implications of trypanosomiasis and tsetse control with particular reference to the possible involvement of private sector tour operators and local communities i.e., the primary beneficiaries (Mullins et al. 1997a, b).
- A preliminary study to develop environmental guidelines for staff handling insecticides or working in environmentally sensitive areas such as the Okavango Delta. A booklet on environmental guidelines was produced in English and Setswana (Grossman and Johnson 1997).

The eradication of tsetse is the current government policy but there is no indication that this will be achieved with the current control programme; indeed few, if any, other countries have actually eradicated *Glossina morsitans* using targets only. An internationally funded team under the auspices of the IAEA does appear to have eradicated *Glossina austeni* from Zanzibar Island after a programme of cattle dipping, target deployment and, finally, SIT.

There have been no cases of sleeping sickness or nagana in Botswana since the mid-1980s and provided that the tsetse distribution does not expand dramatically, there is every possibility that these diseases will remain under control. Thus the continuing presence of low fly densities will present nothing more serious than a "nuisance" problem.

However, maintenance of the *status quo* still requires a recurrent expenditure in excess of US$1.5 million per year. This may be partially defrayed by involving the private sector but full privatisation is certainly not envisaged at this stage. There are, nevertheless, compelling entomological and financial arguments for continuing the search for some means of eradicating tsetse from Botswana.

Should targets ultimately fail to achieve the government's objectives, there is a strong case for supplementing the odour-bait technique with non-residual aerial spraying. This did not achieve eradication between 1972 and 1991 when it was essentially the sole technique employed. But technical advances have been made in recent years - notably with navigation aids - and in combination with targets, it may prove effective. There would be an environmental cost in the world's largest inland delta although past monitoring might suggest that any effects would be short-term only (Douthwaite 1992, Douthwaite et al. 1981, Everts et al. 1983, SEMG 1986, SEMG 1992). There might also be a shift in support from private sector tour operators who currently provide considerable assistance to the Division but who, in general, oppose a return to aerial spraying.

It remains the policy of the government and the Division to retain all options for the eventual elimination of tsetse flies from Botswana. This includes ground and aerial spraying, thermal fogging and SIT. On environmental grounds, the latter has considerable appeal.

In many ways, Botswana presents an ideal situation for the use of SIT to control or eradicate tsetse flies:

- There is only one species, which is at the limit of its natural distribution.
- The infested area is relatively small and discrete.
- After 20 years of aerial spraying, densities are very low in most areas.
- Possible lines of re-invasion from Angola, Namibia or Zambia are tenuous and should be easily defendable.
- The area is flat and therefore conducive to the aerial release of sterile males.
- Tsetse flies currently infest a cattle free area so there is no likelihood of large numbers of sterile flies temporarily raising the endemicity of trypanosomiasis.

On the down side, Botswana would need assistance from the IAEA but is not a member and membership would probably cost in excess of the recurrent TCD budget. There is currently no large colony of *Glossina morsitans centralis*.

## REFERENCES

Douthwaite, R.J. 1992. Non-target effects of insecticides used in tsetse control operations. World Animal Review, FAO, Rome.

Douthwaite, R.J., P.J. Fox, P. Matthiessen and A. Russell-Smith. 1981. The environmental impact of aerosols of endosulfan applied for tsetse control in the Okavango Delta, Botswana. Overseas Development Administration, London. 141 pp.

Everts, J.W., K. van Frankenhuyzen, R. Roman and J.H. Koeman. 1983. Side effects of experimental pyrethroid applications for the control of tsetse flies in a riverine forest habitat (Africa). Arch. Environ. Contamin. Toxicol. 12(1): 91-97.

Grossman, D. and P. Johnson. 1997. Manual on environmental guidelines and procedures for tsetse control operations in the Okavango Delta. Tsetse Control Division, Maun, Botswana. 60 pp.

Mullins, G., R. Allsopp, P. Nkhori, M. Kolanyane and T.K. Phillemon-Motsu. 1997. The effect of tsetse fly and tsetse control on tourism in the Okavango Delta of Botswana. Presented to the 24th meeting of the International Scientific Council for Trypanosomiasis Research and Control (ISCTRC), 29 September - 3 October 1997, Maputo, Mozambique.

Mullins, G., P. Nkhori and M. Kolanyane. 1997. Socio-economics of tsetse fly control in northern Botswana. Department of Animal Health and Production, Gaborone, Botswana.

Scientific Environmental Monitoring Group. 1986. Annual Report to the Regional Tsetse and Trypanosomiasis Control Programme for Southern Africa EC Delegation, Harare, Zimbabwe.

Scientific Environmental Monitoring Group. 1993. Impact of aerial spraying and odour-baited targets on ecosystems. Report 1987-1990. Environmental monitoring of tsetse control operations in Zambia and Zimbabwe. EC Delegation, Harare, Zimbabwe.

# Potential for Area-Wide Control or Eradication of Tsetse Flies in Africa

*J.P. Kabayo*[1] *and U. Feldmann*[2]

[1] Department of Biochemistry, Makerere University, Kampala, Uganda
[2] Insect and Pest Control Section, Joint FAO/ IAEA Division, P.O. Box 100, A-1400Vienna, Austria

## INTRODUCTION

Tsetse flies *(Glossina)* are found in Africa over an area, estimated by various authors, of 7-11 million sq. km. The northern limit of this area corresponds closely to the southern edges of the Sahara and Somali Deserts, running along 14°N and extending across the continent from Senegal in the west to Somalia in the east. The southern limit of tsetse distribution corresponds closely to the northern edges of the Kalahari and Namibian Deserts in the west and runs generally at 20°-30°S to the east of the continent (Ford and Katondo 1977). This tsetse fly belt covers the following 38 countries (listed below) in which the tsetse flies spread African trypanosomiasis, a severe disease that affects man and his domestic livestock, and is among the factors responsible for limiting the pace and extent of development in those countries:

| | | | |
|---|---|---|---|
| Angola | Equ. Guinea | Liberia | Sierra Leone |
| Benin | Eritrea | Malawi | Somalia |
| Botswana | Ethiopia | Mali | South Africa |
| Burkina Faso | Gabon | Mozambique | Sudan |
| Burundi | Gambia | Namibia | Tanzania |
| Cameroon | Ghana | Niger | Togo |
| Cent. Afr. Repub. | Guinea Bissau | Nigeria | Uganda |
| Chad | Ivory Coast | Rwanda | Zambia |
| Congo (Bra.) | Kenya | Senegal | Zimbabwe |
| Congo (Kin.) | Lesotho | | |

The disease is of a major economic importance. Throughout the affected countries within the fly belt, areas that are heavily infested by the tsetse fly are virtually devoid of cattle and other species of domestic livestock. Distribution of livestock in all countries on the African continent where densely infested foci exist is almost exactly the reverse of the distribution of the fly (Finelle 1974, Brunhes et al. 1994).

Attempts to control African trypanosomiasis date back to the beginning of this century. Several different methods of control, some aimed at the disease-causing organism and other aimed at the vector, were employed (Nagel 1995, Jordan 1986). Until after the Second World War, when insecticides became available for use in tsetse control campaigns, the most widely used control measure against tsetse flies was habitat destruction (involving felling trees and bush-clearing), the elimination of host animals (involving killing of wild game) and, to a certain extent, the use of various trapping devices to catch the flies. The tsetse control campaigns mounted in the 40s, 50s and 60s were invariably extensive *"roll up the country"* type of operations, planned and executed military style. Tsetse control campaigns in many countries, including Nigeria, Uganda, Zambia, Zimbabwe, Rwanda and Kenya, succeeded in eliminating tsetse flies from large expanses of land.

The events that punctuated the political history of the African continent throughout the 60s and 70s, characterised by independence and liberation movements, took their toll on the gains and successes achieved earlier in tsetse eradication campaigns. The pre-occupation with the changes from colonial administration to self-rule over much of Africa affected the subsequent budgets, emphasis and management of tsetse control programmes. Where the tsetse control programmes had been managed by a single colonial authority, the situation changed at independence. The government of each country subsequently took charge of its own independent programme, making co-ordination of programmes and control methods difficult. Soon after many countries attained their political independence, variations in the priority each government attached to tsetse control began to emerge. Regional tsetse control authorities and research institutions which were shared between several countries were destroyed. Human and material resources were divided and spread thin in a new trend of prestige, pride and political expediency to create new national facilities and, in effect, to "nationalise" tsetse flies.

## PRELUDE TO RE-INVASION

Countries which attained political independence and became involved in tsetse and trypanosomiasis control activities pursued an approach which stressed political boundaries and national sovereignty lost the opportunity to merge the fragmented efforts of individual countries. The differences which existed in the affected countries at independence reflecting the approach followed by the colonial governments also became "nationalised" and became preserved within the respective political boundaries. For example, many of the countries which were colonised by the British used trypanocidal drugs widely, but conducted large vector control programmes as well, while the countries which were colonised by the French relied exclusively on the use of trypanocidal drugs to control trypanosomiasis (Jordan 1986).

Then came a period, leading to the present times, which was characterised by political instability and other civil disturbances. These combined to precipitate the present critical situation of increased levels of tsetse fly infestation and high disease incidence. Over the years the confident, vigilant efforts and emphasis aimed at eradication of the disease were replaced with supposedly more "realistic" goals of control. This control was subsequently modified and rephrased as integrated control, shifting to containment, then, more recently, sinking to the resigned objective of simply management of the disease. In spite of national efforts and injections of donor support in every country, little success has been recorded and it is increasingly becoming clear that a new approach needs to be designed.

## AREA-WIDE APPROACH

As in campaigns against other insect-borne diseases, the major attraction to the approach to stop transmission of the disease by eradication of *Glossina* spp. is its economic advantages over the control approach. Whatever the cost of eradication may be, eradication is a time-limited, once-and-for-all cost, while the costs involved in control measures recur indefinitely. As stated above, eradication had been achieved in large parts of the continent, but most of these became re-infested, because the areas had not been sufficiently protected from re-invasion and lacked the necessary post-eradication surveillance. The lesson that can be drawn from this experience, however, is that it is inappropriate to initiate a control campaign in the middle of a tsetse-infested area. In countries, such as Zimbabwe, where an effective tsetse control capability exists and where large areas have been reclaimed, the long border perimeter involved demand regular treatments to control re-infestation.

Experience has shown that protection of even small areas, e.g., ranches, stock farms and settlement areas located within a major fly belt, by regular application of insecticides or by any other method or combination of methods is uneconomical, especially with increasing costs of labour and materials. The only viable approach, therefore, is to tackle large areas, preferably covering the entire infestation or in sections which are isolated from each other, e.g. by natural boundaries where no re-invasion is possible. This area-wide approach to tsetse control would represent a major departure from current practice where control is confined within the political boundaries of individual countries. There have been several attempts to control various insect-borne diseases on an area-wide basis, examples of which would illustrate the wisdom and advantages involved. Probably the biggest, albeit unsuccessful, programme ever mounted on the African continent was the attempt in the 1960s to eradicate the malarial mosquito. More recently, the West African onchocerciasis control programme brought together 11 countries in a collective fight against black flies which transmit onchocerciasis and succeeded to reduce drastically its incidence and threat to the population. In 1991, government representatives from Brazil, Paraguay, Uruguay, Bolivia, Argentina, Chile and Peru adopted a resolution calling for action to eradicate *Triatoma infestans*, the insect which spreads *Trypanosoma cruzi*, and the parasite causing Chagas disease (Kingman 1991).

The seven South American countries set up an intergovernmental commission for the Chagas disease, to raise and administer funds and co-ordinate the implementation of the eradication plan. The plan was to spray every home in all areas known to be infested with *T. infestans* in a big programme covering parts of Argentina, Brazil, Chile, Bolivia, Uruguay, Paraguay and Peru and included follow-up inspections and evaluation to ensure that the insect did not return. In another area-wide control programme, the USA and the countries of South America employed the sterile insect technique over the entire area to eradicate the New World screw-worm fly. Earlier eradication programmes in the USA alone had failed because of re-invasion from infested areas in the neighbouring countries.

## NEW FORMS OF ORGANISATION

There is need to create an African-wide organisation similar in structure and function to the South American Commission on the Chagas disease. The following can then be studied:

- The differences currently existing in the level of the priority placed on tsetse and trypanosomiasis control in different countries.
- The need for co-ordination and synchronisation of control operations and the rationale behind the need to exploit natural boundaries in the context of preventing re-infestation of cleared areas.

- The need for an African-wide framework of handling trans-boundary infestations.
- The need to pool resources, avoid duplication and relate to questions of economies of scale.
- The need to agree on methodologies and choice of control strategies.
- The need to better marshal and manage resources for control programmes with emphasis on coherence and accountability.
- The need to direct all available resources with emphasis on the execution of actual control programmes with dates and deadlines.
- The need to preserve a level of continuity to counter the effects of various disruptions on control programmes.
- The need to translate African political rhetoric on unity to the reality of united action to solve a common problem.
- The need, generally, to create a mechanism that will emphasise and enhance the true nature of tsetse and trypanosomiasis as an Africa-wide problem requiring an Africa-wide approach.

The Organisation of African Unity should take up the challenge to mediate in the formation and rationalisation of the proposed Africa-wide Commission on tsetse and trypanosomiasis.

## CONCLUSION

The tsetse and trypanosomiasis problem is an Africa-wide problem which needs an Africa-wide approach.

The most viable form of organisation successfully adopted to handle similar problems elsewhere in the world is the creation of a single organisation to manage and co-ordinate future tsetse and trypanosomiasis control programmes.

## REFERENCES

Brunhes, J., D. Cuisance, B. Geoffroy, J.-P. Hervy and J. Lebbe. 1994. Lles Glossines ou mouches tsé-tsé.Logiciel d'identification. Glossine Expert. Manuel illustré d'utilisation. - Editions d'ORSTOM, Paris, 160 pp.

Finelle P. 1974. African Trypanosomiasis Part IV. Economic Problems, World Anim. Rev. 10, 15-18.

Ford, J. and K.M. Katondo. 1977. Maps of tsetse fly (*Glossina*) distribution in Africa according to sub-generic groups. Bull. Anim. Health Prod. Afr. 15, 187-193.

Jordan, A.M. (1986) Ch. 6 & 7. *In* Trypanosomiasis and African development. Longman Singapore Publ., ISBN 0 582 46356 4, 357 pp.

Kingman, S. 1991. South America declares war on Chagas disease. New Scientist 132 (October 1991), pp.16-17.

Nagel P. 1995. Environmental Monitoring Handbook for Tsetse Control Operations. The Scientific Environmental Monitoring Group (SEMG) (ed.). Weikersheim: Markgraf, ISBN 3 8236 1249 2, 323 pp.

# Screw-worm Eradication in the Americas -- Overview

*John H. Wyss*

USDA, APHIS, International Services, Panama City, Panama

## INTRODUCTION

Screw-worms [*Cochliomyia hominivorax,* (Coquerel)] are found only in the Americas, and are known, therefore, as the New World Screw-worm (NWS). The larval stages of the fly feed on the living flesh of their host. A screw-worm infestation can kill an adult animal in 7-10 days if not treated. All warm-blooded animals are affected including man.

Although screw-worms had long been recognised as a severe pest of animals in the southwestern United States, they had never been detected east of the Mississippi River before 1933. In July 1933, screw-worms were transported on infested cattle to Georgia and became established east of the Mississippi River. Screw-worms spread quickly in the southeastern United States and were able to overwinter in southern Florida. Being new to the region, they were quickly recognised as a severe pest with a tremendous economic impact on livestock production. The livestock owners in the southeastern United States immediately noticed an increase in the number of animal deaths and increased costs of insecticides, veterinary medicines, veterinary services, inspection and handling. At the same time, they observed a decrease in animal weights and in milk production. Due to these observations, the livestock industry in the southeastern United States requested help in controlling screw-worms. Because of these requests, the research community became interested in control and eradication measures for this pest.

Early work by Crushing and Patton in 1933 recognised that *C. hominivorax* was an obligatory animal parasite and different from the secondary blowfly, *Cochliomyia macellaria* (Fabricius). In 1934, the United States Department of Agriculture (USDA), Agricultural Research Service (ARS) opened a research station in Valdosa, Georgia, and E.W. Laake and E.F. Knipling were assigned to work there. In September 1935, R.C. Bushland was hired by ARS to do research related to screw-worms at an ARS Research Laboratory in Dallas, Texas. Melvin and Bushland in 1936 developed artificial media and techniques for screw-worm colonisation. In 1937, Knipling and Bushland began discussing an autocidal theory of screw-worm population suppression. These workers and others were studying screw-worms and developing treatment and control measures, but all work was suspended in 1937 because of the Second World War. Following the war, in 1947, interest once again turned to screw-worm control and eradication, and the development of what would become the sterile insect technique (SIT). Muller in 1950 described the effects of X-ray exposure on the reproductive system in *Drosophila.* Bushland and Hopkins in 1951 demonstrated that adult screw-worms of both sexes were sexually sterilised by irradiating the pupal stage of the insect. This promising work was field tested twice between 1951 and 1953 on Sanibel Island. Sanibel Island is a small island with an area of only 36 km$^2$ located 3 km off the southeast coast of Florida. In these first tests, 38 sterile males were used per km$^2$ on a weekly basis. The sterile flies were produced in the ARS Laboratory located at

Kerrville, Texas. The results of the last field test were very encouraging. Sterile egg masses were detected the first week and within several weeks, 80 percent of the egg masses collected were sterile and the number of egg masses declined. The experiment was continued for several months, but screw-worms were not eradicated. This discouraging outcome was explained by the premise that the migration of fertile flies from the mainland was causing continuous re-infestations of the island. In 1954, this premise was tested in Curaçao, West Indies. Here was an opportunity to test the sterile insect technique (SIT) on a 440 $km^2$ island located 65 km from the mainland of South America. On this island, they doubled the number of sterile males released from 38 to 76 per $km^2$ on a weekly basis. The sterile flies used in this experiment were produced at the facility at Bithlo, Florida. The results of this field test were very good. Sterile egg masses were detected the first week; within several weeks, 84 percent of the egg masses collected were sterile and the number of egg masses was declining. Within several months, 100 percent of the collected egg masses were found to be sterile. This experiment was a complete success; the island of Curaçao was determined to be screw-worm-free in 1954.

## Screw-worm Eradication Programme in the Southeastern United States (1957-1959)

The success of the screw-worm eradication experiment on the island of Curaçao in 1954, with the findings of Lindquist in 1955, that even under good conditions, there were no more than a few hundred screw-worm flies per square mile, led people to believe that screw-worm eradication was indeed feasible. In the spring and summer of 1957, livestock owners experienced the worst losses from screw-worms since their introduction into the southeast in 1933. This same summer, Bushland and his group of researchers conducted a pilot field test with sterile screw-worm flies covering 5,180 $m^2$ near Bithlo, Florida. In this field test, there was a 70 percent reduction in the number of cases in the centre of the test area. This result led to the implementation of an eradication programme in the southeastern United States. The programme began in 1957 and got off to a good start because the winter of 1957-1958 was one of the coldest and wettest in the history of Florida. To take advantage of the severe winter, sterile fly production at the Bithlo, Florida, facility was increased from 2 million per week to 14 million per week. The new sterile fly facility located near Sebring, Florida, was officially opened on 10 July 1958, and within 3 months, was producing over 50 million per week. The capacity of this new facility was 75 million per week. The last two cases detected in the southeastern United States occurred on 19 February 1959 and 17 June 1959. Sterile fly dispersal continued over the area until November 1959. No more cases were detected and except for imported non-replicating infestations, the southeastern United States has been free of screw-worms since that time.

## Screw-worm Eradication Programme in the Southwestern United States (1962-1966)

Livestock producers in the southwestern United States watched the eradication efforts in the southeastern United States with much interest. The Southwest Animal Health Research Foundation (SWAHRF) was formed to obtain support for a screw-worm eradication programme in the southwestern United States. From 1960 to 1962, SWAHRF struggled to get a screw-worm programme authorised and to raise producer-contributed funds. In February 1962, authorisation was obtained for the programme.

Once authorised, the Animal Disease Eradication Division of the USDA expanded an existing rearing facility in Kerrville, Texas. Between February and June 1962, SWAHRF constructed a new screw-worm rearing facility at Mission, Texas. Vice-President Lyndon B. Johnson dedicated the new Mission, Texas, Screw-worm Rearing Facility on 16 June 1962. The new Mission facility had a capacity of 200 million sterile flies per week. With the facility now operational, the screw-worm eradication programme was ready to begin in the southwestern United States. The states of Texas and New Mexico were declared screw-worm-free in 1964, and the eradication programme was then expanded into Arizona and California. The entire United States was declared screw-worm-free in 1966. The plan at that time was to maintain a sterile fly biological barrier, by carrying out weekly dispersals of sterile flies along much of the border between the United States and Mexico, to prevent the migration of fertile flies from Mexico back into the United States. In addition, animals were inspected and dipped before entering the United States from Mexico. Despite these efforts, cases continued to occur in the United States. In 1972, there was a screw-worm outbreak in the United States that approached pre-eradication levels. There were also some cases in Texas in 1976 and 1978. The last autochthonous screw-worm case reported in the southwestern United States occurred in August 1982.

## Screw-worm Eradication Programme in Mexico (1972-1991)

Due to these continued outbreaks and the interest of Mexican livestock producers in extending the eradication programme into Mexico, it was decided to move the barrier south to the Isthmus of Tehuantepec in Mexico. This location would be more economical (a 190 km width as compared to 2,400 km at the United States-Mexico border) to maintain. In addition, a barrier farther from the US border would afford more protection for the United States. An agreement was signed on 28 August 1972, to form the Mexico-United States Commission for the Eradication of Screw-worms (Commission). The objectives of the Commission were to construct and operate a sterile screw-worm fly production plant, and to eradicate screw-worms southwards to the Isthmus of Tehuantepec, in southern Mexico. The eradication programme got underway in 1972, using sterile flies produced in the plant at Mission, Texas. The Commission production plant was constructed in Tuxtla Gutierrez, Chiapas, Mexico, and was inaugurated on 28 August 1976, by the Secretaries of Agriculture from the United States and Mexico. The Tuxtla Gutierrez plant had the capacity to produce 500 million sterile flies per week. The eradication programme continued using sterile flies from both the Mission and Tuxtla Gutierrez plants until the Mission plant was closed in 1982. The Commission's objective, of eradicating screw-worms to the Isthmus of Tehuantepec, was achieved in 1984. However, the barrier at the Isthmus of Tehuantepec divided the country, with livestock producers to the south claiming the Mexican government was showing favouritism to producers north of the barrier. Further studies showed that Panama was a much better site for a permanent biological barrier. A barrier extending from the Panama Canal to the border with Colombia would require only 40 million sterile flies per week, compared to 150 million per week needed at the Isthmus of Tehuantepec in Mexico. In addition, in the eastern half of Panama, the movement of livestock is significantly less, when compared to the animal movements at the Isthmus of Tehuantepec in Mexico.

## Screw-worm Eradication Programme in Central America

Following indications of interest in screw-worm eradication from all the Central American countries and Panama, a plan was developed in 1985 to extend the Screw-worm Eradication Programme through Central America and establish a permanent biological barrier in the eastern half of Panama. The first step in expanding the eradication programme to Central America was to enlarge the programme in Mexico to cover all of the country. Mexico was declared screw-worm-free on 25 February 1991. After eradication, Mexico had two screw-worm outbreaks, one in January 1992, and another in June 1993. The last screw-worm case in Mexico was collected on 17 June 1993, and no screw-worm cases have been detected since.

### Guatemala (1988-1994)

The Commission signed an agreement with the Guatemala Ministry of Agriculture, Livestock and Food (MAGA), on 10 December 1986, to eradicate screw-worms from Guatemala. A programme office was established in Guatemala City in 1988. Aerial dispersal of sterile flies over the large northern state of Peten started in September 1988. The area dispersed was extended until the entire country was covered by January 1991. At the peak of dispersal, approximately 115 million flies per week were being dispersed over Guatemala. Case numbers reached a maximum of 10,573 in 1988 and decreased each year afterwards. The numbers are biased on the low side because field surveillance activities did not commence in the southeastern part of the country until well after the number of cases in the north had already started dropping because of sterile fly dispersals. The last autochthonous screw-worm infestation in Guatemala was recorded on 10 May 1992. Eight introduced infestations were found later in 1992 at the Playitas Inspection Station, one of a number of inspection stations situated near the borders with Honduras and El Salvador on main livestock transportation routes. Two additional imported cases, one in February and one in April 1993, were found at the same inspection station. The number of flies dispersed was decreased starting with the northern state of Peten until, from January through June 1993, an average of only 41 million sterile flies per week was dispersed. The number of flies dispersed continued to decrease until all dispersals ceased at the end of December 1993. An official declaration of freedom from screw-worms was held on 20 May 1994. The programme office was closed on 30 June 1994.

### Belize (1988-1994)

The Commission signed an agreement with the Belize Ministry of Agriculture and Fisheries (MAF), on 2 August 1988, to eradicate screw-worms from Belize. The programme office in Belmopan was dedicated on 25 August 1989. Dispersal of sterile flies began over the northern 40 percent of the country on 26 August 1989. Total coverage of the country with sterile flies was achieved on 10 April 1990. Screw-worm case numbers reached a high, after the programme began, of 172 cases in 1989, and declined each year thereafter. These numbers are also biased on the low side, because sterile fly dispersal started in the north, where the majority of cattle are found, nearly 3 months before all field inspectors were hired, trained and provided with transportation. Interviews with livestock producers indicated that much higher numbers of screw-worm infestations had occurred prior to the start of dispersal. At the peak of dispersal, an average of 24 million sterile flies per week was dispersed. The last autochthonous screw-worm infestation was collected on 1 July 1991. No cases have been found since then. Dispersal of sterile flies over the northern half of the country ceased on 1 June 1992 and all dispersals ceased on 31 December 1992. The country was officially declared screw-worm free on 22 May 1994. The programme office was closed 30 June 1994.

**El Salvador (1991-1995)**

The USDA signed a bilateral agreement with the El Salvador Ministry of Agriculture and Livestock (MAG), on 24 July 1991, to eradicate screw-worms from El Salvador. A programme office had already been opened in San Salvador and field surveillance activities initiated, resulting in an early start on obtaining data on the number of screw-worm infestations. Aerial dispersal of sterile flies over the entire country started in October 1991. At the peak of dispersal, an average of 24 million flies per week was being dispersed. A maximum of 800 cases in 1 month occurred in December 1991, and then declined each month thereafter. A small number of infestations was detected after March 1993, and only two infestations occurred in 1994. One, in January, was autochthonous and another, in May, was an imported case. No cases have occurred since. All dispersal of sterile flies ended in July 1994. An official declaration of freedom from screw-worms was held on 19 June 1995. The programme office closed on 30 June 1995.

**Honduras (1991-1996)**

The USDA signed a bilateral agreement with the Honduras Secretariat of Natural Resources (SRN), on 26 July 1991, to eradicate screw-worms from Honduras. A programme office was opened in September 1991, in Tegucigalpa. Field surveillance activities were initiated in the southwestern half of the country shortly afterwards. Sterile fly dispersal over the southwestern half of the country began in November 1991. The area of dispersal was increased until 100 percent of the country was being dispersed by June 1993. At the peak of dispersal, 120 million flies per week were dispersed. The area of dispersal began decreasing from the southwestern part of the country in June 1993 at about the same time that dispersal started in the eastern-most part of the country. Only about 33 percent of the country was dispersed in the first half of 1995. Beginning in July 1995, the dispersal area was further decreased to about 16 percent, dispersing only in areas adjacent to the border with Nicaragua. All dispersal of sterile flies in Honduras was terminated on 30 September 1995. The number of screw-worm infestations found peaked in 1992, and decreased each year thereafter. The number of cases detected was also biased on the low side, because country-wide surveillance did not commence until well after the number of cases in the southwestern half of the country had markedly decreased because of programme activities. The last case in the country was collected on 10 January 1995, and Honduras was declared screw-worm-free on 6 August 1996. The programme was essentially closed down with the exception of a small contingent to maintain screw-worm prevention and surveillance activities.

## Nicaragua (1992-1998)

The USDA signed a bilateral agreement, on 26 November 1991, with the Nicaragua Ministry of Agriculture and Livestock (MAG) to eradicate screw-worms from Nicaragua. A programme office was opened in May 1992, in Managua. Countrywide field surveillance began in July 1992. The number of screw-worm infestations reported peaked in 1993 and declined each year thereafter. The largest number of cases occurred in June 1993, in which 3,595 were reported. Dispersal of sterile flies over the northern half of the country began in July 1993, and 100 percent coverage was achieved in July 1994. Approximately 120 million flies per week were required to disperse over the entire country. Following the beginning of sterile fly dispersal, the number of infestations countrywide dropped markedly. Only 137 cases were reported in March 1995. Dispersal over 100 percent of the country continued until January 1996, when dispersal in the northern one-third of the country was stopped. The last case in Nicaragua was reported on 1 June 1997. The plan is for Nicaragua to be declared screw-worm-free before the end of 1999. Programme activities will continue at a much lower level to maintain a critical level of screw-worm prevention and surveillance activities and provide dispersal centre services for Costa Rica.

## Costa Rica (1995-1999)

The USDA signed a bilateral agreement with the Ministry of Agriculture and Livestock (MAG) on 29 November 1993, to eradicate screw-worms from Costa Rica. The screw-worm programme office opened in June 1995, in San José. Sterile fly dispersal over the northern part of the country began in April 1996, using flights operating out of Nicaragua. Dispersal of sterile flies, all from Nicaragua, over 100 percent of the country was achieved in October 1996. Approximately 60 million sterile flies per week were needed to cover the entire country at the programmed rate. Field surveillance activities began in April 1997. Because of the delay in beginning field operations, the number of reported screw-worm cases is relatively low. Currently (April 1999), the last reported case occurred on 18 March 1999.

## Panama (1997-2000)

The USDA signed a bilateral agreement with the Ministry of Agriculture and Livestock Development (MIDA), on 11 February 1994, to eradicate screw-worms, construct and operate a sterile fly production facility and establish a permanent biological barrier in Panama. The regional screw-worm programme office in Mexico was relocated to Panama in July 1997, and field surveillance activities began in June 1998. Dispersal of sterile flies over approximately 12 percent of the western part of country began in July 1998, using over-flights out of Nicaragua. In October 1998, 50 percent of the country was dispersed from a new dispersal centre located at Tocumen International Airport near Panama City. Approximately 80 million sterile flies per week will be needed to cover the entire country at the programmed rate of 3,000 sterile flies per $km^2$ dispersed in 3.2 km swath widths.

Plans call for the construction of a new sterile fly production plant east of Panama City near the Pacora River on the land of the old Filipillo sugar refinery. A new plant would permit the current plant near Tuxtla Gutierrez, Chiapas, Mexico, to be closed. That plant has been in continuous operation 24 hours per day, 365 days per year since 1976. The costs of operating and maintaining that plant are steadily increasing.

The cost of aerial transportation of pupae from Tuxtla Gutierrez, Mexico, to dispersal centres in countries where the programme is operating is increasing as the distance from the plant to dispersal centres increases. The danger of fertile flies escaping from the plant in Mexico that is located so far inside the eradicated area is an ever-increasing concern.

## ECONOMICS OF SCREW-WORM ERADICATION

Although screw-worm eradication is an expensive undertaking, the benefits of eradication justify the costs. The economics of living with screw-worms and the benefit to cost ratio of eradication are important factors in selling an eradication programme to foreign officials.

In 1996, the annual producer benefits in the United States, Mexico, and Central America were 796 million, 292 million, and 77.9 million US dollars respectively. These benefits were derived from decreases in death losses, decreases in veterinary services, veterinary medicines, insecticides, inspections, and handling costs and increases in meat and milk production. These producer benefits have an additional positive effect on the general economy of each country, and the region in general, due to the forward and backward linkages in the economy. This multiplier effect is estimated to be 3.5. The estimated total effect on the general economy of producer benefits is, therefore, estimated to be US$2.8 billion for the United States, US$1.0 billion for Mexico, and US$272.6 million for Central America. In addition, it is estimated that consumer benefits are equal to the effects on the general economy of the producer benefits. Adding this to the annual producer benefits results in a significant beneficial overall effect on the economy of all the countries that are now screw-worm-free.

The benefits to perpetuity for the annual producer benefits were calculated at 3, 6, and 8 percent discount rates. While the 8 percent rate represents the true cost of money, it is customary to use 3 percent for public-good programmes such as screw-worm eradication. The benefits to perpetuity for the producer benefits are US$26.5 billion for the United States, US$9.7 billion for Mexico, and US$2.6 billion for Central America, using the 3 percent discount rate.

The benefits to perpetuity are used to calculate the benefit to cost ratio for the screw-worm eradication programmes. The benefit to cost ratios for the eradication programmes have ranged from an average of 12.2 for Central America to 18 for the United States and Mexico. Even the smaller number is very supportive of an eradication programme.

In addition, screw-worm eradication has a significant human and wildlife health component that has not been included in the above figures.

## FUTURE SCREW-WORM ERADICATION PROGRAMMES

All Central America is expected to be free of screw-worm infestations by the end of the year 2000. The Caribbean region is the next most logical area to eradicate this pest. Of all the Caribbean Islands, only Cuba, Hispanola (Haiti and Dominican Republic), Jamaica, and Trinidad-Tobago are infested with screw-worms. Their proximity, with the exception of Trinidad-Tobago, to the United States, Mexico, and Central America makes them a possible source of re-infestation for the screw-worm-free countries. The feasibility of expanding the eradication programme to the Caribbean is very realistic. The Food and Agriculture Organisation (FAO) has done considerable work in the Caribbean and South America on the distribution of screw-worms. The International Atomic Energy Agency (IAEA) has been working with Jamaica for the past two years on developing a screw-worm eradication project. In May 1998, IAEA, Jamaican officials, and representatives of Mexico and the United States, individually, and the Mexico-United States Screw-worm Commission met and agreed to enter into a Memorandum of Understanding to cooperate on the Jamaican eradication project.

In South America, the economic benefits to livestock producers and consumers, and the general effect on the economies of the countries, would be very significant and would strongly support the idea of a South American screw-worm eradication programme. I personally believe that such an eradication programme is feasible and realistic. However, such a large undertaking would require a strong long-term political and financial commitment on the part of all the South American countries, and would necessitate coordination and an extreme dedication to work in harmony. It would not be easy; if it were, it would have already been done.

## ACKNOWLEDGEMENTS

I would like to acknowledge the support and work of all the government officials, Commission and programme employees, livestock owners, and the general public of all the countries involved. The screw-worm programme, to be the success it has been, required the cooperation of all of these individuals and groups within a country, as well as cooperation between all the countries involved.

## REFERENCES

Bram, R.A. 1985. Future screw-worm eradication programmes. Misc. Publ. Entomol. Soc. Am. No. 62, pp. 37-40.

Bushland, R.C. 1985. Eradication programme in the southwestern United States. Misc. Publ. Entomol. Soc. Am. No. 62, pp. 12-15.

Knipling, E.F. 1985. Sterile insect technique as a screw-worm control measure: the concept and its development. Misc. Publ. Entomol. Soc. Am. No. 62, pp. 4-7.

Meadows, M.E. 1985. Eradication programme in the southeastern United States. Misc. Publ. Entomol. Soc. Am. No. 62, pp. 8-11.

Snow, J.W., C.J. Whitten, A. Salinas, J. Ferrer and W.H. Sudlow, 1985. The *screw-worm (Cochliomyia hominivorax)* (Diptera: Calliphoridae) in Central America and proposed plans for its eradication south to the Darien Gap in Panama. J. Med. Entomol. 22: 353-360.

# The New World Screw-worm as a Pest in the Caribbean and Plans for its Eradication from Jamaica and the Other Infested Caribbean Islands

*George H. Grant[1], J. Wendell Snow[2] and Moises Vargas Teran[3]*

[1] Veterinary Services, P.O. Box 309, Hope Gardens,Kingston, Jamaica
[2] International Atomic Energy Agency, P.O. Box100, A-1400, Vienna, Austria
[3] FAO Division, Ministry of Agriculture, P.O. Box 309, Hope Gardens, Kingston, Jamaica

## INTRODUCTION

The screw-worm, *Cochliomyia hominivorax* (Coquerel) (NWS), was eradicated from the Caribbean island of Curacao in 1964 (Baumhover et al. 1955). This programme was considered as a test of the SIT principle. In 1959, the pest was eradicated from Florida with the concept fully established as a sound and novel entomological principle (Baumhover 1966). In 1962, a similar programme was initiated in the southeastern United States with a barrier established along the Mexican-United States of America border (Bushland 1975). In 1975, the pest was eradicated from the island of Puerto Rico, the United States of America (US) and the British Virgin Islands (Williams et al. 1977). In 1981, the pest was totally eradicated from the US and in 1986 from all of Mexico.

It has since been eradicated from Belize, Guatemala, Honduras, El Salvador and Nicaragua. The eventual goal of the programme is to eradicate the pest from Costa Rica and Panama down to the Derail Gap where a sterile fly barrier will be maintained.

In the Caribbean, an estimated 86% of the land mass is considered infested by the New World screw-worm. Jamaica, Hispaniola, Cuba, Trinidad and Tobago are the countries known to be infested. Despite this fact, no comprehensive plans have ever been made for its eradication from these countries which together have a total livestock population of well over 16 million (Table 1). However, based on the great success of the programme elsewhere and recent interest shown by various international organisations and governments in countries which are infested, the situation is changing rapidly. Currently, the country which is most prepared for an eradication programme is Jamaica where government officials have long shown an interest in eradicating the pest. For example, in 1959, a group of Jamaican livestock owners visited officials associated with the Florida eradication programme. Since this date, serious consideration has been given and several attempts made by Jamaica to implement an eradication programme but without success. Real progress was not made until 1997, based principally on efforts by the Jamaican Veterinary Services Division (VSD) when cooperation was established with the International Atomic Energy Agency (IAEA), the Food and Agriculture Organisation (FAO) and the Animal and Plant Health Information Service of the United States Department of Agriculture (USDA/APHIS/ARS) with respect to the preparation for an eradication programme. Preliminary organisational activities for this programme have been started with the expectation that the first sterile fly releases will be made in January, 1999 to initiate what is projected to be a 3-year eradication campaign.

With respect to the rest of the Caribbean, FAO currently has a project in Cuba which is aimed at determining the extent of the problem and its economic impact.

Concurrently, IAEA is developing a "thematic" plan for the entire Caribbean as well as South America.

Table 1. Current livestock population of the NWS endemic countries in the Caribbean ('000').

| Country | Beef Cattle | Dairy Cattle | Horses | Donkeys /Mules | Sheep | Goats | Water Buffaloes | Pigs | TOTAL |
|---|---|---|---|---|---|---|---|---|---|
| Cuba | 4,020 | 480 | 580 | 37 | 310 | 95 | - | 1550 | 7072 |
| Dominican Republic | 1,836 | 250 | 329 | 279 | 135 | 570 | - | 950 | 4349 |
| Haiti | 1,055 | 145 | 400 | 290 | 85 | 910 | - | 360 | 3245 |
| Guyana | 163 | 27 | 2 | 1 | 130 | 79 | - | 50 | 452 |
| Jamaica | 347 | 53 | 4 | 33 | 4 | 440 | - | 210 | 1091 |
| Trinidad & Tobago | 48 | 7 | 1 | 4 | 14 | 52 | 9 | 48 | 183 |
| Suriname | 89 | 10 | - | - | 9 | 9 | 1 | 38 | 156 |

Source: FAO 1996 Report
Data Banks of Trinidad and Tobago, Guyana, Jamaica

## CUBA

This is the largest NWS infested country in the Caribbean with a landmass of approximately 114,525 km$^2$ and a human population of 10,870,000 people. An eradication programme for Cuba will require about 150 million sterile flies per week for two years. Similarly, in the case of this country, a full year would most likely be needed to organise the programme, two years for actual eradication and about one year for verification of results. It is possible that the eradication programme for the country could be carried out in two sections which would take longer but requiring fewer sterile flies at one time. The cost of eradication would likely to be somewhat higher if the 2-phase approach were to be selected. A one-step approach would incur additional costs based on the need for more emergence chamber space for flies. On the other hand, it would reduce the time needed for the eradication while at the same time reducing the need for quarantine and excluding the need for a buffer zone.

The cost-benefits from eradication have not yet been determined for Cuba but this work is currently being undertaken through the sponsorship of FAO. The current information is that sufficient data have already been collected to establish that the NWS is a serious problem in the country and that the insect is widely distributed throughout and affects all warm blooded species while being active at all seasons of the year. The cost of implementing an eradication programme for Cuba is being estimated at some US$54 million (Table 2).

Table 2. Estimated cost of eradicating the NWS from Cuba.

| ITEM | ESTIMATED COST US$ |
|---|---|
| Sterile Flies (estimated at 1,700/million for 2 years) | 26,000,000 |
| Chilled Fly Chambers for Fly Emergence | 3,000,000 |
| Information Campaign | 4,000,000 |
| Quarantine Campaign | 3,000,000 |
| Administration | 3,000,000 |
| Dispersal Centre Operation | 12,000,000 |
| Miscellaneous Costs | 4,000,000 |
| Field Operations | 7,000,000 |
| **TOTAL** | **54,000,000** |

## HISPANIOLA

Hispaniola, which comprises both Haiti and the Dominican Republic, is considered an infested area. Both countries occupy an equivalent landmass of 10,710 $km^2$ each and with human populations of 6,764,000 (Haiti) and 7,471,000 (Dominican Republic). The populations of both countries are reported to be fast growing with an expected increase of 1.5 million within the next 10 years. The best estimates of NWS damage at this time (FAO) are US$16 million and US$10 million annually in the Dominican Republic and Haiti respectively. In both countries, there is currently no organised effort to control the pest. While the NWS is well known to local people, the problem is believed to be either ignored or neglected by the local authorities. It is suspected that once the relevant impact studies are conducted, the greatest losses to the livestock industry and the biggest human health problem resulting from NWS infestation anywhere may well be found in these two countries.

Similarly, despite the current difficulties, these two countries may well be the most important for eradication to take place at this time in the region. It is being suggested that the best approach to the implementation of an eradication programme for these two countries is the use of a combined programme utilising a single distribution centre for both.

The estimated requirements for such a programme is about 95 million sterile flies weekly and at a total cost of almost US$36 million (Table 3). The time frame would be similar to that for Cuba.

## SOUTH AMERICA AND TRINIDAD AND TOBAGO

The landmass associated with the NWS in South America is very large. The literature contains numerous reports of infestations in Columbia, Venezuela, Suriname, Guyana, French Guyana, Ecuador and Paraguay. There are literature reports of human infestations in Uruguay, Peru and Bolivia. The temperatures in the southern regions of Argentina and Chile, as well as high elevations of the Andes Mountains, are notably too cold for the survival of the NWS. The situation is similar for parts of the Brazilian and

Guyana highlands. An estimate of the areas of South America which are continuously NWS infested is placed at 50% with another 30% of the land area invaded each year but only to be eliminated by cold weather at the end of the warm season. This is only an estimate of the situation and data have to be collected in all of the countries in order to determine the true situation.

The twin island state of Trinidad and Tobago is also NWS infested. Given its geographical location with its almost contiguous border with South America an eradication effort at this time would best be considered in association with this area and not as a part of the Caribbean. This may well be so unless new information to the contrary becomes available.

Table 3. Estimated cost of eradicating the NWS from Hispaniola.

| ITEM | ESTIMATED COST US$ |
|---|---|
| Sterile Flies (estimated at 1,700/million for 2 years) | 17,000,000 |
| Chilled Fly Chambers for Fly Emergence | 1,000,000 |
| Information Campaign | 2,000,000 |
| Quarantine Campaign | 1,000,000 |
| Administration | 1,500,000 |
| Dispersal Centre Operation | 9,050,000 |
| Miscellaneous Costs | 1,000,000 |
| Field Operations | 3,000,000 |
| **TOTAL** | **35,550,000** |

## JAMAICA

Jamaica lies in the Caribbean Sea 145 km south of the southernmost extremity of Cuba. The greatest length of the island is 235 km and its greatest width 82 km. The topography consists mainly of coastal plains around the island separated by a central mountain range running from the east with hills and a limestone plateau occupying the central and western areas of the interior. The land area is estimated at 11,422 km$^2$. The island has a tropical climate which is modified by the influence of the sea, the trade winds, and to a lesser extent, by land and sea breezes. There are four seasons distinguishable mainly by the differences in rainfall but conditions are not uniform over the island and vary considerably according to altitude and location.

Usually, the major rainy season starts in August and reaches a peak in May. However, periods of heavy rainfall and drought may occur at any time during the year. The lowest temperature occurs in January or February; the peak temperature usually occurs in July or August. In coastal areas, the average daily temperature ranges from about 23-28°C. However, the temperature often rises to about 30°C during the afternoon and may fall as low as 18°C in the early mornings during the cool season.

Screw-worm infestation is widespread in Jamaica and regardless of seasonal variations, altitude or ecological conditions. All types of livestock operations are affected irrespective of size and management practice. Vo (2000) estimated that the annual benefits from eradication would be between US$5.5 million and US$7.7 million. Benefits were defined as losses avoided due to: 1) mortality, 2) additional expenses for labour associated with surveillance, prevention and treatment of infested wounds, and 3) loss in productivity of infested animals. Assuming an eradication cost of US$9.0 million, she calculated net savings after 3 years as ranging from US$4.2 million to US$13.5 million. She further estimated net benefits to be between US$25 million and US$43 million after 10 years.

Active infestations are likely to occur in any season but appear to be related to the wetter periods. However, this pattern is modified by traditional production schemes for calving, branding etc. In most instances, wounds are treated by the owner and not reported to veterinarians. Snow et al. (1976) reported that peak occurrence was in October during the major rainy season, and that there was a smaller peak in February, several months before the minor rainy season. They reported 210 cases in their paper as cattle (151), swine (20), sheep (11), goats (23) and horses (5).

Cattle were by far the principal economic host, followed by swine and goats. They reported dogs were more likely to be the single most important host of the screw-worm in Jamaica. Private veterinarians have reported to the senior author that 15-30% of clinically treated dogs were for screw-worm infestation. Dogs are most heavily infested during the mating seasons when they stray from home for days at a time and become wounded while fighting.

Unlike cattle, dogs are rarely treated until after infestation. Rawlins and Sang (1984) reported pigs as the most important host in Jamaica and that screw-worms occurred in all parishes. Table 1 shows data taken from this paper where they reported that infestation was the most prevalent in the umbilicus of neonates, bites and barbed wire cuts. Tick bites, castration wounds and branding scars were of lesser significance.

The occurrence of screw-worm infestation in Jamaica is such that potentially all wounds occurring at any time of the year will become infested. We estimate that 80% or more of all untreated wounds would become infested. Most producers have become so accustomed to living with the screw-worm that they take prophylactic actions without considering the cost involved.

The livestock operations range from large-scale to medium-size commercial smaller backyard and "down-the-road" type operations. Cattle production comprises the largest and most important component of livestock industry in Jamaica with approximately 350,000 heads, beef (67%), dairy (15%) and dual purpose and draft animals the other 18%. Currently, Jamaica is not self-sufficient in beef production but hopes to attain this goal in the next 10 years. The other livestocks of importance include goats (440,000), pigs (210,000), equines (33,000) and sheep (4,000). Most of these animals, particularly goats, are found in small-type operations of fewer than 5 animals. However, all operations, regardless of size, have some methods of controlling screw-worm infestation.

No wild animals capable of supporting infestations such as the white-tailed deer, rabbits, opossum or peccaries, are present in Jamaica. The only likely candidates are a few wild pigs in the eastern region of the island and the mongoose, an animal introduced in the last century to destroy rats and snakes in the sugar plantations. It has not been determined if the mongoose serves as a reservoir host for the NWS.

Reports of human myiasis have been made by both public and hospital officials who considered this a minor problem. However, anecdotal information suggests that NWS myiasis in humans is a significant problem on the island. It should be noted that a major problem in terms of determining the true prevalence rate is that of non- or under-reporting of cases, The cases reported are for the most part observed in children, the senile, the mentally retarded and individuals not receiving adequate medical attention or those experiencing substandard levels of personal hygiene. Usually, infestations are found in the legs, toes, facial sores and nasal cavities.

## THE ERADICATION PROGRAMME

The plan is to eradicate the NWS from Jamaica by the use of the sterile insect technique (SIT) which involves the sequential aerial dispersion of adequate numbers of radiation-sterilised NWS over the entire island. An estimated twenty million pupae per week will be obtained from the Mexican-US Screw-worm Commission in Mexico. These pupae will be flown weekly to emergence facilities in Jamaica. Once emerged, pupae will be subjected to aerial release 4 days of each week. An estimated 15 million sterile flies are expected to emerge from the 20 million pupae each week, thereby making for a 1,200 sterile flies / $km^2$ aerial release over the entire country.

It is estimated that this eradication programme will require three years for completion. This includes six months for the programme organisation, two years of sterile fly releases and another six months for free-status verification. The cost of the programme will be approximately US$9 million. Most of these funds will be used for the purchasing of flies and the aerial release component of the programme. The necessary funds for implementation have been secured through budgetary allocations by the Jamaican government and with financial assistance from the International Atomic Energy Agency (IAEA). Additional support will be given through a "cess" on the slaughter of cattle agreed to by the local livestock association. Technical and "in kind" assistance will also be provided by the USDA/ARS and the US-Mexican Screw-worm Eradication Commission (Table 4).

Table 4. Estimated cost of eradicating the NWS from Jamaica.

| ITEM | ESTIMATED COST |
|---|---|
| Sterile Flies (estimated at 1,700/million for 2 years) | 3,536,000 |
| Chilled Fly Chambers for Fly Emergence | 500,000 |
| Information Campaign | 800,000 |
| Quarantine Campaign | 50,000 |
| Administration | 1,214,000 |
| Dispersal Centre Operation | 1,500,000 |
| Miscellaneous Costs | 300,000 |
| Field Operations | 1,100,000 |
| **TOTAL** | **US$ 9,000,000** |

To facilitate the economic and effective use of SIT and to ensure an early positive impact, implementation will be supported by ground-based activities aimed at reducing the local wild fly population. Intensive animal inspection measures and wound treatment regimes will be initiated in collaboration with local livestock owners. These activities will involve the regular inspection of all domestic animals throughout the island and will include prophylactic and curative treatment with selected insecticides. Larvae found in wounds will be collected, recorded, preserved and identified. The data obtained will be used as the basis for estimating the density and distribution of the screw-worm and for monitoring the progress made towards final eradication.

Finally, to prevent the spread of infestation, all livestock movement will be effectively controlled through the strengthening of quarantine and other existing regulatory measures. Similarly, all animals will be inspected and treated prior to their leaving the country and with strict import entry for all animals regarding screw-worm-free status. Animals with infested wounds or with abrasions susceptible to infestation will be quarantined and treated. At some point in the future, we believe that following a successful Jamaican programme, similar eradication programmes will be planned and be implemented for the other Caribbean islands.

In summary, the available information suggests that NWS myiasis is well recognised and acknowledged to be economically devastating to the livestock sub-sector of the Caribbean. Despite this fact, in most of these countries, competing national priorities for scarce budgetary allocations may suggest that such eradication programmes might not be seen as being expedient at this time. However, based on known experience, the immediate and long-term returns from a NWS-free status will more than off-set the cost of any eradication programme. More importantly, the eradication of the NWS has repeatedly been shown to contribute greatly to the alleviation of rural poverty and the promotion of an orderly development of integrated crop and livestock production systems in those countries which have undertaken such a programme.

A successful Jamaican eradication programme would not only serve as a model for future eradication efforts in the other countries of the Caribbean but should also serve as an important barrier against re-infestation of already free areas of both the Caribbean and mainland America.

## REFERENCES

Baumhover, A.H. 1966. Eradication of the screwworm flies. J. Am. Med. Assoc. 196:240-8.

Baumhover, A.H., A.J. Graham, A. Bitter, D.E. Hopkins, H.D. Dew, F.H. Dudley and R.C. Bushland. 1955. Screw-worm control through release of sterile flies. J. Econ. Entomol. 48:462-6.

Bushland K.C. 1975. Screwworm research and eradication. Bull Entomol. Soc. Am. 21:23-6.

Rawlins, S.C. and D.B. Barnett. 1983. Internal Human Myiasis. W.I. Med. J. 32:184-86.

Rawlins, S.C. and A. Mansingh. 1987. A review of ticks and screwworms affecting livestock in the Caribbean. Insect Sci. Applic. 8:259-67.

Rawlins, S.C. and J.C. Sang. 1984. Screwworm Myiasis in Jamaica and proposals for its eradication. Trop. Pest. Manag. 30:125-29.

Vo, T.T. 2000. Economic impact of eradicating the New World Screw-worm (*Cochliomyia hominivorax*) from Jamaica. pp 113-116. *In* Area-Wide Control of Fruit Flies and Other Insect Pests. K.H. Tan (ed.). Penerbit Universiti Sains Malaysia, Penang, Malaysia.

Wendell, S.J. 1976. A report on the screwworm situation in Jamaica. Unpublished ARS report, 77 pp.

Wendell, S.J., H.C. Hofmann and A.H. Baumhover. 1977. The screwworm as a pest on the island of Jamaica and the feasibility of eradication of the Sterile Insect Method. Southwestern Entomol. 2:202-206.

William, D.L., S.C. Gartman, J.L. Hourrigan. 1977. The eradication of screwworms from Puerto Rico and the Virgin Islands. World Animal review FAO. Number 21-1.

# Insurance Against an Old World Screw-worm Fly Invasion of Australia

*Neil Tweddle*[1] *and Rod J. Mahon*[2]

[1] Emergency Disease Strategies Section, National Office of Animal and Plant Health, GPO Box 858, Canberra, ACT 2601, Australia
[2] CSIRO Entomology, GPO Box 1700, Canberra, ACT 2601, Australia; CSIRO, Institute Haiwan, P.O. Box 520, 86009, Kluang, Johor, Malaysia

Australia is fortunate that neither the New World screw-worm fly *Cochliomyia hominivorax* (Coquerel) (NWSWF) nor the Old World screw-worm fly, *Chrysomya bezziana.* Villeneuve (OWSWF), have become established within our country, although much of the northern areas are environmentally suitable (Suthurst et al. 1989). The OWSWF is a substantial threat as it is prevalent in the neighbouring countries of Papua New Guinea and Indonesia. Perhaps more importantly, it is also present in many of our trading partners in Southeast Asia. The export of live cattle from northern Australian ports to Southeast Asian nations has become an important and rapidly expanding trade. The fly is also present near ports in the Middle East which receive considerable numbers of live sheep exported from various Australian ports. In both situations, there is the ever-present opportunity for the screw-worm fly (SWF) to return to Australia as larvae, pupae or adults on stock carrying vessels. While this means of transport probably represents the most likely route for the pest to gain access to Australia, accidental transport on aircraft and active myiases on humans or companion animals remain possibilities.

The introduction of the NWSWF into Australia is considered unlikely, but not impossible. Indeed, in 1992, an Australian tourist returning from South America carried live NWSWF larvae into Australia in a neck wound. Fortunately, the diagnosis was made early and the wound was treated, preventing any larvae evacuating. This was in southern Australia in May, where climatic conditions are unlikely to favour survival at that time of the year, but it demonstrates the potential for inadvertent introduction.

Models of the impact of the OWSWF indicate that the cost of an invasion would be high. It has been estimated by Anaman et al. (1993), that the annual cost in an average climate year to beef cattle, sheep and dairy producers of an endemic establishment would be A$281 million. These costs would be trivial in comparison with the societal cost (McKelvie et al. 1993). Extensive cattle grazing is the dominant industry throughout much of the northern pastoral areas of Australia. It is likely, based on USA experience, that extensive cattle production, as practised in northern Australia, would not be viable if SWF became established.

The extensive grazing industries could theoretically maintain herd numbers in the presence of the OWSWF; however, the labour costs of doing so would be crippling. In areas where the fly is endemic, e.g., Malaysia, beef cattle must be inspected as frequently as every three days in order to deal with new myiasis cases. This is in stark contrast to practices in the Australian extensive pastoral zone, where cattle are inspected rarely, perhaps as infrequently as once or twice each year. In the presence of the OWSWF, this would result in high mortality among adult cattle and extensive mortality among newborn calves, causing serious reduction in recruitment. A characteristic site of

OWSWF infestations is the navel of new-born calves which if untreated, quickly cause the death of the calf. The failure of the livestock industries would in turn impact severely on the small to medium towns servicing the industries and, as there are few or no viable alternative business opportunities, many could die.

Australia's native fauna is naive to this pest and it is inevitable that some impact on the fauna would occur if an incursion occurred, although it is impossible to predict the extent. It is of concern to note that the NWSWF finds the pouch of the American opossum an attractive oviposition site, and if the OWSWF was similarly attracted, the impact on our diverse marsupial fauna could be severe. Human cases of myiases also occur, although the incidence is poorly documented (Searson et al. 1992). An exception is the records from Iraq where numerous human myiasis cases have occurred following an invasion of the OWSWF into that country in 1996.

The Department of Primary Industries and Energy (DPIE) Australia has developed contingency plans to respond to a number of exotic diseases, in collaboration with state governments and the livestock industries. The OWSWF is considered one of the most serious among them. The Australian Veterinary Emergency Plan (AUSVETPLAN 1996) contains a strategy for the control and eradication of the screw-worm fly if ever it gains a foothold on our continent. The policy is to eradicate the pest in the shortest possible period, while limiting economic impact using a combination of strategies including:

- The *sterile insect release method (SIRM)* to control and eradicate the fly.
- *Quarantine and movement controls* in declared areas to prevent the movement of infested animals.
- *Decontamination and disinfestation* of larvae/pupae-contaminated areas and things.
- *Tracing and surveillance* to determine the source and extent of the infestation, and to provide proof of freedom from the disease.
- *Treatment* of individual animals and groups, to prevent or cure infestation especially before movement.
- *Zoning* to define infected and disease-free areas.
- *A public awareness campaign* to encourage rapid reporting of suspected infestations, and to facilitate cooperation from industry and the community.

A fundamental plank in the eradication plan is the implementation of SIRM. Models prepared for DPIE using the best available data (unfortunately largely from NWSWF sources) indicate that an 8:1 benefit-cost ratio would be achieved through undertaking an eradication programme (Anaman et al. 1993). The modelling indicates that large areas of tropical and sub-tropical Australia are suitable for year round survival of the SWF, with further southern extensions in summer that would recede after frosts (Anaman et al. 1993). Incursions in southern areas would not persist over winter, but incursions or spot outbreaks in the south from northern endemic areas could be active over spring, summer and autumn.

The OWSWF has great dispersal ability (Spradbery et al. 1995) and it is anticipated that by the time a SIRM programme can be mounted, the fly will have colonised 500,000 km$^2$. Under this scenario, eradication would take nearly two years. It is envisaged that a facility capable of producing 200 or 250 million flies per week would be required to achieve eradication. Such a facility would be extremely expensive to construct and there would be intense pressure to complete it as quickly as possible to prevent further spread of the fly and to minimise economic losses. Under such circumstances, it would be advantageous to know exactly what should be constructed, and perhaps even have blueprints for tried and tested systems on the shelf ready to implement. This is the rationale for a experimental mass rearing facility, the Institut Haiwan Screw-worm Fly Laboratory in Kluang, Malaysia (Mahon and Ahmad, in this volume), where we hope to apply production engineering methods to the rearing of the OWSWF. In addition, a demonstration that mass reared and sterilised OWSWF were fit under field conditions is considered important in order to confirm and provide confidence in the efficacy of a SIRM programme before embarking on such a large and expensive exercise. This is a collaborative project between DPIE and the Department of Veterinary Services, Malaysia (DVS). DVS has provided the land and a number of staff, while DPIE has provided the capital for the construction of the facility, and Australian producer research and development funding bodies (the Meat Research Corporation, the Australian Wool Research and Promotion Organisation and the Dairy Research and Development Corporation) are providing the funds for the research and development (R & D) programme.

At the completion of this project, a design brief will be produced that would facilitate the construction of a 250 million sterile flies per week facility within Australia if ever it were required. Models indicate that there is merit in construction of the facility and mothballing it until required, though funding of this option may prove difficult. Nevertheless that option will be presented to producers to consider. Insurance policies against insects are expensive!

A multi-species sterile insect production facility was another attractive option evaluated by Anaman et al. (1993). The multi-insect facility concept would see the plant being used for production of sterile insects for control or eradication of endemic pests (e.g., the Queensland fruit fly, the Australian sheep blowfly) until an incursion by an exotic insect pest. The already-operational plant would then be fairly quickly diverted (in full or part) to the production of sterile SWF or other exotic insect horticultural pests. Since production could start earlier in an outbreak, a smaller capacity would be required, as the pest distribution would still be restricted. The estimated benefit-cost ratios for a range of options, based on producer losses only (Anaman et al. 1993) or economic surplus changes for producers and consumers (McKelvie et al. 1993), are very favourable. A full feasibility study of a multi-insect facility is yet to be undertaken.

SIRM could be ineffective if *Chrysomya bezziana* populations from different geographic locations proved to be a complex of sibling species (Strong and Mahon 1991). To explore this potential limitation, a study was initiated to determine the extent of inter-population variability within what we know as *C. bezziana*. Collections were made from as many localities as possible throughout the species' geographic range, and included South Africa and Zimbabwe in southern Africa, Oman in the Middle East, Malaysia, Indonesia and Papua New Guinea in Southeast Asia. A range of techniques previously shown to indicate the presence of sibling species and/or variability within insect populations was applied to the collection. The allozyme study (Strong and Mahon 1991) indicated that intra-population variation was remarkably limited, despite the extensive geographical range of the species. From this result, Strong and Mahon (1991) postulated that as gene flow from the extremes of the range is highly unlikely, it is likely that the species has "recently" expanded its range. Perhaps the range expansion occurred with human aid. Hybridisation tests (Spradbery, unpublished) indicated that adults from different populations would mate, and viable and fertile offspring could be produced. Cytogenetic studies (Bedo et al. 1994) and biochemical profiles of the cuticular hydrocarbons also indicate that while minor variability occurs, there is no indication that *Chrysomya bezziana* consists of a complex of sibling species. Thus, while it would be ideal if the colony used to breed sterile flies for SIRM is from the same location as the source of an incursion into Australia, probably any colony would be effective.

There are other components of the AUSVETPLAN designed to both minimise losses and supplement a SIRM eradication programme. An early warning system has been established as part of the North Australian Quarantine Strategy (NAQS), which is based on enhanced quarantine surveillance, education, and a regular trapping programme using "Swormlure", an attractant developed for NWSWF. A review of the NAQS SWF monitoring programme was undertaken in 1991 (Nunn et al. 1991). As a result of the report, the trapping sites and monitoring programme were modified to improve the chances of early detection. Particular care is taken to prevent the entry of the SWF with returning livestock vessels. Educational programmes have also been implemented to make livestock owners and residents of northern Australia and the Torres Strait more aware of the SWF. Kits have been provided and promoted to encourage the submission of larvae from strikes on animals.

Despite the surveillance, depending on the location of an incursion of the OWSWF, it might be several months before an incursion is detected. Unless a SIRM facility was in existence, a further delay of between one to two years would occur before a SIRM programme could be implemented.

In the interim, restrictions on stock movements would be applied within the infested area in order to limit the spread of the pest though transfer of infected stock. A restricted area will be declared enclosing the infested area and movement of animals within or out of the restricted area will be allowed, subject to inspection, treatment and permit. Treatment with avermectin, a systemic drug, will be mandatory to all animals prior to movement, with movement permitted 7–14 days after treatment.. Animals showing clinical evidence of SWF infestation will be sampled and treated with approved insecticidal smears/pressure packs at the time of the systemic insecticide treatment and inspected again before moving. A larger Control Area will be declared around the Restricted Area. Movement of animals within, or out of, the control area will be by inspection and permit, without treatment, unless required.

As natural dispersal of the adult fly and dispersal via myiases on native fauna will be inevitable, barriers around known infested areas will be forced to continually retreat until the SIRM programme can be implemented. Surveillance for the presence of the pest in as yet uninfested areas will be important, in order to identify new outbreaks and erect new barriers to limit stock movements.

Ivermectin and avermectin are effective systemic parasiticides against larval stages of the OWSWF (Spradbery et al. 1991) and would be used for the treatment of infected animals. Topical treatment of struck wounds with smears or pressure pack insecticides, followed by a systemic insecticide for at-risk animals, is recommended. Ivermectin prophylaxis is preferable for animals wounded in the course of normal husbandry procedures. Wounds are still attractive to the SWF after ivermectin treatment; however, larvae hatching from egg masses laid by female flies will be killed.

Treatment of all myiases located would help reduce the population of the OWSWF, and thereby perhaps, the incidence of new myiases, and should reduce the rate of dispersal. Native and feral animals are a problem, since they cannot be effectively treated. The new chemical, Moxidectin (Cyanamid), has recently been released onto the Australian market. Moxidectin has similarities to the ivermectin group but may be less damaging to the environment. This drug has not been assessed for efficacy against larval OWSWF. An assessment of the efficacy of Moxidectin and new formulations of ivermectin against the OWSWF is proposed.

Public awareness and early reporting of suspect myiases are given major emphasis both prior to and in response to an outbreak. A video, *Recognising Exotic Livestock Disease No. 7: Screw-worm Fly*, has been produced as part of a series to alert veterinarians and other health professionals, to the threat of SWF. It instructs on the epidemiology and clinical signs of the disease. The SWF is also included in general awareness material published from time to time, as well as in feature articles, radio interviews and other materials. Specimen collection kits are supplied to producers (and health centres) in northern Australia to encourage submission of larvae from myiasis strikes when seen. A diagnostic manual has been prepared (Spradbery 1991) and training given in identifying and differentiating the SWF from other species. The NAQS education programme in northern Australia also has a significant emphasis on the SWF.

In conclusion, the approach Australia has taken to provide insurance against invasion by the SWF, has been to undertake preventative and early recognition measures, to identify and quantify the economic and public health threats, and finally, to develop the science for control and eradication that should enable an effective eradication response to be implemented.

## REFERENCES

Anaman, K.A., M.G. Atzeni, D.G. Mayer, M.A. Stuart, D.G. Butler, R.J. Glanville, J.C. Walthall and I.C. Douglas. 1993. Economic assessment of the expected producer losses and control strategies of a screwworm fly invasion of Australia. Queensland Department of Primary Industries, Project Report Series QO93016.

AUSVETPLAN 1996. The Australian Veterinary Emergency Plan, Edition 2.0. Canberra, Department of Primary Industries and Energy.

Bedo, D.G., J.P. Spradbery and R.J. Mahon. 1994. Cytogenetic variation in natural populations of the Old World screwworm fly *Chrysomya bezziana* (Diptera: Calliphoridae). Genome 37: 390-398.

McKelvie, L., K. Hamal and R. Reynolds. 1993. Producer and consumer welfare effects of an invasion of screwworm fly in the Australian livestock sector. ABARE Report to the Queensland Department of Primary Industries, Brisbane, Australia.

Nunn, M.J., S. Smith and J.P. Spradbery. 1991. A review of screw worm fly surveillance in Northern Australia. Bureau of Rural Resources, Canberra.

Searson, J., L. Sanders, G. Davis, N. Tweddle and P. Thornber. 1992. Screwworm fly myiasis in an overseas traveller – case report. Communicable Disease Intelligence, 16: 239-240.

Spradbery, J.P. 1991. *A Manual for the Diagnosis of Screw-worm Fly*. Department of Primary Industries and Energy, Canberra.

Spradbery, J.P., R.J. Mahon, R. Morton and R.S. Tozer. 1995. Dispersal of the Old World screw-worm fly *Chrysomya bezziana*. Med. Vet. Ent. 9: 161-168.

Spradbery, J.P., R.S. Tozer, N. Drewett and M.J. Lindsey. 1991. The efficacy of ivermectin against larvae of the screw-worm fly (*Chrysomya bezziana*). Aust. Vet. J. 62:311-314.

Strong, K.L. and R.J. Mahon. 1991. Genetic variation in the Old World screw-worm fly, *Chrysomya bezziana* (Diptera: Calliphoridae). Bull. Entomol. Res. 81: 491-496.

Suthurst, R.W., J.P. Spradbery and G.F. Maywald. 1989. The potential geographical distribution of the Old World screw-worm fly, *Chrysomya bezziana*. Med. Vet. Entomol. 3: 273-280.

# Incidence of Old World Screw-worm Fly, *Chrysomya bezziana* in Iraq

*Ayad A. Al-Taweel[1], Mohammed A.J. Al-Izzi[1] and Fadhil A. Jassim[2]*

[1] Entomology Department, IAEC, P.O. Box 765, Baghdad, Iraq
[2] Faculty of Veterinary Services, Ministry of Agriculture, Baghdad, Iraq

## INTRODUCTION

The Old World screw-worm fly (OWSWF), *Chrysomya bezziana* Villenuve, is a member of the insect family Calliphoridae and is an obligate parasite of warm-blooded animals in the tropics and sub-tropics (Norris and Murray 1964). Flies lay their eggs on the edge of wounds or body orifices; the resulting larvae invade the host tissues and produce lesions and infertility if the genitals become infested (Humphrey et al. 1980). Recorded hosts include cattle *(Bos indicus)*, sheep *(Ovis aries)*, goats *(Caprus hircus)*, dogs *(Canis familiaris)*, cats *(Felis domesticus)* and man *(Homo sapiens)* (Patton 1920, 1922, Stoddar and Peck 1962, Norris and Murray 1964). This investigation describes the incidence of myiasis caused by *C. bezziana* in Iraq from September 1996 to March 1998.

## MATERIALS AND METHODS

On 9 September 1996, Iraqi veterinary officers noticed an unusual myiasis case presentation along the river bank of the Tigris river within the Baghdad governorate. The infestation caused a hole inside the infested tissues which resulted in rotten areas, accompanied by filtration of liquid serum and distortion of the tissue with larvae of different stages. The veterinary officers were advised to collect a sample of at least ten larvae and treat wounds with available insecticidal preparation. These larvae were sent to a main governmental veterinarian clinic in Baghdad for identification with the collaboration of the Natural History Museum/Baghdad University following Zumpt (1965) and Spradbery (1991). This method has been followed in all governorates where such natural infestation is detected. On the other hand, egg masses and larvae were reared in the entomology laboratory/IAEC on a meat diet consisting of:

| | |
|---|---|
| Minced beef | 58% |
| Whole blood | 15% |
| Distilled water | 27% |

Blood was obtained from the jugular venepuncture of cattle or from slaughterhouses and EDTA (1%) was added to prevent clotting. These egg masses and larvae were put on the meat based diet in small containers and maintained at a temperature of 37°C in an incubator. The diet was changed regularly. At about six days, the larvae migrated from the food, or were transferred to trays containing corncob grits for pupation. The pupae were maintained at room temperature for about 6-8 days; then the young flies were collected for identification following the Spradbery manual (1991) or were kept for rearing.

## RESULTS AND DISCUSSION

During the 19 months of the census, 58,063 cases of myiasis were recorded (Table 1) with 13.9% occurring during 1996 from the time the first case was reported on 9 September 1996, 80.7% occurring during 1997 and 5.4% occurring during the first three months of 1998. This was in spite of the use of available insecticides such as ectopor, diazinon, nagasunt and coumaphos. Furthermore, 22 human cases were also recorded with 54.5%, 22.7%, 9.1%, 9.1% and 4.5% in the Diyala, Babil, Karbala, Basrah and Wassit governorates respectively.

Table 1. Number of Old World screw-worm fly myiasis in Iraq distributed according to the governorates.

| Governorate* | Date of first record | Myiasis cases in animals** | | | Total |
|---|---|---|---|---|---|
| | | Up to 31/12/96 | 1/1/97 31/12/97 | 1/1/98 - 31/03/98 | |
| Baghdad (C) | 09/09/96 | 1802 | 9327 | 737 | 11866 |
| Diyala (E) | 12/09/96 | 1604 | 899 | 204 | 2707 |
| Karbala (W) | 17/09/96 | 2935 | 10422 | 254 | 13611 |
| Wassit (S) | 23/09/96 | 119 | 1089 | 95 | 1303 |
| Babil (W) | 25/09/96 | 823 | 17827 | 461 | 19111 |
| Qadesiya (S.W.) | 29/09/96 | 53 | 2129 | 69 | 2251 |
| Anbar (N) | 14/10/96 | 11 | 1863 | 506 | 2380 |
| Najaf (W) | 27/10/96 | 693 | 2673 | 742 | 4108 |
| Muthanna (S) | 02/01/97 | - | 463 | 7 | 470 |
| Tikrit (N) | 01/10/97 | - | 30 | 0 | 30 |
| Thi-Qar (S) | 11/11/97 | - | 148 | 56 | 204 |
| Missan (S) | 20/11/97 | - | 3 | 0 | 3 |
| Basrah (S) | 14/12/97 | - | 1 | 18 | 19 |
| Total | | 8040 | 46874 | 3149 | 58063 |

* C = Centre; E = East (relative to C); W = West; S = South and N = North

** Weekly data collected at government veterinarian clinics

The monthly incidence of myiasis in relation to prevailing meteorological conditions suggested that myiasis was more common during the cooler seasons of the year and rare during the hot, dry summer (Table 2). Furthermore, this table also shows that 4.6% of cases occurred during autumn 1996, 21.1% in winter 1996-1997, 2.2% in spring 1997, 1.5% in summer 1997 and 25.7% and 44.7% in autumn 1997 and winter 1997-1998 respectively.

Table 2. Number of Old World screw-worm fly myiasis cases recorded in Iraq monthly.

| Month | Total Myiasis Cases | Temperature | | |
|---|---|---|---|---|
| | | Average | Min. | Max. |
| September 1996 | 266 | 30 | 22 | 48 |
| October 1996 | 433 | 25 | 13.5 | 43 |
| November 1996 | 1959 | 16 | 13 | 30 |
| December 1996 | 5386 | 11 | 5 | 27 |
| January 1997 | 4361 | 10 | -3 | 23 |
| February 1997 | 2418 | 11.5 | -3 | 22 |
| March 1997 | 999 | 16 | 11.5 | 26.5 |
| April 1997 | 180 | 22 | 7 | 33 |
| May 1997 | 113 | 26 | 18 | 41 |
| June 1997 | 179 | 32 | 21 | 44 |
| July 1997 | 254 | 35 | 24 | 45 |
| August 1997 | 431 | 34 | 25 | 46 |
| September 1997 | 766 | 30 | 21 | 44 |
| October 1997 | 3123 | 25 | 15 | 39 |
| November 1997 | 11042 | 16 | 13 | 25 |
| December 1997 | 23004 | 12.5 | 9 | 24 |
| January 1998 | 2462 | 9 | 4 | 16 |
| February 1998 | 353 | - | - | - |
| March 1998 | 334 | - | - | - |

Finally, it can be seen from Table 3 that the cases of OWSWF myiasis in the four governorates of the central region of Iraq fluctuated throughout the nineteen months of the survey in spite of the use of insecticides. For example, in Baghdad, the lowest number recorded by governmental veterinarian clinics was during April 1997 when the minimum and maximum temperatures were 7°C and 33°C respectively, while the highest number recorded was during December 1997 when the minimum and maximum temperatures were 9°C and 24°C respectively.

Table 3. Number of Old World screw-worm fly myiasis in the central region of Iraq.

| Month | Baghdad | Diyala | Karbala | Babil |
|---|---|---|---|---|
| September 1996 | 153 | 71 | 36 | 2 |
| October 1996 | 82 | 165 | 107 | 75 |
| November 1996 | 520 | 431 | 739 | 178 |
| December 1996 | 1047 | 937 | 2053 | 568 |
| January 1997 | 752 | 102 | 2075 | 124 |
| February 1997 | 229 | 139 | 404 | 137 |
| March 1997 | 108 | 0 | 324 | 66 |
| April 1997 | 6 | 0 | 1 | 20 |
| May 1997 | 21 | 4 | 1 | 1 |
| June 1997 | 29 | 1 | 24 | 2 |
| July 1997 | 88 | 3 | 44 | 0 |
| August 1997 | 109 | 0 | 170 | 5 |
| September 1997 | 119 | 0 | 108 | 289 |
| October 1997 | 588 | 35 | 1014 | 988 |
| November 1997 | 972 | 508 | 1306 | 6172 |
| December 1997 | 6306 | 107 | 4951 | 8077 |
| January 1998 | 622 | 0 | 254 | 312 |
| February 1998 | 85 | 0 | 0 | 124 |
| March 1998 | 30 | 204 | 0 | 25 |

It is concluded from these data that fast action is needed to control this pest because using insecticides alone is not enough. Therefore, we believe that the sterile insect technique (SIT) is the only successful method for controlling it.

## REFERENCES

Humphrey, J.D., J.P. Spradbery and R.S. Tozer. 1980. Exp. Parasitol. 49:381.

Norris, K.R. and M.D. Murry. 1964. CSIRO Division of Entomology. Technical paper number 6. Canberra, Australia.

Patton, W.S. 1920. Ind. J. Med. Res. 8: 17.

Patton, W.S. 1922. Ind. J. Med. Res. 9: 635.

Spradbery, J.P. 1991. A Diagnostic Manual for Screw-worm Fly. CSIRO, Australia Division of Entomology, 64.

Stoddard, H. and E.F. Peck. 1962. Vet. Rec. 74: 591.
Zumpt, F. 1965. Myiasis in man and animals in the old world. London, Butterworth, XV, 267.

# Cuticular Hydrocarbons of *Glossina austeni* and *Glossina pallidipes*: Similarities Between Populations and Activity as Sex Pheromones

*D.A. Carlson*[1], *B.D. Sutton*[2] *and U.R. Bernier*[1]

[1] USDA, Agricultural Research Service, Center for Medical Agricultural and Veterinary Entomology, P.O. Box 14565, Gainesville, FL 32604, USA
[2] Department of Agriculture and Consumer Affairs, State of Florida, Gainesville, FL 32611, USA

## INTRODUCTION

Tsetse flies are a hazard to the health of humans and domestic animals because they spread trypanosomiasis, also known as nagana. *Glossina austeni* Newstead and *Glossina pallidipes* Austen are important vectors of this disease in East Africa. Sex pheromones were shown to be present in the surface or cuticular hydrocarbon waterproofing waxes of female of several species of the tsetse fly (Huyton et al. 1980). The pheromones identified in *Glossina morsitans morsitans* Westwood (Carlson et al. 1978) and *G. pallidipes* (Carlson et al. 1984, McDowell et al. 1985) have been shown to consist of species-specific, long-chain, high molecular weight hydrocarbons with several methyl branches, present with at least 20 other hydrocarbon compounds in the surface waxes (Nelson and Carlson 1986, Nelson et al. 1988, Sutton and Carlson 1997). The assignment of KI (Kovacx Index) narrows the range of possible methyl-branch configurations in cases of ambiguous or insufficient EI (electron impact) spectra (Carlson et al. 1998). We used gas chromatography/mass spectrometry to demonstrate that different populations of tsetse flies (Carlson et al. 1993) are closely related by investigating these patterns of surface hydrocarbons.

## *G. austeni*

### Chemical separation

We attempted to isolate and identify sex pheromones found in female flies, by open-column chromatography. Bioactivity was found only in the surface hydrocarbons extracted from females. Normal phase column chromatography on silica gel and argentation chromatography indicated that the hydrocarbon fraction contained both alkanes and alkenes (unsaturated hydrocarbons with one double bond) (Carlson and Langley 1986). Gas chromatography (GC) of the cuticular hydrocarbons of individual specimens and pooled samples was supported by GC-mass spectrometry (GC-MS) to identify the hydrocarbons found in each GC peak. The major alkanes from females were consistent with previously published data on this species, which demonstrated that they possess two homologous isomers, 13,17-dimethyltritriacontane and 13,17-dimethylpentatriacontane. The two major alkene peaks in females also contained previously unknown compounds, namely 13,17-dimethyltritriacont-1-ene, and 13,17-dimethylpentatriacont-1-ene, structures determined by GC-MS after derivation with deuterium or dimethyldisulfide (Carlson et al. 1989) (Figure 1).

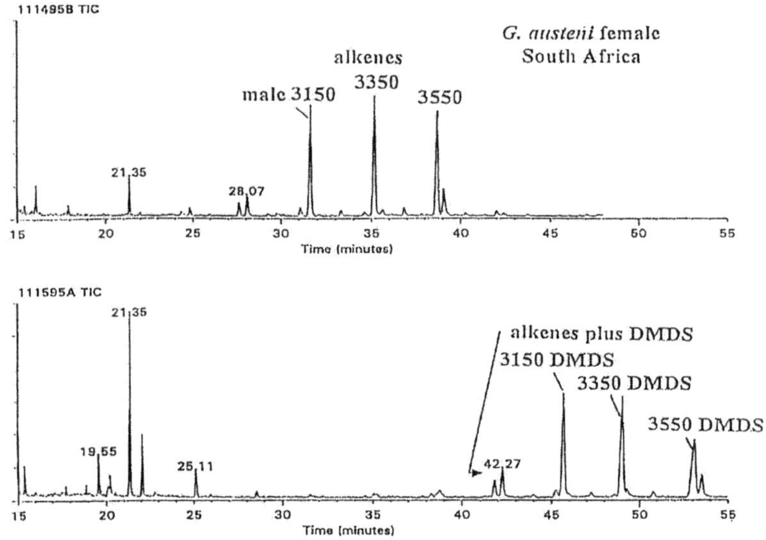


Figure 1. Gas Chromatographic analysis of alkenes from mated *G. austeni* females before and after formation of derivatives.

## Bioassays

Contact sex stimulant activity was studied in bioassays in Austria and Tanga, Tanzania, using tests similar to those used in previous studies with other species (Carlson et al. 1978, 1984). Test showed that the alkanes were not active, but that the unsaturated fraction was active. Dose-response data showed $ED_{50}$ at ca. 5 μg per decoy, using solvent-washed female flies of different species as decoys. This is the first example in a tsetse fly that sexual activity was initiated in males by alkenes, including most of the mounting behaviours released by contact with conspecific females. So far as is known, these alkenes were structurally the same among all specimens examined. Whether one or both of these compounds are responsible for this behaviour is not known.

## Analysis of other specimens

The alkane fractions of conspecific females from seven locations were analysed to determine differences and similarities. They appeared to cluster into three groups: 1) Bristol laboratory, ILRAD/Kenya/Bristol, Maisons-Alfort/Bristol, Vienna, Tanga laboratory, 2) Mozambique wild, Natal-Zululand-SA wild, and 3) Kenya wild (Figure 2). The significance of these differences is not known, since the alkanes are not known to possess biological activity.

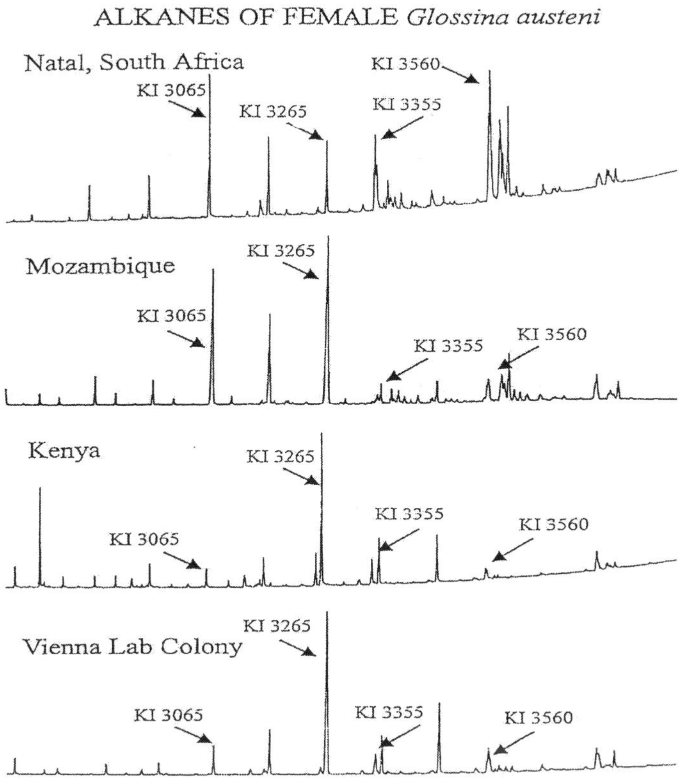


Figure 2. Gas Chromatographic analysis of alkanes from *G. austeni* females from different locations.

The alkenes of these specimens are under investigation, but appear universally the same by GC. There appear to be only two compounds present in the active fraction from unmated females. The male-produced compound is structurally similar, but its biological activity is unknown.

## *G. pallidipes*

### Chemical separation

The sex stimulant pheromone of this species consists of female-produced 13,21-dimethylpentatriacontane, with smaller amounts of chemical homologues. The alkanes of conspecific females from several locations were analysed to determine differences and similarities with published work on the sex pheromone of the species. The surface lipids were obtained by extracting individual specimens and/or pooled specimens in hexane, then concentrating the extracts before normal phase chromatography and argentation chromatography. The alkane fractions were analysed by GC (Figure 3) and GC-MS.

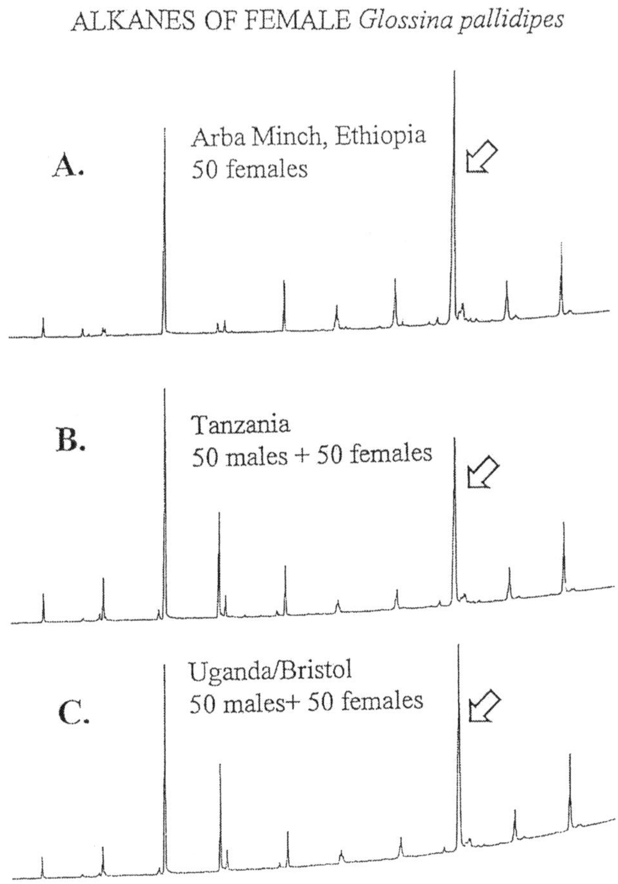


Figure 3. Gas Chromatographic analysis of alkanes from *G. pallidipes* females from different locations.

**Bioassays**

No bioassays have been conducted recently.

**Analysis of other specimens**

New specimens and several older samples were re-examined recently. All females analysed contained a major component consisting of 13,21-dimethyl C35, a minor 11,21-dimethyl C35 isomer and much smaller amounts of homologous 13,21-dimethyl C36 and 37 carbon alkanes. These recent results, and data published previously (Table 1), were compared. The quantities of the major C35 pheromone found were similar, at 7 to 13 μg per female, although the old laboratory females contained more. The results appeared to cluster the females into two groups: 1) Amsterdam, ICIPE/Kenya, Kenya, Tanzania wild, Uganda/Bristol, Arba Minch/Ethiopia, 2) Zimbabwe wild (not shown). The biological significance of these differences is not known.

Table 1. Sex pheromone components recovered from female *G. pallidipes**.

| Origin of females | 11,21-/13,23-$Me_2C_{35}$** proportions found (year) | Quantity in µg/female (age in days) |
|---|---|---|
| Bristol | 1/5 | 7.4 |
| | 1/3 (1978) | |
| Mozambique | Trace/1 | |
| Zambia | Trace/1 | |
| Uganda/Amsterdam | 1/8 | |
| Zimbabwe wild | 1/6, trace/1 | 13 |
| Kenya/Austria colony | 1/6.6 | 11 (8 d) |
| Kenya/Austria colony | | 26 (60 d), 18 (120 d) |
| Kenya (McDowell, ICIPE) | | 81+ |
| Arba Minch, Ethiopia | 1/3 (1998) | |
| Uganda | 1/3 (1998) | |

** Most estimates from GC-MS results, 1982-1985
+ Quantity of 13,17-dimethylpentatriacontane estimated
* From Carlson et al. 1984

## DISCUSSION

The technical information presented here may be useful in several circumstances. First, specimens of tsetse flies collected from different populations can be chemically analysed to ensure that the cuticular chemistry and the sex pheromone, at least, are correct before a new scheme using the sterile insect technique (SIT) begins. Second, knowledge of a species-specific sex pheromone may help in modern biocontrol efforts against this disease vector by ensuring that a competitive strain of fly is used in large-scale sterile male releases. For quality control, colony females can be analysed to see that they produce such a material. Third, sexual behaviour in reared males intended for mass release can be checked against a standard. Fourth, a synthetic pheromone might be used to increase the very slow rate of reproduction in laboratory-reared flies.

The activity of the racemic synthetic sex pheromone was shown conclusively in Wageningen and ICIPE *G. pallidipes* males in 1984-86 tests (Carlson et al. 1984), although the compound was less active in bioassays against Zimbabwe males (unpublished data). This compound was present in all female *G. pallidipes* examined recently, in about the same amount, which suggests that the flies are very similar.

After the success of SIT against *G. austeni* in Zanzibar, plans have been advanced to establish SIT schemes in other locations. There have been questions about the relatedness of widely-separated populations of these flies, for example: 1) Are flies from distant populations still the same species, have they become subspecies, or strains? 2) Do they have different sexual behaviours? 3) Will males from one location recognise females from another location as conspecifics? 4) Do females from different populations possess the same sex stimulant pheromones recognised by males? 5) Do the differences found between populations by molecular techniques (RFLP[restriction fragment length polymorphism], isozymes) have meaning if the sex pheromones between the populations are the same? 6) Are the compounds detected possible kairomones for parasitoid wasps that prefer to attack tsetse flies? 7) Can some elements of quality control for colonised flies be addressed using GC?

We believe that these and other important questions may be addressed by the use of information on cuticular hydrocarbons and sex pheromones in tsetse flies.

## ACKNOWLEDGEMENTS

We (DAC) thank J. Hendrichs, A. Robinson and Udo Feldmann of the International Atomic Energy Agency for the opportunity to pursue this effort in Austria at the Seibersdorf laboratory and for travel funds to attend the FAO/IAEA Conference. We thank Andrew Parker, then at TTRI, Tanga, Tanzania for his assistance. We thank M. Hosack for her technical assistance.

## REFERENCES

Carlson, D.A. and P.A. Langley. 1986. Tsetse alkenes: Appearance of novel sex-specific compounds as an effect of mating. J. Insect Physiol. 32: 781-790.

Carlson, D.A., U.R. Bernier and B.D. Sutton. 1998. Elution patterns of methyl-branched alkanes by capillary GC. J. Chem. Ecol. 24: 845-865.

Carlson, D.A., P.A. Langley and P.M. Huyton. 1978. Sex pheromone of the tsetse fly: isolation, identification and synthesis of contact aphrodisiacs. Science 201: 750-753.

Carlson, D.A., S.K. Milstrey and S.K. Narang. 1993. Genetic classification of the tsetse flies using cuticular hydrocarbons. Bull. Entomol. Soc. 83: 507-515.

Carlson, D.A., D.R. Nelson, P.A. Langley, T.W. Coates, T.L. Davis and M. Leegwater-Vander Linden. 1984. Contact sex pheromone in the tsetse fly *Glossina pallidipes* (Austen); Identification and synthesis. J. Chem. Ecol. 10: 429-450.

Carlson, D.A., C.-S. Roan, R.A. Yost and J. Hector. 1989. Dimethyl disulfide derivatives of long chain alkenes, alkadienes, and alkatrienes for gas chromatography/mass spectrometry. Analyt. Chem. 61: 1564-1571.

Huyton, P.M., P.A. Langley, D.A. Carlson and M. Schwartz. 1980. Specificity of contact sex pheromones in tsetse flies, *Glossina* spp. Physiol. Ent. 5: 253-264.

McDowell, P.G., A. Hassanali and R. Dransfield. 1985. Activity of the diastereomers of 13,23-dimethylpentatriacontane, the sex pheromone of *Glossina pallidipes,* and comparison with the natural pheromone. Physiol. Entomol. 10: 183-190.

Nelson, D.R. and D.A. Carlson. 1986. Cuticular hydrocarbons of the tsetse flies *Glossina morsitans, G. austeni* and *G. pallidipes.* Insect Biochem. 16: 403-416.

Nelson, D.R., D.A. Carlson and C.L. Fatland. 1988. Cuticular hydrocarbons of the tsetse flies, II: *G.p. palpalis, G.p. gambiensis, G. fuscipes, G. tachinoides* and *G. brevipalpis.* J. Chem. Ecol. 14: 963-987.

Sutton, B.D. and D.A. Carlson. 1997. Cuticular hydrocarbons of the *Glossina* (Diptera: Glossinidae): subgenera *Glossina* and *Nemorhina.* J. Chem. Ecol. 23: 1291-1320.

# Economic Impact of Eradicating the New World Screw-worm *(Cochliomyia hominivorax)* from Jamaica

*Trang T. Vo*

Policy Analysis and Development, Policy and Program Development, Animal and Plant Health Inspection Service, U.S. Department of Agriculture, Riverdale, Maryland, USA

## INTRODUCTION

The purpose of the study is to assess the economic feasibility of eradicating the New World screw-worm (NWS), *Cochliomyia hominivorax* (Coquerel), from Jamaica. The endemic presence of the NWS in Jamaica has caused the livestock sector to incur recurrent economic costs and losses. Eradication of the pest utilising the environmentally-benign sterile insect technique (SIT) has proved technologically feasible on other islands and other parts of the world. Based on these successful experiences, the proposed project to eradicate the NWS from Jamaica is expected to be similarly effective in eliminating the pest from Jamaica in a relatively short period of time. The elimination of the pest from the Caribbean would lessen the significant risk of re-infestation of eradicated areas and pest-free countries in the region.

## METHOD

In order to assess the feasibility of eradicating the NWS, information is required on the economic losses resulting from the presence of the pest in the Jamaica livestock population. The benefits of eradication is then considered to be the losses that are avoided from having the programme. Benefits are then compared with programme costs in order to determine the economic worth of eradication.

The study addresses the impacts only on livestock producers in terms of an accounting of costs and losses; impacts on consumers of livestock products, human health and pets are not considered. Information on pest incidence is obtained from a survey of 114 producers of cattle, goats, pigs and sheep. Results from the survey questionnaires are then applied to the total livestock population to obtain aggregate impacts. Five major categories of costs and losses are quantified. These include increased cost of production associated with:

- insecticides and medicine
- veterinary services
- additional time to finish animals
- additional labour for surveillance, treatment and prevention of wounds
- loss in value due to animal mortality.

## RESULTS

Because of the differences in magnitude of the findings, loss estimates due to mortality are based on two sources: one on the survey and the other from a 1984 study conducted by Rawlins and Sang (1984). These rates, along with the average infestation rates of young and adult animals as shown from the survey, are presented in Table 1.

Table 1. Infestation and mortality rates of screw-worm-infested livestock in Jamaica.

| Type of Livestock | Infestation Rates (%) | Mortality Rates from the Survey (%) | Mortality Rates from Rawlins and Sang Study (%) |
|---|---|---|---|
| Dairy cattle | 9.8 | 8.8 | 1.8 |
| Beef cattle | 15.1 | 6.0 | 1.8 |
| Goats | 18.3 | 27.5 | 1.8 |
| Pigs | 11.8 | 27.6 | 1.8 |
| Sheep | 12.6 | 4.8 | 1.8 |

The annual value of loss due to screw-worm deaths is estimated to be between US$624,000 (utilising the Rawlins and Sang mortality rate) and US$3 million using the rates from the recent survey.

The total increase in the cost of production of the Jamaican livestock industry of living with the screw-worm is estimated to be nearly US$5 million annually (Table 2). Expenditures for labour required for surveillance and treatment of animals comprise 74 percent of the additional production costs. Costs of insecticides and medicine represent over 23 percent of the screw-worm-related costs. These farm-level control measures that have been incorporated into livestock production practices have been the necessary tradeoffs towards minimising productivity losses.

In sum, the total losses due to mortality and increases in the cost of production are estimated to be between **US$5.5 million** and **US$7.8 million** annually. Once eradication is achieved, these costs could be avoided and can therefore be thought of as the benefits of the eradication programme.

Table 2. Increase in cost of production due to screw-worm flies.

| Production Cost Increase | US$ | | % of total cost |
|---|---|---|---|
| Total Chemical Costs | | 1,111,511 | 23.3 |
| - Cost of Insecticides | 1,004,616 | | |
| - Cost of Medicine | 106,895 | | |
| Cost of Veterinary Services | | 119,022 | 2.4 |
| Cost of Increased Sale Time | | 13,254 | 0.3 |
| Total Labour Cost | | 3,584,199 | 74.0 |
| - Surveillance and Prevention | 3,255,988 | | |
| - Treatment | 328,211 | | |
| Total increase in costs | | 4,827,986 | |

The eradication programme of the NWS from Jamaica is projected to take three years. Eradication is expected to be achieved by the end of the second year of initiating aerial dispersion of sterile screw-worm flies. The third year efforts would be dedicated to surveillance and monitoring activities. The project is estimated to cost between **US$4.9 million** and **US$9 million**.

Benefits and programme costs of eradication are simulated over ten years. The present values of the stream of benefits and costs, discounted at a rate of 11 percent, are then compared. Comparisons of the two ranges of benefits and programme costs are provided in Tables 3 and 4. Scenario 1 assumes a programme cost of US$4.9 million, while scenario 2 assumes a cost of US$9 million.

Table 3. Comparison of discounted benefits and costs at the end of year 3 and year 10, assuming benefits of US$5.5 million.

| Discounted Value of | After Year 3 | | After Year 10 | |
|---|---|---|---|---|
| | Scenario 1 (US$ million) | Scenario 2 (US$ million) | Scenario 1 (US$ million) | Scenario 2 (US$ million) |
| Benefits | 12.9 | 12.9 | 33.7 | 33.7 |
| Programme Costs | 4.5 | 8.2 | 4.5 | 8.2 |
| Net Savings | 8.4 | 4.7 | 29.2 | 25.5 |
| Benefit-Cost Ratio | 2.9:1 | 1.6:1 | 7.5:1 | 4.1:1 |

Table 4. Comparison of discounted benefits and costs at the end of year 3 and year 10, assuming benefits of US$7.8 million.

| Discounted Value of | After Year 3 | | After Year 10 | |
|---|---|---|---|---|
| | Scenario 1 (US$ million) | Scenario 2 (US$ million) | Scenario 1 (US$ million) | Scenario 2 (US$ million) |
| Benefits | 18.3 | 18.3 | 48.0 | 48.0 |
| Programme Costs | 4.5 | 8.2 | 4.5 | 8.2 |
| Net Savings | 13.8 | 10.1 | 43.5 | 39.8 |
| Benefit-Cost Ratio | 4.1:1 | 2.2:1 | 10.7:1 | 5.9:1 |

## CONCLUSIONS

Under both estimates of benefits, the eradication programme realises net savings after three years, ranging from **US$4.7 million** to **US$13.8 million**, depending upon the budget scenario chosen. Net savings are predictably higher after ten years, as benefits continue to accumulate beyond the targeted three-year eradication phase. Net benefits are estimated to range between **US$25.5 million** and **US$43.5 million** at the end of ten years. Benefit-cost ratios for all scenarios and time periods are correspondingly greater than one indicating the economic feasibility of the project investment under varying assumptions of benefits and programme costs.

In sum, benefits to producers would more than offset the programme cost almost immediately, given that full benefits ranging from US$5 million to US$8 million annually would be realised by the end of the second year of eradication. Out of the five categories of loss attributable to screw-worm estimated in this study, the greatest savings would be realised in terms of the reduction in the cost of livestock production. The labour required for surveillance and the use of insecticides to control screw-worms have resulted in high recurrent costs. These costs are necessary tradeoffs toward minimising production losses. With eradication of the pest, these additional resource expenditures could be redirected to other investment activities. Based on increases in the cost of production of about US$5 million, the *additional cost per animal* is estimated to be **US$5** for inspection and treatment-related costs.

Benefits from screw-worm eradication have important implications for Jamaica, in the light of the large distribution of subsistence farmers dependent upon small-animal holdings for their food source. The gains that could be potentially achieved in Jamaica serve as a compelling argument for the extension of the eradication initiative to the remaining endemic islands of the Caribbean. In addition to lessening the risk of re-infestation to pest-free areas, countries such as Haiti and the Dominican Republic could realise significant gains proportionate with their large livestock populations.

## REFERENCE

Rawlins, S.C. and J.C. Sang 1984. Screw-worm myiasis in Jamaica and proposals for its eradication. Trop. Pest Manag. 30(2): 125-129.

# The Vector Control Operations in the Onchocerciasis Control Programme in West Africa

*Jean-Marc Hougard*

Institut Francais de Recherche Scientifique pour le Developpement et Cooperation, Vector Control Unit, Onchocerciasis Control Programme, P.O. Box 549, Ouagadougou, Burkina Faso

## INTRODUCTION

Onchocerciasis is a dermal filariasis transmitted to man by a blood sucking blackfly belonging to the *Simulium* genus. The most serious manifestations of the disease are blindness and debilitating skin lesions. Africa is by far the most affected continent both in terms of distribution and severity of the clinical manifestations of the disease. That is the reason why an ambitious regional onchocerciasis control project, the Onchocerciasis Control Programme in West Africa (OCP), was launched in 1974 (Molyneux 1995). The objective is to eliminate onchocerciasis as a public health problem and as an obstacle to socio-economic development and to ensure that the countries are in a position to maintain these achievements. Seven countries were concerned at the beginning of the programme (Figure 1), delimiting the "initial area" (Bénin, Burkina Faso, Côte d'Ivoire, Ghana, Mali, Niger and Togo). In 1988, the OCP began operations in the "western extension", an additional four countries in the West (Guinea, Guinea Bissau, Senegal and Sierra Leone) and extended operations into the "southeastern extension" (south Bénin, Ghana and Togo). The rationale for these extensions related to findings that the vectors were able to migrate and hence re-invade controlled areas over several hundred kilometres (Garms et al. 1979). Until 1989, in the absence of a non-toxic drug which could be used on a wide scale to kill the adult worm, the vector control strategy was the only method to interrupt the transmission of the blinding form of the parasite until the adult worm in the human body was eliminated (the maximum duration of the adult worm is estimated to be about fourteen years). In the late 1980s, ivermectin, a microfilaricide which is the only drug available to date, became an integral part of the OCP control strategy (Webbe 1992). In the extension areas, larviciding is still going on with satisfaction, combined with the distribution of ivermectin. In pursuing this combined therapeutic and vector control strategy, the whole of the basins treated should be freed from blinding onchocerciasis by the year 2002 at the latest, which is the end point of the programme activities.

## THE VECTOR CONTROL ACTIVITIES

Vector control activities consist of treating with insecticides the rivers where larval stages of the vector, *Simulium damnosum* Theobald s.l., breed. Taking into account that the development of the aquatic stage which, from egg to pupae, hardly exceeds one week, the spraying is done on a weekly basis. In addition to this, the large number of breeding sites (actually over 15,000 km of river in the rainy season) and the difficulties to reach some of these sites by ground, explain why the aerial spraying method has been chosen. Vector control has encountered three major obstacles which have been successfully overcome. First, it was clearly established very early that the border of the treated area (654,000 km$^2$ spread over seven countries) was re-invaded by infective blackflies from regions outside the programme area. In order to protect permanently the original area and also clean the basins at the source of the re-invasions, the incriminated hyperendemic regions were identified and then put under larvicidal treatment. The second obstacle was the emergence, five years after the beginning of the programme, of a resistance of *S. damnosum* s.l. to temephos (Guillet et al. 1980). The compound (organophosphate) was the only insecticide used at the beginning of the programme. It was selected because of its efficacy, the distance over which it remains effective, its lack of impact on non-target fauna and also its acceptable cost. When the development of resistance to another organophosphate (chlorphoxim) was discovered thereafter, the programme hence adopted a strategy of rotational use of insecticides, where possible, from different chemical groups, with different modes of action, to slow down and to suppress the appearance of new cases of resistance (Hougard et al. 1993).

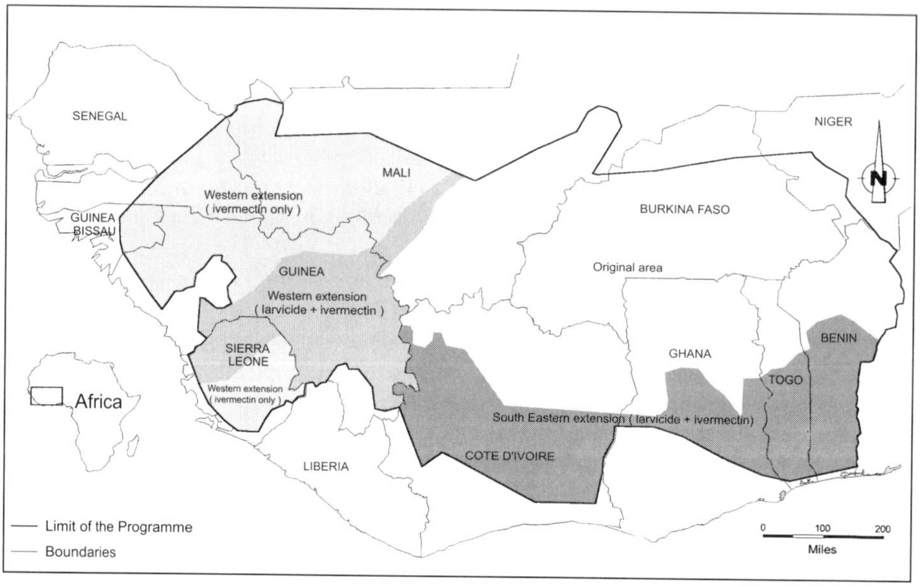


Figure 1. Onchocerciasis Control Programme in West Africa.

Seven insecticides are currently in use. Six of them are chemicals formulated as emulsifiable concentrates (temephos, phoxim, pyraclofos, permethrin, etofenprox and carbosulfan) while the seventh is a liquid concentrate of a biological insecticide (*Bacillus thuringiensis* H-14 or *B.t.* H-14). This rotational use of insecticides has been so effective that, currently, there is very little resistance left to the organophosphates in use while the susceptibility of the *Simulium* population to other compounds remains unchanged. The third obstacle to be dealt with was of environmental concern (Lévêque 1989). A blackfly larvicide, no matter how effective it is, could not be accepted for use by OCP if it had a short, an intermediate or a long-term deleterious effect on the environment. Temephos and *B.t.* H-14 are considered harmless to the environment and can be used without any particular restriction. On the other hand, pyraclofos may, in case of accidental overdose, show some toxicity against the non-target fauna, namely fish, and for that reason, it is not for use under water discharge of 15 $m^3$/s. Permethrin and carbosulfan should also be used with precaution, never below 70 $m^3$/s and, if possible, for not more than six weeks per year on the same river stretch. As the toxicity of etofenprox for fishes and crustaceans is lower than permethrin, its use is allowed to discharges above 15 $m^3$/s but without any restriction on the number of annual cycles. In general, all chemical insecticides should be used in such a way as to avoid any overdose that is likely to cause harm to the environment. For this reason, the programme has set up a satellite transmission network for recording water discharge allowing for in-time management of hydrological data in the treatment of the rivers (Servat and Lapetite 1990). In addition to this, treatment helicopters are equipped with a spraying system that allows for dosing of quantities of insecticides with an accuracy of around one centilitre.

## THE STRATEGY IN THE INITIAL AREA

Before the launching of the programme, the initial area was plagued mainly with the most severe form of onchocerciasis, the "savanna" type, which is characterised by a high incidence of blindness, and is transmitted by the group of "savanna-dwelling" species of the *S. damnosum* complex. In its principle, the vector control strategy in the initial area was thus simple to implement as it consisted in arresting transmission of the parasite regardless of its pathogenicity, through the elimination of its vector regardless of the species involved (Le Berre et al. 1979). From 1990, the first decisions to definitively stop larviciding were made for the basins in which the situation seemed to be satisfactory, both from the epidemiological and entomological points of view. Now, onchocerciasis is no longer a public health problem in the whole of the initial area. Nevertheless, despite the long duration of vector control activities, a few foci of infections remain, even if the clinical signs of the disease have totally disappeared (Boatin et al. 1997). Although these foci represent a negligible percentage of the oncho-free area, their persistence remains a cause for concern. Indeed, a risk of contamination of the adjacent basins is always possible and there is no guarantee that an alternative control strategy will succeed where the current strategy has partially failed. Therefore, the programme and the participating countries are paying particular attention to these foci so as to identify the factors which have hindered the success of control operations. To date and in the present state of our investigations, a succinct analysis of the residual onchocerciasis foci remaining in the original area allows for the identification of three main obstacles to a good achievement of vector control operations. One is a seasonal contamination by wind assisted migrating infective blackflies from hyperendemic areas. The second is linked to migrations of infected human populations coming from prevailing endemic areas. Another factor is the difficulty of accessibility to the area under control, resulting in an incomplete treatment of the breeding sites or in an inadequate entomological assessment network. The discovery of the characteristic features of these foci has made it possible to take a number of corrective actions principally based on the use of ivermectin. Indeed, a long-term control of morbidity through community directed distribution of ivermectin is the most reasonable solution. This method of distribution, which was retained by the African Programme for Onchocerciasis Control (APOC), directly involves the exposed communities (Dadzie 1997). In the present socio-economic context, this approach is probably the best way of ensuring the optimal and durable protection of the populations after the OCP ends.

## THE STRATEGY IN THE EXTENSION AREAS

The extension zones under larviciding are characterised by different types of vegetation ranging from the Sudanese and Guinean savanna to degraded forests. This diversity of landscape favours the presence of various habitats conducive to the development of all species of the *S. damnosum* complex, from the savanna-dwelling vectors to the main forest-dwelling ones. At the time of institution of the larvicide treatments, the idea of using a selective strategy exclusively targeted at the savanna-dwelling blackflies had been considered. However, the mechanism of transmission under natural conditions was not clearly understood because of the inability to distinguish between the strains of the parasite and the difficulty in accurately determining the identity of the adult infected females. This option was therefore discarded in favour of a less targeted strategy taking into account, thanks to the cytotaxonomic tools available (Crosskey 1987), the seasonal abundance of the savanna blackflies in their larval habitat. To date, the technique of morphological identification of adult blackflies has improved, making it now possible to differentiate the savanna species from the others at the same catching point (Wilson et al. 1993). The use of DNA probes allows for the differentiation of the savanna strain from the forest strain of the parasite and another technique allowing for the identification of the adult flies is being developed. These new methods are shedding further light on the mechanisms of transmission under natural conditions, showing that the link of savanna-dwelling vectors with the blinding strain of the parasite is not as close as suggested by the results of the xenodiagnostic studies conducted in the context of experimental transmissions (Toé et al. 1997). However, many other observations made in several basins of the extension areas suggest that the current strategy can be maintained without jeopardising the achievements of vector control. These differing observations led to uncertainties regarding the threat the "blinding" strain of the parasite represents in the extension areas. To attempt to know better, research is oriented in two directions. The first orientation, which is under way, is attempting to improve the performance of the molecular tools (microsatellites, heteroduplex) to allow for a still finer identification of the vector and parasite populations. The second orientation, which is at the planning stage, will call upon immunodiagnostic techniques. It will try to determine whether there exist different levels of pathologenecity in the parasite or, in man, any predispositions to develop blinding onchocercal lesions or not. Pending these results, and four years to the definitive cessation of larvicide treatments, the current vector control strategy which is a realistic compromise between an "overselective" option and, at the other extreme, a too "global" option, should be pursued without any major change until the conclusion of the programme. For the time being, vector control, combined with ivermectin distribution, is continuing in almost all of the basins of the extension areas. However, in a few river basins where the blinding strain of the parasite holds sway, transmission sometimes remains difficult to control despite continued larviciding and the distribution of ivermectin. As in the cases observed in the initial area, unsuspected transmission may subsist here and there and the possibility of a few residual foci of "savanna" onchocerciasis subsisting after the end of the programme cannot be excluded. In the absence of vector control measures and since a macrofilaricide is not suitable for mass treatment, ivermectin will also remain the only means of control of these residual foci after the year 2002.

## THE TRANSFER OF RESIDUAL ACTIVITIES

In less than five years, the residual activities of the Onchocerciasis Control Programme in West Africa will be fully in the hands of the participating countries. A few residual foci will still subsist both in the original and in the extension areas. However, their total size will remain fairly small compared to the 25 million hectares of fertile valleys which will be made free from the disease by the end of the programme. Nevertheless, the emergence of new onchocerciasis foci after 2002 in the areas considered to be freed cannot be excluded. Whatever the situation, the countries must be provided with the means to detect any onchocerciasis focus early before it becomes too difficult to control. As far as entomological surveillance is concerned, the necessary tools and know-how will be made available to the national health services to enable them to detect any transmission. For that purpose, a simple and cheap method of detection of the infective larvae of the parasite in blackflies is in the process of being operationalised (Yaméogo et al. 1997). Thanks to this method, the countries will be able to mobilise themselves even before a recrudescence of the infection in man, which is more difficult to control, is detected, through the various techniques currently available or in the process of becoming so. Finally, as far as vector control is concerned, all the activities will have probably ceased (Hougard and Sékétéli 1998). However, larviciding from the ground with non-toxic and non-persistent insecticides to control blackfly nuisance could increase, at least in those sites of socio-economic interest, where an efficient and sustainable management of such treatments can be ensured (Hougard et al. 1997).

## ACKNOWLEDGEMENTS

All my thanks goes to Sammy Sowah, a professional staff member of the OCP, in reading the English version of this manuscript.

## REFERENCES

Boatin, B., D.H. Molyneux, J.-M. Hougard, O.W. Christensen, E.S. Alley, L. Yaméogo, A. Sékétéli and K.Y. Dadzie. 1997. Patterns of epidemiology and control of onchocerciasis in West Africa. J. Helminthol. 71: 91-101.

Crosskey, R.W. 1987. A taxa summary for the *Simulium damnosum* complex, with special reference to distribution outside the control areas of West Africa. Ann. Trop. Med. Parasitol. 871: 181-192.

Dadzie, K.Y. 1997. Onchocerciasis control: the APOC strategy. Africa Health, March 1997: 13-15.

Garms, R., J.F. Walsh and J.B. Davies. 1979. Further studies on the reinvasion of the Onchocerciasis Control Programme in the Volta River Basin by *Simulium damnosum* s.l. with emphasis on the south-western areas. Trop. Med. Parasitol. 30: 345-362.

Guillet, P., H. Escaffre, M. Ouedraogo et D. Quillévéré. 1980. Mise en évidence d'une résistance au téméphos dans le complexe *S. damnosum (S. sanctipauli* et *S. soubrense)* en Côte d'Ivoire (zone du programme de lutte contre l'onchocercose dans la région du bassin de la Volta). Cah. ORSTOM, Sér. Entomol. Méd. Parasitol., 18(3): 291-299.

Hougard, J.-M., H. Agoua, L. Yaméogo, K.L.B. Akpoboua, A. Sékétéli and K.Y. Dadzie. 1997. Blackfly control: what choices after onchocerciasis? World Health Forum, 18: 1-7.

Hougard, J.-M., P. Poudiougo, P. Guillet, C. Back, L.K.B. Akpoboua and D. Quillévéré. 1993. Criteria for the selection of larvicides by the Onchocerciasis Control Programme in West Africa. Ann. Trop. Med. Parasitol. 87(5): 435-442.

Hougard, J.-M. and A. Sékétéli. 1998. Combatting onchocerciasis in Africa after 2002: the place of vector control. Ann. Trop. Med. Parasitol. 92: 165-166.

Le Berre, R., R. Garms, J.B. Davies, J.F. Walsh and B. Philippon. 1979. Displacements of *Simulium* and strategy of control against onchocerciasis. Phil. Trans. R. Soc. Lond. B. 287: 277-288.

Lévèque, C. 1989. The use of insecticides in the Onchocerciasis Control Programme and aquatic monitoring in West Africa. Scope, 317-335.

Molyneux, D.H. 1995. Onchocerciasis Control Programme in West Africa: Current Status and Future of the Onchocerciasis Control Programme. Parasitol. Today, 11: 399-402.

Servat, E. and J.-M. Lapetite. 1990. Hydrologie, Informatique et Satellite, un refort à la lutte contre l'onchocercose. Orstom Actualités, 29: 22-26.

Toé, L., J. Tang, C. Back, C.R. Katholi and T.R. Unnasch. 1997. Vector-parasite transmission complexes for onchocerciasis in West Africa. Lancet, 349: 163-166.

Webbe, G. 1992. The Onchocerciasis Control Programme. Trans. Royal Soc. Trop. Med. Hyg. 86: 113-115.

Wilson, M.D., R.J. Post and L.M. Gomulski. 1993. Multivariate morphotaxonomy in the identification of adult females of the *Simulium damnosum* Theobald complex (Diptera: Simuliidae) in the Onchocerciasis Control Programme area of West Africa. Ann. Trop. Med. Parasitol. 87: 65-82.

Yaméogo, L., L. Toé, J.-M. Hougard, B. Boatin and T. Unnasch. 1997. Use of a simple entomological method for detection of onchocerciasis infection in the oncho-freed zones. Suppl. Am. J. Trop. Med. Hyg. 57(3): 285.

# Progress in the Eradication of *Amblyomma variegatum* Fabricius, 1794 (Ixodoidea, Ixodidae) from the Caribbean


*R.G. Pegram[1], E.F. Gersabeck[2], D.D. Wilson[2] and J.W. Hansen[3]*

[1] Caribbean *Amblyomma* Programme, Food and Agriculture Organization, Barbados
[2] United States Department of Agriculture, Animal and Plant Health Inspection Service, 4700 River Road, Riverdale, Maryland 20737,USA
[3] Food and Agriculture Organization, Animal Health Service, Viale delle Terme di Caracalla, 1-00100 Rome, Italy


## INTRODUCTION

*Amblyomma variegatum*, commonly known as the tropical Bont tick, was introduced from West Africa to the islands of Guadeloupe and Antigua in the 19th century. The tick spread only to two other islands in the French West Indies during the following 70 to 80 years. Since the mid-1970s, however, it became widely distributed in the eastern Caribbean islands (Figure 1). There is now increasing evidence that migratory birds, especially the cattle egret, *Bubulcus ibis* Linnaeus (Ciconiformes: Ardeie) disseminated larvae and nymphs, as the egret itself has been spreading to new islands over the last 20 years (Corn et al. 1993, Barre et al. 1995). Subsequently, the tropical Bont tick, through its association with the bacterial skin disease known as dermatophilosis, caused the loss of large numbers of animals. On Nevis, for example, cattle numbers were decimated from 5,000 to 500 in less than 10 years. Elimination of this tick and its associated diseases could help to alleviate the deficit in animal protein in the region which imports about US$100 million worth of livestock products per annum.

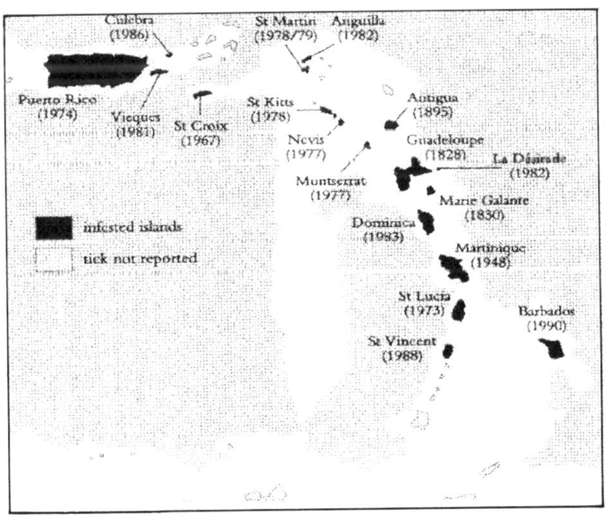


Figure 1. Distribution of Tropical Bont Tick – Eastern Carribean Islands.

Risk analysis, feasibility and cost-benefit studies were carried out to assess the importance of these problems (Alderink and McCauley 1988, Gersabeck 1994). The US Department of Agriculture estimated potential losses of US$762 million annually if the tick and its associated diseases became established in mainland countries. These studies led to the formulation of the Caribbean *Amblyomma* Programme in the late 1980s (Barre and Garris 1989, CARICOM/FAO/IICA 1995, Wilson 1996). During the early 1990s, potential donors were concerned about the proposed tick control strategy whereby government veterinary teams would be responsible for treatment of all livestock. Based on practical experience in the region, it was considered particularly unrealistic for teams to treat 600 - 1,000 animals per day under the current livestock management practices. An alternative, or complementary, method proposed the use of mobile crush-pens which would also be difficult to implement. Who would erect them and where? Once erected, who would bring the animals to them? Most livestock in the region are owned by part-time farmers who tend their livestock early in the morning or late in the evening or on weekends. Consequently, they are rarely available to assist veterinary personnel in handling their livestock during normal working hours. This unique system of animal ownership and husbandry practices in the Caribbean would, therefore, due to very high operational costs, almost prohibit conventional delivery of tick control by veterinary services personnel.

## A NEW APPROACH

For the tick eradication process to be cost-effective and sustainable, the programme regional coordinating unit considered it necessary to modify the methodology originally proposed, and to impose upon livestock owners the responsibility of treating their own animals. Fortunately, the eradication process is based upon a simple biweekly application of the pour-on acaricide Bayticol for a minimum of two years (Barre et al. 1993). There remained, however, some concern regarding farmer compliance as it had been noted that "... whilst the technology was available and eradication was considered to be feasible, the most important obstacle in attempts to eradicate the tropical Bont tick would be the human factor". Thus, the revised approach would require Caribbean livestock owners to adopt new animal husbandry practices, and in order to keep them informed and motivated to follow the treatment schedules, an intensive public awareness campaign was implemented based upon experiences in other animal health programmes (Villet 1994).

## IMPLEMENTATION AND FINANCE

The multi-donor Caribbean *Amblyomma* Programme became operational at the end of 1994. It is a regional Caribbean Community (CARICOM) programme jointly implemented by the Food and Agriculture Organization (FAO) and the Inter-American Institute for Cooperation on Agriculture (IICA). Overall policy guidance is provided by the *Amblyomma* Programme Council, whose membership is comprised of donor representatives, technical agencies, CARICOM and government representatives.

The overall estimated total cost of the tropical Bont tick eradication programme is approximately $37 million for both the CARICOM and the French territories. The programme will take about five years to complete and includes the following components:

- National projects in the CARICOM region, with an overall cost of US$15.6 million, are supported by direct farmer inputs to the value of US$8.2 million, government funds (US$2.9 million) and donor contributions (US$4.5 million).
- The CARICOM regional coordination project includes responsibility for training and production of promotional materials for use in the national projects as well as expenditure for training, travel, support costs and overall coordination through the *Amblyomma* Programme Council (APC). Estimated cost is about US$5.5 million.
- The cost of the French West Indies programme, which is managed independently, is estimated at about US$17 million, which is secured through funds from the French government and the European Union.

A summary of current donor funding is shown in Table 1. USDA continues its long-term commitment to the programme with a pledge for an additional year to 2001. FAO continues to source additional trust funds and donor funds, but has continued its commitment through the FAO Technical Cooperation Programme and the IFAD technical assistance grant. National governments, through CARICOM and FAO, are now negotiating with the European Union (EU) for additional financial support, under the Lome IV programme for 1998-2000, to purchase tick control chemicals and support the continuation of the extension, training and communications component.

Table 1. Summary of Current Donor Funding (in US$ million).

| Donor/Government | Committed | Requested | Total |
|---|---|---|---|
| IICA/USDA | 2.20 | 0.90 | 3.10 |
| IICA | 0.18 | 0 | 0.18 |
| IICA/USDA Sub-total | 2.38 | 0.90 | 3.28 |
| FAO/TF:ODA/GTZ | 0.75 | 0 | 0.75 |
| FAO/TF: USDA/SECNA | 0.20 | 0 | 0.20 |
| FAO:TF/Other | 0 | 0.57 | 0.57 |
| FAO/TF:BEL/ITA (APO Staff) | 0.20 | 0 | 0.20 |
| FAO/TF:NET (APO Staff) | 0.32 | 0.08 | 0.40 |
| FAO/TF: IFAD | 1.00 | 0 | 1.00 |
| FAO/TCP (Regional Communications) | 0.29 | 0 | 0.29 |
| FAO/TCP (Barbados) | 0.30 | 0 | 0.30 |
| FAO Sub-total | 3.06 | 0.65 | 3.71 |
| EU/CARIFORUM (FAO) | 0.90 | 2.10 | 3.00 |
| Total | 6.34 | 3.65 | 9.99 |

A regional coordinating unit assists in overall management of the programme and coordinates the four main components:

- Administrative management.
- Eradication and surveillance.
- Extension, training and communication.
- Adaptive research.

The main tasks of the regional coordinating unit are to support national efforts in tick control in the infested countries in the following areas:

- Providing policy advice to governments on legislation and quarantine.
- Strengthening veterinary services (livestock census and animal registration).
- Purchasing and distributing acaricide and other equipment and supplies.
- Training veterinary personnel in monitoring and surveillance.
- Providing and maintaining a regional CAP database.
- Increasing public awareness and motivation among livestock owners.
- Training in public information and communication techniques.

Government veterinary staff are responsible for the following activities:

- Registering all livestock owners.
- Registering, and where appropriate, tagging all animals.
- Maintaining a national data base.
- Training farmers in the correct use of Bayticol.
- Distribution of Bayticol.
- Monitoring the eradication process.
- Carrying out surveillance for the tropical Bont tick.

These activities are supported by extension and communications staff in order to reinforce compliance with the treatment schedules.

## ADAPTIVE RESEARCH

An important, new adaptive research component, supported by the International Fund for Agriculture Development (IFAD), aims to evaluate alternative, and adjunct, cost-effective and environmentally friendly control technologies. Equally important during the post-treatment follow-up phase, is the availability of improved surveillance methods that do not jeopardise the level of eradication achieved on each island. The adaptive research work plan includes the following field studies to determine the efficacy of various acaricide formulations and novel methods of application, and biological agents to control the tropical Bont tick:

- ACATAK®, a growth development inhibitor.
- Duncan and other self-medicating applicators.
- Pheromone/acaricide-impregnated decoy ear and tail tags.
- Biological control using parasitoids (*Ixodiphagus hookeri*) and "myco-insecticides" (*Beauveria bassiana* (Balsamo) Vuillemin and *Metarhizium anisopliae* (Metchnikoff) Sorokin).

## RESULTS AND DISCUSSION

Despite the uncertainty of long-term funding, encouraging results have been achieved since the launch of eradication activities in Anguilla, St. Kitts and Nevis (1995), Montserrat and St. Lucia (1996), and Antigua and Barbados (1997). The number of livestock registered and under treatment is summarised in Table 2. The active participation of the livestock owners on all islands has been encouraging. Both tick populations (Table 3) and the prevalence of dermatophilosis (Figure 2) have been reduced markedly over the past two years.

Table 2. Status of Registration and Treatment (Targeted) by Country (as of February 1998).

| Country [Owners] | Cattle | | Sheep and Goats | | Comments |
|---|---|---|---|---|---|
| | Estimated | Targeted | Estimated | Targeted | |
| Anguilla [597] | 100 | 100 | 6,500 | 6,500 | Suspended in June 1997 |
| Antigua [1436] | 12,000 | 11,094 | 54,400 | 38,039 | Started in August 1997 |
| Barbados [364] | 12,500 | 1,600 | 44,000 | 3,100 | Only partially infested |
| Dominica [?] | 3,400 | 1,090 | 15,100 | 1,900 | Only partially infested |
| Montserrat [?] | 1,000 | | 1,300 | | Under surveillance |
| St. Kitts [761] | 2,700 | 2,700 | 9,500 | 9,500 | To be completed in 1998 |
| Nevis [742] | 541 | 541 | 27,900 | 27,900 | To be completed in 1998 |
| St. Lucia [458] | 13,000 | 1,050 | 22,500 | 1,360 | Only partially infested |

Table 3. Qualitative and Quantitative Surveillance for *A.variegatum* (TBT) in 1997.

| Country [owners] | Animal Populations, Number Examined and Tick Sightings | | | | | | Comments |
|---|---|---|---|---|---|---|---|
| | Cattle | | | Sheep and Goats | | | |
| | Population | Examined | Positive | Estimated | Examined | Positive | |
| Anguilla [597] | 100 | 8 | 2 | 6,500 | 707 | 3 | Suspended |
| Antigua [1436] | 12,000 | | | 54,400 | | | Too early |
| Barbados [364] | 12,500 | | 2 (sites) | 44,000 | | | Qualitative |
| Dominica [?] | 3,400 | | 2 (sites) | 15,100 | | | Qualitative |
| Montserrat [?] | 1,000 | | | 1,300 | | | No TBT |
| St. Kitts [761] | 2,700 | 318 | 13 | 9,500 | 127 | 0 | Quantitative |
| Nevis [742] | 541 | 0 | 0 | 27,900 | 1,701 | 20 | Quantitative |
| St. Lucia [458] | 13,000 | | 1 (site) | 22,500 | | | Qualitative |

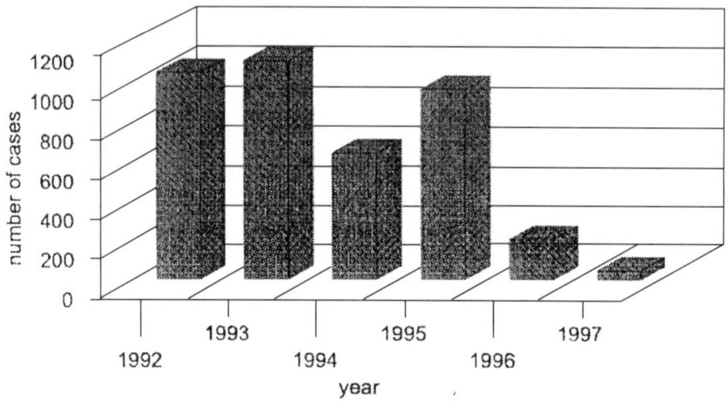


Figure 2. Prevalence of dermatophilosis.

It is now imperative for the remainder of the programme that tick control activities be carried out simultaneously on all islands for at least one of the two years of mandatory treatment. The northern islands which are now being freed of the tick must not become re-infested via tick movement on migratory birds. For the *A. variegatum* eradication programme to be successfully concluded, therefore, it is essential that these efforts be sustained through continued and increased support for a further two years.

## CONCLUSIONS

The mid-term review/evaluation report (George et al. 1997) concluded that incredible progress had been made towards the achievement of the goal, despite resource, technical, and personnel constraints. The approach which relied on farmers for the treatment of animals was highly commended. The public information and communication strategy reinforced the technical veterinary component of the programme. It was further concluded that the approach was cost-effective and sustainable and will contribute to the greater involvement of farmers in future livestock improvement programmes.

The review emphasised that the progress would only be sustained if additional funding be secured to finance the eradication of *A. variegatum* from the entire Caribbean. It recommended that an updated, coordinated Caribbean plan for the eradication programme be prepared and approved with input from all participating countries. A project proposal was prepared and forms the basis of the application to the European Union for further support.

## ACKNOWLEDGEMENTS

Technical and scientific support of both national and regional coordination unit staff is gratefully appreciated. The authors acknowledge the technical and financial assistance provided by the national governments of Anguilla, Antigua, Barbados, Dominica, Montserrat, St. Kitts and Nevis and St. Lucia. Financial support provided by international donors (the USDA, IFAD, IICA, the EU, FAO, and the governments of Belgium, Germany, Italy, the Netherlands, and the United Kingdom) is gratefully acknowledged. Additional sponsorship was kindly provided by Bayer, AG Germany, and various sponsors in the CARICOM countries, particularly in St. Lucia.

## REFERENCES

Alderink, R.J. and E.H. McCauley. 1988. The probability of the spread of *Amblyomma variegatum* in the Caribbean. Prev. Vet. Med. 6: 286-198.

Barre, N. and G.I. Garris. 1989. Biology and ecology of *Amblyomma variegatum* in Guadeloupe and its importance in eradication strategy. *In* The Eradication of Ticks. FAO Animal Production and Health Paper 75: 230 - 245.

Barre, N., G. Garris and R. Aprelon. 1993. Acaricides for eradication of the tick *Amblyomma variegatum* in the Caribbean. Rev. Elev. Med. Vet. Pays Trop. 46: 349-354.

Barre, N., G. Garris and E. Camus. 1995. Propagation of the tick *Amblyomma variegatum* in the Caribbean. Rev. Sci. Tech. Off. Int. Epiz. 14: 841-855.

CARICOM/FAO/IICA. 1995. Eradicating the Tropical Bont Tick from the Caribbean. Words and Publications. Oxford, UK. pp. 7.

Corn, J., N. Barr, B. Theihot, R. Creekmore, G.I. Garris and V.F. Nettles. 1993. The potential role of cattle egrets, *Bubulcus ibis* (Ciconiformes: Ardeie), in the dissemination of *Amblyomma variegatum* (Acari: Ixodoidae) in the eastern Caribbean. J. Med. Entomol. 6: 1029-1037.

George, J., S. Farez, B. Francois and M.T. Sebrechts. 1997. Mid-term Review of the Caribbean *Amblyomma* Program. Unpublished Report, FAO/IICA/USDA, Barbados.

Gersabeck, E.F. 1994. The tropical Bont tick in the Western Hemisphere. pp. 121-132. *In* Comprehensive Reports on Technical Items Presented to the International Committee. O.I.E., Paris.

Villet, J. 1994. The Pan-African Rinderpest Campaign: Where communication is the key: World Anim. Rev. 80/81: 53-62.

Wilson, A. 1996. Appropriate strategies for the control of ticks and tickborne diseases. *In* Vector-borne Pathogens: International Trade and Tropical Animal Disease. Ann. N.Y. Acad. Sci. 791: 554-63.

# Area-Wide Control of Chagas Disease Vectors in Latin America

*C.J. Schofield*

ECLAT Coordinator*, LSHTM, London WC1 E7HT, United Kingdom

Chagas disease (American trypanosomiasis) is now ranked by the World Bank as the most serious parasitic disease of the Americas, with a medical and economic impact far outranking even the combined effects of other parasitic diseases such as malaria and schistosomiasis (World Bank 1993). The infection is virtually impossible to cure and the disease is difficult and costly to treat. In contrast, transmission can be halted by eliminating the domestic insect vectors – large blood sucking reduviids of the subfamily Triatominae – and by improved screening of blood donors to minimise the risk of transfusional transmission (WHO 1991).

Improved screening of blood banks requires appropriate legislation backed by a well-developed system of reference laboratories and standardised procedures, although to a large extent, this can be implemented in a progressive way from local to national levels (Schmunis 1991). By contrast, the key to success in Chagas disease vector control lies in the implementation of large-scale regional or international programmes coupled with long-term community-based vigilance. This is a classic intervention model beginning with a vertical intervention, the attack phase, in which all infested houses are sprayed by trained professionals, progressively backed by a more horizontal community-based system where householders themselves can report the presence of any residual infestations for retreatment where necessary. Elimination of domestic vectors of Chagas disease is facilitated by their slow reproductive rates and limited genetic variability, but is hampered by the ease of passive transport of the insects from one house to another, even across state and international boundaries (Schofield 1994). For this reason, international collaboration is particularly important in Chagas disease vector control.

Since early trials in the 1940s, there have been many local and regional campaigns designed to control domestic populations of Triatominae, especially in Argentina, Brazil and Venezuela. Without exception, the results followed a similar course, with a rapid decline in house infestation rates and a reduction in the incidence of human infection. Generally however, it was impossible to consolidate the initial successes due to changes in intervention policy and a corresponding lack of programme continuity. Although domestic populations of Triatomine bugs can be cleared from individual houses, or from particular localities, the houses remain vulnerable to re-infestation by bugs brought in from untreated regions. In Brazil, the national programme to eliminate the primary vector, *Triatoma infestans* (Klug), was launched in 1983. It was highly successful, but – as with previous local campaigns – there were problems of sustainability and re-infestation of many of the treated localities (Dias 1987).

*The European Community – Latin America research network on the biology and control of Triatominae (ECLAT) is supported by the European Commission and the AVINA Foundation of Switzerland.

To address this problem, governments of the six southern cone countries (Argentina, Bolivia, Brazil, Chile, Paraguay and Uruguay) in 1991 set up a joint initiative to control Chagas disease through the elimination of all domestic and peridomestic populations of the triatomine vectors (Kingman 1991). With the addition of Peru to this programme in 1996, the initiative now covers the entire geographic distribution of the primary vector, *T. infestans*. The Southern Cone Initiative was planned as a 10-year programme to eliminate all domestic and peridomestic populations of *T. infestans*, with concurrent suppression of other vector species of local importance. Now just past its mid-point, the programme has already met with considerable success with Chagas disease transmission interrupted over vast areas of Argentina, Chile, Uruguay, central Brazil, eastern Paraguay and southern Bolivia (Schmunis et al. 1996, WHO 1996, 1997a, 1998). Other domestic vectors in the region – such as *Triatoma brasiliensis* Neiva in northeast Brazil – are also being controlled and the World Health Organisation (WHO) now predicts complete interruption of Chagas disease transmission by the year 2000 (WHO 1997b). Programme costs were originally estimated at US$190-350 million over the 10-year period, with most of the investment to be committed during the initial attack phase over the first 3 years. On this basis, the internal rate of return of the programme was predicted to be just over 14% (Schofield and Dias 1991,1999). In fact, investment to date has been just over US200 million and recent studies in Brazil and Argentina have revealed economic rates of return in excess of 30% and 60%, respectively (Akhavan 1996, Basombrio et al. 1998).

The Southern Cone Initiative has also spurred other regions to plan similar programmes, notably the recently announced Central American Initiative to eliminate *Rhodnius prolixus* Stahl from El Salvador, Guatemala, Honduras and Nicaragua and the Andean Pact Initiative for the control of *R. prolixus* and other vectors in Colombia and Venezuela. In Central America at least, *R. prolixus* seems to be exclusively domestic with a very restricted genetic repertoire and is considered a particularly vulnerable target for control (Dujardin et al. 1998). In Andean Pact countries however, the status of silvatic *R. prolixus* remains unclear and urgent studies have been initiated to assess rates of gene flow and population dispersal between silvatic and domestic habitats (Schofield and Dujardin 1997).

With successful control or elimination of *T. infestans* and *R. prolixus* and control of sympatric vectors, a figure approaching 70% of Chagas disease transmission will be interrupted, while large-scale improvements in blood bank screening will interrupt a further 10-15%. But maintenance of this success and further progress, will depend very largely on improved systems of entomological surveillance and these in turn will depend crucially on maintaining poltical continuity for the control and surveillance programmes. Although technologically simple, the administrative scale of the regional Chagas disease control initiatives has revealed major complexities and intervention coverage has not been entirely uniform. In each country the programmes have faced difficulties primarily of an administrative nature, especially as each country has simultaneously sought to decentralise its intervention services – producing inevitable disruption to the chain of operational decisions and lapses in entomological surveillance. Already there are indications of little-known species of Triatominae adapting to new domestic habitats and our understanding of this evolutionary process is still in its infancy. It should not be forgotten – especially by those responsible for budgetary priorities – that the domestication of Triatominae and initiation of Chagas disease transmission is a relatively recent phenomenon. Experience shows that we can now combat this, but if we relax vigilance, it could happen again.

## REFERENCES

Akhavan, D. 1997. Análise de custo-efetividade do programa de controle da doença de Chagas no Brasil. Report to the Ministerio da Saúde (FNS), Brasilia, Brazil. 28 pp.

Basombrio, M.A., E.C. Del Rey, E.C. Rojas and C.J. Schofield. 1998. A cost-benefit analysis of Chagas disease control in northwest Argentina. Trans. R. Soc. Trop. Med. Hyg. 92: 137-143.

Dias, J.C.P. 1987. Control of Chagas disease in Brazil. Parasitol. Today 3: 336-341.

Dujardin, J.P., M. Mu-oz, T. Chavez, C. Ponce, J. Moreno and C.J. Schofield. 1998. The origin of *Rhodnius prolixus* in Central America. Med. Vet. Entomol. 12: 113-115.

Kingman, S. 1991. South America declares war on Chagas disease. New Sci. (19 October) 16-17.

Schmunis, G.A. 1991. *Trypanosoma cruzi*, the etiologic agent of Chagas disease: status in the blood supply in endemic and non-endemic countries. Transfusion 31: 547-557.

Schmunis, G.A., F. Zicker and A. Moncayo. 1996. Interruption of Chagas disease transmission through vector elimination. Lancet 348: pp.1171.

Schofield, C.J. 1992. Eradication of *Triatoma infestans*: a new regional programme for southern Latin America. Ann. Soc. Belg. Méd. Trop. 72: (suppl.1) 69-70.

Schofield, C.J. 1994. Triatominae – Biology and Control. Eurocommunica Publications, West Sussex, UK. 80 pp.

Schofield, C.J. and J.C.P. Dias. 1991. A cost-benefit analysis of Chagas disease control. Mem.Inst. Oswaldo Cruz 86: 285-295.

Schofield, C.J. and J.C.P. Dias. 1999. The Southern Cone Initiative against Chagas disease. Adv. Parasitol. 42: 1-27.

Schofield C.J. and J.P. Dujardin. 1997. Chagas disease vector control in Central America. *Parasitol. Today* 13: 141-144.

WHO 1991. Control of Chagas Disease. WHO Technical Report Series 811, World Health Organization, Geneva. 95 pp.

WHO 1996. Chagas disease. Progress towards elimination of transmission. Weekly Epidemiological Record 12 January 1996. 12 15.

WHO 1997a. Chagas disease. Interruption of transmission. Weekly Epidemiological Record 3/10 January 1997. 1-5.

WHO 1997b. Chagas disease, leprosy, lymphatic filariasis, onchocerciasis: prospects for elimination. Document no. TDR/Gen/97.1, World Health Organization, Geneva. 35 pp.

WHO 1998. Uruguay declared free of Chagas disease transmission. TDR News 56:6.

World Bank 1993. World Development Report 1993. Investing in Health. Oxford University Press, New York. 329 pp.

# Large-Scale Control of Mosquito Vectors of Disease

*C.F. Curtis and M.H. Andreasen*

London School of Hygiene and Tropical Medicine, London WC1E 7HT, United Kingdom

## MOSQUITO BORNE DISEASE

By far the most important vector borne disease is malaria transmitted by *Anopheles* mosquitoes causing an estimated 300-500 million clinical cases per year and 1.4-2.6 million deaths, mostly in tropical Africa (WHO 1995).

The second most important mosquito borne disease is lymphatic filariasis, but there are now such effective, convenient and cheap drugs for its treatment that vector control will now have at most a supplementary role (Maxwell et al. 1999a). The only other mosquito borne disease likely to justify large-scale vector control is dengue which is carried in urban areas of Southeast Asia and Latin America by *Aedes aegypti* L. which was also the urban vector of yellow fever in Latin America. This mosquito was eradicated from most countries of Latin America between the 1930s and 60s but, unfortunately in recent years, it has been allowed to re-infest and cause serious dengue epidemics, except in Cuba where it has been held close to eradication (Reiter and Gubler 1997).

In the 1930s and 40s, invasions by *Anopheles gambiae* Giles *s.l.*, the main tropical African malaria vector, were eradicated from Brazil (Soper and Wilson 1943) and Egypt (Shousha 1947). It is surprising that greatly increased air traffic has not led to more such invasions of apparently climatically suitable areas, e.g., of Polynesia which has no anophelines and therefore no malaria.

The above mentioned temporary or permanent eradications were achieved before the advent of DDT, using larvicidal methods (of a kind which would now be considered environmentally unacceptable) carried out by rigorously disciplined teams.

## THE ATTEMPT TO ERADICATE MALARIA

Between the end of the Second World War and the 1960s, the availability of DDT for spraying of houses allowed eradication of malaria from the Soviet Union, southern Europe, the USA, northern Venezuela and Guyana, Taiwan and the Caribbean Islands, apart from Hispaniola. Its range and intensity were also greatly reduced in China, India and South Africa and, at least temporarily, in Sri Lanka. In several Latin American countries much progress was made, but this has been reversed following the abandonment of DDT without any replacement being brought into use (Roberts et al. 1997). After eradication from the Soviet Union in the 1960s, malaria epidemics are now returning to Azerbaijan and Tadjikistan following the collapse of the health system and the descent into civil war (Nikolaeva 1996).

In a few instances, unlooked-for eradication has been claimed to have occurred locally as a result of DDT house spraying of species which are strongly endophilic, i.e., with a strong tendency to rest in houses. Examples are of *Anopheles funestus* Giles from Mauritius and possibly from Zanzibar and Pemba (Gillies and de Meillon 1968). There had been plans to eradicate the vector population from Sardinia (Logan 1953), but it was eventually found to be unfeasible and also unnecessary for the eradication of the disease. As emphasised by MacDonald (1957), a moderate reduction in the probability of daily survival may be enough to reduce almost to zero the proportion surviving the 12 days required for completion of development of malaria parasites in the mosquito, and thus malaria transmission may die out. This view was borne out by the collapse of malaria in, for example, Italy, when DDT was brought to bear on it (Merzagora et al. 1996). On this basis, many (but not all) experts advised WHO to go ahead with a world-wide attempt at malaria eradication. It is usual to ascribe its failure in most parts of the tropics to insecticide resistance, and this did make some contribution. However, more important factors were lack of resources and political will and the saturating levels of malaria transmission in many parts of lowland Africa, leading to the requirement for very high percentage reductions in transmission rate if any real progress was to be made. This also means that if a small percentage of individuals in the vector population are consistent in choosing not to enter and rest in houses, this fraction which cannot be killed by domestic insecticides may be enough to maintain an almost unchanged level of malaria prevalence (Molineaux and Gramiccia 1980).

Another aspect of this problem, which was raised in the early discussions about the feasibility of eradication of malaria in Africa and has recently been raised again (Trape and Rogier 1996, Snow et al. 1997), is the negative feedback effect of immunity. In areas of high transmission, the human population only survives by acquisition early in life of a moderate degree of immunity, at the expense of suffering much severe illness and much infant mortality. It has been argued that a moderate reduction of transmission, by reducing the rate of acquisition of immunity, would delay the onset of malaria attacks until later in life. This could reduce the net lifetime benefit of the vector control efforts and could even render them counterproductive if, as some believe, the risk of cerebral malaria is highest not in infants, but in children without adequate acquired immunity. This vitally important issue is difficult to settle within the bounds of medical ethics (Molineaux 1997). Theoretically, the problem could be avoided by replacing the missing immunity by a vaccine, when it becomes available, or by eradicating the vector population so that the missing immunity could do no harm. The possibility of eradication of an *Anopheles* population by SIT will be discussed below.

## IMPREGNATED BEDNETS

Most malaria infective bites by *Anopheles* mosquitoes occur at hours when people are in bed (88% of such bites between 2200 and 0500 hrs according to a recent study in Tanzania by Maxwell et al. 1998). Bednets without chemical treatment provide some degree of protection against malaria infection (about 33% reduction in incidence of re-infection in Tanzanian children cleared of their pre-existing infection by drug treatment (Maxwell et al.1999b). This can be boosted to about 83% protection with either of two pyrethroid treatments of the nets (Maxwell et al. 1999b). In fact, treatment of nets can be considered a more targeted way of using insecticide than spraying it on walls and ceilings and hoping that mosquitoes will rest on it. Mosquitoes are attracted to bednets by the body odour and/or carbon dioxide emitted by the occupants. Pyrethroids are currently the only class of insecticide used on bednets, but in view of the threat of pyrethroid resistance in *Anopheles* (Vulule et al. 1996, Darriet et al. 1997, Curtis et al. 1998b), it is important that satisfactory non-pyrethroid alternatives are found. Despite the cheapness of DDT per unit weight, impregnated bednets are preferable to DDT spraying because:

- they are more effective against malaria (Kere et al. 1996),
- at the dosages of each type of insecticide which are needed, pyrethroid treatment of nets is cheaper per family protected (Li 1990, Kere and Kere 1992, Xu et al. 1998),
- pyrethroids are biodegradable but DDT is not, and residues of it derived from house spraying have been found in human breast milk (Bouwman et al. 1990), though whether this is harmful remains to be proved, and
- DDT intended for anti-malaria use may illicitly find its way into agriculture, and may produce residues in crops which make them unacceptable to importers (Curtis 1994).

A comparison between pyrethroids used for bednet impregnation and for house spraying in an area of intense transmission in Tanzania (Curtis et al. 1998a) showed very similar effects of each form of vector control on the mosquito populations, incidence of malaria infection and prevalence of anaemia (one of the main routes by which malaria causes death or chronic ill health). However, because of the six times greater amount of a relatively expensive insecticide required for providing house spraying for a given community, it was concluded that it would be cheaper to provide nets and their treatment. House spraying has always been considered part of the public health service and the urge to try to make poor people pay for their health care has fortunately not yet spread to this form of disease prevention. However, many wish to try to make individual householders pay for nets and insecticide, though switching from free provision of insecticide to selling it (at cost price) led to a drastic fall in percentage coverage of beds by treated nets in Gambia (Cham et al. 1997).

Apart from the effect of treatment of nets in improving the personal protection which they provide, studies in Tanzania (Curtis et al. 1998a) have shown that a major part of the benefit of a community-wide programme with treated nets is the attraction and killing of a large percentage of the vector population so that few survive to an age at which they can become infective. Thus, it is in the interests of the rich to help the poor to acquire treated nets (e.g., via taxation and a public health system) as it would reduce the number of mosquitoes biting the unprotected poor, picking up infection and surviving to threaten the rich if they ever venture out from their own nets during the night.

Extensive trials in four African countries have shown significant reductions in all-cause child mortality as a result of providing treated nets or the insecticide to treat people's own nets (Lengeler et al. 1996). These reductions are comparable to what were achieved in the 1950s with house spraying using dieldrin (Bradley 1991).

Numerous small projects are springing up all over the tropical world on impregnated nets, but so far only in China and Vietnam are treated nets numbered in millions, i.e., on a scale comparable to the heyday of house spraying.

## STERILE INSECT TECHNIQUE

There was much enthusiasm for SIT for mosquitoes in the 1960s and early 70s but it went into eclipse, largely for political reasons (Anonymous 1975). In the 70s, it was shown in various species of mosquito that chemically sterilised males, or males carrying translocations and a meiotic drive factor or cytoplasmically incompatible with the local population, could compete reasonably well for mates as shown by induction of sterility in the eggs laid by wild females (Lofgren et al. 1974, Grover et al. 1976a, b). An unreplicated report (Bracken and Dondale 1974) that spiders, fed a diet consisting only of chemosterilised mosquitoes, were sterilised has made radiation sterilisation a more acceptable option. However, there is reduced competitiveness if pupae are irradiated and treatment of adults inactivated by chilling would probably be necessary (Smittle and Patterson 1974).

The problem of immigration of females already inseminated with fertile sperms, and unwilling to mate again, was found to be a major obstacle to progress with SIT against *Culex quinquefasciatus* Say in India (Yasuno et al. 1976). Theoretically, the problem could be solved by large-scale rolling programmes of release. However, finding the capital for this seems unlikely for a programme directed against *Anopheles gambiae* Giles, which extends over huge areas of rural Africa and threatens the lives of the children of the poor, but not cash crops which accountants see as a worthwhile investment.

A more achievable target would seem to be eradication of urban mosquito populations where there is little chance of re-invasion from the surrounding countryside which is occupied by different species ill-adapted to urban life. This argument was first proposed with respect to *Aedes aegypti* in Indian towns (Reuben 1974). With a perceptible decline in confidence in legally enforced or community motivated control of its breeding, there may be a revival of interest in the use of SIT against it. Methods of mass rearing, sex separation to avoid release of biting females and release in urban environments were all developed 25 years ago (Singh et al. 1974).

Among malaria vectors, *Anopheles stephensi* Liston (type form) is the species which maintains urban malaria as an important fraction of India's total malaria problem because this mosquito is well adapted to breeding in water tanks, urban wells and pools of water on building sites. Another form, *An. stephensi mysorensis* Liston, has been described from rural areas (Ramachandra Rao 1984). However, the numbers of ridges on the eggs, which was the original way of defining the two forms, have been found to intergrade (Subbarao et al. 1987). Molecular genetics or chromosome inversions may provide evidence on whether there is gene flow between rural and urban *An. stephensi* and hence, whether rural immigrants would interfere with an urban eradication programme. Urban *An. stephensi* populations can be suppressed (but not eradicated) by rigorous enforcement of larval control by screening water tanks, etc. Any SIT programme against *An. stephensi* should be integrated with such an environmental management campaign, the role of the sterile males being to eliminate the residual population left after the maximum feasible reduction in urban breeding sites.

It is important to avoid release of biting female mosquitoes, and in anophelines (unlike culicines), this unfortunately cannot be done on the basis of pupal size. Genetic sex separation was developed more than 20 years ago for several *Anopheles* species based on translocation of dominant genes for insecticide resistance onto the Y chromosome (Curtis et al. 1976, Seawright et al. 1978). This proved difficult, but was achieved, for *An. stephensi* (Robinson 1986). With most of the translocations there were some crossovers, which broke the male sex linkage of the resistance, but in *Anopheles albimanus* Wiedemann, this was suppressed with a fortunately positioned inversion (Seawright et al. 1978). To select a suitable inversion in a more systematic way, an attempt is in progress to use *in situ* hybridisation of a probe of the cloned dieldrin resistance gene (ffrench-Constant 1993) to locate where, on the polytene map, this gene is located and hence, where the inversion (induced or naturally occurring) needs to be. If released males were not completely sterile and the population failed to be eradicated, releases might spread the resistance gene. This probably would not matter in the case of resistance to dieldrin as this insecticide is now considered too toxic for anti-mosquito work.

## MAKING MOSQUITOES GENETICALLY HARMLESS

Theoretically, the problem for genetic control of continuity between rural populations of species such as *An. gambiae* could be solved (and even become an advantage) if, instead of trying to eradicate with SIT, one aimed to drive genes making mosquitoes harmless through wild populations. Unlike SIT, a gene driving system would not necessarily require large capital investment in a rearing plant nor would it be swamped by a moderate level of immigration. The obvious candidate for a mosquito gene for harmlessness to man would be one for non-susceptibility (refractoriness) to *Plasmodium*, but another possibility would be the gene(s) in *An. quadriannulatus* Theobold (a member of the *An. gambiae* complex) which cause it to bite animals and not humans and therefore prevent it carrying malaria. Several cases of refractoriness have been selected in anophelines (e.g., Collins et al. 1986, Feldmann and Ponnudurai 1989), but all have been due to at least two genes, and it would thus be difficult to link these reliably to driving mechanisms. The recent identification by Billker et al. (1998) of the *Plasmodium* gametocyte activating factor as xanthurenic acid, a product of the metabolic pathway to Dipteran eye colour pigments, suggests that a single dominant gene for refractoriness may be obtainable, but it seems that abnormalities in this pathway would have severe fitness costs.

Assuming that a fully fit, single dominant gene for refractoriness is eventually found, the real problem to applying this method remains that of finding a reliable gene driving system. There are precedents in *Drosophila* for the spontaneous spread of certain transposons and of *Wolbachia* symbionts and there is some understanding of under what conditions they would spread (Kidwell and Ribeiro 1992, Sinkins et al. 1997). However, attempts to inject *Wolbachia* into *Anopheles* have so far been unsuccessful and there are precedents for breakdown of the linkage of transposons to other genes with which they had been "loaded" and for selection to favour the transposons alone.

Tight linkage of the gene driving system to the gene to be driven is essential if the combined system is to spread from a small "seeding" release and counter the effects of immigration. It is necessary to avoid even a limited rate of recombination of the driver and the factor to be driven because the same arguments about the danger of loss of anti-malaria immunity from incomplete control with bednets apply to genetic control which fails to go to completion.

## REFERENCES

Anonymous. 1975. Oh, New Delhi, Oh, Geneva. Nature 256: 355-357.

Billker, O., V. Lindo, M. Panico, A.E. Etienne, T. Paxton, A. Dell, M. Rogers, R.E. Sinden and H.R. Morris, 1998. Identification of xanthurenic acid as the putative inducer of malaria development in the mosquito. Nature 392: 289-292.

Bouwman, H., R.M. Coopan, A.J. Reinecke and P.J. Becker. 1990. Levels of DDT and metabolites in breast milk from Kwa-Zulu mothers after DDT application for malaria control. Bull. W.H.O. 68: 761-768.

Bracken, G.K. and C.D. Dondale. 1974. Fertility and survival of *Achaearanea tepidariorum* on a diet of chemosterilised mosquitoes. Can. Entomol. 104: 1709-1712.

Bradley, D.J. 1991. Morbidity and mortality at Pare-Taveta, Kenya and Tanzania, 1954-66: the effect of a period of malaria control. pp. 248-263. *In* Disease and Mortality in Sub-saharan Africa. R,G.Feachem and D.T. Jamieson (eds.). World Bank and Oxford University Press.

Cham, M.K., B. Olaleye, U. D'Alessandro, M. Aikins, B. Cham, N. Maine, L.A. Williams, A. Mills and B.M. Greenwood. 1997. The impact of charging for insecticide on the Gambian National Impregnated Bednet Programme. Health Policy and Planning 12: 240-247.

Collins, F.H., R.K. Sakai, K.D. Vernick, S. Paskewitz, D.C. Seeley, L.H. Miller, W.E. Collins, C.C. Campbell and R.W. Gwadz. 1986. Genetic selection of a *Plasmodium* refractory strain of the malaria vector *Anopheles gambiae*. Science 234: 607-610.

Curtis, C.F. 1994. Should DDT continue to be recommended for malaria vector control? Med. Vet. Entomol. 8: 107-112.

Curtis, C.F., J. Akiyama and G. Davidson. 1976. A genetic sexing system in *Anopheles gambiae* species A. Mosq. News 36: 492-498.

Curtis, C.F., C.A. Maxwell, R.J. Finch and K.J. Njunwa. 1998a. A comparison of use of a pyrethroid for house spraying or bednet treatment against malaria vectors. Trop. Med. Int. Health 3: 619-631.

Curtis, C.F., J.E. Miller, M.H. Hodjati, J.H. Kolaczinski and I. Kasumba. 1998b. Can anything be done to maintain the effectiveness of pyrethroid impregnated bednets against malaria vectors? Phil. Trans. R. Soc. London 353: 1769-1775.

Darriet, F., P. Guillet, F. Chandre, R. N'Guessan, J.M.C. Doannio, F. Rivière and P. Carnevale. 1997. Présence et évolution de la résistance aux pyrethrinoides et au DDT chez deux populations d'*Anopheles gambiae s.s.* d'Afrique de l'Ouest. WHO document WHO/CTD/VBC/97.1001.

Feldmann, A.M. and T. Ponnudurai. 1989. Selection of *Anopheles stephensi* for refractoriness and susceptibility to *Plasmodium falciparum*. Med. Vet. Entomol. 3: 421-52.

Ffrench-Constant, R.H. 1993. The molecular and population genetics of cyclodiene insecticide resistance. Insect Biochem. Mol. Biol. 24: 335-345.

Gillies, M.T. and B. de Meillon. 1968. The Anophelinae of Africa south of the Sahara. S. Afr. Inst. Med. Res., Johannesburg.

Grover, K.K., C.F. Curtis, V.P. Sharma, K.R.P. Singh, K. Dietz, H.V. Agarwal, R.K. Razdan and V. Vaidyanathan. 1976a. Competitiveness of chemosterilised and cytoplasmically incompatible translocated males of *Culex pipiens fatigans* in the field. Bull. Entomol. Res. 66: 469-480.

Grover, K.K., S.G. Suguna, D.K. Uppal, K.R.P. Singh, M.A. Ansari, C.F. Curtis, D. Singh, V.P. Sharma and K.N. Panicker. 1976b. Field experiments on the competitiveness of three genetic control systems for *Aedes aegypti*. Entomol. Exp. Appl. 20: 8-18.

Kere, J.F. and N.K. Kere. 1992. Bednets or spraying? Cost analysis of malaria control in the Solomon Islands. Health Policy Planning 7: 382-386.

Kere, N.K., A. Arabola, B. Bakote'e, O. Qalo, T.R. Burkot, R.H. Webber and B.A. Southgate. 1996. Permethrin-impregnated bednets are more effective than DDT house-spraying to control malaria in the Solomon Islands. Med. Vet. Entomol. 10: 145-148.

Kidwell, M.G. and J.M.C. Ribeiro. 1992. Can transposable elements be used to drive disease refractoriness genes into vector populations? Parasitol. Today 8: 325-329.

Lengeler, C., J. Cattani and D. de Savigny. eds. 1996. Net Gain. I.D.R.C., Ottawa and WHO, Geneva.

Li Zuzi. 1990. *In* Appropriate Technology in Vector Control, ed.Curtis, C.F., C.R.C., Boca Raton, Florida, pp. 42.

Lofgren, C.S., D.A. Dame, D.E. Breeland, D.E. Weidhaas, G. Jeffrey, R. Kaiser, H.R. Ford. M.D. Boston and K.F. Baldwin. 1974. Release of chemosterilised males of *Anopheles albimanus* in El Salvador, 3. Field methods and population control. Amer. J. Trop. Med. Hyg. 23: 288-297.

Logan, J.A. 1953. The Sardinian project. Johns Hopkins Press, Baltimore.

Macdonald, G. 1957. The Epidemiology and Control of Malaria. Oxford Univ. Press, Oxford.

Maxwell, C.A., K. Mohammed, U. Kisumku, and C.F. Curtis. 1999a. Can vector control play a useful supplementary role against bancroftian filariasis? Bull. W.H.O. 77: 138-143.

Maxwell, C.A., J. Myamba, K.J. Njunwa, , B.M. Greenwood and C. F. Curtis. 1999b. Comparison of bednets impregnated with different pyrethroids for their impact on mosquitoes and on re-infection with malaria after clearance of pre-existing infections with chlorproguanil-dapsone. Trans. R. Soc. Trop. Med. Hyg. 93: 4-11.

Maxwell, C.A., J. Wakibara, S. Tho and C.F. Curtis. 1998. Malaria infective biting at different hours of the night. Med. Vet. Entomol. 12: 325-327.

Merzagora, L., G. Corbellini and M. Coluzzi. 1996. L'Altra Battaglia di Cassino: contro la malaria a cinquant'anni dall'epidemia della Valle del Liri. Soc. Ital. Parassitol.

Molineaux, L. 1997. Malaria and mortality: some epidemiological considerations. Ann. Trop. Med. Parasitol. 91: 811-825.

Molineaux, L. and G. Gramiccia. 1980. The Garki project: research on the epidemiology and control of malaria in the Sudan savanna of West Africa. WHO, Geneva.

Nikolaeva, N.V. 1996. Resurgence of malaria in the former Soviet Union. Soc. Vector Ecol. Newsletter 27: 10-11.

Ramachandra Rao, T. 1984. The Anophelines of India. Monogr. Malaria Research Centre, Delhi.

Reiter, P. and D.J. Gubler. 1997. Surveillance and control of urban dengue vectors, pp. 425-462. *In* Dengue and Dengue Hemorrhagic Fever. D.J. Gubler and G. Kuno (eds.). CAB International, Wallingford.

Reuben, R.1974. Factors in the biology of *Aedes aegypti* of importance to a genetic control programme. J. Communic. Dis. 6: 117-120.

Roberts, D.R., L.L. Laughlin, P. Hsheih and L.L. Legters. 1997. DDT, global strategies and a malaria crisis in South America. Emerg. Infect. Dis. 3: 295-302.

Robinson, A.S. 1986. Genetic sexing of *Anopheles stephensi* using dieldrin resistance. J. Amer. Mosq. Contr. Assoc. 2: 93-95.

Seawright, J.A., P.E. Kaiser, D.A. Dame and C.F. Lofgren. 1978. Genetic method for the preferential elimination of female *Anopheles albimanus*. Science 200: 1303-1304.

Shousa, A.T. 1947. Species eradication: the eradication of *Anopheles gambiae* from Upper Egypt 1942-1945. Bull. W.H.O. 1: 309-353.

Singh, K.R.P., G.D. Brooks and M.A. Ansari. 1974. Mass rearing of mosquitoes. J. Communic. Dis. 6: 121-126.

Sinkins, S.P., C.F. Curtis and S.L. O'Neill. 1997. The potential application of inherited symbiont systems to pest control, pp. 155-175. *In* Influential Passengers. S.L.O'Neill, A.A. Hoffman and J.H. Werren (eds.). Oxford University Press, Oxford.

Smittle, B.J. and R.S. Patterson. 1974. A container for irradiation and mass transport of adult mosquitoes. Mosq. News 34: 406-408.

Snow, R.W., J.A. Omumbo, B. Lowe, C.S. Molyneux, J.O. Obiero, A. Palmer, M.W. Weber, M. Pinder, B. Nahlen, C. Obonyo, C. Newbold, S. Gupta and K. Marsh. 1997. Relation between severe malaria morbidity in children and level of *Plasmodium falciparum* transmission in Africa. Lancet 349: 1650-1654.

Soper, F.L. and D.B. Wilson. 1943. *Anopheles gambiae* in Brazil 1930-1940. Rockefeller Foundation, New York.

Subbarao, S.K., K. Vasantha, T. Adak, V.P. Sharma and C.F. Curtis. 1987. Egg-float number in *Anopheles stephensi*: ecological variation and genetic analysis. Med. Vet. Entomol. 1: 265-271.

Trape, J.F. and C. Rogier. 1996. Combating malaria morbidity and mortality by reducing transmission. Parasitol.Today 12: 236-240.

Vulule, J.M., R.F. Beach, F.K. Atieli, D.L. Mount, J.M. Roberts and R.W. Mwangi. 1996. Long-term use of permethrin-impregnated bednets does not increase *Anopheles gambiae* permethrin tolerance. Med. Vet. Entomol. 10: 71-79.

Xu Bozhao, Xiao Xianqi, R.H. Webber and J.D. Lines. 1998. Comparisons of the effect of insecticide-treated bed nets and DDT residual spraying on the prevalence of malaria transmitted by *Anopheles anthropophagus* in China. Trans. R. Soc. Trop. Med. Hyg. 92: 135-136.

World Health Organisation. 1995. Vector control for malaria and other mosquito-borne diseases. WHO Tech. Rep. Ser. 857.

Yasuno, M., W.W. MacDonald, C.F. Curtis, K.K. Grover, P.K. Rajagopalan, L.S. Sharma, D. Singh, K.R.P. Singh, H.V. Agarwal, S.J. Kazmi, P.K.B. Menon, R. Menon. R.K. Razdan, D. Samuel and V. Vaidyanathan. 1978. A control experiment with chemosterilised male *Culex pipiens fatigans* in a village near Delhi surrounded by a breeding free zone. Jap. J. Sanit. Zool. 29: 325-343.

# The French National Programme of Bovine Hypodermosis Eradication

*Isabelle Amouroux*

FNGDS, 149 rue de Bercy, 75012 Paris, France

## INTRODUCTION

The National Federation of Sanitary Defence Organisations (FNGDS) is a professional organisation which brings together more than 95% cattle breeders within its departmental structures (GDS). Working closely with breeders within their local structures, the GDS maintains a high level of efficiency in communicating and informing farmers about sanitary programmes as well as rally them to participate in these programmes.

FNGDS has built a solid cooperation with the veterinary administration and private veterinarians in the fight against the most serious animal diseases. In the 1980s, new sanitary guidelines regarding export emphasised the necessity of fighting against the warble fly. As parasitic hypodermosis causes a major decrease in zootechnic performances, as well as is an important immuno depression, the GDS from different regions began the fight against it. Taking into account the fly biology, it was deemed necessary to establish a concerted fight in order to avoid recontamination in their regions as well as improve sanitary conditions.

In 1989, FNGDS proposed and implemented a national scheme for the eradication of hypodermis in collaboration with other professional and industrial organisations, scientists, veterinarians and the Ministry of Agriculture. In 1994, a law against the warble fly was enacted. This law required every region to fight against the fly beginning 1 July 1998. Before this deadline, all regions were encouraged to begin regional plans on a voluntary basis, which, nevertheless, had to be approved by a national committee.

Under the national scheme which was based on parasitic biology, each region had to build its own programme. It had to be in charge of the details of operation, e.g., regarding topography and the presence of *Hypoderma lineatum* Villers. Cooperation between animal health partners was greatly encouraged.

## DEVELOPMENT CYCLE OF HYPODERMA

There are two species that infect cattle: *Hypoderma bovis* L. and *H. lineatum*. The adult stage is entirely a reproductive and dispersal phase. The infected cattle is dead within few days after larvae leave the body. The eggs hatch in about four days into larvae which immediately penetrate the host's skin, often causing irritation and exudation. The larvae (L1) spend 8 months migrating through the inter-muscular connective tissue to the subcutaneus tissue of the back. They spend a significant period around the spinal cord *(H. bovis)* or in the oesophageal submucosa *(H. lineatum)*. As the larvae reach the back of the infected animal, they attain the second stage: L2. Soon afterwards at the larger final stage, they reach the length of about 25 mm (L3). After 4 weeks the larvae emerge from the breathing hole and fall to the ground. The pupal stage lasts about a month. Then, the adults emerge, and the cycle goes on.

Warble flies appeared on the back of the cattle from March to August. Some observations have indicated that the highest sensitivity of visual controls can be obtained from 15 May to 15 June (Argente 1995).

A couple of hypoderma is able to parasite a hundred animals, 5 kilometres around, in two years. A collective fight is, therefore, necessary to eradicate hypodermosis.

## ORGANISATIONAL PLAN

### Administrative Organisation of the Plan

The territorial units to put the programme into practice are the administrative regions. To put the programme into practice, it must be agreed by more than 60% of breeders. Then each unit of cattle must be identified and registered in a list to be made accessible to the GDS. The GDS must also obtain information about the movement of each animal in a particular department.

At both the regional and national levels, the GDS established a plan describing what will be done for the next five years in each department of the region. This provisional plan regroups technical and financial provisions. Regional committees, formed by the GDS, which is the leader of these committees, veterinarians, the regional administration and professional organisations, meet twice per year. The GDS examines the result of the past campaigns, choose the orientation of the next campaign and validate it. The report sent to the national level is examined at two levels. First, the National Technical Commission examines programme results and compares the regional programme to national schemes from a technical point of view. This committee is composed of regional GDS, national veterinarians, scientists, other professional organisations and an administration representative. Second, the report is officially validated by the National Committee. This is the political committee, comprising all the professional organisations and the administrative representative. The GDS has to carry out the regional plan validated by the National Committee.

Each stage of the plan comprises systematic treatments, tactic treatments, introduced animal treatments, curative treatments and visual control. At each stage, the person in charge (breeder, veterinarian, etc.) will notify the GDS in writing the nature of the treatment that has been carried out. The GDS computerises all information. The organisation, at central and regional levels, is therefore provided with a reliable recording and reporting system. Two reports, representing regional results and plans, are sent in July and in September to the National Technical Committee to be examined.

### Technical Organisation of the Plan

A region is usually divided into three areas. This programme is extended progressively. Usually three campaigns of systematic treatments are necessary in areas where the prevalence of infested cattle is 5%. Current methods to detect hypoderma infestation are the visual observation or palpation of the grubs when the larvae get to the back of the cattle, usually during spring. A cattle is declared infested when only one larva is seen on one animal.

As the prevalence of infested cattle becomes less than 5%, systematic treatments are no more obligatory. Tactical treatment however, continues. This treatment concerns each head of cattle.

An area is declared "cleansed" when the prevalence of infested cattle is below 5% for two consecutive years. Breeders can then benefit from the notice of "varron: zone assainie" under their bovine sanitary document (ASDA).

## Treatment

First of all, the initial prevalence is calculated by visual control. Systematic treatments are preventive ones; they are conducted in autumn, in order to avoid the parasite incidence. The principle of systematic treatments is to treat each head of cattle in a determined area.

The breeders can treat only their young animals. Veterinarians have to be in charge of the treatment of all registered adult animals.

A list of products to be used in treatments is available and authorised in the national scheme. These products contain avermectine and milbemycine. The use of microdose ivermectin for the adults and for dairy cattle is recommended by the Technical National Committee. Only veterinarians can administer this type of treatment. The microdose ivermectin is 0.1 ml /animal, that is, corresponding to one hundredth of the approved dosage.

Treatments against the warble fly is done in the same time as the prophylactic visit. It is recommended that animals be treated as soon as possible and before 30 December, especially if they are infested by *H. lineatum.*

As soon as visual control indicates that the infested cattle rate is below 5%, tactical treatments are only required. These apply to cattle infested during the previous campaign and its neighbourhood 5 kilometres around and cattle where there is no information about the treatment of introduced animals.

## Treatment for Introduced Animals

As the introduction of animals is the most likely way of re-infestation (two out of three heads) in a cleansed area, the national guideline dictates that each animal has to be treated except those coming from an officially cleansed area. The recommended treatment is microdose ivermectin, all year long, except when the animal is infected. In the latter case, as soon as larvae L2 or L3 are seen under the back of the animal, a curative treatment is conducted. A decree will soon oblige farmers to declare an infested animal in cleansed areas.

## Control

The national guideline recognises a clinical infestation warble grub survey conducted in spring, as the common method.

With spring controls, it was possible to evaluate the efficiency of treatments conducted in autumn. It was imperative that treatments were effective because the management of treatments for the next campaign depended on their results.

In areas where systematic treatments were conducted, 20% of cattle had to be controlled. There are two types of visual controls:

- "Aleatory": Lots were drawn on cattle (Snedecor and Cochran 1971). Those picked were used to calculate prevalence of infestation.
- "Oriented": Cattle suspected of infestation were picked. This included cattle infested during the last spring, neighbouring cattle for which no proof of preventive treatment had been registered and cattle for which no information existed about introduced animal treatment, cattle owned by breeders who refused treatment, etc. Through oriented controls, breeders in the programme are instructed to look for warbles. This is very important to decrease the infestation.

"Aleatory" controls must be realised from 15 May to 15 June. However, depending on climatic variations, they can be extended from 1 May to 30 June.

An area is declared cleansed when the prevalence of infested cattle is below 5% for two consecutive years. To reach this certification of a "cleansed area", the number of cattle, determined on a statistical table, has to be controlled. Each following year, such controls are put into practice to validify the certification for the area.

At the same time, some visual controls are also conducted at the cattle market. In 1997, an information network between the regions of the GDS was created. This provided information to the GDS regarding the source of an infested animal. Further action by the GDS can then be taken.

As soon as an animal is considered infested by visual control (aleatory or oriented), a curative treatment is inspected by a veterinarian or in his presence.

## RESULTS

### Technical Aspects

Progression from region to region has been slow in some cases but the eradication plan is beginning to cover the entire national territory of 21 regions. Indeed, in the 1997/98 campaign, the five previous regions, which had originally opposed the plan, had agreed to it. For the first time in 1998, all French regions put into practice a regional programme of eradication.

Before the programme commenced, the prevalence of hypodermosis in French cattle had varied greatly with the regions, from 30% to 90%. While some regions achieved the 5% level in only two years, others took three or more years to do so.

Figure1 shows the rates of regional infested cattle after the implementation of the eradication programme. There is still however, high infestation rates in some areas, as indicated by the National Technical Committee.

The actual situation is presented in Fiugre 2. Nine regions contain at least one cleansed area.

In the 1996/1997 campaign, 11 million animals were covered, corresponding to the ownership by 200,000 cattle breeders. During this period, nearly 3.5 million animals were treated in the systematic way while the other two-thirds were treated in a tactical way. That means that nearly 9 million animals were located in areas where the infestation rate was below 5%.

For the 1997/1998 campaign, 14 million animals in the ownership of 250,000 cattle breeders were involved.

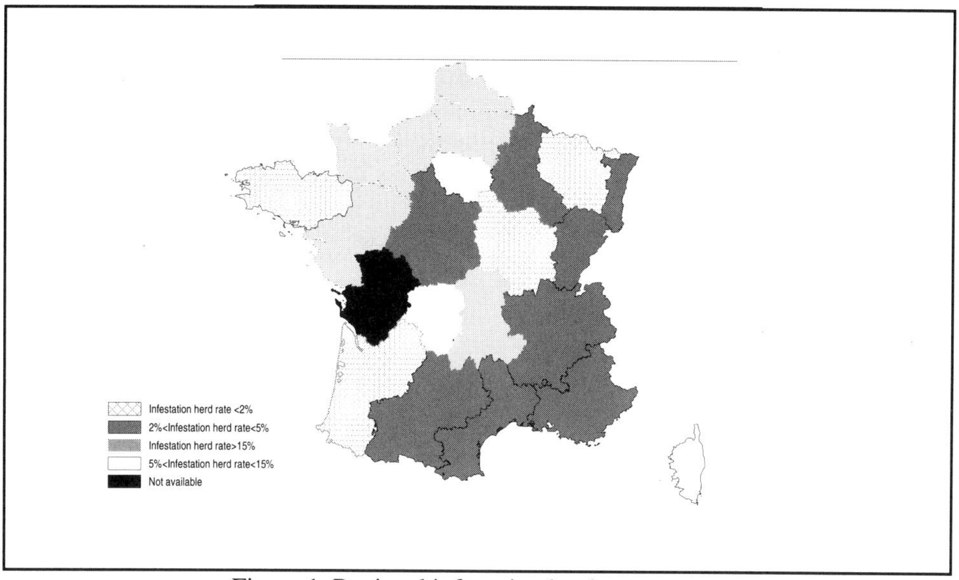


Figure 1. Regional infestation herd rate.

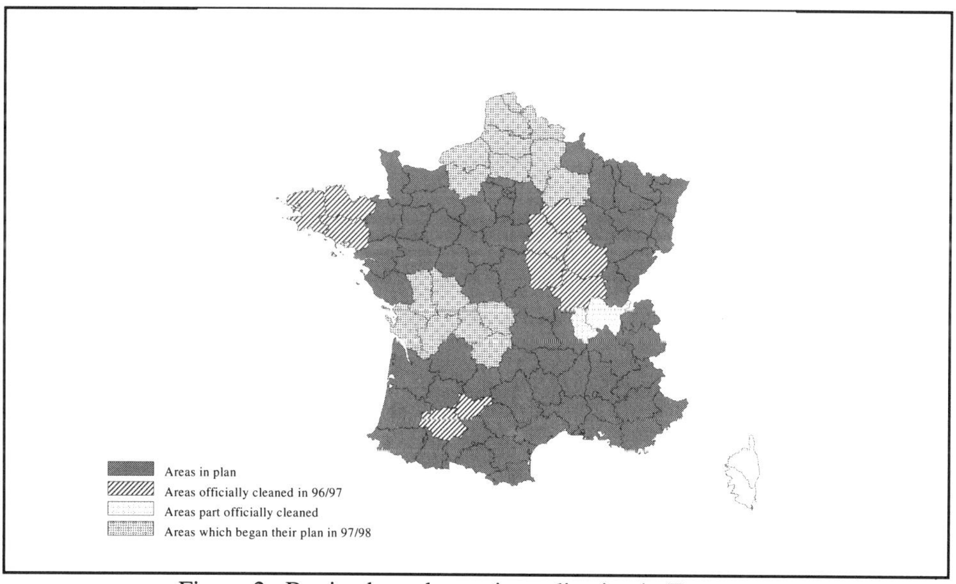


Figure 2. Bovine hypodermosis eradication in France

## Financial Aspects

The eradication programme for the 1996/1997 campaign cost about 60 million francs. Treatments cost the greatest amount. The cost was less than 5 francs on an average per animal, but this varied depending if an animal was in an area where systematic treatments were conducted (about 15 francs/animal), or in a cleansed area (about 2 francs/animal). Furthermore, costs varied even within regions.

Breeders and GDS bore more than 70% of the total costs. There was some financial support from national organisations such as the Ministry of Agriculture, which helped financially with visual controls. OFIVAL (Office National Interprofessionnel des Viandes, de l' Aviculture – National Interprofessional Meat. Livestock and Poultry Bureau) assisted by providing the breeders information while INTERBEV (L'Association Nationale Interprofessionnelle du B• tail et des Viandes – French Meat and Livestock Association) assisted the breeders when an animal died following preventive or curative treatment. There was also some assistance at the European level and local level.

## DISCUSSION

After nearly ten years of experience, we can establish the following points with regard to the management of the national eradication plan.

First of all, it is necessary to have solid collaboration between the different partners at the national and the local levels. Furthermore, to succeed in an eradication programme, there must be motivation among farmers from the beginning, as they have to be prepared to take action.

There were problems with ecologist breeders. A number of them did not want to treat their animals even with microdosed ivermectine which was, from FNGDS' point of view, an "ecologist treatment", considered acceptable by the administration, scientists, veterinarians and ecologist breeders. Using this method, breeders had to control each head of cattle every week during the entire emergence period of warbles. They had to extract larvae manually with oxygenated water. This method was time consuming, but was necessary because of the warble biology. Its implementation meant that there was no risk of recontamination of cattle which had already been treated. Furthermore, chemical products were not used.

During the previous ten years, we had to face some technical difficulties. One of the main problems was that no treatment for dairy cows was available. The recommended dose of ivermectine could not be used on dairy cattle and an interval of 42 days between treatment and slaughter was required to prevent the persistence of residues in milk and tissues. In fact, some studies have pointed out that microdosed ivermectine was highly efficient against the first larvae L1 of *H. lineatum* an *H. bovis* (Argente and Hillion 1984). Furthermore, other studies have shown that ivermectine residues in milk from lactating dairy cows which have been treated with microdoses cannot be detected by usual methods (Alvinerie et al. 1994). An official note published by Ministry of Agriculture fixed a milk LMR (Limit of Maximum Residue), which officially allowed the use of microdosed ivermectin, which does not have a AMM (licensed dosage).

The use of microdose ivermectin, however, has some major advantages: it is very low in cost, and causes minimal harm to both animal and environment. From the scientist's point of view, such a treatment does not increase the risk for selecting drug resistant nematodes (Bauer 1994). Nevertheless, this treatment does not protect against side effects related to the destruction of larvae around the spine or the oesophagus and consequently the timing of treatment remains a priority.

However, it is crucial to use a "microdosage" and not the so called "minidose" (below).

The use of immunological diagnosis has to be recognised in the national guideline. It has already been put into practice in some areas. Indeed, visual controls are effective in a problem area, but as the area becomes cleansed, other methods should be used. Epidemiological surveys are carried out using clinical detection for warbles, but with one inspection only about 50% of the real prevalence is expected. The use of immunological diagnosis on the ground has pointed out some major difficulties in serological interpretation (Petit 1994). In fact the two methods – visual control and serological diagnosis – do not measure the same indicators. The warble count reflects the total evolution of the parasite and the real risk of parasite pressure for the future. The immunological diagnostic indicates a contact between the parasite and the host which remains a proof of the presence of the disease in the herd within the previous year. Some studies have been realised, the effectiveness of a previous campaign to determine characteristics of this method: its sensitivity and its detectability. This method will be necessary to certificate cleared areas, corresponding to areas from which *Hypoderma* have been completely eliminated.

At the same time, an information system has been developed between GDS in order to warn each other about the possibility of infestation of cattle through an introduced animal which may have been infested. More generally, we have to develop epidemiological surveillance in order to be beware of infestation and to detect any residual infestation in a cleansed area.

## CONCLUSION

In order to operate an eradication scheme successfully, public awareness and goodwill are essential. Farmers must be aware of the seriousness of problems and the benefits of eradication. That is why the implementation of the departmental structures with the GDS is an advantage. Indeed a lot of breeder meetings are organised each year and regularly, articles are published in national or local newspapers.

In the course of ten years, the plan is now at a strategic time of development. The entire French territory is covered and many regions have a very low infestation level. However, even if systematic treatment is implemented in most parts of France, the fight itself is far from ending. Protection of cleansed areas needs specific control measures that we have to develop: epidemiological surveillance and, of course, serological testing.

During the coming years, the different partners at the national and at the regional level will have to be determined and united to ensure that the programme is a national success.

We have to aim for the objective of the complete absence of the warble in the year 2000.

## REFERENCES

Alvinerie, M., J.F. Sutra, P. Galtier and P.L. Toutain. 1994. Microdose d'ivermectine chez la vache laitière: concentrations plasmatiques et résidus dans le lait; Revue de Médecine Vétérinaire, Tome 145, No. 10, pp. 761-764.

Argente, G. 1995. Choix d'une périodicité pour les comptages de varrons; COST 811, European cooperation in the field of scientific and technical research, pp. 153.

Argente, G. and E. Hillion. 1984. Utilisation de petites doses d'ivermectine pour le traitement préventif de l'hypodermose bovine; Le Point Vétérinaire, 16, pp. 614-618.

Bauer, C. 1994. Does the use of microdosed ivermectin to control bovine hypodermosis increase the risk for selecting drug resistant nematodes?; COST 811, European cooperation in the field of scientific and technical research, pp. 145.

Petit, E. 1994. Interprétation des résultats sérologiques au cours du plan d'éradication du varron en Bourgogne; COST 811, European cooperation in the field of scientific and technical research, pp. 83-104.

Snedecor, G. and W. Cochran. 1971. Planification et Analyses des sondages. Methods Statistiques, pp. 560-600.

# PART II.

## APPLICATIONS OF THE AREA-WIDE CONCEPT TO NON-FRUIT FLIES

---

## B) PLANT PESTS

# Eradication of the Cotton Boll Weevil *(Anthonomus grandis)* in the United States – A Successful Multi-Regional Approach

*Gary L. Cunningham and William J. Grefenstette*

USDA-APHIS, 4700 River Road, Riverdale, Maryland 20737, United States

## INTRODUCTION

The cotton boll weevil, *Anthonomus grandis* Boheman, is believed to have entered the United States of America (US) from Mexico and was first detected in South Texas in 1892. Since that time, the pest has spread throughout most of the nation's cotton-producing areas and has become the industry's number one nemesis (Figure 1).

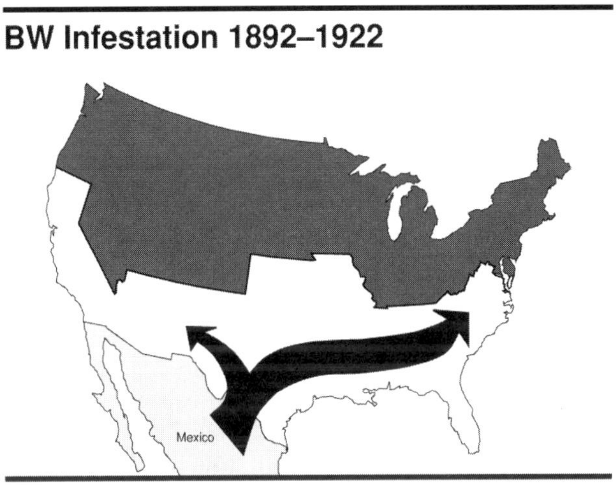


Figure 1. Boll weevil spread throughout most of the US within 30 years and led to major changes in agricultural practices.

More than US$13 billion in economic losses have occurred since its introduction, with recent annual expenditures of more than US$300 million for control costs alone. Although the weevil has been eradicated from over 4 million acres, its presence in non-programme areas continues to dictate production practices within the mid-south, Texas and Oklahoma.

## INITIAL ERADICATION EFFORTS

The National Cotton Council, representing the US cotton industry, began efforts in 1958 to accelerate boll weevil research. As a result, a pilot eradication project began in portions of the mid-South in the early 1970s and led to successful eradication of the boll weevil in Virginia and North Carolina by 1980. Following grower-approved referenda, the programme expanded in 1987 into Georgia, Florida, and Alabama in the southeast. Eradication efforts also began in the southwest US in Arizona and California. Both the southeast and southwest portions of the country, as well as a portion of northwest Mexico, are now weevil-free. Building upon these successes, the programme continues to expand rapidly throughout most of the remaining infested areas.

It is important to note that this programme has always had full support from the industry itself. This is not a United States Department of Agriculture (USDA)-mandated initiative that requests industry involvement. In fact, it is just the opposite. It is a grower-driven and grower-supported programme which coordinates implementation activities through USDA's Animal and Plant Health Inspection Service (APHIS) and a number of other federal and state agencies such as the Cooperative State Research, Education and Extension Service, the Farm Service Agency and State Departments of Agriculture. The programme's success continues to be contingent upon strong grower involvement and backing, as well as the industry's ability to organise itself in a united effort.

## PROGRAMME COMPONENTS

The programme uses three principle techniques on a large, area-wide basis to eradicate the boll weevil. They include pheromone trapping, chemical treatment and cultural practices. These techniques have not changed significantly since the programme's inception – primarily because they continually prove to be the most effective, least expensive and most environmentally suitable.

The main function of the boll weevil pheromone trap is to detect weevil populations throughout the season. Before any eradication treatments are applied within the programme area, all cotton acreage is identified, mapped, and surveyed. The survey is done with pheromone traps, which are placed around all cotton fields at a density of approximately 1 trap per 50-75 acres. Traps are installed in mid-summer and serviced for several weeks prior to the onset of diapause. These surveys indicate large areas within the overall programme increment that are likely to support significant populations of overwintered weevils each season. The traps are also used to help monitor treatment efficacy. In subsequent years, during active eradication, traps are placed in the spring at a density of 1 trap per 1-5 acres, and they are monitored for the entire season.

Pesticide treatments, usually malathion at ultra-low volumes (10-16 ounces per acre; 0.75-1.2 lb. active ingredient per acre), are carefully applied in response to trapping results, and only infested fields are treated. In those areas where aerial application is not practical, such as sensitive areas near dwellings or along streams and ponds, high-clearance ground equipment or truck-mounted mistblowers provide for a more precise placement of pesticides. In addition, an extensive environmental monitoring plan is implemented to comply with federal and state pesticide laws and to protect human health along with threatened and endangered plant and animal species.

During the first full season of the programme, after the initial series of fall treatments, the average number and frequency of insecticide applications vary across the Cotton Belt. In most infested areas, 6-8 applications generally will be required during the growing season. In a few heavily infested fields, as many as 15 treatments may be required. During the second full season, the number of fields and the number of chemical treatments applied are usually reduced by 40 to 60 percent. The third season is used to isolate and eliminate lingering weevil populations.

Growers are encouraged to follow good cultural practices that limit early- and late-season food supply for the weevil. These practices include uniform delayed planting, use of short-season cotton varieties, early harvesting and destroying stalks immediately after harvest. Other cultural practices which have been identified as helpful include keeping field borders clear and accessible, and not planting cotton next to environmentally sensitive areas, such as schoolyards, churches, and bodies of water.

The eradication phase is considered complete when spring-trapping indicates 99 percent of the programme acreage is free of infestation. The final phase confirmation concentrates on the remaining infested acres and lasts until boll weevils are no longer detected in traps. Post-eradication activities continue until the risk of re-infestation becomes insignificant and include annual field mapping, trapping, and response to any weevils captured. In most programme areas, boll weevil populations will be eliminated in an average of two years, with an additional year of moderate-density trapping to confirm eradication.

## ORGANISATIONAL STRUCTURE AND FUNDING

The continued success of the boll weevil eradication programme is due in large part to an organisational structure which places the effected commodity group (cotton growers) in the central leadership role. Each state has a Boll Weevil Eradication Foundation, consisting of local growers and a representative of the State Department of Agriculture to help guide activities and assist in policy decisions which may arise periodically. Before a state programme is started, a referendum must be passed (usually by two-third majority) which then requires 100 percent participation by all cotton growers in the proposed zone.

Federal assistance, which until recently provided up to 30 percent of the programme's funding, is no longer guaranteed. The US Department of Agriculture's Farm Services Agency, however, initiated a loan programme in 1997. Loans are made to officially recognised boll weevil eradication foundations to help make the programme more affordable. These loans, which totalled US$40 million in 1997, usually involve modest interest rates and a term of seven or eight years. This financing helps to defray the high cost of starting a programme, spreads the cost over a few more years, and makes the growers' annual programme cost more acceptable. Depending on the particular programme area, state funds may be used to offset grower assessments, in some instances by more than 50 percent. The APHIS federal funds to the belt-wide programme are down significantly from previous years, providing only 13 percent of overall costs in 1998.

## CURRENT PROGRAMME STATUS

Within the 13 million acres of cotton grown nation-wide, the boll weevil has been eradicated from more than four million acres, spread across eight states in the southeast and southwest portions of the country (Figure 2). Approximately two million acres are currently in active eradication in the mid-south and Texas. Another 700,000 acres within four new zones in New Mexico, Oklahoma, Mississippi, and Tennessee will start with a series of diapause treatments this fall. Other zones, especially in Texas, Louisiana, and Mississippi, are preparing to hold referenda and could begin as early as the fall of 1999.

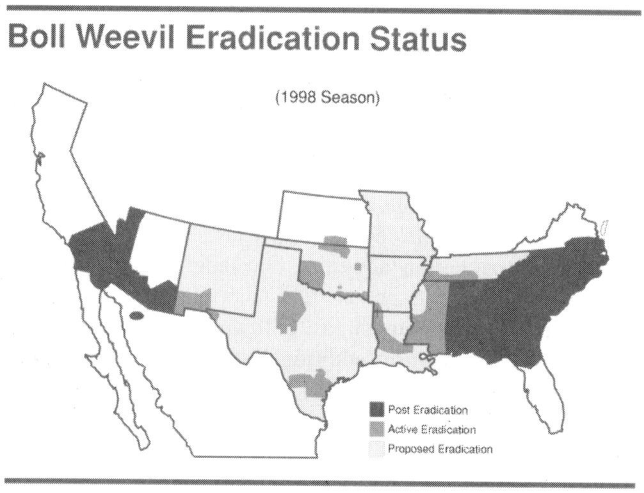


Figure 2. Current status of the boll weevil eradication programme in the US.

## BENEFITS

Recent farm legislation and the resulting decline in federal price supports make it imperative that growers compete successfully in international markets. For this to occur, there must be significant increases in yield and/or substantial reduction in production costs. Boll weevil eradication accomplishes the latter and has also dramatically increased yields in many areas. The National Cotton Council (Anonymous 1994) conservatively projects a 12:1 benefit-to-cost ratio for nation-wide eradication. Agricultural economists in Louisiana recently projected a benefit-to-cost ratio of 40:1 for boll weevil eradication in their state. These ratios are derived, in part, from eliminating the dollar value of yield lost to weevils and the resultant benefits pumped into the local economy. New jobs in ginning, warehousing, seed crushing and textile manufacturing have occurred as a direct result of eradication. In Georgia alone (Haney et al. 1996), insecticide applications have decreased more than 60 percent, pest management costs are 30 percent lower and overall crop damage is down by nearly 70 percent. Farm land values have also increased.

Environmentally, integrated pest management strategies for controlling secondary pests have a much better chance of succeeding in the absence of the boll weevil. And just as important, eradication provides a cleaner environment as a result of the 40- to 90-percent reduction in the amount of pesticides used each year on cotton.

## FUTURE PROSPECTS

These are exciting times for the boll weevil eradication programme. Momentum is building within the industry to hasten the weevil's demise, in part, because growers now realise that they can no longer compete effectively with those who have rid themselves of the pest. In the southeast US alone, where the weevil has been eliminated, there has been a 3.8-fold increase in cotton acreage during the past ten years, while other areas such as the mid-south and Texas have barely maintained their acreage. Increased use of *Bacillus thuringiensis* (Bt) cotton has added to growers' awareness of boll weevil pressure. With Bt cotton, less insecticides are now needed to control lepidopterous pests. As a result, boll weevil populations which used to be suppressed indirectly by worm treatments, are now free to increase. Areas which were not previously aware of weevil problems are now realising the need to take action.

Numerous challenges remain. As the programme moves northward where weevil populations are less intense, there will be likely less interest by affected growers to finance an eradication effort because their losses to the pest are less severe. But an unusually warm winter throughout most of the Cotton Belt in 1997-1998 greatly increased the weevils' ability to overwinter. This could result in much higher than normal populations during the 1998 season. Increased urbanisation of farmland and the presence of sensitive areas such as churches, schools, and hospitals, as well as organic cotton fields where no pesticides can be sprayed, require the consideration of alternative techniques such as bait tubes, biological control with *Catolaccus grandis* Burks, and cotton-free buffer zones. As the weevil is eradicated along the US border, cooperative efforts with Mexico for suppression or eradication on their side will have to be maintained. Such is already the case in the northern portions of the Mexican states of Baja California and Sonora.

Although not insurmountable, these situations will undoubtedly challenge programme leaders and cause some delay in programme implementation. The industry, however, maintains its strong commitment and support for boll weevil eradication, and remains undaunted in its determination to rid the country of this pest within the next 5-10 years.

## REFERENCES

Anonymous. 1994. Boll weevil eradication: a national strategy for success. National Cotton Council of America. Memphis, Tennessee.

Haney, P.B., W.J. Lewis, and W.R. Lambert. 1996. Cotton production and the boll weevil in Georgia: history, cost of control, and benefits of eradication. University of Georgia Research Bull. 428.

# Corn Rootworm Area-Wide Management Across the United States

*Laurence D. Chandler[1], James R. Coppedge[2], C. Richard Edwards[3], Jon J. Tollefson[4] and Gerald E. Wilde[5]*

[1] USDA-ARS Northern Grain Insects Research Laboratory, Brookings, South Dakota, USA
[2] USDA-ARS College Station, Texas, USA
[3] Purdue University, W. Lafayette, Indiana, USA
[4] Iowa St. University, Ames, Iowa, USA
[5] Kansas St. University, Manhattan, Kansas, USA

## INTRODUCTION

The western corn rootworm, *Diabrotica virgifera virgifera* LeConte, northern corn rootworm, *D. barberi* Smith and Lawrence, and Mexican corn rootworm, *D. virgifera zeae* Krysan and Smith are among the most economically and environmentally important pests of United States maize (*Zea mays* L.) production systems (Metcalf 1986). Annually, 8 to 10 million hectares of maize are treated with soil applied insecticides to protect the crop from larval feeding damage. Crop rotation, however, is also widely used to minimise the need for soil insecticide applications. Insecticides for adult rootworm management are also frequently used.

Numerous problems are currently associated with corn rootworm management approaches. Soil insecticides are normally used to protect maize roots from larval feeding damage. However, they are ineffective in controlling the management of corn rootworm populations (Gray et al. 1992, Sutter et al. 1991). It is not uncommon for large numbers of rootworms to develop within treated fields. Thus, when maize is grown in the same field year after year (continuous cropping), soil insecticide applications must be used to protect the plant. These applications are generally made without knowledge (prophylactic) of the rootworm population levels within the field due to the difficulty of sampling for immature life stages.

Crop rotation, where maize is planted every other year following soybean (*Glycine max* (L.) Merr.), is used throughout the United States Corn Belt as a primary rootworm management tool. In the southwestern United States, maize is also rotated with grain sorghum (*Sorghum bicolor* (L.) Moench). Rootworm species have altered their behaviour to adapt to these rotational schemes. Portions of northern and western corn rootworm populations throughout the Corn Belt have developed extended diapause traits which allow eggs to overwinter for more than one year, thus allowing larvae to hatch when maize is again planted (Krysan et al. 1986). A significant portion of the western corn rootworm population in the midwestern United States (east central Illinois, northern Indiana, southern Michigan, and western Ohio) has developed an affinity for ovipositing eggs into soybean (Edwards et al. 1996). These eggs hatch the following season when maize is again planted resulting in significant economic damage and management costs. A similar phenomenon has been observed with Mexican corn rootworms laying eggs in grain sorghum in Texas (J. R. Coppedge, personal communication).

Western corn rootworm resistance to chlorinated hydrocarbon insecticides has been extensively documented (Ball and Weekman 1962). Recently, two distinct populations of western corn rootworms in Nebraska were found to be resistant to carbaryl and methyl parathion which are commonly used for adult control (Meinke et al. 1998). Although the occurrence of resistance has not spread outside of these areas, the potential for increased tolerance of western corn rootworm populations to carbamate and organophosphate insecticides across the region does exist.

In response to many of the management problems discussed above, scientists with the USDA Agricultural Research Service and cooperating north central United States universities developed a semiochemical insecticide-bait targeted at adult corn rootworms (Sutter and Hesler 1993). The bait, composed of cucurbitacins, a minute amount of carbaryl, and a non-toxic edible carrier, can be applied by conventional aerial or ground sprayers (Hoffmann et al. 1996, Chandler and Sutter 1997, Chandler 1998). The bait adheres to plant surfaces and stimulates rootworm adult feeding, resulting in high levels of mortality and little impact on secondary pests or beneficial arthropods. The bait is applied when females predominate in the field and before significant oviposition begins. By targeting females at this critical developmental stage, egg laying can be reduced and thus, economic larval infestations can be avoided in the following growing season. Unlike soil insecticides, where maize root protection is the primary goal, this control tactic manages rootworm populations.

The development of this bait along with improved rootworm monitoring techniques and better understanding of rootworm biology/ecology, forms the basis for the corn rootworm area-wide management programme. Management of rootworm populations using the bait concept is best accomplished when conducted over a large area. Thus, a regional or area-wide approach may effectively reduce rootworm populations, resulting in significant economic savings to growers and improved environmental stewardship through the reduction of insecticide use throughout maize production areas. The information presented discusses the development of the corn rootworm area-wide programme, the initiation of the operational component of the programme, and the results of the first year of semiochemical-bait applications.

## MATERIALS AND METHODS

### Programme Development

In 1995, the Corn Rootworm Area-Wide Management Ad Hoc Committee was formed to guide development and implementation of the programme. The Committee is currently composed of representatives from several USDA agencies (Agricultural Research Service (ARS), Cooperative States Research, Education, and Extension Service and the Economic Research Service) and individuals from cooperating universities. The Committee developed the following mission statement as a guide to the overall programme, this being the successful establishment and implementation of an area-wide demonstration programme that:

- is the result of a partnership of growers, private consultants, applicators and suppliers, research and extension personnel, and local, state and federal agencies who have a stake in the development and adoption of improved crop management technologies,

- clearly demonstrates advantages of enhanced grower profits, reduced risks, enhanced environmental compatibility, and superiority of Integrated Pest Management (IPM) approaches compared to current pest control approaches.

The goals of the programme are:

- to demonstrate an area-wide IPM concept for the control of corn rootworm and other pests of maize such that voluntary adoption will occur throughout all maize growing regions,
- to develop a partnership of federal, state, local, and private interests which will be involved in the programme from conception to adoption.

The mission and goals of the programme were first presented publicly at a stakeholders' meeting held in St. Louis, Missouri in the fall of 1995. This meeting included both supporters and critics of the concept resulting in open and frank discussion of the merits of a corn rootworm area-wide management programme. The group recommended that evaluation of the area-wide concept at multiple locations throughout the United States maize production area was needed, and that expansion of the programme should begin in 1996. It was immediately recognised that the extensive size of the maize production area in the United States presented numerous differences in environmental conditions and resulting crop and pest management tactics. Therefore, the multiple location demonstration concept was devised to encompass as many of the major maize production differences encountered across the country as possible.

The Ad Hoc Committee recommended that the programme be implemented in three phases: 1) Phase I - evaluation and selection of study sites, development of background information, and education – 1996; 2) Phase II - full scale implementation of the programme – 1997-1999; and 3) Phase III - technology transfer to growers and interested parties – 2000.

The Agricultural Research Service provided approximately $550,000 for Phase I funding in the fiscal year 1996. Funding for Phase II was approximately US$1.6 million in 1997 and is targeted at approximately US$1.5 million for 1998. Funds were distributed to cooperators using established Agricultural Research Service intra-agency fund transfers and specific cooperative agreements with cooperating universities.

**Site Locations**

Five sites across the United States are currently being used for the area-wide management study/demonstration. Each site has a companion "normal" maize production area for comparison of area-wide vs. conventional management practices. The sites are:

**Cooperator Site A**

This site is located in eastern Iroquois Co., Illinois, and western Benton Co. and Newton Co., Indiana (41km$^2$ in size). This programme targets western corn rootworm within a corn/soybean rotation system and is located in the heart of the region experiencing significant western corn rootworm behavioural changes resulting in oviposition in soybean. Purdue University manages the site in cooperation with the University of Illinois.

### Cooperator Site B

This site is located in Clinton Co., Iowa (41 $km^2$ in size). This area-wide programme targets western and northern corn rootworms in a primarily continuous maize production area. Some soybean is grown in rotation. Iowa State University manages the site. Staff from the University of Minnesota, South Dakota State University, and the Agricultural Research Service Corn Insects Laboratory in Ames, Iowa cooperate on various aspects of the research activities.

### Cooperator Site C

The site is located in Republic Co., Kansas (41 $km^2$ in size) near the town of Scandia. This programme targets western corn rootworm in a furrow irrigated continuous maize production area. Kansas State University manages the activities within the site.

### ARS Site A

This site is located in Bell Co., Texas (21 $km^2$ in size) near the town of Little River. This management area targets Mexican corn rootworms in a maize/grain sorghum production system. The Agricultural Research Service Area-wide Pest Management Research Unit in College Station, Texas manages the site in cooperation with the Texas Agricultural Extension Service and Texas A&M University.

### ARS Site B

This study area is located in Brookings Co., South Dakota (41 $km^2$ in size) near the town of Aurora. This management area targets western and northern corn rootworm in a maize/soybean rotation production area. Some continuous maize is grown using a centre pivot irrigation system. The Agricultural Research Service Northern Grain Insects Research Laboratory in Brookings, South Dakota, manages the site in cooperation with South Dakota State University.

### General Conduct of the Programme

Background information was gathered at each study site during 1996 and the first few months of 1997. That information was used to predict the severity of corn rootworm populations and to determine the need for soil insecticide applications to reduce larval feeding damage during the first growing season (Phase II) of the programme. Soil insecticide applications were made where necessary to assist maize production before full fledged corn rootworm population management was introduced at the sites.

Two methods were selected to monitor corn rootworm populations during the implementation phase of the programme. Pherocon® AM (yellow sticky traps) traps were selected as the primary means for tracking insect populations and initiating semiochemical insecticide-bait applications in maize at Cooperator Sites B and C and ARS Site B. Adult counts from maize plants were used to trigger bait applications to maize at Cooperator Site A and ARS Site A. Pherocon AM traps were used to initiate bait applications in soybean at Cooperator Site A. Pherocon AM traps were monitored at all sites on at least a weekly basis. Cumulative counts of 5 to 6 beetles/trap/day for 7 consecutive days were used as a treatment threshold in maize, while counts of 2 beetles/trap/day for 7 consecutive days were used to trigger applications in soybean. Additionally, these applications were made only when gravid females were present. Maize plant counts in Illinois and Indiana triggered bait applications when counts of 0.5 beetles/plant or more were observed and when gravid females were present. Re-application of bait was made as necessary if corn rootworm populations reached treatment thresholds.

Adult emergence cages were also placed in several fields within the centre most area (core) of each area-wide management site to determine the timing and pattern of male/female emergence. These data are being used to assist decisions in bait applications and to help ground truth predictive models.

Root ratings using the Iowa 1-6 scale (Hills and Peters 1971) will be conducted in 1998-2000 at each site to assess the success or failure of the previous years semiochemical-bait applications. Ratings will be conducted in a sampling of core and companion area fields.

Secondary arthropod pests and beneficial insects will be monitored weekly at each site using visual observations, trapping methods, sweep nets, etc. Secondary pest outbreaks will be handled on a case by case basis and management will be based upon the recommendation of the site managers and cooperating crop consultants.

Each year, all sites will develop GIS/GPS maps of each field within the area-wide and companion sites. These maps will assist overall analysis of the programme and provide useful information on insect movement and population dynamics, crop phenology and production, soil types, etc. GPS will be used by aerial applicators to locate fields for treatment.

### Semiochemical-bait Applications

Corn rootworm adults will initially be managed using the semiochemical insecticide-bait SLAM® (MicroFlo Co., Lakeland, FL). Bait applications will be made using either high clearance ground sprayers or aircraft. SLAM will be applied at 230 to 561 gm of product/ha in 9 to 19 litres of water per ha. An additive will be used to improve rain-fastness of the bait on plant leaves. Rates selected will be dependent upon corn rootworm population pressure. Finished product droplet size should average 600 microns in diameter.

### Economic Assessment

Agricultural economists at Purdue University will determine the impacts of the corn rootworm area-wide management programme on farm profits and will develop spreadsheet budget models to analyse programme and non-programme farms for each site. Additionally, a study to determine the implications of area-wide management on input supply firms (agricultural chemicals and associated pest control/ consulting services) will be conducted.

### Sociological Assessment

Rural sociologists at Iowa State University will conduct studies to assess sociological barriers and opportunities for education, understanding, and successful promotion of corn rootworm area-wide management at all study sites.

### Insecticide Resistance Management

Entomologists at the University of Nebraska in cooperation with the Agricultural Research Service Northern Grain Insects Research Laboratory, will monitor corn rootworm susceptibility to carbaryl in all area-wide management study sites. Development and validation of carbaryl diagnostic concentrations will occur. Factors responsible for the development of insecticide resistance and the biochemical and molecular mechanisms of resistance in existing resistant populations will also be investigated.

## RESULTS AND DISCUSSION

### Cooperator Site A

The Illinois/Indiana site enlisted the partnership of 45 growers and approximately 4,600 ha of corn (94 fields) and soybean (73 fields). An additional 14 maize and 11 soybean fields (a total of approximately 700 ha) comprised the untreated companion area. Western corn rootworm sampling began on 21 July and ended on 3 September.

Western corn rootworm beetle populations were extremely high in the study site and companion area. First applications of SLAM were made on 28 July to two maize fields. In the 3 days following, 63 maize fields and 5 soybean fields were treated. By 22 August, 39, 50, and 5 maize fields, and 40, 21, and 6 soybean fields had been treated with SLAM once, twice, or three times, respectively. Several fields were sprayed after 22 August with Sevin XLR Plus due to economic considerations.

Final statistics showed that 75 fields were treated once with SLAM, 65 twice, and 11 three times. Twenty-eight fields were treated with Sevin XLR Plus. Initial efficacy from the SLAM applications was good. However, a continuous wave of western corn rootworms that inundated the area-wide site resulted in the need for numerous additional applications. Despite the large numbers of corn rootworms encountered, it is reasonable to expect that there will be fewer larvae and fewer beetles to manage in 1998.

### Cooperator Site B

In 1997, the Iowa area-wide management site was composed of 118 maize fields (approximately 2500 ha) of which 90% were considered continuous growing of maize. Forty growers participated in the activity during the year.

Western corn rootworm beetle emergence began on 12 July, about 10 days later than normal due to the cooler weather in May and June. Beetle populations reached treatment threshold beginning 28-29 July. SLAM applications began on 5 August. About 10 days of residual activity was obtained with the SLAM treatments. Continued beetle emergence brought beetle numbers to the re-treatment level by 22 August.

Beetle populations were very high in nearly all the monitored fields. A total of 2270 ha (about 93% of the total) were sprayed with SLAM once. An additional 1223 ha of maize was re-treated. Beetle kill following applications was effective.

**Cooperator Site C**

Thirty-six growers were partners with the area-wide management site located near Scandia, Kansas. These growers had 98 fields of maize encompassing approximately 1720 ha in total size. An additional 11 growers provided maize fields (23 fields with a total size of 529 ha) in the control area.

First western corn rootworm beetle emergence was observed on 4 July. The last adults were collected from emergence cages on 4 August within the management area. Peak emergence occurred from 10-21 July, with SLAM applications occurring during this time.

Forty-four of the 98 fields (1179 ha or 68.5% of the total maize hectares) required a SLAM application. An additional 316 ha required a second treatment due to rootworm numbers re-approaching threshold or because of product (SLAM) problems (unreliability or droplet size/placement problems). After adjusting some of the initial application problems, SLAM appeared to provide excellent ($\geq$ 95%) control in all treated fields.

**ARS Site A**

The 1997 activities at this site were a continuation of full-scale area-wide management initiated at this location in 1996. A total of 8 growers participated in this study. A total of 791 ha of maize and 334 ha of grain sorghum was included within the area-wide site and an additional 4 maize fields were used as controls.

Mexican corn rootworms first emerged in mid-May with 90% emergence complete by the third week of June. SLAM applications were initiated on 24 June and were completed by 26 June. Root ratings taken in 1997 from fields treated with SLAM in 1996 indicated there was significantly less root damage in those fields compared to control fields and thus fewer beetles emerging (95% reduction in the SLAM treated fields compared to the control fields) throughout the area.

A total of 209 ha of maize was treated with SLAM in 1997. No fields required re-treatment. SLAM applications significantly reduced (>95% control) Mexican corn rootworm adult numbers and appeared effective in managing the population.

**ARS Site B**

In 1997, 20 growers had 56 maize fields (1383 ha) within the area-wide management site east of Brookings, South Dakota. Continuous maize was grown in 12 fields (316 ha). Five continuous maize production fields outside the area were used as untreated controls.

Northern and western corn rootworm emergence began during the second week of July. Peak emergence occurred during the period from 25 July through 8 August. SLAM applications were initiated the week of 4 August and continued through the first week of September. No re-treatments were required.

Eighteen maize fields (535 ha) were treated with SLAM during the season. Seven continuous (233 ha), 10 first year (238 ha), and 1 mixed (both first and continuous corn - 65 ha) field were treated. In all cases, SLAM applications substantially reduced corn rootworm adult numbers below treatment thresholds.

## GENERAL DISCUSSION AND CONCLUSIONS

The initial results from the first year of Phase II corn rootworm area-wide management were encouraging. Corn rootworm beetle numbers were reduced at all study sites. Initial carbaryl resistance surveys did not detect any development of increased levels of beetle tolerance to the insecticide-bait. Thus, the bait should remain a viable management tool for the foreseeable future. Economic and sociological surveys are being conducted to determine the fiscal effectiveness and general attitude towards conducting corn rootworm management across the Corn Belt. It is too early to tell how successful 1997 was in managing corn rootworm populations. Likewise, it is also too early to predict when, and if, adoption of this regionalised approach to rootworm management will occur.

Cooperation among all participants has been outstanding. The management structure of each site has evolved and is positioned to handle the transfer of the area-wide technology if further assessment indicates this to be a feasible approach. We recognise that area-wide management of corn rootworms may not work well in every maize production area throughout the United States. However, the data we expect to gather in 1998 and beyond will provide us with much needed information to more accurately structure area-wide programmes to the needs of local clientele and to determine feasibility of expanding the approach.

## ACKNOWLEDGEMENTS

The corn rootworm area-wide management programme would not have been possible without the immense support provided by numerous individuals. Although we cannot list all cooperators as authors, we wish to gratefully acknowledge the following individuals for their participation, in and support of, the corn rootworm area-wide management programme: Larry Bledsoe, Wayne Buhler, Larry Buschman, Jesse Cocke, Stan Daberkow, Jerry DeWitt, Ric Dunkle, Robert Faust, Billy Fuller, Corey Gerber, Mike Gray, Deb Hartman, Randy Higgins, Clint Hoffmann, Denise Hovland, Gretchen Jones, Buddy Kirk, Les Lewis, Scott Lingren, Mickey McGuire, Marshall Martin, Lance Meinke, Tim Nowatski, Eldon Ortman, Ken Ostlie, Steve Padgitt, Randy Pingell, Walt Riedell, Rich Roehrdanz, Mike Scharf, Roxanne Shufran, Blair Siegfried, Phil Sloderbeck, Kevin Steffey, Harry Stockdale, Jeff Whitworth, Dave Woodson, Robert Wright, and Kun Yan Zhu. Additionally, the efforts of our administrative support staff at each cooperating institution, industry partners (BASF Inc., MicroFlo Co., Trece Inc.) and numerous local agricultural businesses are much appreciated. Finally, the programme could not have gone forward without the cooperation and enthusiasm of the many growers who actively embraced the concept and who have supported the activities of the many research groups involved.

## REFERENCES

Ball, H.J. and G.T. Weekman. 1962. Insecticide resistance in the adult western corn rootworm in Nebraska. J. Econ. Entomol. 55: 439-441.

Chandler, L.D. 1998. Comparison of insecticide-bait aerial application methods for application of corn rootworm (Coleoptera: Chrysomelidae). Southwest. Entomol. 23: 147-159.

Chandler, L.D. and G.R. Sutter. 1997. High clearance sprayer methods for application of corn rootworm (Coleoptera: Chrysomelidae) semiochemical-based baits. Southwest. Entomol. 22: 167-178.

Edwards, C.R., L.W. Bledsoe and J.L. Obermeyer. 1996. The dramatic shift of the western corn rootworm, *Diabrotica virgifera virgifera* LeConte (Coleoptera: Chrysomelidae), to maize in rotation with soybeans in Indiana. Proc. XX Internat. Cong. of Entomology, Florence, Italy, paper no. 15-082, pp. 469.

Gray, M.E., A.S. Felsot, K.L. Steffey and E. Levine. 1992. Planting time application of soil insecticides and western corn rootworm (Coleoptera: Chrysomelidae) emergence: implications for long-term management programmes. J. Econ. Entomol. 85: 544-553.

Hills, T.M. and D.C. Peters. 1971. A method of evaluating post-plant insecticide treatments for control of western corn rootworm larvae. J. Econ. Entomol. 64: 764-765.

Hoffmann, W.C., P.S. Lingren, J.R. Coppedge and I.W. Kirk. 1996. Corn rootworm control with a semiochemical-based insecticide. 1996 NAAA/ASAE Joint Tech. Session, Paper No. AA96-007. ASAE, St. Joseph, MI.

Krysan, J.L., D.E. Foster, T.F. Branson, K.R. Ostlie and W.S. Cranshaw. 1986. Two years before the hatch: rootworms adapt to crop rotation. Bull. Entomol. Soc. Am. 32: 250-253.

Meinke, L.J., B.D. Siegfried, R.J. Wright and L.D. Chandler. 1998. Adult susceptibility of Nebraska western corn rootworm (Coleoptera: Chrysomelidae) populations to selected insecticides. J. Econ. Entomol. 91: 594-600.

Metcalf, R.L. 1986. Foreword, pp. vii-xv *In* (J.L. Krysan and T.A. Miller, eds.) Methods for the study of the pest *Diabrotica*, Springer-Verlag, New York.

Sutter, G.R., T.F. Branson, J.R. Fisher and N.C. Elliott. 1991. Effects of insecticides on survival, development, fecundity, and sex ratio in controlled infestations of western corn rootworm (Coleoptera: Chrysomelidae). J. Econ. Entomol. 84: 1905-1912.

Sutter, G.R. and L.S. Hesler. 1993. New biorational and environmentally safe technology for management of insect pests in maize production systems, pp. 303-306. *In* Proc. Stress Symposia: Mechanisms, Responses, Management. College Agric. and Biol. Sci., South Dakota State Univ., Brookings, South Dakota.

# Area-Wide Pest Management of Locusts and Grasshoppers: The Striking Similarities of Problems and Solutions in Africa and The United States

*Jeffrey A. Lockwood*

Entomology Section, Department of Renewable Resources, University of Wyoming, Laramie, Wyoming, USA

## INTRODUCTION: OPENING A DIALOGUE

Grasshoppers and locusts are among the most devastating pests of human agriculture. These insects cause serious damage to crops and forage on every arable continent, and their depredations have become the basis for legends, myths, and (in recent times) complex political and economic programmes. No pest problem spans such immense areas, with up to 8 million ha treated for rangeland grasshoppers during outbreaks in the US and 16 million $km^2$ prone to outbreaks of the Desert locust, *Schistocerca gregaria* Forskål, alone. The traditional management approach has involved extensive, regionalised control programmes, but recent trends suggest a decentralised future for grasshopper and locust management. Hence, we have a dynamic situation that presents the opportunity for a comparative analysis of the costs and benefits of an area-wide approach to pest management at different scales.

As political, cultural, and communication barriers between scientists dissolve, the possibility of learning from one another's experiences (both failures and successes) promises to dramatically accelerate the rate of innovation, progress and discovery in pest management. For example, the recent advances in Reduced Agent-Area Treatments (RAAT, in which insecticide is applied to swaths, separated by untreated swaths or buffers) for management of rangeland grasshoppers in the US (Lockwood and Schell 1997) are based on the adaptation of tactics developed by African, Australian, Asian, and European scientists (Rachadi and Foucart 1996, Musuna and Mugisha 1997, Scherer and Célestin 1997, Wilps and Diop 1997, Launois and Rachadi 1997).

The key to successful adaptation of management methods must begin with intellectual modesty and nationalistic humility so that the insights of non-scientists and experts from outside one's country are given respect and serious consideration. It is subsequently necessary to recognise the essential similarities and differences between land use systems and understand the political and socioeconomic contexts in which strategies have developed. In the light of these considerations, I shall briefly review the comparative ecologies, economics, and politics of area-wide locust and grasshopper management in Africa and the US and derive a set of common concerns and approaches to resolving the emergent issues.

I cannot claim to fully understand the complexities of locust management in Africa, as it has taken more than a decade for me to gain insight regarding the biological and cultural subtleties of grasshopper management on rangeland in the US. However, I am confident that we have a great deal to gain from an open dialogue, and this paper is an attempt to open such a discussion with apologies for what are sure to be simplifications and errors of interpretation in summarising an enormously complicated and sophisticated system of locust management.

## COMPARATIVE ECOLOGY: GRASSHOPPERS *VS.* LOCUSTS

### Similarities: A Common Set of Essential Characteristics

For the purposes of the analysis, I shall use the term "grasshoppers" to refer to rangeland acridids in the US (understanding that this definition in no way reflects the tremendous diversity of temperate and tropical grasshoppers and their attendant management systems) and the term "locusts" to refer to locusts in Africa (and primarily the Desert locust, *S. gregaria*, which remains the dominant pest species but represents only a portion of the remarkable range of locusts found around the world). In essence, locusts are simply those species of grasshoppers that exhibit behavioural and physiological "phase changes" associated with migration or under crowded conditions. This capacity to exhibit a gregarious phase is a continuous, rather than discrete, biological characteristic with some species being more "locust-like" than others. In this context, grasshoppers and locusts have three critical similarities that allow a basis for common concerns and approaches to pest management. First, both are in the Family Acrididae, and thereby share a number of fundamental physiological, behavioural, and ecological similarities, including: hemimetabolous development (eggs, nymphs, and adults), egg pods which are usually buried in the soil, susceptibility to similar pathogens, efficient water conservation, capacity for rapid population growth, etc. (Uvarov 1966, 1977).

Second, all pest species of acridids are native to the habitats in which they occur. Although movements may be extensive, there are no exotic pest species in Africa or the US. Hence, these organisms are not, by definition, targets of classical biological control programmes (the use of exotic agents to control co-evolved, exotic pests). Moreover, given that these species have evolved within the ecosystems where they occur, eradication is an ecologically risky and economically untenable management strategy. Considerable evidence suggests that the extirpation or chronic suppression of native species from core areas of their range, particularly those that appear to be so deeply integrated into immense, native ecosystems, is ill-advised (Lockwood 1993a, b).

Third, the outbreak dynamics of grasshoppers and locusts are the manifestation of natural processes in which the organisms exploit ephemeral resources that arise and disappear as a function of erratic weather conditions. These opportunistic resource trackers (Kemp 1992), may have the frequency, duration or scale of their population dynamics modified by human activities (Lockwood et al. 1988), but the essential capacity for catastrophic outbreaks is an evolved strategy to exploit unpredictably abundant resources. This life history strategy allows locust and grasshopper outbreaks to encompass immense spatial scales, which necessitates an "area-wide" approach to their detection and management.

### Differences: The Necessity for Careful Extrapolation

Although locusts and grasshoppers have essential ecological similarities, there are a number of important differences that require careful adaptation of management approaches from one system to the other (El-Gammal et al. 1995). These differences limit the direct extrapolation of methods, but with thoughtful modifications, some approaches may be even more viable in the novel setting than in the original context. In essence, there are five factors that must be considered as we compare and exchange management tactics and philosophies between Africa and the US.

First, grasshopper species are typically univoltine (one generation per year), while locusts are almost invariably multivoltine, at least during the development of outbreaks (Uvarov 1966, 1977, Pedgley 1981). Hence, locusts have an intrinsically greater capacity for population increase on an annual basis, so the optimal mortality of a locust control programme is nearly 100%, and a lag in our response to a growing infestation has serious consequences. Conversely, 80-90% control is the goal of a grasshopper control programme, and the timing of intervention is much more critical, as there is typically only a 2 to 4 week window during which treatments are efficacious in any given year (DeBrey et al. 1993).

Second, grasshopper outbreaks are often composed of many species, with 2 to 5 species typically comprising the majority of the infestation and another 5 to 15 species "coat-tailing" as predators are swamped by the enormous population densities (Joern and Gaines 1990, Lockwood 1997). Thus, management strategies should consider a wide range of feeding, reproductive, and related behaviours. Consequently, target-specific approaches are rarely employed (e.g., the use of bran bait formulations has been largely abandoned, as many common pests species do not consume baits, and viral biological control agents have been dismissed as being too narrow in their host range to warrant the cost of development (Onsager et al. 1990-1993)). Locust outbreaks are usually comprised of a single species, so the opportunity for host-specific cultural and biological control approaches is greatly enhanced (Lomer and Prior 1992).

Third, during an outbreak, grasshoppers are usually more sedentary than locusts. At high densities, locust nymphs travel long distances in tightly cohesive bands (a phenomenon virtually unknown in the grasshoppers). Although a few grasshopper species can exhibit swarming behaviour, this is an unusual behaviour (Pfadt 1988). The greater mobility of locusts has profound management implications. For example, much larger areas must be incorporated into a locust control programme, and tracking the movement of swarms becomes a critical element of a campaign. Indeed, their movements may necessitate the repeated treatment of a re-invaded area. However, there are also logistical advantages of a mobile pest. Strips of insecticides spaced at >1 km can serve as barriers during control programmes for nymphal bands, while effective control of grasshopper populations requires strips of insecticide at 8 to 70 m intervals (Rachadi and Foucart 1996, Lockwood and Schell 1997).

Fourth, in an area-wide context, grasshoppers are pests of rangelands (and hence, animal production) in the US. Although crop damage can occur when rangeland infestations move into agricultural lands, the primary focus of large-scale survey and treatment programmes has been the native grass- and shrublands of the western states (Hewitt and Onsager 1983, Schell and Lockwood 1997a,b). Locust outbreaks may also originate in uncultivated lands, but the singular focus of control is the prevention of crop damage as swarms migrate into agricultural areas (Pedgley 1981). Hence, concerns for residues on food crops are particularly relevant in locust control programmes.

Fifth, the management of grasshoppers in the US has primarily been a matter of economic concern. That is, the decision support systems and infrastructure of management (Berry et al. 1994, Hastings et al. 1996) are concentrated on the question of economic thresholds, benefit:cost ratios and similar financial measures of a programme's success. Recently however, the National Grasshopper Management Board (1995, 1998) has taken the more expansive position that the goal of grasshopper management is to keep good stewards on the land. This re-perception of purpose has brought grasshopper management much closer to the philosophy of locust management. In Africa, the fundamental concern is food security, not economic efficacy, and keeping productive and sustainable agriculturalists on the land is a critical aspect of food security. This convergence of motivations serves as a common point of departure for discussions of how US and African acridologists and pest managers may learn from one another as the parameters of area-wide programmes are changing.

## MANAGEMENT ISSUES: A CONVERGENCE OF FUTURES

In developing a comparative analysis of area-wide grasshopper and locust control programmes, it is valuable to consider the following geopolitical premise: the relationships between states and the federal government in the US are remarkably similar to the relationships between nations and international agencies in Africa. In the US, the primary responsibility and source of funding for grasshopper control have been with the US Department of Agriculture's Animal and Plant Health Inspection Service (USDA-APHIS). The states have not had to coordinate, fund, or conduct surveys, environmental documentation, area-wide control programmes, or training and research efforts associated with grasshopper management. Similarly, the United Nations' Food and Agriculture Organization (UN-FAO) has been the primary source of logistical resources and the coordinating body for utilisation of bilateral and multilateral funds for control operations. The other key component of a sound comparative analysis is the common vision of decentralisation, which appears to be a dominant trend for both APHIS and FAO (Husnick 1995, FAO 1998). In turn, decentralisation is a function of several, interacting factors which include declining funding, which may, itself, reflect the general paucity of evidence that large-scale pest management has been effective in reducing the severity, frequency or duration of grasshopper or locust outbreaks in the US and Africa. Indeed, some evidence suggests that large-scale, long-term use of broad-spectrum insecticides may exacerbate grasshopper population dynamics (Lockwood et al. 1988). Thus, the combination of the geopolitical conditions and the localisation of programmes underlies the converging, fundamental pest management issues defining the future of grasshopper and locust control, which include: the scale of the management programmes, the purpose of intervention, the sources of funding for control, the timing of the response to an outbreak and the constraints on the type of programme that emerges.

### Scale of Management: Regional vs. Local

The traditional scale of grasshopper and locust pest management has been an extreme version of the area-wide concept. This approach has been perhaps the single most important factor in defining the other elements of the pest management programmes. When intervention occurs over millions of hectares, the cost, timing, and constraints are essentially dictated. However, in recent years, there appears to be a trend towards localisation of grasshopper and locust management as an element of a general decentralisation. As grasshopper and locust programmes become more localised and site-specific, the other parameters of these efforts will inevitably change as well.

### Cost of Management: External vs. Internal

As programmes are reduced in size, the rationale for federal (US) or international (Africa) funding is correspondingly diminished. The geopolitical unit of management becomes the determinant of funding, so that smaller-scale programmes are viewed as the responsibility of more localised governments (states in the US or nations in Africa).

To a certain extent, economics are both an effect and a cause of management localisation. That is, as funding to federal (US) and international (FAO) agencies declines, decentralisation becomes a necessity as much as a "strategy". The systematic effort to reduce federal spending in the US has undermined APHIS' capacity to sustain its grasshopper management programme, and presumably the US failure to maintain its payments to the UN may have affected FAO's ability to support international, locust control programmes.

### Timing of Pest Management: Reactive vs. Proactive

Large-scale pest management programmes must overcome financial and logistical momentum in order to respond to an outbreak, and the attendant lag in mobilising such efforts can result in serious delays and agricultural losses. Although early warning systems are possible in theory, history suggests that the intensity of survey declines following the recession of major outbreaks, such that the next outbreak often begins without notice (Lockwood et al. 1988). Thus, the reduction in spatial scale may allow a shift from a reactive to a proactive or preventative programme that detects and suppresses incipient loci or "hot spots" of infestation. Again, cause and effect become confounded, as FAO's Emergency Prevention System (EMPRES 1998) was developed as a means of taking proactive measures on a more localised, but still multinational, basis. Indeed, it can be argued that a philosophical transition from treatment to prevention has played an important role in precipitating the downsizing of grasshopper and locust management.

### Constraints of Pest Management: Political vs. Economic

With declining spatial, economic, and temporal scales of grasshopper and locust management, the limiting factor in developing and sustaining a pest management system is shifting from political to economic considerations. Rather than political pressure driving the availability of external resources, the matter is rapidly becoming a function of local economics (states in the US or nations in Africa). In the US, there has been a series of recent efforts to develop analytical and decision-support tools necessary to assess the economic benefit-cost ratios of control programmes, and reducing the cost of control has become a driving factor in the development of alternative management strategies (Lockwood and Schell 1997). In Africa, economic models have not been developed to assess the viability of local programmes, but it is clear that the rekindled interest in barrier treatments is, at least in part, motivated by the economic benefits of this strategy (Rachadi and Foucart 1996, Musuna and Mugisha 1997, Scherer and Célestin 1997, Wilps and Diop 1997, Launois and Rachadi 1997).

### Purpose of Pest Management: Expediency vs. Sustainability

As the scale of pest management declines, the purpose of grasshopper and locust control is being reassessed. Again, it could be argued that the changing philosophy of pest management has been not only a consequence but a cause of this trend towards decentralisation. In any case, it is evident that the pest management systems that are developing for grasshoppers and locusts may become less influenced by political expediency. Although it is yet to be seen if the immense public pressure and consequent political repercussions associated with outbreaks of acridids in the US and Africa will be sufficient to once again re-establish the crisis-driven, reactionary pest management programmes of APHIS and FAO, some evidence suggests that the pattern and philosophy of pest management may be changing. Rather than the periodic disassembly and reassembly of immense programmes in response to grasshopper and locust outbreaks, with assistance often arriving only after serious losses have occurred, the purpose appears to be shifting to developing sustainable, ongoing pest management systems under local control (e.g., the "hot spot" control programme (Lockwood and Schell 1995) and the EMPRES programme (EMPRES 1998, FAO 1998)). Indeed, the growing interest in preventative programmes involving cultural and biological control (Lockwood and Schell 1997) reflects this underlying change in philosophy. Perhaps the most explicit evidence of such a transition is the statement by the US National Grasshopper Management Board (1995, 1998) that the purpose of control is to "keep good stewards on the land".

## DECENTRALISATION: A COMPLEX OF COST AND BENEFITS

The rapidly changing conditions of area-wide pest management make the documentation of dynamics extremely difficult. As such, I shall rely heavily on my experiences with the effects of decentralisation of grasshopper management in Wyoming, with the hope that the reader will forgive my reliance on narrative evidence. To the extent possible, the case study of Wyoming will be related to apparent changes in Africa, but the nature of political communications and diversity of policies makes such extrapolations rather speculative.

### Costs of Localising Pest Management

There are a number of costs associated with reducing the scale of grasshopper and locust control programmes, and without very careful and proactive leadership, these disadvantages may well lead to an ineffective array of disconnected efforts that fail to provide economic returns, food security or land stewardship.

Perhaps the most obvious cost of localisation is the loss of various economies of scale. The cost of insecticide, biological control agents, aerial application, etc., can all be expected to increase as the scale of programmes is reduced. In a related manner, there are likely to be inefficiencies with the fragmentation of management programmes. The border areas will be dramatically increased, so the integration of survey and management will be greatly complicated. For example, because surveyors are rarely located in the geographic centre of new, arbitrarily (politically, rather than ecologically) decentralised management units, these individuals are likely to terminate surveying at borders that are relatively proximate to their point of origin in one direction but travel to relatively distant borders in another direction (that would be much more efficiently sampled by an individual in the adjacent management unit). Indeed, we are seeing such logistical inefficiencies arising in Wyoming during the early development of a state-based survey programme.

The fragmentation associated with decentralisation may lead not just to inefficiencies but to actual conflicts of programmatic goals and strategies across an expanding network of boundaries. Differences in state-based surveys that are arising as a function of different perceptions of the purpose of grasshopper management and the consequent value of surveys will preclude any regional integration of survey data in the US. Conflicting programmatic goals are a particularly serious matter in the context of a highly mobile pest, such as the Desert locust in Africa. If one nation decides to wait until swarms develop in order to organise a response (reactive or curative tactics) and an adjacent country has adopted a hopper band management programme (proactive or preventative tactics), then the success of the latter effort may be undermined by the strategy of the former. This concern can be conceptually expanded into the maxim that discord will arise when costs are individualised but the benefits are collectivised. That is, when early, preventative actions are taken in one locale, the adjacent areas may avoid the losses associated with an expanding outbreak (in the US) or a mobile swarm (in Africa). Hence, sustaining a decentralised mosaic of programmes in which we localise the costs and regionalise the benefits will be extremely difficult. This concern has been raised in the context of "hot spot" management in the US (Lockwood and Schell 1995), and it is evident that considerable energy and resources will be needed to assure that decentralised, smaller-scale programme are cohesive – or at least non-antagonistic – networks and to avoid the creation of area-wide pest management anarchy, in which fragmented units parasitise the benefits of prevention or pursue conflicting objectives across borders.

Finally, the current transition from large- to small-scale area-wide grasshopper and locust management can create considerable uncertainty and indecision. For example, the virtual elimination of the APHIS survey left states with no means of assessing the scale and intensity of ongoing grasshopper infestations as there was no systematic transfer of technology, infrastructure, or training. Hence, the sudden departure from federal to state responsibility has created a serious gap in our capacity to assess or respond to grasshopper outbreaks. In Africa, it appears that FAO has made a conscious effort to address the decentralisation of locust management through training and technology transfer, but it has yet to be demonstrated that nations or regions are prepared to respond to an outbreak. The history of external support via rescue efforts by APHIS or FAO entering a region when the grasshopper or locust situation becomes severe and politically untenable may undermine development of localised, management infrastructures. There are those who expect that when an acridid outbreak becomes sufficiently serious, the external agencies capable of imposing a system of area-wide control will be drawn back into the affected region. However, experience also suggests that this approach generates suboptimal solutions, as the lag time in mobilising large-scale interventions invariably permits extensive losses and precludes preventative strategies.

## Benefits of Localising Pest Management

It would be misleading to suggest that decentralisation is necessarily a negative trend with respect to pest management. Reducing the scale of pest management has a number of possible and apparent benefits. As grasshopper and locust control programmes are localised, there may be greater participation by agriculturalists, as the immediate relevance of their efforts becomes apparent and their assistance becomes necessary. In Wyoming, we have seen a resurgence of interest by local (county) Weed & Pest districts with the diminishment of federal programmes. FAO efforts in Africa also have the potential to enhance local "ownership" of management programmes. Furthermore, as local agencies take on greater portions of the programmatic responsibilities, we can expect enhanced accountability and responsiveness. The Weed & Pest districts in Wyoming are under the control of local boards and funded by county taxes, so there is intense interest and scrutiny with respect to fiscal efficacy. In particular, the traditional overuse of insecticides that was possible through federal cost-sharing may soon become an obsolete strategy, as local funds are managed to optimise economic benefits (Berry et al. 1994, Hastings et al. 1996). The consequent reduction in insecticide usage translates into not only increased economic returns but decreased environmental impacts as well (Lockwood and Schell 1997). When the distance between the source and the expenditure of funds is reduced via decentralisation, it is reasonable to expect that the attention to the efficient use of the funds will be correspondingly increased.

Interestingly, localised oversight has stimulated, rather than inhibited, innovation and risk-taking with respect to testing, adopting, and advocating new methods in Wyoming. This openness to novelty may be due to the realisation that the large-scale programmes of APHIS cannot simply be down-sized and forced into the local conditions. Our experience in Wyoming suggests that the adoption of the RAAT method and discussions of county-level surveys using the most recent innovations in sampling (Legg et al. 1993, 1996, Legg and Lockwood 1995) have been made possible (and necessary) by decentralisation; the old methods are simply inappropriate for the new scale. Indeed, the first studies of the RAAT method in the US were funded by, and conducted in cooperation with, two county-level Weed & Pest districts. Perhaps the frequency of, and apparent interest in, training programmes for locust management in Africa also reflects a willingness to explore new approaches (FAO 1998).

A related advantage of decentralisation is the potential for higher efficiency via site-specificity. It is clear that different habitats, landscapes, and production practices all generate localised ecological and economic situations, the response to which cannot be optimised by broad, federal policies and regionalised strategies. For example, the "hot spot" management method of suppressing outbreak by treating incipient loci of infestation is most effective when integrated with intense, site-specific knowledge of local pest species and habitats (Lockwood and Schell 1995). Indeed, FAO has concluded that, "... countries concerned with EMPRES are witnessing more efficient preventive control of pests. Response times to members' requests are shortening, and FAO staff in the field are in closer contact with the countries' needs" (FAO 1998).

## CONCLUSIONS

The benefits and costs of an area-wide approach to grasshopper and locust management suggest that there may be no "ideal" scale for all aspects of a programme. It is safe to say that the maxim of "bigger is better" fails to adequately address the complexities of pest management, but the notion that "small is beautiful" oversimplifies the situation. Rather, it seems that different elements of a sustained area-wide pest management (as distinguished from the simpler objectives of a pest eradication programme) are optimised at different spatial scales. At this time, it appears that survey may be the sort of activity that is most efficiently conducted and coordinated over very large areas, due to economies of scale and with respect to logistics, consistency of training, standardisation of methods, uniformity of reporting and integration of information. Although there are also economies of scale with respect to treatment, it appears that site-specific or localised treatment programmes may yield greater economic and environmental returns, as well as more appropriate and timely responses to particular conditions. No matter what pattern of management emerges in the course of decentralisation, it will be critical to communicate and integrate across scales. While a mixture of management scales may be more complex that the historical, regionalised efforts of APHIS and FAO, it appears that there is considerable promise that a systematically integrated network of small- and large-scale responsibilities will result in a more effective strategy than uniform scaling (either localised or centralised) without regard to the scale-dependent opportunities and difficulties of each management component.

## REFERENCES

Berry, J.S., W.P. Kemp and J.A. Onsager. 1994. Hopper: decision support system for rangeland grasshopper management (software and manual), USDA-APHIS, Boise, ID.

DeBrey, L.D., M.J. Brewer and J.A. Lockwood. 1993. Rangeland grasshopper management. Wyoming Agric. Exp. Stn. Bull. B-980, 8 pp.

El-Gammal, A.A.M., J.A. Lockwood, J.A. Onsager and M.P. Pener. 1995. The future of grasshopper and locust pest management. J. Orthop. Res. (panel discussion transcript edited by J.A. Lockwood) 4: 15-17.

Emergency Prevention Systems (EMPRES). 1998. World wide web site. http://www/fao.org/WAICENT/FaoInfo/Agricult/AGA/AGAH/EMPRES/Info.htm.

Food and Agricultural Organization (FAO) 1998. World wide web site. http://www/fao.org/default.htm.

Hastings, J.D., L.K. Branting and J.A. Lockwood. 1996. A multiple paradigm system for rangeland pest management. Comp. Electron. Agric. 16: 47-67.

Hewitt, G.B. and J.A. Onsager, 1983. Control of grasshoppers on rangeland in the Unites States – a perspective. J. Range Manage. 36: 202-207.

Husnik, D.F. 1995. Short-term actions for grasshopper/Mormon cricket programme delivery. Memo from the Deputy Administrator, USDA-APHIS-PPQ, December 28, 1995.

Joern, A. and S.B. Gaines. 1990. Population dynamics and regulation in grasshoppers, pp. 415-482. *In* R.F. Chapman and A. Joern (eds.), Biology of grasshoppers. Wiley, New York.

Kemp, W.P. 1992. Rangeland grasshopper (Orthoptera: Acrididae) community structure: a working hypothesis. Environ. Entomol. 21: 461-470.

Launois, M. and T. Rachadi. 1997. The problem of a replacement for dieldrin, pp. 247. *In* New strategies in locust control. S. Krall, R. Peveling and D. Ba Diallo (eds.). Birkhäuser, Basel, Switzerland.

Legg, D.E. and J.A. Lockwood. 1995. Estimating densities of grasshopper (Orthoptera: Acrididae) assemblages: an extension of the binomial sampling technique. J. Kansas Entomol. Soc. 68: 178-183.

Legg, D.E., J.A. Lockwood and M.J. Brewer. 1996. Variability in rangeland grasshopper (Orthoptera: Acrididae) counts when using the standard, visualised sampling method. J. Econ. Entomol. 89: 1143-1150.

Legg, D.E., J.A. Lockwood, W.P. Kemp and M.A. Nolan. 1993. Estimating densities of grasshopper (Orthoptera: Acrididae) assemblages using binomial sampling. Environ. Entomol. 22: 733-742.

Lockwood, J.A. 1993a. The benefits and costs of controlling rangeland grasshoppers with exotic organisms: the search for a null hypothesis and regulatory compromise. Environ. Entomol. 22:904-914.

Lockwood, J.A. 1993b. Environmental issues involved in the biological control of rangeland grasshoppers (Orthoptera: Acrididae) with exotic agents. Environ. Entomol. 22: 503-518.

Lockwood, J.A. 1997. Rangeland grasshopper ecology, pp. 83-102. *In* Bionomics of grasshoppers and their relatives. S.K. Gangwere, M.C. Muralirangan and M. Muralirangan (eds.). CABI, London.

Lockwood, J.A. and S.P. Schell. 1995. Rangeland grasshopper outbreak dynamics: Gradient, eruptive, both, or neither? J. Orthop. Res. 4: 35-48.

Lockwood, J.A. and S.P. Schell. 1997. Decreasing economic and environmental costs through reduced area and agent insecticide treatments (RAATs) for the control of rangeland grasshoppers: Empirical results and their implicaitons for pest management. J. Orthop. Res. 6: 19-32.

Lockwood, J.A., W.P. Kemp and J.A. Onsager. 1988. Long-term, large-scale effects of insecticidal control on rangeland grasshopper populations. J. Econ. Entomol. 81: 1258-1264.

Lomer, C.J. and C. Prior (eds.). 1992. Biological control of locusts and grasshoppers, Redwood Press, Melksham, United Kingdom, 394 pp.

Musuna, A.C.Z. and F.N. Mugisha. 1997. Evaluation of insect growth regulators for the control of the African migratory locust, *Locusta migratoria migratorioides* (R. & F.), in Central Africa, pp. 137-142. *In* New strategies in locust control. S. Krall, R. Peveling and D. Ba Diallo (eds.). Birkhäuser, Basel, Switzerland.

National Grasshopper Management Board. 1995. Proceedings and resolutions of the 1995 meeting of the National Grasshopper Management Board, January 10-12, Rapid City, South Dakota.

National Grasshopper Management Board. 1998. Proceedings and resolutions of the 1998 meeting of the National Grasshopper Management Board, January 6-8, Denver, Colorado.

Onsager, J.A., D.A. Streett and S.A. Woods. 1990-1993. Grasshopper pathogen field evaluation: virus. *In* USDA-APHIS Grasshopper Integrated Pest Management Project 1990-1993 Annual Reports, Boise, Idaho.

Pedgley, D. 1981. Desert locust forecasting manual. Volume 1, Centre for Overseas Pest Research, London, United Kingdom.

Pfadt, R.E. 1988. Field guide to common western grasshoppers. Wyoming Agric. Exp. Stn. Bull. 912 and supplements.

Rachadi, T. and A. Foucart. 1996. L'efficacité du fipronil en traitement en barrières contre les bandes larvaires du Criquet pèlerin, *Schistocerca gregaria* (Forskål, 1775) en conditions réelles d'opérations antiacridiennes. CIRAD Document 538, Montpellier, France.

Schell, S.P. and J.A. Lockwood. 1997a. Spatial analysis of ecological factors related to grasshopper population dynamics in Wyoming. Environ. Entomol. 26: 1343-1353.

Schell, S.P. and J.A. Lockwood. 1997b. Spatial characteristics of rangeland grasshopper population dynamics in Wyoming: implications for pest management. Environ. Entomol. 26: 1056-1065.

Scherer, R. and H. Célestin. 1997. Persistence of benzoylphenylureas in the control of the migratory locust *Locusta migratoria capito* (Sauss.) in Madagascar, pp. 129-136. *In* New strategies in locust control. S. Krall, R. Peveling and D. Ba Diallo (eds.). Birkhäuser, Basel, Switzerland.

Uvarov, B. 1966. Grasshoppers and locusts: a handbook of general acridology, Volume 1. Cambridge University Press, London, 481 pp.

Uvarov, B. 1977. Grasshoppers and locusts: a handbook of general acridology, Volume 2. Centre for Overseas Pest Research, London, 613 pp.

Wilps, H. and B. Diop. 1997. Field investigations on *Schistocerca gregaria* (Forskål) adults, hoppers and hopper bands, pp. 117-128. *In* New strategies in locust control. S. Krall, R. Peveling and D. Ba Diallo (eds.). Birkhäuser, Basel, Switzerland.

# The Sterile Insect Technique for Commercial Control of the Onion Fly

*M. Loosjes*

De Groene Vlieg, Duivenwaardsedijk 1, Nieuwe Tonge, NL-3244LG, The Netherlands

## INTRODUCTION

Since 1981, the sterile insect technique (SIT) has been applied commercially for the control of the onion fly *(Delia antiqua* Meigen, Diptera: Anthomylidae) in the Netherlands, by a private company called de Groene Vlieg. This paper describes the practical application and problems encountered.

## DESCRIPTION OF THE METHOD

The larvae of the onion fly are a severe pest in onions in temperate regions. In the Netherlands, onions are grown by individual growers on fields of approximately 2-6 hectares each. Onion fields are distributed among fields with other crops, and in areas where onions are grown, they account for about 5-10% of the surface area. The dispersal activity of the released flies is limited with the majority of sterilised flies remaining in the field where they were released. Therefore, it is possible for each individual farmer to use the SIT control method just as he would chemical control and this is independent of the types of control his neighbours apply. The application of SIT in this type of ecosystem is in direct competition with normal chemical control.

SIT is generally used as a population control method by reducing pest numbers or even causing local and temporary eradication. As low population levels are very difficult to monitor and immigration or passive transport of flies is rather common with the onion fly, eradication is not a feasible goal. Moreover, releasing a standard low number of sterile flies is more economical than trying to sustain barriers and monitoring for possible immigrants, at least for the time being.

Onion flies are mass reared by de Groene Vlieg. A new strain is regularly collected from the field, in or near the area where the releases take place. Flies which are released have been reared in the facility for 4-10 generations. Continuous rearing is carried out as young pupae can be stored for up to one year at 3°C. This ability to store pupae is extremely important for the implementation of the programme. The rearing capacity of the facility is over 400 million pupae per year and is used to treat about 2,600 hectares of onions, i.e., about 16% of the total Dutch onion crop area.

Onion fly pupae are sterilised shortly before emergence with gamma radiation from a Co-source. A few years ago, the Dutch government decommissioned a very appropriate radiation plant for pupal irradiation. Pupae have now to be irradiated in Belgium and this is expensive and inconvenient. The source is not really suitable for this purpose or to irradiate the volume of pupae required.

At emergence, the flies are marked with a fluorescent powder, which they retain in their retracted ptilinum. The flies are stored for a short period at 3°C and then loaded into the release containers by weight. Release containers consist of two gauze screens which form a cage of about 40 x 60 x 4 cm, each containing up to 40,000 sterilised flies.

The flies are transported to onion fields by car. They are released by hand, the required number of release containers being emptied along the borders of the onion fields. Three traps are placed in every field and they are sampled weekly. During the first years of the programme, releases were made mainly from aircraft at a height of 20 m and a speed of 180 km/h. This method was abandoned for logistic reasons and for the fact that information on the distance between release site and trapping site could not be obtained.

Onion flies are removed from the traps and checked for the presence of dye. Flies without dye have to be taxonomically certified to be onion flies as there are several very similar fly species. The trap catch data are used to adjust the next release. This causes a considerable time pressure as releases are generally done weekly. Both sexes have to be released and while females have no effect on the technique, data from treatment can improve the information to be derived from the trap catches.

## PRACTICAL PROBLEMS

The main problem is that the targeting of the fly population in an area-wide approach is only partially achieved. This means that the main economic advantage of SIT in area-wide application cannot be realised. The reason is the limited participation of growers due to their lack of confidence in the technique and the selfish behaviour of some growers.

The lack of confidence is due to the fact that most farmers are accustomed to easy application and the visible effects of chemical control. They put more trust in coloured powders (insecticides) applied by themselves all over the field, than they do in flies released onto their fields by other people. These flies can also fly to their neighbours' fields!

The initial cost to the farmer for the application of SIT was at first about one-third cheaper than chemical control. However, this wide difference in price turned out to be another reason for distrust – the farmers could not understand that a cheaper technology could give them the required level of control. Remarkably, the number of customers increased when the price was raised to about the level of the costs of chemical control. At present, the prices charged do not cover costs but making SIT more expensive than chemical control would lower participation much more.

During the first two years of its introduction, SIT was financially supported by the Dutch government. The government often reacts positively to new developments of benefit to the environment, but in the present case when the "newness" had worn away, the government retreated and left matters to market forces. This has been taken as a negative sign by the farmers and they are not clear why governmental support has been withdrawn.

Another cause of distrust in the farmers' eyes is that SIT "doesn't always work". This is due to the fact that at high population levels, SIT is much less effective than chemical control. The reason for this and the action taken are as follows. To reduce populations by a certain factor, the number of sterile flies to be released has to be related to the size of the wild population. However, growers are not at all interested in population levels and are only concerned with the prevention of damage, i.e., a certain low population level. So the higher the population density the stronger the reduction has to be. To achieve this, the number of flies to be released will have to be exponentially dependent on the number of wild flies present. Thus, costs are also exponentially dependent on the numbers of wild flies to be treated.

Chemical control, on the contrary, is based on prescribed treatment sufficient to protect every single plant treated, irrespective of it being the only plant attacked or whether, in fact, all plants being attacked. Treatment costs are independent of population size.

This density dependent character of SIT makes competition with chemical control difficult and as long as high population levels occur, it is at a disadvantage. Once populations are reduced, SIT becomes very effective in comparison with chemical control.

In the onion fly, variation of pest densities from field to field is considerable; differences may reach a factor of 100. In silverskin onions and onion sets, population levels can be especially high with the result that the number of sterilised flies required easily exceeds the rearing capacity or financial constraints of the programme. Where these crops are concerned, the flies are treated in an integrated approach. As soon as the number of sterile flies required becomes difficult to produce, the grower is advised to spray. About equal fractions of both fertile and sterile flies are thus killed, and releases of sterile flies a few days later will result in better sterile to fertile ratios again. However, growers are not concerned about the economics of SIT and they just observe that "it doesn't always work".

Second, selfish behaviour by growers is a considerable problem. Difficulties arise due to the growers who do not use SIT and who constitute some 40% of the onion growers in the regions where it is marketed. Some growers prefer to use a cheaper chemical control or, in fact, practise no control at all but simply hope to benefit from sterile flies that are released in their neighbours' field. This selfish behaviour, generally, does not work very well as the number of flies they receive will always be low and their wild population will increase.

Every year, onions are grown in different fields. In spring, the flies emerging from diapausing pupae, have to search for onions. This leads to a redistribution of populations among onion fields. So, fields where flies have been insufficiently controlled, or not controlled at all, have a major negative effect on the implementation of SIT.

An overall reduction in pest population levels has only occurred in regions where few onion sets or silverskin onions are grown, and where SIT has been applied for more than ten years to over half of the fields.

In summary, the success of SIT against the onion fly in the Netherlands is limited, due to the growers' trust in chemical control, a lack of governmental support, the growers' interpretation of integrated control at high population densities as a sign that SIT does not always work, and the selfish behaviour of some growers.

## FUTURE PROSPECTS

Many growers who believe in, and use, chemical control tend to see SIT as a convenient insurance policy in the event that chemicals would no longer be effective due to the development of resistance. However, this is a dangerous option because when resistance develops, the size of the population to be controlled will far exceed the number of sterile flies that can be reared and released. Only strict and regional management of onion production can provide a solution, but in the Netherlands, such a degree of regulation is in practice impossible.

So at the moment, SIT can only be applied locally until elsewhere in the Netherlands the onion fly develops resistance to insecticides. This would give the opportunity to target nearly 100% of the onions in the present area under SIT treatment and for a better price. The other choice is to terminate the programme immediately and save any future losses.

# The Sterile Insect Technique in the Integrated Pest Management of Whitefly Species in Greenhouses

*M. Calvitti, P.C. Remotti and U. Cirio*

ENEA, C.R. Casaccia - Division of Agriculture and biotechnologies via Anguillarese 301, I-00060 S.M. di Galeria, Rome, Italy

## INTRODUCTION

Insect pests commonly known as whiteflies are Hemiptera belonging to the family of Aleyrodidae *Trialeurodes vaporariorum* Westwood (greenhouse whitefly) and the B-biotype of *Bemisia tabaci* Gennadius (=*Bemisia argentifolii* Bellows and Perring) are pests whose economic importance is constantly increasing within the European agriculture. The B-biotype of *B. tabaci*, in particular, has become more problematic by causing damage over a wide range, from the temperate climates of Californian squash fields to European greenhouses and field crops. In the absence of valid alternatives, many growers have resorted to intensive application of insecticides to control these pests, creating a severe environmental and health hazard. Several new environmentally safe technologies are currently available and have opened up new opportunities in the integrated pest management (IPM) of whiteflies under greenhouse conditions. In particular, biological or biologically-based control means, including a number of fungi, insects, and compounds have been recently developed. However, the limitation of whitefly population outbreaks in greenhouses is a problem that needs to be solved.

The idea to extend the use of sterile insect technique (SIT) to a confined environment against whitefly species is novel, and especially when we consider that the target species undergo arrhenotoky (unfertilised females generate only male progenies). The possibility to join this approach to the IPM of the whitefly species in the greenhouse may open new perspectives in the safe application of nuclear technology for pest control. The present work reviews recent advances in research and practice related to the development of SIT for the control of whiteflies in greenhouses. Explanations on whitefly radiation biology, with data on *Bemisia* spp. radio-sterilisation, methods for whitefly mass rearing and collection, and the definition of a complete SIT procedure tested against the greenhouse whitefly on greenhouse tomatoes, are discussed in detail.

## RADIATION BIOLOGY OF WHITEFLY SPECIES

### *Trialeurodes vaporariorum*

Genchev (1986) first demonstrated the feasibility of sterilising males and females of the greenhouse whitefly by treating pupae or adults at the dose range 50-60 Gy. He found that γ-rays induced sterility in oocytes and as a result of lethal mutation transferred by spermatozoa, combined to partial sperm inactivation. These findings were later confirmed by Huang and Cirio (1989) who studied the radio-sensitivity of pupae of this species testing a 1-100 Gy dose range, 1-3 d before adult eclosion.



**Area-Wide Control of Fruit Flies and Other Insect Pests,**
ed. K.H. Tan. Penerbit Universiti Sains Malaysia, Penang, 2000.

The results obtained from a successive series of laboratory tests (Calvitti et al. 1997) led to the following conclusions: 1) *T. vaporariorum* may be completely sterilised by treating adults with γ-rays at the dose of 50 Gy for females and 60 Gy for males, 2) sterilisation procedures do not affect insect longevity, 3) sterile males preserve a sufficient sperm transfer capacity, resulting in the transmission of dominant lethal mutations to progeny of untreated females. This capacity is highest in the first week after irradiation, and 4) sperm precedence, investigated with a computing $P_2$ value, occurred for the sperm of the most recent copulation (Calvitti et al., unpublished data). We considered these results to be a valid basis for an extension of our studies on the greenhouse environment.

***Bemisia tabaci***

The same experimental procedure used to study the radio-sensitivity of the greenhouse whitefly (Calvitti et al. 1997) has been used to investigate the response of *B. tabaci* (B-biotype) to γ-rays. The results are summarised in Figures 1 and 2. Complete female sterility was achieved at 70 Gy. Males, tested over a 10-d period after irradiation, were less radio-sensitive. In fact, at 80 Gy, they were still able to fertilise a low percentage of eggs (0.01%). Furthermore, a higher resistance to irradiation of the silverleaf whitefly compared with the greenhouse whitefly, was observed. The analysis of the viability and fertilisation status of eggs laid by untreated virgin females that were mated with sterilised males, at various γ-doses (Figure 2), had the highest values of sterile male induced egg mortality (up to 51%) at 70 Gy. These findings confirmed that sterile silverleaf whitefly males produce spermatozoa still able to penetrate the eggs, generating non-viable embryos. This may play a key role in reducing the reproductive capacity of untreated females mated with treated males.

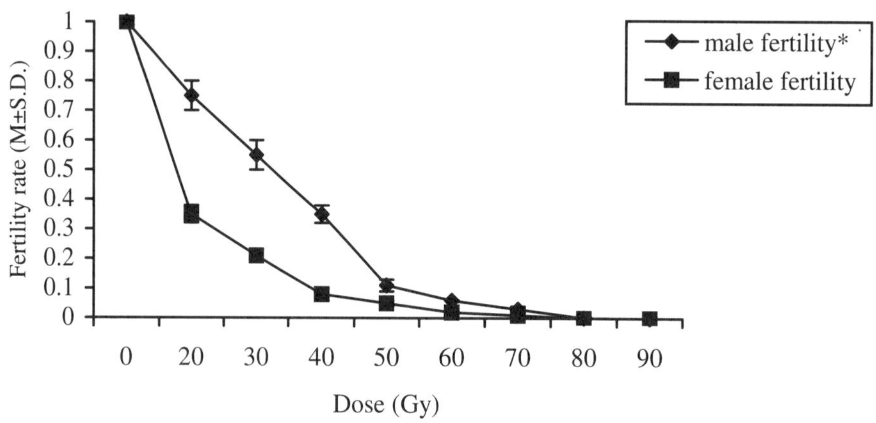


Figure 1. Effect of gamma irradiation on silverleaf whitefly fertility rate.
* Male fertility was considered 100% when crosses between virgin females and treated males produced at least 60% of female offspring (see Figure 2; 0 Gy).

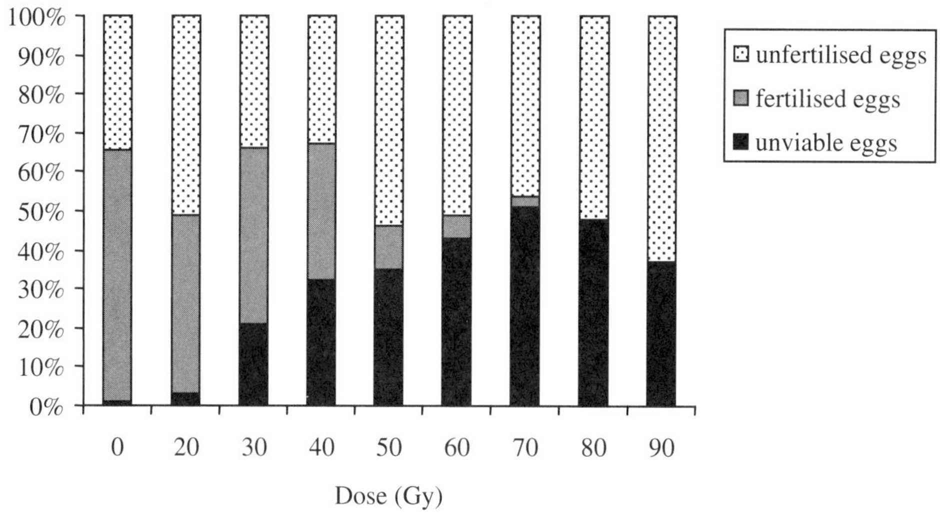


Figure 2. Distribution of unviable, fertilised (resulting in females) and unfertilised (resulting in males) eggs laid by untreated virgin females mated with males irradiated at different doses.

Because of the possible transmission of plant disorders (silverleaf induction) and viruses by the sterile whitefly adults, open field and greenhouse experimentation of SIT is being delayed until the capacity of treated insects to act as vectors for viruses is fully clarified. However, Yokomi et al. (1990) demonstrated that the transmission of the squash silverleaf disorder is due to a toxicogenic factor associated with nymphal feeding.

## EXTENSION OF SIT TO COMMERCIAL GREENHOUSES

### Criteria for Selection in Experimental Greenhouses

An experimental application of SIT on a greenhouse scale, was set up in spring 1997 in two areas located in the province of Latina, central Italy. In this province, protected crops are widespread and *T. vaporariorum* is the dominant whitefly species. Extensive research previously published on the relationship between whiteflies and their natural plant hosts (Calvitti and Remotti 1998) contributed to additional information on the distribution and dynamics of the whitefly species in areas characterised by a high concentration of greenhouses. Whiteflies infest a greenhouse crop by 3 common routes: 1) acquired plant material or from older plantings of the same crops, 2) from weeds or other host plants occurring in the greenhouse, and 3) from host plants in the surrounding area.

According to the presence of whitefly susceptible crops, or weeds occurring in the areas around the greenhouses, two areas were identified for the study. The first one (area A), located at a sandy shore in a natural indentation, was characterised by a low concentration of greenhouses. In this area wild erbaceous plants growing around the greenhouses are generally removed. The second (area B), was characterised by widespread greenhouses and abundant weed hosts for whiteflies. In both areas, we selected 16 unheated greenhouses of about 500 m$^2$, half of which were provided with a 50-mesh screen along the sides, to prevent migratory insect flow. For each area, eight greenhouses were used for SIT trials (four screened and four opened). The remaining eight greenhouses were not subjected to any treatment against whiteflies and were used as controls. In all greenhouses, tomato (*Lycopersicon esculentum* Miller) cultivar "Monica" plants were transplanted at the beginning of March 1997.

## Whitefly Mass Rearing, Collection and Irradiation

Mass rearing of the greenhouse whitefly is feasible at an acceptable cost as shown by the commercial production of its parasitic wasp *Encarsia formosa* Gahan (Hymenoptera Aphelinidae). A mass rearing facility for the whitefly sterile insect campaign was developed at the ENEA-Casaccia research centre (Rome). An isolated heated (T= 25 ± 5°C) greenhouse of ca. 120 m$^2$ was filled with the same proportion of eggplant (*Solanum melongena* L.), tomato and zucchini (*Cucurbita pepo* L.) plants, grown in pots and placed on four benches of 12 m$^2$ (60 plants per bench). Plants were irrigated and fertilised abundantly. Four weeks prior to starting the campaign, some hundred adult whiteflies were released and allowed to spread and multiply in this greenhouse. Insects were captured in a greenhouse where the plants did not show any symptoms of pathology. All plants were soon colonised with enough insect offspring for the latter experiments. The scheme of whitefly production required the replacement of damaged plants not able to sustain high whitefly population. It was possible to collect about 300 adults per plant per day. The adults, normally migrating to the upper part of the plants, were collected with a modified aspirator into petri dishes that had black cheese cloth on the bottom. The collected insects were stored at 4-6°C and counted with an image analyser. Each petri dish contained approximately 1,000 to 2,000 adults. The dishes were successively irradiated at 55 Gy and transported in a cool box to the greenhouses.

## Sterile Insect Release

SIT protocol was defined according to the availability of insects and to the presumed reproductive capacity of the whiteflies. Sterile insects were released into greenhouses at 10-d intervals in order to ensure a continuous supply of sexually active sterile males, for a total of six successive releases. The number of sterile whiteflies (sex ratio 1:1) released, varied from 10 (first release) to 50 insects per plant in each of the last two releases. The complete sterile insect release schedule is shown in Figure 4. We avoided overloading the young plants with sterile insects, although it was demonstrated that sterile whiteflies were not able to cause direct damage to plants due to their feeding activity, confining the effective crop injury to nymphs of the native population (Calvitti et al. 1998). About 200 sterile whiteflies per plant (ca. 120,000 per greenhouse) were released over the experimental period. In each of the two areas, we applied SIT protocol both on the four screened and four opened greenhouses. The same programme of sterile insect release was applied to all the greenhouses selected for SIT trials.

## Sampling Procedure

Prior to the release of the sterile whiteflies, the initial population density of the greenhouse whiteflies was estimated. Since the plants were monitored in the first week after transplant, only the adult stage was monitored. All plants were checked early morning because the adults are less mobile due to the lower temperatures. Since we considered the application of SIT against whiteflies as a preventive measure, we did not select for our purpose, greenhouses in which the adult whitefly density exceeded the mean value of 1 to 2 pairs of adult whiteflies per plant. Yellow sticky were applied to traps around the greenhouse (one trap every 10 m) in order to acquire additional information on the whitefly migratory flow. Traps were replaced at 10-d intervals and the adults trapped were counted (Figure 3).

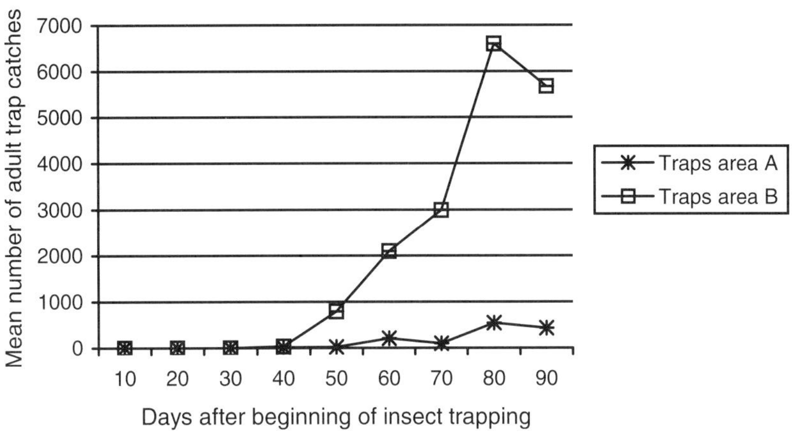


Fig. 3. Mean 10-d whitefly trap catches in A and B areas.

The greenhouse whitefly population was analysed over a 90-d period. Due to the presence of sterile adults, we monitored the density of nymphs (L3 and L4) and pupae by a random sampling of 10% of plants located in the greenhouses. We used the leaf as the sample unit by removing three leaves from the upper, middle and lower sections of plants. All the greenhouses were subjected to the same programme of chemical treatment as any commercial crop except that no insecticides were applied. At the end of the experimental period, all mature fruits were harvested from the different classes of the greenhouses and examined to determine the mean percent of fruits coated with honeydew and mould formation.

**SIT Evaluation**

The application of SIT treatment resulted in a slower population increase in all four different environmental conditions (Figure 4). Absolute population reduction was not achieved, nor could this be expected, due to the parthenogenesis of these insects.

It should be emphasised that after 90-d, the differences in whitefly density, between treated and untreated crops, were remarkable, particularly in the greenhouses in area A, characterised by low presence of whitefly fliers in the outside environment (Figures 3, 4a and b). In this area, the application of 50-mesh screens did not result as an essential measure.

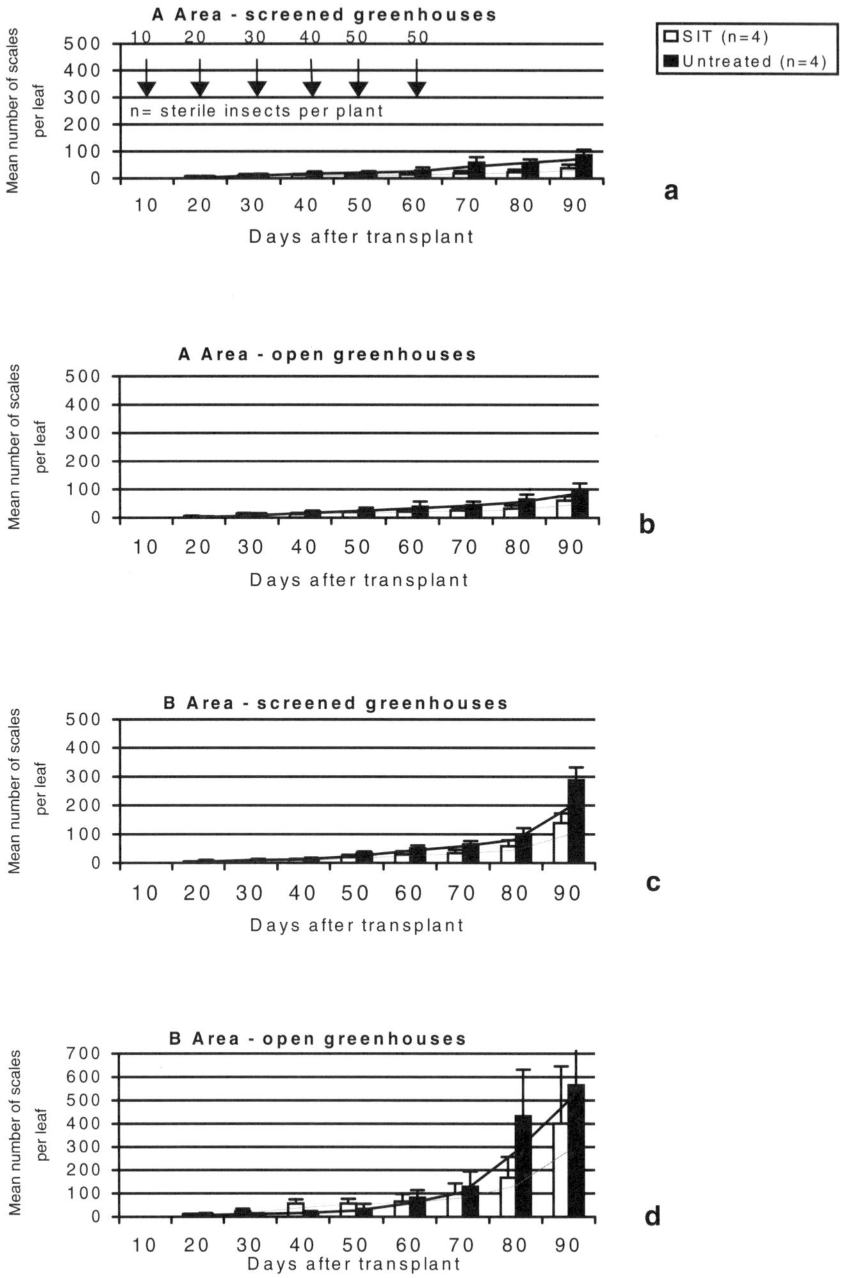


Fig. 4. Density trends of whitefly scales (3rd and 4th instars) and pupae in greenhouses of area A (a-b) and area B (c-d). Arrows and numbers show the releases of sterile insects valid for all SIT greenhouses.

Area B was characterised by abundant fliers in and outside the greenhouses (as shown by the adult mass trapping) (Figure 3). In greenhouses lacking a screen along the sides, the efficacy of SIT application resulted in a decrease. In fact, the delaying effect on the natural drift of whiteflies to population outbreaks (Figure 4c and d) was not remarkable. Because of the constant input of insects from the outside, the sterile insect technique did not work sufficiently well to achieve a significant retardation of the population increase. Whiteflies could develop up to the mean density of 400 scales per leaf in spite of the slowing effect of SIT. The isolation of greenhouses achieved with net-screens proved a very useful measure. In fact, this study has demonstrated that the joint use of screens and sterile insects retarded the whitefly population increase by as much as ca. 50%.

The qualitative analysis of tomato fruits, performed at the 90$^{th}$ d, showed that the differences found in the population densities did not consistently result in a different quality of fruits. In area A, whiteflies did not cause any damage to plants in treated or in untreated greenhouses, thereby confirming that the release of sterile whiteflies could not cause damage to the crop.

In area B, we observed honeydew and mould on 12.2 ± 3.8% of harvested fruits from untreated and unscreened greenhouses. The application of net-screens as a unique measure to prevent whitefly input reduced the mean value to 8.6 ± 2.1%. Approximately, the same percentage of contaminated fruits was recorded when SIT was applied in greenhouses lacking screens (7.6 ± 2.0). The most notable results in terms of whitefly population growth reduction and fruit quality (1.5 ± 1.1% of fruits with honeydew deposits) were achieved when SIT was coupled with the application of 50 mesh-screens.

Application of SIT against whitefly species, under greenhouse conditions, may work well as a preventive measure by delaying significantly the occurrence of population outbreaks, which causes relevant economical damage. This role may become of primary importance in relation to short period greenhouse crops.

The sterile insects do not cause damage to plants as previously shown on zucchini plants (Calvitti *et al.* 1998), and the results obtained on tomatoes have confirmed this. We may presume that since sterilised females lay fewer eggs only in the first week after irradiation, they probably require less food. Research is in progress to verify this, which is fundamental to determine whether or not sterile whiteflies completely preserve the capacity as viral vectors.

The results obtained on greenhouse tomatoes suggest that SIT worked better in greenhouses not exposed to high migratory pressure. This pressure depended on the ecological characteristics of the area and was reduced by applying net-screens. Nevertheless, we believe that SIT may be considered a useful tool in supporting the augmentative biological control of whiteflies. SIT is not limited either by adverse climatic conditions or the use of agro-chemicals. The only limiting factor now is incomplete knowledge of the physiological effects on plants associated with the excessive pressure of sterile whiteflies.

## REFERENCES

Calvitti, M., C. Govoni, M. Buttarazzi and U. Cirio. 1997. Use of induced sterility in adult *Trialeurodes vaporariorum* (Homoptera: Aleyrodidae) treated with ionising radiation. J. Econ. Entomol. 90: 1022- 1027.

Calvitti, M. and P.C. Remotti. 1998. Host preference and demographic parameters of *B. argentifolii* on weeds. Environ. Entomol. 27 (6): 1350-1356.

Calvitti, M., P.C. Remotti, A. Pasquali and U. Cirio. 1998. First results in the use of the sterile insect technique against *Trialeurodes vaporariorum* in greenhouse. Ann. Entomol. Soc. Am.. 91 (6) 513-517.

Genchev, N. 1986. Radiation sterilisation of the greenhouse whitefly (*Trialeurodes vaporariorum* Westw., Homoptera: Aleyrodidae). I. Lethal radiation effects on puparia and imagoes. Soil Sci. Plant Prot. 21: 125-133.

Huang, Y.G. and U. Cirio. 1989. Tecnica dell'insetto sterile per il controllo del *Trialeurodes vaporariorum* Westw. Inf. tore fitopatologico 6: 63-66.

Yokomi, R.K., H.A. Hoelmer and L.S. Osborne. 1990. Relationships between the sweet potato whitefly and the squash leaf disorder. Phytopatol. 80: 895-900.

# Area-Wide Integration of Lepidopteran $F_1$ Sterility and Augmentative Biological Control

*James E. Carpenter*

Insect Biology and Population Management Research Laboratory, Agricultural Research Service, United States Department of Agriculture, Tifton, Georgia, USA

## INTRODUCTION

Area-wide pest management (APM) and integrated pest management (IPM) originated from two different efforts to combine two or more control techniques into programmes in which each method could synergise the effectiveness of others and thus create a level of pest control that was greater than that of a single technique (Perkins 1982). Since then, the concept of APM has evolved to include many aspects of IPM and often is now referred to as area-wide IPM. Still, the element of total population management is central to this approach of insect pest management. In support of APM, Knipling (1998) stated that of the insect pests that were of major concern to agriculture before the newer classes of insecticides were available, most are still pests today, the major exceptions being the screw-worm fly and the boll weevil in the southeastern US cotton growing region. Knipling also noted that both of these pest species were subjected to area-wide suppression programmes. In response to the USDA IPM Initiative (USDA 1993, 1994) which seeks to achieve the national goal of having 75% of the crop acres under IPM by the year 2000, the Agricultural Research Service developed an Area-wide IPM Programme. This programme combines environmentally-sound pest control techniques with the advantages of APM and develops partnerships with other federal, state, local and private sector entities. Technologies such as the integration of lepidopteran $F_1$ sterility and augmentative biological control may be considered for future programmes.

## EFFICACY OF $F_1$ STERILITY AND AUGMENTATIVE BIOLOGICAL CONTROL

The ability of the $F_1$ sterility technique to impact upon seasonal populations of highly mobile lepidopterans was demonstrated on *Helicoverpa zea* (Boddie) (Lepidoptera: Noctuidae). Carpenter and Gross (1993) conducted a pilot test in small mountain valleys in North Carolina to assess the influence of released, substerilised (100 Gy) males on wild *H. zea* populations and to measure the infusion rate of inherited sterility into the wild population. Results from this study revealed that the number of wild males captured per hectare was positively correlated with the distance from the release site of irradiated males. Analyses of seasonal population curves of wild *H. zea* males calculated from mark-recapture data suggested that seasonal increases of wild *H. zea* males were delayed and/or reduced in mountain valleys where irradiated males were released. The incidence of larvae with chromosomal aberrations (progeny of irradiated, released *H. zea* males (Carpenter 1991)) collected from the test sites during the growing seasons indicated that irradiated males were very competitive in mating with wild females and were successful in producing $F_1$ progeny which further reduced the wild population. Overall, release ratios averaged <5:1 but reduced the wild population of *H. zea* by more than 70%.

Augmentative releases of parasitoids also have the potential to moderate the seasonal increase of lepidopteran pest populations. For example, a 3-yr test was conducted to examine the ability of augmentative releases of the tachinid parasitoid *Archytas marmoratus* (Townsend) (Diptera: Tachinidae) to control early season populations of *H. zea* (Proshold et al. 1998). Percentage parasitism of *H. zea* was increased to 42% in non-isolated fields of whorl-stage corn and >90% in isolated fields following inundative releases of *A. marmoratus*. In a contiguous corn growing area, there was a positive correlation between density of *H. zea* larvae and percentage parasitism within 0.8 km of the release field. The field with the greatest larval density and percentage parasitism of *H. zea* larvae was the one farthest from the release site, indicating good host finding capability by *A. marmoratus*.

The potential benefit of combining sterile insects with conventional pest control methods was recognised by Knipling (1964). Later, researchers studied the idea of combining sterile insect releases with the release of parasitoids to improve the efficiency and efficacy of pest control (Barclay 1987, Wong et al. 1992). According to Knipling (1992) and Barclay (1987), combining inundative releases of natural enemies and sterile insects should yield synergistic effects. Although natural enemies and the sterile insect technique have different modes of action, the effectiveness of the sterile insect technique increases the ratio of natural enemies to adult hosts and the effectiveness of natural enemies increases the ratio of sterile to fertile insects. However, greater suppression could be expected if parasitoid releases were combined with the $F_1$ sterility technique (Carpenter 1993). Not only is $F_1$ sterility theoretically more effective than full sterility in reducing population increases (Carpenter et al. 1987), the $F_1$ sterility technique produces sterile $F_1$ larvae that would provide an increased number of hosts for the parasitoids. As a result, the number of parasitoids produced should increase even if the rate of parasitism remained the same (host density independent) and whether or not additional parasitoids were released. Although population models that independently consider augmentative releases of parasitoids (Knipling 1992) and $F_1$ sterility (Carpenter et al. 1987) suggest both tactics are highly efficacious, integrating lepidopteran $F_1$ sterility and augmentative biological control results in synergistic effects (Carpenter et al. 1996b). Therefore, the full potential of $F_1$ sterility and of augmentative releases of parasitoids as area-wide control tactics for lepidopteran pests may be realised only when the two suppression methods are integrated.

Population models may provide some insight into how different control strategies could be combined for greatest efficiency. Although the effectiveness of $F_1$ sterility continues to increase as the ratio of irradiated to non-irradiated insects increases, the efficiency decreases quickly once a 10:1 ratio has been obtained. A similar loss of efficiency occurs in the parasitoid release technique when the parasitoid to host ratio increases above 5:1 (Carpenter 1993). According to these models, the economic benefit of combining $F_1$ sterility and parasitoid augmentation techniques would be greatest when the ratio of irradiated to non-irradiated is <10:1 and the ratio of parasitoid to host is $\leq$5:1.

The model presented in Figure 1 demonstrates that population suppression is increased when $F_1$ sterility and parasitoid releases are combined and that the percentage reduction in normal pest population growth is greater when parasitised host produce adult parasitoids than when no parasitoids are produced. However, the number of parasitoids produced from parasitised hosts following an inundative release of parasitoids is difficult to predict. As the parasitoid:host ratio increases, super-parasitism and, consequently, parasitoid mortality may also increase. During 1993 when *A. marmoratus* were released against *H. zea* larvae in whorl-stage corn, the rate of parasitism was very high and super-parasitism was quite common (Proshold et al. 1998). Consequently, few flies were produced from the parasitised *H. zea* collected from the field.

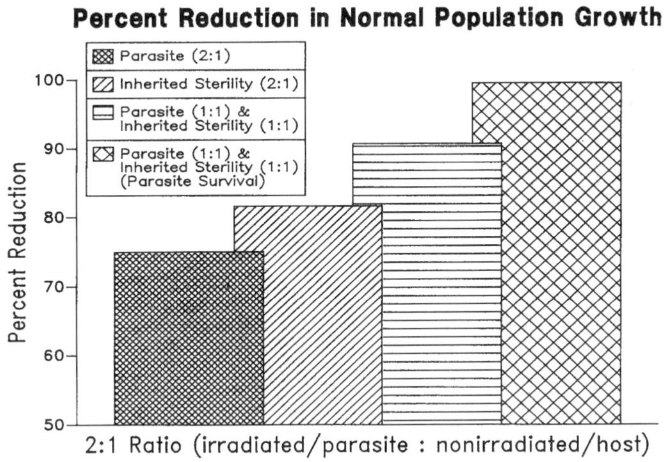


Figure 1. Comparison of the projected reduction in normal population growth when only parasitoids are released, when only irradiated (100 Gy) male moths are released, when both parasitoids and irradiated males are released and no parasitised host produces parasitoids and when both parasitoids and irradiated males are released and all parasitised hosts produce a parasitoid. (Carpenter et al. 1996b).

## COMPATIBILITY OF $F_1$ STERILITY AND AUGMENTATIVE BIOLOGICAL CONTROL

Fully successful integration of $F_1$ sterility and parasitoid augmentation into a management approach can occur only if parasitoid strategies do not negatively impact irradiated insects and their progeny more than that of the wild population and if $F_1$ sterility does not negatively impact upon the efficacy of parasitoids. Knowledge of any negative impact of $F_1$ sterility on parasitoids would be important before an APM programme using $F_1$ sterility was implemented. For example, if parasitoids that attacked $F_1$ were unable to develop normally and most of the host larvae present were $F_1$ larvae, there could be a negative impact on subsequent parasitoid populations. Conversely, if parasitoids were able to develop normally on $F_1$ eggs, larvae and pupae, the greater number of hosts available should allow for a subsequent increase in the parasitoid population. Because many hosts of the $F_1$ generation would die before they reached the adult stage, any parasitoids developing on these hosts would result in a positive and synergistic increase in the efficacy of the APM programme.

Field, greenhouse and laboratory studies compared the acceptability and suitability of progeny from irradiated (100 Gy) and non-irradiated *Spodoptera exigua* (Hubner) (Lepidoptera: Noctuidae) males as hosts for the larval parasitoid *Cotesia marginiventris* (Cresson) (Hymenoptera: Braconidae) (Carpenter et al. 1996a) and progeny from irradiated (100 Gy) and non-irradiated *H. zea* males as hosts for *A. marmoratus* (Mannion et al. 1994, 1995). Results from these studies revealed that progeny of irradiated male and non-irradiated female moths were acceptable and suitable hosts for parasitoid development. Female parasitoids showed no oviposition preference for progeny from females paired with either irradiated or non-irradiated males. Other researchers studying different lepidopteran pests also are considering the possibility of combining $F_1$ sterility and parasitoid augmentation into an effective management approach (Anonymous 1996).

Carpenter (1993) described several different scenarios in which $F_1$ sterility could be integrated with natural enemies to control pest lepidopteran populations. The release of partially sterile male and female moths would produce large numbers of $F_1$ eggs and larvae that could be field-reared on early season weeds, or reared on crop plants that could tolerate some feeding damage (e.g., whorl-stage corn) by the larvae. Natural enemies (native and/or released) could use the $F_1$ eggs and larvae as hosts and, thereby, substantially increase the natural enemy population for the next generation of the pest insect. Also, surviving $F_1$ larvae would produce sterile adults that would negatively impact the next generation of the pest insect. If the economic injury level of cultivated host plants indicated that the additional $F_1$ larvae were undesirable, the dose of radiation administered to the target pest could be increased to a level that would reduce or eliminate the number of progeny from irradiated females, or releases could be limited to irradiated males.

Although $F_1$ sterility is compatible with synthetic organic insecticides, parasitoids and/or predators are not generally compatible with these products. If insecticides are needed to reduce pest infestations, insect growth regulators, biologicals, or other formulations that are compatible with natural enemies should be considered.

Another management scheme could be to establish host plants for the lepidopteran pest in insecticide-free areas adjacent to insecticide-treated crops. Host plants could be artificially infested with pest larvae to provide natural enemies (native and/or released) with an adequate supply of hosts. If the pest larvae used in the artificial infestations (nursery crops) were sterile (progeny of irradiated parents), then non-parasitised larvae would not contribute to the increase of the wild population, but would produce sterile adults that would negatively impact the next generation of the pest insect.

## CHALLENGES AND CONSTRAINTS

Although many studies have demonstrated that $F_1$ sterility is much more effective than full sterility in producing competitive, irradiated male moths and in reducing populations of the target pest (Anisimov et al. 1989, Anisimov 1993, Carpenter et al. 1983, Proverbs et al. 1978), researchers have been reluctant to recommend substerilising doses because of concerns that hatching $F_1$ larvae from irradiated females would cause unacceptable or economic damage to the cultivated crop. These concerns are valid, especially when the crop has a high cash value and can tolerate little or no damage and/or when release ratios of irradiated: wild are >10:1.

In an attempt to better understand the potential damage to plants by $F_1$ larvae, a model is being developed to predict the influence of different parameters on larval populations resulting from increasing release ratios of partially sterile : wild moths. The general form of the model is as follows:

$$\mathbf{M_j fesph/d = l_j} \quad (1)$$

when f = number of female moths/area; d = number of plants/area; e = number of eggs laid/female; s = % survival of larvae to adulthood excluding any effect of augmentative biocontrol; p = % surviving the effects of augmentative biocontrol; h = % of eggs hatching; $M_j$ = probability of mating for each type of male moth present; and $l_j$ = number of larvae/plant resulting from a specific cross. If irradiated (I) males and females have been released into a wild (W) population, then $l_j$ must be calculated for each of the 4 possible crosses (i.e., $l_{WW}$, $l_{WI}$, $l_{IW}$, $l_{II}$). Therefore, the sum of all $l_j$ = the total number of larvae/plant. Also, $\mathbf{M_I + M_W = 1.0}$ for each female type (W&I) when $\mathbf{M_I}$ is calculated as the number of "I" males present divided by the number of all males present and $\mathbf{M_W}$ is calculated as the number of "W" males present divided by the number of all males present.

When the number of larvae/plant needed to reach the economic threshold ($l_t$ ) is known, the number of females/area ($f_t$) required to reach the economic threshold can be calculated by modifying equation (1) as follows:

$$\mathbf{f_t = dl_t/esph} \quad (2)$$

Although this model is incomplete in its present form, it provides a general estimate of how different release ratios of irradiated moths will affect the population of larvae in the field when subjected to simultaneous releases of a natural enemy such as a parasitoid. The data represented in Figure 2 were summarised from radiation biology studies (Anonymous 1993, Carpenter and Malakrong, unpublished data) of the diamondback moth, *Plutella xylostella* (L.) (Lepidoptera: Yponomeutidae), irradiated at 150 Gy. The economic threshold, plant densities and expected parasitism of released *Cotesia plutellae* Kurdjumov (Hymenoptera: Braconidae) were estimated from Biever et al. (1994) and data provided by the Ministry of Agriculture, Mauritius. The number of wild females/area needed to produce an economic threshold of 3 larvae/plant was estimated at 850 and the sex ratio was set at 1:1. Other estimates included: e = 100; d = 10,000; s = 0.7; p = 0.5; $h_{WW}$ = 100%; $h_{WI}$ = 50%; $h_{IW}$ = 10%; $h_{II}$ = 10%. This model estimates that the total number of larvae resulting after a release of irradiated male and female moths will not exceed the number of larvae that would occur naturally, as long as the release ratio of irradiated : wild does not exceed 4:1 (Figure 2) . If the economic threshold were lower and thereby requiring fewer larvae, adjustments would be necessary such as releasing only irradiated males, releasing fewer females, or increasing the dose of radiation. The % of larvae that are $F_1$ sterile increases as the release ratio of irradiated : wild increases (Figure 2). At a release ratio of 4:1, approximately 84% of all larvae in the field would be $F_1$ sterile larvae. The effect of the rate of survival (s) on the number of larvae present at different release ratios is presented in Figure 3. The model predicts that an increase in the rate of survival from s = 0.7 to s = 0.9 would have almost no effect on the number of larvae present.

## SUMMARY

Augmentative biological control and $F_1$ sterility in Lepidoptera have emerged as promising control strategies for lepidopteran pests. The potential for combining these two pest management tactics has been examined by laboratory, greenhouse and field studies. Results from these studies have revealed that progeny of irradiated Lepidoptera and progeny from non-irradiated Lepidoptera are equally acceptable and suitable as

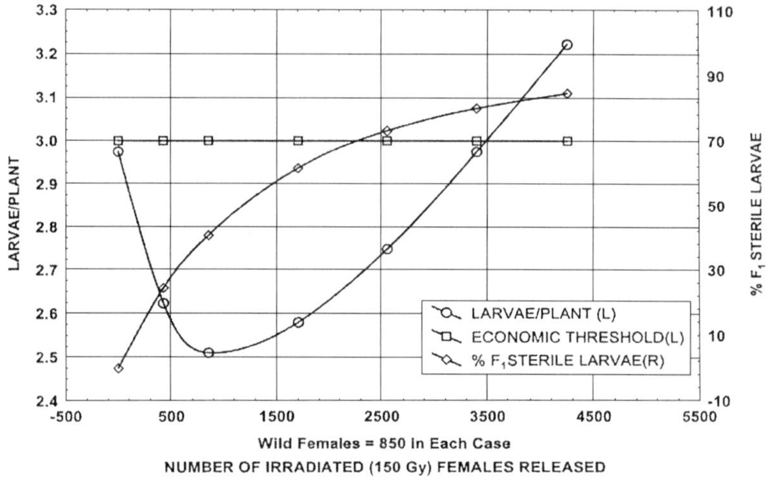


Figure 2. Model showing the effect of the number of released, irradiated DBM on the number of larvae per plant and the percentage of those larvae that are progeny of irradiated parents ($F_1$ sterile larvae). The economic threshold is set at three larvae per plant. Data points on the graph represent values calculated at different release ratios (0:0:1:1; 1:1:2:2; 1:1:1:1; 2:2:1:1; 3:3:1:1; 4:4:1:1; and 5:5:1:1) of irradiated males : irradiated females : wild males : wild females.

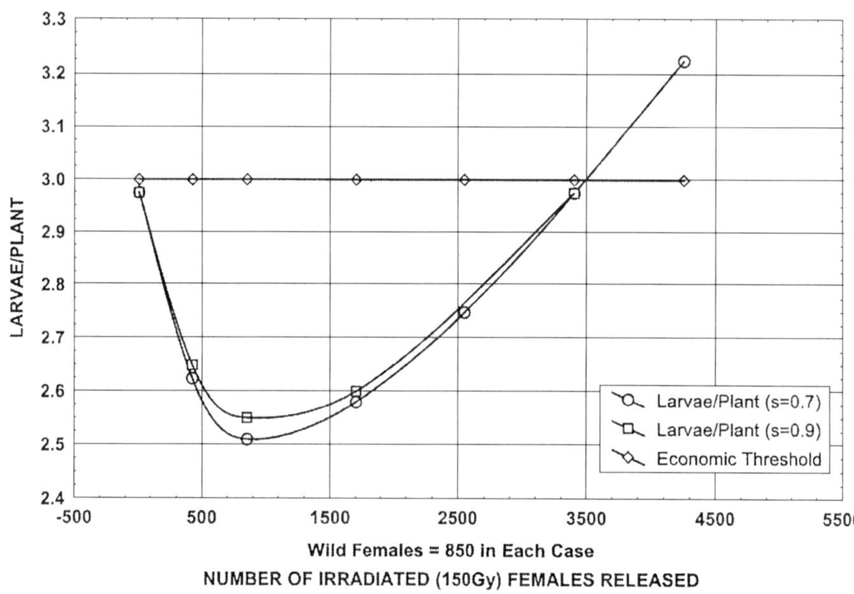


Figure 3. Model (as in Figure 2) comparing the number of larvae per plant for two different survival rates (70% and 90%).

hosts for natural enemy development. Integration of $F_1$ sterility and augmentative releases of natural enemies should yield synergistic effects, especially at release ratios of <10:1. Even lower release ratios of natural enemies have proved to impact pest populations and lower release ratios of irradiated moths would reduce the probability that $F_1$ sterile larvae would cause economic damage to crops.

## REFERENCES

Anisimov, A.I. 1993. Study of the mechanism and possibilities of using $F_1$ sterility for genetic control of codling moth. In Radiation Induced $F_1$ Sterility in Lepidoptera for Area-Wide Control, IAEA-STI PUB/929, Vienna, Austria. pp. 162.

Anisimov, A.I., N.V. Lazurkina and A.N. Shvedov. 1989. Influence of radiation-induced genetic damage on the suppressive effect of inherited sterility in the codling moth (Lepidoptera: Tortricidae). Ann. Entomol. Soc. Am. 82: 769-777.

Anonymous. 1993. Proc. Intl. symposium on management of insect pests: nuclear and related molecular and genetic techniques. Jointly organised by the Intl. Atomic Energy Agency and the Food and Agric. Organization of the U.N., Vienna, 19-23 October 1992. IAEA-SM-327/35.

Anonymous. 1996. Proc. of FAO/IAEA First Research Co-ordination Meeting on Evaluation of population suppression by irradiated lepidoptera and their progeny, Jakarta, Indonesia, 24-28 April 1995. IAEA-D4-RC-561.

Barclay, H.J. 1987. Models of sterile insect releases for populations under attack by parasitoids. Ecol. Model. 36: (3-4) 155-169.

Biever, K.D., D.L. Hostetter and J.R. Kern. 1994. Evolution and implementation of a biological control - IPM system for crucifers: 24-year case history. Am. Entomol. 40: 103-108.

Carpenter, J.E. 1991. Effect of radiation dose on the incidence of visible chromosomal aberrations in *Helicoverpa zea* (Lepidoptera: Noctuidae). Environ. Entomol. 20: 1457-1459.

Carpenter, J.E. 1993. Integration of inherited sterility and other pest management strategies for *Helicoverpa zea*: status and potential. pp. 363 370. *In* Proceedings, Intl. Symposium on Management of Insect Pests: Nuclear and Related Molecular and Genetic Techniques. Jointly organised by the Intl. Atomic Energy Agency and the Food and Agric. Organization of the U.N., Vienna, 19-23 October 1992. IAEA-SM-327/35.

Carpenter, J.E. and H.R. Gross. 1993. Suppression of feral *Heliothis zea* (Lepidoptera: Noctuidae) populations following the infusion of inherited sterility from released substerile males. Environ. Entomol. 22: 1084-1091.

Carpenter, J.E., Hidrayani and W. Sheehan. 1996a. Compatibility of $F_1$ sterility and a parasitoid, *Cotesia marginiventris* (Hymenoptera: Braconidae), for managing *Spodoptera exigua* (Lepidoptera: Noctuidae): acceptability and suitability of hosts. Fl. Entomol. 79(3): 289-295.

Carpenter, J.E., C.M. Mannion and Hidrayani. 1996b. Potential of combining inherited sterility and parasitoids for managing lepidopteran pests. Proc. of FAO/IAEA First Research Co-ordination Meeting on Evaluation of population suppression by irradiated Lepidoptera and their progeny, Jakarta, Indonesia, 24-28 April 1995. IAEA-D4-RC-561, pp. 273-283.

Carpenter, J.E., A.N. Sparks and H.L. Cromroy. 1987. Corn earworm (Lepidoptera:

Noctuidae): influence of irradiation and mating history on the mating propensity of females. J. Econ. Entomol. 80: 1233-1237.

Knipling, E.F. 1964. The potential role of the sterility method for insect population control with special reference to combining this method with conventional methods. US Dept. Agric. Agric. Res. Serv. Ser. 33-98.

Knipling, E.F. 1992. Principles of insect parasitism analysed from new perspectives: practical implications for regulating insect populations by biological means. Agricultural Handbook No. 693, United States Department of Agriculture, Washington, DC.

Knipling, E.F. 1998. Sterile insect and parasite augmentation techniques: unexploited solutions for many insect pests problems. Fl. Entomol. 81: 134-160.

Mannion, C.M., J.E. Carpenter and H.R. Gross. 1994. Potential of the combined use of inherited sterility and a parasitoid, *Archytas marmoratus* (Diptera: Tachinidae), for managing *Helicoverpa zea* (Lepidoptera: Noctuidae). Environ. Entomol. 23: 41-46.

Mannion, C.M., J.E. Carpenter and H.R. Gross. 1995. Integration of inherited sterility and a parasitoid, *Archytas marmoratus* (Diptera: Tachinidae), for managing *Helicoverpa zea* (Lepidoptera: Noctuidae): Acceptability and suitability of hosts. Environ. Entomol. 24:1679-1684.

Perkins, J.H. 1982. Insect, experts and the insecticide crisis. Plenum Press, New York.

Proshold, F.I., H.R. Gross and J.E. Carpenter. 1998. Inundative release of *Archytas marmoratus* (Diptera: Tachinidae) against the corn earworm and fall armyworm (Lepidoptera: Noctuidae) in whorl-stage corn. J. Entomol. Sci. 33: 241-255.

Proverbs, M.D., J.R. Newton and D.M. Logan. 1978. Suppression of codling moth, *Laspeyresia pomonella* (Lepidoptera: Olethreutidae), by release of sterile and partially sterile moths. Can. Entomol. 110: 1095-1102.

US Department of Agriculture. 1993. Three agency release, presidential announcement regarding IPM adoption. Office of Communications, June 23.

US Department of Agriculture. 1994. *USDA announces national plan to increase use of integrated pest management,* Release Number 0943.94; and *Backgrounder: USDA's IPM Initiative,* Release Number 0942.94, Office of Communications, USDA, Washington, D.C.

Wong, T.T.Y., M.M. Ramadan, J.C. Herr and D.O. McInnis. 1992. Suppression of a Mediterranean fruit fly (Diptera: Tephritidae) population with concurrent parasitoid and sterile fly releases in Kula, Maui, Hawaii. J. Econ. Entomol. 85(5): 1671-1681.

# Pink Bollworm Integrated Management Using Sterile Insects Under Field Trial Conditions, Imperial Valley, California

*M.L. Walters[1], R.T. Staten[1] and R.C. Roberson[2]*

[1] USDA, APHIS, PPQ, Phoenix Plant Protection Center, Phoenix, AZ, USA
[2] California Department of Food and Agriculture, Sacramento, CA, USA

## INTRODUCTION

The pink bollworm moth (*Pectinophora gossypiella* Saunders) feeds almost exclusively on cotton (*Gossypium* spp.) and causes economic loss (Pfadt 1978). The pink bollworm (PBW) is often the key pest of cotton in Arizona, southern California, and northwestern Mexico. The larvae (immature stages) bore into the developing cotton fruit, where they feed on the cotton lint and seeds, causing significant damage and dramatically reducing the yield of cotton lint (Pfadt 1978). The PBW is difficult to control with conventional means (insecticides) because it spends the destructive larval phase inside the cotton boll where it is well protected from control measures. Cultural controls, such as a short growing season, have successfully decreased the population in the Imperial Valley (Chu et al. 1992) to the point where eradication may be possible using sterile insects and genetically engineered cotton. Because the PBW is an introduced insect, with few plant hosts other than cultivated cotton, its eradication from continental USA is a desirable and economically attractive alternative to the continued use of pesticides and/or further loss to the pest.

Mass releases of sterile insects began in earnest in 1970 in the San Joaquin Valley, California, in order to inhibit normal reproduction and to eradicate the pest in an environmentally responsible manner. Sterile release involves mass production and sexual sterilisation using irradiation (20 krad for PBW adults). This was accomplished by building a rearing facility in Phoenix, AZ. The facility has 6,410 square metres of permanent laboratories, rearing and irradiation chambers and insect packing rooms. The facility operates the year round but with a variable production rate, that is, maximal during the cotton growing season (May through September).

Sterile insect technology is based on the monitoring of the native and sterile populations in the field and the subsequent release of appropriate numbers of sterile insects in order to inundate the native population and drastically reduce native to native mating. Sterile release has protected the San Joaquin Valley from developing a reproducing PBW population but cannot prevent wind-borne immigration from other infested areas.

## THE IMPERIAL VALLEY STERILE RELEASE PROJECT

The overall objective of this programme was to test the effectiveness of sterile insects, used in combination with pheromones, in a generally infested, optimum season, area. Insect populations were monitored and population data were collected using ca. 200 pheromone baited Delta traps (Foster et al. 1977) at no less than two traps per cotton field. Four traps were used in fields of more than 28 hectares. Sterile and native male moths were attracted by the lure and became stuck in the trap. Twice per week, field technicians checked the traps, counted native and sterile moths captured, and cleaned or replaced the traps. They recorded the trap catches on a preprinted, automated data entry sheet. Project management used the field data to regulate the rates of the release of sterile PBW moths and application of pheromones, as described below.

In the years 1994-1996, moths were aerially released six days a week (Table 1), from the time before cotton fruit was available until after fruit was available or until we could no longer hold a beneficial sterile to native ratio. Specifically, sterile release began in all fields on or before four true leaves were found on plants, and at the same rate in all fields. There were four release rates available. The rate was determined by the risk, calculated from native and sterile moth trap data, and limited by moth availability. When plants reached the 6th leaf stage, if a 60:1 ratio was not obtained, the Mitsubishi® PBW rope pheromone formulation was applied at 400/acre (Staten et al. 1987). During mid-season, if a good ratio of native:sterile moths could not be obtained at the highest release rate (4x), a sprayable pheromone formulation was used with three applications (Ecogen MEC® at 5 grams a.i./acre). "Pheromone" refers to an artificial sex pheromone that mimics the chemical mixture of substances secreted by female PBW moths. The female uses the natural pheromone to signal to and attract a male for mating purposes. The artificial pheromone has a similar effect on the male but since it is distributed all over the cotton field, it confuses or incapacitates the male so that he cannot find a female and thus reproduction is suppressed. Pheromone applications integrate well with sterile releases and pheromones are also used to bait the sticky traps used for population monitoring.

A Cessna 206 aircraft was used as the delivery aircraft. The seat rails were used to fasten the insect delivery machine and refrigeration equipment to the aircraft body. A drop tube installed through the aircraft skin allowed the insects to be metered out into the slipstream. Modifications to the air conditioning improved efficiency and, subsequently, increased insect quality. A new auger system, in conjunction with a stepping motor drive, improved calibration and allowed rate changes during the flight. Video card camera technology provided direct visual monitoring of insect discharge. The camera assured the pilot that the system was operating and that all insects had been discharged before the termination of the flight.

The geographic coordinates associated with a sample, such as trap captures, were input into a geographic information system (GIS) and correlated with other data layers. The total GIS has multiple functions: it serves as the basis for data analysis and output of recommendations.

Table 1 describes the control measures used from 1994-1996. In 1995, we experimented with a low-level sterile release in an effort to find a very affordable, yet effective level of release. We found this level insufficient to control the PBW population and so returned to a higher level in 1996, with good results.

In 1997, due to the widespread use of genetically engineered cotton, moths were released three days a week from mid-May through July at a rate of 40 insects/hectare/ day. Note in Figure 1 that although local reproduction was heavily suppressed by genetically engineered cotton, sterile insects and pheromones, late season moth numbers were similar to those of previous years. These catches seem to be correlated more to migration from cotton acreage to the south than to actions taken in the Imperial Valley. Migration is indicated by increasing numbers of PBW intercepted in traps located between the two growing areas. This evidence supports the need for expansion of the PBW management area.

Table 1. SIT and Pheromone Parameters for 1994-1997.

| | 1994 | 1995 | 1996 | 1997 |
|---|---|---|---|---|
| Sterile moths/hectare | 100 | 70 | 140 | 40 |
| Pheromones available | Rope & MEC | Rope & Fibre | Rope and Fibre | Rope |

Imperial Valley, 1997

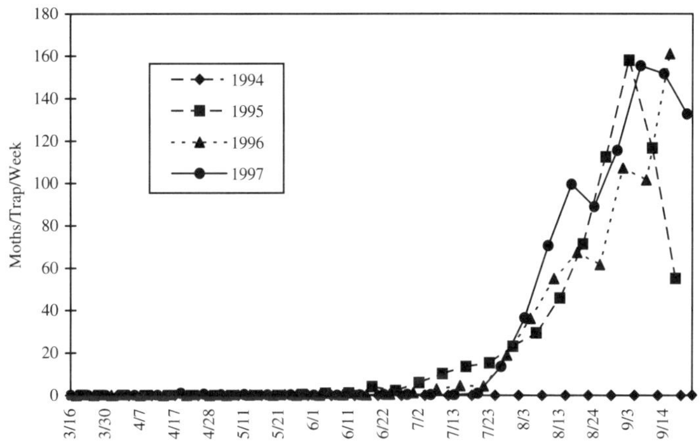


Figure 1. Wild PBW males trapped in the Imperial Valley, California, 1994-1997.

The cotton planted in the Imperial Valley in 1997 consisted of 81% genetically engineered cotton, 17.5% conventional (non-genetically engineered, susceptible) treated with pheromone rope and 1.5% conventional untreated refuge. Defoliation began in mid-August. Bolls sampled after that time were taken from a diminished boll population and were under heavy pressure from increasing PBW activity. Late season samples were heavily biased toward high PBW levels, as reflected in the boll data (Table 2).

## 1998 AND BEYOND

The objective of eradication is being pursued as an expanding, multi-season project. In 1998, sterile insect coverage was extended to the California Blythe/Palo Verde Valleys. Widespread use of genetically engineered cotton in these two growing areas began in 1997.

Genetically engineered cotton plants produce the *Bacillus thuringiensis* (Bt) toxin. Most Lepidoptera cannot survive ingestion of the toxin and thus cannot reproduce on these cottons. With genetically engineered cotton, we expect high larval mortality in early instar, a few surviving larvae found in the 1-3% non-Bt producing plants found within the cultivar, and a drastic reduction of PBW biotic potential. However, the use of genetically engineered cotton may be short-lived due to resistance. By late season, 95 %

of remaining bolls will have eggs laid on them and have 1st instar penetration. This increases resistance pressure. Genetically engineered cotton alone will not eradicate the pest since the producing company requires a certain percentage of cotton to be planted in conventional, susceptible cotton varieties. This is to protect against the rapid development of resistance to the toxin.

Table 2. Boll data for 1997, Imperial Valley, California.

| 100 % Genetically Engineered cotton | | | Conventional Cotton | | | | Weighted Average | |
|---|---|---|---|---|---|---|---|---|
| | | | Rope | | Refuge | | | |
| Week of | Bolls Sampled | Larvae/ 100 Bolls | Bolls Sampled | Larvae/ 100 Bolls | Bolls Sampled | Larvae/ 100 Bolls | Bolls Sampled | Larvae/ 100 Bolls |
| 6/22/97 | 2000 | 0.00 | 1000 | 0.00 | 900 | 0.00 | 1808.50 | 0.00 |
| 7/6/97 | 2000 | 0.00 | 1000 | 0.10 | 1000 | 0.00 | 1810.00 | 0.02 |
| 7/20/97 | 1990 | 0.00 | 975 | 0.00 | 800 | 0.00 | 1794.53 | 0.00 |
| 8/3/97 | 1998 | 0.00 | 995 | 0.10 | 1000 | 0.00 | 1807.51 | 0.02 |
| 8/17/97 | 1383 | 0.00 | 280 | 19.29 | 900 | 0.00 | 1182.73 | 3.38 |
| 8/31/97 | 1148 | 0.00 | 606 | 38.28 | 400 | 1.75 | 1041.93 | 6.73 |
| 9/14/97 | 322 | 0.31 | 519 | 102.12 | 100 | 3.00 | 353.15 | 18.17 |
| 9/28/97 | 200 | 0.00 | 203 | 44.83 | 100 | 1.00 | 199.03 | 7.86 |

The combined effect of sterile insects and genetically engineered cotton may make eradication possible in a relatively short period and the potential for resistance forces a short time frame. We use the term "eradication" to indicate zero detection with an extensive survey for three years following completion of programme. The first two years of the eradication effort entail the highest costs (Table 3). Since the PBW is a highly mobile insect, the programme must encompass large contiguous areas. Suppression (control) must be intensive for at least two years and must start where the pest has maximum biological potential. The programme must include a large area such as southern California, Baja Norte, northern Sonora and western Arizona. Cultural controls must be used including early plough down, a host free period and clean seeds. The monitoring of native and wild populations and good information management are critical to making optimal resource allocations. Mating disruption with several forms of synthetic pheromones greatly enhance the effectiveness of the programme, especially when wild moth population pressure increases. Pheromone rope formulations must be used at adequate levels and with proper timing (6 true-leaf stage). Pheromone fibre may be applied later in the season to thwart immigrating populations. The release of sterile PBW is a key component to the overall success of the programme and is recommended for active eradication/control areas. The final component is the partnering of cotton growers with state and federal agencies.

Table 3. Estimated costs for a sample eradication scenario, first 2 years.

| Treatment | Hectare | Seasonal Cost |
|---|---|---|
| 125 sterile insects/day * 60 days = 7,500 insects/hectare @ US$30/ hectare | 80,000 | US$ 2,400,000 |
| Genetically engineered cotton @ US$80/hectare | 72,000 | US$ 5,760,000 |
| High rate pheromone @ US$70/hectare | 8,000 | US$ 560,000 |
| Mid-season pheromone application @ US$25/hectare | 8,000 | US$ 200,000 |
| Monitoring costs @ US$12.50/hectare | 80,000 | US$ 1,000,000 |
| Total costs per growing season @ US$124/hectare | 80,000 | US$ 9,920,000 |

## PROGRAMME RESULTS AND CONCLUSIONS

Sterile pink bollworm moths have been released in the San Joaquin Valley, California, USA, since 1970. Population suppression has been achieved and cotton protected by an integrated programme. In 1994, the programme was extended to the generally infested Imperial Valley (1994-1996 was in the pre-genetically engineered cotton era). In 1998, we extended the programme to the Blythe/Palo Verde Valley, California, USA. Plans include the addition of all southwestern cotton states, Arizona and New Mexico (USA) and Baja Norte and Sonora (Mexico) to the programme during the next five years.

Geographic display and analysis have been, and will continue to be, very important for data analysis and as decision implementation tools. The development of this entire system has been evolutionary. The programme has been well received by cotton producers in northern Mexico and southern California.

## ACKNOWLEDGEMENTS

Our thanks to Amit Patel for computer engineering, GIS expertise, expertise on insect release operations, and forms development. Our deepest gratitude goes to David Pierce for ongoing support and vast knowledge in the areas of GIS, GPS, sterile insects, general equipment use and maintenance.

## REFERENCES

Chu, C.C., R.C. Weddle, R.T. Staten, T.J. Henneberry, S.L. Birdsall and J.R. Carson. 1992. Pink bollworm: populations two years following initiation of a short-season cotton system in the Imperial Valley, CA, pp. 804-806. *In* Proceedings, Beltwide Cotton Conferences, Nashville TN, January 5-9.

Foster, R.N., R.T. Staten and E. Miller. 1977. Evaluation of traps for pink bollworm. J. Econ. Entomol. 7: 289-291.

Pfadt, R.E. 1978. Insect pests of cotton, pp. 369-403. *In* Fundamentals of Applied Entomology. R. E. Pfadt (ed.), Third Edition. MacMillan Publishing Company, Inc., New York, New York.

Staten, R.T., H.M. Flint, R.C. Weddle, E. Quintero, R.E. Zarate, C.M. Finnell, M. Hernandes and A. Yamamoto. 1987. Pink Bollworm (Lepidoptera: Gelechiidae): Large-scale field trials with a high-rate gossyplure formulation. J. Econ. Entomol. 80: 1267-1271.

# SIT for Codling Moth Eradication in British Columbia, Canada

*Kenneth A. Bloem*[1] *and Stephanie Bloem*[2]

[1] OK-CM SterileInsect Release Program, Box 1080, Osoyoos, BC, V0H 1V0, Canada
[2] USDA-ARS YARL, 5230 Konnowac Pass Rd., Wapato, WA 98951, USA

## INTRODUCTION

The codling moth (CM), *Cydia pomonella* L. (Lepidoptera: Tortricidae), is considered the key pest of apples and pears in the fruit growing regions of south central British Columbia. This region includes about 18,000 acres of commercial production, as well as several urban centres with abundant backyard fruit trees and ornamental crab apples. Now, after 30 years of research and planning, an eradication programme using the sterile insect technique (SIT) has been implemented against CM. This article reviews the progress that the programme has made and how well reality has met expectations in key areas. Proverbs (1982) and Proverbs et al. (1982) reviewed the techniques for mass rearing, sterilising and releasing CM, DeBiasio (1988) developed the initial implementation plan and Dyck et al. (1993) reviewed the history and development of the programme up to 1992 when it became operational.

## GEOGRAPHY AND TOPOGRAPHY

The fruit growing regions of southern British Columbia occur along isolated arid valleys that are surrounded by high mountains. The areas generally receive less than 300 mm of precipitation per year. Fruit is grown under irrigation and closely follows the lakes and rivers in the valley bottoms. The eradication programme encompasses all commercial apple and pear production in four such valleys, the Okanagan, Similkameen, Creston and Shuswap Valleys. The Okanagan Valley lies about 500 km east of Vancouver and the Pacific Coast. It runs northsouth and extends into the state of Washington in the US, where it ties into the Columbia River system. The smaller adjacent Similkameen Valley extends west from the south end of the Okanagan Valley, and the Creston Valley is located about 200 km east of the Okanagan on the US/Canada border above the state of Idaho. The Shuswap Valley joins and extends the Okanagan Valley to the north. The topographical features of these valleys (steep mountain sides, arid terrain) and the fact that CM fly relatively short distances, particularly females, all help to isolate the area and reduce the risk of re-infestation once eradication is achieved.

The programme is currently operating in Zone 1, which includes the south Okanagan Valley from the US border near Osoyoos to Summerland, as well as the Similkameen and Creston Valleys. Zone 2 covers the area from Summerland north to Winfield and includes the largest urban centre, Kelowna. Zone 3 extends from Winfield north and east to Vernon and Salmon Arm.

## CONSTRUCTION AND OPERATION OF THE REARING FACILITY

Capital funds in the amount of Cdn$7.7 million were approved by the federal and provincial governments for construction of the CM mass rearing facility and the purchase of operating equipment. Ground breaking began in January of 1992 in Osoyoos, BC, and construction was completed in March 1993. The project finished Cdn$300,000 under budget.

Although the reproductive colony was established and a small number of release trials were conducted in 1993, budget constraints did not permit full-scale testing of either rearing or release equipment until the first operating field season in 1994. Unfortunately, equipment failures soon became a chronic problem and threatened programme delivery. Some of the problems included undersized gear boxes for the diet pumps, micro-switches on diet lines that were not water or dust proof, insufficient cooling capacity to maintain proper rearing temperatures, humidity control problems, and unreliable electrical systems on the moth dispensing units.

Start-up problems are not uncommon for a facility of this type. Given that operational and equipment problems should be expected, it is critical to allow time for sufficient testing of all systems under actual operating conditions (while under warranty) and before demands for rearing and releasing become critical. This is particularly true when new and/or "improved" technologies are being introduced.

Despite these initial problems, the mass rearing facility has proved to be very well suited to rearing CM. From an initial production goal of 5.3 million CM per week (Dyck et al. 1993), actual production has increased from 8.7 million per week in 1994 to 14.2 million CM per week in 1997. This increase in production has largely been the result of improvements in process control, such as better humidity and temperature control, more uniform diet dispensing, and a more consistent and optimal number of eggs being placed on each tray of diet.

## RELEASES

CM releases begin around 1 May and continue for 20 weeks until mid-September. Small 4-wheeled all-terrain vehicles (ATVs) outfitted with coolers on the back and moth dispensing units on the front, similar to those described in McMechan and Proverbs (1972), are used to release CM. Adult moths are packaged by volume into plastic petri dishes (700-900 per dish), sterilised with 25 krad of gamma radiation, loaded into coolers on the ATVs and transported to the field. Ice packs keep the CM chilled and inactive. When a driver is ready to release them, a petri dish is emptied into the hopper on the dispensing unit and chilled CM are gently blown under the trees as the ATV is driven up and down the orchard rows. Three or four passes are made through each acre of orchard. The programme releases sterile CM twice per week and the release rate varies with production levels. In 1994, the rate was ca. 450 moths per acre or 900 CM per acre per week. In 1997, increased production allowed for a release rate of ca. 750 moths per acre or 1,500 CM per acre per week.

The programme monitors the wild population, overflooding ratios and uniformity of releases on a weekly basis with wing-type pheromone traps baited with red rubber septa containing 1 mg codlemone, placed at a density of 1 per 2.5 acres during clean-up and 1 trap per 12.5 acres during active releases. A random examination of fruit is made yearly in approximately 600 or 1/3 of the orchards treated just prior to harvest to determine percent fruit damage by CM.

## WILD POPULATION AND OVERFLOODING RATIOS

Overflooding ratios for a given species are calculated based on field experience and computer models that predict population rates of increase and/or decrease. Influencing factors determining overflooding ratios include the size of the wild population, the reproductive potential of the female, the potential impact of weather conditions, the quality and competitiveness of the sterile insects, cost constraints, and the time in which eradication is to be achieved. For CM, it was estimated that the release of enough sterile moths to provide an initial overflooding ratio of 40 sterile:1 wild would bring about eradication in 2-3 years.

Because most of these factors are difficult to measure with accuracy, calculations of production needs should err on the high side and calculations of production capabilities should be conservative. Miscalculations resulting in over-production may increase costs but should bring about eradication more quickly, whereas miscalculations on the low side may jeopardise a programme's success. Once a facility has been built and production capabilities established, the only recourse for dealing with low overflooding ratios is to decrease the size of the treatment area or combine releases with other control tactics to lower the wild population to a manageable level.

In 1992, the programme assumed that it would be possible to reduce the CM wild population to one that resulted in an average trap catch of less than two moths per trap in any given week. Taking into account trap density and efficiency, this number of CM was then multiplied by 40 in order to arrive at the production requirements for the facility. Unfortunately, when releases began in 1994, trap captures of wild CM reached 8 moths per trap per week during first generation. Although production estimates were conservative and output has steadily increased, the programme had to initially struggle with poor overflooding ratios, less than adequate control, and cost and timeline overruns.

## INSECT QUALITY

Producing good quality insects is one of the most important components of a SIT programme (Huettel 1976). Rearing managers are continually pressured to reduce costs, improve efficiency and increase production numbers. Sterile to wild ratios, after all, are what drive SIT programmes. However, the requirement for success is not only to produce large numbers of sterile insects, but large numbers of insects that are competitive. Unfortunately, many steps in the rearing and release processes can have a negative impact on field performance and mating competitiveness (quality) of mass reared insects.

Insect quality issues should be addressed whenever possible during facility design. It is critical, therefore, that the engineering firm in charge of construction be flexible and willing to work alongside scientists so that new technologies are introduced and adapted to provide a better rearing environment, not rearing procedures adapted to fit an engineer's idea of "improvements" in the new facility. For example, the CM facility has a unique adult collection system where adults emerge directly into rearing rooms and are attracted by UV-lights attached to vacuum hoods mounted in the ceiling. This system only collects CM capable of flying to the UV/hoods and thus insures that good flight ability is selected for, and maintained in, the reproductive colony.

Two quality issues of concern to the CM programme have been the adverse effects of handling the moths prior to release and poor CM activity (relative to that of wild CM) in cool weather. With respect to handling, we discovered that it is important to minimise the time that the CM are kept in cold storage. In 1994, turn-around time between adult collection and field release was 36-48 hours. In 1995, a third worker shift was implemented to irradiate CM immediately after collection. As a consequence, sterile CM are now no more than 12 hours old when released and field quality has improved significantly. It was also determined that bouncing of the petri dishes in the coolers on the ATVs had a negative impact on CM quality. A recommendation was made instructing release drivers to carry no more than half of their day's supply of CM at any one time. With respect to poor competitiveness when CM are released in cool weather, we found that fluctuating the temperature during larval rearing only marginally increases adult flight activity in the field. Unpublished data by Judd at the AAFC-PARC, Summerland, BC, suggest that the problem may be more related to the way the CM are released (cool moths onto cool ground) than to the quality of the moths. We did find that mass reared CM that are induced into diapause (Bloem et al. 1997) and later released, perform significantly better in cool weather than CM reared under constant temperatures (Bloem et al. 1998). Other unpublished research also demonstrates that CM treated with lower doses of gamma radiation are significantly more competitive.

## BYLAW ENFORCEMENT AND URBAN AREAS

Possibly, the greatest barrier to geographic eradication of CM in southern BC is the intimate interface between viable commercial orchards, abandoned orchards, hobby farms, and urban subdivisions. Unfortunately, all of the research for CM-SIT was conducted in, and developed for, areas of commercial production. However, urban areas, with relatively few host trees that are widely spaced and may harbour CM infestations of 50-100%, present a much different scenario. Under such conditions, it is difficult to imagine that the uniform distribution of sterile CM at standard orchard release rates would result in sufficiently high numbers of CM arriving at infested trees and bringing about any meaningful level of control. Releases in urban centres such as Penticton, BC, during 1994 bore this out.

In 1995, the programme adopted a policy of "zero tolerance" for CM infested fruit in all urban and non-commercial orchard properties. The preferred method for achieving "zero tolerance" is for property owners to remove all fruit from their trees before the end of the first or spring generation (during eradication years) or to remove the trees entirely. Incentive programmes were developed where those who strip their trees receive a discount on replacement fruit at packing houses and those who remove their trees receive a discount on replacement non-host trees at garden centres. Other methods of CM control such as removal of only infested fruit and the use of pesticide sprays are acceptable, but only if the same "zero tolerance" level is maintained throughout the season.

The programme employs 15 urban monitors to make inspections and enforce "zero tolerance" on roughly 5,000 properties in Zone 1, and 35 monitors to make inspections on over 12,000 properties in Zone 2. When a property is found to have any level of CM damage, it is issued with a control order for complete fruit removal. The failure to identify the requirements for urban clean-up in the implementation plan has been a major factor in the higher than anticipated costs for the programme. Since 1995, the programme has seen trap captures in the urban centres of Zone 1 drop from an average of 18 moths per trap per week to two moths per trap per week in 1997. However, strict enforcement of the programme's zero tolerance policy remains a critical and a controversial issue, and many doubt that the effort is aggressive enough to achieve eradication.

## PUBLICITY

Eradication programmes cannot be conducted in an information vacuum. They require the active support and participation of various levels of government, the grower community and the general public. To effectively solicit and maintain this support requires an aggressive, ongoing communications campaign.

Probably the biggest failing of the programme in 1994 was insufficient communication with orchardists and homeowners. The insufficient pre-release publicity that was done left people with high expectations and little understanding of how SIT works. As a result, many orchardists stopped all control measures for CM, did not realise they had wild infestation levels that were much too high for SIT to control, and sustained damage well above the accepted economic injury level (0.5% at harvest).

In order to improve public awareness and perception of the programme, a communications firm was hired in 1995 and the budget for communications increased from Cdn$5,000 to $100,000 per year. The strategy was to inform the public of the importance of CM eradication and appeal to their sense of responsibility, not unlike campaigns for community recycling programmes. Commercial growers now receive direct mail four times a year updating them on the progress of the programme. In addition, they are reminded of compliance and spray requirements through regular news releases, radio announcements, the Ministry of Agriculture's information network, and weekly programme updates that are available through packing houses and local fieldmen. Urban and non-commercial property owners also receive direct mail stressing the "zero tolerance" policy and compliance dates. Other methods used to inform the public include extensive radio and newspaper advertising, presentations about CM-SIT to school and community groups, and information booths at shopping malls, garden centres and agricultural fairs.

## FIELD PROGRESS IN ZONE 1

Releases of sterile CM began in 1994. Unfortunately, the wild population was still unacceptably high (trap captures of wild CM reached an average of eight moths per trap per week during peak spring emergence) and expectations of success were equally unrealistic. As the year developed, start-up problems, poor communication with growers, high temperatures, heavy crop losses due to hail damage, and low fruit and juice prices combined to produce high CM damage and a large overwintering population at the end of the season.

Things began to change in 1995 when growers were convinced to combine an aggressive spray programme with the release of sterile CM. Since then, the resulting decline in the wild CM population has been dramatic (Figure 1). The average trap catch has been reduced from 13 CM per trap per week during peak first generation and 2.5 moths during peak second generation to an average of 1.2 and <0.5 moths per trap per week for first and second brood, respectively, in 1997. The desired 40 sterile:1 wild overflooding ratio was achieved in all but two weeks in the spring and it was greater than 500:1 throughout the last half of the season in 1997. The majority of growers applied only one cover spray for CM in 1997, essentially no damage was reported, and it is likely that many growers will not be required to spray for CM in the south Okanagan in future years.

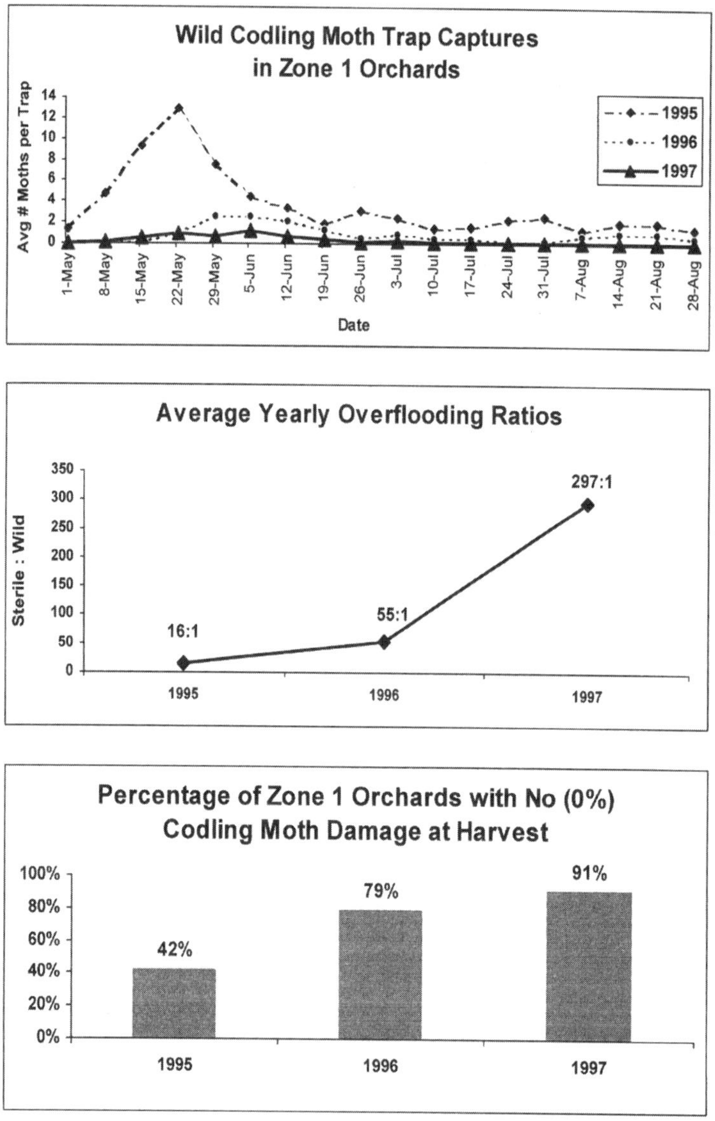


Figure 1. Sterile Inesct Release Program field results for 1995–1997, including average trap captures, overflooding ratios and apple damage at harvest.

## CURRENT TIMELINE

The Sterile Insect Release (SIR) Mandate Area has now been divided into three (rather than two) treatment zones, the south (Zone 1), central (Zone 2) and north (Zone 3) Okanagan, with approximately 8,500, 6,250 and 3,250 acres of commercial apple and pear production, respectively (SIR Programme Strategic Plan 1996). Each zone will undergo three years of pre-release sanitation using conventional methods to lower the wild population and eliminate wild trees and derelict orchards. This will be followed by three years of CM releases to achieve eradication. Eradication in Zone 1 is anticipated by the end of 1999. The clean-up phase for Zone 2 began in 1997, with releases scheduled to begin in 2000 and eradication anticipated by the end of 2002. Eradication in Zone 3 is anticipated by 2005.

## BUDGET

Federal and provincial governments paid the initial capital costs (Cdn$7.4 million), while the five municipal governments agreed to levy a CM-SIT parcel tax on commercial apple and pear growers and a mil rate tax on property owners to fund the yearly operating costs of the programme. According to the implementation plan, operating costs were projected to be Cdn$2.0-2.5 million per year for eight years (1992-1999). It was proposed that growers would pay Cdn$40 per acre of apples and pears (roughly the yearly cost of CM control for an average grower based on 2.5 applications of azinphosmethyl at 1.4 kg/2.5 acres). The mil rate tax on property owners would vary between Cdn$0.13 – Cdn$0.26 per Cdn$1,000 of property value or about Cdn$6.00 per household.

It soon became obvious, however, that the programme had significantly underestimated the effort and costs involved, particularly to deliver the programme to urban areas and to remove wild trees and derelict orchards (SIR Programme Strategic Plan 1996). As a result, payment for growers in Zone 1 increased to Cdn$70 per acre in 1994, Cdn$72.10 in 1995, Cdn$74.26 in 1996 and Cdn$80.00 in 1997. Although property owners were only being charged Cdn$0.15-0.20 per Cdn$1,000 of assessed land value, municipal government representatives argued that because property values had increased they were actually paying more than double (Cdn$12-15 per household) what they had originally agreed to. After an intense lobby effort by the programme's governing board in 1997, the federal and provincial governments agreed to each contribute an additional Cdn$2 million to stabilise the annual cost of the programme to growers at Cdn$80 per acre of apples and pears and to homeowners at Cdn$0.195 per Cdn$1,000 of assessed land value. In 1997, the total operating budget was Cdn$3.2 million. The operating budget for 1998 is Cdn$3.7 million.

## CONCLUSION

The SIR programme has faced many challenges since its inception and it continues to struggle with the concept of eradication (vs. area-wide control) and the associated long-term financial, political and operational demands. Despite this, the SIR programme has been making steady progress for the past three years and each year, more of the end users are satisfied with the results. This is evidenced by: the additional funds contributed by the federal and provincial governments in 1997, the willingness of growers to pay higher than expected yearly programme costs, and the cooperation by homeowners who not only remove the fruit or the host trees from their properties but also help the programme to find CM infestations in the surrounding neighbourhoods. As conventional pest control methods become more expensive, less effective and less tolerated, it is hoped that export of the technology for CM-SIT will be possible to other apple growing areas around the world.

## REFERENCES

Bloem, S., K.A. Bloem and L.S. Fielding. 1997. Mass rearing and storing codling moth larvae in diapause: a novel approach to increase production for sterile insect release. J. Entomol. Soc. B. C. 94: 75-81.

Bloem, S., K.A. Bloem and A.L. Knight. 1998. Assessing quality of mass reared codling moths (Lepidoptera: Tortricidae) using field release-recapture tests. J. Econ. Entomol. 91: 1122-1130.

DeBiasio, D. 1988. Codling moth sterile insect release study. Project no. 15001, Crane Management Consultants Ltd., prepared for the British Columbia Fruit Growers Association.

Dyck, V.A., S.H. Graham and K.A. Bloem. 1993. Implementation of the sterile insect release programme to eradicate the codling moth *Cydia pomonella* (L.) (Lepidoptera: Olethreutidae) in British Columbia, Canada, pp. 285-297. *In* Management of Insect Pests: Nuclear and Related Molecular and Genetic Techniques, Proc. Intl. Symp., Vienna (1992), IAEA-SM-327/29.

Huettel, M.D. 1976. Monitoring the quality of laboratory-reared insects: a biological and behavioral perspective. Environ. Entomol. 5: 807-814.

McMechan, A.D. and M.D. Proverbs. 1972. Equipment and procedures for release of sterile codling moths. Can. Agric. Engineering. 14: 42-45.

Proverbs, M.D. 1982. Sterile insect technique in codling moth control, pp. 85-99. *In* Sterile Insect Technique and Radiation in Insect Control, Proc. Intl. Symp. Neuherberg (1981), IAEA, Vienna.

Proverbs, M.D., J.R. Newton and C.J. Campbell. 1982. Codling moth: a pilot programme of control by sterile insect release in British Columbia. Can. Entomol. 114: 363-376.

Sterile Insect Release Programme Strategic Plan. 1996. Wayside Press, SIR Board, 9848 Aberdeen Rd., Vernon, BC, Canada, V1B 2K9. 41 pp.

# Area-Wide Population Suppression of Codling Moth

*C.O. Calkins[1], A.L. Knight[1], G. Richardson[2] and K.A. Bloem[3]*

[1] USDA, Agricultural Research Service, Yakima Agricultural Research Laboratory, Wapato, Washington, WA 98951 USA
[2] USDA, ARS, Yakima Agricultural Research Laboratory, Oroville, Washington, USA
[3] USDA, APHIS, PPQ, Central Biological Control at Florida, A&M University, Tallahassee, Florida, USA

## INTRODUCTION

The area-wide pest population control concept began with E.F. Knipling (1979) in the 1970s. Control of a pest population on individual fields does little to control the overall pest population because only a portion of the population is being affected. Expanding control tactics beyond individual farms tends to suppress the population on a wider scale and frequently results in suppression of the population for more than one year. The Agriculture Research Service (ARS) believes that this concept has not been addressed with the focus and support that it deserves. The ARS Administration made a conscious decision in 1994 to create a series of area-wide programmes funded out of ARS-based funds that had previously been used for pilot tests. These programmes involve a coordinated effort among ARS and university scientists, growers, and fieldmen for agriculture supply centres and fruit packing houses.

The first area-wide programme supported by ARS was the codling moth (CM), *Cydia pomonella* L. (Lepidoptera: Tortricidae) suppression programme. The codling moth is the key pest of pome fruit throughout the western United States (Beers et al. 1993). About half of the insecticides applied on these crops are directed toward this pest. A non-insecticidal control technique, mating disruption (MD), is available to replace the organophosphates. Removal of the hard pesticides directed against this pest would do the most to allow natural enemies to survive and reproduce in the orchards, which in turn would have the effect of reducing secondary pests. Elimination of the pesticides would also remove much of the health risks to workers and would minimise buildup of pesticide resistance.

The objectives of the Codling Moth Area-wide Program are to enhance the efficacy of the non-pesticide approach, to demonstrate that mating disruption will work if conducted properly, to develop biological technology to lower costs of control that complement mating disruption, to implement effective biological control systems, to improve monitoring systems for all pests and their principal natural enemies, to improve worker safety and to improve the perception that fruit production is safe for consumers, especially for children and infants.

## INITIATION OF THE AREA-WIDE PROGRAMME

Prior to the inception of this programme, several large growers in Washington were using mating disruption for the codling moth routinely and getting satisfactory control. However, small growers seemed to have frequent failures. They and others concluded that the main problem was from neighbouring orchards that, by not using the technique, were a source of mated females that were flying into their MD orchards. Other complaints were that control could not be maintained on hilly topography, on orchard perimeters, and near bin, prop and brush piles. The complaints from growers, who had tried the technique and failed, discouraged others from using it. The problem at the outset was to convince growers to try the idea of creating large contiguous blocks under the same management strategy by joining together and coordinating the timing of applying mating disruption dispensers and cover sprays.

To participate in the area-wide programme, certain criteria were required. These were: 1) the production area and cultural practices should be typical of good fruit growing areas, 2) populations of key pests should occur consistently so that meaningful data could be collected to measure the effects of reduced pesticide use, 3) producers within the test areas should be willing to cooperate and share costs, and 4) the test area should have the organisational structure to support and establish the enhanced integrated pest management (IPM) systems in the local community.

The initial task was to set up the test areas for the pilot projects. The sites selected were: one in California, one in Oregon and three in Washington. These sites, referred to as the Codling Moth Area-wide Management Project (CAMP), were composed of apple and pear orchards owned by interested growers whose properties were fairly contiguous. To ensure that participating growers would remain in the programme, a partial subsidy of US$125 per hectare was granted to each participant for three years. There was no subsidy during the last two years of the programme. The initial effort was for the growers to install 1,000 pheromone dispensers per hectare in the upper one-third of the tree canopy. The typical orchard in the Pacific Northwest has about 500 trees per hectare, thus two dispensers per tree were most commonly recommended. Because mating disruption is most efficient at low moth populations (Beers et al. 1993, Vickers and Rothschild 1991), the growers were also encouraged to apply at least one cover spray of azinphosmethyl to lower the initial codling moth population. Cover sprays of azinphosmethyl are recommended at 250 degree days after biofix, about 15 days before peak codling moth egg hatch occurs. Biofix is the date when the first consistent emergence of codling moths are detected in the spring. This usually happens just before "Red Delicious" trees bloom. Traps baited with 10 mg codlemone lures were placed throughout the blocks at the rate of one trap per hectare shortly after biofix to monitor the population. A threshold of five moths per trap per generation was established for deciding whether to apply sprays for the second generation.

The amount of fruit damaged by the codling moth was assessed through fruit examinations during the larval phase of the first generation and at harvest. Traps were also maintained for leafrollers and constant monitoring for other secondary pests was conducted. Other types of monitoring consisted of determining percentages of parasitism of leafhoppers, leafminers, and aphids in the individual orchards.

## MATING DISRUPTION AND THE STERILE INSECT TECHNIQUE

One of the CAMP sites, Lake Osoyoos, consisted of about 180 hectares on the east and west sides of Lake Osoyoos near the town of Oroville and abutting the Canadian border to the north. A codling moth eradication programme using the sterile insect technique (SIT) is being conducted immediately north of the border in Canada. The codling moth sterile insect release programme (SIR) was started in 1992 to eradicate this pest from 7,500 ha of orchards in south central British Columbia (Dyck et al. 1993). A modern rearing facility costing US$5 million was constructed just outside the town of Osoyoos, British Columbia. In 1997, it produced an average of 14 million sterile codling moths per week. The first sterile moths were released in 1994, and the programme initially encountered major problems due to the unexpectedly high levels of wild codling moths present in many orchards. In 1995, growers were required to supplement the SIR programme with multiple applications of organophosphate insecticides (OP). However since 1995, the number of sprays has been steadily decreasing and control has been improving. The SIR programme was deemed successful in 1997. As most growers applied less than one OP spray per season, the catch of wild moths averaged < 2 per trap per season, the sterile:fertile moth ratio in traps soared to 297:1, and 91% of the orchards had no discernible codling moth fruit injury at harvest (Bloem and Bloem 2000).

Adjacent pome fruit orchards are present south of the Canadian -US border along both sides of Lake Osoyoos. The Canadian SIR programme envisioned creating a moth-free buffer within the US to minimise continual reintroductions into Canada. Through close cooperation with the SIR Board, ARS organised a similar SIR programme in 1995 for 160 ha of orchards extending 1 km into the US. The initial Lake Osoyoos CAMP site included 14 growers farming 64 orchards (91% apples, 5% pears, and 4% cherries). One grower did not participate during the first two years of the study and his orchards remained a major source of wild moths during those years. All other apple and pear orchards were treated with 1,000-ISOMATE-C+ dispensers per hectare for mating disruption of codling moths. In addition, orchards were treated with cover sprays of insecticides as needed for the codling moths and leafrollers.

Sterile codling moths were provided by the Canadian SIR programme. During the first two years of the project, moths were released only during the second generation flight period from 1 July to 15 September. During 1997, chilled moths were released twice weekly at the rate of 1,120 moths per hectare during both flight periods from 1 May to 15 September. Releases were made from all-terrain vehicles using a special moth release blower device mounted on the front of the vehicles (McMeacha and Proverbs 1972). Moths were released along prescribed transects spaced 30 m apart through all of the MD orchards. The cost of the programme, including labour and equipment, has been ca. Cdn$38 per hectare but does not include the cost of moths or trap monitoring. Canadian growers are charged Cdn$150 per hectare but the programme is also supported by taxes on homeowners. The full cost of the SIR programme is estimated to be nearly Cdn$500 per hectare.

All orchards were monitored with sex pheromone-baited traps (10 mg lures) at a density of 1 trap per hectare. Two species of leafrollers, oblique-banded, *Choristoneura rosaceana* Harris, and pandemis, *Pandemis pyrusana* Kearfott, were monitored with sex pheromone-baited traps at a density of one per 4 hectares. All orchards were routinely scouted to visually assess pest and natural enemy populations. Ten orchards were extensively sampled for all the major pests and their natural enemies during the season. Six to eight orchards treated with conventional pesticides outside the project served as controls and were monitored similarly each year to provide an evaluation of the programme's impact.

Baseline data on pest populations from this site were collected from ca. 80 hectares in 1994 prior to the start of the area-wide project. Growers were applying an average of 5.5 sprays of azinphosmethyl to control the codling moth. Traps loaded with 1 mg lures caught an average of 60 fertile and 14 sterile Canadian moths during that season. These sterile moths were released on the Canadian side of the border. Fruit injury from the codling moths averaged 0.3%, but ranged from 0.05 to 3.0% among the orchards. Fruit injury by leafrollers averaged 1.1%. Surprisingly, growers were unaware of the presence of leafrollers and this fruit injury had been previously misidentified as codling moth damage in the packing houses.

The results of the Lake Osoyoos CAMP project have been evaluated both in terms of the baseline data and by comparison with the conventional programmes used by the growers outside the project. Since 1994, the growers in the project have reduced their use of broad spectrum organophosphate insecticides from 5.5 to 0.2 sprays per season. Fruit injury due to codling moths has fallen from 0.30% to 0.06%. Growers in the project have had 70% less injury than those outside the project during these three years. During 1995 and 1996, growers in the project used 70% less organophosphate insecticides than growers outside the project. Then in 1997, growers at the Lake Osoyoos CAMP site nearly eliminated the use of these insecticides (Table 1).

Table 1. Summary of results for codling moth from the Lake Osoyoos CAMP, during the baseline year (1994) and the first three years, 1995 – 97.

| Year | % CM injury | Mean wild moth catch/trap/season | OP applications per season | Sterile/Fertile moth ratio |
|---|---|---|---|---|
| 1994 | 0.30 | 60.0 | 5.5 | 0.3:1 |
| 1995 | 0.22 | 4.29 | 2.2 | 31:1 |
| 1996 | 0.04 | 0.82 | 1.8 | 181:1 |
| 1997 | 0.06 | 0.31 | 0.2 | 369:1 |

Leafrollers have been a major concern of growers during the three years of the programme. Control was not improved in the first year when compared with the baseline data and was somewhat higher than in the outside orchards (1.0% vs. 0.75%). During the last two years, however, fruit injury by leafrollers has been 70% lower within versus outside the project (0.36% vs. 1.0%). Leafrollers inside the project have been managed by an increase in the use of *Bacillus thuringiensis* insecticides.

Other secondary pests have not been a major concern during the three years of this project and few additional insecticidal sprays have been applied. During 1997, injury from true bugs (Hemiptera: Heteroptera) averaged 0.21% inside versus 0.48% in the comparison blocks. Cutworm (Lepidoptera: Noctuidae) injury was somewhat higher in the project than outside (0.8 vs. 0.04%).

ARS will terminate their role in the area-wide programme on 30 September 1999. This programme was never intended to be a perpetual government programme; rather it was designed as a self-help programme whereby ARS would demonstrate the strategy and effectiveness of mating disruption and SIT for small growers. The growers are expected to continue the coordinated programme on their own and from all indications, it appears that this is the case with the Lake Osoyoos programme.

## REFERENCES

Beers, E.H., J.F. Brunner, M.J. Willet and G.M. Warner. 1993. Orchard pest management: a resource book for the Pacific Northwest. Good Fruit Grower, Yakima, Washington. 276 pp.

Bloem, K.A. and S. Bloem. 2000. SIT for codling moth eradication in British Columbia, Canada. pp 207-214. *In* Area-Wide Control of Fruit Flies and Other Insect Pests. K.H. Tan (ed.). Penerbit Universiti Sains Malaysia, Penang, Malaysia.

Dyck, V.A., S.H. Graham and K.A. Bloem. 1993. Implementation of the sterile insect release programme to eradicate the codling moth, *Cydia pomonella* (L.) (Lepidoptera: Olethreutidae), in British Columbia, Canada. pp. 285-294. *In* Management of Insect Pests: Nuclear and Related Molecular and Genetic Techniques, IAEA-SM-327/29, Vienna.

Knipling, E.F. 1979. The basic principles of insect suppression and management. Agriculture Handbook No. 512. USDA, SEA, Washington, D.C. 659 pp.

McMecha, A.D. and M.D. Proverbs. 1972. Equipment and procedures for release of sterile codling moths. Can. Agric. Engineering 14:42-45.

Vickers, R.A. and G.H.L. Rothschild. 1991. Use of sex pheromones for control of codling moth. pp. 339-354. *In* Tortricid Pests, Their Biology, Natural Enemies and Control. L.P.S. van der Geest and H.H. Evenjuis (eds.). Elsevier Science Publishers B.V., Amsterdam, The Netherlands.

# Use of Nuclear Techniques in Biological Control

*Patrick D. Greany*[1] *and James E. Carpenter*[2]

[1] Center for Medical, Agricultural and Veterinary Entomology, ARS-USDA, Gainesville, FL, USA
[2] Insect Biology and Population Management Research Laboratory, ARS-USDA, Tifton, GA, USA

## INTRODUCTION

As pointed out by Benbrook (1996), pest management is at a crossroads, and there is a great need for new, biointensive pest management strategies. Among these approaches, biological control is a keystone. However, because of increasing concerns about the introduction of exotic natural enemies of insect pests and weeds (Howarth 1991, Delfosse 1997), the overall thrust of biological control has moved toward augmentative biological control, involving releases of established natural enemy species (Knipling 1992). This in turn has created a need to develop more cost-effective mass rearing technologies for beneficial insects. Nuclear techniques could play an especially important role in augmentative biological control, not only in facilitating mass rearing, but in several other ways, as indicated below.

Recognising the potential value for use of nuclear techniques in biological control, the Insect and Pest Control Section of the Joint FAO/IAEA Division of Nuclear Techniques in Food and Agriculture, International Atomic Energy Agency, sponsored a Consultants' Group Meeting on this subject in April 1997. The Group produced a document entitled *Use of Nuclear Techniques in Biological Control: Managing Pests, Facilitating Trade and Protecting the Environment.* The consultants included the authors of this paper as well as Ernest Delfosse (at that time, with the USDA-APHIS National Biological Control Institute), Garry Hill (Intl. Institute for Biological Control), Sinthya Penn (Beneficial Insectary), and Felipe Jeronimo (USDA-APHIS PPQ, Guatemala). The remarks presented in this paper reflect the thoughts presented by these consultants and other participants at the IAEA-sponsored meeting.

Several potential uses for nuclear techniques were identified by the Consultants' Group, including: 1) improvements in rearing media (either artificial diets or natural hosts/prey), 2) provision of sterilised natural prey to be used as food during shipment, to ameliorate concerns relating to the incidental presence of hitchhiking pests, 3) provision of supplemental food or hosts in the field, to increase the initial survival and buildup of released natural enemies, and 4) reproductive sterilisation of weed-feeding insects that are candidates for biological control, for use in open field trials.

## APPLICATIONS OF IONISING RADIATION

Ionising radiation offers a reliable means to achieve: 1) developmental arrest of hosts/prey for use in *in vivo* rearing, 2) microbial pasteurisation or sterilisation of artificial media and possibly even natural hosts/prey (e.g., to kill *Nosema* spp.), and

3) reproductive sterilisation of hosts/prey to prevent release of viable pests along with beneficial insects.

## Developmental Arrest

Gamma radiation has been used to inhibit development of Caribbean fruit flies, *Anastrepha suspensa* Loew, that escape parasitisation by *Diachasmimorpha longicaudata* (Ashmead) (Sivinski and Smittle 1990). This was done to preclude the inadvertent release of fertile adult fruit flies along with parasitoids in inoculative and inundative release programmes in Florida. In this case, the fruit fly larvae were irradiated using ca. 4 kR (40 Gy) during the third instar, prior to exposure to parasitoids. This was a more useful application of gamma radiation than that of Ramadan and Wong (1989), who exposed pupae of the Oriental fruit fly, *Bactrocera (Dacus) dorsalis* (Hendel) to gamma radiation *after* having already exposed the larvae to parasitisation by *Diachasmimorpha (Biosteres) longicaudata*, resulting in sterility of the adult parasitoids. In the studies of Sivinski and Smittle, the dose of gamma radiation (from a $^{137}$Cesium source) prevented adult eclosion of non-parasitised caribflies, but it did not prevent these larvae from serving as viable hosts for *D. longicaudata*. This allowed the investigators to safely set out puparia from larvae exposed to parasitoids without fear of releasing fertile flies into the area. Their work parallelled the earlier studies of Morgan et al. (1986), who used gamma radiation (50 kR or 500 Gy) to inhibit development of pupae of *Musca domestica* L. that were then used as hosts for the parasitoid *Spalangia endius* Walker. Similar benefits were obtained by Roth et al. (1991), who used irradiated horn fly pupae as hosts for hymenopterous parasitoids. Morgan et al. (1986) also found that irradiated housefly pupae could be held successfully for an extended period (ca. 10 weeks) prior to parasitisation by storing them at a low temperature (4.4°C), as long as adequate humidity was maintained.

Another application for ionising radiation that has promise is to inhibit the cellular and/or humoural defense reactions of host insects that might otherwise serve as optimal factitious hosts for beneficial insects. This approach was tested as a means of inhibiting encapsulation of the parasitoid *Microplitis croceipes* Cresson in a candidate factitious host, *Galleria mellonella* L. (S. Ferkovich, personal communication). This parasitoid usually attacks larvae of *Heliothis* and *Helicoverpa* spp., but it will oviposit into *Galleria* larvae, and ca. 20% routinely escape encapsulation and complete development in this laboratory host, although the *Microplitis* adults derived from *Galleria* larvae were considerably smaller than those reared in *Heliothis/Helicoverpa* larvae (Gupta et al. 1996). It was felt that by using gamma radiation, it might be possible to substitute *Galleria* larvae for *Heliothis/Helicoverpa* larvae and thereby economise considerably in the rearing of *Microplitis*. Unfortunately, at no radiation dose (up to 100 Gy) was it possible to achieve an increase in success in the use of *Galleria* (S. Ferkovich, personal communication). Although this approach was not successful in this instance, the principle should be kept in mind for other host-parasitoid systems, to enable more economical *in vivo* rearing.

An untested, but promising, application for gamma radiation is to inhibit the behavioural resistance of hosts/prey, so that they can be made more suitable for attack by parasitoids and/or predators that may otherwise be injured by their hosts. Similarly, it may be possible to prevent other behaviours that diminish the suitability of candidate hosts/prey, such as bothersome web spinning by *G. mellonella* larvae.

**Sterilising/Pasteurising Media/Hosts/Prey**

An excellent application for ionising radiation is for use in microbial pasteurisation and sterilisation of artificial media and even natural hosts/prey for rearing parasitoids and predators. Gamma radiation, as well as X-rays, provides a means of killing bacteria, fungi, viruses, and protozoa that may impair growth and development of insect parasitoids and predators without nutritional damage to the artificial media, and it can be used to dramatically increase the medium's shelf life upon storage. Irradiation is ideally suited for use in insect rearing for several reasons. First, it is easy to achieve repeatable doses. Second, there is no thermal degradation of the diet components, in contrast to the use of steam or dry heat sterilisation, which changes the physical properties of media due to denaturation of proteins, for example. In addition, thermolabile enveloping membranes and diet packaging films such as Parafilm®, are generally unaffected by gamma radiation and X-rays at the rates used for these purposes. Ionising radiation can also be easily employed to sterilise materials that cannot be filter sterilised, such as thick, viscous media with numerous particles present. One of the greatest virtues of this approach is that it can be used to sterilise media *after* packaging (described as "terminal sterilisation" in the pharmaceutical industry). This helps ensure long shelf life during storage, and helps prevent rapid microbial contamination of media when presented to beneficial insects.

As recently as 1984, little success had been made in the development of artificial media for entomophagous insects (King and Leppla 1984). However, in the past ten years or so, great advances have been made in developing artificial media that are suitable for a myriad of insect natural enemies (see Anderson and Leppla 1992, Grenier et al. 1994 and references therein). In many of these instances, it probably would have been extremely helpful to use ionising radiation to improve the rearing success on artificial media. This approach has been of great value in our own studies on a variety of predators and ectoparasitoids that we are rearing on an artificial medium ("DI-Diet") free of any insect components (Greany and Carpenter 1996, Carpenter and Greany 1998). Our approach is to prepare the medium under clean (but not sterile) conditions, and then to encapsulate it before subjecting the final product to gamma radiation, using a dose of ca. 1.7 kGy. This achieves a high degree of microbial control which, along with storage at 4°C, allows the product to be kept for at least several weeks. Use of this system is illustrated in Figure 1, showing feeding of *Podisus maculiventris* nymphs upon our medium encapsulated in Parafilm.

We are currently working jointly with a scientific and engineering research firm (Analytical Research Systems, Inc., of Gainesville, FL) to develop a sophisticated encapsulation process for our medium which will also include sterilisation by gamma radiation. This combination of high volume diet packaging and sterilisation should significantly reduce the cost of rearing a variety of beneficial insects and simultaneously improve the success of the use of the artificial medium. The approach may be very useful to complement the excellent artificial diet and rearing system developed by Rojas et al. (1996) for the boll weevil parasitoid, *Catolaccus grandis*. It also provides a novel means for presenting control agents such as pathogenic microorganisms to insect pests, for example, fire ants (the response of fire ants to diet-filled, polymer-coated capsules is illustrated in Figure 2).

Finally, it may also be possible to use ionising radiation to kill disease organisms such as the microsporidian *Nosema* spp. that are often present in host organisms intended for parasitoid/predator rearing, and which can be transmitted from infected hosts to their natural enemies. For example, Undeen et al. (1984) showed that spores of *Nosema algerae* could be killed by gamma radiation in excess of 3 kR (30 Gy).

Figure 1. Fourth instar nymphs of *Podisus maculiventris* feeding on "DI-diet" encapsulated in Parafilm, and irradiated after encapsulation using gamma radiation.

Figure 2. Imported fire ants investigating diet-filled capsules prior to consuming contents.

## Reproductive Sterilisation

Gamma radiation has been used to cause inherited sterility among progeny of radiation-exposed insects (i.e., $F_1$ sterility), so that these reproductively incompetent insects can be safely field-released for use as hosts/prey for indigenous natural enemies. This subject will be covered in detail by J. Carpenter elsewhere, and therefore will not be described further here.

Use of ionising radiation for reproductive sterilisation could be a great assistance in providing the safe international shipment of biological control organisms by equipping these organisms with natural prey that had been reproductively sterilised prior to packaging. For example, it would allow shipment of predaceous phytoseiid mites with prey mites (for use as food during shipment) without fear that the spider mites could become a problem upon release of the predatory mites.

Another promising application of ionising radiation would be to allow testing of agents intended for weed control outside quarantine conditions if these agents are first sterilised reproductively. This could enable them to be released into the field to determine whether they will feed upon the target weed and any other non-target plants without fear that they could become established and perhaps cause inadvertent harm.

## IMPLEMENTATION OF IONISING RADIATION IN BIOLOGICAL CONTROL

Considering the demonstrated benefits associated with the use of nuclear techniques for these purposes, it is logical to ask, "Why isn't this approach being used to greater advantage?" One of the reasons is the relative scarcity of irradiators, but it is, in part, due to fear of irradiation and a lack of understanding of the effects of this process. While a number of USDA-ARS, APHIS and university laboratories have a gamma radiation source readily available, relatively little use has been made of these facilities for purposes relating to rearing beneficial insects. Commercial firms have had even less opportunity to avail themselves of this promising technology because of difficulty in obtaining access to irradiators.

Some persons mistakenly fear that the use of a radioactive source, or even a linear accelerator or X-ray machine, will cause the exposed materials to become radioactive. Another common misconception is that this process will destroy the nutritional value of the irradiated materials, or will cause the formation of an abundance of free radicals of oxygen or other radiolytic products that will render the foodstuff toxic. The comfort level of potential users might be increased through an educational programme to enlighten them about the safety of this technology; an excellent booklet on this topic was published by the IAEA (Anonymous 1995).

One of the genuine "hassles" that inhibits potential users from taking better advantage of this technology is the abundance of regulatory agency requirements that must be met for acquisition and maintenance of a radioactive source, such as a Cesium or Cobalt source. Along with this, there is a high initial cost for purchase of even a small (ca. 0.5 litre volume) gammacell (in the order of US$100,000). The need to ultimately dispose of the radioactive waste also constitutes at least a minor problem. Fortunately, X-ray machines and linear accelerators for use in food irradiation are being developed. These devices are subject to fewer regulatory constraints than $^{137}$Cesium or $^{60}$Cobalt sources, and they may prove much more user-friendly for the insect rearing community. Relatively low-cost cabinet X-ray machines are being developed that will suffice for many small-scale users.

## SUMMARY

Overall, there is great promise for the use of ionising radiation in support of the development of improved mass rearing methods to be used for augmentative biological control. The advent of more and more artificial media that are proving suitable for numerous beneficial insects is providing an impetus for development of appropriate sterilisation regimes. Similarly, ionising radiation may also be used to great advantage to improve conventional *in vivo* rearing strategies for many parasitoids and predators.

Finally, the regulatory climate is becoming much more stringent, and radiation techniques may help facilitate international, interstate, and even intrastate shipments of insect natural enemies, by preventing the accidental release of reproductively viable pest organisms along with their natural enemies (Delfosse 1997, Hill 1997), and it could enable improved testing of agents intended to be used for biological control of weeds prior to field release by first releasing reproductively-sterilised individuals for field trials.

## REFERENCES

Anderson, T.E., and N.C. Leppla. 1992. Advances in insect rearing for research and pest management. Westview Press, Boulder. 519 pp.

Anonymous. 1995. Facts about food irradiation. Int. Atomic Energy Agency. 38 pp. *In* Pest Management at the Crossroads. C.M. Benbrook (ed.). 1996. Consumers Union, New York. 272 pp.

Bushland, R.C. and D.E. Hopkins. 1951. Experiments with screwworm flies sterilised by X-rays. J. Econ. Entomol. 44: 725-731.

Carpenter, J.E. 1997. Development and integration of alternative management strategies using inherited sterility and natural enemies to control lepidopteran pests. Report for IAEA Consultants Group on Use of Nuclear Techniques in Biological Control; held 14-18 April, 1997, Vienna.

Carpenter, J.E. and P.D. Greany. 1998. Comparative development and performance of artificially-reared vs. host-reared *Diapetimorpha introita* (Cresson) (Hymenoptera : Ichneumonidae) wasps. Biological Control 11:203-208.

Delfosse, E. 1997. International issues in relation to biological control regulation, coordination and accountability. Report for IAEA Consultants Group on Use of Nuclear Techniques in Biological Control; held 14-18 April, 1997, Vienna.

Greany, P.D. and J.E. Carpenter. 1996. Culture Medium for Parasitic and Predaceous Insects, U.S. patent filing; s/n 08/692,565 - docket no. 0010.96, filed Aug. 6, 1996.

Grenier, S., P.D. Greany and A.C. Cohen. 1994. Potential for mass release of insect parasitoids and predators through development of artificial culture techniques. pp. 181-205. *In* Pest Management in the Subtropics: Biological Control - a Florida Perspective, D. Rosen, F.D. Bennett and J.L. Capinera, (eds.). Intercept, Andover. 737 pp.

Gupta, P., C.R. Dillard and S.M. Ferkovich. 1996. Potential of an unnatural host, *Galleria mellonella* (Lepidoptera: Galleriidae), in rearing the corn earworm endoparasitoid *Microplitis croceipes* (Hymenoptera: Braconidae). Ann. Entomol. Soc. Am. 89:103-108.

Hill, G. 1997. The potential for the integration of nuclear techniques in arthropod biological control. Report for IAEA Consultants Group on Use of Nuclear Techniques in Biological Control, 14-18 April, 1997, Vienna.

Howarth, F.G. 1991. Environmental impacts of classical biological control. Annu. Rev. Entomol. 36:485-509.

King, E.G. and N.C. Leppla. 1984. Advances and challenges in insect rearing. U.S. Govt. Printing Office: 1983-769-037:10.

Knipling, E.F. 1992. Principles of insect parasitism analysed from new perspectives: practical implications for regulating insect populations by biological means. USDA ARS Agriculture Handbook No. 693. U.S. Govt. Printing Office, 337 pp.

Morgan, P.B., B.J. Smittle and R.S. Patterson. 1986. Use of irradiated pupae to mass culture the microhymenopterous pupal parasitoid *Spalangia endius* Walker (Hymenoptera: Pteromalidae). I. *Musca domestica* L. (Diptera: Muscidae). J. Entomol. Sci. 21:222-227.

Ramadan, M.M. and T.T.Y. Wong. 1989. Effect of gamma radiation on *Biosteres longicaudatus* (Ashmead) (Hymenoptera: Braconidae), a larval parasitoid of *Dacus dorsalis* Hendel (Diptera: Tephritidae). Proc. Hawaii. Entomol. Soc. 29:111-113.

Rojas, M.G., J.A. Morales-Ramos and E.G. King. 1996. *In vitro* rearing of the boll weevil (Coleoptera: Curculionidae) ectoparasitoid *Catolaccus grandis* (Hymenoptera: Pteromalidae) on meridic diets. J. Econ. Entomol. 89:1095-1104.

Roth, J. P., G.T. Fincher and J.W. Summerlin. 1991. Suitability of irradiated or freeze-killed horn fly (Diptera: Muscidae) pupae as hosts for hymenopteran parasitoids. J. Econ. Entomol. 84: 94-98.

Sivinski, J. and B. Smittle. 1990. Effects of gamma radiation on the development of the Caribbean fruit fly (*Anastrepha suspensa*) and the subsequent development of its parasite *Diachasmimorpha longicaudata*. Entomol. Exp. Appl. 55: 295-297.

Undeen, A.H., R.K. Vandermeer, B. J. Smittle and S.W. Avery. 1984. The effect of gamma radiation on *Nosema algerae* (Microspora: Nosematidae) spore viability, germination and carbohydrates. J. Protozool. 31:479-482.

# Biological Control Against the Carob Moth *Ectomyelois ceratoniae* in Oases and in Packing Houses in Tunisia

*M.H. Dhouibi[1], N. Hawlitsky[2], H. Zaaraoui[3], T. Krisaane [3], T. Cheikh[1], M. Cherni [1] and I. Ben Moussa[1]*

[1] Institut National Agronomique de Tunisie, 43 Avenue Charles Nicolle cité Mahrajène, 1082 Tunis Mahrajene, Tunisia
[2] INRA Versaille France
[3] Groupement Interprofessionnel de Dattes de Toseur,Tunisia

## INTRODUCTION

The carob moth, *Ectomyelois ceratoniae* Zeller is abundant in the Mediterranean countries. It attacks various dry fruit in cultures or in stored products, notably pomegranate, *Punica granatum* L.; date palm, *Phoenis dactylifera* L. plantations; citrus, *Citrus* spp., apricot, *Prunus armeniaca* L. and pistachios, *Pistachio vera*. We can find *E. ceratoniae* in the north as well as in the south of Tunisia, especially in central zones and Saharan areas where caterpillar infestations can reach 90% of pomegranate fruit and 20% of dates (Dhouibi 1991). To reduce this damage, several control methods have been experimented. Chemical control is the most effective means of control against pests. However, against this species, insecticides seem to be difficult and randomly used, due to the endophytic behaviour of the pyralid and the position of the fruit on the pomegranate tree. Moreover, this method has very ominous repercussions on biological cadence. Besides, it is necessary to look for other control means to allow the preservation of the ecosystem. In Tunisia, several efforts have been directed at biological control, by using local parasitoids and through usage of the bio-insecticides mainly *Bacillus thuringiensis* (Dhouibi 1992, 1994, Dhouibi and Jemmasi 1993). In order to substitute the chemical control and to strengthen the integrated control, other possibilities can be envisaged, for example, the genetic method or the autocidal control, that is, based on mass rearing and the substerile male releases into the natural population. For the purpose, it provokes the sterility to ulterior generations and evaluates the impact of irradiation on the different biological parameters of emerged adults from treated nymphs and their competitiveness. Dhouibi and Omran (1995) and Dhouibi and Tijani (1996) have studied the mass rearing of the carob moth pyralid on an artificial diet and the effect of different irradiation doses, especially a substerilising dose, on *E. ceratoniae* pupae.

## MATERIALS AND METHODS

Since 1982, research has been undertaken to study the possibility of integrated control against date pyralids by the judicious combined use of all possible control methods. For this purpose we have realised the following:

## Mass Trapping in Packing Houses

Following the identification of the sex pheromone of females in 1989 (Dhouibi 1989), different types of traps have been tested in packing houses. The two types of traps are:

- the **Delta System** – a trap constructed from plastic-coated cardboard (universal)
- **bottle with transparent plastic** (locally made) - four traps each lured with sex attractant of the pyralid were used in each packing house. The experiment was repeated 4 times.

## Control with a Bio-insecticide *Bacillus thuringiensis*

*Bacillus thuringiensis* formulations (Bactospeine XLV to 13,000 U.I/mg A.K., Biobit XL 9,000 U.I/mg A.K, Ecotek pro 24,000 U.I/mg T.n.) are used in terrestrial or aerial treatments. The terrestrial treatment is conducted with the Bactospeine powder in several trials. The treatment of plantations on very high grounds is done through the use of lengthened tubes. Aerial treatment is likewise made using Bactospeine XLV. For this purpose, the Ministry of the Defense has put at our disposal, two equipped helicopters, to ensure a regular application of 5 litres/ha (1.5 litres of the product added with 3.5 litres of water). The homogenity of the application has been verified with a sensitive paper (18.75 $cm^2$) to oil and water (Teejet) placed horizontally and vertically in cardinal points of trees.

## Utilisation of Native Parasitoids

Since 1993, we have started the rearing of two local parasitoids:

- *Phanerotoma flavitestacea* Fischer: a ovolarval parasitoid, which is laid in eggs of the host and developed inside the larvae. The experimental releases have been conducted in an orchard of SODAD (100 ha) in Tozeur during these last three campaigns to study the dispersion, to determine time of release and the parasitoïd quantities per unit area.
- *Habrobracon hebetor* Say: a larval parasite, the whole development of which occurs on the body of the host. We have multiplied this parasite under laboratory conditions on a substitute host *Ephestia kuehniella* Zeller. Experimental releases have been conducted in Mornag (suburbs of Tunis), the experimental station of Ksar-Chriss (SIDI Bousid) and in Chakmou (Tozeur).

## Impact of Fruit Protection Using Different Materials on Infestation and Fruit Quality

Different protection materials have been used. The experimental site is divided into blocks and each block contains several treatments: protected regimes with the blue plastic, white plastic, yellow plastic of the same thickness, kraft paper, screens and the check in 5 replicates. This operation made regularly each year aims to protect regimes against the rain. This study is conducted as we are looking for the type of material to use, its impact on physico-chemical qualities of fruit and its effect on date infestation.

### Utilisation of Substerile or Sterile Males

We have achieved in the laboratory the mass rearing of the date pyralids *E. ceratoniae* on an artificial diet. We then studied the impact of substerilising doses on adult emergence from old pupae treated with 20, 25 and 30 krads doses and their competitiveness. The source of irradiation is variable with different debits (Salah asaes, Instiut Pasteur, Inrat and Tajoura: Nuclear centre in Libya).

## RESULTS AND DISCUSSION

We present here biological control methods used to control the carob moth in oases and in packing houses (parasitoids, *B. thuringiensis*, alone or in combination, the mass trapping and irradiated insects).

### Control with a Bio-insecticide by the Use of *Bacillus thuringiensis*

The *B. thuringiensis* variety Kurstaki is a bacteria that acts on larvae of *E. ceratoniae* by ingestion before their penetration in dates. The efficacy of the formulation of the *B. thuringiensis* 16,000 I.U.A.K/mg by the terrestrial way ensures the maintenance of a level of infestation at least lower than that before application. Similar results have been found by aircraft application using the following formulations: Bactospeine XLV to 13,000 I.U.A.K/mg and Biobit XL 9,000 I.U.A.K/mg and Ecotek pro to 24,000 I.U./T.n./mg (Figures 1 and 2). The terrestrial treatment is hardly used in the old plantations because of the excessive heights, the important palm tree density and the lack of appropriate application devices. To this end and for rapid interventions, the use of *B. thuringiensis* ULV formulations is recommended. Currently a national campaign is made regularly by aircraft in all the oases of the Djérid.

### Utilisation of Native Parasitoids

The multiplication and experimentation of the two local parasitoids, *H. hebetor* and *Phanerotoma flavitestacea,* have been repeated during these last five years and have given very interesting results in limiting the pest population.

### Release of *Habrobracon hebetor*

*Habrobracon hebetor* is an ectoparasite found exclusively on fallen dates and in pomegranates looking for *E. ceratoniae* caterpillars. The release of *H. hebetor* spaced over 10 days in orchards of pomegranates is recommended especially in oases where pomegranate trees and the palm trees are planted together.

The combined use of *B. thuringiensis* and *H. hebetor* appeared particularly promising to limit populations of the date pyralid especially in packing houses (Figures 1 and 2).

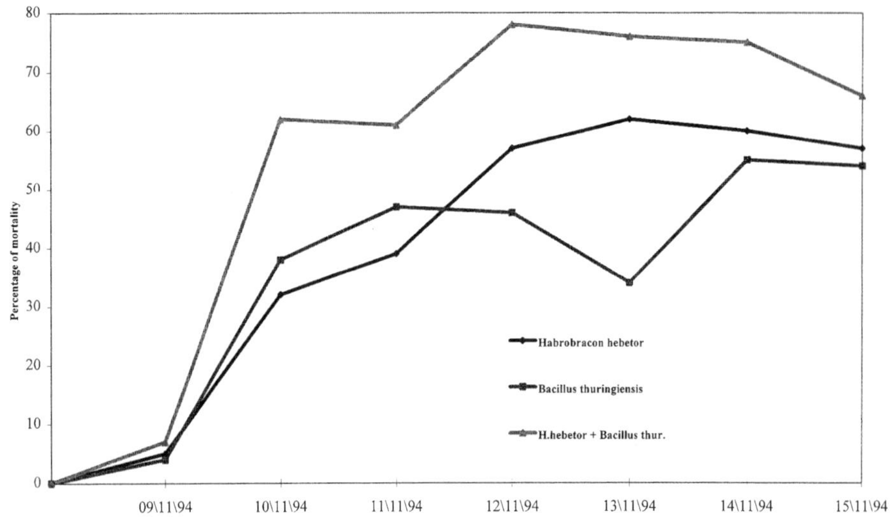


Figure 1. Effects of *Bacillus thuringiensis* and *Habrobracon hebetor* on mortality of *Ectomyelois ceratoniae* in a packing house.

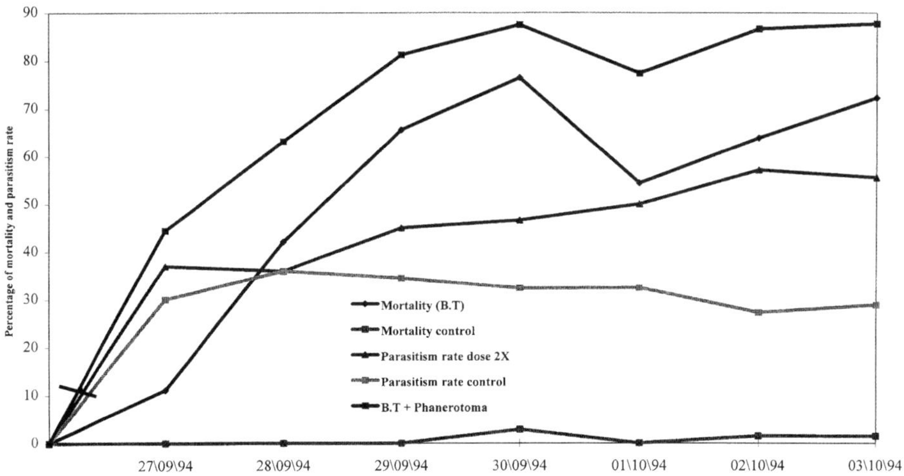


Figure 2. Percentage of mortality and parasitism rate in oasis treated with *Bacillus thuringiensis* and *Phanerotoma flavitestacea* (separately and in combination) and their respective controls.

## Utilisation of *Phanerotoma flavitestacea*

*Phanerotoma flavitestacea* is an hymenoptera endoparasite, frequently present in the oases. It is found in the eggs of its host and develops inside the larvae. In Tunisia, this braconid is present naturally in the oases with a parasitism rate of 40% (Dhouibi 1993). The releases of *P. flavitestacea* from 4 points per ha in an oasis of 100 ha increased the natural parasitism rate to reach 95% (Figure 3). In the oases, the combined use of *P. flavitestacea* release and the aerial release of *B. thuringiensis* ULV formulations has positively increased the larval mortality (Figure 2).

## Mass Trapping of Pyralids in Date Packing Houses

The mass trapping of the date pyralid with sex attractants (Dhouibi 1993) as well as other phycitine moths has resulted in the removal of almost all adults from packing houses and consequently has decreased the infestation level during the test period (Figure 4).

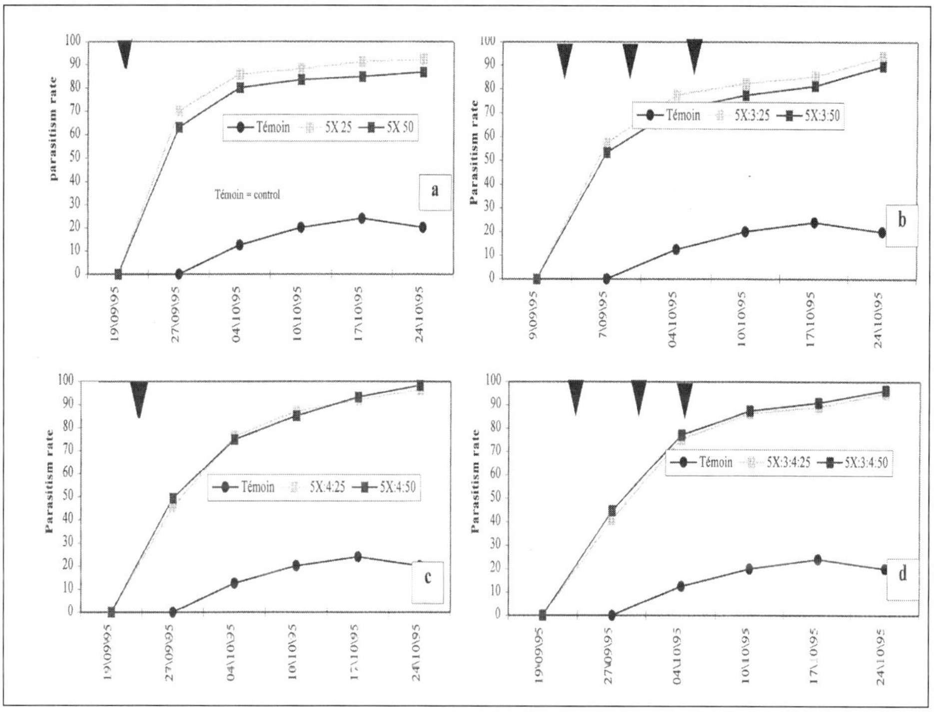


Figure 3. Parasitism rate of *Phanerotoma flavitestacea* at 25 and 50 individuals per release point on carob moth. Graph a) released one at one point/ha; b) released thrice at one point/ha; c) released once at four points/ha; and d) released thrice at four points/ha.

## Utilisation of Substerile Males

To be able to use the sterile male technique or substerile insects, it is necessary to succeed in mass rearing of the carob moth *E. ceratoniae*, followed by a study of substerile dose impact on emerged adults from old irradiated pupae by substerile doses: 20, 25 and 30 krads and to determine the competitiveness of these adults (Figure 5). The results obtained in this field are very interesting. The main objective is to use substerile males in isolated areas in combination with the parasitoids to reduce natural population of the carob moth.

## CONCLUSION

The study has focused on the multiplication in the insectarium of the natives and introduced parasitoids of the date pyralids and their release in different sites of pomegranate orchards, in the oases or in orchards where the palm tree is associated with the pomegranate tree. The results of research during the five consecutive years have shown an increase of the parasitism rate in orchards (monoculture or polyculture).

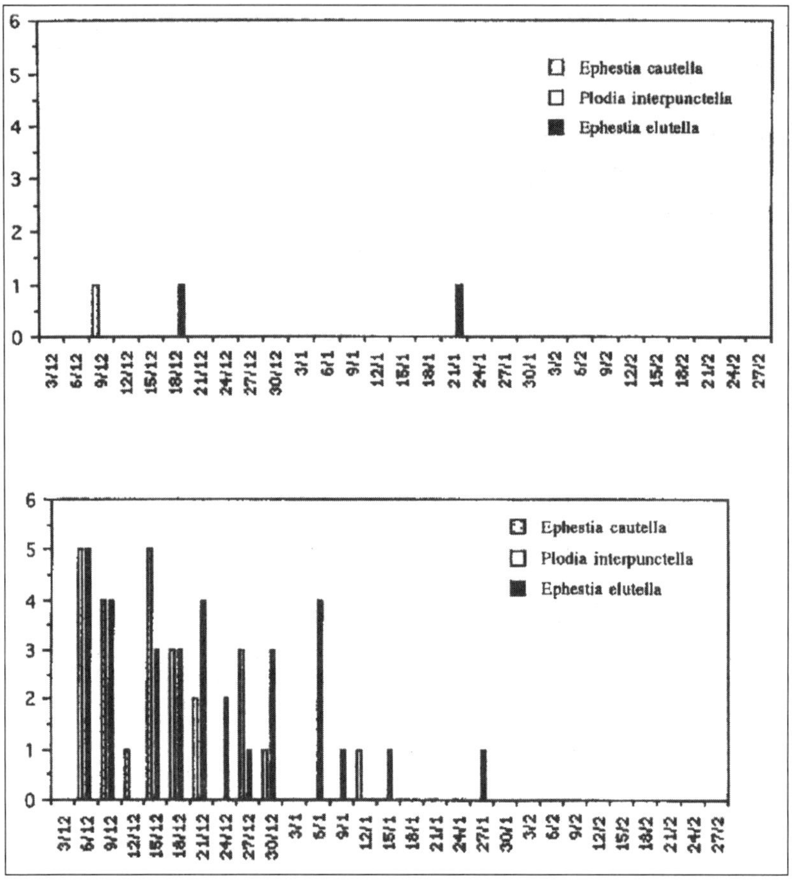


Figure 4. Numbers of phycitine moths caught in packging houses.
(Top - effect of mass trapping; Bottom - control.)

The treatment of oases since the beginning of fruit infestation, operations during several campaigns using different *B. thuringiensis* formulations, has currently been applied to all the oases of the Djerid using appropriate equipment. The combined use of these biological methods improves their efficacy appreciably.

Autocidal control, first involving the multiplication in sufficient quantities in the laboratory of the carob moth on a semi-artificial diet, then the irradiation of produced pupae for the production of substerile and competitive males, has been tested. It can be promising especially when we have to deal with isolated areas. To undertake a strategy of integrated biological control, it is necessary to continue these research works by regulating the behaviour of the parasitoids in the oases, quantities of parasitoid released and the released periods in relation to the level of the moth population. There are also possibilities of combinations between treatment with the *B. thuringiensis* and the parasitoid releases as well as mass trapping and the impact of sterile males in isolated oases and in packing houses.

The integration of all these methods will result in a decrease in the level of the pest population in the field and in packing houses without having to seek recourse to chemical products and fumigants. The quality of fruit production will then improve.

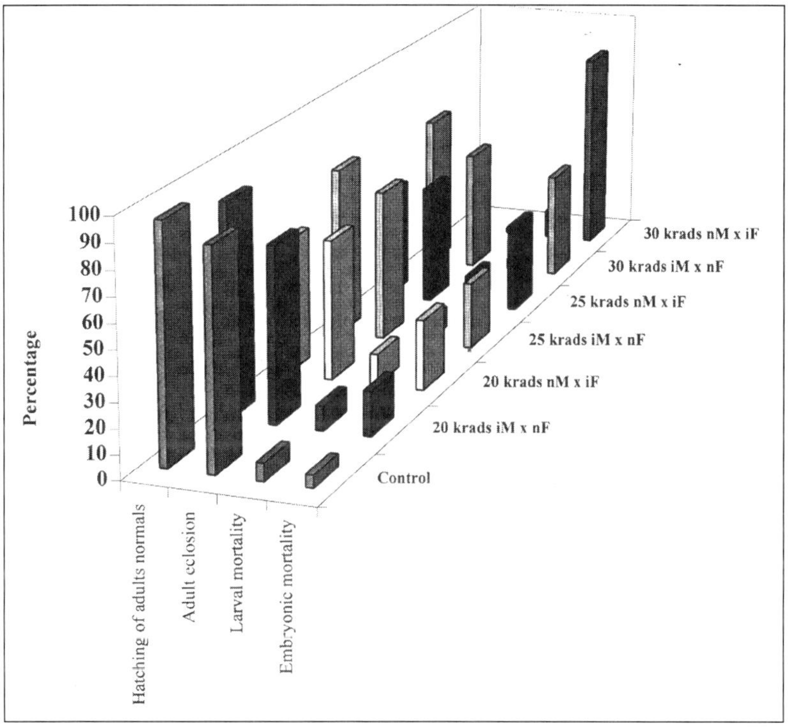


Figure 5. Irradiation effect on the F1 progeny of *Ectomyelois ceratoniae*.

## REFERENCES

Abderrahmane, C.T. 1997. Elevage de la pyrale des dattes (Ectomyelois ceratoniae Zeller 1881) et effet des doses substérilisantes d'irradiation sur les nymphes et la compétitivité des adultes. DEA D'Ecologie Animale Faculté des Sciences de Tunis. 75 pp.

Calkin, C.O. 1996. Multifaceted Control Strategies for Areawide Suppression of codling Moth Populations. Second Research Co-ordinated Meeting on Evaluation of Population Suppression by Irradiated Lepidoptera and Their Progeny, Vienna 2-6 September 1996.

Dhouibi, M.H. 1992. Effet de la Bactospeine XLV sur la pyrale des dattes *Ectomyelois ceratoniae* Zeller (Lepidoptera: Pyralidae). Med Fac. Landbouww. Univer. Gent, 57/2b,505- 514.

Dhouibi, M.H. and A. Jemmasi. 1993. Lutte biologique contre la pyrale des caroubes *Ectomyelois ceratoniae* (Lepidoptera: Pyralidae) par *Habrobracon hebetor* (Hymenoptera: Braconidae) en vergers de grenadiers. Med. Fac. Landbouww. Univ. Gent.58/2a, 427-436.

Dhouibi, M.H. and M. Omrane. 1995. Effect of gamma radiation at different ages on the carob moth *Ectomyelois ceratoniae* Zeller pupae (Lepidoptera: Pyralidae). First research Co-ordinated meeting: Evaluation of population suppression by irradiated Lepidoptera and their progeny. Jakarta 17- 28 April 1995.

Dhouibi, M.H. and A.C. Tijani. 1996. Mating competitiveness of irradiated males and females of the carob moth *Ectomyelois ceratoniae* (Lepidoptera: Pyralidae). Second research Co-ordinated meeting on Evaluation of Population Suppression by Irradiated Lepidoptera and Their Progeny, Vienna 2-6 September 1996.

Mansour, M. 1996. Evaluation the potential of Suppressing Codling moth, *Cydia pomonella* L., Field population by the use of the Sterile Insect Release Method. Second Research Co-ordinated Meeting on Evaluation of Population Suppression by Irradiated Lepidoptera and Their Progeny, Vienna 2-6 September 1996.

Omran, M. 1996. Contribution à l'étude de quelques caractéristiques écophysiologiques de la pyrale des dattes Ectomyelois ceratoniae Zeller 1881 (Lepidoptera: pyralidae) et effet des radiations gamma sur ses nymphes. DEA D'Ecologie Animale Faculté des Sciences de Tunis. 127 pp.

# PART II.

## APPLICATIONS OF THE AREA-WIDE CONCEPT TO NON-FRUIT FLIES

## C) MOLECULAR BIOLOGY AND GENETICS IN RELATION TO AREA-WIDE INSECT CONTROL

# Polytene Chromosome Maps and RAPD Polymorphisms in *Glossina austeni*

*A. Gariou-Papalexiou*[1]*, G. Yannopoulos*[1]*, A.S. Robinson*[2] *and A. Zacharopoulou*[1]

[1] Department of Biology, Division of Genetics, Cell and Developmental Biology, University of Patras, Greece
[2] Entomology Unit, FAO/IAEA Agriculture and Biotechnology Laboratory, A-2444 Seibersdorf, Austria

## INTRODUCTION

A combined methodology of cloned RAPD (random amplification of polymorphic DNA) polymorphic bands and *in situ* hybridisation to polytene chromosomes is an efficient way to initiate construction of a physical and genetic map of insect disease vectors (Dimopoulos et al. 1996, Mutebi et al. 1997). The studies presented here are the first step in developing this approach in tsetse flies. This technology will be used to support tsetse sterile insect technique (SIT) programmes by providing tools with which population structure and isolation can be assessed and genetic markers that can be used to differentiate released flies from wild flies identified. An added benefit is their possible use in unravelling epidemiological complexity and problems regarding speciation (Besansky et al. 1997).

Polytene chromosomes of Diptera have been shown to be excellent material for the study of chromosome structure and function as well as for an understanding of the genetics of natural populations (Lefevre 1976). They provide a means for the accurate mapping of chromosome rearrangements and the precise localisation of genes, using both rearrangement analysis and *in situ* hybridisation. Previous reports on the cytology of the tsetse flies ( Riordan 1968, Maudlin 1970, 1979, Southern et al. 1972, Southern and Pell 1973, Davies and Southern 1976, Southern 1980) have described the basic mitotic karyotype in several *Glossina* species, and demonstrated the presence of well banded polytene chromosomes in pupal trichogen cells (Southern and Pell 1974, 1981, Pell and Southern 1976). Polytene chromosomes were described for *G. austeni* Newstead, *G. morsitans morsitans* Westwood, *G. pallidipes* Austen and *G. fuscipes fuscipes* Newstead, but these descriptions are difficult to work with as they are drawings of polytene chromosome elements. In this paper, the photographic chromosome maps of pupal scutellar bristles of *G. austeni* are presented. They show that these chromosomes can be used with much greater ease than standard cytological material in the interpretation of any population differences.

The RAPD analysis was used to analyse three different populations of *G. austeni* and several primers were able to discriminate between these populations. RAPD analysis provides an easy, although sometimes problematic, access to naive genomes (Williams et al. 1990) through the use of commercial oligonucleotide primers. The pros and cons of the use of this polymerase chain reaction (PCR) based technique have been well documented (Black IV 1993) but there is now a considerable body of evidence testifying to their use in several areas of applied entomology e.g., the development of linkage maps (Hunt and Page 1995) and genetic fingerprinting (Brown et al. 1997).

These studies are the first attempt to develop a set of cytological and molecular tools which can aid in the improvement of SIT intervention measures for tsetse flies. The development of a basic set of genetic tools has proved to be very useful in other insect pest species. In the Medfly, (*Ceratitis capitata* Wied.), strains have been synthesised which produce only males for release in SIT programmes (Robinson et al., 2000).

## MATERIALS AND METHODS

### Polytene Chromosomes

#### *Preparation*

Polytene chromosomes were prepared from pupal trichogen cells associated with apical scutellar bristles. Male pupae have four long scutellar macrochetae while females two long and two short ones. *G. austeni* flies were reared in the Entomology Unit at Seibersdorf, Austria, and pupae were shipped to Patras for analysis. Upon receipt, the pupae were maintained at 18° C and chromosome preparations were carried out on 17-30 day old pupae according to the following procedure.

Pupae were removed from their pupal case and fixed for at least 24 hours in ethanol:acetic acid (3:1). The first cuticular layer was then removed and the scutellum isolated and transferred to a drop of 45% glacial acetic acid on a clean coverslip. Following the removal of the second cuticular layer, the remaining tissue was manipulated by a pair of needles and a drop of glacial acetic acid: $H_2O$: lactic acid (3:2:1) was added. 1-2 minutes later, a drop of lactic-acetic orcein was added. After 1-2 minutes staining, the tissue was picked up by placing a clean slide over the coverslip. The slide was then inverted and the coverslip was gently moved in order to spread the polytene chromosomes out of the nuclei. Finally, the slide was lightly blotted to remove the excess stain. Slides were then examined under low power phase contrast microscopy for the presence and quality of chromosomes. Gentle tapping and squashing were performed to obtain adequate spreading and separation of chromosomes.

#### *Construction of photographic maps*

Well spread chromosomes with clear banding patterns were photographed under X100 magnification using the Agfapan-25 black and white film. Selected photographs from each chromosomal region were used for the construction of the composite chromosome maps, by putting together pieces of several different nuclei. These maps were scanned at 1200 dpi with Adobe Photoshop 4.0. Finally, the maps were printed on Agfa Accuset 1000 Film at 2400 dpi and then on Agfa Proof ABL 45.

## RAPD Analysis

### *DNA preparation*

DNA was extracted according to a modified protocol of Bender et al. (1983). Wild flies from the field which had been preserved in 90% alcohol were placed in several changes of $ddH_2O$ before DNA extraction. The head and thorax were removed from each fly and homogenised in 250 µl of buffer (0.1 M NaCl, 0.2 M sucrose, 0.1 M Tris, 0.5 M EDTA and 0.5% SDS) and the pestel rinsed with a further 250 µl of buffer. The solution was incubated at 65°C in a water bath for 30 min. 70 µl of 8 M potassium acetate were added, the solution well agitated and incubated on ice for 30 min. The solution was centrifuged at high speed for 15 min and the supernatant transferred to a new tube. 2 vols of 100% EtOH were added and the solution incubated at room temperature for 5 min followed by centrifuging for 15 min. The supernatant was removed, the pellet washed in 70% EtOH, spun and then dried. The dry pellet was resuspended in 50 µl of TE.

### PCR Reaction

All reactions were performed in 25 µl as follows: 1.0 µl DNA, 1.0 µl primer(0.05 µg/µl stock), 2.5 µl dNTP's (2 mM stock), 2.5 µl PCR Buffer, 1.0 µl $MgCl_2$ (50 mM stock), 0.2 µl Taq polymerase (1 unit) and 16.8 µl of sterile $ddH_2O$. The reactions were run in a Thermal cycler with the following programme: 1 min. at 94°C, 1 min. at 36°C and 2 min. at 72°C for 45 cycles. The amplified product was run on 1.6% agarose gels. Primer set # 100/1 from UBC was used.

### Fly Strains

*G. austeni* was used in all the studies. The cytology was performed on a laboratory strain which is maintained at the Entomology Unit, Seibersdorf. The strain was established from a pupal collection made on Unguja Island, Zanzibar, in 1964 (Nash et al. 1966). *G.austeni* from mainland Tanzania and Unguja Island were trapped on sticky panels and preserved in 90% alcohol. *G. austeni* from S. Africa were collected in normal tsetse traps and then placed in 90% alcohol.

## RESULTS AND DISCUSSION

### Polytene Chromosomes

### *General description*

*G. austeni* polytene nuclei contain three long polytene elements, representing the X chromosome and the two autosomes, $L_1$ and $L_2$. In the polytene maps presented here, we followed the designation, as well as the number of divisions for each arm, used by Southern and Pell (1974), who were the first to obtain well banded polytene chromosomes of tsetse flies. However, we adopted the numbering system used for *Drosophila* polytene maps to standardise these maps with those from other insects. The major divisions were numbered 1-100 from the telomere of XL (instead of XR in

Southern and Pell 1974) to the telomeric region of $L_2R$ (instead of $L_2L$, as in Southern and Pell 1974). Divisions 1-15 were reserved for XL, 16-36 for XR, 37-53 for $L_1L$, 54-72 for $L_1R$, 73-85 for $L_2L$ and 86-100 for $L_2R$. Each major division usually commences to the left of a marker band and was further subdivided from left to right into a number of minor zones, A, B, C or D, depending on its size. Because of the numbering system used, the band sequence within a major or minor division does not correspond to that reported by Southern and Pell (1974) in the maps of their line drawings.

Polytene chromosome maps provide a means to pinpoint the precise location of a cloned DNA, a chromosome breakpoint or a rearrangement, i.e., they can be used to produce a physical map. In order that data from polytene chromosome analysis can be compared between laboratories, it is important that any particular band or region seen under the microscope can be correlated with the chromosome map. If the maps are represented only by drawings of chromosomes, the investigator has to translate any observed chromosome band seen under the microscope into the lines on a drawn map. This creates confusion and wastes time. Photographic maps minimise these difficulties, since the individual chromosome arm is assembled by piecing together selected pictures of overlapping chromosomal regions. It is for these reasons that photographic chromosome maps of the *G. austeni* polytene chromosomes have been developed.

The current maps have been based on previous work (Southern and Pell 1974), though the band sequence within a major or minor division on the line drawn maps does not correspond to those of the current photographic maps. These differences are due to: 1) the adoption of a numbering system used for the construction of polytene chromosomes of Drosophila (Bridges 1935, Lefevre 1976) in which the numbering of the 100 sections begins at the free left end of the X chromosome and finishes at the telomere of the $L_2R$ chromosome arm, and 2) difficulties encountered in translating any observed chromosome region into the line drawing maps constructed by Southern and Pell (1974). The constructed photographic maps are shown in Figure 1. A brief description of the principal chromosome markers that could serve as diagnostic landmarks for each chromosome arm is given below.

### *X chromosome (sections 1-36)*

The X chromosome of *G. austeni* is intermediate in size among the three polytene elements. The two arms differ slightly, with the left arm being shorter than the right. Close to the left end, in regions 1C-2A, three consecutive landmarks are found, the second of which occupies a slight constriction. Section 3 is characterised by two dark bands in A and C and the constriction at the B subdivision. An important diagnostic landmark of this arm is region 7A-8B. Two dark bands at 7A are followed by a puff at 7B, by three consecutive swellings at 7C, 7D, 8A and by a second characteristic puff at 8B. The most characteristic diagnostic sections of the XL arm are the 12AB regions where four dark bands are followed by a constriction at the beginning of 12C with the two puffs at 12C and 13A, separated by a constriction consisting of a dark band. The centromeric region, 15C, is also identified by a pair of weak bands followed by a dark dot.

The XR arm is identified easily by its unique chromosome region at 24-26. Three sharp bands at 24AB are followed by a diffused puff at 24C and a constriction

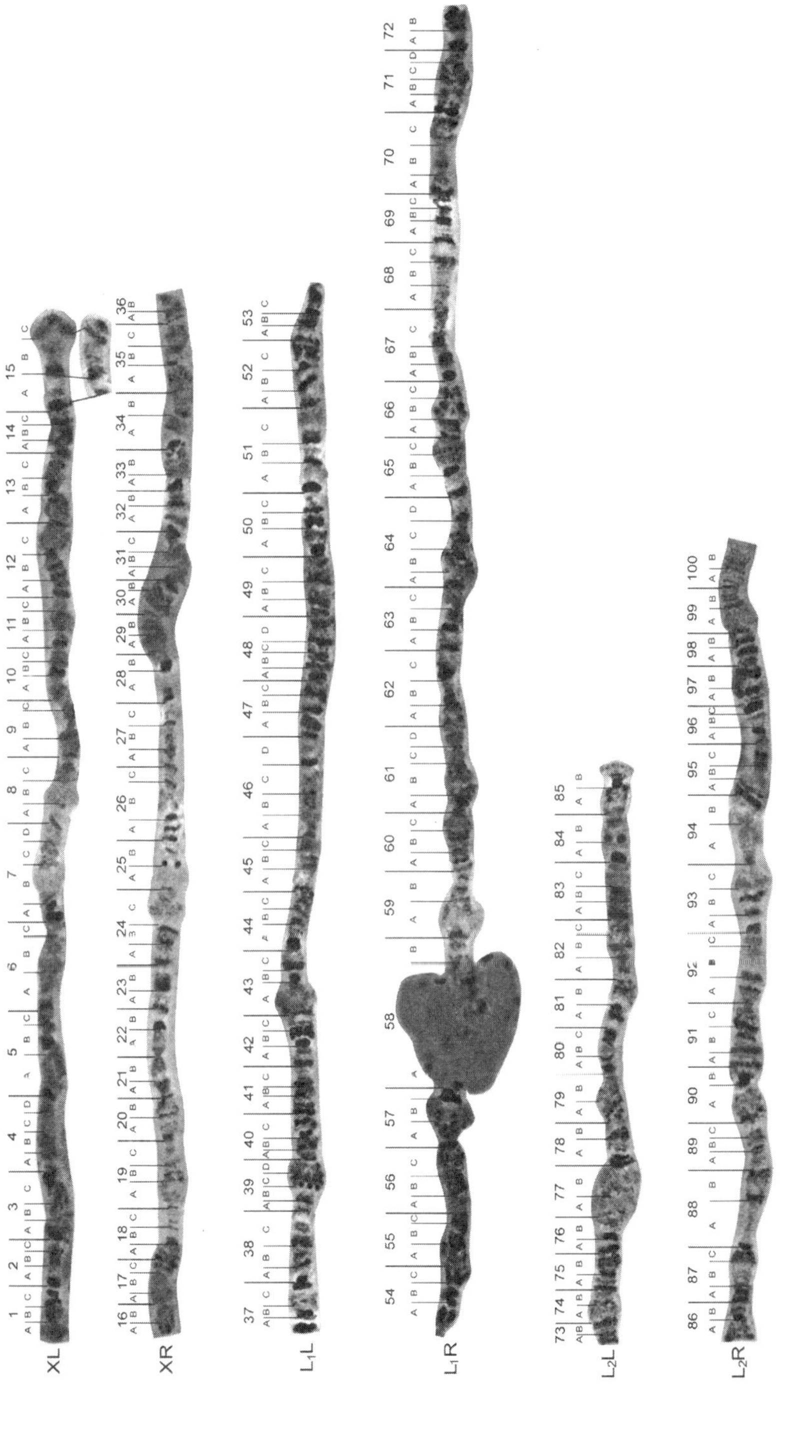


Figure 1. Polytene chromosome maps of *Glossina austen.*

consisting of a pair of light bands at the 24/25 boundary. The two dark dots at 25B, just after the puff at 25A, are followed by a series of four bands, the first of which is lighter than the others. Additional landmarks of this arm are the centromeric area, 16A, with its dot-like appearance and the narrow free end, 36B, capped by a slightly dark band.

### *$L_1$ chromosome (sections 37-72)*

$L_1$ is the longest chromosome element and the most distinctive in the polytene complement. The left end is easily identified by two dark bands in 37AB and a constriction consisting of two heavy bands at the boundary of 37/38. The 39D region is characterised by a diffused swelling consisting of four dark dot-like bands. An additional recognition site is at region 43 with the expanded area 43A and two dark bands at B and C, the first of which also constitutes a characteristic constriction. Numerous heavy bands from 47-52 are characteristically separated by small puffs or diffused expanded areas. The two heavy dark dots at 52C are followed by two thin dot-like bands. The centromeric region of this arm, 53C, is always thin and consists of a heavy band with two heavy dark dots at 54A and two dark bands at 53B. The unique recognition site is the 58A region where the nucleolus is always associated with the constriction at the 57/58 boundary. The arm is usually broken at this site. This arm is also characterised by heavy dark bands, expanded areas and puffs (62C, 63B, 65A, 65B, 66B, 67AB, 69AB). The $L_1R$ tip is easily identified by the presence of a dark band at 72B and a constriction at 72A.

### *$L_2$ chromosome (sections 73-100)*

The $L_2$ chromosome is the shortest polytene element. Important landmarks of this arm are the two tips and the centromeric region. The left tip is narrow and round and shows a series of light bands, followed by a small puff at 74A. The principal marker of this arm is region 76-77 where a large puff, 77AB, is bordered by two pairs of dark bands at 76A and 78A. The three pairs of dark bands in region 80-81A are also characteristic. Additional diagnostic sites are a dark band at 84A, two heavy dark dots at 84B and a dark band at 85B close to the centromere, consisting of a unique constriction.

The $L_2R$ tip is characteristically flared with numerous light bands. Three narrow areas at 88A, 92B and 94A constitute useful landmarks. Additional recognition sites are in sections 97-98, which contain a series of dark bands, some of which are double (region 95A and region 90B-91), where a constriction in 90B is followed by an expanded region at 91 with four distinct dark bands.

### *Heterochromatic Network*

In addition to the polytene elements described above, a heterochromatic network is usually observed (Figure 2), associated with the XR centromeric region. According to Southern and Pell (1974), it may represent a heterochromatic segment of the X chromosome which pairs with the Y chromosome at male meiosis. However, we cannot exclude the possibility that the whole structure, or at least a part of it, represents the heterochromatic supernumerary chromosomes which are not polytenised.

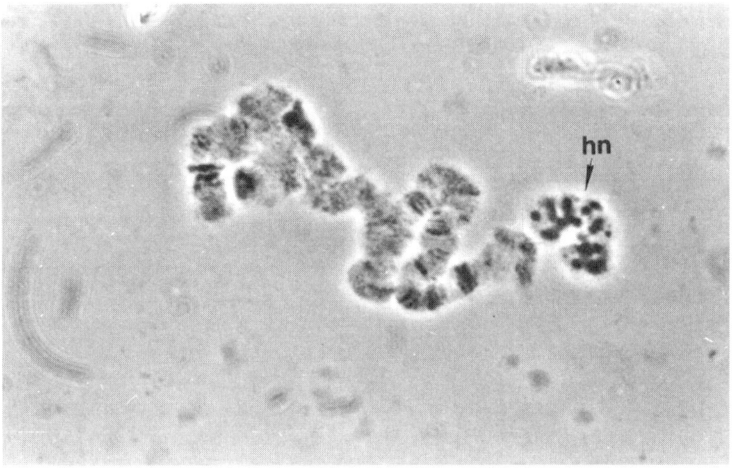


Figure 2. The heterochromatic network (hn) observed in pupal trichogen cells associated with the centromere of the X chromosome in *Glossina austeni.*

### *G. austeni from Zanzibar and mainland Tanzania*

As reported in these proceedings, there has been a successful SIT programme to eradicate *G. austeni* from Unguja Island, Zanzibar, and this was the reason that this species was chosen. As in all SIT programmes, the degree of movement of flies into and out of the treatment area should be as low as possible and, therefore, islands are ideal ecological areas to develop this technique. Unguja Island is 35 km from the mainland where *G. austeni* is also present. It was considered prudent to try to develop diagnostic markers which would help to distinguish these two populations. The generally accepted view is that the *G. austeni* population on Unguja is a descendent of the one which was trapped on the island when it separated from mainland Africa. This scenario should have provided sufficient time for significant genetic differentiation to occur. Included in this comparison were flies from the strain which was being released on Unguja and which was established about 30 years ago with material from the same island.

A total of 48 primers was tested on panels of DNA containing samples from Unguja Island, Seibersdorf and mainland Tanzania. An equal number of males and females for each strain was analysed and no sex specific differences were observed. In general, there was not a large degree of genetic differentiation among the three populations although a complete analysis has not yet been made. However, there were many polymorphisms detected which will be used to develop *in situ* hybridisation to the polytene chromosomes. In general, the Unguja Island (Zanzibar) and the Seibersdorf flies (Laboratory) produced similar profiles with the flies from the mainland (Tanga) being the most different. Only two primers, #76 (Figure 3) and #84 (Figure 4) produced profiles which could be used to differentiate Unguja flies from mainland flies. The possible diagnostic bands are indicated by arrows. The lack of genetic differentiation between the two field populations is surprising and perhaps reflects a different history than the one currently accepted i.e., that the *G. austeni* was introduced into Unguja Island more recently than previously thought.

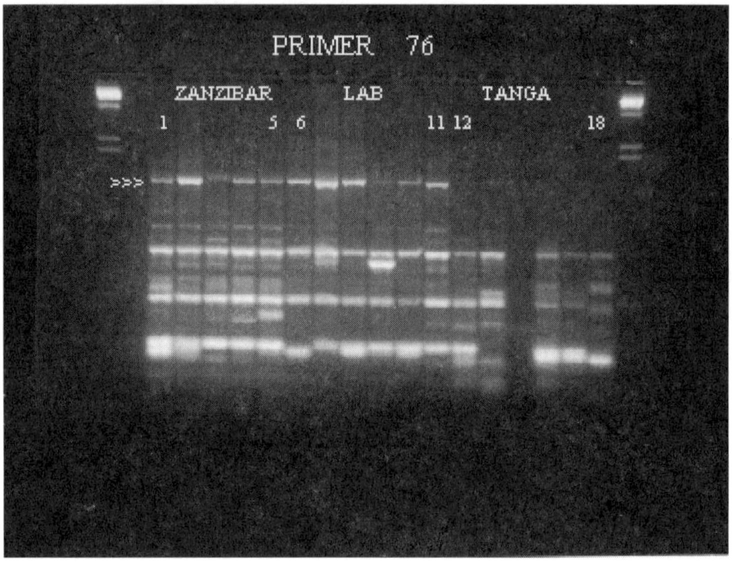


Figure 3. RAPD analysis of 3 populations of *Glossina austeni* using primer 76. Lanes 1-5, males from Unguga Island, Zanzibar; lanes 6-11, males from a laboratory colony at Seibersdorf; lanes 12-18, males from Tanga, Tanzania. Outermost lanes are DNA markers. The arrow indicates a possible diagnostic band.

### *G. austeni from Zanzibar and S. Africa*

The population of *G. austeni* in S. Africa could be a target for future SIT as it is geographically isolated from other *G. austeni* fly belts. However, it has been given sub-specific status based on morphological and ecological parameters (Travassos Santos Dias 1987) and its true status will only be resolved when hybridisation experiments are carried out. At the moment, there is no laboratory colony of the South African population available and the question was raised as to whether the mass reared *G. austeni* colony at Tanga could be used to supply flies to South Africa. With no colony available for hybridisation studies, an initial RAPD analysis was made on flies from South Africa and those from the mass reared colony. Flies from Zanzibar were also included in the analysis.

Twenty-two primers were tested on panels of DNA from the three populations, South Africa (Zululand), Seibersdorf (Laboratory) and Tanzania (Tanga). Only one primer #32 (see Figure 5) gave a clear diagnostic band, although several others were informative and will be further analysed. If the South African population is indeed sub-species of *G. austeni,* the degree of genetic differentiation is low. Attempts are now being made to establish a colony from South Africa to clarify the taxonomic status of this population.

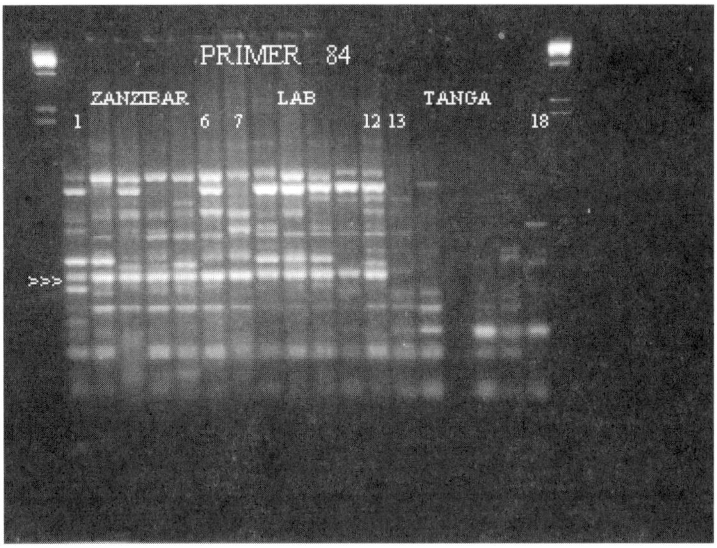


Figure 4. RAPD analysis of 3 populations of *Glossina austeni* using primer 84. Lanes 1-6, males from Unguga Island, Zanzibar; lanes 7-12, males from a laboratory colony at Seibersdorf; lanes 13-18, males from Tanga, Tanzania. Outermost lanes are DNA markers. The arrow indicates a possible diagnostic band.

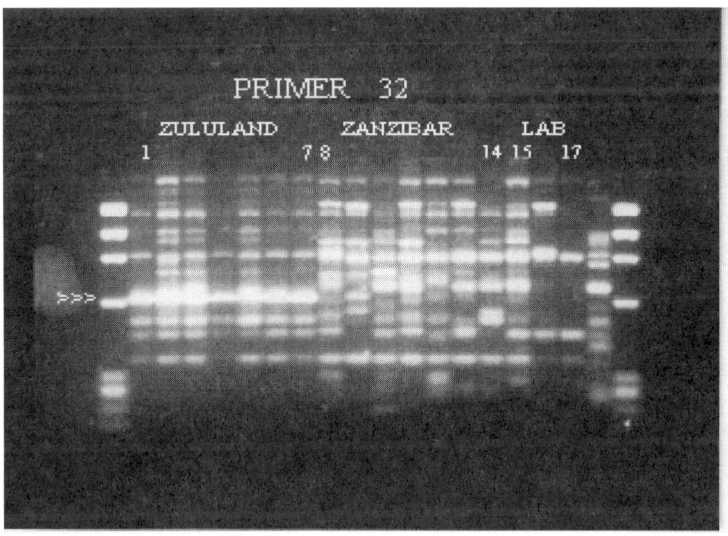


Figure 5. RAPD analysis of 3 populations of *Glossina austeni* using primer 32. Lanes 1-7, males from Zululand, South Africa; lanes 8-14, males from Unguga Island, Zanzibar; lanes 15-17, males from Tanga, Tanzania. Outermost lanes are DNA markers. The arrow indicates a possible diagnostic band.

## CONCLUSIONS

A combination of RAPD mapping and *in situ* hybridisation is a powerful tool to construct genetic and physical maps in Dipteran species of agricultural and public health importance. The results presented in this paper are a first step to develop better maps for tsetse. Fluorescent *in situ* hybridisation (FISH) to mitotic chromosomes has already been carried out in tsetse using ribosomal DNA probes (Willhoeft 1997). Some genetic fingerprinting were also carried out (Blanchetot and Gooding 1994). Photographic maps are now available and RAPD polymorphisms identified for *G. austeni*. The next phase will include the characterisation of the polymorphic bands followed by *in situ* hybridisation to the polytene chromosomes. The reproductive biology of tsetse makes traditional Mendelian genetics very difficult and protein analysis has revealed low levels of genetic differentiation in field populations although isozyme analysis has been successfully used both to generate genetic maps (Gooding 1984) and to analyse hybridisation experiments (Gooding 1997). Isozyme analysis, however, presents logistic problems as it is difficult to obtain suitable material from the field. DNA analysis of ethanol-preserved individuals will enable many questions to be answered regarding the epidemiological significance of tsetse populations in various ecological situations. Chromosomal polymorphisms, ideally visualised by polytene analysis, have been very important to an understanding of the epidemiology of malaria (Coluzzi et al. 1979). In tsetse, nothing is known about chromosomal polymorphisms in the field, despite the fact that there are clear population differences in many important traits e.g., susceptibility to infection (Tarimo Nesbitt et al. 1990) and response to traps (see discussion in Bayliss and Nambiro 1993). Gooding (personal communication) has some preliminary genetic evidence that inversions are present in some populations of *G. morsitans morsitans*.

## ACKNOWLEDGEMENTS

AG-P, GY and AZ gratefully acknowledge the financial support of the IAEA.

## REFERENCES

Baylis, M. and C.O. Nambiro. 1993. The responses of *Glossina pallidipes* and *G. longipennis* (Diptera: Glossinidae) to odour-baited traps and targets at Galana Ranch, south-eastern Kenya. Bull. Entomol. Res. 83: 145-151.

Bender, W., P. Spierer and D. Hogness. 1983. Chromosomal walking and jumping to isolate DNA from the ACE and rosy loci and the bithorax complex in *Drosophila melanogaster*. J. Mol. Biol.168: 17-33.

Besansky, N.J., T. Lehman, G.T. Fahey, D. Fontenille, L.E.O. Braack, W.A. Hawley and F.H. Collins. 1997. Patterns of mitochondrial variation within and between African malaria vectors, *Anopheles gambiae* and *An. arabiensis*, suggest extensive gene flow. Genetics 147: 1817-1828.

Black IV, W.C. 1993. PCR with arbitrary primers: approach with care. Insect Mol. Biol. 2: 1-6.

Blanchetot, A. and R.H. Gooding. 1994. Genetic variability and segregation analysis in *Glossina morsitans morsitans* (Diptera: Glossinidae) using DNA fingerprinting. Genome 37: 289-295.

Bridges, C.B. 1935. Salivary chromosome maps. J. Hered. 26: 60-64.

Brown, R.J., C.A. Malcolm, P.L. Mason and R.A. Nichols. 1997. Genetic differentiation between and within strains of the saw-toothed beetle, *Oryzaephilus surinamensis* (Coleoptera:Silvanidae) at RAPD loci. Insect Mol. Biol. 6: 285-289.

Coluzzi, M., A. Sabatini, V. Petrarca and M.A. Di Deco. 1979. Chromosomal differentiation and adaptation to human environments in the *Anopheles gambiae* complex. Trans. R. Soc. Trop. Med. Hyg. 73: 483-487.

Davies E.D.G. and D.I. Southern. 1976. Giemsa C-banding within the genus *Glossina (Diptera:Glossinidae).* Genetica 46: 413-418.

Dimopoulos, G., L. Zheng, V. Kumar, A. Della Torre and F.C. Kafatos. 1996. Integrated genetic map of *Anopheles gambiae*: Use of RAPD polymorphisms for genetic, cytogenetic and STS landmarks. Genetics 143: 953-960.

Gooding R.H. 1984. Tsetse genetics: a review. Quaest. Entomol. 20: 89-128.

Gooding R.H. 1997. Genetic analysis of hybrid sterility in crosses of tsetse flies *Glossina palpalis palpalis* and *Glossina palpalis gambiensis* (Diptera: Glossinidae). Can. J. Zool. 75: 1109-1117.

Hunt, G.J. and R.E. Page. 1995. Linkage map of the Honey Bee, *Apis mellifera,* based on RAPD markers. Genetics 139: 1371-1382.

Lefevre, G. 1976. A photographic representation and interpretation of the polytene chromosomes of *Drosophila melanogaster* salivary glands. pp 31-36. *In* the Genetics and Biology of *Drosophila.* Vol.1a. M. Ashburner and E. Novitski, (eds.). Acad. Press, London.

Maudlin, I. 1970. Preliminary studies on the karyotypes of five species of *Glossina.* Parasitology 61: 71-74.

Maudlin, I. 1979. Chromosome polymorphism and sex determination in a wild population of tsetse. Nature 277: 300-301.

Mutebi, J-P., W.C. Black IV, C.F. Boslo, W.P. Sweeny Jr. and G.B. Craig. 1997. Linkage map for the Asian Tiger mosquito (*Aedes (Stegomyia) albopictus*) based on SSCP analysis of RAPD markers. J. Hered. 88: 489-494.

Nash, T.A.M., R.J. Kernaghan and J.A. Boyle. 1966. The large scale rearing of *Glossina austeni* (Newst.) in the laboratory. 1. The use of pregnant and non-pregnant goats as hosts. Ann. Trop. Med. Parasitol. 60: 39-47.

Pell P.E. and D.I. Southern. 1976. A cytogenetical study of *Glossina fuscipes fuscipes* including a comparison of the polytene chromosome maps with those of *Glossina austeni (Diptera: Glossinidae).* Genetica 46: 511-528.

Riordan, K. 1968. Chromosomes of the tsetse fly *Glossina palpalis R-D.* Parasitology 58: 835-838.

Robinson, A.S., G. Franz and K. Fisher. 2000. Genetic sexing strains in the Medly, *Ceratitis capitata*: Development, Mass Rearing and Field Application. Res. Trends Entomol. (in press).

Southern, D.I. 1980. Chromosome diversity in tsetse flies. *In* Insect Cytogenetics. Symposium of the Royal Entomological Society of London. R. Blackman (ed.). Blackwell Scientific Publications.

Southern, D.I., T.A. Graig-Cameron and P.E. Pell. 1972. The meiotic sequence in *Glossina morsitans morsitans.* Trans. R. Soc. Med. Hyg. 66: 145-149.

Southern, D.I. and P.E. Pell. 1973. Chromosomal relationships and the meiotic mechanisms of certain *morsitans* group tsetse flies and their hybrids. Chromosoma 44: 319-334.

Southern, D.I. and P.E. Pell. 1974. Comparative analysis of the polytene chromosomes of *Glossina austeni* and *Glossina morsitans morsitans*. Chromosoma (Berl.) 47: 213-216.

Southern, D.I. and P.E. Pell. 1981. Cytogenetical aspects of *morsitans* tsetse flies with particular reference to *Glossina pallidipes* (Diptera: Glossinidae). Cytobios 30: 135-152.

Tarimo Nesbitt, S.A., R.H. Gooding and B.M. Rolseth. 1990. Genetic variation in two field populations and a laboratory colony of *Glossina pallidipes* (Diptera: Glossinidae). J. Med. Entomol. 27: 586-591.

Travassos Santos Dias, J.A. 1987. Contribuicao para o estudo da sistematica do genero *Glossina* Wiedemann, 1830 (Insecta, Brachycera, Cyclorrapha, Glossinidae) Proposta para a criacao de um novo subgenero. Garcia de Orta, Ser. Zool., Lisboa, 14: 67-78.

Willhoeft, U. 1997. Fluorescence *in situ* hybridisation of ribosomal DNA to mitotic chromosomes of tsetse flies (Diptera: Glossinidae: *Glossina*). Chromosome Res. 5: 262-267.

Williams, J.G.K., A.R. Kubelik, K.J. Livak, J.A. Rafalski and S.V. Tingey. 1990. DNA polymorphisms amplified by arbitrary primers are useful as genetic markers. Nucleic Acids Res. 18: 6531-6535.

# Arriving at the Age of Pest Insect Transgenesis

*Peter W. Atkinson*[1] *and David A. O'Brochta*[2]

[1] Department of Entomology, University of California, Riverside, CA, 92521, USA
[2] Center for Agricultural Biotechnology, University of Maryland Biotechnology Institute, College Park, MD, 20742, USA

## INTRODUCTION

Technologies that enable the stable genetic transformation of insects other than the vinegar fly, *Drosophila melanogaster* Meigen, have been sought since *D. melanogaster* was initially transformed using the *P* transposable element (Rubin and Spradling 1982). *D. melanogaster* transformation can now be achieved by using Type II eukaryotic transposable elements such as *P*, *hobo*, *Hermes*, *mariner*, *Minos* and *piggyBac* (Blackman et al. 1989, Lidholm et al. 1993, Loukeris et al. 1995a, O'Brochta et al. 1996, Rubin and Spadling 1982, A. M. Handler, personal communication). The success of this strategy led to many attempts to extend it into non-drosophilid insects and this approach has recently been successful with the use of four different transposable elements to transform two non-drosophilid insect species, the Medfly, *Ceratitis capitata* (Wied.) and the yellow fever mosquito, *Aedes aegypti* (L.) (Coates et al. 1998, Handler et al. 1998, Jasinskiene et al. 1998, Loukeris et al. 1995b). The generation of these transgenic insects has, in part, arisen through the adoption of two approaches. One has been the isolation of new transposable elements from non-drosophilid insects. The second has been the implementation of mobility assays that have quickly enabled the mobility properties of these new elements in the target pest species to be determined. The success of these approaches will most likely be extended to other pest insect species over the next five years and will increase our ability to use modern genetic techniques to develop new strategies to control pest insects.

## GENETIC TRANSFORMATION OF *AEDES AEGYPTI* BY THE *HERMES* AND *MARINER* ELEMENTS; TRANSFORMATION OF *CERATITIS CAPITATA* BY THE MINOS AND *PIGGYBAC* ELEMENTS

### *Hermes*

The *Hermes* transposable element was identified from the house fly, *Musca domestica* L., initially on the basis of the *hobo* element instability in this species (Atkinson et al. 1993). The instability of plasmid-borne *hobo* elements injected into *M. domestica* led to speculation that an endogenous *hobo*-like element, now known to be the *Hermes* element, was perhaps responsible for this instability (Atkinson et al. 1993). The full length 2.749kb *Hermes* element was subsequently isolated and interplasmid transposition mobility assays used to determine whether or not *Hermes* was mobile in non-host species. These assays are simple and quick to perform provided that a satisfactory technique for the introduction of DNA into insect embryos is available. Microinjection of plasmid DNA into developing insect embryos has been the physical delivery method used. These assays revealed that *Hermes* is mobile in a wide range of insect species such as *D. melanogaster*, *M. domestica*, *Bactrocera tryoni* Froggatt, *C. capitata*, *Lucilia cuprina* Wied., *Ae. aegypti* and *Helicoverpa armigera* Hübner (Pinkerton et al. 1996, Sarkar et al. 1997a and b). This broad host range soon led to *Hermes* being successfully employed as a genetic transformation vector in *D. melanogaster* (O'Brochta et al. 1996). The transformation frequencies achieved with *Hermes* in this species routinely exceed those obtained by using the *P* transposable element which has been the standard vector for *Drosophila* transgenesis for over 15 years.

*Hermes* element transposition assays showed that *Hermes* could transpose accurately in the yellow fever mosquito, *Ae. aegypti*, indicating that it can also be used to genetically transform this species . In order to successfully identify transgenic adults, an eye pigmentation marker gene from *D. melanogaster* encoding the enzyme kynurenine hydroxylase was cloned into the *Hermes* element. This gene, called *cinnabar* (*cn*) is involved in ommochrome biosynthesis and it is the ommochromes that produce the purple colour seen in the ommatidia of adult mosquitoes. The *cn* DNA fragment subcloned into *Hermes* contained all the genomic sequences of this gene, including the intronic and promoter sequences of *cn*. This genetic marker had been shown, using transient expression assays, to complement a specific white-eyed mutation in the Bhalla *Ae. aegypti white* mutant strain, indicating that this strain was deficient in kynurenine hydroxylase (Cornel et al. 1997).

A *Hermes* element containing the entire *cn* gene was co-injected into developing Bhalla *Ae. aegypti w* embryos along with a helper plasmid which supplied active *Hermes* transposase – the enzyme that mediates the transpositional recombination of the *Hermes* element. This *Hermes* element was capable of genetically transforming *Ae. aegypti* at a transformation frequency of approximately 8% when measured as the number of matings producing transgenic offspring divided by the total number of fertile matings (Jasinskiene et al. 1998). Molecular analysis of transformed lines indicated that *Hermes* had integrated into the *Ae. aegypti* chromosomes. Furthermore, the inheritance of the *Hermes*[*cn*] element in transgenic lines was clearly Mendelian as would be expected for a gene that had integrated into the chromosomes. The other remarkable result of this experiment was that *D. melanogaster cn* gene was correctly expressed in *Ae. aegypti*. This indicates that there is sufficient conservation of promoter and RNA splicing signals between *Drosophila* and *Aedes* to allow the expression of at least some heterologous genes.

*Hermes* can transpose in at least seven insect species (see above), two of which have now been transformed. It is reasonable to expect that *Hermes* will also be able to be harnessed as a genetic transformation vector in at least these other five species as well. Experiments aimed at introducing *Hermes* into *C. capitata* are currently underway. The high *Hermes* transposition frequency observed in this species and another tephritid, *B. tryoni*, may result in relatively high transformation frequencies in these economically important species.

The ability to successfully transform *Ae. aegypti* has immediately led to experiments aimed at examining gene expression in the salivary glands of this insect. Genetic transformation enables this research to be performed in the original host species, rather than a model species, like *D. melanogaster*. The outcomes of this work will increase our knowledge of gene expression in this tissue and can be expected to lead to strategies that will prevent the transmission of viruses and parasites through the salivary glands of this, and other, mosquito species. This is just one example of the benefits of a transgenic based technology in pest insect species.

### *Mariners*

The *mariner* element has also been used as a transformation vector in *Ae. aegypti*. *Mariner* elements are members of the *mariner/Tc1* superfamily of transposable elements (Robertson 1995). Members of this superfamily are found in a tremendous number and variety of animal species and the ancestry of these elements can be traced back to the IS3 sequences of prokaryotes. The abundance and wide distribution of *mariners* have enabled the phylogenetic lineage of these elements to be constructed. The most striking feature of these phylogenies is that, in several clear cases, the phylogenies based on the analysis of *mariners* is in strong disagreement with the accepted host species phylogenies (Robertson 1995). That these discrepancies are most likely due to the ability of *mariners* to move horizontally between species is now generally accepted. A corollary of this is that some *mariners* may be harnessed as gene vectors in a number of insect species.

The *MOS* element has been used successfully as a genetic transformation vector in *D. melanogaster* albeit with a low transformation efficiency relative to the *P*, *hobo* and *Hermes* elements. Furthermore, the integrated *MOS* element was refractory to further mobilisation by *MOS* transposase which was in contrast to the behaviour of integrated *P*, *hobo* and *Hermes* elements in transgenic lines of *D. melanogaster* following the re-introduction of the relevant transposase. These results suggested that the development of *mariners* as transformation vectors may be somewhat problematic. However, recent results strongly indicate that *mariners* may be adaptable as genetic transformation vectors.

Following the same strategy that was developed for *Hermes*, mobility assays showed that the *MOS* element from *D. mauritiana* was capable of carrying several kilobases of cloned DNA and transposing in at least several non-drosophilid insect species, including *Ae. Aegypti* (Coates et al. 1998, Coates et al. 1997). These results contradicted other data that indicated that *MOS* was limited in its ability to carry extra DNA and that this inability may have contributed to previous failed attempts to utilise *mariners* as genetic transformation vectors in insects. While the reasons for the *mariners'* relative immobility in *Drosophila* remain unclear, the transformation of *Ae. aegypti* using the *MOS* element carrying the *cn* genetic marker is unequivocal (Coates et al. 1998). Transformation was achieved at a frequency of 4% and, as for *Hermes*, the transformed lines displayed Mendelian inheritance. Integration by transpositional recombination was confirmed by Southern blots and determination of the DNA target sequence flanking the *MOS*[*cn*] insertions (Coates et al. 1998). As expected with *mariner* transposition, insertion was at a TA site and a TA site duplication was created as a consequence of transposition. The *MOS* element can be used as a genetic transformation vector in *Ae. aegypti* and it is reasonable to predict that it will find use as a gene vector in a number of insect species as well.

Other evidence that *mariners* will be developed as transformation vectors comes from experiments in which transposition of another member of the *mariner* element family has been demonstrated *in vitro*. The *Himar* element is a consensus *mariner* element based on the sequence of genomic copies of *mariners* from the horn fly, *Haemoatobia irritans* L. (Lampe et al. 1996). *In vitro* transposition of *Himar* requires only the *Himar* transposase and the *Himar* transposable element (acting as the substrate) and magnesium (Lampe et al. 1996). These minimal requirements for transposition are perhaps not surprising since, as mentioned above, it is accepted that *mariners* have been capable of being transmitted horizontally between species and genera. The recent demonstration that the *MOS* element can be used as a transformation vector in the protozoan *Leishmania major* Yakimoff and Schokhor (Gueiros-Filho and Bevereley 1997) shows that *mariners* can actually transpose across kingdoms, further supporting the hypothesis that *mariners* can truly function as autonomous transposable elements with little or no requirements for host encoded factors.

### *Minos*

The *Minos* transposable element of *D. hydei* Sturtevant is another Type II transposable element. It was discovered serendipitously within the transcribed spacer of the ribosomal DNA region of the *D. hydei* genome (Franz and Savakis 1991). The distribution of *Minos* in the *D. hydei* genome suggested that *Minos* might be a mobile transposable element. Subsequently, *Minos* was harnessed as a genetic transformation vector in *D. melanogaster* although comparable transformation frequencies were in the order of 6% – some 4-5 fold lower than those obtained using the *D. melanogaster P* or *hobo* elements or the *M. domestica Hermes* element (Loukeris et al. 1995a). Nevertheless, the mobility properties of *Minos* soon led to it being tested in *C. capitata* which it was also able to transform at a frequency of approximately 1-2% (Loukeris et al. 1995b). This constituted the first report of the genetic transformation of a non-drosophilid insect species by transpositional recombination. One of the significant features of these experiments was the use of the *C. capitata white* cDNA gene as the genetic marker. The recipient strain was deficient for this gene product, thereby enabling transformants containing the functional copy of this gene to be easily recognised. The success with the *white* marker gene placed in the *Minos* transposable element further confirmed the importance of selectable genetic markers as integral parts of any insect transformation technology.

### *PiggyBac*

The *piggyBac* transposable element from the lepidopteran *Trichoplusia ni* (Hübner) is a Type II eukaryotic transposable element and is unrelated to any of the other insect Type II transposable elements so far isolated and characterised. *PiggyBac* was initially isolated on the basis of its mobility. It was detected during the routine passage of *Galleria mellonella* L. nucleopolyhedrosis baculovirus though *T. ni* tissue culture cells (Fraser et al. 1993). *PiggyBac* was demonstrated to be capable of accurate excision and transposition using *in vivo* mobility assays but, most significantly, it has been recently and repeatedly used as a genetic transformation vector in *C. capitata* (Elick et al. 1996, Fraser et al. 1995, Handler et al. 1998). Moreover there is every reason to believe that higher genetic transformation frequencies of *C. capitata* using this transposable element can be achieved.

*PiggyBac* transformation of *C. capitata* has been achieved on at least eight independent occasions in two different laboratories (Handler et al. 1998 and personal communication). The transformation frequency has been approximately 2% and most of these lines appear to be stable although analysis of all of them is still in the preliminary stages. Other than the high degree of repeatability of these experiments, two other aspects of them are noteworthy. First, the eye colours recovered were quite strong relative to the pale eye colours obtained from similar transformation experiments using *Minos* as the gene vector. For both vectors, the same genetic marker was used – the cDNA of the *C. capitata white* gene that had been placed under the control of the hsp70 promoter of *D. melanogaster*. This may indicate that the *piggyBac* element may have a predisposition to insert in transcriptionally active regions of the *C. capitata* genome – at least in comparison to the *Minos* element. Second, the *piggyBac* helper transposase that was used contained the *piggyBac* promoter. While the activity of this promoter in various tissues has not been analysed, other insect Type II transposable element promoters are typically weak and, in order to increase the transcriptional activity from them, the transposase gene is placed under the control of a strong inducible promoter like the hsp70 promoter from *D. melanogaster*. Thus, the ability to generate *piggyBac*-mediated transformants of *C. capitata* using the *piggyBac* promoter is significant and it is reasonable to predict that the construction of a *hsp70piggyBac* transposase helper should result in an increased transformation frequency. Indeed, in recent experiments, this *hsp70piggyBac transposase* has increased the *piggyBac* transformation frequency of *D. melanogaster* by five-fold relative to the transformation frequency obtained using the original unmodified transposase (A. M. Handler, personal communication). It is reasonable to expect that a similar enhancement of *C. capitata* transformation will ensue upon the use of this modified transposase in future transformation experiments.

The ability of a lepidopteran element – *piggyBac* – to genetically transform two higher dipteran species, *D. melanogaster* and *C. capitata*, bodes well for the widespread use of *piggyBac* as a genetic transformation vector in a range of insect species. *PiggyBac* has recently been shown, using transposition mobility assays, to accurately transpose in developing embryos of the pink bollworm, *Pectinphora gossypiella* Saunders and the genetic transformation of this important pest species can now be anticipated (S. Thibault, personal communication).

The four examples of transposable element mediated non-drosophilid transformation are listed in Table 1.

Table 1. The four cases of non-drosophild transformation mediated by transpositional recombination (Cc indicates that the genetic marker came from *C. capitata*; Dm indicates that the genetic marker came from *D. melanogaster*. In the case of the white gene, the structural gene came from *C. capitata,* however, it was placed under the control of the *D. melanogaster* hsp70 promoter).

| Species | Tranposon | Normal host of TE | Genetic Marker | Total Size of TE/insert | | Reference |
|---|---|---|---|---|---|---|
| *C. capitata* | *Minos* | *D. hydei* | *White* (Cc/Dm) | 5.4/3.6kb | 1-2% | Loukeris et al. 1995b |
| *C. capitata* | *piggyBac* | *T. ni* | *White* (Cc/Dm) | 6.1/3.6kb | 1-2% | Handler et al. 1998 |
| *Ae. aegypti* | *Hermes* | *M. domestica* | *Cinnabar* (Dm) | 5.9/4.7kb | 8% | Jasiniskiene et al. 1998 |
| *Ae. aegypti* | *MOS1* | *D. mauritiana* | *Cinnabar* (Dm) | 6/4.7kb | 4% | Coates et al. 1998. |

## OTHER SYSTEMS FOR THE INTRODUCTION OR MANIPULATION OF GENES IN INSECTS

Four other systems are currently available for the introduction and/or manipulation of genes in insects. However, none as yet have led to the production of transgenic insects. While they will only be considered briefly here, each will most likely play significant roles in insect molecular biology either alone or in combination with transposable elements.

### Sindbis Virus

Sindbis virus is an alphavirus and has a single stranded RNA genome. Sindbis viruses occur naturally and some varieties have been genetically engineered as expression vectors. These can be grown to high titres in mammalian cells and the virus particles produced used to infect either mosquitoes or mosquito cell culture lines (Higgs et al. 1993). Infected cells do not lyse; rather, the infection is sustained and this enables the analysis of strategic genes carried into the insect by the recombinant virus. The most dramatic demonstration of the power of Sindbis-based technology has been the elegant demonstration that *Ae. aegypti* adults, infected with a genetically engineered Sindis virus expressing an antisense form of a dengue virion assembly protein, are refractory to dengue transmission (Olson et al. 1996).

### Pantropic Retroviral Vectors

These viruses are pseudotyped with the G envelope protein from vesicular stomatitis virus (VSV-G), thereby enabling them to interact with a number of generic components of cell membranes. As a consequence, these viruses have the ability to infect a broad range of species. They are incapable of further transmission since their genomes lack the information needed for the synthesis of VSV-G. Once inside the target cells, the retroviral DNA can insert into the host genome. Pantropic retroviral vectors have been used to genetically transform zebrafish, surfclams and mosquito cell cultures (Lin et al. 1994, Lu et al. 1996, Matsubara et al. 1996). However, they have yet to be successfully used to genetically transform insects.

### Insect Endosymbionts

Molecular biological techniques, such as the polymerase chain reaction, have increased the sensitivity with which bacterial endosymbionts of insects can be detected (O'Neill et al. 1992). In particular, this approach has dramatically expanded our knowledge of the distribution of members of the genus *Wolbachia*. This, in turn, has rekindled efforts aimed at understanding and exploiting *Wolbachia* biology with one aim being to use *Wolbachia* to introduce strategic genes into economically important insect species. These paratransgenic insects would not be transgenic in the classical sense of carrying the transgene in their genomes. However, their phenotype would be caused through their carrying of the transgene in their cells infected with *Wolbachia*. One advantage of this approach is that the ability of endosymbionts to spread through populations would also help drive the transgene through the population.

A recent demonstration of the power of this approach has been the genetic engineering of *Rhodococcus rhodnii* Goodfellow and Anderson, which is an endosymbiotic bacterium inhabiting the gut of the reduviid bug *Rhodnius prolixus* Stahl (Durvasula et al. 1997). *R. prolixus* is the vector of *Trypanosoma cruzi* Chagas which is the causative agent of Chagas disease. A genetically engineered plasmid containing the cecropin A gene was introduced into *R. rhodnii* by genetic transformation and transformed bacteria were then introduced into the gut of *R. prolixus*. Cecropin is a naturally occurring peptide which comprises part of the humoral immunity response insects. Production of cecropin in the gut of *R. prolixus* is toxic to *T. cruzi* without affecting the viability of the insect vector in any detectable manner. The feeding behaviour of *R. prolixus* in which, it feeds on fecal droplets, provides a convenient and effective way that the engineered endosymbiotic bacteria can be spread through the vector population, thereby decreasing the transmission rate of Chagas disease.

### Other Recombination Systems

The FLP-FRT recombinase system from yeast has been shown to function correctly in both *D. melanogaster* and *Ae.aegypti* (Golic and Lindquist 1989, Morris et al. 1991). This system enables the targeting of insertions to FRT sites and also enables the deletion of intervening DNA sequences located between FRT sites. In a similar fashion, the cre-lox recombinase system from bacteriophage P1 has been shown to function in mosquito embryos (Carlson et al. 1995). The ability to introduce genes into insects using transposable elements now opens up the possibilities of using systems like the FLP-FRT and cre-lox recombinase systems to modify transformed insect genomes in very precise ways. Transposable element vectors can be inactivated following integration to ensure stability of the introduced element. Strains containing FLP or cre recombinase can be created by introducing this gene into the genome through transposable element mediated recombination. Subsequent manipulations can be achieved by crossings with strains containing transposable element introduced FRT or lox sites.

## CHALLENGES AND FUTURE DIRECTIONS

Non-drosophilid transformation is now a reality and the number of insect species that will be able to be genetically transformed will increase over the next several years. With the successful development of this technology, now comes the challenge of both improving and harnessing it to achieve practical outcomes for human health and agriculture. Generally speaking, development will occur in two areas. Strains will be developed for use only in the laboratory and these strains will, in turn, enable the production of genetic strains for applied use. The second area of development will be the use of genetic strains in the field. These two areas of development are not mutually exclusive and the need to develop genetic strains for applied use will continue to be the driving force for the generation of sophisticated laboratory based genetic strains.

### Transformation Technology – in the Laboratory

Sophisticated genetic technologies based on transposable element-mediated genetic transformation have been in use in *D. melanogaster* biology for many years. Significantly, these are not necessarily dependent on the availability of pre-existing genetic mutations (or the corresponding genes) but can be developed for any insect species, provided a convenient method for tracking the transposable element in successive generations exists. In *Drosophila,* this is normally provided by using an eye pigmentation gene as the genetic marker and this strategy will be applicable in other insects, such as *Ae. aegypti*, *Anopheles gambiae* and *C. capitata* in which analogous eye pigmentation mutations already exist. Indeed, the ability of the *D. melanogaster cn* gene to rescue the corresponding mutation in *Ae. aegypti* and the ability of the *C. capitata white* gene to do the same in *D. melanogaster* indicate that many potential marker genes can be expressed in heterologous systems. Failing this, new genetic markers that are dominant and can be selected in living organisms must be sought. Perhaps the most promising candidate is the green fluorescent protein(GFP) from the jellyfish, *Aequorea victoria* Murbach et Shearer (Chalfie et al. 1994).

Gene tagging and enhancer trapping are two sophisticated technologies that we can anticipate being applied to pest insects over the next five years. Transposon tagging provides a way by which target genes can be identified by virtue of the mutant phenotype arising from their inactivation through the insertion of a transposable element. All that is required is that the likely mutant phenotype be predictable so that flies displaying this phenotype can be recognised. The target gene can then be simply cloned by cloning the transposable element that has inserted into it.

Enhancer trapping is similar to gene tagging in that it requires the integration of the transposable element in, or near to, the target genes. In this case, the transposable element carries an indicator gene, such as the β-galactosidase, β-glucuronidase or GFP gene placed in frame with the 5' end of the element. Insertion of this element near the promoter of a gene expressed in a particular tissue (or at a particular time) will lead to the expression of the indicator gene in that tissue (or at that developmental time). Thus, genes can be cloned on the basis of their tissue specific or temporal expression. Genes expressed in response to environmental stress (such as that induced by chemicals) or to viral or parasite infection can also be cloned using this strategy. The ability for two transposable elements – *piggyBac* and *Minos* – to transform *C. capitata* and the ability of *Hermes* and *mariner* to transform *Ae.aegypti*, allows us to contemplate the construction of genetically engineered strains that can be used for gene tagging and enhancer trapping screens in these important pest species.

The outcome of these approaches will be the identification, isolation and characterisation of genes involved in key physiological, behavioural, developmental and pathogenic pathways in the pest insect. These genes could then be directly used in field applications, such as the sterile insect technique. Alternatively, the basic knowledge gained from the analysis of these genes, and their gene products, should enable the development and formulation of new pest management strategies or chemicals which are species specific and environmentally compatible. In particular, the comparison of homologous strategic genes between pest and non-pest insect species would be expected to provide valuable information concerning the basic molecular basis of the pestiferous behaviour of these species.

## Transformation Technology – in the Field

The ability to introduce genes into pest insects already allows us to embark on projects that will generate genetically tagged insects for use in SIT. For *C. capitata,* the use of GFP as a genetic tag for mass reared and mass released strains might provide a means by which these strains are easily recognisable in the field and will also provide an efficient approach to quality control in the mass rearing facility. Genetically tagging the sperm of the SIT strain (for example, with GFP) would enable the mating competitiveness of the SIT strain to be evaluated in the field. It should be possible to develop new genetic sexing strains in quicker times than required for conventional genetic manipulations. Indeed, the relative speed with which genetic engineering of strains can be accomplished will decrease the amount of time that populations need to be reared in the laboratory during strain development. The amount of wild type genetic material that can be maintained in these genetically engineered strains will also be increased since the procedures used in genetic engineering are precise at the nucleotide level and only target very small regions of the genome. Consequently, strains produced or augmented by genetic engineering might be predicted to be efficient competitors for field females following release.

The strategies outlined above will enable the isolation and manipulation of genes involved in sex determination and make possible the generation and maintenance of male-only strains suitable for use in SIT. The *tsl* gene could also be cloned and the molecular basis of its phenotype resolved. Knowledge of the basic molecular mechanism responsible for *tsl* might enable the generation of new *tsl* alleles that are even more effective in SIT. The generation of cold sensitive "*tsl*-type" strains could also be envisaged.

## CONCLUSIONS

Genetic manipulation of two important insect pest species, *C. capitata* and *Ae. aegypti*, is now a reality. These experiments not only now enable sophisticated genetic technologies to be exploited in these two species but also serve as proof of principle for the use of transposable elements as gene vectors in insect pest species. More examples of insect pest transformation can be expected in the near future. Based on transposable element mobility assays, these would be expected to include bactrocerids, calliphorids, anophelines, noctuids and muscids. The outcomes of these experiments will provide new knowledge and new approaches that can be incorporated into ongoing efforts to minimise the devastating agricultural and medical effects of insect pests.

## REFERENCES

Atkinson, P.W., W.D. Warren and D.A. O'Brochta. 1993. The *hobo* transposable element of *Drosophila* can be cross-mobilised in houseflies and excises like the *Ac* element of maize. Proc. Natl. Acad. Sci. USA 83:9693-9697.

Blackman, R.K., M. Macy, D. Koehler, R. Grimaila and W.M. Gelbart. 1989. Identification of a fully functional *hobo* transposable element and its use for germ line transformation of *Drosophila*. EMBO *J.* 8:211-217.

Carlson, J., K. Olson, S. Higgs and B. Beaty. 1995. Molecular genetic manipulation of mosquito vectors. Annu. Rev. Entomol. 40:359-388.

Chalfie, M., Y. Tu, G. Euskirchin, W.W. Ward and D. Prasher. 1994. Green Fluorescent Protein as a marker for gene expression. Science 263:802-806.

Coates, C.J., N. Jasinskiene, L. Miyashiro and A.A. James. 1998. *Mariner* transposition and transformation of the yellow fever mosquito, Aedes aegypti. Proc. Natl. Acad. Sci. USA 95:3748-3751.

Coates, C.J., C.L. Turney, M. Frommer, D.A. O'Brochta and P.W. Atkinson. 1997. Interplasmid transposition of the *mariner* transposable element in non-drosophilid insects. Mol. Gen. Genet. 253:728-733.

Cornel, A.J., M.Q. Benedict, C. Salazar-Rafferty, A.J. Howells and F.H. Collins. 1997. Transient expression of the *Drosophila melanogaster cinnabar* gene rescues eye colour in the white eye (WE) strain of *Aedes aegypti.* Insect Biochem. Mol. Biol. 27:993-997.

Durvasula, R.V., A. Gumbs, A. Panackal, O. Kruglov, S. Aksoy, R.B. Merrifield, F.F. Richards and C.B. Beard. 1997. Prevention of insect-borne disease: An approach using transgenic symbiotic bacteria. Proc. Natl. Acad. Sci. USA 94:3274-3278.

Elick, T.A., C.A. Bauser, and M.J. Fraser. 1996. Excision of the *piggyBac* transposable element *in vitro* is a precise event that is enhanced by the expression of its encoded transposase. Genetica 98:33-41.

Franz, G.and C. Savakis. 1991. *Minos*, a new transposable element from the *Drosophila hydei*, is a member of the *Tc1*-like family of transposons. Nucleic Acids Res. 19:6646.

Fraser, M.J., G.E. Smith and M.D. Summers. 1993. The acquisition of host cell DNA sequences by baculoviruses: relation between host DNA insertions and FP mutants of *Autographa californica* and *Galleria mellonella*. Virology 47: 287-300.

Fraser, M.J., L. Cary, K. Boonvisudhi and H.G.H. Wang. 1995. Assay for movement of lepidopteran transposon IFP2 in insect cells using a baculovirus genome as target DNA. Virology 211:397-407.

Golic, K.G. and S. Lindquist. 1989. The FLP recombinase of yeast catalyses site-specific recombination in the *Drosophila* genome. Cell 59:499-509.

Gueiros-Filho, F.J. and S.B. Beverley. 1997. Trans-kingdom transposition of the *Drosophila* element *mariner* within the protozoan Leishmania. Science 276:1716-1719.

Handler, A.M., S.D. McCombs, M.J. Fraser and S.H. Saul. 1998. The lepidopteran transposon vector, *piggyBac,* mediates germ line transformation in the Mediterranean fruit fly. Proc. Natl. Acad. Sci. USA 95: 7520-7525.

Higgs, S., A.M. Powers and K.E. Olson. 1993. Alphavirus expression systems: Applications to mosquito vector studies. *Parasitol. Today* 9:444-452.

Jasinskiene, N., C.J. Coates, M.Q. Benedict, A.J. Cornel, C. Salazar-Rafferty, A.A. James and F.H. Collins. 1998. Stable, transposon-mediated transformation of the yellow fever mosquito, *Aedes aegypti*, using the *Hermes* element from the house fly. Proc. Natl. Acad. Sci. USA 95:3743-3747.

Lampe, D.J., M.E.A. Churchill and H.R. Robertson. 1996. A purified *mariner* transposase is sufficient to mediate transposition *in vitro.* EMBO J. 15:5470-5479.

Lidholm, D., A. Lohe and D. Hartl. 1993. The transposable element *mariner* mediates germ line transformation in *Drosophila melanogaster*. Genetics 134:858-868.

Lin, S., N. Gaiano, P. Culp, J.C. Burns, T. Friedmann, J. Yee and N. Hopkins. 1994. Integration and germ line transmission of a pseudotyped retroviral vector in zebrafish. Science 265:666-669.

Loukeris, T.G., B. Arca, I. Livadaras, G. Dialektaki and C. Savakis. 1995a. Introduction of the transposable element *Minos* into the germ line of *Drosophila melanogaster.* Proc. Natl. Acad. Sci. USA 92:9485-9489.

Loukeris, T.G., I. Livadras, B. Arca, S. Zabalou and C. Savakis. 1995b. Gene transfer into the Medfly, *Ceratitis capitata*, using a *Drosophila hydei* transposable element. Science 270:2002-2005.

Lu, J.K., T.T. Chen, S.K. Allen, T. Matsubara and J.C. Burns. 1996. Production of transgenic dwarf surfclams, *Mulinia lateralis*, with pantropic retroviral vectors. Proc. Natl. Acad. Sci. USA 93:3482-3486.

Matsubara, T., R.W. Beeman, H. Shike, N.J. Besansky, O. Mukabayire, S. Higgs, A.A. James and J.C. Burns. 1996. Pantropic retroviral vectors integrate and express in cells of the malaria mosquito, *Anopheles gambiae*. Proc. Natl. Acad. Sci. USA 93:6181-6185.

Morris, A.C., T.L. Schaub and A.A. James. 1991. FLP-mediated recombination in the vector mosquito, *Aedes aegypti*. *Nucleic Acids Res.* 19:5895-5900.

O'Brochta, D.A., W.D. Warren, K.J. Saville and P.W. Atkinson. 1996. *Hermes*, a functional non-drosophilid gene vector from *Musca domestica*. Genetics 142:907-914.

Olson, K.E., S. Higgs, P.J. Gaines, A.M. Powers, B.S. Davis, K.I. Kamrud, J.O. Carlson, C.D. Blair and B.J. Beaty. 1996. Genetically engineered resistance to Dengue-2 virus transmission in mosquitoes. Science 272:884-886.

O'Neill, S.L., R. Giordano, A.M. Colbert, T.L. Karr and H.M. Robertson. 1992. 16S rRNA phylogentic analysis of the bacterial endosymbionts associated with cytoplasmic incompatibility in insects. Proc. Natl. Acad. Sci. USA 89:2699-2702.

Pinkerton, A.C., D.A. O'Brochta and P.W. Atkinson. 1996. Mobility of *hAT* transposable elements in the Old World American bollworm, *Helicoverpa armigera*. Insect Mol. Biol. 5:223-227.

Robertson, H.M. 1995. The *Tc1-mariner* superfamily of transposons in animals. J. Insect Physiol. 41:99-105.

Rubin, G.M. and A.C. Spadling. 1982. Genetic transformation of *Drosophila* with transposable element vectors. Science 218:348-353.

Sarkar, A., K. Yardley, P.W. Atkinson, A.A. James and D.A. O'Brochta. 1997a. Transposition of the *Hermes* element in embryos of the vector mosquito, *Aedes aegypti*. Insect Biochem. Mol. Biol. 27:359-363.

Sarkar, A., C.J. Coates, S. Whyard, U. Willhoeft, P.W. Atkinson and D.A. O'Brochta. 1997b. The *Hermes* element from *Musca domestica* can transpose in four families of cylorrhaphan flies. Genetica 99:15-29.

# *Drosophila* as a Model for the Study of Sex Determination in *Anopheline* and *Aedine* Mosquitoes

*A. Pannuti, T. Kocacitak and J.C. Lucchesi*

Department of Biology, Emory University, Atlanta, Georgia, USA

## INTRODUCTION

Sterile insect technique control strategies consist of releasing laboratory produced male insects that have been sterilised by irradiation. These strategies require the production of massive quantities of males. Population-replacement strategies rely on the genetically engineered interruption of that portion of the malaria parasite's life cycle that occurs in the mosquito. This could be achieved by the inundative introduction of transformed males or the more limited introduction of males carrying an infective agent capable of driving a parasite-inhibiting transgene into the vector population. Once again, the release of genetically engineered males would require genetic systems for their mass production.

Mass production of males can be accomplished most effectively through genetic sexing techniques. Genetic sexing can be achieved by identifying the key steps in the genetic regulation of sex differentiation and by modifying one or more of these steps so that their execution would result in sex-specific lethality. As the necessary and seminal first step towards this goal, we set out to identify and isolate a gene whose primary transcript is processed differently in males and females of *Anopheles gambiae* Giles.

A survey of sex determination among insects reveals a vast array of different mechanisms. Our understanding of these mechanisms consists only of information derived from classical cytological and genetic studies. Using the knowledge derived from the study of *Drosophila*, it has been possible to discern a fundamental pattern in the sex determining mechanisms of many diverse insect species (Nöthiger and Steinmann-Zwicky 1985). The challenge now, is to determine if there has been an evolutionary conservation of the genes responsible for the fundamental pattern, i.e., if the molecular mechanisms that underlie sex determination in *Drosophila* are the same in other insects of interest or if in these insects, the apparent fundamental pattern is achieved by completely different genetic functions and molecular mechanisms.

Our experimental approach is based on the extensive molecular genetic information available on the regulation of sex differentiation in *Drosophila* (Cline and Meyer (1996) and references therein). In this organism, the primary sex determining signal is the number of X chromosomes in relation to the number of sets of autosomes in the genome, the so-called X:A ratio. The products of several X-linked "numerator" genes and of at least one autosomal "denominator" gene, interact with maternal gene products to determine the function of a master regulatory gene (*Sex lethal* (*Sxl*)), that is responsible for the different morphology as well as behaviour of the two sexes. Although *Sxl* is transcribed in both chromosomal male and female embryos, differential splicing results in a functional product only in females where it initiates the regulatory pathway of somatic female differentiation.

In the initial phase of our work on the mosquito genetic sex determination pathway, we asked whether a homologue of the *D. melanogaster* Meigen "master regulatory" *Sxl* gene exists in mosquitoes and whether it is regulated in a sex-specific manner. Using the *D. melanogaster Sxl* cDNA sequence as a guide, we amplified a fragment of *An. gambiae* genomic DNA with an open reading frame that encodes a conceptual polypeptide 75% similar to the *Drosophila* gene product. We used this fragment to isolate a 6 kb genomic fragment. Following sequencing, several probes were constructed and used for hybridisation to Northern blots of poly(A)$^+$ RNA from *An. gambiae* males or females. These experiments failed to show any evidence of sex-specific processing. This is consistent with similar observations derived from other Diptera: the blowfly *Chrysomya rufifacies* Macquart (Müller-Holtcamp 1995), the phorid fly *Megaselia scataris* (Sievert et al. 1997), the common house fly *Musca domestica* L. (Meisen et al. 1998) and the Medfly *Ceratitis capitata* Wied. (Saccone et al. 1998).

In light of the above results, we have turned our attention to the *Drosophila* "switch gene" *double-sex* (*dsx*). In *Drosophila, Sxl* initiates differentiation in the female mode by regulating the splicing of the *transformer* (*tra*) gene in a sex-specific manner. In turn, this gene product (together with the product of another gene, *transformer2* (*tra2*), present in both sexes) directs the splicing of the *dsx* primary transcript to yield a regulatory protein that acts to repress the male-specific realiser genes, thus achieving female sexual differentiation (by "realiser genes", we mean those genes and their products that are responsible for the actual differentiation of cells into the tissues and organs that are involved in sexual reproduction). In male embryos, an alternate mode of splicing of the *dsx* transcripts occurs by default and leads to a product that represses the female-specific realiser genes and results in male sexual differentiation. As described below, we have obtained evidence indicating that a homologue of this gene exists in *An. gambiae* and that its primary transcript is spliced in a sex-specific manner.

## RESULTS

In *D. melanogaster,* alternative splicing of the *dsx* transcript results in the production of two proteins, *dm*DSXM in males and *dm*DSXF in females, which share a common amino-terminal domain but have unique sex-specific carboxy-terminal domains (Figure 1). By comparing the *D. melanogaster* and *D. virilis* Sturtevant DSX protein sequences, we identified a region of high similarity and used it to design partially degenerate primers with which we amplified a fragment of 97 bp from a *An. gambiae* cDNA (Besansky et al. 1995) library (derived from male and female RNA primed with oligo dT and random primers). This fragment had a high degree of identity with the 3' end of exon 3 (common) and the 5' end of exon 4 (female-specific) of the *dm*DSXF protein. Screening of the cDNA library with this fragment resulted in the isolation of five independent clones which can be separated into two classes (Figure 2). The longest cDNA from class A contains an open reading frame (ORF) encoding 236 aa. A blast analysis reveals that the N-terminus of the predicted protein is most similar to the common DSX region of *D. melanogaster* while the C-terminus is most similar to the female-specific *dm*DSXF. The presence of a long 5'UTR and of a polyA tail suggests that this clone includes the full-length ORF of the *An. gambiae* female-specific double-sex protein (*ag*DSXF).

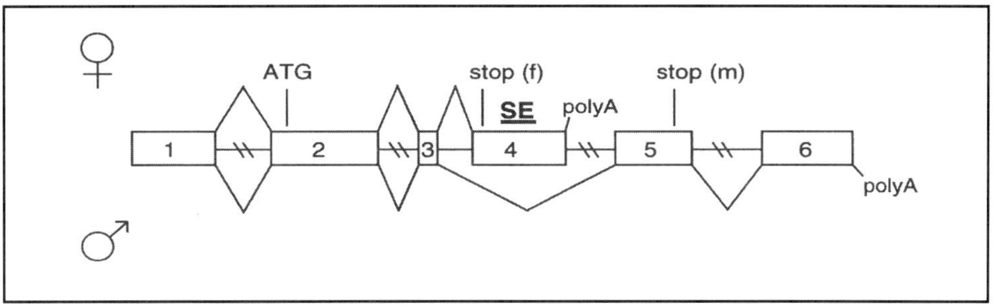


Figure 1. The sex-specific pattern of *dsx* pre-mRNA alternative splicing in *D. melanogaster*.
Boxes 1, 2 and 3 represent common exons, exon 4 is the female-specific exon, exon 5 and 6 are male-specific exons. Lines between the boxes represent introns, and the lines above and below the pre-mRNA illustrate the female- and male-specific patterns of alternative splicing. Sites of cleavage and polyadenylation are labelled polyA. In females, binding of factors like TRA and TRA-2 to the dsx splicing enhancer region (SE) activates the weak female-specific 3' splice site. The start (ATG) and end (stop) of the female and male ORFs are indicated.

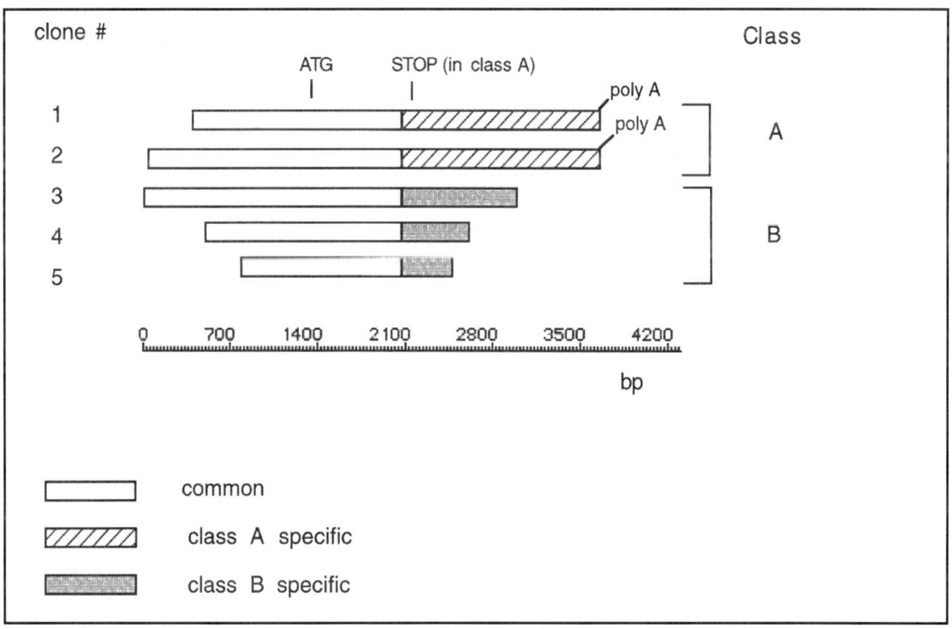


Figure 2. Schematic representation of five *An. gambiae* cDNA clones similar to *D. melanogaster* dsx transcripts. The clones can be separated into two classes (A and B). Common and specific regions are indicated.

### *An. gambiae* dsx cDNA clones

The three clones in Class B, have a 5' terminal sequence which is identical to the DNA sequence of the 5' end of the class A cDNA. This identity stops at a point beyond which the ORF of the clone is very different from the *an*DSXF ORF. The point of divergence corresponds to the site where, in *D. melanogaster*, the two sex-specific transcripts diverge due to sex-specific splicing (i.e., the 3' end of exon 3, in Figure 1). This suggests that these clones are partial clones of the *An. gambiae* male-specific double-sex protein (*an*DSXM).

To estimate the length of the transcript(s) corresponding to the class B cDNAs, we performed Northern blot analysis of male or female poly(A)$^+$ RNA using as a probe a 600 bp fragment from the unique 3' end of class B cDNA. The results, presented in Figure 3, show the presence of a single male-specific transcript of approximately 4 kb.

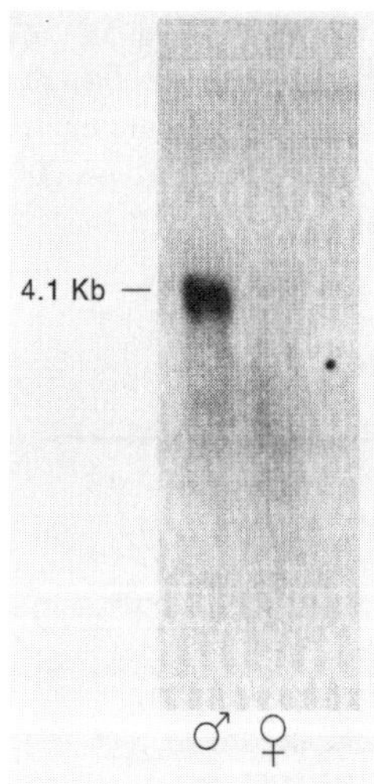


Figure 3. Northern blot of male and female *An. gambiae* poly(A)$^+$ RNA using as a probe a 600 bp fragment from the unique 3' end of class B cDNA.

To determine whether the two classes of cDNA discussed above are derived from sex specific transcripts, we performed RT-PCR with male or female poly(A)$^+$ RNA using primers from the common region in combination with class A or B specific primers. The results, presented in Figure 4, show the expected sex-specific amplified fragments.

## EXPERIMENTAL GOALS

We plan to finish the characterisation of the *An. gambiae* homologue of the *Drosophila dsx* gene and to clone and characterise the *Ae. aegypti* L. homologue as well. An important reason for proceeding with *Ae. aegypti* is the valuable information that can be gained by comparing the sequence and structural features of the same gene in two different and distantly related species of the same family. An equally important reason is based on the practical consideration that an effective technique of germ line transformation has recently been developed in this species in the Frank Collins (University of Notre Dame) and Anthony James (University of California at Irvine) laboratories. Since it is reasonable to expect that the sex-specific regulation of *dsx* splicing depends on the function of specific regulatory factors, as it does in *Drosophila,* a finite probability exists that these factors may not be able to regulate the differential splicing of the *dsx* transcript from a different species. Therefore, we believe that we should perform our experiments within the species that can be transformed. Nevertheless, given the great importance of *An. gambiae* as the principal malaria vector, and the likelihood that germ line transformation will be available in this species in the near future, it seems obviously worthwhile to us to carry out certain experiments, in parallel, in this species.

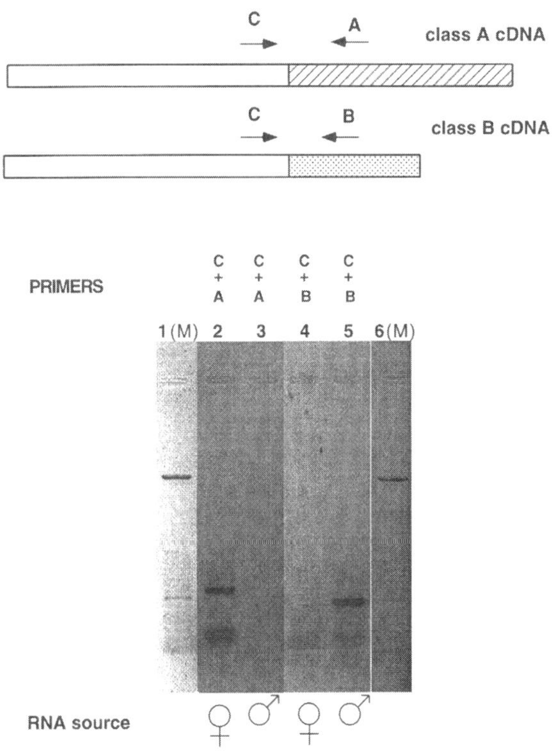


Figure 4. RT-PCR of poly(A)$^+$ RNA from female or male *An. gambiae* adults. First strand cDNA was synthesised by AMV-RT priming the reaction with random hexamers, and then PCR amplified using primers from the common region (C) of *(ag)dsx* clones in combination with class A or B specific primers.

The main goal of our research is to engineer recombinant genes that would lead to a sex-specific phenotype resulting from alternative splicing. As an important preliminary step, we will synthesise transgenes, modified only to the extent necessary to distinguish their products from those of the endogenous gene, introduce them by transfection into *Aedes* cultured cells or by transformation into *Ae. aegypti*, and monitor their products in Northern blots or by RNA protection. We will then engineer *(an)dsx* or *(ae)dsx* constructs containing reporter or cell-lethal genes that can be specifically spliced out in one sex or the other. Transcription of the constructs containing such genes should be under the control of an inducible promoter and the gene's ORF should be interrupted by an intron whose sex-specific removal would yield the active gene product in a sex-specific manner. Therefore, it is particularly important that the promoter used be as strictly conditional as possible, i.e., that it has as little uninduced activity as possible and that its induction yields a very high level of transcription.

## CLOSING REMARKS

Adam Wilkins (1995) has put forth a hypothesis to account for the evolution of the genetic regulation responsible for sex differentiation in diploid, multicelluar organisms. He has proposed that the final step in a complex sex-differentiation pathway is the oldest and that, as one moves back along the pathway, the genes (steps) found in one type of organism will be replaced by other genes, as one surveys more distantly related species. To date, a number of experimental observations provide support for this hypothesis. In *Drosophila,* activation of the *Sxl* gene represents the first step in sex differentiation; although this gene is conserved among a wide variety of Dipteran *genera (Drosophila, Ceratitis, Anopheles, Chrysomia, Megaselia* and *Musca),* it has not been recruited for sex differentiation. Presumably, in these groups, the role played by *Sxl* in *Drosophila* is played by other, as yet unidentified or uncharacterised genes. In contrast, *dsx*, the last gene in the regulatory pathway, appears to have retained its sex-specific regulation in *Ceratitis* and *Anopheles.* These considerations leave unanswered the question of the evolutionary conservation of the genes that are upstream of *dsx.*

The research that we are carrying out is not targeted to a particular type of control of mosquito vectors; we believe that it will be equally valuable to the different strategies that ultimately require the release of massive quantities of males, foremost among which is the sterile insect technique.

Finally, the eventual practical application of our results has guided our experimental planning. We have chosen promoters whose induction does not require the massive administration of complex and expensive chemical agents. We emphasise the potential of killing females during their early development, thereby avoiding the cost of rearing them to adulthood and then eliminating them, when only males are needed. We also have chosen cytotoxic genes that would not pose an environmental or health problem if organisms carrying the cytotoxic constructs were to accidentally escape or if other organisms were to ingest escapers on purpose or accidentally.

## REFERENCES

Besansky, N.J., J.A. Bedell, M.Q. Benedict, O. Mukabayire, D. Hilfiker and F.H. Collins. 1995. Cloning and characterisation of the white gene from *Anopheles gambiae*. Insect Mol. Biol. 4: 217-231.

Cline, T.W. and B.J. Meyer. 1996. Vive la difference: males vs. females in flies vs. worms. Annu. Rev. Genet. 30:637-702.

Meisen, M., D. Bopp, D. Hilfiker-Kleiner, A. Dubendorfer, C. Brunner and R. Nothiger. 1998. *Sex-lethal*, the master sex-determining gene in *Drosophila,* is not sex-specifically regulated in *Musca domestica*. Development 125: 1487-1494.

Müller-Holtkamp, F. 1995. The *Sex-lethal* gene homologue in *Chrysomya rufifacies* is highly conserved in sequence and exon-intron organisation. J. Mol. Evol. 41:467-477.

Nöthiger, R. and M. Steinmann-Zwicky. 1985. A single principle for sex determination in insects. Cold Spring Harb. Symp. Quant. Biol. 50:615-621.

Saccone, G., I. Peluso, D. Artiaco, D. Bopp and L.C. Polito. 1998. The *Ceratitis capitata* homologue of the *Drosophila* sex-determining gene *Sex-lethal*: a case of high sequence conservation but different regulation and possible divergent function. Development 125: 1495-1500.

Sievert, V., S. Kuhn and W. Traut. 1997. Expression of the sex determining cascade genes *Sex-lethal* and *double-sex* in the phorid fly *Megaselia scataris*. Genome 40: 211-214.

Wilkins, A. 1995. Moving up the hierarchy: a hypothesis on the evolution of a genetic sex determination pathway. BioEssays 17:71-77.

# Molecular Approaches to the Modification of Insect Pest Populations

*J.M. Crampton*[1,2], *S. Stowell*[1], *R. Parkes*[1], *M. Karras*[3] *and R.E. Sinden*[3]

[1] Liverpool School of Tropical Medicine, Liverpool L3 5QA, United Kingdom
[2] The University of Liverpool, Liverpool, L69 7ZB, United Kingdom
[3] Imperial College of Science, Technology and Medicine, London SW7 2BB, UK

## INTRODUCTION

After considerable research effort over the last decade or more, the ability to routinely introduce specific genes and other DNA constructs (such as linked promoter:gene cassettes) into a range of pest insect genomes at high efficiency using transgenic approaches is fast becoming a reality. The critical issue that now needs to be addressed is how best to incorporate these techniques into Sterile Insect Technique (SIT) in order to improve its effectiveness or efficiency. Manipulation of insect pest genomes using transgenic approaches may be used in two ways. It may be used as an analytical tool, or to introduce or modify either endogenous or heterologous genes and their expression in the pest insect of choice. In this way, new strains may be generated with a set of desired characteristics beneficial to SIT. In order to realise the full potential of the technology, a number of issues and research areas is being explored and progress to date is reviewed below. Specific examples are drawn from work on mosquito systems in order to illustrate the approaches available to identify genes and promoters of interest and the potential applications to SIT.

## WHAT TARGET GENES ARE AVAILABLE OR REQUIRED FOR INTRODUCTION OR MANIPULATION?

A range of different gene systems is being investigated or of interest for potential incorporation into SIT. These systems include those involved in: programmed cell death; sex determination; insect/pathogen interactions (including refractory/susceptibility mechanisms); insect immunity; blocking pathogen transmission; use as simple tags to distinguish released from wild type insects to monitor SIT release programmes; controlling insect behaviour (including blood feeding, or stinging of fruit); enhancing the competitiveness of released insects over endogenous populations or the ability of released males to attract females. A number of these is briefly discussed below.

### Insect/Pathogen Interactions

If the interactions that occur between insect and parasite can be defined at a molecular level, it may then be possible to modify or utilise either the gene itself or its expression to produce an insect which is unable to support parasite development. For the malaria vector *Anopheles gambiae* Giles, the recent development of an integrated genetic and cytogenetic map should greatly facilitate the cloning of such genes and the characterisation of their response at a molecular level. Using this approach, three linkage groups have been identified, corresponding to the three chromosomes with a total of 131



**Area-Wide Control of Fruit Flies and Other Insect Pests,**
ed. K.H. Tan. Penerbit Universiti Sains Malaysia, Penang, 2000.

microsatellite markers (46 on the X chromosome, 57 on chromosome 2 and 28 on chromosome 3). In addition, a biochemical selectable marker was identified, *Dieldrin resistance* (*Dl*) on chromosome 2, together with five visible markers, *pink eye* (*p*), *white* (*w*) on the X, *collarless* (*c*) and *lunate* (*lu*) on the second and *red-eye* (*r*) on the third (Zheng et al. 1993, 1996).

Genetically selected refractory strains of mosquitoes exist that block parasite development, either through lysis of the early ookinete stage, during penetration of the midgut epithelium (Vernick 1995) or by melanotic encapsulation of the ookinete after penetration, as they transform into oocysts on the basal side of the epithelium (Collins et al. 1986). A laboratory strain of the *An. gambiae* mosquito has been selected which is capable of mounting a melanotic encapsulation response against several species of *Plasmodium* parasites (Zheng et al. 1998). Quantitative Trait Linkage (QTL) analysis identified three genetic loci, *Pen* 1-3, for the encapsulation of *P. cynomolgi* B oocytes. Genetic mapping revealed one major and two minor QTLs, with 76% of the total response attributed to the combined action of these three loci (Zheng et al. 1997). Further work has also demonstrated that the enzymes involved in melanisation include prophenoloxidases (proPO) (Lee et al. 1998) and serine proteases that activate proPOs. Three serine protease genes have been cloned that are very similar to the *Drosophila* gene *easter* and appear to be upregulated in response to immune challenges. These genes map cytogenetically to division 14 which has been implicated as a minor locus (*Pen* 3) involved in *P. cynomolgi* B melanisation (Paskewitz 1998).

In the mosquito *Ae. aegypti* L., susceptibility to the avian malaria parasite *P. gallinaceum* is largely determined by a dominant allele of a locus on chromosome 2 (Thathy et al. 1994). This single locus, *pgs* [2, IF98], was found to be responsible for more than 50% of the refractory phenotype observed, with an additional locus, *pgs* [3, MalI] contributing about 10% (Severson et al. 1995). Refractoriness of *Ae. aegypti* to the nematode *Brugia malayi* has been shown to be associated with a major and a minor locus on chromosomes 1 and 2 respectively (Macdonald and Ramachandran 1965, Severson et al. 1994). This situation differs from *An. gambiae*, however, in that the QTLs identified in *Ae. aegypti* control the intensity of parasite infection.

In each of these cases, molecular mapping techniques coupled with genetic and physiological research of refractory mechanisms will eventually lead to the molecular definition of a range of genes which controls aspects of the ability of insects to transmit pathogens to vertebrate hosts. Once these genes are well defined, the genes themselves or their expression may be manipulated in strains employed in SIT to reduce the problems posed by pest insects.

## Insect Immunity

Insects are known to mount strong cellular and humoral innate immune reactions in response to infection by bacteria, fungi and parasites. A number of antimicrobial peptides has now been identified that are involved in such responses together with the transcriptional activation mechanisms controlling their production. Several immune markers have now been identified in *An.gambiae* including *Gambif*-1 (Barillas-Mury et al. 1996) which shows homology to the *Drosophila* immune-response transcription factor *Dorsal* and *Defensin* (Richman et al. 1996), an antibacterial peptide. Most recent has been the identification of a number of immune markers induced specifically within the mosquito midgut (Dimopoulos et al. 1997). Two such markers have been cloned and characterised and their transcripts shown to be induced upon bacterial challenge. One, AgGNBP, isolated by PCR using degenerate primers, encodes a homologue of GNBP of the moth *Bombyx mori* L. (Lee et al. 1996). This peptide also shows strong similarities to glucanase, including a putative polysaccharide-binding domain. The other, ISPL5 (Immune-related Serine Protease-like Sequence 5) shows features of serine proteases involved in haemolymph clotting, innate immunity and development, but lacks two residues of the catalytic triad that are necessary for enzymatic activity. Experiments have demonstrated that these two markers are induced by bacteria and bacterial cell wall components, as are three other sequences, *def*, encoding defensin, the infection-responsive putative serine protease ISP13 (Dimopoulos et al. 1996) and the putative infection-responsive galactose lectin IGALE20. Using these markers, it was demonstrated that significant reactions are elicited by the malaria parasite, both locally in the midgut and systemically in the rest of the body. Defensin and AgGNBP participate strongly in both responses; IGALE20 is predominantly induced locally and ISP12 at a distance. ISP13, which is constitutively expressed at high levels in the midgut, shows no more than marginal induction. The significance of the observed immune reaction to the parasite has still to be determined, together with an understanding of the physiological roles of these markers.

Expression of two midgut-specific defensin sequences, *Smd*1 and *Smd*2, exclusively localised in the anterior midgut, has also been observed in the fly *Stomoxys calcitrans* L. (Lehane et al. 1997). However, Northern analysis has demonstrated that both of these molecules are tissue specific, being produced by the insects anterior midgut tissue and not the fat body or haemocytes, unlike other insect defensins.

Currently, there is a large range of antibacterial and immune response proteins which have been defined in a broad spectrum of insect pest species. There are a number of ways in which manipulation of insect immunity genes or their expression could be utilised to benefit SIT and related approaches. An enhancement of the insect's ability to mount an immune response to infection would presumably enhance the life expectancy of the sterile insect in the environment. This could clearly be of benefit in that released flies would persist longer, fewer would need to be released, so that unit costs would be reduced. Enhancement of specific elements of the insect immunity system (either endogenous or heterologous) which are known to be involved in pathogen susceptibility would reduce the transmission potential of manipulated insects. Alternatively, impairing the ability of an insect to mount an immune response to infection through molecular approaches would reduce the overall fitness of the manipulated insect in a specific situation. Thus, if such a manipulation were undertaken on a population-wide scale on a specific pest insect, it may have the overall effect of suppressing the population size.

## Blocking Pathogen Transmission

One strategy in disease control is to identify those genes involved in blocking transmission of the pathogen. These genes can then be incorporated into the germline of the transgenic insect, the products of which would effectively block transmission either directly in the insect or in the vertebrate host. It is well documented that antibodies raised to proteins expressed on the surface of the *Plasmodium* parasite can bind to the parasite, thereby inhibiting further development and hence blocking transmission. One example of such a protein is Pbs21, a sexual stage surface protein of the rodent malaria parasite *P. berghei* (Matsuoka et al. 1996). A very exciting possibility, therefore, is to introduce the genes coding for such antibodies into the mosquito genome, thus directly conferring the transmission blocking phenotype to the insect. In this case, a transgenic mosquito would be created, incorporating an antibody gene expressed in the insect midgut in response to a blood meal and which therefore blocks transmission of malaria. In its simplest form, this approach would involve introducing what is, essentially, a monoclonal antibody into the mosquito. This would not be ideal for a number of reasons and some way of introducing polyclonal transmission blocking antibody coding genes would be required. However, for the present, the introduction of a single transmission blocking antibody gene, under the control of a gut specific, blood meal inducible promoter, is being attempted as a model to assess the system. As an alternative, the gene coding for the parasite antigen could be introduced into the genome of a blood sucking insect such as a mosquito in such a way that the vaccine protein was produced in the saliva of the insect. To assess this concept, work is currently underway to place the cloned gene encoding Pbs21 under the control of a salivary gland-specific promoter and introduce this promoter/gene cassette into the germline of the mosquito *Ae. aegypti* (Stowell et al. 1998). It is envisaged that recombinant protein will be present as a component of the insect's saliva and will be delivered to the host when the female takes a blood meal, thereby eliciting the production of antibodies capable of blocking parasite transmission in the mosquito vector.

## Tags

A number of genes currently available would perhaps lend themselves to early incorporation into SIT. The particular need here is to introduce a defined gene which would act as a simple tagging system so that all released insects would be quickly, easily and cheaply identified and separated from wild flies. One possible candidate is a gene isolated from the jellyfish *Aequorea victoria* Murbach et Shearer encoding the autonomous *Green Fluorescent Protein (GFP).* This gene has already been used as a dominant selectable marker in *Drosophila melanogaster* Meigen and is also likely to be of use in non-drosophilid transgenesis, as a marker in a wide range of dipteran and lepidopteran species. GFP has already been expressed in both cultured insect cells and larvae (Laukkanen et al. 1996, Cha et al. 1997) and experiments are now underway to construct chimeric genes in which GFP is placed under the control of *actin* promoters and the resulting plasmids injected into blastoderm embryos of non-drosophilid insects. GFP expression is then monitored in $G_0$ individuals as a way of detecting putative transformants (Michel et al. 1998). If this proves successful, the technique may rapidly be applied to improve the efficiency of SIT.

## Sex Determination

Genetic sexing in insects is achieved by linking the wild type allele of a selectable marker, via a reciprocal translocation, to the Y chromosome. Large-scale separation of homozygous females and heterozygous males allows the possible release of only males in SIT programmes. The disadvantages of such a system are twofold. Firstly, occurrence of unbalanced gametes leads to semi-sterility and secondly, recombination in heterozygous males, although rare, leads to the breakdown of the system (Franz et al. 1994). It is envisaged that sexing systems developed using molecular approaches will overcome such problems. The aim is to develop a system which will enable the inducible production of only males during mass rearing. One approach would be to combine a conditional promoter with a sex-specific promoter to which a lethal gene could be expressed to kill one sex only. One genetic sexing strain has been developed for the Mediterranean fruit fly (Medfly) based on a temperature sensitive lethal mutation which allows female killing at an early embryonal stage (Hendrichs et al. 1995), thereby allowing the mass rearing and release of only males. This approach and the research undertaken to date is reviewed elsewhere in this volume.

## Enhancing Competitiveness and Modifying Pest Insect Behaviour

Mass reared insects are usually less competitive for mates and more susceptible to predation than wild type insects. Identification and characterisation of genes involved in the production and release of pheromones may allow genetic manipulation to enhance the competitiveness of released insects, or their ability to attract females. In either case, such a manipulation may greatly increase the efficiency of the SIT programme.

Recently, much progress has been made in extending our understanding of the genes involved in insect behaviour and learning. Much of this research has come through investigations in *Drosophila* (e.g., Meller and Davis 1996, Rendahl and Hall 1996). As yet, there has been no move to undertake similar studies in other insects, although there is an appreciation that related genes are likely to be found in these insects. The potential of using genes involved in influencing behaviour is, at this stage, speculative. The aim would be to manipulate specific genes so as to influence the behaviour of the insect in such a way as to reduce detrimental determinants, or enhance beneficial ones. A simple example of this concept derived from simple genetic approaches is based on the knowledge that some mosquitoes bite humans preferentially, whilst others bite animals. In instances where sibling species exhibit these divergent feeding habits and where they can mate to produce viable offspring, there is the possibility of identifying and selecting the gene influencing this behavioural pattern and introducing it into specific species or strains. Clearly, if such a gene could be identified and characterised, it may represent an attractive target for manipulation or introduction in pest insects.

In both instances discussed above, little progress has been made towards identifying the genes involved in pest insects. There are a number of approaches which could and, no doubt, will be utilised to identify such genes. These will include using related genes from *Drosophila* where these are available to identify their homologues in the pest insect of interest. Where the genes are currently not available in *Drosophila*, an efficient approach is to use the extensive genetic and molecular information available for this insect to identify mutants with the required characteristics so that the genes may then be identified. Once this is achieved, finding related genes in the pest insect may be undertaken with more confidence. However, the recent availability of an efficient germline transformation system for non-drosophilid insects now makes it possible to consider the use of transposon tagging to functionally clone the genes of interest in a particular insect pest species. The introduction and insertion of a mobile genetic element at or near a particular locus can cause that allele to mutate, producing a structural or developmental effect. In *Drosophila* in particular, transposable genetic elements (TGEs) have been used as mutagens in order to clone genes or gene clusters of interest via transposon tagging (Bingham et al. 1981, Searles et al. 1982). In essence, the TGE is introduced into the germ line of the insect by microinjection of the embryo and the progeny scored for mutants in the phenotype of interest. Subsequently, cloned TGE probes are used for *in situ* hybridisation to chromosomes of mutant and wild type individuals. This identifies a TGE "newly" integrated at or near the genetic locus of interest. DNA clones are then retrieved from a genomic library prepared from the mutant stock, using the TGE DNA as a probe. DNA sequences adjacent to the TGE in such clones represent the gene for the locus of interest. This approach is an extremely powerful application of the technology, as it allows the cloning of genes purely on the basis of their function.

## WHAT PROMOTER SYSTEMS ARE AVAILABLE OR DESIRABLE FOR CONTROLLING THE EXPRESSION OF THE INTRODUCED GENES?

Promoters of interest and/or currently under investigation include: those induced or repressed by external agents or environmental factors; sex-specific promoters; tissue (e.g., gut or salivary gland specific) or developmental stage-specific promoters. For example, both the insect midgut and the salivary glands play important roles in pathogen transmission, as well as providing targets for manipulation and pest population suppression. This is well illustrated by the *Plasmodium* parasite which differentiates from ookinete to oocyst as it passes through the mosquito midgut wall and may reside in the salivary glands for considerable periods of time as sporozoites before being delivered to a vertebrate host when the female mosquito takes a blood meal. Therefore, it is envisaged that the isolation and characterisation of female-specific salivary glands or blood meal inducible gut promoters will be of value for tissue, sex-specific expression of anti-parasite genes. Another example would include characterisation of fat body specific promoters. The fat body synthesises components involved in oogenesis and the insect immune system, and may therefore prove a useful target for manipulation. The search for promoters specific to these tissues is briefly discussed below.

### Salivary Gland

The salivary glands of the yellow fever mosquito *Ae. aegypti* express a number of genes encoding a variety of proteins. Research is currently aimed at isolating and characterising the upstream regions of these genes with a view to their possible use as putative promoters to express exogenous DNA. One such gene, Apyrase, encodes an ATP-diphosphohydrolase, a secreted enzyme that facilitates blood feeding by preventing platelet aggregation in the host. Apyrase expression is temporal, sex and tissue-specific, being limited to the cells of the distal-lateral and medial lobes of the adult female salivary glands (Smartt et al. 1995). Recently, a 1.66kb upstream region from this gene was used to express the luciferase reporter gene in transformed *Ae. aegypti* mosquitoes. Luciferase assays performed on isolated salivary glands and carcasses indicated that the majority of luciferase expression was occurring in the salivary glands and that this expression could be localised to the distal-lateral and medial lobes (C.J. Coates and A.A. James, personal communication). Similar results have also been achieved using an upstream region from another *Ae. aegypti* salivary gland gene, MalI (Coates et al. 1998).

### Midgut

In the mosquito, considerable research has gone into the isolation and characterisation of the trypsin genes, some of which are induced by blood meal and the promoters regulating their transcription have now been well characterised. In *An. gambiae*, the gut trypsin genes are arranged as a tightly clustered gene family consisting of seven related coding sequences, *Antryp1-7* (Muller et al. 1993). Each of these genes is transcribed in a sex, tissue and temporal-specific manner. Two members of this family, *Antryp1* and *2* (late trypsins) are induced in the gut upon blood feeding whereas genes *Antryp3,4,5,6* and *7* (early trypsins) are all expressed constitutively (Muller et al. 1995). Sequence analysis has recently revealed the presence of a highly conserved 12 nucleotide motif upstream to the TATA box of all trypsin genes, designated the Putative Trypsin Regulatory Element (PTRE). This is thought to be involved in the control mechanism regulating transcription of the trypsin genes. Transient transformation experiments of mosquito cells using luciferase reporter constructs fused to deleted or mutagenised forms of the PTRE region indicated that this PTRE motif consists of 2 *cis*-acting elements (Giannoni et al. 1998).

### Fat Body

The mosquito fat body is responsible for the production of massive amounts of yolk protein (YP) precursors. The powerful YP promoters directing this response could also be utilised for expression of foreign genes in a tissue-specific manner, in response to a blood meal. To elucidate the hormonal regulatory cascade that controls high-level expression of vitellogenin (Vg) and vitellogenic carboxypeptidase (VCP) in the mosquito fat body, the ecdysteroid receptor (AaEcR) and several genes mediating ecdysone response have been cloned (Raikhel 1998). Three isoforms of the mosquito homologue of the *Drosophila* E75 early gene are induced at the onset of vitellogenesis in the mosquito fat body, suggesting their involvement in the regulation of vitellogenesis. The action of ecdysone on the target vitellogenic genes is indirect. Putative binding sites to two early ecdysone inducible gene proteins, E75 and E74, are present in the promoter regions of VCP and Vg. Several binding sites for transcription factors determining tissue-specific transactivation, including hepatocyte nuclear factors (HNF) 3 and 4, are also present in the promoters of both genes. The mosquito HNF 4 homologue is highly expressed in the fat body during vitellogenesis. When transformed into *Drosophila*, the promoter regions of both VCP and Vg containing these tissue-specific and hormone response elements confer fat body-specific expression of a reporter gene.

Thus, using a range of techniques which has generally involved the identification of tissue or stage- specific gene expression followed by gene cloning, some potentially useful promoters are now available. In some cases, the difficulty has been to establish the true functionality of these promoters. The use of *Drosophila* P element transformation with a reporter gene linked to the putative promoter sequence has, in general, proved successful as an approach. However, this may not always be feasible. A particular example would include the blood meal inducible, gut specific promoters where clearly this approach is limited and an alternative has to be found. The availability of an effective germline transformation system now overcomes this problem.

## ONCE NEW STRAINS OF INSECTS ARE PRODUCED WITH THE DESIRED CHARACTERISTICS, WHAT OTHER FACTORS NEED TO BE CONSIDERED?

Once transgenic vectors have been created, the long-term stability of the introduced genes will need to be investigated, initially by determining the copy number and location of the integrated promoter/gene cassette within the genome. The stability of this cassette over a number of generations will have to be ensured. The impact of the introduced gene on the fecundity and fitness of the insect will also be determined in comparison to wild type insects. Once these issues have been addressed, the conditions for effective mass rearing may then be considered. It may be possible to directly apply novel strains created by transgenic technology to mass rearing techniques. However, where this is not the case, it will be necessary to consider the problems likely to be faced. For example, such a situation may disrupt the normal adaptive process and therefore be opposed by natural selection.

In addition to using transgenic techniques for the improvement of SIT, there may also be situations where the release of non-sterile insects may prove beneficial for the suppression of insect pest populations or reducing their ability to cause health or economic problems. In this case, how the desired genes may be driven through the insect populations in such a way that the desired gene becomes fixed in the population is clearly an important issue. Three types of "drive" mechanisms have been proposed. One is meiotic drive, where a given chromosome is transmitted to more than the expected 50% of offspring. Any desirable genes linked to the driven chromosome would eventually approach fixation even with the release of relatively few individuals. There is experimental evidence to support the use of meiotic drive in *Ae. aegypti*. This mechanism, driven by the $M^D$ locus, has been used to force the marker gene *re* (red eye) into a laboratory cage population (Wood et al. 1977). Interestingly, meiotic drive also occurs during hybrid dysgenesis and it might, therefore, also be possible to exploit this phenomenon by using Type II transposable elements such as P (Kidwell and Ribeiro 1992). It is widely known that transposable elements can spread rapidly in natural populations. However, very little is known about whether these elements will spread in a similar way if they have been "loaded" with additional DNA sequences. The success of such a mechanism may also depend on the initial absence of that transposable element family from the population or species.

The second type of mechanism is the exploitation of genetic traits that reduce heterozygote fitness (Curtis and Graves 1988). For example, the gene to be driven could be introduced into a translocation chromosome such that viable and fertile homozygotes were formed, whereas heterozygotes would display reduced fertility or viability. In this way, translocations, pericentric inversions, inter-racial hybrid sterility, cytoplasmic incompatibility and compound chromosomes all have potential since, in each case, hybrids have reduced fitness. Such mechanisms require larger release numbers since there is no exponential increase in the frequency of the driven chromosome as with meiotic drive. However, fixation of desirable genes would occur more quickly than with meiotic drive because of the reduced fitness of heterozygous combinations. Therefore, there is a wide number of issues which must be addressed before the use of such a mechanism to drive foreign genes into insect populations, not least of which are the environmental impact and safety issues surrounding any proposed release. These are critical and require early consideration.

## CONCLUSIONS

There is now real potential to improve SIT and broaden its application to a wider range of insect pests by using transgenic techniques. Much still needs to be done to identify the genes and promoter systems which can then be employed to allow the potential application of this technology to SIT to be assessed. At that stage, it will then be possible also to determine the environmental and safety issues which will need to be addressed before genetically manipulated insects can be released into the environment as part of a SIT programme.

## ACKNOWLEDGEMENTS

This work was funded by the Wellcome Trust and the Medical Research Council.

## REFERENCES

Barillas-Mury, C., A. Charlesworth, I. Gross, A. Richman, J.A. Hoffman and F.C. Kafatos. 1996. Immune factor GAMBIF1, a new REL family member from the human malaria vector *Anopheles gambiae*. EMBO J. 15: 4691-4701.

Bingham, P.M., R. Levis and G.M. Rubin. 1981. The cloning of DNA sequences from the *white* locus of *Drosophila melanogaster* using a novel and general method. Cell 25: 693-704.

Cha, H.J., M.Q. Pham, G. Rao and W.E. Bentley. 1997. Expression of green fluorescent protein in insect larvae and its application for heterologous protein production. Biotech. Bioeng. 56: 239-247.

Coates, C.J., N. Jasinskiene, L. Miyashiro and A.A. James. 1998. Mariner transposition and transformation of the yellow fever mosquito, *Aedes aegypti*. Proc. Nat. Acad. Sci. USA. 95: 3748-3751.

Collins, A.F., R.K. Sakai, K.D. Vernick, S. Paskewitz, D.C. Seeley, L.H. Miller, W.E. Collins, C.C. Campbell and R.W. Gwadz. 1986. Genetic selection of a *Plasmodium* refractory strain of the malaria vector *Anopheles gambiae*. Science 236: 607-610.

Curtis, C.F. and P.M. Graves. 1988. Methods for replacement of malaria vector populations. J. Trop. Med. Hyg. 91: 43-48.

Dimopoulos, G., A. Richman, A. della Torre, F.C. Kafatos and C. Louis. 1996. Identification and characterisation of differentially expressed cDNAs of the vector mosquito, *Anopheles-gambiae*. Proc. Nat. Acad. Sci. USA 93: 13066-13071.

Dimopoulos, G., A. Richman, H.M. Muller and F.C. Kafatos. 1997. Molecular immune responses of the mosquito *Anopheles gambiae* to bacteria and malaria parasites. Proc. Nat. Acad. Sci. USA. 94: 11508-11513.

Franz, G., E. Gencheva and P. Kerremans. 1994. Improved stability of genetic sex-separation strains for the Mediterranean fruit fly, *Ceratitis capitata*. Genome 37: 72-82.

Giannoni, F., J. Vizoli, F. Catterucia, H.M. Muller, F.C. Kafatos and A. Crisanti. 1998. Cis and trans-acting promoter elements regulating the transcription of the *Anopheles gambiae* trypsin genes. Keystone Symposia on Molecular and Cellular Biology: Toward the Genetic manipulation of Insects No. 028;28.

Hendrichs, J., G. Franz and P. Rendon. 1995. Increased effectiveness and applicability of the sterile insect technique through male-only releases for control of mediterranean fruit flies during fruiting seasons. J. App. Entomol. 119: 371-377.

Kidwell, M.G. and J.M.C. Ribeiro. 1992. Can transposable elements be used to drive disease refractoriness genes into vector populations? Parasitol. Today 8: 325-329.

Laukannen, M.L., C. Okerblom and K. Keinanen. 1996. Secretion of green fluorescent protein from recombinant baculovirus-infected insect cells. Biochem. Biophys. Res. Comm. 226: 755-761.

Lee, W.J., J.D. Lee, V.V. Kravchenko, R.J. Ulevitch and P.T. Brey. 1996. Purification and molecular cloning of an inducible gram-negative bacteria-binding protein from the silkworm, *Bombyx mori*. Proc. Nat. Acad. Sci. USA. 93: 7888-7893.

Lee, W.J., A. Ahmed, A. della Torre, A. Kobayashi, M. Ashida and P.T. Brey. 1998. Molecular cloning and chromosomal localisation of a prophenoloxidase cDNA from the malaria vector *Anopheles gambiae*. Insect Mol. Biol. 7: 41-50.

Lehane, M.J, D. Wu and S.M. Lehane. 1997. Midgut-specific immune molecules are produced by the blood-sucking insect, *Stomoxys calcitrans*. Proc. Nat. Acad. Sci. USA 94: 11502-11507.

Macdonald, W.W. and C.P. Ramachandran. 1965. The influence of the gene $f^m$ (filarial susceptibility, *Brugia malayi*) on the susceptibility of *Aedes aegypti* to seven strains of *Brugia, Wuchereria* and *Dirofilaria*. Ann. Trop. Med. Parasitol. 59: 64-73.

Matsuoka, H., J. Kobayashi, G. Barker, K. Miura, Y. Chinzei, S. Miyajima, A. Ishii and R.E. Sinden. 1996. Induction of antimalarial transmission blocking immunity with a recombinant ookinete surface antigen produced in silkworm larvae using a baculovirus expression vector system. Vaccine 14 120-126.

Meller, V.H. and R.L. Davis. 1996. Biochemistry of insect learning - lessons from bees and flies. Insect Biochem. Mol. Biol. 26: 327-335.

Michel, K., J. Watson, A.C. Pinkerton, S. Thibault, D.A. O'Brochta and P.W. Atkinson. 1998. Expression and utilisation of green fluorescent protein as a genetic marker in insects. Keystone Symposium on Molecular and Cellular Biology : Toward the Genetic Manipulation of Insects No. 410:51.

Muller, H.M., J.M. Crampton, A. della Torre, R. Sinden and A. Crisanti. 1993. Members of a trypsin gene family in *Anopheles gambiae* are induced in the gut by blood meal. EMBO J. 12: 2891-2900.

Muller, H.M., F. Catteruccia, J. Vizioli, A. della Torre and A. Crisanti. 1995. Constitutive and blood meal-induced trypsin genes in *Anopheles gambiae*. Exp. Parasitol. 81: 371-385.

Paskewitz, S. 1998. Biochemical and genetic analyses of *Plasmodium*-mosquito interactions. Keystone Symposium on Molecular and Cellular Biology No. 014:23.

Raikhel, A. 1998. Integration of hormonal and tissue-specific control in the expression of yolk protein precursors in the mosquito fat body. Keystone Symposium on Molecular and Cellular Biology: Toward the Genetic Manipulation of Insects No. 018:25.

Rendahl, K.G. and J.C. Hall. 1996. Temporally manipulated rescue of visual and courtship abnormalities caused by a nona mutation in *Drosophila*. J. Neurogen. 10: 247-256.

Richman, A.M., P. Bulet, C. Hetru, C. Barillas-Mury, J.A. Hoffmann and F.C. Kafatos. 1996. Inducible immune factors of the vector mosquito *Anopheles gambiae* - biochemical purification of a Defensin antibacterial peptide and molecular-cloning of preprodefensin cDNA. Insect Molec. Biol. 5: 203-210.

Searles, L.L., R.S. Jokerst, P.M. Bingham, R.A. Voelker and A.L. Greenleaf. 1982. Molecular cloning of sequences from a *Drosophila* RNA polymerase II locus by P element transposon tagging. Cell 31: 585-592.

Severson, D.W., A. Mori, Y. Zhang and B.M. Christensen. 1994. Chromosomal mapping of two loci affecting filarial worm susceptibility in *Aedes aegypti*. Insect Mol. Biol. 3: 67-73.

Severson, D.W., V. Thathy, A. Mori, Y. Zhang and B.M. Christensen. 1995. Restriction fragment length polymorphism mapping of quantitative trait loci for malaria parasite susceptibility in the mosquito *Aedes aegypti*. Genetics 139: 1711-1717.

Smartt, C.T., A.P. Kim, G.L. Grossman and A.A. James. 1995. The Apyrase gene of the vector mosquito *Aedes aegypti* is expressed specifically in the adult female salivary glands. Exp. Parasitol. 81: 239-248.

Stowell, S.L., R.E. Sinden and J.M. Crampton. 1998. Expressing parasite proteins in the mosquito: a model system to examine the possibility of using haematophagous insects to deliver vaccines. Keystone Symposium on Molecular and Cellular Biology: Toward the Genetic Manipulation of Insects No. 216:40.

Thathy, V., D.W. Severson and B.M. Christensen. 1994. Reinterpretation of the genetics of susceptibility of *Aedes aegypti* to *Plasmodium gallinaceum*. J. Parasitol. 80: 705-712.

Vernick, K. 1995. Molecular aspects of *Plasmodium* resistance in *Anopheles gambiae*. J. Cell. Biochem. S21A SIA:189.

Wood, R.J., L.M. Cook, A. Hamilton and A. Whitelaw. 1977. Transporting the marker gene *re* (red eye) into a laboratory cage population of *Aedes aegypti* (Diptera: Culicidae), using meiotic drive at the $M^D$ locus. J. Med. Entomol. 14: 461-464.

Zheng, L.B., F.H. Collins, V. Kumar and F.C. Kafatos. 1993. A detailed genetic map for the X chromosome of the malaria vector, *Anopheles gambiae*. Science 261: 605-608.

Zheng, L.B., M.O. Benedict, A.J. Cornel, F.H. Collins and F.C. Kafatos. 1996. An integrated genetic-map of the African human malaria vector mosquito, *Anopheles gambiae*. Genetics 143: 941-952.

Zheng, L.B., A.J. Cornel, R. Wang, H. Erfle, H. Voss, W. Ansorge, F.C. Kafatos and F.H. Collins. 1997. Quantitative trait loci for refractoriness of *Anopheles gambiae* to *Plasmodium cynomolgi* B. Science 276: 425-428.

Zheng, L.B., A.J. Cornel, R. Wang, H. Erfle, H. Voss, W. Ansorge, F.C. Kafatos and F.H. Collins. 1998. Genetic and molecular approaches to refractoriness of *Anopheles gambiae* to *Plasmodium cynomolgi* B. Keystone Symposium on Molecular and Cellular Biology: Toward the Genetic Manipulation of Insects No. 004:20.

# PART II.

## APPLICATIONS OF THE AREA-WIDE CONCEPT TO NON-FRUIT FLIES

## D) SUPPORTIVE TECHNOLOGIES FOR AREA-WIDE CONTROL

# The Use of Semiochemical-Based Devices and Formulations in Area-Wide Programmes: A Commercial Perspective

*Owen T. Jones and Enzo D. Casagrande*

AgriSenseBCS Ltd., Treforest Industrial Estate, Pontypridd, S. Wales, CF37 5SU, UK.

## INTRODUCTION

It is now almost 40 years since the first insect pheromone was isolated and identified and many authors have referred over the years to their commercial potential. Although much progress has been made during those years in terms of scientific and technological advances related to semiochemicals, the "progress" that has actually been made in the commercial exploitation of pheromones and other semiochemicals has been rather limited when viewed in terms of their share of the agrochemical market. Based on published data and unpublished information which has been made available to the authors, semiochemical-based products are estimated to have world-wide sales in the US$70-80 million range at the manufacturers' level. This compares with a world-wide insecticide market of nearly US$8 billion in 1995 so that it therefore constitutes less than 1% of that market. In terms of the biopesticide market (bacteria, viruses, botanical insecticides, entomopathogenic nematodes and beneficial insects) on the other hand, semiochemicals constitute a much higher percentage (ca. 30%) and probably are third in importance after bacterial and botanical products. The industry has taken over 20 years to achieve its current size and has been pioneered mostly by small to medium-sized enterprises (SMEs).

Several technical and biological factors have played a part in determining the rate at which semiochemical-based technology has been adopted commercially in the field but one factor in particular has greatly influenced this technology transfer process, namely, area-wide programmes. These have often been government-backed programmes that have demonstrated the effectiveness of the semiochemical-based technique being implemented and have convinced the farmers and growers of the benefits that these novel techniques can bring.

Several strategies have been developed where semiochemicals have been used either alone or in combination with other control technologies but three in particular stands out.

### Mating Disruption

Most of the efforts made in using semiochemicals for controlling insect pests over the last two decades have been in the field of mating disruption using the sex pheromones of Lepidopteran pests. Several problems have had to be overcome in order to make the technique viable and cost effective, such as developing good low cost synthetic routes to the pheromone active ingredients, developing reliable controlled release dispensers, agreeing with the regulatory requirements for such products and actually getting them registered. It was shown early on in the development of these technologies that results were often poor if trials were done in small plots or where the treated plots were not isolated from surrounding crop not treated with mating disruption formulations. Area-wide mating disruption programmes overcame both these problems and such programmes have played a very important role in the gradual acceptance of this technique for controlling certain key Lepidopteran pests. Area-wide mating disruption programmes have been very successful in top-fruit. The Oriental fruit moth (*Grapholitha molesta* Busck) has been successfully controlled in a 1,200 ha mixed stonefruit area of the Tulbagh valley in South Africa (Barnes and Blomefield 1997). Programmes against the codling moth *(Cydia pomonella L.)* in over 10,000 ha of orchards in the Pacific Northwest of the USA (Thomson 1997), and in 3,500 ha of orchards in both South Africa and northern Italy are also being undertaken. In other crops too, control of *Chilo suppressalis* Walker in a 2,500ha rice growing area of Spain (Jones et al. 1990) and the control of the pink bollworm in over 250,000 ha of cotton in Egypt every year since 1995 shows what can be achieved when area-wide programmes are undertaken.

## Mass Trapping

Although the concept of mass trapping is rather simple, in practice it has been difficult to apply successfully, primarily for the following reasons:

- the lack of attraction of females by the attractant source used,
- the lack of highly efficient traps,
- the problem of high insect populations and trap saturation, and
- the need for a high density of traps per unit of surface area which makes the technique too costly.

The boll weevil, *Anthonomous grandis* Boheman, is a good example of an important pest where mass trapping has been applied successfully. Traps for the weevils are baited with grandlure which acts both as a sex pheromone attracting females and as an aggregation pheromone in early and late season when it attracts both sexes. Numerous attempts at mass trapping the boll weevil were conducted during the 1970s with different formulations and traps (Hardee 1982). However, low trap capacities and low trapping efficiencies meant that supplemental insecticide applications had also to be used and mass trapping was almost abandoned as a means of combating this pest. Later in the same decade, however, an area-wide boll weevil eradication project was initiated in a number of the southern states of the USA. The system used involved area-wide insecticide treatments to reduce boll weevil populations together with pheromone traps for monitoring and suppression of very low density populations. This programme has been highly successful in both the southeastern and southwestern USA (Ridgway et al. 1990). In 1987, the southeastern boll weevil eradication programme was expanded through Florida, most of Georgia and a major portion of Alabama and similar successes were recorded. In the southwestern programme that started in 1985, the boll weevil had been eradicated from California, southwestern Arizona and adjoining areas of Mexico by 1987 through a similar system of trapping and insecticide applications. The expanded southwestern programme was completed in 1990 and a buffer zone or containment programme is now in operation to prevent re-infestation (Ridgway and Inscoe 1992). It is clear, therefore, that mass trapping can have a role where problems of high populations and trap saturation are overcome through supplementary spraying with insecticides and particularly where an area-wide programme is undertaken.

Mass trapping of insect pests has also been successfully applied against certain bark beetle pests such as the spruce bark beetle, *Ips typographus* L. which causes widespread damage to pines in Europe and certain parts of Asia. It uses an aggregation pheromone to orchestrate mass attacks against susceptible trees and with the identification of the chemicals used in this process, new powerful lures became available for baiting traps to catch this insect. Such lures were approved for use in drainpipe traps by the pesticide boards of the Ministry of Agriculture in Norway and Sweden for an integrated pest management (IPM) programme against *I. typographus* in those two countries (Bakke and Lie 1989). During the 1970s, there were severe outbreaks of this pest in Scandinavia with significant losses of good timber. Mass trapping was initiated in 1979 in Norway as part of an IPM package which involved changes to cultural practices which also disfavoured the beetle. In addition to mass trapping, felling and removal of beetle-infested trees were also an important components of the IPM approach.

Over 600,000 traps were used in this programme with a total of over 3 billion beetles caught per annum. Trap catches peaked in 1980 at 4.5 billion and by 1982, the incidence of tree deaths had reduced dramatically. Since the IPM measures taken against *I. typographus* were many and varied, it is difficult to know what contribution to the overall effect was made by the mass trapping efforts. The incidence of the beetles in Scandinavia has not risen to epidemic levels since the early eighties but indications are that this might happen again in the not too distant future. It will be interesting to see whether the authorities concerned will again mount such an ambitious programme which covered 2.0 to 2.5 million hectares of forest the previous occasion.

In the olive fly, *Bactrocera oleae* (Gmelin), olfactory attractants have been described for both males and females. The possibility of combining this insect's visual attraction to yellow sticky traps with the response of the males to the female-produced sex pheromone (1,7 dioxaspiro [5,5] undecane) and the female's response to ammonium salts, has been investigated as a population suppression mechanism (Broumas et al. 1982). Positive results were obtained in these early experiments which covered areas with 10,000 olive trees. The system suffered, however, from a number of the limitations noted earlier. These included the need for a high density of traps, their limited catching capacity especially in areas with high populations, and in addition, concerns about the non-selectivity of the traps when it comes to beneficial insects, such as hymenopteran parasites and egg predators, e.g., *Chrysoperla carnea* Steph. (Neuenschwander 1982).

Many of these problems were overcome by abandoning sticky traps and turning instead to "target" devices which were treated with insecticides to kill the attracted insects once they made contact with them. In this way, problems of trap saturation were overcome because the insect, once having picked up a lethal dose of insecticide from the target device, flies or walks away from it until the toxic effects of the insecticide manifest themselves. Also, by using targets with grey, green or brown colours, the attraction of beneficial insects was reduced thus conserving their populations in the olive grove.

**Lure and Kill**

The use of olfactory attractants in combination with insecticides has been a well-established practice since the 1950s in the case of Tephritid fruit flies such as the Mediterranean fruit fly, *Ceratitis capitata* (Wied.), and the olive fly *(B. oleae)*. The use of a protein hydrolysate-insecticide mixture in a "lure and kill" strategy allows the operator to spray a reduced surface area of crop relying on the attractant to bring the insect to the treated area. Protein hydrolysates are used in many fruit-growing areas of the world for fruit fly control and have undoubtedly led to reduced environmental contamination by insecticides. However, being a proteinaceous material, the hydrolysates are highly attractive to most insects, both good and bad, and as a consequence, their use can significantly reduce natural enemy populations in the crops thus treated. The availability of species-specific semiochemical attractants, however, has overcome this problem in many cases and some examples of their successful application are given below.

## "Lure and Kill" Strategies for Tephritid Fruit Flies

"Lure and kill" strategies for Tephritid fruit flies fall into two groups: those that employ some form of *target device* and those that rely on attracting the insect onto a natural surface, e.g., host tree foliage, which has been treated with an attractant/insecticide mixture; this second technique will be referred to as *sprayable formulations*.

## Target Devices for Male Annihilation

Of all the parapheromones that have been developed for Tephritid fruit flies, methyl eugenol stands out as being the most potent. Fibrous blocks containing methyl eugenol and an insecticide such as Naled (dibrom) have been used as target devices for male Oriental fruit flies, *Bactrocera dorsalis* (Hendel), in several eradication programmes. Steiner et al. (1965) made the first successful attempt at eradication of *B. dorsalis* by male annihilation on the island of Rota in the Marianas. They deployed 5% by weight of an insecticide in methyl eugenol which was used to saturate 5 cm fibreboard squares so that each held 24 g of the mixture. These fibreboard squares were then thrown out of an aircraft over uninhabited areas at given rates and re-application intervals to achieve eradication in about six months.

This technique was then adopted by the Japanese government in an ambitious eradication campaign against the Oriental fruit fly in the Ogasawara Islands. Over a period of ten years, they were able to eradicate the fly from all the islands in the archipelago from Amami in the north to Okinawa in the south (Koyama et al. 1984).

The success of this technique was undoubtedly based on the fact that methyl eugenol is very attractive indeed to *B. dorsalis* and is capable of attracting a sufficiently high percentage of males to the insecticide-treated target devices to leave the females unfertilised. None of the other parapheromones has been shown to produce the same effect when used alone as the lure; in most cases, some other attractant has also been used that attracts females.

## Target Devices for Male and Female Annihilation

The olive fly, *B. oleae*, was mentioned above as a species that can be controlled, at low population levels at least, using yellow sticky traps baited with sex pheromones and food attractants. However, such traps suffer from problems of trap catch saturation and detrimental effects on beneficial insects. Work in Greece over the last 15 years has been aimed at overcoming these problems through the development of target devices that carry an insecticide for killing the attracted flies instead of adhesives. The target devices used in most of the large-scale trials undertaken in Greece consisted of plywood rectangles (15 x 20 x 0.4 cm) dipped into appropriate concentrations of the pyrethroid insecticide Deltamethrin for a sufficiently long period of time to saturate the wood with the insecticide solution. These devices were baited with ammonium salt dispensers (food attractant) and one target in 3 or 5 was also baited with a sex pheromone dispenser. Fewer sex pheromone dispensers than food attractants were thought necessary as their distance of attraction was shown in earlier experiments to be 60-80 m while that for the food attractants was only 15-20 m.

Great logistical problems had to be overcome during the installation period of the devices in June and July, especially in years where over 2,000,000 trees were treated. As the controlled release devices for the food and sex attractants became more advanced, it was possible to install the target devices during early summer and reasonably expect them to last until late autumn when olive fly populations reduced in importance through decreasing temperatures and the olives were harvested. The effectiveness of the target devices was assessed by laboratory bioassays throughout their period of field use. Olive fly populations were monitored by the use of traps and through taking samples of olive fruit for periodic examination of damage levels.

The results over five years from the area-wide application of these target devices can be summarised as follows: fly populations as measured by McPhail traps were consistently lower in target-device treated areas compared with conventionally treated controls. The average number of bait sprays that had to be used in target treated areas during the early years of the programme (1984/1985) were 1 as opposed to 2.5. No supplementary bait sprays were required in later years in the target-device treated areas. In most years, fruit infestation was lower than, or equal to, that in the controls where bait sprays were applied (Haniotakis et al. 1991). The target device method of controlling *B. oleae,* therefore, was very effective as a method of eliminating insecticide bait sprays and significant increases in beneficial insect numbers were observed in target-device treated areas (Paraskakis 1989). However, for the method to work to its greatest effect, it has to be applied on a large area. In small plots, large-scale adult movements over short distances can significantly over-ride the effects of the devices. Similarly, when the system fails to contain pest populations, complementary measures is almost invariably required, significantly affecting the cost effectiveness of the technique.

In Spain, similar target device technology has been tested on a relatively large scale since 1992. The device is similar to that used in Greece but it is made out of cloth that is then soaked in Deltamethrin rather than wooden boards. The devices have been baited with long life baits such as those used in Greece or have been baited repeatedly with sprayable protein and pheromone baits. The latter need to be replenished several times during the season but are very low cost in materials yet labour-intensive. With the long life lures, the up-front costs are greater but require very little maintenance thereafter during the rest of the season. In areas as large as 400 ha, very satisfactory results have been obtained in most years.

This method of managing *B. oleae* populations, although more labour-intensive generally, will nevertheless be pursued in most olive growing countries since legislative and environmental pressures will eventually restrict the broad-scale use of bait sprays.

## Sprayable Formulations for *B. oleae*

The wide-scale use of protein/insecticide bait sprays for controlling the olive fly, *Bactrocera oleae*, has become well established in most Mediterranean olive growing countries, together with several disadvantages to its use. Probably, the most important problem with this technique is its lack of selectivity. Many important insect predators and parasites are known to be attracted by the protein hydrolysate component of the bait spray mix and this often leads to substantial reductions in their populations with continued use of this technique.

With the isolation and identification of the sex pheromone of *B. oleae*, a selective attractant became available which could substitute the protein hydrolysate. Trapping experiments showed the 1,7 dioxaspiro [5.5] undecane major component to be strongly attractive to the males (Jones et al. 1983). Observations made during attempts to disrupt the mating of *B. oleae* using techniques similar to those used for Lepidoptera, showed that instead of producing mating disruption, the wide-scale treatment of experimental olive groves with the pheromone produced, in some instances, only large immigration of *B. oleae* adults, *both male and female.* It appeared therefore that the pheromone, in addition to attracting the males in a clearly directed manner as seen with monitoring traps, was also attractive to females but not in such a strongly directed way. Attention was therefore moved from trying to achieve mating disruption with the pheromone to using the pheromone as a substitute attractant in bait sprays that were much more selective.

A sprayable formulation of the pheromone was therefore required which would slowly release it over a period of time consistent with the effective life of the insecticide once applied in the field. The pheromone was micro-encapsulated in Poly-urea type micro-capsules (5-10 micrometer) and in polymer entrapped micro-beads (5-10 micrometer) (Polycore SKL™) containing 20 g spiroacetal per litre, similar to those used for Lepidopteran mating disruption. The formulated pheromone was then tank mixed with either malathion or dimethoate and applied aerially or from the ground.

In aerial sprays, the plane delivers a 20 m wide swathe every 100 m of grove so that only 20% of the crop is treated. From the ground, the pheromone/insecticide mixture is applied either from a tractor mounted sprayer which applies the mixture to the south side of each row of trees or, if a knapsack sprayer is used, only 1 square metre of foliage, again on the south side of each tree, needs to be treated. Trials carried out over many years in southern Spain have shown that both application methods give consistently good results (Montiel 1989). Indeed, this technique is now being used on large areas of environmentally sensitive national parks in southern Spain totalling 13,800 ha in 1996 (Spanish Ministry Data 1996).

## STERILE INSECT TECHNIQUE AND SEMIOCHEMICALS

Sterile insect techniques by definition have to be carried out over large areas, and area-wide eradication programmes such as those for screw-worm and the Mediterranean fruit fly in North and Central America are well known examples of the very successful application of this technique. SIT is now, however, being attempted in a number of countries as a conventional pest management strategy as opposed to pest eradication. In this way, selected areas of infested crops are treated such as whole valleys which are contiguous with other neighbouring infested areas. With such a strategy, buffer zones have to be established at the extremities of the SIT areas that prevent the ingress of gravid females. Semiochemical-based traps, targets and sprayable formulations are ideal candidates for inclusion in such programmes and work done with Medfly in Europe illustrates the ways in which the two technologies of SIT and semiochemicals can be integrated.

## Reduction of Natural Pest Populations within SIT Areas Prior to the Release of Sterile Insects

In many SIT programmes, before the release of sterile insects, natural populations of Medflies have been reduced through the use of protein bait sprays. Protein hydrolysates are very good attractants for adult tephritids and, when mixed with a suitable killing agent such as malathion, can reduce natural populations substantially and therefore increase the chances of success for the SIT programme. Protein hydrolysates are, however, highly non-selective in the insects which they attract. In addition to attracting non-target species such as muscids, they have been found to attract beneficial parasite or predator species such as *Chrysopa* spp.

In an attempt to develop a more selective sprayable attractant for Medflies, the para-pheromone trimedlure was entrapped in a particulate polymeric matrix for controlled release where the particle sizes ranged from 0.1 to 2 microns in diameter. This is used instead of the protein hydrolysate in tank mixes with conventional insecticides and sprayed in bands from the air or onto one metre square areas of foliage on the south side of each tree in citrus orchards.

Trials with this formulation were carried out in a 5 ha orchard with 400 citrus trees per ha. Each tree was treated with a mixture of malathion at manufacturers' recommended rates and the polymer-entrapped trimedlure emulsion at a rate of 20 g a.i. per ha but treating only one metre square of every tree to the south side. Each tree was treated with 0.125 g of trimedlure. Fly populations were measured before and after treatment with yellow sticky traps baited with trimedlure dispensers.

Trap catch data showed that the Medfly numbers were reduced to almost 10% of pre-spraying population levels immediately following treatment and they remained at about 15% for up to 20 days. In unsprayed control orchards, no corresponding reductions in populations were observed. Although the majority of the reduction observed was of the male population, there were also reductions in the female numbers suggesting that trimedlure is also mildly attractive to the females.

In the trials carried out in Egypt with the same trimedlure polymer entrapped formulation, substantial reductions in adult males were observed following treatments in both peach and guava orchards. In the guava orchards, the percentage infestation was also reduced by between 30% and 40%, compared with untreated control plots. The polymer-entrapped trimedlure formulation lasts about 7-10 days in the field once applied and therefore can be followed by sterile male releases after about 10 days without fear of them being killed by a selective bait insecticide mixture.

## Reduction of Natural Populations Surrounding SIT Release Areas

In the buffer zones surrounding SIT area, both traps and target devices have been evaluated. In buffer zones where populations are relatively small, mass trapping can be a solution provided the traps used work for the length of the sterile insect release period and do not require a lot of servicing. Ideally, such traps should catch both males and females and the recent advance by Heath et al. (1998) in the development of selective ammonium and amine salt attractants for Medflies is a very important development in this direction. Indeed, if the lures were totally selective for female Medflies, the traps could be used within the sterile male release areas.

If traps cannot be used because of catch saturation or excessive servicing requirements, insecticide treated target devices can be used instead. In trials done in Greece, the targets have consisted of inverted yellow cups coated on the inside with a mixture of sugar and methomyl and baited with a trimedlure dispenser – these attract and kill male Medflies and have given good control when used by themselves. However, they give much better control when used side-by-side with target devices baited with protein hydrolysates that attract both males and females. The trimedlure baited target increases significantly the attractiveness to females of a protein-baited target placed at a distance of about 0.5 metres away. The resultant attraction of females was found to be very selective in terms of the sex of the responding insect and in terms of the non-target species. This paired target concept was tested as a "lure and kill" technique for Medflies both on Crete and mainland Greece and proved to be very effective in trials. Very low Medfly numbers were caught in monitoring traps placed in buffer zones thus treated with target devices during the whole of the season-long SIT release programme on Crete.

## REFERENCES

Bakke, A. and R. Lie. 1989. Mass Trapping. pp. 67-87. *In* Insect Pheromones in Plant Protection. A.R. Jutsum; R.F.S.Gordon (eds.). John Wiley & Sons, Chichester.

Barnes, B.N. and T.L. Blomefield. 1997. Goading growers towards mating disruption: the South African experience with *Grapholitha molesta* and *Cydia pomonella* (Lepidoptera, Tortricidae). pp. 45-56. *In* Technology Transfer in Mating Disruption, IOBC wprs Bulletin 20.

Broumas, T., C.M. Liaropoulos, P. Katsoyannos, C. Yamvrias, G. Haniotakis and F. Strong. 1983. Control of the olive fruit fly in a pest management trial in olive culture. pp. 584-592. *In* Fruit Flies of Economic Importance. R. Cavalloro (Ed.). Proceedings of the CEC/IOBC International Symposium, Athens, Greece, 16-19 November 1982.

Haniotakis, G., M. Kozyrakis and T. Fitsakis. 1991. An effective mass trapping method for the control of *Dacus oleae* (Diptera: Tephritidae). J. Econ. Entomol. 84, 564-569.

Hardee, D.D. 1982. Mass trapping and trap cropping of the boll weevil *Anthonomus grandis* Boheman. pp. 65-72. *In* Insect Suppression with Controlled Release Pheromone System (Vol II), A. F. Kydonius and M. Beroza (eds.). CRC Press, Boca Raton, Florida.

Heath, R.R., N.D. Epsky, J. Hendrichs, A. Bakri, B. Barnes, H. Camacho, F. Jeronimo, B.I. Katsoyannos, P. Montoya, R. Pereira, J.P. Ros, Y. Rossler, S.I. Seewooruthun, L.A. Vasquez, E.M. Vattuone and A. Zumreoglu. 1998. Evaluation of female Mediterranean fruit fly (Ceratitis capitata) synthetic attractant systems for trapping and sterility assessment in thirteen countries. Fifth International Symposium on Fruit Flies of Economic Importance. 1-5 June, 1998. Penang, Malaysia.

Jones, O.T., S.R. Jenkins, E. Casagrande, A. Olmos, V. Aznar, C. Fabregues, A. Perdiguer, D.R. Hall and J.L. Smith. 1990. The mating disruption control of the rice stem borer (*Chilo suppressalis*) through the use of sex pheromones. pp. 943-944. ANPP – Deuxieme Conference Internationale Sur Les Ravageurs En Agriculture, 4-6 Decembre, 1990.

Jones, O.T., J.C. Lisk, C. Longhurst, P.E. Howse, P. Ramos and M. Campos. 1983. Development of a monitoring trap for the olive fly, *Dacus oleae*, (Diptera: Tephritidae) using a component of its sex pheromone as lure. Bull. Entomol. Res., 73: 97-106.

Koyama, J., T. Teruya and K. Tanaka. 1984. Eradication of the oriental fruit fly (Diptera: Tephritidae) from the Okinawa Islands by male annihilation. J. Econ. Entomol. 77: 468-472.

Montiel, A. 1989. Control of olive fly by means of its sex pheromone. pp. 443-453. *In* Fruit Flies of Economic Importance. R. Cavalloro (ed.). Proceedings of the CEC/IOBC International Symposium, Rome, Italy, 7-10 April, 1987.

Neuenschwander, P. 1982. Beneficial insects caught by yellow traps used in mass trapping of the olive fly, *Dacus oleae*. Entomol. Exp. Appl. 32: 286-296.

Paraskakis, H.I. 1989. Results from the biological control of *Saissetia oleae* Oliv. on olive trees in Crete, Greece. pp. 274-284. *In* Proceedings of the 2nd panhellenic congress of entomology, 11-13 November, 1987, Athens Greece.

Ridgway, R.L. and M.N. Inscoe. 1992. Insect behaviour-modifying chemicals: practical applications in the United States. *In* Insect Pheromones and other Behaviour-Modifying Chemicals: Applications and Regulation. pp. 19-28. *In* Proceedings of a Symposium held at the Brighton Crop Protection Conference – Pests and Diseases, 19 November, 1990. R.L.Ridgway, M.Inscoe & H. Arn (eds.). BCPC Monograph No 51.

Ridgway, R.L., M.N. Inscoe and W.A. Dickerson. 1990. Role of the boll weevil pheromone in pest management. pp. 437-471. *In* Behaviour-Modifying Chemicals for Insect Management: Applications of Pheromones and Other Attractants. R.L. Ridgway, R.M. Silverstein and M.N. Inscoe (eds.). Marcel Dekker, Inc., New York.

Steiner, L.F., W.C. Mitchell E.J. Harris, T. T. Kozuma and M. S. Fujimoto. 1965. Oriental fruit fly eradication by male annihilation. J. Econ. Entomol. 58: pp. 961-964.

Thomson, D. 1997. Confusion amongst codling moth fellows continues: a commercial perspective on the implementation of codling moth mating disruption in North America. pp. 57 –63. *In* Technology Transfer in Mating Disruption, IOBC wprs Bulletin 20.

# Environmentally-Safe Pest Control Using Novel Bioelectrostatic Techniques: Initial Results and Prospects for Area-Wide Usage

*P.E. Howse and K.L. Underwood*

School of Biological Sciences, The University, Southampton SO16 7PX, UK

## INTRODUCTION

The aim of our work is to provide an alternative to broad-scale pesticide spraying for crop protection. The need for such a development, in order to protect the environment, wildlife and human health, has never been more urgent. The project was conceived in 1992 deriving from concepts which are the subject of patent applications in 60 countries throughout the world. Since then, feasibility studies have been completed and links have been made with other organisations keen to evaluate and develop the technique. A parallel project, using similar electrostatic powder technology, was the development of an environmentally-friendly cockroach trap, which received a Prince of Wales Award for Innovation in 1997. In order to hasten the commercialisation of the technology, a new company, *ExoSect Ltd*, has been set up in association with the University of Southampton.

The technique is based on the discovery that certain inert powders that can be electrostatically charged will adhere to the insect cuticle. The cuticle appears to be permanently electrically polarised (i.e., it functions as an electret). Powders of opposite polarity to the surface will therefore attach readily and it is extremely difficult for the insect to dislodge itself. In contrast, powders which do not charge easily or which rapidly lose their charge, quickly fall off, or can be groomed off by the insect. Male insects are lured to a source of charged powder by odour attractants (e.g., sexual pheromones). They then contaminate themselves with the powder. After leaving the trap, they pass on the powder to females in mating attempts. An insecticidal material formulated with the powder can thus be delivered to the females. The insecticide must be slow-acting in order for this transfer to take place. The components of the method are the following:

### Carrier Particles

There are various inert or biodegradable materials with appropriate electrostatic characteristics which can be used as carriers for other biologically active materials. Long-lasting charge retention is perhaps the most important factor.

### Pesticide

Slow-acting approved insecticides in a dry powder formulation can be applied to an insect long enough for it to be killed. The knock-down time can be varied between one and three days for certain synthetic insecticides, or over four days for biological insecticides, during which time the insect behaves normally. Spores of the entomopathogenic fungus *Metarhizium* can be used as biological insecticides, for example. Dry spores of the fungus can be formulated with suitable electrostatically chargeable particles.

**Particle Transfer**

There is a high rate of loss of particles during the first 48 hours, particularly from hairs and other projections of the insect (after which the loss is very low). This means that particles are readily transferred to females in mating attempts.

**A Selective Attractant**

Pheromones or parapheromone attractants for males are available for almost all of the major insect pest species. Control by trapping males alone, however, is generally not a viable method because over 90% must be trapped to ensure that sufficient eggs in the next generation are infertile.

**A Dissemination Station**

Bait stations have been developed which retain formulated powders, minimising their loss by wind and facilitating transfer to insects (Patent applied for).

This method mimics a natural epidemic infective process (such as a sexually-transmitted disease), with the following advantages:

- insecticides do not come into contact with the crop or soil,
- extremely small amounts of insecticide are used,
- the method targets the pest species only, and others (beneficial insects etc.) are unaffected ,
- materials are all low-cost,
- unskilled labour is required only for placing devices around the crop,
- does not preclude the use of other methods that might be used in integrated pest management, and
- the way is open to using a range of pesticides to which insects have not previously been exposed and to which they will not have started evolving resistance.

One of our main research programmes is into the chemical ecology of fruit flies species such as the Mediterranean fruit fly, melon fly and olive fly. All are of enormous economic importance in fruticulture which is of vital importance to the exports of many developing countries. We have therefore chosen these as species for initial trials and carried out small-scale trials in fruit orchards in Mallorca (Balearic Is.) and in Mauritius.

It has been possible to show that populations of fruit flies in Mauritian villages decline rapidly when exposed to either a synthetic insecticide, or to a strain of the entomopathogenic fungus, *Metarhizium anisopliae* (Metchnikoff) Sorokin (Figure 1).

A difficulty encountered is that of recording the number of females in the population. Protein baited traps which can be used for monitoring both males and females are far less sensitive than pheromone traps for males and catches of females in Mauritius were too low to show clear trends. However, trials in Mallorca with the Mediterranean fruit fly showed that the female catch declined when compared the male catch during treatments (Figure 2).

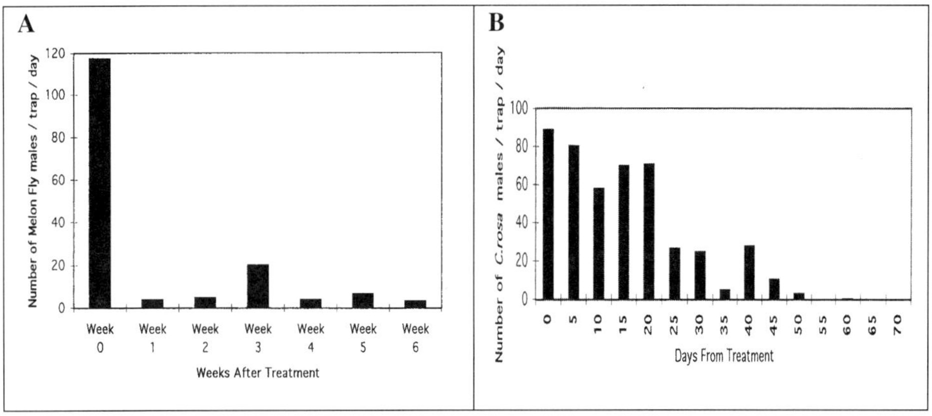


Figure 1. Fruit fly trials in fruit orchards, Mauritius 1996. (A) Trap catches of melon fly males after introduction of synthetic insecticide bait stations. (B) Trap catches of *Ceratitis rosa* Karsch males after introduction of entomopathogenic fungal bait station.

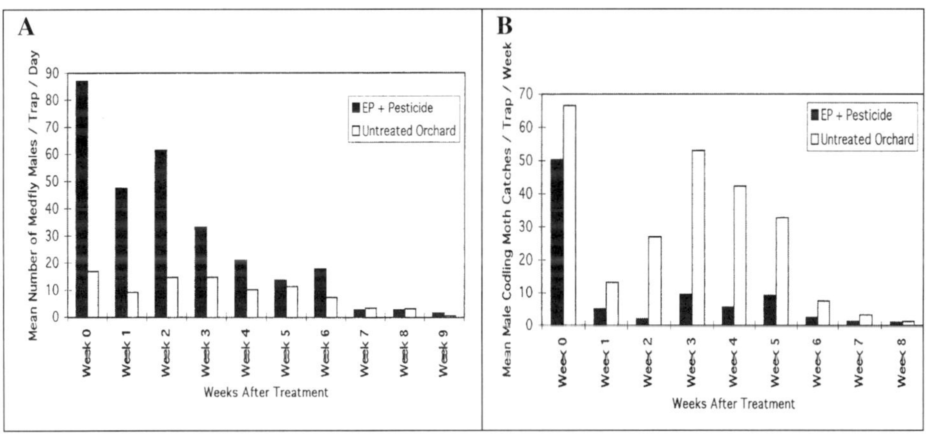


Figure 2. Medfly insecticide bait station trials and control in citrus orchards, Mallorca 1997. (A) Mean number of Medfly males/trap/day (N=2). (B) Number of Medfly females/trap/day (EP - electrostatic powder).

Small-scale trials have been carried out against fruit flies (*Bactrocera* and *Ceratitis* species in Mauritius and the Balearic Is. and in mainland Spain), and Lepidoptera (*Cydia pomonella* L. in Britain, and *Plutella xylostella* L. in Mauritius), using slow-acting insecticides or entomopathogenic fungal spores (*M. anisopliae*) as active ingredients. The results have all been consistent, and indicate that a high level of population reduction can be achieved with the use of several orders of magnitude less of the active ingredient than would be possible with crop spraying. A mating disruption system, which could have considerable cost benefits, has also been partially evaluated in the field.

## CODLING MOTH

Treatments of codling moth-infested orchards in southern Britain were carried out with bait stations placed at 20 - 40 /ha. Synthetic pheromone was used as an attractant and the active ingredient was 50 mg chlorpyrifos. The population changes were measured using sticky delta traps (Agrisense BCS) baited with 10 mg of pheromone. In one orchard, the pest population was reduced by approximately 70% when compared with an untreated area, and in another orchard, pest levels were below those of a control area that had been sprayed two weeks before the start of the trial (Figure 3). The total amount of pesticide used was 10 g /ha. This represents the maximum available, and the amount taken up by insects must be less.

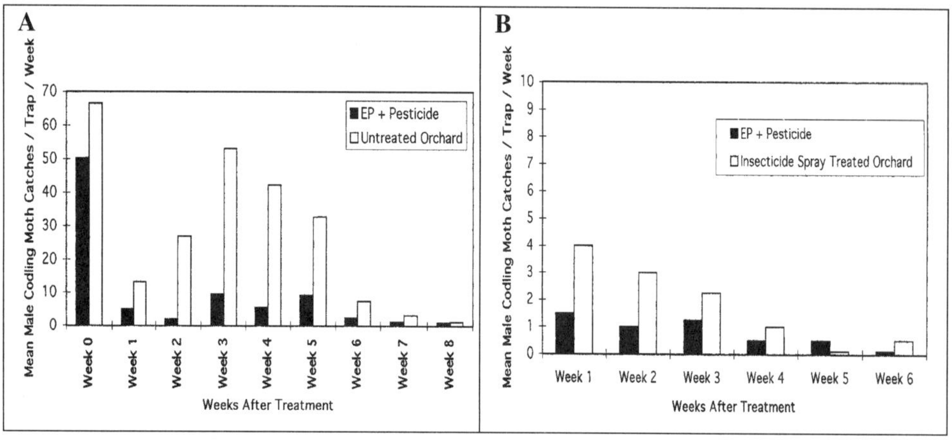


Figure 3. Codling moth trials in southern Britain, 1997. Mean number of male codling moth caught per trap per day (N=4). (A) Insecticide bait station trial and untreated control. (B) Insecticide bait station trial and treated control. (EP - electrostatic powder).

Mating disruption is a method of population reduction which does not involve synthetic insecticides. Various slow-release formulations of sex attractant pheromone have been developed commercially over the past 20 years, ranging from microcapsules to hollow plastic fibres and "sandwich" plastic flakes.

Mating disruption involves saturating the air in the crop with pheromone to such an extent that the males are unable to detect live females and may be drawn to false trails from dispensers. A total of about 320k hectares of cotton is under treatment by mating disruption at US$33/ha. The key cost element is the price of the pheromone, which is used at ~70 g per hectare. For codling moth (in apples), 15k hectares are treated by mating disruption at a cost of US$320/ha. About 200 g of pheromone are used per hectare (Howse et al. 1998).

Our trials on the codling moth involve use of charged powders formulated with pheromone. Trials in two orchard areas indicate that control can be achieved with 100 mg/ha, thus three orders of magnitude less than used in conventional treatment. A rapid decline in trap catches similar to that seen in insecticide trial was observed. It should therefore be possible to reduce the cost of control and at the same time considerably magnify the profit margin on treatments with charged powders. The main issue to be considered is how to distribute bait stations. Use of manual labour is likely to be a disincentive in cotton, but possibly not so in orchards and small plots.

These are only two examples of a large number of pests that could be controlled by our technique. This could make organic farming of many fruit and vegetable crops viable. No insecticide is sprayed onto the crop, and insecticide use is zero when a mating disruption approach is used.

## BROWN-TAIL MOTH *(EUPROCTIS CHRYSORRHOEA* L.)

This insect is an amenity pest in southern coastal Britain, on the northwest coasts of Europe and in limited areas of the east coast of the USA. Urticating hairs produced by the larvae can produce severe allergies. The moth finds its way into urban areas through green corridors and then becomes extremely difficult to control by spraying or cultural methods.

This situation offers a good opportunity for evaluation of charged powder technology. The area to be treated involves a heavily urbanised peninsula of the city of Portsmouth (southern England adjacent to a national nature reserve (Farlington Marshes)). Bait stations will be placed out in the reserve containing only pheromone formulation. There will thus be zero impact on the other fauna. In the urban area, a total of 365 hectares representing the main centres of infestation, identified on the basis of public response to allergies in previous years, will be treated with bait stations (10/ha) containing a pheromone attractant and an insecticide-powder formulation. The population in both areas will be monitored independently with sticky pheromone traps containing pheromone.

In the urban area, the maximum amount of insecticide which could find its way into the environment is 1 g /ha. Field tests indicate that less than 1 % of the powder is lost from the bait station due to weathering. Residues in the environment are therefore likely to be undetectable.

The electrostatic powder technique outlined here shows great potential as an alternative control strategy, based on the results to date. Future trials, such as the present Brown-tail moth project, will help to modify and refine the technique.

## REFERENCE

Howse, P.E., I.D.R. Stevens and O.T. Jones. 1998. Insect pheromones and their use in pest management. Chapman & Hall, London, UK. 369 pp.

# Synthetic Analogues of Natural Semiochemicals as Promising Insect Control Agents

*István Ujváry[1], Miklós Tóth[1] and Patrick Guerin[2]*

[1] Plant Protection Institute, Hungarian Academy of Sciences, Budapest, Hungary
[2] Institute of Zoology, University of Neuchâtel, Neuchâtel, Switzerland

## INTRODUCTION

After decades of research and development, insect pheromones and other semiochemicals became indispensable tools of ecologically based agricultural pest and disease vector management programmes with main uses as: 1) detection and population monitoring of emerging and migrating insects, 2) mass trapping of insects, 3) combined formulation of semiochemicals and insecticides ("lure-and-kill"), and 4) mating disruption with specially formulated pheromone components. In spite of their demonstrated safety and biodegradability, the direct application of these semiochemicals for pest control has not fulfilled initial expectations. Nonetheless considerable field experience has been accumulated (Cardé and Minks 1995). Evidently, two important factors limit the practical potential of these substances: 1) inherent in their particular mode of action, semiochemicals, especially pheromones, are effectively cleared by specific enzymes in the insect antennae, and 2) some of these compounds contain labile functional moieties that are prone to degradation (oxidation, isomerisation and polymerisation) under field conditions. Appropriate chemical modifications of these natural compounds, however, can circumvent these problems by providing synthetic analogues (sometimes also called parapheromones or antipheromones; for early studies, see Roelofs and Comeau 1971, Payne et al. 1973) which in ideal cases are not only more potent and environmentally acceptable but more economical as well. It should also be mentioned that many effective attractants have been discovered through the empirical screening of synthetic chemicals, some of which have actually turned out to be structural relatives of natural semiochemicals of the particular insect.

In this paper, selected case studies of analogues of sex pheromones and kairomones will be presented. The examples from our work include nitrile bioisosteres of labile aldehyde pheromone components of the cranberry girdler moth, *Chrysoteuchia topiaria* Zeller; analogues of (*Z*)-11-hexadecenyl acetate, a principal component of several important lepidopteran species and analogues of 1-octen-3-ol and 3-*n*-propylphenol, two kairomonal cattle odour components attractive to tsetse flies (*Glossina* spp.).

## MEDFLY ATTRACTANTS

Trimedlure, a powerful attractant of the Mediterranean fruit fly, *Ceratitis capitata* (Wied.), one of the most successful parapheromones used in agriculture, was discovered by USDA chemists during the routine screening of hundreds of compounds (reviewed by Cunningham 1989, Metcalf and Metcalf 1992). The synthetic product is a mixture of isomeric *tert*-butyl 4- and 5-chloro-*cis*- and *trans*-2-methylcylohexane-1-carboxylates from which the most active component is the (1*S*,2*S*,4*R*) isomer (**1**, Figure 1). Warthen et al. (1995) have recently investigated the structure-biological activity relationship of the various stereoisomers of **1**, while recent molecular modelling studies by Flath et al. (1994) have demonstrated close structural complementarity of **1** and α-copaene (**2**), a sesquiterpene identified by Fornasiero et al. (1969) as the principal component responsible for the attractiveness of angelica seed oil to Medflies. Compound **2**, also present in orange oil and other essential oils, is believed to be the original plant kairomone with which *C. capitata* and other Ceratitinae coevolved. Ceralure, a recently developed highly attractive and persistent analogue of trimedlure (McGovern and Cunningham 1988) has improved persistency and appears to be a promising tool for male annihilation techniques. Of the eight isomeric enantiomer pairs of ceralure, **3** (Figure 1) is the most active isomer (Warthen et al. 1994).

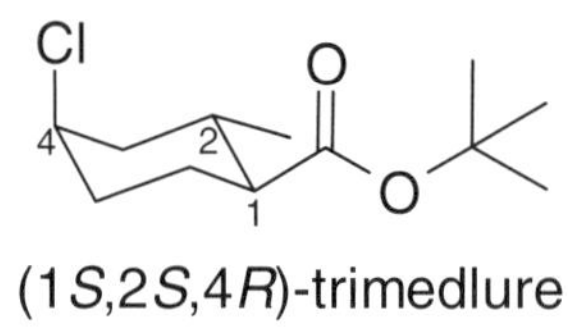


**1**

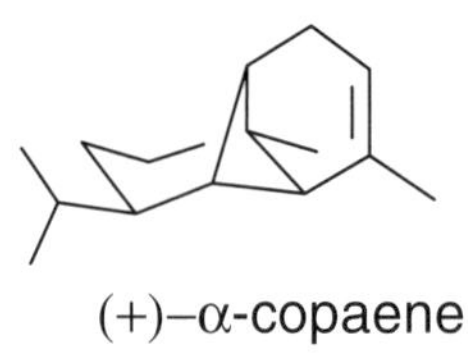


**2**

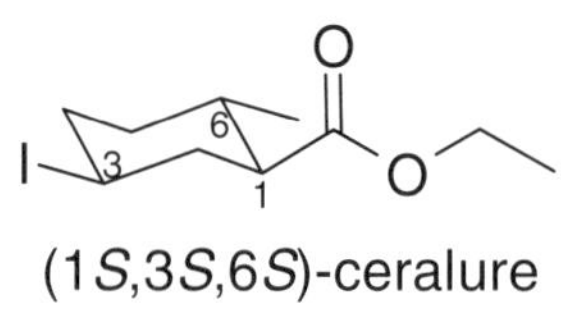


**3**

Figure 1. The most active stereoisomers of natural and synthetic of Medfly attractants.

## ALDEHYDE BIOISOSTERES

Sex pheromonal and kairomonal aldehydes are widely used in practice either in traps or as behavioural disruptants. These compounds, however, readily undergo oxidation and polymerisation resulting in biologically inactive decomposition products (Shaver and Ivie 1982). Various attempts have been made to find structural analogues that are more stable under field conditions. The aldehyde group in various lepidopteran and coleopteran sex pheromones has successfully been replaced by alkane (Kanno et al. 1980), alkene (Carlson and McLaughlin 1982) and formate (Mitchell et al. 1975, Beevor and Campion 1979, Beevor et al. 1981, McLaughlin et al. 1981) functionalities. Metcalf and Lampman (1989) have recently found the nitrile group mimicking the aldehyde moiety of *p*-methoxycinnamaldehyde, an effective kairomonal lure for the adult western corn rootworm, *Diabrotica virgifera virgifera* LeConte.

During our recent studies with aldehyde mimics, a series of pheromone analogues in which the nitrile group replaces the labile aldehyde functionality, has been prepared (Ujváry et al. 1993). The results with the cranberry girdler moth, *C. topiaria*, an important pest of grass grown for seed production and fir tree seedlings in nurseries and cranberries in North America, are briefly summarised here. The main component of the sex pheromone produced by the females of the species is (*Z*)-11-hexadecenal (**4**) the activity of which can be synergised by a low percentage of (*Z*)-9-hexadecenal (**5**) (McDonough and Kamm 1979). Comparative field experiments with 100:3 mixtures of the aldehydes **4** and **5** and with their bioisosteric (*Z*)-11-hexadecenonitrile (**6**) and (*Z*)-9-hexadecenonitrile (**7**) (Figure 2) indicated that the analogue mixture was attractive to *C. topiaria* males although it was not as active as the aldehyde blend (Ujváry et al. 1993).

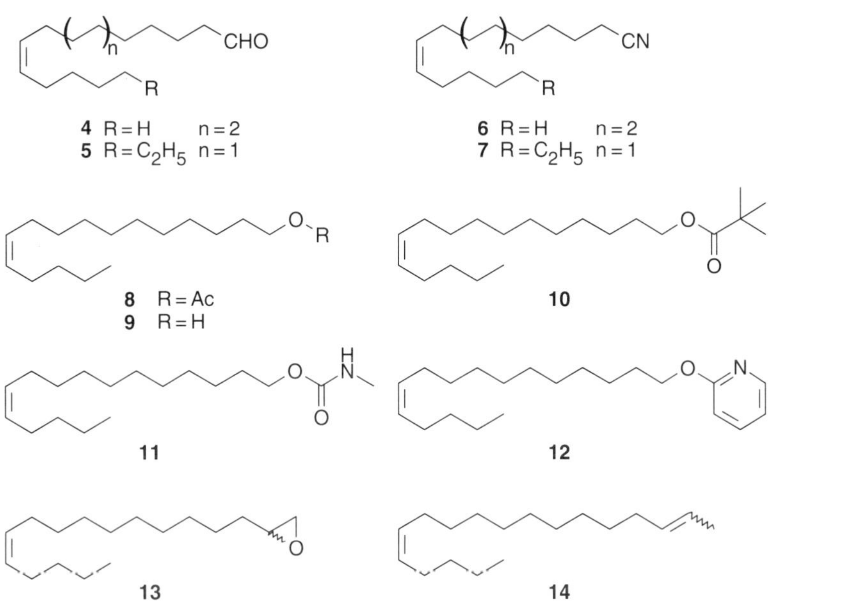


Figure 2. Structure of the test compounds. Natural pheromones **4**, **5**, **8**, and **9**; analogues **6**, **7**, and **10-14**.

## ANALOGUES OF PHEROMONAL ACETATES

We have recently prepared a number of analogues in which the acetate group of (*Z*)-11-hexadecenyl acetate (**8**), a common pheromone component of Noctuidae, was replaced by bioisosteric and other functionalities (for a recent study with acetate bioisosteres, see Gustavsson et al. 1997). The following analogues were studied: (*Z*)-11-hexadecenyl pivalate (**10**), (*Z*)-11-hexadecenyl *N*-methylcarbamate (**11**), 2-[(*Z*)-11-hexadecenyloxy]pyridine (**12**), (±)-(*Z*)-1,2-epoxy-12-heptadecene (**13**), and (2*E*/*Z*,13*Z*)-2,13-octadecadiene (**14**) (Figure 2). The effect of these compounds alone or in combination with the sex pheromone of the clover cutworm, *Discestra trifolii* Hufnagel, was tested (Figures 3 and 4). The sex pheromone blend of this insect contains **8** and (*Z*)-11-hexadecenol (**9**) as a minor synergistic component (Underhill et al. 1976). The traps were baited with rubber tubes impregnated with 100 μg of the 100:2 mixture of **8** and **9**,

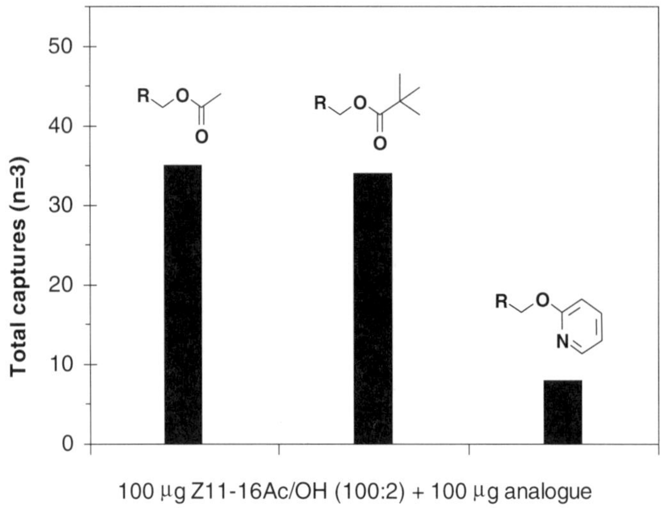


Figure 3. Effect of analogues (100 µg) on the total captures of *Discestra trifolii* in traps baited with 11Z-16Ac and 11Z-16OH (100 µg, 100:2 ratio). 27 July – 29 August 1995, vineyard near Dömsöd, Hungary, 3 replicates; R = (*Z*)-$CH_3(CH_2)_3CH{=}CH(CH_2)_9$.

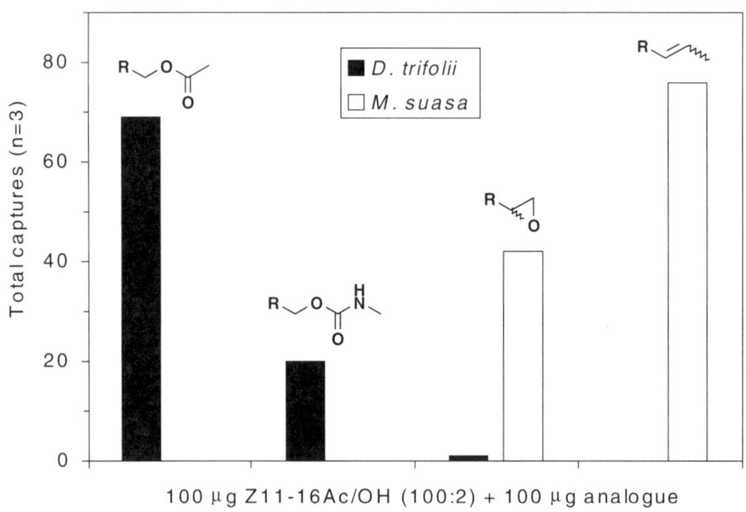


Figure 4. Effect of analogues (100 µg) on the total captures of *Discestra trifolii* and *Mamestra suasa* in traps baited with 11Z-16Ac and 11Z-16OH (100 µg, 100:2 ratio). 22 July – 8 August 1996, vineyard near Dömsöd, Hungary; 3 replicates; R = (*Z*)-$CH_3(CH_2)_3CH{=}CH(CH_2)_9$.

the optimal attractant for the males of this insect, and with 100 μg of the appropriate analogue. Field tests showed that none of the synthetic analogues were attractive alone or in combination with the minor component **9** (data not shown). In assays for synergistic or inhibitory activity, the pivaloyl derivative **10** lacked any influence on the attractivity of the sex pheromone mixture at the dosage tested (Figure 4). However, the carbamate analogue **11** that was designed (Ujváry and Tóth 1987) on the basis of related carbamate pheromone analogues (Albans et al. 1984) reduced trap captures (Figure 3). The novel pyridyl ether **12** was also inhibitory (Figure 3). Interestingly, the epoxide **13** and the olefin **14** analogues not only inhibited the *D. trifolii* captures by the pheromone blend but were also attractive to sympatric *Mamestra suasa* Denis & Schiffermüller males. Since the main sex pheromone component of *M. suasa* is the acetate **8** which is synergised by (*Z*)-11-hexadecenal (**4**), a minor pheromone component of the species required for optimal and selective attraction (Tóth et al. 1986), these two synthetic compounds evidently mimic the aldehyde **4**. Moreover, the absence of the males of the former species in traps containing **8** and **13** or **14** can be explained by these compounds being analogues of aldehyde **4**, a known pheromonal inhibitor for *D. trifolii*. However, the conversion of these synthetic compounds to aldehyde **4**, via decomposition similar to the one described recently for hydrocarbon dienes by Bartelt and Jones (1983), cannot be excluded.

## ANALOGUES OF TSETSE KAIROMONES

Tsetse flies (*Glossina* spp.) are the principal vectors of the African trypanosomiasis diseases of humans and domestic livestock. One of the environmentally benign tsetse control methods relies on baits containing kairomonal attractants. Analysis of the odours of host animals identified several chemicals which are now widely used in traps of various designs (Green 1994, Brightwell et al. 1997). For *G. morsitans morsitans* Westwood and *G. pallidipes* Austen, the attractive components of ox breath are $CO_2$, acetone (Vale 1980) and 1-octen-3-ol (**15**) (Hall et al. 1984, Vale and Hall 1985). The attractivity of buffalo and ox urine to these *Glossina* species is due to a phenolic fraction (Hassanali et al. 1986) that contains 3-*n*-propylphenol (**16**) as one of the active principles (Bursell et al. 1988). Several research groups have recently studied the structure-activity relationship of these tsetse kairomones (Saini et al. 1989, Van Der Goes Van Naters et al. 1996).

To improve the efficiency of the presently used traps and also to find compounds that are attractive to other important *Glossina* species, e.g., members of the *palpalis* and *fusca* groups that respond poorly to the presently known odours, we decided to synthesise and test in the laboratory a series of kairomonal analogues such as two new cycloaliphatic analogues **17** and **18** that were designed to mimic the structure of (*R*)-1-octen-3-ol (**15**) and 3-*n*-propylphenol (**16**) (Figure 5).

The isomeric mixtures of (±)-3-*n*-propylcyclopentanol (**17**) and (±)-3-*n*-propylcyclohexanol (**18**) were synthesised and tested in electrophysiological experiments (Ujváry et al. 1998). Gas liquid chromatography coupled with electroantennographic detection using antennae of *G. brevipalpis* Newstead showed that both the *cis* and *trans* isomer of **17** elicited an average 54% EAG response relative to (±)-1-octen-3-ol, its open chain congener. Of the related isomeric mixture of **18**, only the major isomer of the synthetic mixture was active (25 % response). The natural host odour component 3-*n*-propylphenol (**16**) elicited a 27 % response in this assay. Preliminary field experiments with **17** and **18**, however, were not equivocal (Oloo, personal communication) indicating that the stimulation of chemoreceptors observed for these compounds in the EAG test, is not a reliable indicator of behavioural responses. Nevertheless, the structural similarity noted for 1-octen-3-ol (**15**) and 3-*n*-propylphenol (**16**) and the observed EAG activity for the cyclic analogues warrant further studies.

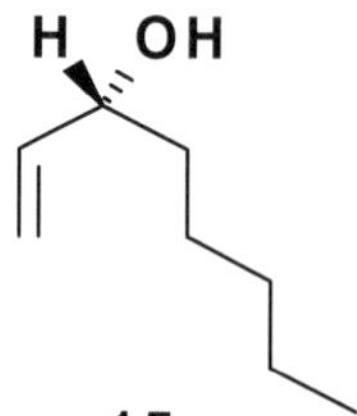


**15**

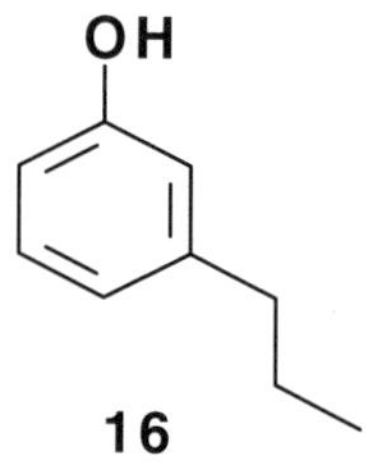


**16**

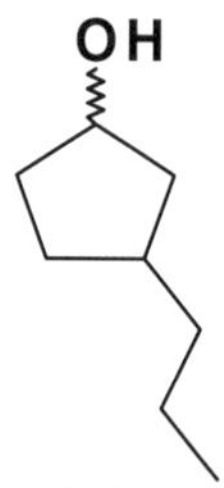


**17**

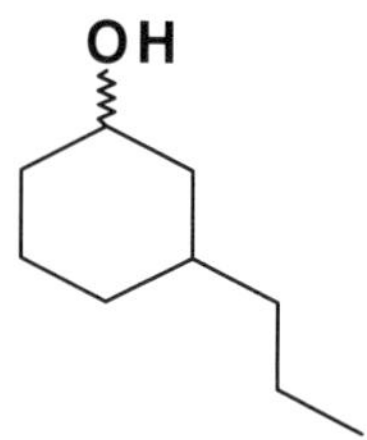


**18**

Figure 5. Structure of tsetse kairomones (*R*)-1-octen-3-ol (**15**) and 3-*n*-propylphenol (**16**) and alicyclic analogues **17** and **18**.

## SUMMARY

Many synthetic insecticides were designed by chemical modifications of natural products affording analogues that, in many cases, bear little resemblance to the original lead compounds (e.g., ether type pyrethroids and juvenoids, and the ecdysone agonist diacyl-hydrazines). Natural semiochemicals are highly specific and apparently safe (Inscoe and Ridgway 1992), thus ideally suited to environmentally sound insect control programmes. Due to their particular mode of action and, in some cases, scarce availability, their wide-scale field application is often limited. It is believed that synthetic analogues could solve these problems. The commercially successful analogues should be: 1) inexpensive (trimedlure vs. α-copaene), 2) effective and stable enough for extended periods (e.g., aldehyde analogues), and 3) safe, that is non-toxic to non-target species. Future research and developmental work in this field will rely on computer-aided molecular modelling techniques as well as behavioural, biochemical and receptor studies and will lead to synthetic alternatives of natural semiochemicals, pheromones in particular. In this paper, we have described several structural variations of kairomones and pheromones giving analogues which, in some cases, possessed unique biological properties that might help to provide an understanding of the intricate chemistry of insect olfaction and provide new prototype compounds for further research.

## ACKNOWLEDGEMENTS

This work was supported by the International Atomic Energy Agency (Research Contract No. 302-D4-HUN-8342) and by the Hungarian Scientific Research Fund (Grant No. T014343).

## REFERENCES

Albans, K.R., R. Baker, O.T. Jones, A.R. Jutsum and M.D. Turnbull. 1984. Inhibition of response of *Heliothis virescens* to its natural pheromone by antipheromones. Crop Prot. 3: 501-506.

Bartelt, R.J. and R.L. Jones. 1983. (*Z*)-10-Nonadecenal: a pheromonally active air oxidation product of (*Z,Z*)-9-19 dienes in yellowheaded spruce sawfly. J. Chem. Ecol. 9: 1333-1341.

Beevor, P.S. and D.G. Campion. 1979. The field use of 'inhibitory' components of lepidopterous sex pheromones and pheromone mimics. pp. 313-325. *In* Chemical ecology: odour communication in animals. F.J. Ritter (ed.). Elsevier/North-Holland Biomedical Press, Amsterdam.

Beevor, P.S., V.A. Dyck and G.S. Arida. 1981. Formate mimics as mating disruptants of the striped rice borer moth, *Chilo suppressalis* (Walker). pp. 302-311. *In* Management of insect pests with semiochemicals: concepts and practice. E. R. Mitchell (ed.). Plenum Press, New York.

Brightwell, R., R.D. Dransfield, P. Stevenson and B. Williams. 1997. Changes over twelve years in populations of *Glossina pallidipes* and *Glossina longipennis* (Diptera: Glossinidae) subject to varying trapping pressure at Nguruman, south-west Kenya. Bull. Entomol. Res. 87: 349-370.

Bursell, E., A.J.E. Gough, P.S. Beevor, A. Cork, D.R. Hall and G.A. Vale. 1988. Identification of components of cattle urine attractive to tsetse flies, *Glossina* spp. (Diptera: Glossinidae). Bull. Entomol. Res. 78: 281-291.

Cardé, R.T. and A.K. Minks. 1995. Control of moth pests by mating disruption: successes and constraints. Annu. Rev. Entomol. 40: 559-585.

Carlson, D.A. and J.R. McLaughlin. 1982. Diolefin analog of a sex pheromone component of *Heliothis zea* active in disrupting mating communication. Experientia 38: 309-310.

Cunningham, R.T. 1989. Parapheromones. pp. 221-230. *In* Fruit flies. Their biology, natural enemies and control. A. S. Robinson and G. Hooper (eds.). Vol 3A, Elsevier, Amsterdam.

Flath, R.A., R.T. Cunningham, T.R. Mon and J.O. John. 1994. Male lures for Mediterranean fruitfly (*Ceratitis capitata* Wied.): structural analogs of α-copaene. J. Chem. Ecol. 20: 2595-2609.

Fornasiero, U., A. Guiotto, G. Caporale, F. Baccichetti and L. Musajo. 1969. Identificazione della sostanza attrattiva per i maschi della *Ceratitis capitata*, contenunuta nell'olio essenziale dei semi di *Angelica archangelica*. Gazz. Chim. Ital. 99: 700-710.

Green, C.H. 1994. Bait methods for tsetse fly control. Adv. Parasitol. 34: 229-291.

Gustavsson, A.L., M. Tuvesson, M.C. Larsson, W. Wenqi, B.S. Hansson and T. Liljefors. 1997. Bioisosteric approach to elucidation of binding of the acetate group of a moth sex pheromone component to its receptor. J. Chem. Ecol. 23: 2755-2776.

Hall, D.R., P.S. Beevor, A. Cork, B.F. Nesbitt and G.A. Vale. 1984. 1-Octen-3-ol. A potent olfactory stimulant and attractant for tsetse isolated from cattle odours. Insect Sci. Appl. 5: 335-339.

Hassanali, A., P.G. McDowell, M.L.A. Owaga and R.K. Saini. 1986. Identification of tsetse attractant excretory products of a wild host animal, *Synercus caffer*. Insect Sci. Appl. 7:5-9.

Inscoe, M.N. and R.L. Ridgway. 1992. Non-target effects of lepidopteran sex attractant pheromones. pp. 49-59. *In* Insect pheromones and other behaviour-modifying chemicals: applications and regulation. BCPC Monographs No. 51. The British Crop Protection Council, Farnham.

Kanno, H., M. Hattori, A. Sato, S. Tatsuki, K. Uchiumi, M. Kurihara, J. Fukami, Y. Fujimoto and T. Tatsuno. 1980. Disruption of sex pheromone communication in the rice stem borer moth, *Chilo suppressalis* Walker (Lepidoptera: Pyralidae) with sex pheromone components and analogues. Appl. Entomol. Zool. 15: 465-473.

McDonough, L.M. and J.A. Kamm. 1979. Sex pheromone of the cranberry girdler, (*Chrysoteuchia topiaria* Zeller) Lepidoptera: Pyralidae. J. Chem. Ecol. 5: 209-217.

McLaughlin, J.R., E.R. Mitchell and J.H. Cross. 1981. Field and laboratory evaluation of mating disruptants of *Heliothis zea* and *Spodoptera frugiperda* in Florida. pp. 243-251. *In* Management of insect pests with semiochemicals: concepts and practice. E. R. Mitchell (ed.). Plenum Press, New York.

McGovern, T.P. and R.T. Cunningham 1988. Persistent attractants for the Mediterranean fruit fly, the method of preparation and method of use. U.S. Patent 4, 764, 366, Aug. 16. 5 pp.

Metcalf, R.L. and R.L. Lampman. 1989. Estragole analogues as attractants for corn rootworms (Coleoptera: Chrysomelidae). J. Econ. Entomol. 82: 123-129.

Metcalf, R.L. and E.R. Metcalf. 1992. Fruit flies of the family Tephritidae. pp. 109-152. *In* Plant kairomones in insect ecology and control. R. L. Metcalf and E. R. Metcalf (eds.). Chapman and Hall, New York.

Mitchell, E.R., M. Jacobson and A.H. Baumhover. 1975. *Heliothis* spp.: disruption of pheromonal communication with (*Z*)-9-tetradecen-1-ol formate. Environ. Entomol. 4: 577-579.

Payne, T.L., H.H. Shorey and L.K. Gaston. 1973. Sex pheromones of Lepidoptera. XXXVIII. Electroantennogram responses of *Autographa californica* to *cis*-7-dodecenyl acetate and related components. Ann. Entomol. Soc. Am. 66: 703-704.

Roelofs, W.L. and A. Comeau. 1971. Sex pheromone perception: synergists and inhibitors for the red-banded leaf roller attractant. J. Insect Physiol. 17:435-448.

Saini, R.K., A. Hassanali and R.D. Dransfield. 1989. Antennal responses of tsetse to analogues of the attractant 1-octen-3-ol. Physiol. Entomol. 14: 85-90.

Shaver, T.N. and G.W. Ivie. 1982. Degradation products of (*Z*)-11-hexadecenal and (*Z*)-9-tetradecenal, components of a sex pheromone of the tobacco budworm (Lepidoptera: Noctuidae). J. Agric. Food Chem. 30: 367-371.

Tóth, M., G. Szöcs, C. Löfstedt, B.S. Hansson and M. Subchev. 1986. Sex pheromone components of *Mamestra suasa*: chemical analysis, electrophysiological activity, wind tunnel activity and field tests in two European countries. Entomol. Exp. Appl. 42: 291-299.

Ujváry, I. and M. Tóth. 1987. Unpublished study cited by G. D. Prestwich: Chemical studies of pheromone reception and catabolism. pp. 473-527. *In* Pheromone biochemistry. G. D. Prestwich and G. Blomquist (eds.). Academic Press. New York.

Ujváry, I., J.C. Dickens, J.A. Kamm and L.M. McDonough. 1993. Natural product analogs: stable mimics of aldehyde pheromones. Arch. Insect Biochem. Physiol. 22: 393-411.

Ujváry, I., V. Tóth and P. Guerin. 1998. Structure-activity studies of new analogues of the tsetse kairomones 1-octen-3-ol and 3-propylphenol. pp 27-46. *In* First Research Co-ordination Meeting on "Improved Attractants for Enhancing the Effectiveness of Tsetse Fly Suppression Operations and Barrier Systems Used in Tsetse Control/Eradication Campaigns". IAEA-D4-RC-631.

Underhill, E.W., W.F. Steck and M.D. Chisholm. 1976. Sex pheromone of the clover cutworm moth, *Scotogramma trifolii*: isolation, identification and field studies. Environ. Entomol. 5: 307-310.

Vale, G.A. 1980. Field studies of the responses of tsetse flies (Glossinidae) and other Diptera to carbon dioxide, acetone and other chemicals. Bull. Entomol. Res. 70: 563-570.

Vale, G.A. and D.R. Hall. 1985. The role of 1-octen-3-ol, acetone and carbon dioxide in the attraction of tsetse flies, *Glossina* spp. (Diptera: Glossinidae), to ox odour. Bull. Entomol. Res. 75: 209-217.

Van Der Goes Van Naters, W.M., L. Bootsma, C.J. Den Otter and R.G. Belemtougri. 1996. Search for tsetse attractants: a structure-activity study on 1-octen-3-ol in *Glossina fuscipes fuscipes* (Diptera: Glossinidae). J. Chem. Ecol. 22: 343-355.

Warthen, J.D., Jr., R.T. Cunningham, A.B. DeMilo and S. Spencer. 1994. *trans*-Ceralure isomers: differences in attraction for the Mediterranean fruit fly, *Ceratitis capitata*. J. Chem. Ecol. 20: 569-578.

Warthen, J.D., Jr., W.F. Schmidt, R.E. Doolittle and R.T. Cunningham. 1995. Structure-activity relationship observations of *trans*-trimedlure enantiomers. J. Chem. Ecol. 21: 69-79.

# Integration of the Sterile Insect Technique (SIT) and Autosterilisation Lethality in the Eradication of *Glossina fuscipes fuscipes* Newst. (Diptera: Glossinidae) in Buvuma Islands in Lake Victoria, Uganda

*F.P. Oloo*[1], *P.A. Langley*[2], *F. Luyimbazi*[3] *and L.M. Ogwal*[4]

[1] Ministry of Agriculture, Veterinary Department, P. O. Kabete, Nairobi, Kenya
[2] Insect Investigations Ltd., School of Pure & Applied Biology, University of Wales, Cardiff
[3] Ministry of Agriculture, Animal Resources and Fisheries, P. O. Box 102, Kampala, Uganda
[4] COCTU, Entebbe, Uganda

## INTRODUCTION

The years of 1970s to 1990s saw the development of bait technology for the control of tsetse flies from their infancy to their optimum levels. The technology relies on attracting the flies to trapping and killing devices, relying on olfactory and visual cues to artificial and natural objects which the flies probably perceive as their hosts. Despite the present level of improvement, the catching or trapping efficiency is low, averaging 20-35%, and some tsetse species hardly respond to these techniques. This makes sustaining control programmes to a level where economic development can proceed effectively elusive, leading to the re-invassion and collapse of many tsetse control projects. Efforts are therefore being made to incorporate different killing methods into the trapping systems. Lethal insect techniques (LIT) with pathogens, insect growth regulators and other chemosterilants are incorporated into bait technology to amplify their effectiveness as the flies can transfer them to other members of the species which cannot get access to the attractive devices. They are compatible with the sterile insect technique. Prospects for autosterlisation of the tsetse flies have been reviewed by Langley and Coates (1982) who also assessed the incorporation of sex pheromones and bisazir in the field. Encouraging field results were obtained with Pyropnxyfen (Sumitomo Chemical Co.) which is a juvenile hormone mimic that allows the larvae of *G.morsitans morsitans* to be produced normally but on pupation, further development was arrested after twenty days of the thirty day intrapupal period, "effectively" making the females sterile (Hargrove and Langley 1990). However, when triflumuron is applied at doses of 0.5 micrograms per tsetse fly the following reproductive cycle was arrested with no recovery over four reproductive cycles (forty five days). Birth products of the sterilised flies ranged from abortion of eggs to fully grown larvae which formed non-viable puparia. The aim of this trial was to assess the effectiveness of triflumuron on *G. f. fuscipes.*

## MATERIALS AND METHODS

The Buvuma Islands have a total area of about 220 square kilometres. The irregular shape of the main island allows for utilisation of narrow necks which can be reinforced with insecticide impregnated pyramidal traps to make barriers separating the flies in the experimental blocks from the rest of the population on the islands (Figure 1). Insecticide impregnated traps which are widely used to control *G. f fuscipes* in the country are known to make effective barriers. However, in-built collecting cages in the traps are left open as accumulating dead flies attract predators which destroy the traps in the field.

The experimental blocks were identified as Banga (Block A), Buwanzi (Block B), Bugabo (Block C) and Mawanga (Block D). The positions of the narrowest part of the neck on both lake shores in each block were fixed using the Global Positioning System (GPS).

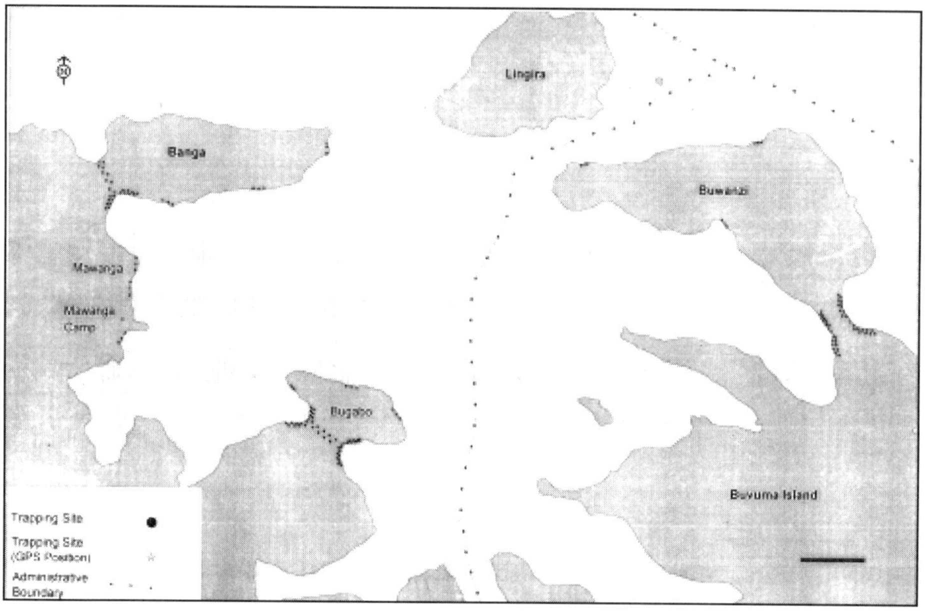


Figure 1. Buvuma Island working area.

## Block A, Banga

The neck separating Banga from the main island is approximately 500 m wide. The experimental block rises to an altitude of 1,215 m above sea level, the lake being approximately 1,145 m. The southern lake shore vegetation is dense and separated from the northern shore by a ridge of relatively high altitude which is more open, especially along the rocky eastern corner. There was apparently no free movement of the flies between the two shores across the ridge. The area of the experimental block was estimated to be 2.5 square kilometres, parts of the lake shore being covered with banana, sweet potato and cassava plantations which give way to swampy shores on the Lukoma and Mawanga sides.

### Block B, Buwanzi

The narrow point of Buwanzi was approximately 350 m wide. The area is well cultivated with stands of forest. Both shores are covered with reeds and papyrus swamps. The block had an area of about 5.0 sq km rising to nearly 1,219 m.

### Block C, Bugabo

The neck of Bugabo is about 600 m wide, comprising dense forest with papyrus swamps along the lake shore. As the neck is relatively wide, the barrier might have a high re-invasion pressure. The experimental block was about 1 sq km rising to 1,158 ft.

### Block D, Mawanga

The block extended from Mawanga Point for about 2 km. The lake shore is intensively cultivated except around Mawanga Point which is rocky. Since this is the untreated control area and is also not a peninsula, barriers were not required.

## Experimental Design

The pilot trial compared the efficacy of insecticide impregnated traps and traps fitted with triflumuron treated contaminating devices (CDs) in reducing *G. f. fuscipes* populations on the islands. Insecticide impregnated traps and triflumuron were placed in the Banga and Buwanzi experimental blocks respectively. Bugabo was a treatment control block where the use of CDs without insect growth regulator (IGR) was evaluated to assess their effect on trap catches. The CDs were used in the Buwanzi experimental block to dispense the IGR. Mawanga was the non-treatment control block in which the tsetse population was monitored without any other treatment. This was to enable the influence of the barriers in the three blocks to be assessed.

The barriers were set up at the narrowest points of the land using pyramidal traps suspended at knee height either from tree branches or provided support before the monitoring transects were established. The traps were impregnated with a 20% deltamethrin suspension concentrate (Glossinex 200 sc Cooper) diluted to 0.3%. At this concentration, a lethal dose will be in the fabrics under field conditions for up to six months when the traps are replaced. At the normally used concentration of 0.1%, the traps need to be re-impregnated after three months.

The barrier was created by two lines of traps 50 m apart, placed along both sides of the lake shores starting from the narrowest point and extending for 500 m to either side. Similarly, two lines of traps were extended in the hinterland to link the narrowest points. The traps were placed at 100 m intervals along the barrier lines, those in the second line being positioned midway between the ones in the first.

Three barriers were set at Banga, Buwanzi and Bugabo using 30, 42 and 48 traps respectively (Figure 1). The traps were impregnated with 0.3% deltamethrin s.c. and were replaced with new ones after 5-6 months during the experiment. A total of 36 monitoring traps were used to establish baseline data in the four experimental blocks from the month of August 1997 to January 1998.

### Setting Monitoring Transects for Baseline Data

A monitoring procedure was established in three areas in each block. These positions were fixed using the GPS. Each area had three trapping sites 50 m apart. Pyramidal traps with top tsetse collecting cages were used to monitor the population. The metal trap support above the ground was greased to stop predators, especially ants, from eating the flies in the cages. Where necessary, the trap positions were adjusted to avoid windy and dark places in order to increase their performance. They were emptied after 24 hours for three consecutive days, the catches being recorded as males and females. All the traps were removed at the end of the monitoring period and monitoring was repeated monthly to provide baseline data before the treatments of the experiment.

## RESULTS

The baseline data show that the tsetse population was generally high on all the four sites with a mean between 0-90 flies per trap per day. There was an increase in the fly population in November and December, apparently due to increased El Nino related rains which occurred generally in the East African region.

## DISCUSSION

The IGR approach to tsetse control is more environment friendly and compatible with the sterile insect technique than the insecticide impregnated traps. Unlike insecticide based suppression methods, the technique can be used simultaneously with, and to complement effects of, the released sterile males to control tsetse populations. The released flies may also be contaminated with the IGR when they pass through the CDs and transfer it to females during mating. The results of the experimental phase will provide the much needed information on additional tools to present tsetse control techniques, most of which rely on instant kill before the fly transfers lethality to others in their habitat. The merits of this approach have been described by Langley and Weidhaas (1986).

## REFERENCES

Hargrove, J.W. and P.A. Langley. 1990. Sterilisation of tsetse (Diptera:Glossinidae) in the field: a successful trial. Bull. Entomol. Res. 80, 397-403.

Langley, P.A. and T.W. Coates 1982. Prospects for autosterilisation of tsetse flies, *Glossina* species (Diptera: Glossinidae), using sex pheromone and bisazir in the field. Bull. Entomol. Res. 72, 319-327.

Langley, P.A. and D. Weidhaas. 1986. Trapping as a means as a means of controlling tsetse, *Glossina* spp. (Diptera: Glosasinidae): the relative merits of killing and sterilisation. Bull. Entomol. Res. 76, 89-95.

# Insect Cold Storage: Using Cryopreservation and Dormancy as Aids to Mass Rearing

*Roger A. Leopold*

US Department of Agriculture, Agricultural Research Service, Biosciences Research Laboratory, Fargo, North Dakota, 58105 USA

## INTRODUCTION

Low temperature storage is a strategy often used for obtaining efficiency and flexibility in the mass rearing of insects for release in bio-control and sterile insect technique (SIT) programmes. Numerous studies have been conducted which have assessed tolerance of various insect and acarid species to subambient temperatures in attempts to gain the benefits of extending shelf life, to accumulate for release, to inactivate or hold during shipment and to preserve indefinitely for other sundry reasons (Leopold 1998). Tolerance to cold temperatures varies greatly between species and between life stages within individual species. Many insects and acarids regularly tolerate long periods of cold or even freezing temperatures by utilising their inherent cold tolerance mechanisms while others die after only short exposures to non-freezing temperatures (Bale 1987, Lee 1989, Morris et al. 1983 and Sømme 1995). This report gives information on two approaches, namely cryopreservation and dormancy, which can be used to enable insects to survive a cold storage period and also reviews research progress using adaptations of these two approaches.

## CRYOPRESERVATION VS. DORMANCY

The use of either of these two approaches in the cold storage of insects and acarids requires that certain knowledge of developmental, physiological and physical aspects of these animals be known and exploited. For example, cryopreservation involves combined chemical and physical manipulations of cells to enable storage at liquid nitrogen temperatures. It is important to determine how cellular integrity can be maintained in the presence of potentially damaging processes such as dehydration, introduction of multi-molar concentrations of protective solutes and freezing temperatures. Likewise, the use of dormancy as a cold storage method requires that the organisms be prepared physiologically and be developmentally competent to survive extended low temperature exposure. Successful dormancy induction requires that precise manipulation of environmental conditions be made during the mass rearing regime. Further, difficulty in making the choice of using either dormancy or cryopreservation is usually not an issue. The approach can be easily ascertained after consideration of the number of insects to be stored, the length of storage period, the stage of development desired for storage, the inherent physiological and developmental capabilities of the organism and the respective logistics involved.

## Cryopreservation

Cryopreservation typically refers to procedures which have been developed for processing cells, tissues or organisms for long-term storage at liquid nitrogen temperature. The general principles of the cryopreservation methodology and some of the barriers to the successful cryopreservation of insects have been described in earlier communications (Leopold 1991, 1993). Briefly, the process of cryopreservation is based on adequate water management. Cryopreservation of cells and multicellular organisms requires that sufficient water be removed to avoid lethal ice formation upon exposure to sub-zero temperatures. A cryoprotectant (CPA) solution is used to replace the lost water in order that intra- and extracellular solutes do not reach toxic levels by keeping them in the aqueous phase during the freezing process. Cryoprotectants can also protect cells on a colligative basis and it has been suggested that these chemicals may help avoid damage by stabilising unfreezable water, changing ice crystal configuration, and by resisting denaturation of essential cellular components during dehydration (Mazur 1984, Merryman 1974, Storey and Storey 1988, Franks 1985, Baust 1973).

Cryopreservation protocols place constraints on the choice of the developmental stage of an insect or acarid that can be cryopreserved. Since it is imperative that the cells of an organism be freely permeable to water loss and the influx of cryoprotectants, only those forms which are already permeable or can be rendered permeable without inflicting significant damage are candidates for treatment with this technique. Immature and adult stages of most insects and mites are equipped with an integument which resists water loss and permeation. Thus, in most cases, the stage of development that will be amenable to treatment with cryopreservation methods is the embryonic stage. This stage is most desirable because of the relative ease of permeabilisation, handling logistics involved with cryopreservation protocols and the intrinsic developmental plasticity that is often displayed during embryogenesis. Further, cryopreservation would not ordinarily be a method of choice for storing insects or acarids in a programme where there was a need for immediate release, the number to be released was frequent and great or the storage period required was relatively short. This method has the greatest potential for use in area-wide release programmes when the need exists for storing founder strains, standardised strains selected for a particular area or behaviour and strains to be used in genetic research.

To date, acceptable recovery from liquid nitrogen storage as cryopreserved embryos has been accomplished with only one insect, the vinegar fly, *Drosophila melanogaster* Meigen (Oregon R strain). Mazur et al. (1992) and Steponkus and Caldwell (1993) were able to gain > 80% hatching and 40-50% viable adults from a protocol which employed multi-molar concentrations of ethylene glycol as the cryoprotectant, ultra fast cooling for storage into liquid nitrogen and fast warming for recovery. This technique has not yet been incorporated into the routine storage of flies at the various *Drosophila* stock centres around the world. A similar lapse of time occurred between the development and the subsequent routine use of the method for liquid nitrogen storage of bovine and murine embryos (P. Mazur, personal communication). Nevertheless, this method requires further testing on strains having less viability or different developmental attributes than the Oregon R strain used in the initial studies.

Attempts at using the *Drosophila* method for cryopreservation of the embryos of several insect pests of livestock has required making several modifications relative to permeabilisation, cryoprotectant loading and unloading and the culture medium. However, these modifications have not as yet resulted in an effective method for preserving these flies. The data in Figure 1 show that there is a stepwise decline in the capability of the embryos to emerge as larvae after each successive phase of the procedure. The most noticeable decline in survival is the concentrating step going from 2.0 M to the 8.0 M ethylene glycol. This step involves raising the level of ethylene glycol in the embryos by removing water osmotically upon a brief exposure to the high concentration of ethylene glycol (Rall 1987). A high internal concentration of cryoprotectant is required to form a vitreous solution rather than ice crystals upon ultra fast cooling. The apparent toxicity of this particular step was not observed by Steponkus' or Mazur's groups in the development of their cryopreservation protocols for *D. melanogaster* embryos. Whether ethylene glycol has a direct toxic effect on the species tested (Figure 1) or certain essential elements are removed from the embryos upon removal of the cryoprotectant and replacement of the lost water is not known. Interestingly, after removal of the 8.0 M ethylene glycol, the embryos will develop to hatching, even after liquid nitrogen exposure, but they are unable to emerge from the vitelline membrane. Currently, other cryoprotectant solutions and culture media are being tested to alleviate this problem.

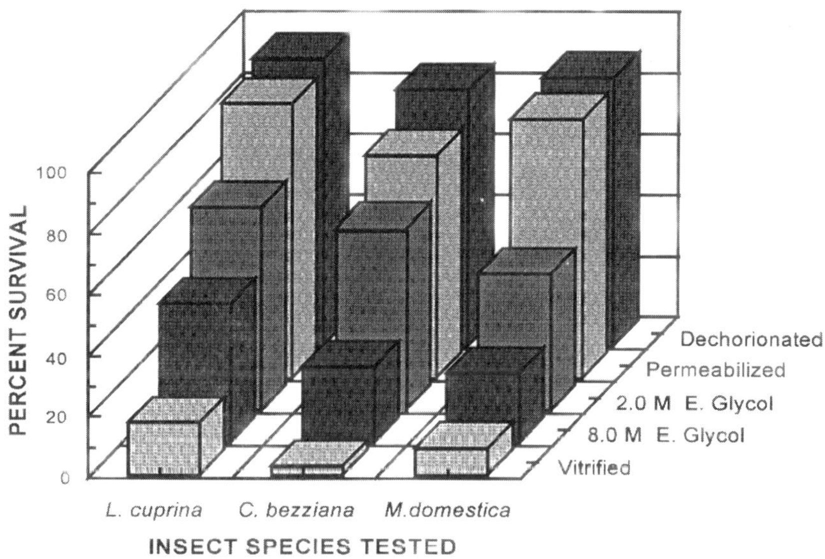


Figure 1. An illustration of larval survival of three species of flies after treatment with a protocol developed for cryopreservation of *Drosophila* embryos *(Lucilia cuprina* Wied., *Chrysomya bezziana* Villenueve and *Musca domestica* L.).

## Dormancy

Of the two strategies that could be employed for storage purposes, the induction of a dormancy mechanism is a far more conservative approach. It is one of the major strategies employed by insects and mites to survive harsh environmental conditions. Dormancy can generally be divided into two main categories, diapause and hibernal quiescence. Diapause can be an obligatory or facultative suspension of the development or activity of an organism. It is usually utilised by the organism to avoid periods of severe environmental stress such as winter, seasonal drought or lack of food sources.

Obligatory diapause is a dormant period that is embedded into the developmental programme and regulation or modification by external means is limited.

Facultative diapause is evoked by a prior programming through change in certain environmental conditions occurring over time. The use of the facultative type of diapause is most amenable for storage situations since it is subject to manipulation by external means. In some insect species, diapause may be elicited by signals occurring over two generations or there can be maternal effects which dictate induction or only modify the threshold of the response (Zaslavski 1988). The main environmental signal involved in regulating facultative diapause is change in day length, although temperature, food composition, relative humidity or rainfall may also act as stimuli for some species. Organisms in diapause may also require additional signals such as a change in day length, temperature and/or moisture to activate a recovery process and this can result in a delay in returning to the normal developmental programme or activity. Diapause is also different from quiescence because it is mediated by the organism's endocrine system and is usually limited to expression during a specific stage of development (Beck 1980, Denlinger 1985 and Tauber et al. 1986).

A dormant period expressed as a hibernal quiescence is usually elicited by temperature extremes but dehydration, anoxia and increased salt concentration can also induce this condition (Sømme 1995). In contrast to diapause, quiescence is more of an immediate response to adverse conditions, and more often, it results in a dramatic lowering of the organism's metabolic rate. Moreover, upon removal of the potentially harmful condition, the organism quickly recovers its ability to carry on normal life processes.

Unfortunately, all insects do not possess the capability to enter dormancy or if they do, the particular life stage having dormancy capabilities may not be the stage desired for storage. Further, there is a metabolic expense for maintaining dormancy which varies between species and with the severity of conditions to which the organism is exposed. Thus, damage can occur to organisms while in a dormant status if the conditions are too harsh or last too long.

Dormancy can be associated with, or modified by, cold hardening. Cold hardening refers to the physical and/or metabolic modifications that occur within an organism which enable it to survive low temperature situations such as overwintering in a cold climate. Like diapause, induction of a cold hardening response is dependent upon reception of an environmental signal which is usually low temperature, but can also be caused by change in day length or lack of water availability (Lee 1991). Cold hardening can be linked to dormancy and share the same environmental cues, occur with dormancy and have different cues or occur independently without induction of dormancy (Denlinger 1991). Unlike diapause, cold hardening is not manifested as an all or none response. It is typically expressed in proportion to the magnitude of the appropriate stimulus received.

It is implicit to remember when using dormancy and/or cold hardening to facilitate or extend a period of cold storage that there are often very precise regulatory cues which must be provided in the correct sequence and magnitude. Failure to correctly incorporate these cues into the mass rearing regime prior to the cold storage period will result in the release of an inferior insect or will reduce the yield of the mass production. Salt (1959) determined with hymenopteran, *Bracon cephi* Gahan, that the signals which induced a cold hardening response followed those which induced diapause. Thus, in such cases, placing diapausing insects under cold storage conditions without the proper cold hardening could produce irreparable damage.

In addition to cold hardening, dormancy/cold storage can also be enhanced or modified by other factors such as storage within a host, supplying sustenance, denying sustenance and providing recurrent recovery periods during storage. Aphid mummies (Whitaker-Deerberg et al. 1994, Liu and Tian 1987, Archer et al. 1973), house fly pupae (Fabritius 1980) and lepidopteran eggs (Vigil 1971, Gautam 1986) have been shown to be convenient and sometimes necessary sites for placing parasitoids into cold storage. Feeding during a pre-acclimation period, during storage and/or prior to shipment have also been used successfully in extending shelf life and increasing the survival of stored insects and mites (Leopold 1998). In contrast, feeding prior to cold exposure can be detrimental to some insects, especially if the storage temperature is to be below freezing. Ingestion of food or water has been associated with loss of cold hardiness (Sømme 1966, Baust and Morrisey 1975, Young and Block 1980, Cannon et al. 1985). Heterogenous gut contents have been attributed as a source of incidental ice nucleators which cause lethal freezing in supercooled, freeze intolerant species (Salt 1936, 1953, Hofsvang and Hågvar 1977, Zachariassen 1985).

Giving insects intermittent recovery periods during cold storage has been shown to increase shelf life. Rudolf et al. (1993) showed that cycling the storage temperature between 3°C and 13°C was substantially better than continuous storage at 9°C for the anthocorid predator, *Orius laevigatus* Fieber. Leopold et al. (1998) provided 2 or 3 hr recovery periods at 28°C every four days for pupae of three fly species stored at 10°C and found that survival was increased over that of continuous storage at the lower temperature (Figure 2). Further, when the recovery period was shorter or was given at longer intervals, the effect was not as great. The effect of providing recurrent recovery periods during cold storage was even greater when two of the blow fly species, *Lucilia cuprina* Wied. and *L. sericata* Meigen, were given only 4 hr/wk. Survival to adult emergence for *L. sericata* was > 75% after 90 days and for *L. cuprina,* it was > 85% after 60 days at 10°C (Leopold et al. 1998). It was hypothesised that the recovery periods were necessary for elimination of toxic products building up during storage and/or for repair of damage caused by chilling.

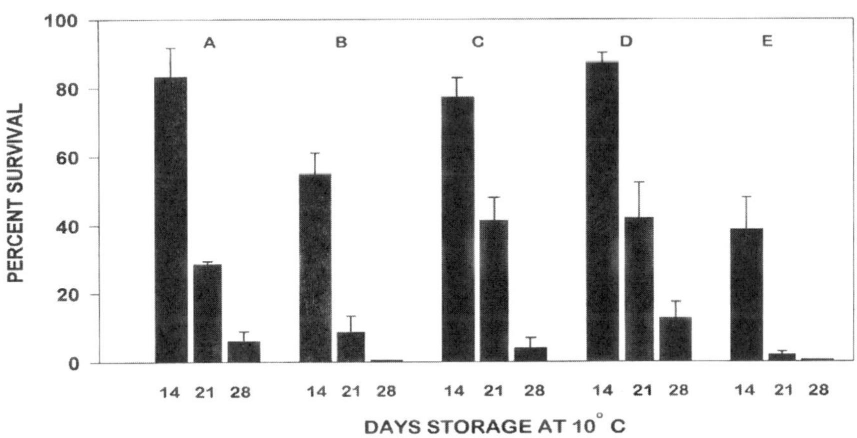


Figure 2. Data showing adult emergence of house fly pupae treated with various recurrent 28°C recovery regimes during a 21 day storage period at 10°C. Recovery regimes: A = 1 hr every other day; B = 1 hr/wk; C = 2 hr every 4th day; D = 3 hr every 4th day; E = no recovery (from Leopold et al. 1998).

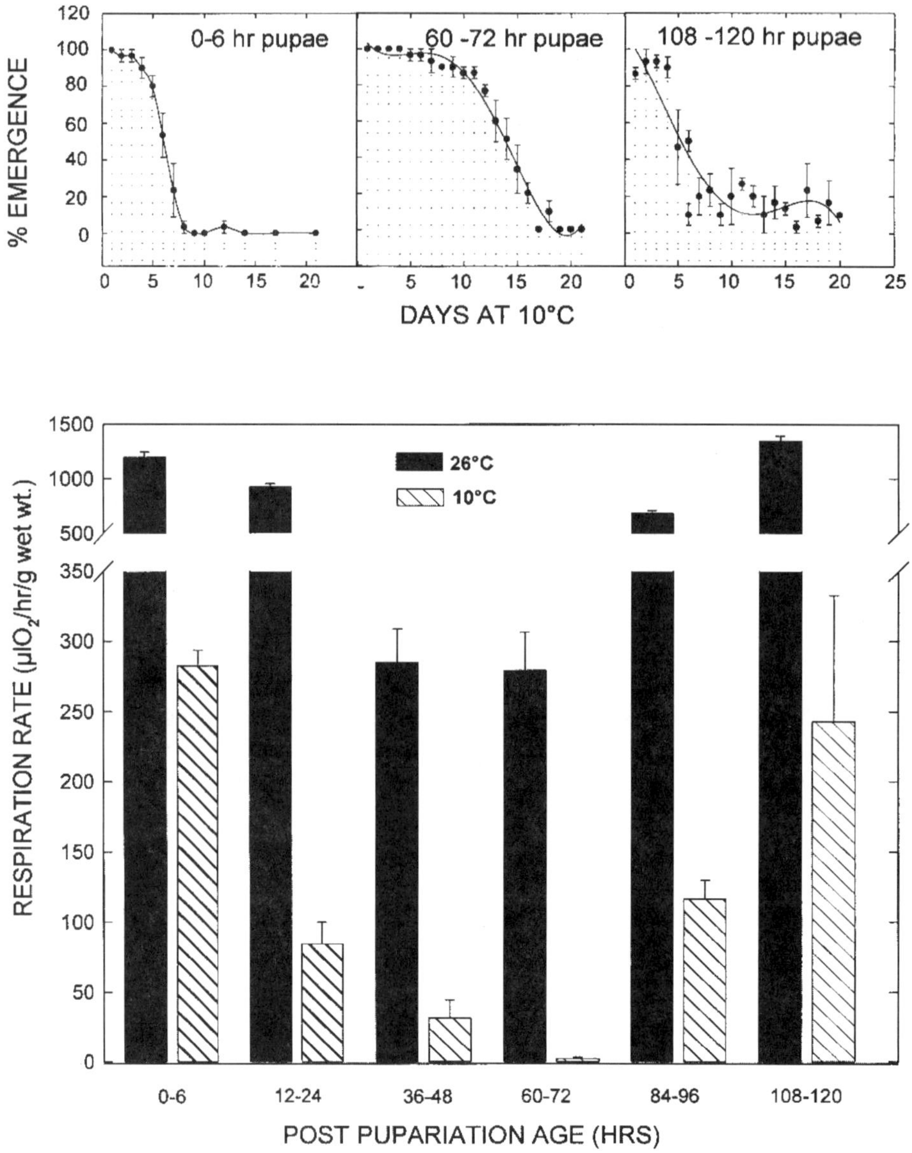


Figure 3. The upper 3 graphs show daily survival data (adult emergence) of house fly pupae placed in 10 °C storage at either 0 - 6, 60 - 72 or 108 - 120 hr after pupariation. The lower graph shows the respiration rates of the house flies measured throughout the pupal stage (modified from Leopold et al. 1998).

This study also showed that survival after storage was correlated to the respiration rate of a particular stage of pupal development. The data presented in the lower graph of Figure 3 show that respiratory rates of house fly pupae followed throughout the pupal stage represent a U-shaped curve. The insects at mid-stage have significantly lower respiratory rates than either the very young pupae or the pharate adults. When the insects are exposed to a storage temperature of 10°C, the respiration rates fall to negligible levels for the ages 36 to 72 hr old. When comparing these data to the upper 3 graphs in Figure 3, it illustrates that survival over time at a storage temperature of 10°C is greatest with those pupae having the lowest rate of respiration. The length of time for 50% survival for pupal ages 0-6 and 108-120 hr was 5.7 and 6.5 days, respectively, while that for the 60-72 hr group was 13.1 days. Thus, damage caused by continuous, prolonged cold storage can be minimised by choosing pupae in a stage of development where the ability to significantly lower respiration is greatest upon low temperature exposure.

In addition, the data in the upper three graphs of Figure 3 show that death of the pupae stored at 10°C begins to occur shortly after exposure to the cold with the 0-6 and 108-120 hr groups while death of the 60-72 hr group is not substantially expressed until the $10^{th}$ or $11^{th}$ day of storage. The 'shoulder-type' contour of this latter survival curve suggests that damage is accumulating with this group before it is latently expressed.

Combining a pre-storage cold hardening period of slow cooling (0.017°C/min) with a regime of recurrent recovery periods further increases the adult emergence of house fly pupae held in storage at 7°C (Leopold et al. 1998). Survival after 28 days of storage was > 55% as opposed to < 5% in the controls. The cold hardening/recurrent recovery treatment regime was effective with only the 12-24 hr age group while the recurrent recovery treatments alone enhanced survival for pupae placed in storage from 12 through 72 hr post pupariation.

## CONCLUSIONS

The two strategies described herein for use in the development of cold storage protocols for mass reared insects and mites have different requirements with corresponding outcomes. Cryopreservation relies on a chemical/physical processing which requires that close attention be given to developmental tolerance, membrane permeabilities, cryoprotectant efficiencies and toxicities and cooling/warming rates. Typically, this procedure would most often be used where there is a need for long-term storage as the medium for storage is usually liquid nitrogen and would not be an economical means for storing large numbers of organisms. In contrast, induction of dormancy generally relies upon some type of environmental manipulation over time, such as length of day and/or temperature. While dormancy induction, like cryopreservation, may have a strict developmental tolerance, some mechanisms such as cold hardiness and hibernal quiescence do not necessarily require that an insect or mite be in a specific stage of development. Further, storage time for dormancy mechanisms can be as short as one or two days and may range up to one year in length. Storage time depends upon individual species tolerance and/or the type of dormancy mechanism induced. These types of procedures would be most amenable to storage of large numbers of organisms at temperatures easily and economically maintained by mechanical refrigeration.

Development of a cryopreservation method for insect embryos has been accomplished thus far for only one species, *D. melanogaster.* Adaptation of this method for insects currently being used in SIT or to be used in future SIT programmes, such as the New World and Old World screwworm flies, has had limited success thus far. The apparent problem lies in the correct introduction of a CPA or in the formulation of a CPA with low toxicity.

The use of dormancy as a strategy for cold storage of large numbers of mass reared insects has had wide usage. Using cold acclimation and recurrent recovery periods during a cold storage period shows promise in combating chilling damage in the more cold intolerant species. The key to using quiescence mechanisms for cold storage lies in choosing the stage of development having the lowest metabolic activity. This presumably would also apply if storage involved the use of anoxic conditions.

It is expected that as more species are cultivated for use in area-wide control programmes, the need for storage, shipping, and maintenance technology for mass rearing these organisms will greatly increase. Dormancy and cryopreservation are two strategies that can help meet these needs.

## REFERENCES

Archer, T.L., C.L. Murray, R.D. Eikenbary, K.J. Starks and R.D. Morrison. 1973. Cold storage of *Lysiphlebus testaceipes* mummies. Environ. Entomol. 2: 1104-8.

Bale, J.S. 1987. Insect hardiness – an ecological perspective. J. Insect Physiol. 33: 899-908.

Baust, J.G. 1973. Mechanisms of cryoprotection in freezing tolerant animal systems. Cryobiology. 10: 197-205.

Baust, J.G. and R.E. Morrissey. 1975. Supercooling phenomenon and water content independence in the overwintering beetle, *Coleomegilla maculata.* J. Insect Physiol. 21: 1751-54.

Beck, S.D. 1980. Insect photoperiodism. 2nd ed. Academic Press, New York. 387 pp.

Cannon, R.J., W. Block and G.D. Collett. 1985. Loss of supercooling ability in *Cryptopygus antarcticus* (Collembola: Isotomidae) associated with water uptake. Cryoletters. 6: 73-80.

Denlinger, D.L. 1985. Hormonal control of diapause. *In* Comprehensive Insect Physiology, Biochemistry and Pharmacology. 8: 353-412. G.A. Kerkut and L.I. Gilbert (eds.). Pergammon Press, Oxford.

Denlinger, D.L. 1991. Relationship between cold hardiness and diapause. pp. 174-198. *In* Insects at low temperature. R.E. Lee and D.L. Denlinger (eds.). Chapman and Hall, New York.

Fabritius, K. 1980. Storage of pupae of *Musca domestica* L. parasitised by *Muscidifurax raptor* Gir. and Sand. (Hymenoptera: Chalcidoidea). Stud. Cercet. Biol. Anim. 32: 83-88.

Franks, F. 1985. Biophysics and biochemistry at low temperatures. Cambridge University Press, Cambridge. 210 pp.

Gautam, R.D. 1986. Effect of cold storage on the adult parasitoid *Telenomus remus* Nixon (Scelionidae: Hymenoptera) and the parasitised eggs of *Spodoptera litura* (Fabr.) (Noctuidae: Lepidoptera). J. Entomol. Res. 10: 125-31.

Hofsvang, T. and E.B. Hågvar. 1977. Cold storage tolerance and supercooling points of mummies of *Ephedrus cerasicola* Stary and *Aphidius colemani* Viereck (Hym. Aphidiidae). Norw. J. Entomol. 24: 1-6.

Lee, R.E. 1989. To freeze or not to freeze. Bioscience 39: 308-13.

Lee, R.E. 1991. Principles of insect low temperature tolerance. pp. 17-46. *In* Insects at low temperature. R.E. Lee and D.L. Denlinger (eds.). Chapman and Hall, New York.

Leopold, R.A. 1991. Cryopreservation of insect germplasm: cells, tissues and organisms. pp. 379-407. *In* Insects at low temperature. R.E. Lee and D.L. Denlinger (eds.). Chapman and Hall, New York.

Leopold, R.A. 1993. Advances in the preservation of insect germplasm. pp. 63-78. *In* Proceedings of FAO/IAEA International Conference on Management of Insect Pests: Nuclear and Related Molecular and Related Genetic Techniques. Vienna, Austria, IAEA-SM-327/27.

Leopold, R.A. 1998. Cold storage of insects for integrated pest management. pp. 235-67. *In* Temperature sensitivity in insects and application in integrated pest management. G.J. Hallman and D.L. Denlinger (eds.). Westview Press, Boulder.

Leopold, R.A., R.R. Rojas and P.W. Atkinson. 1998. Post pupariation cold storage of three species of flies: increasing chilling tolerance by acclimation and recurrent recovery periods. Cryobiology 36: 213-24.

Liu, J.J. and Y. Tian. 1987. Cold storage of *Encarsia formosa* Gahen. Chin. J. Biol. Cont. 3: 4-6.

Mazur, P. 1984. Freezing of living cells: mechanisms and implications. Am. J. Physiol. 247: 125-42.

Mazur, P., K.W. Cole, J.W. Hall, P.D. Schreuders and A.P. Mahowald. 1992. Cryobiological preservation of *Drosophila* embryos. Science 258: 1932-35.

Merryman, H.T. 1974. Freezing injury and its prevention in living cells. Annu. Rev. Biophys. 3: 341-63.

Morris, G.J., G. Coulson, M.A. Meyer, M.R. McLellan, B.J. Fuller, B.W.W. Grout, H.W. Pritchard and S.C. Knight. 1983. Cold shock – a widespread cellular reaction. Cryoletters 4: 179-92.

Rall, W.F. 1987. Factors affecting the survival of mouse embryos cryopreserved by vitrification. Cryobiology 24: 387-402.

Rudolf, E., J.C. Malausa, P. Milot and R. Pralavario. 1993. Influence of cold temperature on biological characteristics of *Orius laevigatus* and *Orius majusculus* (Het.: Anthocoridae). Entomophaga 38: 317-25.

Salt, R.W. 1936. Studies on the freezing process in insects. Tech. Bull. 116, Univ. Minn. Agric. Exper. Stn.

Salt, R.W. 1953. The influence of food on the hardiness of insects. Can. Entomol. 85: 261-69.

Salt, R.W. 1959. The role of glycerol in the cold hardening of *Bracon cephi* (Gahan) Can. J. Zool. 37: 59-69.

Sømme, L. 1966. The effect of temperature, anoxia, or injection of various substances on haemolymph composition and supercooling in larvae of *Anagasta Kuehniella* (Zell.). J. Insect Physiol. 12: 1069-83.

Sømme, L. 1995. Invertebrates in hot and cold arid environments. Springer-Verlag, Berlin. 275 pp.

Steponkus, P. and S. Caldwell. 1993. An optimised procedure for the cryopreservation of *Drosophila melanogaster* embryos. Cryoletters 14: 375-80.

Storey, K.B. and J.M. Storey. 1988. Freeze tolerance in animals. Physiol. Rev. 68: 27-84.

Tauber, M.J., C.A. Tauber and S. Masaki. 1986. Seasonal adaptations of insects. Oxford University Press, New York. 411 pp.

Vigil, B.O. 1971. Laboratory multiplication and release of *Trichogramma sp.* with a view to controlling *Heliothis zea* (Boddie) and *Alabama argillacea* (HB.) in El Salvador (Central America). Coton Fibres Trop. 26: 211-16.

Whitaker-Deerberg, R.L., G.J. Michels, L.E. Wendel and M. Farooqui. 1994. The effect of short-term cold storage on emergence of *Aphelinus asychis* (Walker) (Hymenoptera: Aphelinidae) mummies. Southwest. Entomol. 19: 115-18.

Young, S.R. and W. Block. 1980. Experimental studies on the cold tolerance of *Alaskozetes antarticus*. J. Insect Physiol. 26: 189-200.

Zachariassen, K.E. 1985. Physiology of cold tolerance in insects. Physiol. Rev. 65: 799-832.

Zaslavski, V.A. 1988. Insect development: photoperiodic and temperature control. Springer-Verlag, Berlin. 187 pp.

# Mass Rearing the Old World Screw-worm Fly, *Chrysomya bezziana*

*R.J. Mahon*[1] *and H. Ahmad*[2]

[1] CSIRO Entomology, GPO Box 1700, Canberra, ACT 2601, Australia
CSIRO, Institut Haiwan, P.O. Box 520, 86009 Kluang, Johor, Malaysia
[2] Institut Haiwan, P.O. Box 520, 86009 Kluang, Johor, Malaysia

Many countries within the tropics are afflicted with one of two species of screw-worm, either the New World screw-worm fly *Cochliomyia hominivorax* (Coquerel) (NWSWF), or the Old World screw-worm fly, *Chrysomya bezziana* Villenuve (OWSWF). In nature, larvae of both species are obligate parasites and feed on the living flesh of mammals and to a lesser extent, birds. Female flies lay eggs at the site of a pre-existing wound or near body orifices of the host. First instar larvae feed superficially on the wound; however, larger larvae use their hooked mouthparts to burrow deep into the flesh of the host. Damaged blood vessels provide a steady stream of blood and plasma that typically oozes from the infested wound. The wound also acquires a characteristic odour. Presumably, some of the volatile components emanating from the wound, also provide strong signals to gravid female flies, as once infested, wounds become far more attractive as a site to lay eggs than uninfested wounds (Spradbery 1990, Hall 1995). As the number of larvae increases, the myiasis enlarges and the well-being of the host is threatened. In areas where gravid female SWF are numerous, the wound has little chance to heal, thus death of the host is likely unless the wound is treated and re-infestation prevented.

While New and Old World species are distantly related, they are remarkably similar ecologically and in their biological characteristics. Wherever either species occurs, it is considered a serious pest of livestock (see review by Spradbery 1994). Australia is fortunate that neither species of the screw-worm is present despite evidence that extensive areas appear environmentally suitable (Suthurst et al. 1989). The OWSWF is considered the most serious threat to Australia as it is present in the neighbouring countries of Papua New Guinea and Indonesia.

It is envisaged that the sterile insect release method (SIRM) will be employed to eradicate the OWSWF if it becomes established in Australia. To facilitate such a programme, and to reduce delays in constructing a suitable mass rearing facility, the Department of Primary Industries and Energy, Australia, in collaboration with the Department of Veterinary Services Malaysia, has established an experimental insect rearing facility within the Institut Haiwan, Kluang, Johor. The facility has a nominal capacity of rearing 10 million sterile flies per week. Information gathered while operating the facility will be used to develop plans and protocols for an Australian facility with a capacity of rearing 100 to 200 million per week.

During the period 1973 to 1991, Dr. Philip Spradbery and CSIRO Entomology colleagues developed procedures for mass rearing OWSWF in a facility on the outskirts of Port Moresby, Papua New Guinea (Spradbery 1990). Larval rearing employed the hydroponic system first developed by USDA to rear the NWSWF. Larvae were fed on a diet consisting of water and suspended/dissolved blood, egg and milk powder. During the period 1984-1991, the facility provided Australia with a response capacity if an incursion of the OWSWF occurred. Full production could be achieved within weeks, to supply sterile flies for release in Australia. However, the Australian Department of Primary Industries and Energy (DPIE), the authority responsible for responding to exotic diseases entering Australia, conducted a review of its preparedness strategy, and concluded that except under the most favourable circumstances, the capacity of the PNG facility was insufficient to cope with an incursion into Australia. Furthermore, at that time, the NWSWF facility in Tuxtla had developed a new rearing system using a gelled diet that was superior to the hydroponic system. The production of NWSWF larvae using the gelled diet was found to require less dietary ingredients, and also significant savings could be achieved through a reduction in labour requirements. While DPIE wished to capture the benefits of the gelled diet for the OWSWF, conversion of the PNG facility to one using gelled diet was not considered practical (see Tweddle and Mahon, this procedings). Accordingly, the facility was closed in 1991.

The experimental fly rearing facility established in the Institut Haiwan Kluang, Malaysia, was specifically designed to use gel systems. Diet is pumped from the mixing vat directly to larval trays in measured quantities under the control of a programmable logic controller (PLC). The PLC controls the period that diet is dispensed and as the delivery pump supplies at a constant rate, the volume can be accurately controlled. The system is essentially a replica of the functional Tuxtla one, with the addition of the PLC to permit accurate doses we believe will be required in the experimental facility.

Rearing trays (30 trays per chamber) are located on a rack within environmental chambers. Each rack and associated trays remain within the chamber from the time a small portion of gel is seeded with eggs until larvae have evacuated onto the trays seven days later. Additional aliquots of diet are added on days two, three and four. The addition of diet takes place *in situ* without removal of the racks or trays from the chamber. Fully-fed larvae evacuating onto the tray are gathered via funnels into collecting containers in a compartment below the chamber.

A personal computer (PC) equipped with building management system (BMS) software, controls environmental conditions throughout the facility. The building-specific software was developed from the proprietary programme called SIGNAL (SIEBE Environmental Controls). BMSs are normally employed to control environmental conditions in large buildings. In our mass rearing facility, certain conditions can be specified for constant temperature rooms, laboratories and work areas. The BMS also operates a range of devices as diverse as the operation of pumps in response to signals from water level detectors in tanks, to the management of the sequential activation of essential equipment when the emergency power source is required.

Most importantly, the BMS controls conditions within the rearing cabinets. There are 36 cabinets arranged in groups of three. Each group is independent from the other 11 groups. Eight of the groups are configured to rear larvae, and four for pupariation and pupae maturation. Each group has equipment to cool or heat the air, and to vary the volume of air entering the chambers where it circulates over all the trays. The cabinet control system is also capable of varying the proportion of air recycled, with the remainder being exhausted and replaced with fresh air.

Airflow is measured and reported to SIGNAL at three locations, the first in the exhaust air duct, the second in the fresh air duct and the third immediately before the air enters the cabinets. Signals from airflow stations are used by SIGNAL to control three variable speed devices and the positions of two dampers. The first variable speed device controls the speed of a fan forcing air through the cabinets and can achieve airflows within the range of 50 to 200 litre per min. A second fan, located in an exhaust duct, controls the amount of air exhausted with supplementary control provided through the operation of modulating dampers. Fresh air is regulated in a similar fashion. It is envisaged that for the period that the eggs are hatching and the larvae are in early instars, limiting the amount of exhaust air and recycling the remainder can save energy. However, during latter days of the 7-day rearing cycle, the now large and metabolically active OWSWF larvae produce quantities of ammonia ($NH_3$) and presumably other noxious gasses. Airflow over the rearing trays will be increased to remove the $NH_3$ and the proportion of recycled air will be reduced to flush unwanted gases from the system.

SIGNAL receives temperature data from two temperature probes, one sited in the duct that returns air from the cabinets, the other in the supply-air duct. SIGNAL regulates the temperature within the cabinet on the basis of the return air temperature. If too cool, four separate heating elements within the air-stream may be activated. If the discrepancy between the set-point (the temperature entered into the PC that SIGNAL attempts to achieve and then maintain) and the input from the return air probe is large, all four heaters will be activated. If the discrepancy is smaller, fewer elements will be operated. If required, cooling is achieved as air passes over a chilled water (10°C) heat exchanger. The flow rate of chilled water is regulated by SIGNAL via a modulating valve.

SIGNAL receives data from two humidity probes located adjacent to the temperature probes, i.e., in the supply and return air-streams. Like temperature, SIGNAL regulates the relative humidity within the cabinet on the basis of the return air measurement. Humidity of the air destined to pass to the cabinets may be raised by the injection of steam from a steam generator, with steam production controlled by regulating the current passing through the heating element in the boiler. During the last two days of the rearing cycle, there is considerable evaporative water loss from the rearing trays because the larvae are large and active. In this situation, it may be necessary to remove water vapour from the circulating air. This will be accomplished by operating the chilled water heat exchanger to cool the air, remove the resultant condensate, and then activating the heating elements to re-heat the air before returning it to the cabinets.

In the Tuxtla NWSWF facility, racks containing rearing trays are moved between rooms with different environmental conditions in order to accommodate the changing requirements of the maturing larvae. In our facility, the trays remain in the cabinet but at intervals the BMS set-points are changed (temperature, relative humidity, airflow (through the cabinet) and the proportion of air exhausted). Optimal conditions have yet to be determined although, if the OWSWF mimics the NWSWF in this respect, hot (37.5°C) and humid (90% RH) conditions during the egg-hatch and early instars will be changed in four steps, to cooler (31°C) and drier (50% RH) conditions.

The BMS receives a vast amount of data from approximately 100 remote sensors and also regularly scans the condition of variable speed devices, heating elements, valve positions, etc. If the operator wishes, some or all data may be logged at whatever frequency required, presenting an opportunity to record a detailed historical picture of the operation. This may be particularly useful to determine the conditions experienced by larvae and pupae that resulted in exceptional, or sub-standard, batches of flies.

When in operation, the rearing cabinets will be opened only three times (to add additional diet) during the rearing cycle. Thus, operator exposure to the unpleasant and remarkably smelly conditions experienced by the larvae is minimised. Pupae and adults are also housed in separate environmental chambers. It is hoped that this separation will reduce human contact with insect allergens. Experience has shown that allergy problems can become an important issue when rearing blowflies.

At the time of writing, (September 1998), the facility is complete and is expected to begin operation within a week, some 22 months after the anticipated completion date. The extensive delays have occurred due to a number of unrelated factors. Initially, the gel handling system, including the mixing and dispensing equipment was unable to cope with the high viscosity of the gelled diet. The equipment was designed using viscosity measurements on test samples; however at the time, it was not known that the gel diet acts as a non-Newtonian fluid, and as a consequence, special care must be made when measuring viscosity. We are grateful for the assistance of the USDA/MAG staff in Tuxtla for their assistance in resolving this problem. The second major problem was to achieve the required control over the environment within the rearing cabinets. The first iteration was found to be inadequate and the systems required major re-design. Another problem has undoubtedly been the location. Kluang is remote from large centres where skilled technicians are available. The system is complex and diverse skills were required to make the whole system function as an integrated unit. Now that the hardware and software are functioning, we look forward to a different array of challenges when the OWSWF colony is introduced into the system.

## REFERENCES

Hall, M.J.R. 1995. Trapping the flies that cause myiasis: their responses to host-stimuli. Ann. Trop. Med. Parasitol. 89: 333-357.

Spradbery, J.P. 1990. Australian Screw-worm Fly Unit: Manual of operations. CSIRO Division of Entomology Technical report No.49. pp. 1-241.

Spradbery, J.P. 1994. Screw-worm fly: a Tale of Two species. Agric. Zool. Rev. 6: 1-62.

Suthurst, R.W., J.P. Spradbery and G.F. Maywald. (1989). The potential geographical distribution of the Old World screw-worm fly, *Chrysomya bezziana*. Med. Vet. Entomol. 3: 273-280.

# Incorporation of Diapause into Codling Moth Mass Rearing: Production Advantages and Insect Quality Issues

*Stephanie Bloem*[1], *Ken A. Bloem*[2] *and Carrol O. Calkins*[1]

[1] USDA-ARS YARL, 5230 Konnowac Pass Road, Wapato, WA 98951, USA
[2] Former Director, OK-CM SIR Program, Box 1080 Osoyoos, BC, V0H 1V0 Canada

## INTRODUCTION

The codling moth (CM), *Cydia pomonella* L. (Lepidoptera: Tortricidae) is a widely distributed species and the key pest of apples and pears in orchards of the Pacific Northwest. CM possesses facultative diapause and the mature 5$^{th}$ instar larva is the overwintering stage (Brown 1991). British Columbia, Canada, is attempting to eradicate CM using the sterile insect technique (SIT). The Okanagan-Kootenay Sterile Insect Release (SIR) Program began releasing sterile CM in 1994 (Dyck et al. 1993, Bloem and Bloem 1996) and results in recent years have been excellent. Despite the fact that standard rearing operations have gone extremely well, and weekly CM production has increased from 8.7 million CM per week in 1994 to 14.2 million per week in 1997, the ability to mass rear diapausing CM holds a number of advantages that are currently not available. The ability to stockpile CM larvae in diapause throughout the fall and winter months when the mass rearing facility is underutilised would:

- provide additional CM for release during the spring/summer field season,
- provide back-up material to guard against colony losses due to operational failures and/or pathogen outbreaks,
- allow for a potentially more cost-effective use of the facility by maintaining production year round,
- facilitate the development of mass rearing techniques for key parasitoids that attack the overwintering stage,
- allow for the easy shipment (and sale) of CM to researchers around the world.

Here we outline the current standard CM mass rearing procedures and discuss research that has allowed diapausing CM to be mass reared (Bloem et al. 1997). We also discuss research that suggests diapaused CM are more competitive than standard non-diapaused CM (Bloem et al. 1998).

## STANDARD CM MASS REARING

The SIR facility in British Columbia uses a modified version of the sawdust-based diet developed by Brinton et al. (1969). Fibreglass trays (45x29x2.5 cm) are filled with fresh diet, the surface is scarified with hot steam and trays are sprayed with a fine mist of liquid paraffin to help control desiccation. Diet trays are then placed on aluminum-frame diet carts, 75 trays per cart (3 trays per shelf x 25 shelves). Waxed paper egg sheets with 2,500-3,000 incubated CM eggs are placed (egg side down) on a wire rack on top of the diet. Diet carts are placed in larval rearing rooms (28 carts per room), at constant temperature (27°C), long-day photoperiod (16L:8D) and relative humidity (RH) that starts at 75% and is gradually lowered to 55% by day 21. Egg sheets are removed after 1 week.

At day 22 the diet trays/carts, which now contain late instar larvae/early pupae, are moved to emergence rooms, where conditions are 28°C, 16L:8D, and 30%RH. At day 28, when adult eclosion begins, the photoperiod is changed to 0L:24D and ultraviolet (UV) lights attached to vacuum traps in the ceiling of the rooms are turned on for 50 min every hour. Peak adult emergence occurs at day 35. Emerging CM attracted to the UV lights fly upward into the vacuum traps and are transported through ducts to a cold room (0-2°C). Inside the cold room, cyclone traps cause the moths to fall from the air stream in the ducts and into large plastic bins. The CM are collected once per hour and are either processed for field release or for colony maintenance as required. Normally, 90% of adult production is packaged and sterilised for release, while 10% is kept to maintain the reproductive colony.

## YEARLY REARING SCHEDULE

Standard full-scale production is in effect for 20-22 weeks each year, from the beginning of April until mid-August. During this period, 32 kettles of diet are prepared per day. The yield per kettle is ca. 62 trays, making daily production about 2,000 trays of diet. In 1997 CM production averaged ca. 1,400 adults per tray (sex ratio 1:1).

During the "off" season, in effect from mid-August until late February, 2 kettles of diet are prepared per week in order to maintain the reproductive colony. The switch to full-scale production requires a step-up period that takes 7 weeks, during which 3 kettles are cooked per day (15 per week) for 6 weeks and 1 week in which 16 kettles are prepared per day.

## MASS REARING THROUGH DIAPAUSE

### System 1

CM larvae can be induced into diapause while developing inside trays of artificial diet (Bloem et al. 1997). This is accomplished by altering larval rearing conditions to simulate autumn field conditions, i.e., shortening the photoperiod to 12L:12D and lowering the temperature to 25°C during photophase and 21°C during scotophase. RH follows the same trend as in standard rearing. At day 21, diet trays are covered with C-fluted corrugated cardboard pallets (45x29x1.5 cm), previously heat-treated (at 65°C for 72 h) to avoid the introduction of pathogens into the rearing system. The additional space required to accommodate the pallets reduces the tray-holding capacity of the diet carts by 50% of standard production (or 36 trays per cart).

Pallets remain on the diet for 14 days while CM larvae wander out of the diet and search for locations in which to spin overwintering hibernacula (Bloem et al. 1997). Pallets are removed at day 35 and packaged in dark polyethylene bags, 15 pallets per bag. Bags are stored at 15°C, 0L:24D and ambient RH for ca. 100 days, after which they are moved to cold storage (0-2°C), where they are kept in the dark for a minimum of 30 days and up to 12 months.

Diapause is broken by returning the pallets to an emergence room where temperature is 27-28°C and photoperiod is 16L:8D. Pallets are placed on diet carts as if they were trays of diet (75 pallets per cart). Adults begin to emerge at about day 16 and peak emergence occurs at day 18-19. The amount of time that diapausing CM larvae are stored in the cold has an impact on the synchronicity of adult emergence (i.e., the longer the cold storage period, the more synchronous the adult emergence). Emerging CM are collected, packaged and sterilised as in standard rearing.

Diapause rearing using this system is approximately 73% as efficient as standard production (Bloem et al. 1997), i.e., only 73% as many diapaused CM adults are produced per pallet (ca. 1,000) as are non-diapaused adults produced per tray of diet using standard rearing procedures (ca. 1,400). However, an additional 15% can be "recovered" if the diet trays are returned to long days after the pallets have been removed. This number (15%) includes CM larvae that were either not induced into diapause and pupated in the diet, or larvae that entered diapause, spun hibernacula outside of the pallets and broke diapause when they were returned to long day conditions. Since about 10% of standard production is normally used to maintain the reproductive colony, the adults collected in this fashion could be used for colony maintenance, thus eliminating the need for additional diet preparation. Unfortunately, 12% of production per diet tray is lost during diapause rearing (Bloem et al. 1997). The missing 12% are larvae that migrate out of the diet, spin hibernacula other than in the pallets, and do not break diapause when returned to long day conditions.

## System 2

In an effort to reduce the cost and improve the efficiency of diapause rearing, a second system has recently been developed. CM larvae are induced into diapause while inside trays of artificial diet as described above. However, on day 21, instead of placing pallets on top of the trays, the trays are placed inside fibreglass mesh bags (50x40 cm). The bags still allow air to flow over the diet, but restrict larval wandering to the immediate vicinity of the tray. CM larvae prefer not to spin cocoons directly on the mesh. Restricting the movement of the larvae in this fashion "forces" them into diapause inside the diet or among the frass that has collected on the diet surface during rearing.

After the trays are placed inside mesh bags, the larvae continue to develop for another 14 days as in System 1. At day 35, the diet trays per carts are moved to emergence rooms and follow the same storage conditions as outlined before. Diapause is broken by placing the carts in emergence rooms where temperature is 27-28°C and photoperiod is 16L:8D. The mesh bags are removed when the material is brought out of cold storage for emergence. As in System 1, adults should begin to emerge at about day 16 upon return to long day conditions, and peak adult emergence should occur at day 18-19.

The efficiency of System 2 has not yet been evaluated; however, we expect a similar percentage of larvae will not have been induced into diapause as occurred in System 1 and will be lost. However, in this system there will essentially be no loss of larvae due to wandering.

## ADVANTAGES, DISADVANTAGES AND COSTS OF SYSTEMS 1 VS. 2

From a rearing standpoint, mass rearing through diapause has a number of disadvantages relative to standard rearing, the principal one being cost. In System 1, the cost of the cardboard pallets and the labour to place them and remove them from the diet, as well as to handle the pallets during storage and emergence significantly increase the unit cost of production. In addition, larvae that migrate out of the diet and spin hibernacula on the carts or elsewhere in the rearing room reduce production efficiency and further increase costs, as well as make the cleaning of the carts and rooms more difficult.

In 1997, the SIR Program averaged 1,400 CM per tray of diet, at a cost of about US$2.20 per tray, which included the cost of all diet ingredients, labour and facility use. As such, the cost of producing 1 million standard CM was ca. US$1,572.00. The cost under System 1 would be US$2.20 per tray of diet + the cost of the pallet, which is currently US$2.20. Since rearing through diapause is only 73% efficient (1,000 adults per pallet), the cost of producing 1 million larvae in diapause would be ca. US$4,400.00. However, this figure does not include the additional handling and utility costs for cool and cold storage. System 2 is also more expensive relative to standard production because of the cost of the mesh bags (ca. US$0.50-US$1.00/bag) and because of similar handling and storage requirements.

System 2 has advantages over System 1 in that: 1) the cost of mesh bags is lower than for pallets and mesh bags are re-useable, whereas pallets can only be used once or twice before they must be discarded, 2) mesh bags restrict larvae from wandering away from the diet, which facilitates clean-up, and 3) mesh bags do not have the space requirements of pallets and thus allow for full use of the carts. As a consequence, production of diapausing larvae per rearing room is potentially 50% greater than in System 1. However, System 2 would: 1) require additional diet preparation to maintain the reproductive colony (non-diapausing larvae are not recovered), 2) require additional cold storage space because trays are left on the carts, and 3) not allow for easy shipment and transportation of diapausing CM. As such, rearing through System 2 would be useful mainly to increase CM production for use at the same rearing facility.

## QUALITY OF DIAPAUSED CM ADULTS

Pupal and adult weights for both males and females, adult longevity (with water), female fecundity and fertility and male mating ability have been examined for both diapause reared and standard colony CM. In general, moths that are mass reared through diapause are comparable in quality to CM reared under standard conditions. Fecundity of diapaused females is significantly lower, although the proportion of eggs that hatched is similar for both diapaused and standard females. Standard males were found to be significantly longer-lived than diapaused males (Bloem et al. 1997).

Field quality of diapaused CM was examined using release-recapture tests (Bloem et al. 1998), and higher recapture was interpreted as an indication of higher insect quality. In small-scale releases conducted in the spring of 1995, significantly higher numbers of diapaused CM were recaptured in passive interception traps than standard colony moths reared under either constant or fluctuating temperatures. When both types of sterile CM were released into 1 ha plots in large-scale tests conducted in the summer and fall of 1996 and the spring of 1997, the proportion of recaptured diapaused CM males was always significantly higher than for standard (non-diapaused) moths. This was true for the recapture of CM using passive interception traps, pheromone-baited traps and virgin female-baited traps.

## CURRENT PRODUCTION POTENTIAL FOR DIAPAUSING CM AT THE SIR FACILITY IN CANADA

### System 1

The SIR facility has 16 rearing rooms, and assuming 1 room is filled per day, it would take 16 days to fill all the rooms. Diapause rearing requires 35-36 days and 1 day for emptying and cleaning each room. If the last day of standard rearing is 15 August, such a schedule would accommodate 4 diapause production cycles to be completed before the SIR facility would have to be cleaned and readied for standard rearing activities the following year.

Given the fact that cardboard pallets reduce the space efficiency of the carts/rooms by 50%, 16 kettles of diet would be cooked per day to fill 1 room (28 carts) and in 16 days, 256 kettles of diet would be prepared. Considering that 62 trays of diet can be filled by 1 kettle, and assuming 4 rearing cycles, then 63,488 trays could be prepared in total. Finally, if 1,400 CM can be obtained per tray under standard rearing, and the 73% efficiency of rearing through diapause is factored in, ca. 63.5 million CM could potentially be produced and stockpiled during the "off" season at the SIR facility in Canada using System 1.

### System 2

Calculations for System 2 would be similar in terms of number of days required per cycle; however, 32 kettles of diet could be prepared per day to fill each rearing room. As such, roughly 127,000 trays and 127 million CM could be produced.

## STORAGE SPACE FOR CM LARVAE IN DIAPAUSE

A corrugated cardboard pallet is placed on top of each tray during System 1 rearing. This pallet provides suitable sites in which larvae spin their winter cocoons. The 63,488 pallets that would be used in total would be generated in 4 groups of 15,872 pallets each. Each group would require storage, first for 100 days at 15°C and 0L:24D (cool storage) and then at 0-2°C and 0L:24D (cold storage) for a minimum of 30 days or until the CM are needed.

The pallet dimensions are 45x29x1.5 cm and ca. 240 pallets can fit in a cubic metre ($m^3$) of space. Therefore, ca. 67 $m^3$ would be needed to store the material from each production cycle (and 67 $m^3$ x 4 cycles ca. 268 $m^3$ to store the entire production). The emergence rooms at the SIR facility in Canada could be used as cool storage for the bags of pallets containing diapausing larvae. Although the rooms are maintained at 28°C during standard rearing, it is possible to reduce the temperature to a constant 15°C. The size of the rooms is roughly 6x6x2.50 $m^3$ and there are a total of 9 emergence rooms. One emergence room has ca. 90 $m^3$ of space. Therefore, if space is allowed for walkways between the stacks of pallets, it appears that 1 production cycle could occupy roughly 1 emergence room, and that there would be enough cool storage to accommodate all 4 production cycles.

In terms of the amount of time required in cool storage, which is ca. 100 days for each production batch, there probably would be a conflict of space with the last batch, as the emergence rooms are needed for standard rearing beginning in early April. As a consequence, it would be necessary to secure alternate cool storage for the last production cycle.

Cold storage (0-2°C) at the SIR facility is very limited, and at this time, there is a maximum of 22 $m^3$ of available cold storage, which is not even enough to store a single diapause production cycle. Alternatives for cold storage space might be: 1) purchase of a prefabricated refrigerated unit, 2) rental of a refrigerated trailer, or 3) rental of cold storage space at packing houses. Since a total of 268 $m^3$ is needed to store the 4 production cycles, the rental of cold storage at a local packing house might be the most reasonably priced alternative.

The same storage bottlenecks would be encountered if System 2 diapause rearing were to be implemented, particularly as it pertains to cold storage. However, if this system were adopted as an alternate mass rearing strategy, then perhaps changes in the standard schedule of operations could be made in order to allow for the production of high quality diapausing CM for part or all of the rearing operations.

## CONCLUSION

Our research has demonstrated that diapause can be successfully incorporated into the open tray mass rearing system used in the Canadian SIR Program. Rearing through diapause, however, is not as efficient as mass rearing CM under constant summer-like conditions. This is true, both in terms of insect recovery and the use of available facility space. However, our research has shown that CM reared through diapause are of superior field quality when compared to standard reared colony CM. This significant improvement in quality partially offsets the higher cost of diapause production. At present, other uses for the diapaused strain are also being investigated.

## REFERENCES

Bloem, K.A. and S. Bloem. 1996. Codling moth eradication programme in British Columbia: a review and update. *In* Biology and Control of the Cherry Fruit Flies: A Worldwide Perspective. M.T. AliNiazee and L.E. Long, (eds.). Oregon Agric. Exper. Stat. Special Report.

Bloem, S., K.A. Bloem and L.S. Fielding. 1997. Mass rearing and storing codling moth larvae in diapause: a novel approach to increase production for sterile insect release. J. Entomol. Soc. B. C. 94: 75-81.

Bloem, S., K.A. Bloem and A.L. Knight. 1998. Assessing quality of mass reared codling moths (Lepidoptera: Tortricidae) using field release-recapture tests. J. Econ. Entomol. 91: 1122-1130.

Brinton, F.E., M.D. Proverbs and B.E. Carty. 1969. Artificial diet for mass production of the codling moth, *Carpocapsa pomonella* (Lepidoptera: Olethreutidae). Can. Entomol. 101: 577-584.

Brown, J.J. 1991. Diapause, pp. 175-186. *In* World Crop Pests, Tortricid Pests their Biology Natural Enemies and Control, Vol. 5. L.P.S. van der Geest and H.H. Evenhuis (eds.). Elsevier, New York.

Dyck, V.A., S.H.Graham and K.A. Bloem. 1993. Implementation of the sterile insect release programme to eradicate the codling moth, *Cydia pomonella* (L.)(Lepidoptera: Olethreutidae), in British Columbia, Canada, pp. 285-294. *In* Management of Insect Pests: Nuclear and Related Molecular and Genetic Techniques, IAEA-SM-327/29, Vienna.

# New Systems for the Large-Scale Production of Male Tsetse Flies (Diptera: Glossinidae)

*E. Opiyo, D. Luger and A.S. Robinson*

Entomology Unit, FAO/IAEA Agriculture and Biotechnology Laboratory, IAEA Laboratories A-2444, Seibersdorf, Austria

## INTRODUCTION

Tsetse flies, vectors of trypanosomiasis, infest 36 African countries and their distribution covers approximately 10 million $km^2$. Trypanosomiasis is a debilitating and often fatal disease of domestic livestock and humans and is considered the most important limiting factor for the development of the livestock sector in Africa. Approximately 50 million cattle and scores of millions of small ruminants are at risk of contracting trypanosomiasis. Direct losses in meat production, milk yield and traction power and the cost of control programmes are estimated to amount to more than US$500 million each year (FAO 1994). In addition, 100 million people are at risk of contracting the disease. According to the World Health Organization, about 300,000 new cases of human trypanosomiasis occur annually (WHO 1997). If the lost potential in livestock production is combined with that of crop production through loss of traction power, trypanosomiasis is estimated to cost Africa US$4 billion or more each year (FAO 1994).

The available and environmentally accepted intervention methods for the management of tsetse and trypanosomiasis include parasite control using drugs, the promotion of trypanotolerant livestock and vector control. Parasite control is plagued by the development of resistance to the available drugs and programmes for the development of new drugs are limited. Vector control and eradication involve application of insecticide treated attractive devices on animals including cattle, and the sterile insect technique (SIT).

In spite of the efforts spent on the control of the disease and the vector, tsetse flies remain a threat to agricultural development of the region. Experience indicates that only a combination of several of these intervention methods can effectively support sustainable agricultural systems.

The feasibility of rearing tsetse flies in Africa for use in SIT was first demonstrated in Tanzania (Williamson et al. 1983) where a colony of 60,000 *Glossina morsitans morsitans* Westwood produced a total of 500,000 sterile males. In Burkina Faso, between 1976 and 1984, a colony of 330,000 *G. palpalis gambiensis* Vanderplank and G. *tachinoides* Westwood provided 950,000 sterile males for release into an area of 3,000 $km^2$ (Clair et al. 1990) while during the Bicot project in Nigeria in an area of 1,500 $km^2$, 1.5 million sterile male *G. p. palpalis* Robineau-Desvoidy were released (Olandunmade et al. 1990). Recently, 8.5 million sterile males were released on Unguja Island, Zanzibar, the United Republic of Tanzania in an area of 1,600 $km^2$ produced by a colony of about 600,000 *G. austeni* Newstead (Saleh et al. 1999, Kitwika et al. 1999). This led to the eradication of the tsetse population and a massive reduction in disease incidence in cattle (Saleh et al. 1999).


ed. K.H. Tan. Penerbit Universiti Sains Malaysia, Penang, 2000.

Tsetse fly SIT has been applied on a limited scale because of the inability to provide large numbers of sterile males for release. The present rearing system is labour intensive and too many quality sensitive steps in the mass production system are not sufficiently standardised to transfer the system directly to large-scale production.

Tsetse rearing evolved from feeding on live hosts to an *in vitro* rearing system where blood is fed to flies through a silicone membrane (Feldmann 1994a). At present, cages are small, hold a small number of flies and have to be manually transferred for feeding and then returned for pupal collection. This limits the number of flies that can be handled at any one time. In order to improve these processes, a Tsetse Production Unit (TPU) was developed and evaluated.

During conventional tsetse rearing, flies need to be sexed with the correct number and sex of flies, whether for stocking production cages or for the release of males only. This has to be done by hand on an individual fly basis following the immobilisation of adults at 4°C. A procedure is reported in this paper for the self-stocking of production cages (SSPC) which enables flies to emerge directly into the production cages in the right sex and number. This eliminates the need for manual sex separation before mating and the same procedure can be used to produce only males for release.

For the SIT aerial release in Zanzibar, sterile flies were placed in boxes and dropped from an aircraft along predetermined flight paths. Boxes are expensive and require much space within the aircraft. For fruit flies, pink bollworms and screw-worms, a chilled adult release system was developed and this will also be required for future tsetse SIT programmes. Limited experiments were carried out on the long-term effects of chilling on male tsetse flies.

## MATERIALS AND METHODS

### Tsetse Flies

A *G. austeni* colony was established from pupae collected from Zanzibar in 1986. It has been maintained at Seibersdorf on membranes since 1986 on a diet of fresh frozen irradiated bovine blood.

*G. pallidipes* Austen originated from wild pupae collected in Uganda and colonised in Amsterdam in 1975. It was brought to Seibersdorf from Bristol in 1986, where it has been maintained on a diet of 75% fresh frozen bovine and 25% porcine blood. The flies are maintained under standard rearing conditions of 23-24°C and 80% RH.

### The Tsetse Production Unit

A programmable machine occupying 7 m$^2$ floor space containing 648 cages with a centralised larval and pupal collection and membrane feeding area was designed, installed and evaluated using *G. austeni*. An estimated 225,000 females could be maintained on the machine with cages holding 300 females and 75 males. The cages were held on a conveyor belt system which moved the cages slowly throughout the day, including transport to and from the feeding table and pupal collection station.

The evaluation included cage design development, fly performance, toxicity and stress tests. Pupal quality was also assessed in terms of size class and weight (Feldmann 1994b), using a pupal size sorting machine which separates daily pupal production into five distinct sizes, size A being the smallest and E the largest size. Pupae in each size class were weighed and counted to give the mean weight in each class.

### Self-stocking of Production Cages

The number of pupae required for stocking a cage was based on the overall emergence rate (%) and the female to male ratio required (1:4). The aim was to stock the cages with 100 females and 25 males. Pupae collected daily were incubated in open trays at 23-24°C with 80% RH for 25 days. After this period, the pupae were placed in petri dishes under the standard emergence cage until the first fly emerged. At this time, the pupae were moved to 26.5°C and placed in a single layer under the netting of a production cage. At 24, 48, 72 and 96 hours from the time the pupae were moved to the higher temperature, the number and sex of flies in production cages were noted. Between 48 and 72 hours, the required number and sex of flies were found in the cages. Therefore, in the next series of experiments, the emergence was checked at hourly intervals between 9:00 and 15:00 h between 48 and 72 hours and the pupal incubation time identified at which the required numbers of males and females were counted in the cage. Cages stocked in this way were observed for pupal production which was expressed as pupae per initial female. It was compared to production cages manually stocked with 100 females and 25 males.

### Cold Temperature Treatment

Male tsetse flies were given three daily blood meals and then divided into three groups. One group acted as control and the other two groups were chilled at 4-5°C for 24 and 48 hours respectively. At 24 and 48 hours, flies were allowed to recover at the insectary holding conditions, while dead flies were removed and counted. The survival of flies was monitored for one week. Treated males were used for mating 1,400 colony females and pupal production during a 13-week period was compared to that of a similar number of females mated by untreated males.

### Data Analysis

Data were subjected to analysis of variance using the Tukey's HSD test where indicated.

## RESULTS

### The Tsetse Production Unit

At the start of the evaluation, a high percentage of daily mortality was observed and this led to a number of tests being undertaken to identify the causes. The tests showed that the flies were not stressed, had access to sufficient blood and that the materials that were used in cage construction were not toxic to the flies. However, continuous movement was found to be a major contribution to the mortality observed (Table 1) and movement was then restricted to the period of feeding.

Table 1. Effect of movement on performance of flies in cages with 300 females and 75 males.

| Treatment | % Mean survival | Pupae per initial female (PPIF) | % Daily mortality Week 1 |
|---|---|---|---|
| Movement | 28.33 ± 2.29a | 1.13 ± 0.16a | 2.50 |
| No movement | 48.85 ± 4.47b | 2.16 ± 0.14b | 1.00 |

Data represent performance of 4 replicates per treatment, values followed by the different letter in each column are statistically different ($p < 0.05$, Tukey HSD test).

For a period of twelve months (mid-1996 and mid-1997), five generations of *G. austeni* were maintained at the TPU. A total of 188,743 flies was added to the TPU and 166,928 pupae were produced. The mean daily mortality was 2.89% compared to 0.76% of the standard colony during the same period. The fecundity of flies that survived was 0.69 pupae per female per 10 days (P/F/10d) at the TPU and 0.86 for colony flies. The difference, however, was not statistically significant. There was improvement in pupal quality as measured by size and weight through the generations (Tables 2 and 3). The percentage of size class A pupae, which are the smallest size, decreased through generation F1 to F5. The inability to maintain a self-sustaining colony at the TPU led to its decommissioning.

Table 2. Size classes of *G. austeni* pupae (%) by generations in July 1997.

| Generations | Classes | | | | |
|---|---|---|---|---|---|
| | A | B | C | D | E |
| G0 | 3.2 | 29.9 | 53.5 | 11.8 | 1.6 |
| F1 | 7.4 | 23.5 | 60.3 | 07.4 | 1.5 |
| F2 | 8.6 | 31.4 | 54.3 | 05.7 | 0.0 |
| F3 | 7.7 | 27.5 | 51.6 | 14.1 | 0.0 |
| F4 | 3.3 | 16.7 | 62.6 | 13.2 | 2.2 |
| F5 | 2.8 | 29.2 | 45.8 | 16.7 | 5.6 |
| COLONY | 2.1 | 24.0 | 62.8 | 11.0 | 1.6 |

## Self-stocking of Production Cages with *G. austeni*

At 23-24°C, emergence was completed after six days (Figure 1) while at 26.5°C, the eclosion period is reduced to four days (Figure 2). During the first two days, predominantly females emerge and the last two days, mostly males. Males start to emerge during the second day and 52 hours from the onset of emergence, production cages could be stocked with about 100 females and 25 males, a 1: 4 male to female ratio (Figure 2). Pupae per initial female (PPIF) at week 10 for flies emerged using this system was 3.39 compared to 3.67 for manually loaded cages. Flies that emerged after the production cages were loaded were predominantly males with a sexing error of less than 0.5%.

Table 3. Mean pupal weight (mg) of *G. austeni* by generations.

| Generations | Classes | | | | |
|---|---|---|---|---|---|
| | A | B | C | D | E |
| G0 | 12.9 | 15.6 | 19.9 | 22.0 | 25.6 |
| F1 | 11.3 | 16.3 | 20.3 | 22.8 | * |
| F2 | 14.0 | 16.8 | 20.5 | 23.4 | * |
| F3 | 12.9 | 16.9 | 19.3 | 22.6 | * |
| F4 | 13.1 | 16.5 | 21.0 | 22.2 | 24.0 |
| F5 | 14.0 | 16.2 | 20.3 | 23.3 | 25.7 |
| COLONY | 12.9 | 16.9 | 20.4 | 23.5 | 26.4 |

* = very few or no pupae in the class

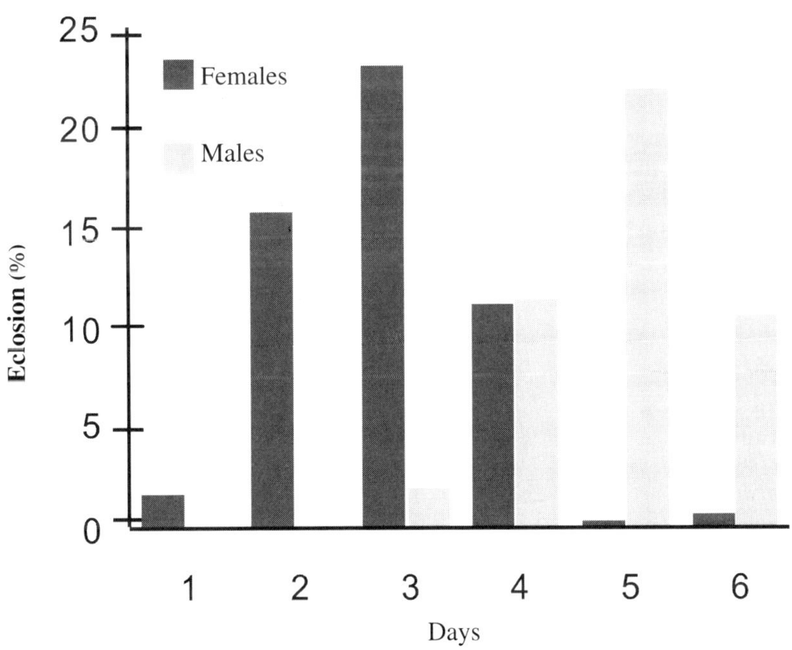


Figure 1. Daily emergence pattern of *G. austeni* females and males at constant temperature of 23-24°C.

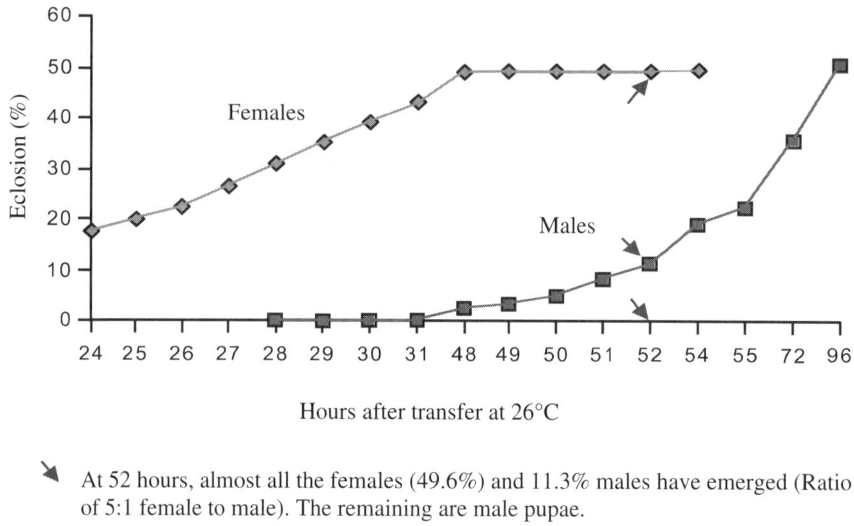


Figure 2. Emergence pattern of *G. austeni* for females and males after transfer to 26.5°C.

## Cold Temperature Treatment of *G. austeni*

The percent survival of the treated and untreated flies is shown in Table 4. The survival of flies kept at 4-5°C for 24 hours immediately after removal from the low temperature and one week later was not significantly different from that of the control group.

After a 48-hour cold temperature treatment, the survival of chilled flies was lower than that of the control group after one week. The fecundity of females mated with males which had a 48-hour cold temperature treatment was lower than that of the control and the group mated with males which had a 24-hour treatment (Table 4).

Table 4. Mean (SD) survival of male *G. austeni* kept at 4-5°C for 24 and 48 hours and pupal production of females when mated with chilled males.

| Treatment | % Survival | % Survival at 1 week | PPIF | P/F/10 d |
|---|---|---|---|---|
| 24 hours | 98.83 ± 2.47 | 85.53 ± 08.09 | 4.34 | 0.72 |
| Control 24 hours | 97.33 ± 2.08 | 87.33 ± 06.66 | 4.71 | 0.89 |
| 48 hours | 90.33 ± 4.22 | 54.33 ± 10.05 | 4.20 | 0.65 |
| Control 48 hours | 95.33 ± 3.06 | 88.67 ± 05.13 | 4.71 | 0.89 |

PPIF = pupae per initial female
P/F/10 d = pupae per female per 10 d

## DISCUSSION

The evaluation of the TPU was carried out using *G. austeni*. A large number of flies was introduced into the TPU but it was not possible to maintain a self-sustaining colony. This was due almost entirely to the high percentage of daily mortality observed. The fecundity of flies that survived was comparable to that of the colony flies. The quality of pupae produced indicated that the feeding of female flies and the transfer of nutrient from mother to the developing larva *in utero* were normal. The acceptable A class pupae in a colony should be below 10% of pupae produced. During the evaluation period, the Class A pupae did not reach 10% of the total pupae produced by flies maintained at the TPU, indicating that the inability of females to survive and not productivity or pupal quality was the key element affecting performance. Cage design and movement were found to be detrimental to fly survival. Even after the modification of internal structures in the cages, the improved performance of the flies still failed to meet the goal of having a self-sustaining colony and without major changes in cage design and holding, the TPU could not be developed into a practical unit for rearing tsetse flies. The TPU did however, demonstrate that the principles of automated feeding and pupal collection are indeed valid.

Sex separation in tsetse flies is possible only in the adult stage and this has been carried out by chilling flies at 4°C, followed by manual sexing. While searching for the optimum conditions for sex separation, it was demonstrated that the adult emergence pattern is temperature dependent with females emerging earlier than males from pupae deposited at the same time (Zdarek and Delinger 1995). At 26.5°C, adult emergence spanned a period of four days; during the first two days, only females eclosed and during the last two days, mostly males. Based on this behaviour, it has been possible to separate the sexes at the time of emergence. Flies were encouraged to emerge directly into production cages and this removed a very tedious step of collecting and manually sexing all newly emerged flies. In addition to removing the need for chilling for sex separation, pupae that are left after the production cages have been stocked, are mainly males with a very low number of females (± 0.5%). In order to use this system, it is essential that constant pupal incubation conditions allow for the synchronised development of pupae.

It will be possible to develop a chilled adult release system for tsetse SIT based on the results presented here. Male flies subjected to chilling for up to 24 hours are capable of successfully inseminating females and they show good survival. The design of the system has still to be completed.

In conclusion, the emergence pattern of male and female tsetse flies can be exploited to stock production cages with flies of the right number and sex. Although the TPU could not maintain a self-sustaining colony, the principle of automatic holding, feeding and pupal collection has been demonstrated to be sound and can lead to the development of large-scale production of tsetse flies.

## ACKNOWLEDGEMENTS

We wish to acknowledge the contribution of D. Nadel and A. Paral to the design and construction of the tsetse production unit, discussion on young fly handling from J. Zdarek and support and helpful discussions from several colleagues especially J. Hendrichs and U. Feldmann.

## REFERENCES

Clair, M., D. Cuisance, H. Politzar, P. Merot and B. Bauer. 1990. Tsetse fly eradication in Burkina Faso and evaluation of traps and targets. pp. 31-43. *In* Proceedings of Final Research Co-ordination Meeting, Vom Nigeria: Sterile Insect Technique for Tsetse Control and Eradication. IAEA/RC/319.3/3.

FAO. 1994. Opening address of Y. Cheneau. *In* A systematic approach to tsetse and trypanosomiasis control; Proceedings of the FAO Panels of Experts held in Rome, Italy, 1 to 3 December, 1993.

Feldmann, U. 1994a. Guidelines for the rearing of tsetse flies using the membrane feeding technique. pp. 449-471. *In* Techniques of Insect Rearing for the development of Integrated Pest and Vector Management Strategies. J.P.R. Ochieng-Odero (ed.). ICIPE Science Press.

Feldmann, U. 1994b. Some quality control parameters used in the rearing of tsetse flies. pp. 13-30. *In* Techniques of Insect Rearing for the development of Integrated Pest and Vector Management Strategies. J. P. R. Ochieng-Odero (ed.). ICIPE Science Press.

Kitwika, W.A., A.R. Msangi, N.E. Kiwia, F. Mramba, I.I. Malele, G. Kasilagila, M.B. Byamungu, A.G. Parker and V.A. Dyck. 1999. Production of *Glossina austeni* Newstead (Diptera: Glossinidae) for SIT in Tanzania. pp. 357-358. *In* 24th Meeting of the International Scientific Council for Trypanosomiasis Research and Control, Maputo, Mozambique. OAU/STRC, Nairobi, Kenya.

Olandunmade, M.A., U. Feldmann, W. Takken, S.O. Tenabe, H.J. Hamann, J. Onah, L. Dengwat, A.M.V. Van der Vloedt and R.E. Gingrich. 1990. Eradication of *Glossina palpalis palpalis* (Robineau-Desvoidy) (Diptera: Glossinidae) from agro-pastoral land in central Nigeria by means of the sterile insect technique. pp. 5-23. *In* Proceedings of Final Research Co-ordination Meeting, Vom Nigeria: Sterile Insect Technique for Tsetse Control and Eradication. IAEA/RC/319.3/1.

Saleh, H.M., M.J.B. Vreysen, S.S. Kassim F.W. Suleiman, K.G. Juma, Z. Zhu, H. Pan and V.A. Dyck. 1999. The successful application of the sterile insect technique (SIT) for the eradication of Glossina austeni (Diptera: Glossinidae) from Unguja Island (Zanzibar). *In* 24th Meeting of the International Scientific Council for Trypanosomiasis Research and Control, Maputo, Mozambique. OAU/STRC, Nairobi, Kenya.

WHO. 1997. Thirteenth Programme Report, UNDP/World Bank/WHO Special Programme for Research and Training. pp. 124-133. *In* Tropical Diseases (TDR). World Health Organization, Geneva 1997.

Williamson, D.L., D.A. Dame, D.B. Gates, P.E. Cobb, B. Bakuli and P.V. Warner. 1983. Integration of insect sterility and insecticides for control of *Glossina morsitans morsitans* (Diptera: Glossinidae) in Tanzania: V. The impact of sequential releases of sterilised tsetse flies. Bull. Entomol. Res. 73: 391-404.

Zdarek, J. and D.J. Delinger. 1995. Changes in temperature, not photoperiod, control the pattern of adult eclosion in tsetse, *Glossina morsitans*. Physiol. Entomol. 20: 362-366.

# Awareness, Flight Guidance and Reliability: Impact of Technology Upon the Sterile Insect Technique

*Kenneth D. Slagell*

K & K Aircraft Inc., Bridgewater, VA, USA

Technology progress impacts upon our lives creating new relationships and opportunities. Development of the microprocessor has been a cornerstone of change and advancing processing capability makes complicated calculations and the activities born of them accessible to increasingly large portions of humanity.

As I consider the impact of technology, I am mindful of the bicycle's invention. The concept is old, dating back to China approximately one thousand years ago. It was not until 1865, however, that a French mechanic added pedals to the large front wheel of a "hobby horse." In the 1870s, the popular bicycle, known as the Pennyfarthery, made its debut, and its popularity initiated some interesting changes.

Bicycles need better roads, and better roads were built in certain cities and towns to accommodate increasing bicycle traffic. Riders needed directions, and signs were installed. Crowded new roads forced development of laws to regulate cyclists and traffic flow.

In 1888, John Dunlop installed the first pneumatic tyres on a bicycle, and in the early 1900s, the bicycle finally came into its own after enhancements by airplane inventors, Orville and Wilbur Wright and others, resulted in a lightweight, economical device suitable for a mass market.

So you can see in something as elementary as bicycles how technology impacted not only transportation but other areas of peoples' lives creating new relationships and opportunities. The microprocessor, as it relates to sterile insect release, has also impacted many aspects of the sterile insect technique.

The microprocessor surpasses the notion of a simple electronic slide rule. For millennia, the only calculating machine was the abacus. Even 30 or 40 years ago, the only machine to add was a device consisting of gears and cogs and mechanical linkages. Having ten rows of ten columns of numbers, selecting the number you wanted to add and pulling a handle was the only way to add numbers together. Multiplication, division, and even subtraction seemed to be way into the future.

In 1964, during my high school days, the state of the art was the dual based log slide rule made by Pickett. This slide rule made quick calculations of multiplication, division, roots, and powers, both to the log ten and natural log, very easy to perform. The only other alternative to multiplying and dividing was to use logarithms by virtue of the fact that addition of numbers achieved multiplication. In those days, the electronic slide rule was only a concept, and at the time, I thought an electronic slide rule would be like a dual-based log slide rule connected to a mechanical device to slide the cursor back and forth to achieve the correct answer. This would have been a purely analogue type of calculating machine.

Of course in those days, the idea of using microprocessor-based switching to perform calculations was futuristic and essentially undreamed of by most people. But today in almost any drug store, for US$4.00 you can buy a basic calculator that uses digital switching to add, subtract, multiply, divide, and to do other mathematical functions. Because it is so commonplace and so inexpensive, the calculator, which is quite special, is very much taken for granted.

Consider turbine engines, in particular, as an example of a technology that has impacted the sterile insect technique. Apart for turbine engines, there are the Global Positioning System (GPS) and the Internet. Let us now consider the reliability of flight technology, then flight guidance and lastly, the awareness of flight technology.

The King Air type aircraft with turbine engines is used in the Medfly programme in Los Angeles and in the Central American Screw-worm Project in Managua, Nicaragua. Turbine engines have largely replaced piston engines because fewer moving parts and stronger, lighter materials make them more reliable than earlier piston engine designs.

On board the aircraft is a microprocessor-based guidance system which guides the pilot and records flight paths and release activities. Sterile insect release projects increasingly rely upon GPS satellite guidance and desktop geographic information systems to identify streets, monitoring activities, treatment zones, and to present flight data for analysis.

Depending upon the specific requirements of the programme manager, aircraft may be guided with or without differential correction and flight activities may or may not be recorded for review. A very simple navigation plan would only indicate waypoints and directions to pilots at predetermined intervals along a flight path.

The GPS guidance system provides managers with the ability to programme vast treatment areas into onboard devices that guide aircraft pilots with precise parallel offsets to evenly cover targeted areas. So as we have looked at technology and reviewed bicycles, electronic slide rules and the reliability provided by gas turbine engines, let us look more in depth now at the GPS flight guidance systems.

If you look at the history of the navigational age, the significance and the impact of GPS on flight guidance become even more apparent. In the early days of world exploration, navigators relied upon celestial observation. It is my understanding that it was relatively easy to determine the latitude of an observer on the face of the earth. The longitude was more difficult to determine because of the time and distance calculations required. Of course, the magnetic compass has been available for many years to give directions and headings, but it did little to determine the actual location of guidance across the face of the earth.

In the early days of aviation, light networks were developed to guide aircraft at night from one point to another. Pilots contracted with farmers to light small fires which were easy to see from the air along the route between Chicago and St. Louis. At least once, a pilot was delayed by an engine problem, and the farmers abandoned their fires. The pilot made his trip anyway, but it was clearly important to develop a reliable flight guidance system.

Networks of high intensity electrical lights were developed across the United States to guide aircraft from point to point. Eventually, ground-based radio broadcast emerged which gave a four course range and provided pilots with a general idea of their location in relation to a predetermined route.

The current system combines the very high frequency omnidirectional range (VOR) and distance measuring equipment (DME) systems. The VOR system gives pilots a signal from a ground-based radio station, but it does not relate the distance to that station. Coupled with the VOR is the DME navigational system which uses a transmitter onboard an aircraft in conjunction with a radio transmitter on the ground. A radio signal is transmitted from the aircraft to the ground station, and distance is determined by calculating the time necessary for returning the signal to the aircraft.

A low frequency radio ranging system (Loran) has existed for a number of years. Because low frequency radio signals are subject to interference, it is easily disrupted. However, it is generally very effective, and the microprocessor makes possible automatic position calculations. In earlier days, intersecting coordinates and factoring time changes from different stations were performed manually. Incorporating the microprocessor into Loran has made it a very reliable, user friendly navigational aid in the past ten years or so.

To overcome the deficiencies inherent in the Loran system, the United States Department of Defense developed and deployed the GPS. This is probably one of the most significant technological advances in the navigation age. It has also impacted other areas of life. GPS offers worldwide coverage, unsurpassed accuracy, and permits highly efficient navigation and transportation.

The GPS system consists of three entities – geosynchronous satellites, ground control stations, and the GPS onboard receiver. Twenty-four satellites in six orbital planes broadcast positing information to portable receivers around the world. Ground control stations monitor the signals from the satellites, and all are linked to a master control station in Colorado which monitors and corrects timing signals from the different satellites.

The master control station relies upon a highly accurate atomic clock that has an estimated error of only two seconds for every seventy years of operation. The receiving units and the satellite transmitters are synchronised, and the transmission time from satellite to receiver is calculated by a microprocessor in the receiving unit. Having three satellites in view by the receiving unit allows precise computation of location and altitude.

Because of security concerns by the United States Department of Defense, an error is introduced into the GPS signals known as selective availability. To overcome this error, differential correctional systems have been developed that use ground based GPS receivers at fixed, known locations. These stationary correcting units read and interpret the introduced errors and calculate and broadcast corrections to the mobile receivers. Differential correction also eliminates a natural error caused by refraction as signals pass from geosynchronous satellites operating at 22,000 miles above the earth to receivers. Differentially corrected GPS is accurate to approximately one metre.

Differentially corrected guidance is provided in real time and flight maps are recorded for quality control purposes. Once the flights are recorded, they can be overlaid in a variety of geographical information or GIS systems. One of the common GIS software packages is MapInfo. Because of the microprocessor and technological developments, programme managers have a greater control over treatments that was not possible a few years ago.

As we look at technology in the GPS system aboard the aircraft, the pilot has a moving map display inside the cockpit to show not only the path of the dispersal plan but where he has dispersed, which swaths he has covered. As he progresses through this swath and predetermined path, a flight guidance is provided him by a Course Deviation Indicator (CDI) on the dashboard of the airplane or as an option, a flight guidance bar is installed outside the aircraft so that, as he is in continual watch of other aircraft and traffic, he can easily see in his peripheral vision exactly where he is in relationship to the prescribed flight line. As he flies across the block, the data are recorded on a computer hard drive which, upon landing, can be delivered to a GIS system for review.

We have looked at technology as it relates to bicycles, how electronic slide rules have made an impact on technology, the reliability that gas turbine engines have provided and the flight guidance of GPS. Finally then, let us look at the awareness that is provided by the Internet. Most of you in this conference have made your arrangements and reservations by use of the Internet. And only a few years ago, no one would have ever thought that the Internet would grow to the extent that it has today. It is a conduit for information shared among organisers, among programme directors, and it allows communications from all parts of the world, which is fast, efficient, and reliable simply because of the advancement of the large-scale integrated circuit microprocessors and the computer age.

So, the impact of technology on sterile insect release is considerable. First of all, it gives us a greater awareness as we gain information and direction through computer communications and the Internet. The GPS and microprocessor based flight guidance systems give us greater accuracy and dispersal that were inconceivable a few years ago. And finally, the reliability that the turbine engines give us makes the whole sterile insect technique dispersal system a fast, reliable way to control Medflies, screw-worm flies, and other pests susceptible to the sterile insect technique.

# PART III.

## APPLICATION OF AREA-WIDE CONCEPT TO FRUIT FLIES

## A) ACTION PROGRAMMES

# The Use of Massive SIT for the Control of the Medfly, *Ceratitis capitata* (Wied.), Strain SEIB 6-96, in Mendoza, Argentina

*Oscar De Longo, Anibal Colombo, Pablo Gomez-Riera and Andrea Bartolucci*

ISCAMen. Boulogne Sur Mer 3050 (5500) Mendoza, Argentina

## INTRODUCTION

The increasing demand for fresh fruits has led to the development of new techniques in farming and in harvest and postharvest areas. Special care has been taken to ensure that the use of pesticides reconciles pest control with a low impact on the environment and human health.

The presence of pests of economic importance, among them, the Medfly (*Ceratitis capitata* Weid.), led to the search for alternative techniques of control. The Medfly was first recorded in the United States in 1901, and in Argentina in 1934.

The use of the sterile insect technique (SIT) began to develop around 1950. Improvement in the technique has produced promising results world-wide, especially in the breeding of sexing strains, which allows the release of sterile males only.

The National Service of Agrifood Quality and Protection (SENASA) of Argentina, coordinates the National Program of Control and Eradication of Fruit Flies (PROCEM). PROCEM has divided Argentina into five different regions. This is according to quarantine regulations, and control and eradication activities to be undertaken. Actually, the Mendoza and San Juan provinces, and the Patagonia Region, have adhered to PROCEM. They have ongoing regional programmes in their respective areas. Mendoza and the Central Valleys of San Juan (Tulum, Ullum and Zonda) are included in the New Cuyo Region, Phase I, where the main objectives are: eradication of the Medfly; verification of the quarantine status; creation of a Medfly free area and exports to Medfly free markets. SIT is the main technique used in Mendoza for Medfly eradication. The Institute of Quality and Protection of Mendoza (ISCAMen) is in charge of the programme in the Province of Mendoza. Mendoza policies for Medfly eradication and the use of SIT, are based on:

- a growing area of 290,000 ha of vegetable and fruits (vineyards, stone, pome and nut orchards) of high quality, with favourable agroecological conditions and geographical isolation,
- exports of fresh fruits for the Northern Hemisphere markets during winter time,
- Mendoza as a crossroad area between the Pacific Ocean markets and Mercosur, the most important Latin American market (US$900 billion of GNP and 200 million people),
- the use of the Pacific Ocean ports near Mendoza as new avenues for exports,
- agroclimatic conditions are not favourable for the commercial production of citrus,
- cold winters with low temperatures often falling below freezing point, which result in a natural suppression of the pest,

- *Ceratitis capitata* is the only species of quarantine importance, and
- the distribution of the irrigated land surrounded by desert with no natural plant hosts, which avoids re-infestation and facilitates quarantine controls.

Taking into account all these considerations, the Mendoza Medfly Program was started in 1990, with pest monitoring for two years, in order to determine the degree of infestation and population cycles. A facility for the mass rearing and sterilisation of the Medfly was also constructed. In the 1992/93 season, flies of the sterile bisexual native strain were released. This activitiy continued through the 93/94 and 94/95 seasons, increasing weekly from 70 to 200 million flies. Results were satisfactory. Suppression achieved was in the order of 98%.

The international and national financial crisis of December 1995 led to a considerable delay in mass rearing and liberation. A severe re-infestation took place then. Medfly native fertile adults returned to the levels of the 90/92 season.

In the winter of 1995, tests were conducted in the rearing of the white pupae sexing strain, developed by the Seibersdorf Laboratory of the IAEA (SEIB 6-96). Taking into account tests performed by Robinson (1996) and Fisher (1997) showing good performance of the sexing strain, the field tests performed in Argentina and the adjustments already made in the mass rearing of the strain, ISCAMen decided to replace the native bisexual strain with the white pupae strain SEIB 6-96 at Bioplant Km8.

The 96/97 season began early in mid-winter (August) with the release of a large number of sterile insects (116 million a week). As a result, the programme achieved the best degree of suppression ever recorded; this has been maintained in the 97/98 season.

The level of re-infestation recorded in the 95/96 season, before the use of the white pupae strain, allowed a statistical comparison with the bisexual native strain. Pre-release Medfly trap day (MTD) indices were also used to compare with the 95/96 season, in the same geographical area.

## MATERIALS AND METHODS

### Geographical Area and Climatic Conditions

The Mendoza province is located between the parallels of 32° and 37°33´ south latitude and the meridians of 66°30' and 70°36' west of Greenwich (Figure 1).

It has an area of 147,800 km$^2$. Its limits are: to the north, the San Juan province; to the south, the La Pampa and Neuquén provinces; to the east, the San Luis province and to the west, the Republic of Chile, across the Andes.

The climate is continental, arid, a desert with low summer rainfalls (90 to 250 mm a year), high sunshine, 15°C of night and day temperature difference, and 220 to 260 days free of agricultural frosts (Figure 2).

There are four oases, covering 690,000 ha (Figure 1). One is in the north, another in the northeast, a third in the centre and the last in the south. These oases are isolated by natural barriers, the most important one, the Andes to the west. The north and east oases are bounded by desert, with low xerophytic vegetation, and no natural Medfly hosts. The central oasis is located in the foothills, and surrounded by desert, with no native plant hosts. It is separated from the north oasis by 40 km of desert, and 160 km from the south oasis. The latter is also isolated in all directions by deserts over 100 km or more.

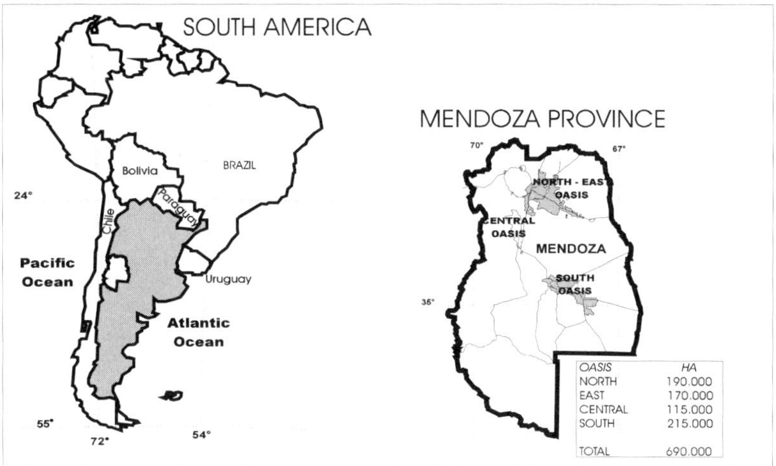


Figure 1. Geographical situation of Mendoza and its Oases.

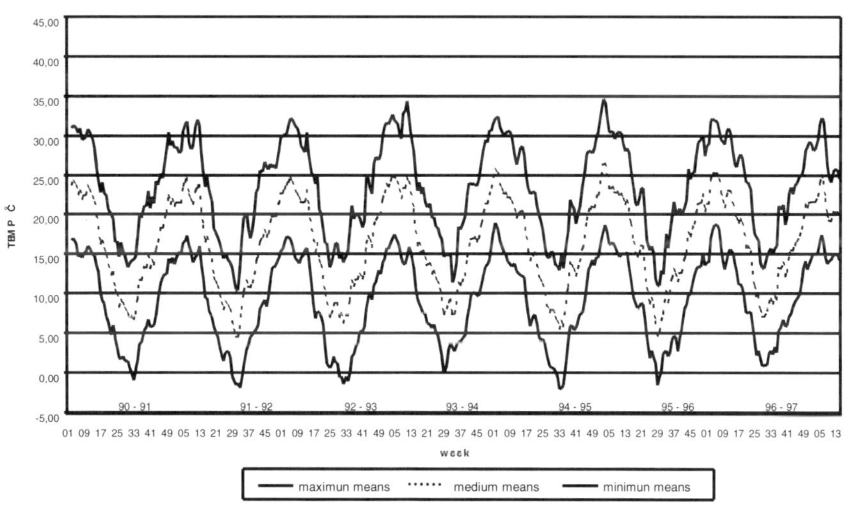


Figure 2. Mean weekly temperatures from 1990-98.

## PRODUCTION

Sterile flies from the white pupae strain are reared at Bioplanta Km8. The mean weekly production is 300 million pupae. 160 million male pupae have been used for the eradication programmes in Mendoza (128 million), San Juan (48 million) and Patagonia (14 million).

Mass production began in August 1996. In the 96/97 campaign, release reached a weekly mean of 116.5 million sterile males, over an area of 219,000 ha (43,000 ha urban and 176,000 ha rural) (Table 1).

Male sterilisation is performed through irradiation with gamma rays from a Cobalt 60 source. This means the use of atomic energy for peaceful purposes. ISCAMen owns a irradiator, IMCO20, which is operated by Nuclear Mendoza SE in accordance with the Argentinian nuclear regulations provided by the ARN (Nuclear Regulatory Authority).

Table 1. Working areas of Mendoza Province.

| AREAS | SURFACE (ha) |
|---|---|
| Province | 14,780,000 |
| Oases | 690,000 |
| Urban | 43,000 |
| Rural | 647,000 |
| Cultivated | 290,000 |
| Under trapping | 546,000 |
| Under Medfly release | 219,000 |
| (a) Urban | 43,000 |
| (b) Rural | 176,000 |

## PACKING

Irradiated pupae are pigmented with fluoresceine, and taken to the packing plant. There, the pupae are packed in paper bags, 30 to 100 cc per bag, according to collectors. This provides an average of 1,400 flying males per bag. Each bag contains a supporting paper to allow the emerging flies to expand their wings, and a paper with sugar to provide food for the flies.

Paper bags are sealed and packed in plastic boxes (12 bags each). Mean production is 5,000 bags per day (2,000 litres of pupae and 35,000 bags per week).

Boxes are stacked in piles under controlled conditions of temperature (25°C ± 1°C) and humidity (65% ± 5%). The bags were allowed to stand for 96 hours after irradiation, after which the pupae are at the optimal time of emergence for release.

### Fly release

Release is done by plane (90%) and by truck (10%). Bags are taken from the packing plant, and distributed, under controlled conditions, in trucks to the various points of release. This is done in the same plastic boxes, in order not to harm the flies.

At the airports, the bags are put into the planes, which can carry from 700 to 800 bags per flight. Planes used are Cessna 180, C. 182 and C. 185. Flights operate at an altitude of 500 to 1,000 feet, depending on whether it is in a rural or urban area. Speed is set at 100 mph. Each flight covers an area of 5,000 ha over strips 300 m wide. Before bags are released through a venturi funnel, they are opened by the operator who ensures that they stay completely so when they reach the ground.

Releases are monitored through the GPS, in the cars and the counted bags recovered on the released areas as well as sterile males trapped in Jackson traps.

Ground releases are made in small areas where it is not economical for aerial releases and when it is necessary to re-inforce aerial releases. Ground releases are done from closed trucks that carry bags of plastic boxes. Bags are released along designated routes.

In the urban areas, fly densities are 900 pupae per ha per week (400 sterile flying males per ha per week), with two weekly releases, each one 3 to 4 days apart. In the rural areas, the mean density is 450 pupae/ha/week (200 sterile flying males/ha/week), with releases done once a week (Figure 3).

In winter months, with frosts occurring, releases are done twice a week, in order to have sterile males in the fields.

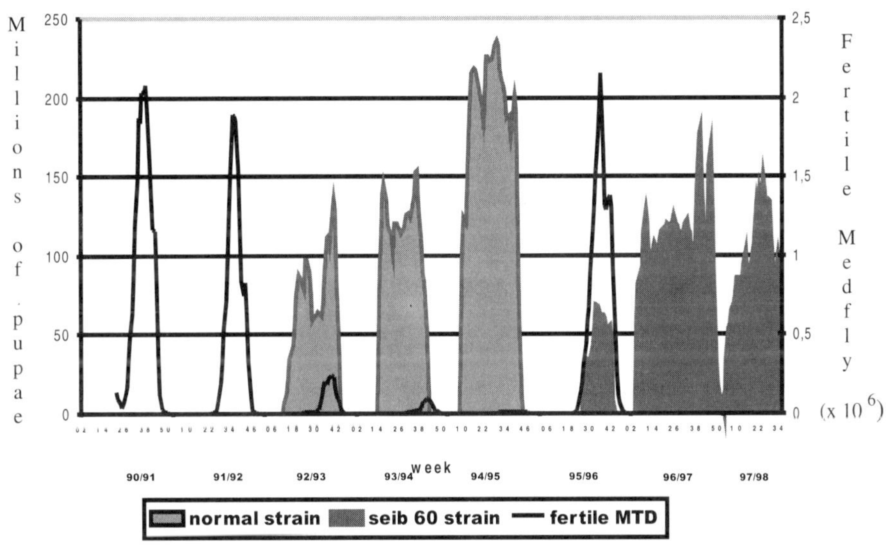


Figure 3. Millions of pupae released and fertile Medfly trapped per week.

## TRAPPING

The trapping net is made up of 4,395 Jackson's traps: 3,059 baited with trimedlure; 111 baited with methyl eugenol; 111 baited with cuelure and 1114 McPhail's baited with torula yeast pellets. All traps are read and monitored on a weekly basis (Table 2).

Table 2. Flies trapped in Mendoza oases.

| OASIS/TRAPS | JACKSON (*) | | | MCPHAIL | TOTAL |
|---|---|---|---|---|---|
| | T | C | M | | |
| NORTH | 803 | 33 | 33 | 291 | 1,160 |
| EAST | 810 | 35 | 35 | 262 | 1,142 |
| CENTRAL | 443 | 16 | 16 | 227 | 1,391 |
| SOUTH | 1,003 | 27 | 27 | 334 | 1,391 |
| **PROVINCE** | **3,059** | **111** | **111** | **1,114** | **4,395** |

* baited with T= Trimedlure
C= Cuelure
M= Methyl eugenol

## IDENTIFICATION

Fly identification is done at three laboratories:

- the Km8 Laboratory, which identifies flies captured on the north and east oases,
- the Uco Valley, which identifies those from the central oasis,
- the Rama Caida, which identifies those from the south oasis.

At the laboratory facilities, sterile males from the Jackson and McPhail traps are counted. Fertile flies from these traps are identified and counted. Mounting and observation of genitalia are carried out to confirm results. Other activities are larvae in fruit samples, to confirm the absence of other fruit flies (in other traps) and to key in data gathered on all the oases into the computer systems.

Six hundred Jackson traps and 320 McPhail traps from 150 to 400 fruit samples, are processed daily. Results are sent to the central field operations office where all data are recorded.

## DATA ANALYSIS

The objective of the work was to compare the effect of using SIT with the bisexual local strain against the white pupae sexing strain (SEIB 6-96). Comparisons were also made against periods of no releases and an intermediate campaign with late release and re-infestation.

The weeks considered were between $18^{th}$ and $45^{th}$, on all campaigns. Each campaign consisted of 52 weeks, the first week of August of each year being week 1. Weeks under consideration ran from the $18^{th}$, this being the first of winter releases and adding up day-degrees until the completion of a biological cycle. In so doing, it was possible to evaluate the effect of the sterile fly without the suppression effect of low winter temperatures.

Comparisons were made using the MTD (Medfly/trap/day) index as the observation variable.

The amount of sterile flies released (millions of pupae per week) was used as a covariant. There were four treatments:

T1: monitoring with no releases, years 1990 to 1992 (53 weeks)
T2: monitoring with releases of the local bisexual strain, years 1992 to 1995 (84 weeks)
T3: monitoring with late releases and re-infestation, years 1995 to 1996 (28 weeks)
T4: monitoring with releases of the white pupae sexing strain SEIB 6-96, from 1996 up to present day (46 weeks considered)

Descriptive statistics were calculated for each treatment. The normality test of Shapiro-Wilks was performed on the data.

MTD was transformed as:

$$TMTD = \ln (MTD + 0.000001)$$

For the bisexual strain, 50% of the volume released was taken into account in order to consider the same pressure of sterile males with respect to the sexing strain.

Analysis of Variance (ANOVA) was performed on a randomised block design, where blocks were the weeks. The Tukey Test was performed to compare means among treatments with unequal numbers of observations (weeks). Also, the Dunnet Test was performed to compare each treatment against a check (T1).

All data were processed using SAS (1982) package.

## RESULTS AND DISCUSSION

Under the Shapiro-Wilks test, T1 showed $p \leq 0.01$. The transformation of MTD could not reject Ho of normality distribution for T2, T3 and T4 (Table 3).

Table 3. Mean MTD (Medfly/trap/day) for the four treatments and significant tests.

| Treat. | Mean MTD | Tukey | Dunnet | std. | N |
|---|---|---|---|---|---|
| T3 | 0.829786 | A | A | 0.702934 | 28 |
| T1 | 0.783308 | A | A | 0.734712 | 53 |
| T2 | 0.035542 | B | B | 0.065234 | 84 |
| T4 | 0.001514 | C | C | 0.001469 | 46 |

Note: Treatments with the same letter do not differ statistically ($p \leq 0.05$) when performed over TMTD. Results are shown in the original data (MTD).

The ANOVA, Tukey Test and Dunnet Test were performed using TMTD, even though results are expressed in the original values. Treatments showed significant differences at $F \leq 0.05$. The Dunnet Test showed significant difference ($p \leq 0.05$) of T2 and T4, with respect to T1. There was no significant difference of T3 with respect to T1 ($p > 0.05$).

The Tukey Test showed significant differences ($p \leq 0.05$) of T2 and T4, with respect to T1, T3 and between them. There was no significant difference of T3 with respect to T1 ($p > 0.05$).

Our results showed that in the Medfly Program of Mendoza, Argentina, the local bisexual strain could be compared, on a statistical basis, against the sexing strain SEIB 6-96. This could be shown, since the 95/96 campaign showed re-infestation levels similar to those of pre liberation dates (Figure 4).

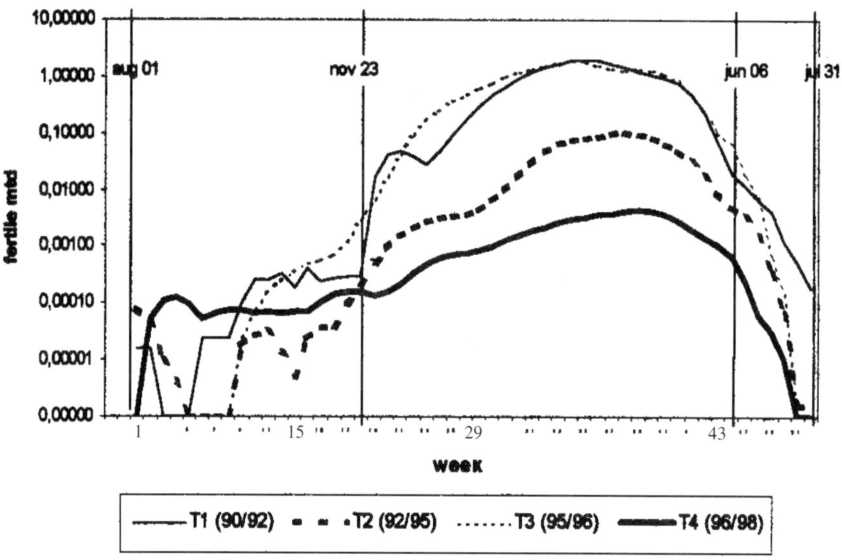


Fig. 4. Female Medfly/ trap/ day (MTD) for four treatments.

The mean comparison of the MTD in each campaign showed that under the bisexual release campaigns of 94/95, the mean MTD was 0.0017. This was the third in a row of campaigns using SIT.

On the other hand, in only one campaign releasing the sexing strain, the mean MTD reached 0.00088. This indicated lower suppression rates than those obtained using the bisexual strain. This could be explained by a better distribution and competitiveness of the sterile males, which are forced to look for wild females, since there are no sterile females in sight.

The sexing strain also showed advantages in management, since the volume for packing and releasing is lower for the same working area and density of release, and with lower costs.

In addition, the absence of sterile females means less risk of fruit punctures. The absence of sterile females could allow the adjustment of releases, forcing sterile males to remain one more day in the bags, with optimal conditions of sexual maturity on reaching the fields, and less predation from natural enemies on non-emerged pupae.

Results obtained confirmed the effectiveness of the sexing strain SEIB 6-96 in the eradication of the Medfly.

## ACKNOWLEDGEMENTS

We would like to thank Edgardo Arlandi and Alberto Tobares for graphs and figure design, and Roberto Crino, for pictures.

## REFERENCES

Fisher K. 1997. Installation of a recombinant filter for the SEIB 6-96 white pupae strain in Mendoza, Argentina. Field Report. 15-28 March. Seibersdorf. Austria.

Hendrichs J., G. Franz and P. Rendon. 1995. Increased effectiveness and applicability of the sterile insect technique through male-only releases for control of Mediterranean fruit flies during fruiting season. J. Appl. Entomol. 119, 317-376.

ISCAMen. 1998. Reportes semanales del Programa de erradicacion de la Mosca del Mediterraneo. Mendoza. Argentina.

Robinson, A.S. 1996. Economic, strategical and operational considerations for the use of Medfly sexing strain in SIT programmes. Consultants meeting. Mendoza, Argentina. Working paper. March. Seibersdorf, Austria.

SAS Institute Inc. SAS User's Guide: Statistics, 1982 Edition. Cary, NC: SAS Institute Inc., 1982. 584 pp.

SENASA-ISCAMen. 1997. Programa de Control y Erradicacion de las Moscas de la fruta en Argentina. Subprograma de erradicacion de la Mosca del Mediterraneo en Mendoza. Segunda revision tecnica del Subprograma Mendoza por el Comite Tecnico Internacional. 21-26 de abril.

# Current Progress in the Medfly Program Mexico-Guatemala

*A. Villaseñor[1], J. Carrillo[2], J. Zavala[3], J. Stewart[4], C. Lira[2] and J. Reyes[5]*

[1] Medfly Commission Mexico-Guatemala 6 Calle 1-36, Zona 10 Edif. Valsari of. 303, Guatemala City, Guatemala.
[2] Programa Moscamed, 6a Calle 1-36, Zona 10, Edificio Valsari, Office 203, Guatemala
[3] Regional Medfly Program, Mexico-Guatemala-United States, Comisión Nacional de Sanidad Agropecuaria, Mexico
[4] Regional Medfly Program, Mexico-Guatemala-United States, USDA-APHIS-IS, Guatemala
[5] Comisión Nacional de Sanidad Agropecuaria (CONASA), Guillermo Perez Valenzuela No. 127, Colonia del Carmen, Coyoacan, D.F. 041000, Mexico

## INTRODUCTION

For twenty years, the Regional Medfly Program in southern Mexico and Guatemala, central America – which is financed by Mexico, Guatemala and USA – has successfully halted the Medfly *(Ceratitis capitata* (Wied.)) dispersion toward Mexico and USA. After the pest eradication in Chiapas, Mexico, in 1982 and some vain efforts to contain it in Guatemala, a strategy has been formed in the construction of a static barrier of containment. However, this has been criticised frequently by the border society which has suffered for a long time because detection and control action sometimes result in big emergency plans as the answer to strong explosions and pest dispersions which have in turn occurred from climatic phenomena, such as El Nino, in 1998.

The Medfly eradication in Guatemala has not been accomplished because the suppression technology used before sterile insect technique (SIT) had been based on malathion aerial bait spray. The aerial bait spray has been prohibited in Guatemala since 1987, following strong complaints from ecological groups and the beekeeping sector, as well as because of financial contraints.

The xanthene dye technology that replaced the use of malathion has given new hopes and possibilities to the old project of pest eradication in Guatemala and Central America. However, moving the barrier from north to south and from east to west is necessary to prevent re-infestations in Mexico and pest-free areas in Guatemala.

The development of new detection and control tools has also strongly supported the project, with the use of more efficient traps, such as the OBDT trap, baited with ammonium acetate, putrescine and trimethylamine (phase IV traps), trimedlure (TML)-laced yellow panel traps and TML-baited Jackson traps.

The use of the improved chilled adult release system, the aerial bait sprays which are guided by the GPS and SATLOC navigation systems, the augmentative release of parasitoids and the artificial biological isolation of the infested areas all help to increase the possibilities of suppression in areas of the control zone.

## AREA-WIDE STRATEGY

To stop the Medfly advance toward the north and east of the region and to protect the free areas of Guatemala and Mexico, we work in large zones, regardless of international borders or political domestic divisions (Figure 1).

### Free Zone

This includes all the areas that have never been infested or those that have been declared eradicated. Here, only preventive trapping is carried out and are protected by quarantines and emergency plans.

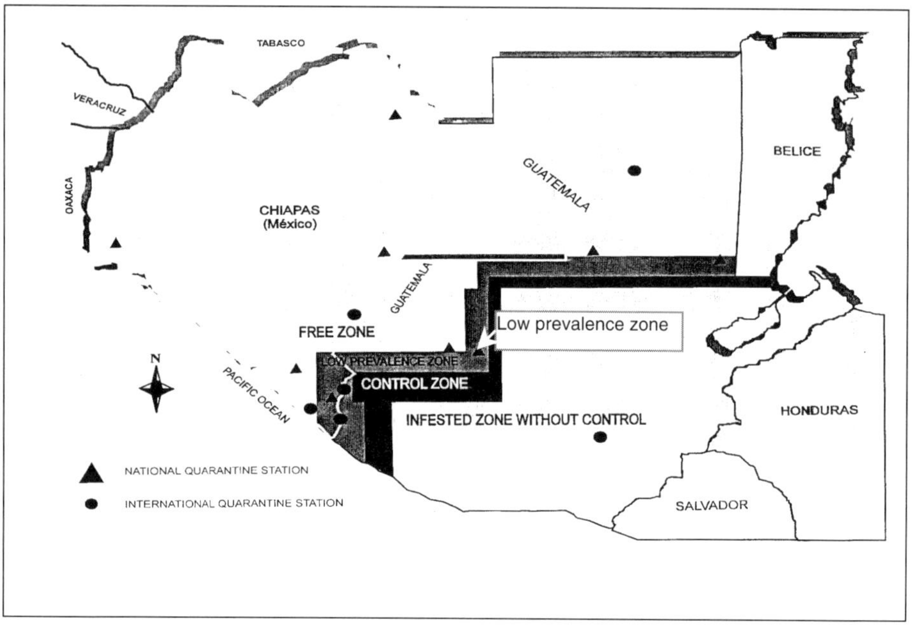


Figure 1. Medfly Regional Program strategies for 1998.

### Low Prevalence Zone

This is an area that has been under eradication activities or where there are few possibilities of pest establishment. Some outbreaks have been detected and they are treated under emergency plans.

### Control Zone

This includes the areas subjected to the most intensive control activities, such as the aerial bait spray applications used to reduce high populations of wild flies, the release of sterile flies at very high densities and the destructions of preferred infested hosts.

### Monitor Zone

This zone conducts pest population dynamics so that future action can be projected and the trapping that is carried out in the control area neighbourhood or into the infested and uncontrolled zones can be studied.

### Infested and Uncontrolled Zone

No control action is carried out in this area.

## ACTIONS AND RESULTS

### Trapping

Jackson traps have been the official traps used in international programmes and detection protocols. However in 1997, this regional programme began the general use of the yellow panel traps and the phase IV traps, resulting in a better detection of the pest and better evaluation of the effects of the control methods.

The free zone of the region operates a total of 16,568 traps, from which 13,960 are Jackson traps and 2,608 yellow traps; checks are done twice a month. In the low prevalence zone, 3,774 traps – 2,923 Jackson, 115 yellow panel traps and 706 phase IV traps – are used, while in the control zone – 2,370 Jackson traps, 100 yellow panel traps and 200 phase IV traps – are used and these are checked once a week.

### Survey

The survey of fruits in the tropical areas is an extremely important tool to enable detection of the presence, distribution and density of the pest population and to assure that it has been excluded from a working area. The detection of the pest in the postharvest coffee periods and dry seasons is made very efficient through the collection of residual coffee berries. However, when these berries are out of season, the pest can only be detected at the larval stage. Furthermore, because of their low-density levels, traps cannot detect adults especially in places with rough topography (common in tropical coffee areas), as it is very difficult to map an ordinary trapping route in such areas. For that reason, sometimes the pest is only reported as larval outbreaks without the detection of adult flies (Figure 2).

### Chemical Control

Malathion ground sprays, formulated from one part of malathion, four parts of hydrolysed protein and 95 water parts, are used to treat isolated pest outbreaks. Apart from this, they were used in a general form in all the zones between the 1988-1997 period when aerial bait sprays were prohibited.

The malathion aerial bait sprays, using the formula of one part of malathion and four or nine parts of hydrolysed protein, were stopped in 1987 because of public complaints and those of ecological groups and beekeepers.The usage of xantene dyes was approved in Guatemala in April 1996 and utilised on a bigger scale from January 1997, when 16,750 hectares were sprayed four times with a dose of 3.5 litres per hectare, each containing 0.69% phloxine B, 0.31% uranine, 20% high fructose, 40%

mazoferm and 39% water. Results similar to those achieved by the malathion bait sprays were reached and between December 1997 and February 1998, 26,000 hectares were sprayed eight times weekly, 16,750 hectares with complete coverage, 5,000 hectares with alternate swaps and 3,250 hectares in "hot spots". The formula used was 0.5% phloxine B, 20% high fructose, 40% mazoferm and 39.5% water.

The suppression and control results of the pest populations were compared with those in adjacent witness areas. Complete coverage and alternate swaps were ensured The suppression and control results of the pest populations were compared with those in adjacent witness areas. Complete coverage and alternate swaps were ensured with the use of SATLOC guidance systems in the fixed wings of the Turbo Trush. "Hot spot" outbreaks in the free zone, the low prevalence zone and the control zone were sprayed by helicopter to reinforce eradication (Figure 3).

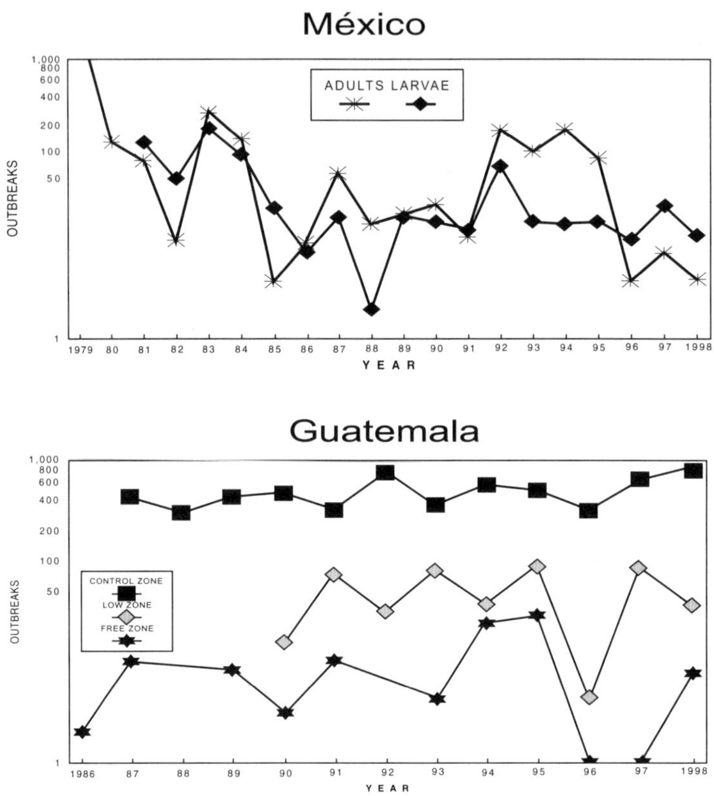


Figure 2. Annual fluctuation of Medfly outbreaks.

## Current Situation up to May 1998

After the aerial sprays, the weather conditions of prolonged drought and high temperatures were in favour of a demographic pest explosion, provoking a big dispersion and pressure on the various areas. The pests passed through the biological barrier zone and aerial spray areas, invading areas in the low prevalence zone. Some outbreaks were also detected in free zones. Altogether, there were 35 isolated outbreaks Chiapas, Mexico, 20 outbreaks in the northwest and 6 in Peten in Guatemala (Figure 4).

## Legal Control

15 quarantine stations have been strategically placed in order to isolate and protect the fly-free zones. However, the region encountered great difficulty following the quarantine regulations because of the military customs of commercial interchange at the border towns as well as a poor culture of cooperation in plant health issues. Nevertheless, the quarantine actions have decreased the appearance of outbreaks in Peten and Huehuetenango, in the same way as the pest introductions in the north of Chiapas and Tabasco, Mexico, have decreased (Figure 1).

## THE STERILE INSECT TECHNIQUE (SIT)

Sterile flies from different strains were targetted for release accordingly. In the control zone after the aerial sprays, only TSL males were released in a density of 1,500 flies per hectare, taking care that the trap captures reflected a minimum 80:1 sterile:fertile ratio. Flies of the normal bisexual strain were released in the rest of the control and low prevalence zones in a gradual decreasing density of 5,000-2,000 flies according to the degree of infestation.

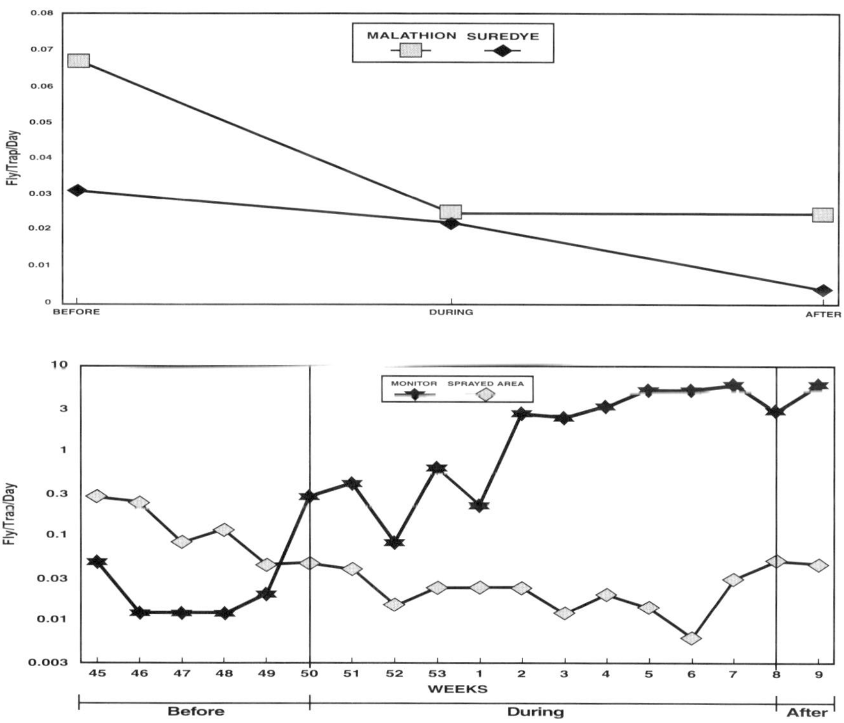


Figure 3. Effect of aerial bait spray for Medfly population in control zone – Guatemala.

Figure 4. Medfly infestation during 1998.

The release activity through the paper bag system has been reinforced since 1997 with the chilled adult release system after improvements were made by the Californian Department of Food and Agriculture (CDFA). Subsequent evaluations showed a better distribution and fly density in the targetted areas. Besides, both systems are guided by the GPS and/or SATLOC navigation systems that assure accuracy in the targeted areas.

**Biological Control**

The parasitoid augmentative release technique was initially used in this programme in 1997. The Metapa facility mass reared 20 million *Diachasmimorpha tryoni* (Cameron) per week and released them in the control zone and in the Rio Ocho infested area in Guatemala. The Aurora facility, in Guatemala, has been rearing 5 million *D. tryoni* per week, these being released in pear orchards in the central part of Guatemala. *D. longicaudata* (Ashmead) and *D. tryoni* are under field evaluation, in order to determine the better species of the two. The package and release system, using paper bags and chilled adults, is being tested in the field.

**Production Facilities**

The regional programme involves two sterile flies production facilities, one in Metapa, Chiapas, Mexico, where 500 million of sterile flies from the bisexual strain are produced per week and another in El Pino, Guatemala, where 80 million of sterile males from the TSL strain are produced. These are for use in the region while 400 million of the bisexual strain are for delivery to California and Florida, USA. The two facilities, nevertheless, have installed capacity to produce 1,400 million flies of the bisexual strain and 800 million of the TSL strain, respectively.

Both faciities have renewed their colony strains in order to assure competitiveness under field conditions. Decision is taken according to the annual field quality control tests and the colony strains produced at the Metapa facility have been changed four times accordingly (Figure 5).

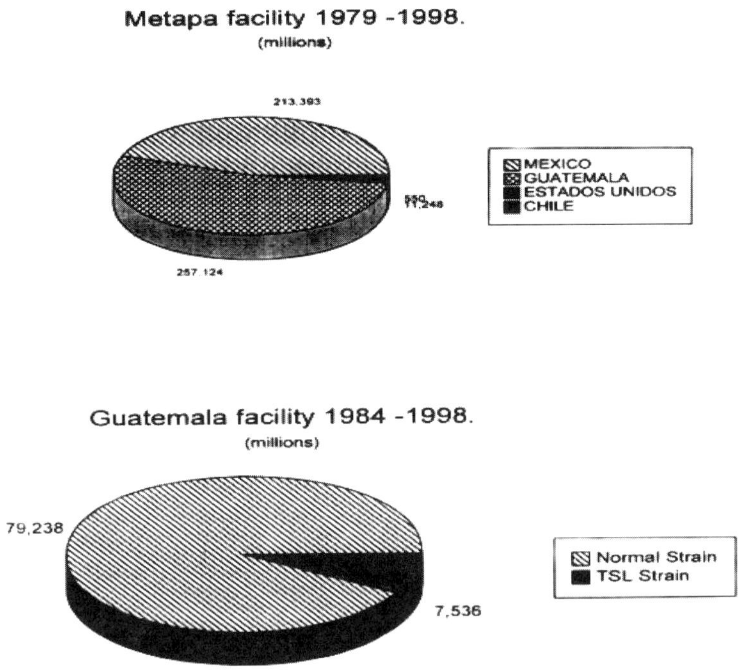


Figure 5. Medfly mass rearing facilities in Mexico-Guatemala.

## EMERGENCY PLANS IN FREE ZONES

Certain areas of the region, such as the Lacandona Jungle in Chiapas (1994) and Rio Ocho in Guatemala (1997), could not be monitored during the political-social conflicts that were conducted by the internal guerrilla groups. As a result, some pest introductions grew without control for generations, until they reached a high level stage of population so that it was necessary to implement strong emergency actions in order to eradicate them.

Pest eradication efforts in these jungle areas were in isolated coffee areas and demonstrated the effectiveness of SIT and new technology in tropical regions.

A maximum infestation of 0.7 flies/trap/day (FTD) was localised in La Selva Lacandona of Chiapas, with the detection of a total of 8,243 fertile detected flies and 50 larvae/kg from 29,749 larvae.

14,500 hectares were sprayed by helicopter, eight times with malathion bait (nine parts of hydrolysed protein and one part of malation with a dose of one litre per hectare). 290 tons of fruits were destroyed; 1,240 million sterile flies were released in a density of 2,000 sterile flies per hectare, achieving pest eradication between March and November 1995. A strict quarantine was applied in order to isolate the infested zone as well as to promote a strong public relations campaign.

In Rio Ocho, Guatemala, a maximum infestation of 0.193 FTD was found with the detection of a total of 903 fertile flies and 3.59 larvae/kg from 1,663 larvae on a coffee area surface of 15,000 hectares, from which 5,500 hectares had been sprayed six times weekly with the formula of 0.069% phloxine B, 0.32% uranine, 20.0% high fructose, 40.0% mazoferm and 39.0% water. In addition, 110 million sterile-only males were released per week at a density of 2,400 flies per hectare after the sprays as well as 1,000 *D. tryoni* per hectare. Eradication was achieved between June and November 1997.

## Financial Sources

The current cost of maintaining the containment barrier is about US$12 million a year (67% contributed by Mexico, 32% by United States of America and 1% by Guatemala). From this budget, US$6.5 million is devoted to Chiapas and US$7.5 million to Guatemala. Expenditure through field operations costs about 52%, sterile flies mass rearing, 38%, and administration, 10%.

The new developed technology, like the TSL strain, xantenes dyes, chilled release adults, augmentative releases of parasitoids, new traps, allows for the strengthening of the containment barrier and its movement into Guatemala territory, but this technology is very expensive and the programme will increase its expenditure if this is used on a large scale.

In order to achieve Medfly eradication from Guatemala and the entire central America as a long-term objective, a cost-benefit study is being prepared with the support of the three governments involved. This study will be fundamental to our attempt to get stronger financial support to carry out such a huge project on Medfly eradication from Central America.

# Mediterranean Fruit Fly Preventative Release Programme in Southern California

*Robert V. Dowell[1], Isi A. Siddiqui[2], Fred Meyer[1] and E. Leon Spaugy[3]*

[1] California Department of Food and Agriculture, 1220 N Street, Sacramento, CA 95814, USA
[2] U.S. Department of Agriculture, 14th & Independence Ave., SW, Washington, DC 20250, USA
[3] Los Angeles County Agricultural Commissioner, 12300 Lower Azusa Rd., Arcadia, Los Angeles, California, USA

California employs several area-wide pest management programmes that use the release of sterile insects to protect its commercial and dooryard agriculture. The first was developed in response to the discovery of the Mexican fruit fly, *Anastrepha ludens*, in Tijuana, Mexico and adjacent areas in San Diego County, California. Initially pesticide sprays of malathion and bait were applied to host plants around each fly find site. Additionally, soil sprays of diazion (0.05 kg per 93 $m^2$) were applied under every host plant around each fly find site. It soon became apparent that this approach was expensive and environmentally damaging. This led the interested parties, the California Department of Food and Agriculture (CDFA), the United States Department of Agriculture (USDA) and the government of Mexico to develop a programme that utilises the release of sterile Mexican fruit flies over the city of Tijuana in order to prevent the establishment of a breeding population of this fly in the city. The belief is that preventing the Mexican fruit fly from breeding in Tijuana will help protect both that city and California. To date, no Mexican fruit fly larvae have been found in Tijuana or the adjacent areas of California.

The second programme was developed in response to the discovery of the pink bollworm, *Pectinophora gossypiella*, in cotton in the Imperial Valley area of southern California. As the pink bollworm spread throughout the cotton growing region of southern California, it became a significant pest that threatened the 405,000 hectares of cotton grown in the San Joaquin Valley to the north. To keep this pest out of the San Joaquin Valley, the CDFA/USDA and California cotton growers use the large-scale releases of sterile pink bollworms in areas in which wild pink bollworms are captured each year. Thus far, the pink bollworm has been prevented from establishing a permanent presence in the San Joaquin Valley and the cotton growers in southern California, Arizona and adjacent areas in Mexico are developing an area-wide pink bollworm eradication programme that will use sterile moths as its primary management tool.

In 1993, CDFA found itself faced with a Medfly situation that required a change in the methodology used to eradicate such Medfly infestations were eradicated from California. Medfly infestations had been found each year since 1987 with the number of wild Medflies detected decreasing only after a large aerial malathion and bait spray programme in 1989-90 (Dowell et al. 1999). Four hundred wild Medfly adults were trapped in 39 cities in five southern California counties in 1993; these finds represented

35 discrete core infestations whose treatment boundaries were merging together as had been seen in 1989-90 (Dowell and Penrose 1995, Penrose 1996).

In response, the USDA formed an international Medfly science advisory panel to help develop a proactive approach to dealing with Medfly invasions in southern California. Based on advice from this panel, the CDFA /USDA/County Agricultural Commissioners (CAC) began a Medfly sterile release programme on 1 March 1994 over a 3,791 km² area of the Los Angeles basin to eradicate extant Medfly infestations. This basin-wide programme used the continuous release of 96,500 sterile Medflies per km² per week for a two-year period. The release rate was the upper limit of that used successfully in Mexico in their Medfly eradication/exclusion programme (Schwarz et al. 1989). Prior to the start of the basin-wide programme, ground sprays of malathion and bait were applied to all hosts within a 200 m radius of all fly finds. The basin-wide programme was designed to eradicate all extant Medfly infestations, and to prevent new infestations from becoming established (Dowell and Penrose 1995). The capture of wild Medflies within the treatment boundaries dropped from 400 in 1993 to no wild Medflies in 1995. In 1996, a single, unusually desiccated, undyed male Medfly was trapped the day after the programme terminated and at a time of programme staff reductions. No further flies were discovered in subsequent delimitation surveys.

Based on the success of the basin-wide programme and advice from the CDFA Medfly Science Advisory Panel, the CDFA/USDA/CAC started a new *proactive* Medfly Preventative Release Program (PRP) over a 5581 km² area (Figure 1) of southern California in July 1996. The goal of the PRP is to prevent or reduce the rate at which new Medfly infestations develop within the PRP boundaries while the USDA/CDFA/CAC identify and begin to close pathways through which Medfly infested produce enters California.

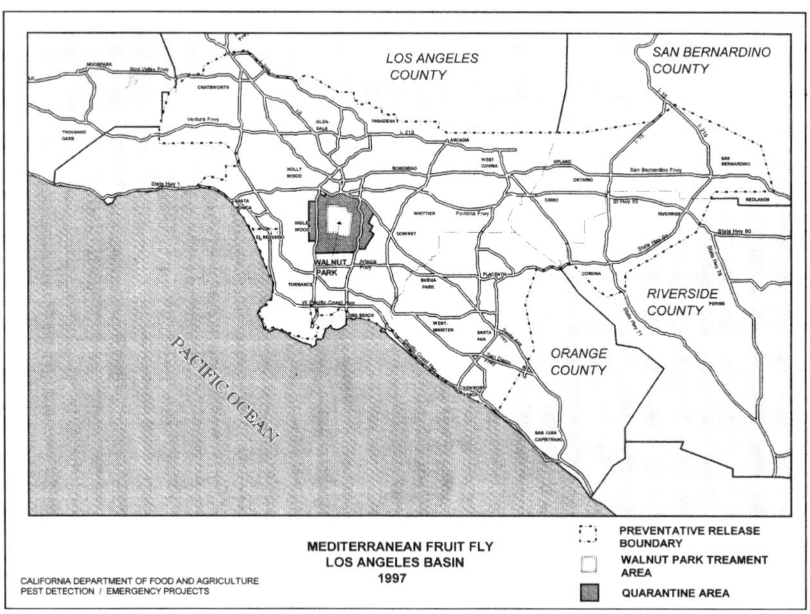


Figure 1. Boundaries of the PRP and location of Walnut Park treatment and quarantine areas.

The PRP will continue until July 2001 at which time the programme will be re-evaluated. The PRP. includes the 1994-1996 basin-wide programme and an additional 1790 km² in Riverside, San Bernardino, and the Orange Counties. The PRP boundaries were drawn to encompass all previous Medfly infestations in the Los Angeles basin and as much of the adjacent contiguous urban area as allowed with the available supply of sterile Medflies.

The PRP is reviewed annually by a Medfly Science Advisory Panel (MEDSAP): Dr. Derrell Chambers (Chair) (USDA retired), Dr. Roy Cunningham (USDA retired), Dr. Richard Rice (University of California), Dr. Jorge Hendrichs (International Atomic Energy Agency, Vienna, Austria) and Dr. Aldo Malavasi (University of Sao Paulo, Sao Paulo, Brazil). The MEDSAP has recommended changes in the PRP which have been implemented by CDFA/USDA/CAC, including a change in the focus of the larval survey, a greater emphasis on using McPhail traps to detect wild Medflies in the fall/winter months and the use of increased release rates of sterile Medflies to decrease the probability of new Medfly infestations developing in high risk areas (CDFA 1997, 1998). MEDSAP recommendations are available in an annual report to the legislature prepared by the CDFA by writing to: Pest Detection/Emergency Projects Branch, CDFA, 1220 N Street, Sacramento, California, 95814 USA.

The PRP has five major components: twice weekly sterile Medfly releases at a minimum rate of 48,250 sterile Medflies per km² over the PRP; detection trapping at the rate of two trimedlure baited Jackson traps and two McPhail traps per km² throughout the PRP; the larval survey of Medfly host fruit; fly identification and data management.

Sterile Medflies for the PRP are supplied by the CDFA and USDA Medfly rearing facilities in Hawaii and from a sterile Medfly rearing facility in Guatemala. The sterile flies are released seven days per week along predetermined flight lines using a Global Positioning System (GPS) satellite navigation guidance and recording system. The sterile fly facility at Los Alamitos, Orange County, receives, incubates and emerges 450 million sterile Medfly pupae per week and evenly disperses the sterile flies over the PRP. Prior to release, the flies from each rearing facility are mixed to prevent any part of the PRP from continually receiving flies from a single genetic stock. This lowers the risk posed by the invasion of a strain of Medfly resistant to a particular strain of sterile Medflies (Iwahashi 1996, McInnis et al. 1996).

The release density is no less than 48,250 sterile Medflies per km² per week in accordance with the recommendation of the MEDSAP. The release rate is the lower limit used in the Mexican Medfly exclusion programme (Schwarz et al. 1989). The twice weekly releases allow for a more even distribution of the sterile Medflies. An estimated 41 billion sterile Medflies were released between 10 July 1996 and June 1999.

Two trimedlure baited Jackson traps and two McPhail traps are deployed per km² throughout the PRP to detect wild Medflies and to help project management monitor project activities. Two trimedlure baited traps per km² is the same density used in the lower risk urban areas of California (CDFA 1995). The CDFA/USDA/CAC felt that this density would provide an adequate degree of ability to detect wild Medfly infestations that might develop under the PRP and that this trap density would not remove sufficent sterile Medflies to endanger the PRP. The McPhail trap density was not changed from that used in the general detection programme in the southern Californian urban areas (CDFA 1995).

The Jackson traps are inspected every one to two weeks and the McPhail traps every week year-round. The total number of traps monitored in the project area is about

15,000. Twenty-three wild Medflies were trapped within project boundaries in 1997 (Figure 2).

In 1996, the larval survey involved a focused effort in areas believed to have a high risk of Medfly introduction i.e., wholesale fruit markets, fruit importers, areas of historical Medfly finds and exotic fruit markets. Host fruits were sampled, cut and inspected for the presence of Medfly larvae weekly throughout the year. Over 1,427 kg of fruit were inspected and no Medfly larvae were found in 1996. In 1997 and 1998, larval surveys were discontinued around historical Medfly find areas and the efforts were concentrated around wholesale markets, fruit importers, and exotic fruit markets. These random larval surveys found no Medfly larvae. Additional fruit cutting was done in the vicinity of the wild Medflies trapped in 1997. These focused, delimitation surveys found 11 properties with Medfly infested fruit.

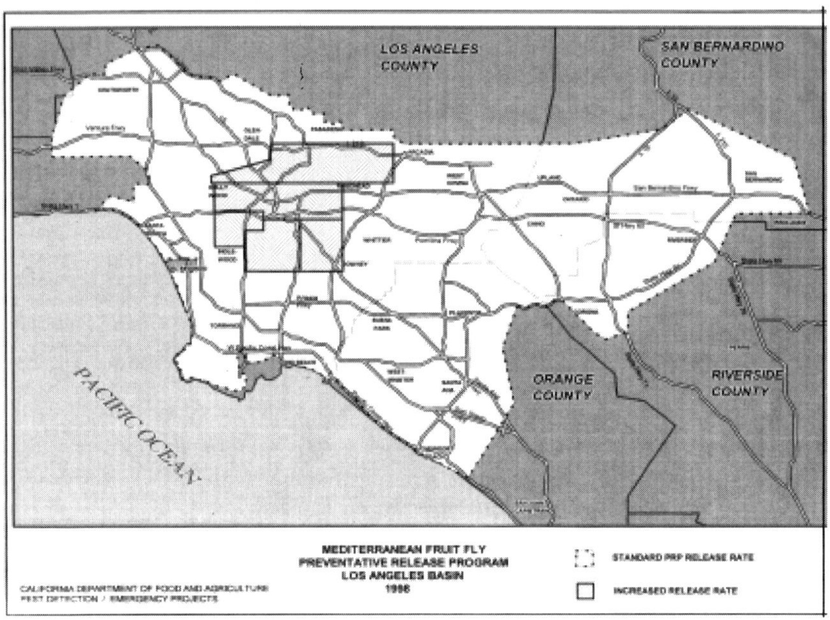


Figure 2. Boundaries of PRP area to receive 200,000 sterile Medflies weekly.

Fly identification determines whether the adult Medflies trapped in the Jackson and McPhail traps in the PRP are marked with a fluorescent pink dye. Undyed flies are dissected by a trained biosystematist who examines their reproductive organs to determine if they are sterile or wild. Some 350,000 to 500,000 dyed sterile Medflies are examined each week with an average of five requiring dissections.

The project maintains an ongoing quality control programme (QC) for the fly identification section. A weekly average of five, undyed, sterile Medflies is placed singly in traps being run through the fly identification laboratory. These "test" flies are placed in both Jackson and McPhail traps in no set pattern. The fly identifiers must find these undyed flies and turn them in for dissection. Failure to find the first test fly results in retraining of the identifier. The second missed test fly results in further retraining and a written letter of warning. The third missed test fly results in the removal of the individual from the fly identification laboratory. Between July 1994 and June 1998, 14 people were removed from the identification laboratory based on the results of these quality control tests. Recovery of the undyed Medflies by identification laboratory staff exceeds 95%.

In addition to maintaining a data base, the PRP publishes a weekly report showing the distribution of sterile Medflies caught in detection traps, and catalogues the quantity and quality of the sterile flies released. These reports are used by programme managers to insure that all areas within the PRP are receiving sufficent sterile Medflies, as indicated by the trap catch of sterile Medflies, and that the quality of sterile Medflies received meets acceptable criteria for emergence, flight and mating propensity. Traps with less than acceptable catches sterile flies are examined to determine the cause and appropriate corrective measures are taken, if needed. The sterile Medfly rearing laboratories are given continual feedback so that they can correct any declines in fly quality.

On 26 September 1997, a wild Medfly was trapped in the Walnut Park area of Los Angeles County (Figure 1). Subsequent delimitation trapping found another 22 adult Medflies. Larval surveys found 11 properties harbouring Medfly larvae in dooryard fruit. The PRP responded to this infestation by applying ground sprays of malathion and bait to host plants within a 200 m radius of each property on which wild Medflies were found. The host fruit from the properties on which Medfly larvae were found and the adjacent properties was stripped and diazion soil sprays were applied under these plants. The release rate of sterile Medflies was increased to 195,000 flies per km² in a 23 km² area around each property on which a wild Medfly was found. The total area receiving the 195,000 flies per km² was 41 km² (Figure 1). These increased release rates were maintained from 3 October 1997 until 16 April 1998, a time interval equal to three Medfly generations based on a temperature driven developmental model using local air and soil temperatures (CDFA 1989). Additionally, the PRP increased the release rate of sterile flies to 98,000 per km² in a 166 km² area buffer area around the eradication zone from 3 October 1997 until 16 April 1998.

The initial Medfly find was in a McPhail trap that had been relocated at the infested area the previous week. Subsequent male and female Medflies were trapped in Jackson and McPhail traps put out as part of the delimitation effort. The infestation was very small, physically encompassing 1.5 km²; 0.03% of the PRP. This is 1/5 of the average size of the infestations discovered during 1988 and 1989. These years were selected for comparison because the infestations occurred after a period of several years free of Medfly infestations similar to that from 1995 to 1996. The small size of the infestation, combined with lack of satellite infestations, suggests that the Walnut Park infestation is new and that the sterile Medflies were able to contain it (CDFA 1998). Nuclear DNA analysis (RAPDs ) of the Walnut Park Medflies found bands not seen in Medflies from previous infestations in southern California which is consistent with the Walnut Park infestation representing a new introduction (Haymer, unpublished data).

Medfly larvae from Walnut Park were sent to the USDA-APHIS laboratory in Hawaii. There the flies were allowed to complete their life cycle and the subsequent females were tested for the mating compatibility against Hawaiian strains of sterile male Medflies. The wild Medflies from California showed no signs of mating resistance to the sterile medfly males (Lance, personal communication).

The CDFA conducted a detailed analysis of the PRP in which the operational aspects of the programme in the Walnut Park area were compared to those from other areas in which wild Medflies were not found. Items examined included: sterile fly quality including emergence, flight and mating propensity, time interval between sterile fly releases, trap catch of sterile flies and the results of quality control tests of the trapping programme. There were no operational aspects of the programme that were unique to the Walnut Park area.

The discovery of the Walnut Park infestation demonstrates that the detection trapping programme in California is able to detect physically small Medfly infestations under a sterile release programme and that the CDFA policy of relocating traps every six weeks is an important component in an effective fruit fly detection programme.

In response to the Walnut Park infestation, the release rate of sterile Medflies is being increased to at least 154,400 flies per km² in a 647 km² area of the PRP (Figure 2). The increased release area will include Walnut Park and other areas of Los Angeles County (South Central Los Angeles, South Gate) and will give additional protection to an area that has had multiple Medfly infestations in the last ten years (Dowell and Penrose 1995, Penrose 1995).

The Walnut Park infestation demonstrates that the release of 96,500 sterile Medflies per km² alone in California cannot stop the development of new Medfly infestations. The CDFA/USDA/CAC must continue to increase their efforts to identify and close the pathways through which illegal fruit is entering the state. The PRP and its predecessor – the basin-wide programme – have reduced the annual rate of detection of wild Medflies from an average of 126.0 per year from 1987 to 1994 to 7.6 from 1995 to 199, a 94% reduction. More importantly, the number of infested areas has decreased from an annual average of 7.5 per year to 0.33 per year over the same time frame, a 95.6% reduction. Using large-scale releases of sterile Medflies has resulted in a substancial reduction in the rate at which wild Medfly adults are being trapped in the greater Los Angeles area and a equally large reduction in the development of Medfly infestations.

## REFERENCES

California Department of Food and Agriculture. 1994. Final programmatic environmental impact report. The exotic fruit fly eradication programme using aerial application of malathion and bait. Clearinghouse number 91043018.

California Department of Food and Agriculture. 1995. Insect Trapping Guide. A.J. Gilbert and R.R. Bingham (eds.).

California Department of Food and Agriculture. 1989. Mediterranean fruit fly eradication action plan.

California Department of Food and Agriculture. 1997. Mediterranean fruit fly preventative release programme. Report to the Legislature.

California Department of Food and Agriculture. 1998. Mediterranean fruit fly preventative release programme. Report to the Legislature.

Dowell, R.V. and R. Penrose. 1995. Mediterranean fruit fly eradication in California 1994-1995. pp. 161-185. *In* The Mediterranean fruit fly in California: defining critical research. J. G. Morse, R. L. Metcalf, J. R. Carey and R. V. Dowell (eds.). Coll. Natur. and Agric. Sci. Univ. Calif. Riverside.

Dowell, R.V., I.A. Siddiqui, F. Meyer and E.L. Spaugy. 1999. Early results suggest sterile flies may protect S. California from Medfly. Calif. Agric. 53 (2): 28-32.

Iwahashi, O. 1996. Problems encountered during the long-term SIT programme in Japan. pp. 391- 398. *In* Fruit fly pests - A world assessment of their biology and management. B. A. McPheron and G. J. Steck (eds.). St. Lucie Press, Delray Beach, Florida.

McInnis, D.O., D.R. Lance and C.G. Jackson. 1996. Behavioural resistance to the sterile insect technique by Mediterranean fruit fly (Diptera: Tephritidae) in Hawaii. Ann. Entomol. Soc. Am. 89:739-744.

Penrose, D. 1995. California's 1993/1994 Mediterranaen fruit fly eradication programme. pp. 551-554. *In* Fruit fly pests. A world assessment of their biology and management. B.A. McPheron and G.J. Steck (eds.). St. Lucie Press, Delray Beach, Florida.

Schwarz, A.J., J.P. Liedo and J.P. Hendrichs. 1989. Current programme in Mexico. Chapter 9.5.3. *In* World Crop Pests. Fruit flies: Their biology, natural enemies and control. A.S. Robinson and G. Hooper (eds.). Volume 3B. Elsevier. 445 pp.

# The Mexican Fruit Fly Eradication Programme

*J. Reyes F., G. Santiago M. and P. Hernández M.*

Comisión Nacional de Sanidad Agropecuaria (CONASA), Guillermo Perez Valenzuela No. 127, Colonia del Carmen, Coyoacan, D.F. 041000, Mexico

The goal of the Mexican Fruit Fly Eradication Programme is to control, suppress or eradicate from Mexico four species of fruit flies of economic and quarantine importance (*Anastrepha ludens* Loew, *A. obliqua* Macquart, *A. serpentina* Wied. and *A. striata* Schiner). These pests cause damage amounting to US$710 millon per year. In addition to this cost, there are other expenses from pest control actions and the loss of international markets, because fruit importing countries have established stringent quarantine measures to restrict the entry of these pests.

For purposes of the programme's implementation, Mexico was divided into three working zones (Figure 1), defined by agro-ecological characteristics, the number of fruit fly species present and the size of fruit growing regions. In addition, a cost:benefit analysis was carried out which indicated that the rate of return, in a 12-year time frame, might be as much as 33:1 in Northern Mexico, and 17:1 in the rest of the country, for an area over 100,000 hectares.

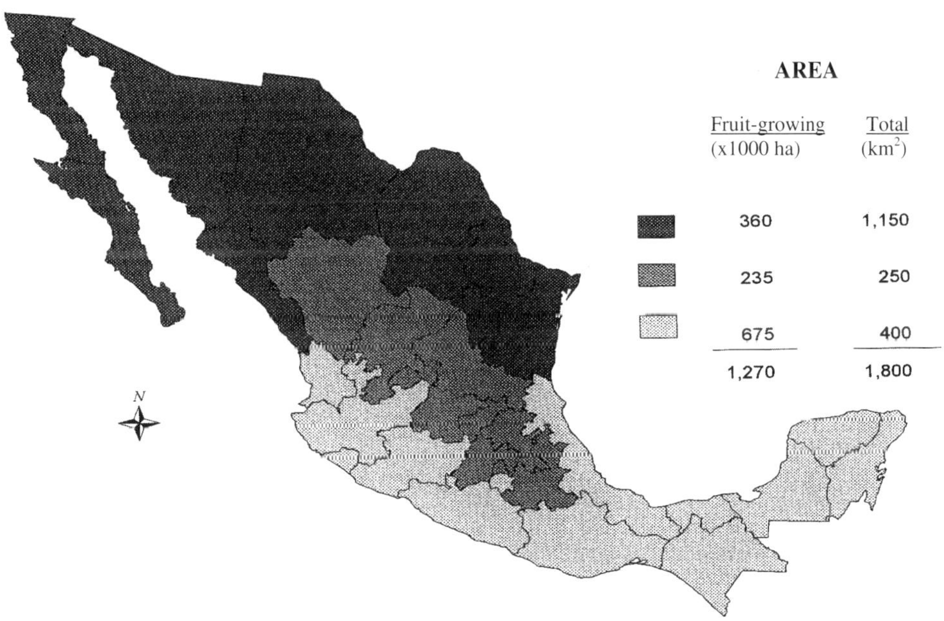


Figure 1. Working zones

Eradication technology involves: 1) surveys of pest populations by trapping and host fruit harvesting to monitor the presence and density of fruit flies, 2) reduction of pest populations applying cultural practices and using selective bait sprays, 3) mass release of sterile flies and augmentative release of parasitoids to eliminate populations and, 4) enforcement of quarantine measures to protect fruit fly free areas. Figure 2 summarises the sequence of these basic eradication actions.

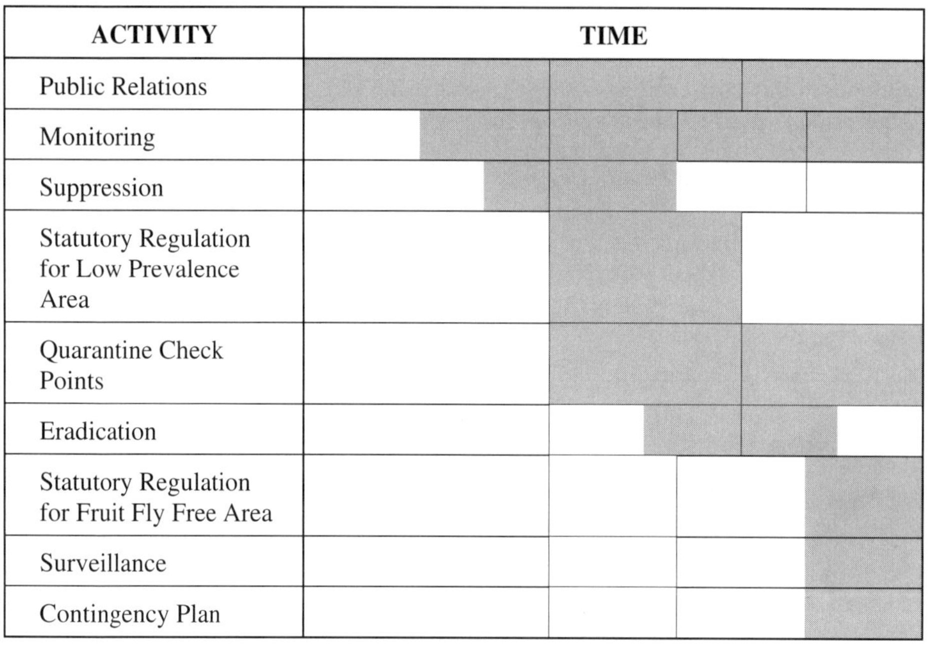


Figure 2. Basic activities on fruit fly eradication.

In 1992, the construction of a sterile fly and parasitiod mass rearing factory was initiated, with an annual production capacity of 300 million sterile flies and 50 million parasitiods. Since 1994, *A. ludens* and the parasitoid *Diachasmimorpha longicaudata* (Ashmead) are being produced on a mass scale. To date, 27 billion sterile flies and 4.4 billion parasitoids have been reared. The mass rearing of *A. obliqua* began in late 1995, and so far, 1.2 billion pupae of this fly species have been produced (Figure 3).

Since 1992, fruit flies have been eradicated from 730,000 $km^2$ of land where there are planted 142,000 hectares of fruit groves, mainly oranges, apples, peaches and mangoes. The elimination of these pests has sped the process of conversion from annual crops to fruit production aimed for export. As a result of the programme, it has also been possible to establish fruit fly low prevalence areas in 150,000 $km^2$, where there are 30,000 hectares of citrus plantations (oranges, grapefruit and tangerines), and 12,000 hectares of mango orchards (Figure 4).

Since 1992, fruit flies have been eradicated from 730,000 $km^2$ of land where there are planted 142,000 hectares of fruit groves, mainly oranges, apples, peaches and mangoes. The elimination of these pests has sped the process of conversion from annual crops to fruit production aimed for export. As a result of the programme, it has also been possible to establish fruit fly low prevalence areas in 150,000 km$^2$, where there are 30,000 hectares of citrus plantations (oranges, grapefruit and tangerines), and 12,000 hectares of mango orchards (Figure 4).

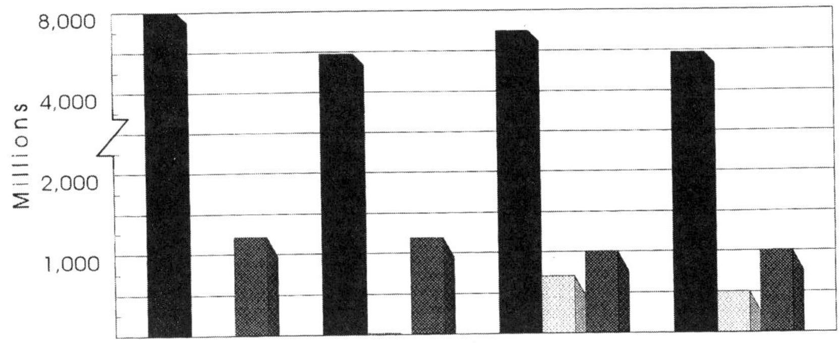


| Spp. / Year | 1994 | 1995 | 1996 | 1997 |
|---|---|---|---|---|
| A. ludens | 8,000 | 6,000 | 7,000 | 6,000 |
| A. obliqua | 0 | 30 | 700 | 500 |
| D. longicaudata | 1,200 | 1,200 | 1,000 | 1,000 |

Figure 3. Amounts of sterile fruit flies and parasitoids produced.

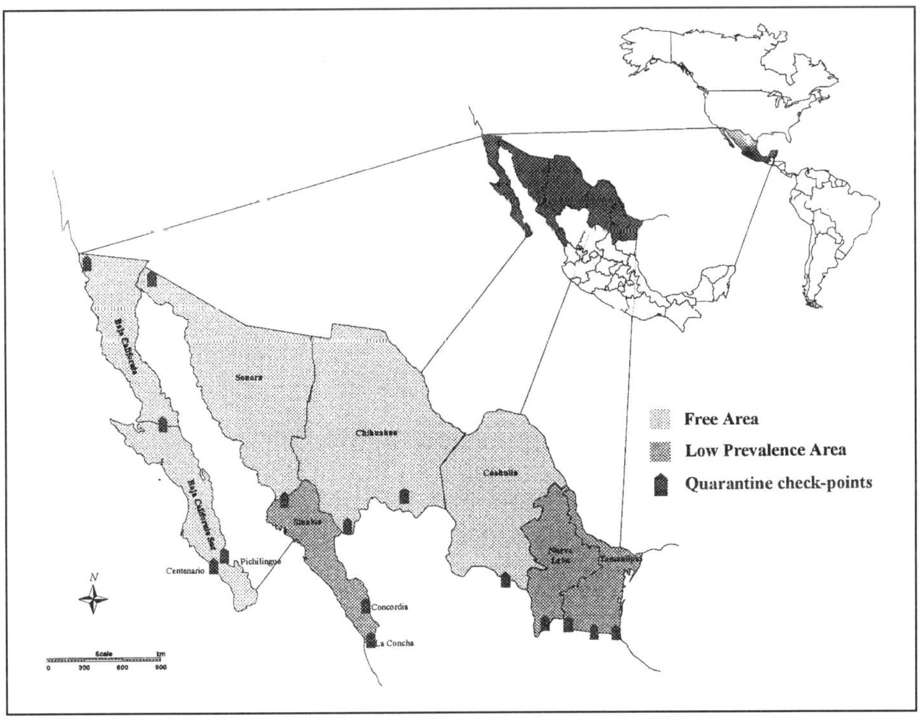


Figure 4. Fruit fly free and low prevalence areas.

To demonstrate the absence of *Anastrepha* in pest free areas, an annual average of 240,000 McPhail trap devices has been deployed and 480,000 devices installed in low pest prevalence areas.

Each year, the quarantine check points seize 300 tons of fruit and intercept an average of 14,000 larvae; 90,000 tons of fruit are fumigated and 2.5 million vehicles are inspected. As a result of enforcing quarantine actions at road check points, airports, sea ports, central markets and fruit warehouses, no fruit flies have been found in the free areas since 1997. All measures taken in quarantine control check points are enforced under the Mexican Plant Protection Law (Figure 4).

As a result of research and development activities, the sterile insect technique (SIT) for area-wide application is now available for *A. ludens*, *A. obliqua* and *A. serpentina*. Studies are still currently in progress to optimise rearing methods in order to reduce production costs. Where the guava fruit fly, *A. striata,* is concerned, substantial progress has been made toward the establishment of bleeding colonies by developing a special ovipositing panel and an artificial diet. The maximum ovipositing period and special environmental conditions for mating have been identified. It is hoped that, in the near future, the methodology for mass rearing of this species will be available.

The process of establishing colonies of native parasitoids of *Anastrepha*, i.e., *Doryctobracon crawfordi* Viereck (Hymenoptera: Braconidae); *Aganaspis pelleranoi* Brethes (Hymenoptera: Eucoilidae); and *Coptera haywardi* Loicano Ogloblin i.l. (Hymenoptera: Diapriidae) has been under way for the past three years and at present, mass rearing methodologies for these flies are well advanced. These methods could soon be incorporated into the programme as an additional strategy for suppression of fruit flies. Research is being conducted in coordination with the Institute of Ecology (Mexico), and the Agricultural Research Service (ARS)-USDA-Gainesville, Florida. Although the mass rearing of the parasitoids *Diachasmimorpha longicaudata* and *D. Tryoni* is now feasible, the current rearing processes can still be improved.

As part of the studies aimed at improving the monitoring systems for *C. capitata* conducted in coordination with the International Atomic Energy Agency (IAEA), work continues aimed at evaluating lures like putrescine, ammonium acetate and trimethylamine, as well as the open-bottom-dry-traps. Results to date indicate that more females than males can be caught with these traps.

Moreover, new chemicals have been discovered as alternatives for malathion, such as cyromazine, borates and Phloxin B. These are currently being validated against *Anastrepha* flies on a large scale in coordination with the ARS-USDA-Weslaco, Texas.

The programme has a cooperative agreement with the Californian Department of Food and Agriculture (CDFA), by which the mass rearing facility can provide sterile flies of any of the above mentioned species for outbreak control in the state of California.

From 1992 to 1997, the cost of running the programme was US$80 million. These funds were provided by the Mexican federal and state governments and by the fruit industry. Of this amount, 50% was invested for the construction of the sterile fly and parasitoid production facility and for its maintenance and operation while the rest was applied to field activities.

# Eradication of *Bactrocera papayae* (Diptera: Tephritidae) by Male Annihilation and Protein Baiting in Queensland, Australia

*D.L. Hancock, R. Osborne, S. Broughton and P. Gleeson*

Queensland Department of Primary Industries, P.O. Box 652, Cairns, Queensland 4870, Australia

## INTRODUCTION

An established population of *Bactrocera papayae* Drew and Hancock (Asian papaya fruit fly) was detected on the Australian mainland near Cairns, north Queensland, on 16 October 1995 (Fay et al. 1997). Coincidentally, the first flies were bred from green papaya, a host not normally utilised in the unripe state by indigenous fruit fly species. *B. papayae* is a member of the Oriental fruit fly (*Bactrocera dorsalis*) complex (Drew and Hancock 1994) and has more than 190 known hosts in Southeast Asia, where it is a major pest species. Consequently, the threat to Australian horticulture posed by the incursion was recognised immediately. By 26 October 1995, breeding populations were confirmed around Cairns, Mossman and Mareeba and a Pest Quarantine Area (PQA) was legislated. By 1 November 1995, following detections south of Innisfail, the PQA boundaries were set at 144°15´E, 19°00´S, covering nearly 78,000 km$^2$. *B. papayae* was ultimately trapped over some 20,000 km$^2$ (Figure 1). A national eradication campaign began in November 1995, undertaken by the Queensland Department of Primary Industries.

## DESCRIPTION OF THE PEST QUARANTINE AREA

The nature of the PQA presented a significant challenge to the goal of eradication of *B. papayae*. Land use within the area comprises a mix of urban development and agricultural production on the coastal plain and Atherton Tablelands, separated by the densely forested mountains and escarpment of the Wet Tropics World Heritage Area. Much of the agricultural land is devoted to sugar cane but there are also widespread plantings of tropical fruits, particularly mangoes, bananas and papayas. The estimated value of horticultural production in the PQA is A$250 million/year. Native fruit fly species in the area include the major pest *B. tryoni* (Froggatt) (Queensland fruit fly) as well as the lesser pest species *B. neohumeralis* (Hardy), *B. frauenfeldi* (Schiner), *B. jarvisi* (Tryon), *B. cucumis* (French) and *B. musae* (Tryon). Only the latter species is attracted to methyl eugenol, the male lure for *B. papayae*.

## STRATEGIES EMPLOYED

Local and international expert advice indicated that eradication based on male annihilation might be possible and a cost-benefit analysis (ABARE 1995) supported this assessment. Consequently, the federal and state governments supported a decision to attempt eradication, bearing in mind the extent of the infestation, the large populations

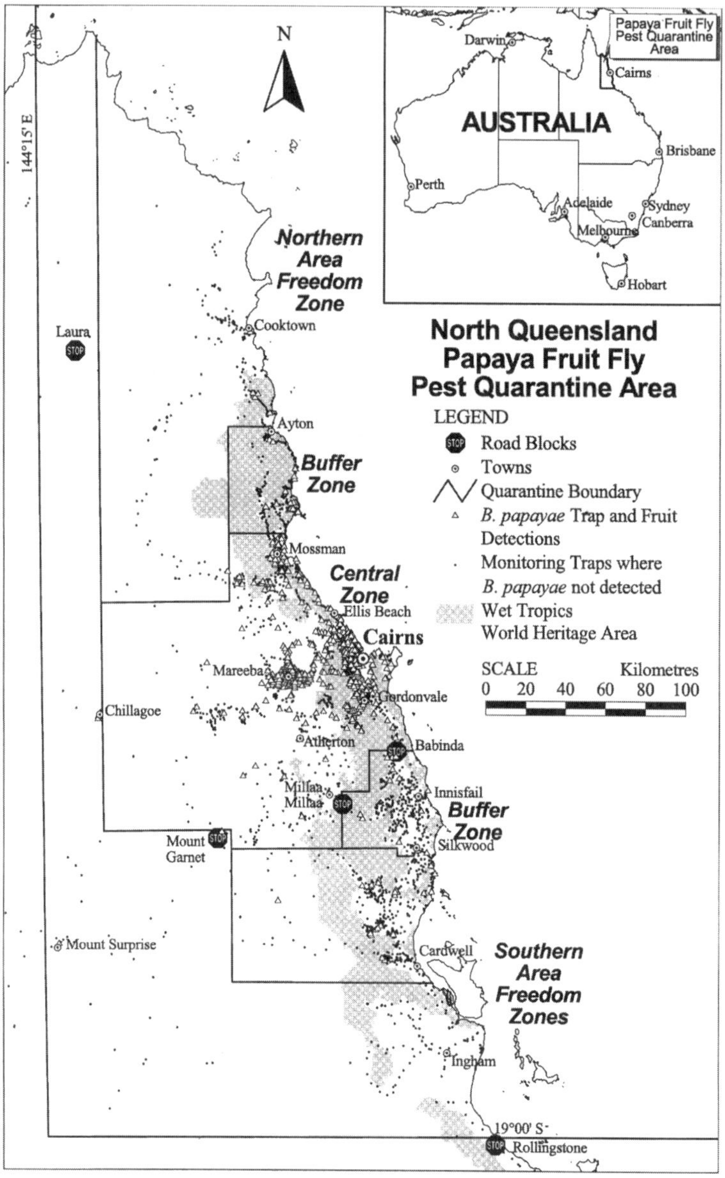


Figure 1. Map of the Pest Quarantine Area for *Bactrocera papayae*.

of *B. papayae* involved and the difficult and environmentally sensitive nature of much of the terrain.

Surveillance for *B. papayae* was primarily by a network of Steiner traps baited with the male lure methyl eugenol. By the end of 1995, 780 traps were in service, the number steadily increasing during the campaign with over 2,600 in service by the end of 1997. The initial use of Lynfield traps was discontinued due to their unsuitability in high rainfall areas. The second survey method was rearing of fruit flies from collected fruit. Two strategies were used, firstly a general fruit-collecting survey in residential, horticultural and abandoned cultivated areas and, secondly, a survey which targeted rainforest fruit to check for evidence of *B. papayae* breeding in the rainforest. Fruit samples were brought to laboratories for fly emergence, a quarantine-secure facility being established for the purpose. As the programme progressed, samples from known hosts and with signs of fruit fly damage were increasingly targeted. Trap catches and emerged adults were identified by trained staff working under guidance of experienced fruit fly taxonomists.

Male annihilation using fibreboard blocks (50 x 50 x 13.4 mm) impregnated with methyl eugenol (13.5-15 ml) and malathion ULV (4.5-5 ml) was the main eradication tool. This was supplemented by protein bait treatment of host trees, particularly in areas of known or suspected breeding activity. An intensive blocking programme saw most of the infested area covered by the end of 1995, normally achieving a coverage rate of 400-600 blocks/km$^2$ in urban and suburban areas. All blocks were applied by hand, being nailed or wired to trees at a minimum height of 1.5 m. Barriers of blocks, about 1 km wide, were initially placed from the coast to the mountain ranges in three separate areas between Babinda and Cardwell, near the southern limit of the PQA; these were superseded by general blocking.

Initially, new blocks were applied every 6 weeks, with the previous blocks left for a further 6 weeks before retrieval. Based on research into the persistence of malathion and methyl eugenol in blocks (Lloyd et al. 1998), the blocking treatment was modified in 1997, with new blocks applied every 8 weeks and old blocks retrieved at the same time. By the end of the blocking campaign in April 1998, a cumulative total of almost 3 million blocks had been used and mostly retrieved. Prior to the establishment of Area Freedom monitoring, every effort was made to remove and destroy any remaining blocks and a publicity campaign conducted to assist in retrieval of any blocks missed.

Protein bait applications (5 litres of protein autolysate plus 2 litres malathion (500g/litres) per 100 litres) were carried out as spot leaf treatments (100 ml/tree). Initially, these were applied on a very limited scale but after their importance in contributing to control of significant infestations of *B. papayae* in coffee was realised, they were used more extensively. The introduction of a "squirt-mobile", a modified vehicle that enabled protein bait to be applied into the canopies of tall host trees such as mango and *Terminalia*, also appears to have been a significant factor in quickly controlling several of the localised outbreaks or "hot spots". Protein bait treatments were applied from the ground and ceased in March 1998. Approximately 156,300 litres of protein bait were applied during the campaign.

## RESULTS

A total of 8,947 *B. papayae* was caught in male lure traps, mostly in the first 5 months of the campaign, after which the number of traps detecting flies and the number of flies within traps declined dramatically. By October 1996, overall trap catches had declined by 99.9% compared with original levels (Figure 2). The most frequent detections were in the Central Zone (Cairns-Mareeba-Mossman area) (Figure 1), with many traps recording more than 50 flies and one recording over 800 in the first 3-month period. Smaller populations were detected as far north as Cooktown, as far south as Cardwell and as far west as Chillagoe and Mount Surprise.

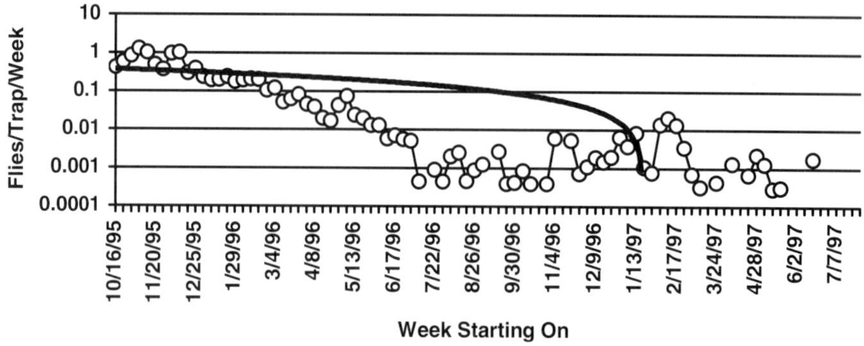


Figure 2. Decline in *B. papayae*/ trap/ week (log scale) covering the period of trap detections, October 1995 - June 1997. Breaks in data series represent weeks of zero trap catches. No trap detections were made after 16 June 1997.

The general fruit survey resulted in 63,950 samples by mid-May 1998, of which 254 produced *B. papayae*, representing 35 plant species from 14 families. Mango (*Mangifera indica* L.), banana (*Musa paradisiaca* L.), sea almond (*Terminalia catappa* L.) and guava (*Psidium guajava* L.) proved to be the major hosts, although citrus (*Citrus* spp.), capsicum/chilli (*Capsicum annuum* L.), papaya (*Carica papaya* L.), coffee (*Coffea arabica* L.), cumquat (*Fortunella japonica* (Thumb.) Swingle) and cherry tomato (*Lycopersicon esculentum* Miller) were also significant. A total of 4,144 *B. papayae* was bred from fruit samples.

In the rainforest survey, 12,500 samples were collected between January 1996 and June 1997, when the survey was terminated. None of the samples contained *B. papayae*. This result confirmed that early catches of *B. papayae* in rainforest traps were of flies dispersing into the rainforest rather than evidence of localised forest-breeding populations.

Although the number of *B. papayae* trapped or reared from fruits steadily declined during 1996, a resurgence was recorded at several foci around Cairns, Mareeba and Mossman during the peak fly season between November 1996 and March 1997 (Figure 2). These were quickly controlled, the last records for *B. papayae* being 6 March 1997 for Mossman and 11 March 1997 for Mareeba. Populations in the Cairns area (northern beaches) lingered on until July, with the last trap detections being 16 June 1997 and the last detection in fruits (*Passiflora suberosa* L.) collected on 7 July 1997.

## Management Strategies in Key Problem Areas

The effectiveness of the monitoring programme in detecting problem areas and the ability of the eradication methods to deal with them swiftly were critical to the success of the campaign. Prior to July 1996, the infestation was widespread and discrete problem areas not evident. As the campaign progressed, some problem areas became more apparent, as discussed below. These often consisted of localised outbreaks "hot spots" and a response protocol was developed to deal with them.

The infestation around Mareeba was notable for its persistence over a wide area. This is largely a rural area with major areas of horticultural production, including coffee and mangoes. A likely cause of the problem was the difficulty encountered in obtaining an adequate coverage with blocks and protein treatments. In March 1996, three *B. papayae* were bred from ripe coffee cherries but the significance of this was not appreciated until July 1996, when 80 specimens were bred from green coffee cherries on another property. This and subsequent rearings or trap detections from other coffee properties suggested that block treatments alone were inadequate and an intensive protein baiting programme was carried out from August to December 1996. This was successful and the threat of heavy population buildup in coffee was averted.

From late January 1997, large numbers of *B. papayae* appeared in traps (135 specimens) and fruit samples (197 specimens) from a number of properties in a localised area along a creek line east of the Mareeba township. Of the bred specimens, 174 came from one cumquat tree. An intensive 10-week campaign was devised for the 43 properties in the affected area and by 11 March 1997, this infestation was effectively controlled. No further detections of *B. papayae* in the Mareeba and surrounding districts have been made since that date.

Localised infestations also appeared near Mossman and in the Cairns-Gordonvale area between November 1996 and March 1997. The Gordonvale infestation was traced to a patch of bananas growing along a riverbank, 200 m from the detecting trap. The infestation near Mossman occurred in mango and guava but was not detected by traps. As in the Mareeba situation, these local infestations were quickly controlled by intensified block and protein bait treatments and host removal.

Ellis Beach-Palm Cove, north of Cairns, presented a rather different problem. This is a narrow strip of coastline bordered by a steep mountainside. *B. papayae* was detected here in November 1995 and, despite treatments, persisted almost continuously until July 1997. From March 1997 treatments were intensified; the protein baiting increased to twice weekly and the blocking density increased to 1,000 blocks/$km^2$. Monitoring activities also increased and, following trap detections in June 1997, Ellis Beach was deblocked and trap density intensified (including canopy traps), in an attempt to determine whether the source of the infestation lay within the area or the hinterland beyond it. A controlled burn off of the undergrowth by the local fire brigade may have contributed to eradication through destruction of a suspected reservoir host, *P. suberosa*, from which specimens were bred in July 1997. However, the intensified treatments finally appeared to have been effective and no further detections were made in traps after 16 June 1997 or from fruit collected after 7 July 1997.

An isolated outbreak of *B. papayae* (8 specimens) was detected outside the PQA at Mt Isa, in western Queensland, in February 1997. Breeding was suspected although not confirmed, despite extensive fruit collections. No further detections were made after a local eradication programme commenced shortly afterwards. Eradication and Area Freedom for the area were achieved by December 1997.

## Quarantine Management and Area Freedom

As much of the PQA is a national and international tourist destination, fruit movement by travellers represented one of the greatest threats to containment of the infestation. A publicity campaign, supplemented by road blocks, was established to restrict movement of infested fruit and guard against its removal by vehicle from within the PQA. Air and rail traffic were also monitored. Later in the campaign, roadblocks were used to protect the southern Area Freedom zone (a major area of horticultural production) and its buffer zone from the remaining infested areas, the roadblock initially established at Cardwell moving to Babinda for this purpose (Figure 1). Certification arrangements were introduced to assist industry with the export of commercial fruit and vegetables.

As a result of the apparent success of the male annihilation programme, Area Freedom was introduced in stages to designated zones of the PQA (Figure 1), following approximately 12 months of nil-detection of *B. papayae* in areas under eradication treatment. After removal of blocks and cessation of protein baiting, 12 weeks of intensive monitoring by trapping and fruit collecting commenced. Traps were placed on a 1 km grid in horticultural and residential areas and at 5 km elsewhere. The southern part of the PQA, from Silkwood southwards, was declared Area Free on 17 October 1997. This was followed by the northern part of the PQA, from Ayton northwards, on 26 January 1998. Monitoring for Area Freedom status begins in the remainder of the PQA on 1 June 1998, with Area Freedom scheduled for declaration on 23 August 1998.

## SIT Planning

Detections of small numbers of *B. papayae* in rainforest traps early in the campaign raised concerns that the fly, known to inhabit rainforests in Southeast Asia, might establish in similar habitats within the PQA. As a result of this, and the ongoing threat of rainforest colonisation and recurring localised infestations in urban and agricultural areas, plans were initiated for the introduction of a sterile insect technique (SIT) programme should such be required. A clear impetus for this programme was recognition that use of widespread chemical-based treatments in protected habitats would be environmentally unacceptable. Scientists at the International Atomic Energy Agency and the US Department of Agriculture (USDA) provided invaluable advice at the planning stage. A contingency plan was developed during 1997, in consultation with environmental authorities, for limited use of blocks, as an interim measure, should small-scale infestations be detected in rainforests. Protein bait treatments were not considered an option.

By July 1997, the number of *B. papayae* detections was so low that it became apparent that SIT would unlikely be required. To facilitate a decision on this issue, a qualitative risk assessment was undertaken in October 1997, taking into account the land use pattern in the PQA to accurately estimate the required release area. Combined with data from the surveillance programme, the risk assessment indicated that the introduction of SIT was no longer needed to eradicate *B. papayae*. Shortly afterwards, the planning process was formally terminated. It is believed that intensive blocking in the PQA successfully prevented the establishment of *B. papayae* in the World Heritage Area rainforests.

## Environmental Implications

Attention to environmental issues was an important part of the campaign. The insecticide used, malathion, is noted for its rapid breakdown upon entering the environment and inability to accumulate through the food chain. Nevertheless, it was desirable to use it as little as possible without jeopardising the campaign. Malathion ULV, being almost water insoluble, was used in the blocks and traps, whilst malathion e.c.500 was used in the protein applications. Chemical analysis of weathered blocks showed that the malathion persisted within them for a considerable period of time (at least 12 months), under field conditions (Lloyd et al. 1998), with no significant loss detectable over the first 28 weeks. Treated blocks were stored and allowed to drip dry for several days before application in the field and under laboratory conditions; only trace amounts of malathion appeared to be capable of entering the environment directly from a block during simulated rainfall. Research into block efficacy (Lloyd et al. 1998) showed that blocks continue to attract and kill significant numbers of flies for at least 10 weeks and that the practice of leaving old blocks alongside new ones was not necessary. This led to an immediate reduction (by half) in the number of treated blocks out at any one time. Further research has suggested that the amount of malathion used in the blocks may also be reduced considerably (to 1 ml) without lessening their effectiveness.

Apart from the target species, *B. papayae*, 14 native species of fruit flies are attracted to methyl eugenol within the eradication treatment area, with two species, *B. amplexiseta* (May) and *B. unirufa* Drew, apparently confined to the PQA. Thus, it was important to ensure that these and other uncommon species were not eradicated along with the target species. Many of the indigenous species, including the two with restricted distributions, are most abundant in or largely confined to rainforests and the lack of eradication treatments therein ensured that these species were not detrimentally affected. This was borne out by continual assessment of their numbers in rainforest monitoring traps, where all specimens collected were identified to species. In areas which attained Area Freedom following extensive eradication treatment, native species populations quickly returned to those comparable with untreated areas. The only native species which declined to undetectable levels in both rainforest and eradication treatment areas was *B. musae*, itself a pest of bananas and a species that appears to be without a substantial refuge population in forests. However, this species also recovered quickly following cessation of eradication treatments in Area Freedom zones.

In contrast to the native species and in particular *B. musae*, *B. papayae* has not been detected in Area Freedom areas. This provides further evidence that eradication in those areas has been achieved. On both practical and environmental grounds, it was not possible to apply widespread eradication treatments in protected habitats and treatments outside those areas failed to impact on species with substantial rainforest populations. This demonstrates that a SIT programme would have been essential to achieve eradication of *B. papayae* had it become established in rainforest habitats.

## REVIEW AND EVALUATION OF CAMPAIGN

The campaign was reviewed twice yearly by a national body of fruit fly specialists (the Scientific Advisory Panel). Two external reviews by international (USDA) and interstate fruit fly experts in May and November 1996 also made recommendations which contributed to the success of the programme. A stringent internal quality assurance programme was implemented to ensure accurate documentation of all monitoring and eradication procedures and establishment of an internal auditing system for both areas of operation. A data management and analysis unit was established to manage the large amount of monitoring and eradication data generated by the programme. This unit was also responsible for generating programme reports and for optimising trap placement requirements, particularly for Area Freedom monitoring, using a geographic information system.

## CONCLUSIONS

Current data demonstrate that the eradication campaign has been successful. Provided no further *B. papayae* are detected, Area Freedom throughout the PQA is scheduled for declaration by the end of August 1998, at which time all roadblocks will be closed and quarantine restrictions lifted. Intensive monitoring with traps will continue throughout the following summer/wet season, until the end of March 1999. This covers the peak period of fruit fly activity, during which any remnant populations of *B. papayae*, should they exist, will become detectable by monitoring traps. Failure of *B. papayae* to re-appear under these conditions will enable the achievement of eradication to be declared by April 1999.

## ACKNOWLEDGEMENTS

We thank the many people involved in the Papaya Fruit Fly Eradication Programme and associated research, plus the support of agricultural producers and local communities, for their contributions to the ongoing success of the campaign.

## REFERENCES

ABARE. 1995. Papaya fruit fly. Cost-benefit analysis of the proposed eradication program. An Australian Bureau of Agricultural and Resource Economics consultancy report to Australian Quarantine and Inspection Service, Canberra, 31 pp.

Drew, R.A.I. and D.L. Hancock. 1994. The *Bactrocera dorsalis* complex of fruit flies (Diptera: Tephritidae: Dacinae) in Asia. Bull. Entomol. Res., Suppl. 2: 1-68.

Fay, H.A., R.A.I. Drew and A.C. Lloyd. 1997. The eradication program for papaya fruit fly (*Bactrocera papayae* Drew and Hancock) in north Queensland. pp.259-261. *In* Management of fruit flies in the Pacific. A.J. Allwood and R.A.I. Drew (eds.). *Aciar Proceedings* 76

Lloyd, A., P. Leach and R. Kopittke. 1998. Effects of exposure on chemical content and efficacy of male annihilation blocks used in the eradication of *Bactrocera papayae* in north Queensland. Gen. Appl. Entomol. 28: 1-8.

# Eradication of an Exotic Fruit Fly from Mauritius

*S. Indira Seewooruthun, Shradanand Permalloo, Bhubaneswar Gungah, A. Rashid Soonnoo and Malini Alleck*

Entomology Division, Ministry of Agriculture, Fisheries and Cooperatives, Reduit, Mauritius

## INTRODUCTION

The Oriental fruit fly, *Bactrocera dorsalis* (Hendel), was first detected in Mauritius in June 1996 (Seewooruthun et al. 1998). A female fly was collected from a McPhail trap baited with ammonium chloride on 5 June 1996, at Camp Carol, a village situated about one kilometre from the airport. The detection of *B. dorsalis* in Mauritius was the first record for the African region which has about 11 species of fruit flies (Allwood and Drew 1997). *B. dorsalis* is presently distributed over Asia, Australia, the Pacific Ocean islands and South America (White and Elson-Harris 1992).

## EARLY DETECTION

An area-wide programme for the control of fruit flies, the National Fruit Fly Control Programme, funded jointly by government of Mauritius and the European Union, has been operational over one-third of Mauritius since 1994 (Soonnoo et al. 1995, Permalloo et al. 1998). This programme has, as one of its prerequisites, an island-wide monitoring system for fruit flies, using male lure traps. Trapping grids for the detection of fruit fly incursions are also operated as part of the quarantine component. Additionally, trapping is used in the context of the research component of the programme. This island-wide trapping system made early detection of the Oriental fruit fly possible. The need for the early detection of exotic pests has been recognised as a very important factor for successful eradication (Allwood and Drew 1997). The speed of detection is also a determining factor in cost, effectiveness and success of eradication.

After the tentative identification of the single female fly suspected to be *B. dorsalis*, confirmation was sought from the International Institute of Entomology.

## INITIAL MEASURES TAKEN

The availability of material, equipment, trained personnel, as well as adequate technical know-how made it possible to intervene immediately at the site of detection. An area of about 1 $km^2$ was subjected to the bait application technique (BAT) and the male annihilation technique (MAT), within 24 hours of detection. These two techniques are currently being used in the area-wide fruit fly programme.

Trapping was intensified around the detection point. Methyl eugenol traps were placed in a radius of 1.5 km from the outbreak centre.

## TARGET AREA

Mauritius is a tropical island of 1860 $km^2$ situated in the southwest of the Indian Ocean, 800 km off the east coast of the Malagasy Republic. Agriculture occupies about 52% of the total area out of which 85% is under sugar cane cultivation. The rest is under vegetable and fruit production.

The eradication programme is conducted in the southern region of the island and covers a total area of about 300 $km^2$ (Figure 1). Fruit fly hosts include fleshy fruits which can be cultivated or wild, vegetables such as cucurbits, tomatoes and chillies. Cultivated fruit fly hosts are found mostly in backyard gardens whereas wild hosts grow mainly along river banks, river estuaries and in natural reserves.

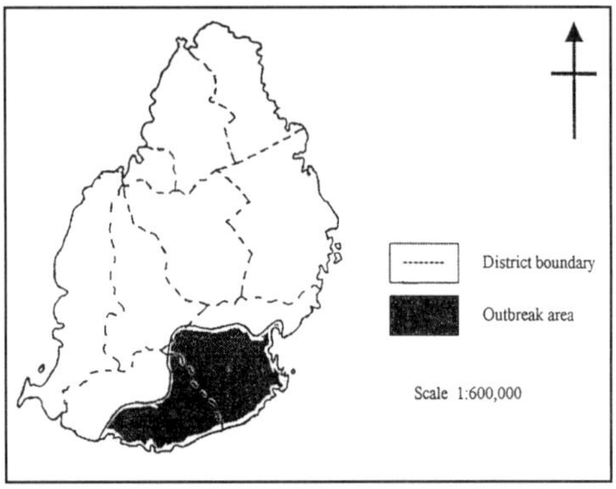


Figure 1(a). Map of Mauritius showing outbreak area.

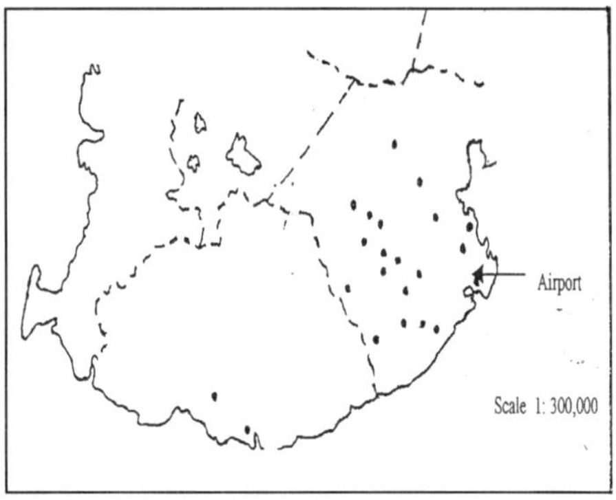


Figure 1(b). Outbreak area showing detection points.

Among fleshy fruits, the main fruit fly hosts are mango (*Mangifera indica* L.), guava (*Psidium guajava* L.), peach (*Prunus persica* (L.) Batsch), loquat (*Eriobotrya japonica* (Thunb.) Lindl.), jujube (*Ziziphus jujuba* Lam.), Indian almond (*Terminalia catappa* (L.) Ridley), ox heart (*Annona reticulata* L.), water apple (*Syzygium samarangense* (Blume) Merry and Perry) and citrus (*Citrus spp.*).

## MEASURES TAKEN FOR CONTAINMENT/ERADICATION

Methods generally utilised for eradication of exotic fruit flies are BAT, MAT and the sterile insect technique (SIT). These are used either singly or in combination. They are also coupled with other actions such as insecticidal sprays, soil treatment, fruit stripping and fruit clean-up. MAT has been used for eradication, either alone (Koyama et al. 1984, Nakamori et al. 1991), together with SIT (Steiner et al. 1970), or combined with BAT (Fay et al. 1997).

After the initial treatments focused in the vicinity of the detection area, trap catches were negative until 6 weeks later, on 18 July, when eight adults were again caught in two traps from a locality about 1 km from the first detection point. Treatments were then gradually extended until a total area of 300 $km^2$ was covered by the eradication programme.

The treatments effected are listed below:

- BAT, using 2% protein hydrolysate and 0.7 % malathion 57 EC.
- MAT, using plywood blocks of 50mm x 50mm x 12mm, each containing 6g of methyl eugenol and 1g of malathion 57 EC.
- Cover spray of trees with ripening fruits.
- Soil drenching under trees with ripening and fallen fruits.
- Fruit clean-up and disposal.

BAT is applied as 40 ml spots at the rate of 200 spots per hectare. Treatment is effected at 7- to 10-day intervals. MAT blocks are nailed onto tree trunks at the rate of 10 to 14 per hectare in buildup areas and regions with wild vegetation, and are renewed every 2.5 to 3 months.

Certain regions, inaccessible by road, and where a large numbers of *T. catappa* trees grow, were subjected to aerial treatment, using an ultra-light motorised (ULM) aircraft. An area of about 50 $km^2$ was treated with BAT and 3000 of MAT blocks were disposed at intervals of 50 m. The blocks were held in pairs by a nylon thread of about 30 cm long. This strengthened the attachment of the blocks to fruit trees while they were being dropped.

## MONITORING FOR EVALUATION

Locally developed methyl eugenol baited traps based on the Steiner model were used for detection of adult flies. These were re-lured every fortnight with 0.1 ml of a methyl eugenol/malathion mixture in the ratio 1:1.

Traps were initially placed around initial detection points and were gradually increased to a total of 367, placed at an average interval of 100 metres. 95 McPhail traps baited with protein hydrolysate, ammonium chloride or Brewer's waste were also set in the area. All traps were monitored daily for a period of 5 months from the first detection, and subsequently checked twice a week.

Regular fruit sampling was effected for determination of presence or absence of *B. dorsalis*. Fruits were held in plastic boxes or trays containing sand. These were kept in a laboratory in the eradication area. Identification of flies was done at the adult stage.

## RESULTS

*B. dorsalis* trapped from 22 villages covering a total area of about 250 $km^2$. A total of 144 adult flies were caught, the majority of which were trapped during a period of three months, between 18 July and 22 September 1996. The maximum number caught from one locality in a single day was 32, with a total of 61 from the same village over the three-month detection period. Trap catches of *B. dorsalis* are summarised in Figure 2. The last trap catch was on 6 May 1997.

*B. dorsalis* was not collected from any of the 250 traps operated within the seven other districts within the purview of the area-wide control programme.

Fruit and vegetables belonging to over 20 species were sampled for detection from the eradication area. Fruit and vegetables kept for observations totalled 70,875 and 5,042, respectively (Table 1). Emergence of *B. dorsalis* was recorded from only two

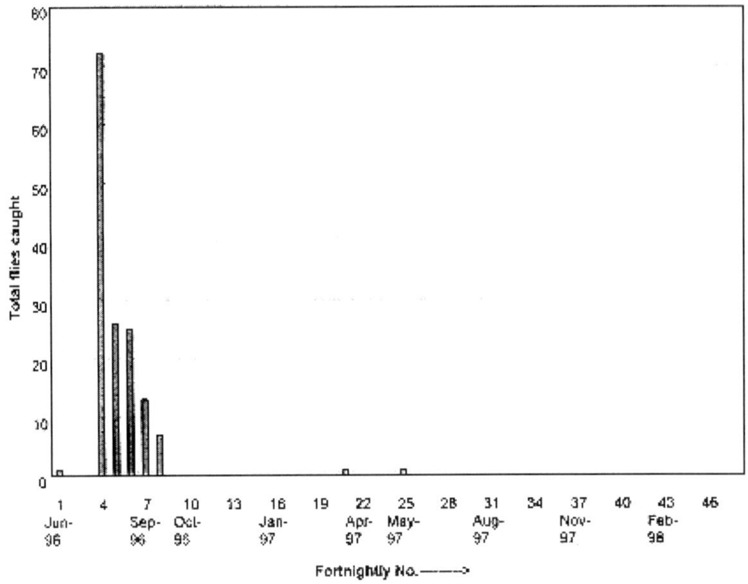


Figure 2. Fortnightly trap catches of *Bactrocera dorsalis,* June 1996 to April 1998.

Table 1. Fruit collection data in the eradication area (July 1996 – April 1998).

| Fruits | No. of fruits collected | No. of fruits kept in breeding | Emergence of fruit flies* | | | | | | |
|---|---|---|---|---|---|---|---|---|---|
| | | | Bd | Bz | Cr | Cc | Bc | Dc | Tc |
| Mango (*Mangifera indica* L.) | 310,289 | 4,794 | 22 | 2,647 | 39 | 114 | 0 | 0 | 0 |
| Indian Almond (*Terminalia catappa* L.) | 3,107,044 | 58,993 | 66 | 20,483 | 11,186 | 3,465 | 0 | 0 | 0 |
| Vegetables | 5,265 | 5,042 | 0 | 0 | 0 | 57 | 11,666 | 176 | 414 |
| Other fruits | 20,424 | 7,088 | 0 | 726 | 259 | 77 | 0 | 0 | 16 |
| Total | 3,443,022 | 75,917 | 88 | 23,856 | 11,484 | 3,713 | 11,666 | 176 | 430 |

* Bd: *Bactrocera dorsalis*
Bc: *Bactrocera cucurbitae*
Bz: *Bactrocera zonata*
Cr: *Ceratitis rosa*
Cc: *Ceratitis capitata*
Dc: *Dacus ciliatus*
Tc: *Trirhithromyia cyanescens*

species of fruit, namely mango and Indian almond. The latter is a non-edible fruit which, apart from growing wild along river banks and coastal areas, is planted for shade. A total of three samples of mangoes and six samples of Indian almond were found to be infested, totalling 88 *B. dorsalis*. The samples originated from six different villages. The last emergence was recorded from a sample of Indian almond fruit collected on 22 April 1997.

Fruit collection outside the eradication area is an ongoing process. About 22,500 fruits and 16,000 vegetables have been collected. The presence of *B. dorsalis* was not detected from any of these samples.

## DISCUSSION

There has been no detection of *B. dorsalis* for a period of one year since May 1997. The time that has elapsed from the last detection of the pest corresponds to about 12 generations of the fly. The possibility of *B. dorsalis* having been eradicated from Mauritius can be considered from now onwards (Nakamori et al. 1991). Eradication treatments will shortly cease. Intensive monitoring through both trapping and fruit sampling will, however, be continued for at least one year.

## ACKNOWLEDGEMENTS

We would like to thank the following for permission to publish this paper and for helping directly or indirectly: the Permanent Secretary, the Chief Agricultural Officer, the Principal Agricultural Officer (Crops) and Mr. P. Jokhun, Mrs. H. Abdoula, and the staff and personnel of the Entomology Division. Our special thanks to the European Union for providing the necessary funds in the initial stages of the eradication, the Imperial College through Dr. J. Mumford and colleagues from different Australian institutions for their valuable advice. We acknowledge the involvement of Mr. A. Joomaye, former Head of the Entomology Division, who took prompt action to start the eradication programme. The identification of *B. dorsalis* by the International Institute of Entomology is also acknowledged.

## REFERENCES

Allwood, A.J. and R.A.I. Drew. 1997. National and regional needs for future activities on fruit flies (Diptera: Tephritidae) in the Pacific region. pp 262-264. *In* Management of Fruit Flies in the Pacific. A.J. Allwood and R.A.I. Drew (eds.). *Aciar Proceedings* 76.

Fay, H.A., R.A. Drew and A.C. Lloyd. 1997. The eradication programme for papaya fruit fly (*Bactrocera papayae* Drew and Hancock) in North Queensland. pp. 259-261. *In* Management of Fruit Flies in the Pacific. A.J. Allwood and R.A.I. Drew (eds.). *Aciar Proceedings* 76.

Koyama, J., T. Teruya and K. Tanaka. 1984. Eradication of the Oriental fruit fly (Diptera: Tephritadae) from the Okinawa Islands by a male annihilation method. J. Econ. Entomol. 77:468-472.

Nakamori, H., M. Nishimura and H. Kakinohana. 1991. Eradication of the Oriental fly, *Dacus dorsalis* Hendel (Diptera: Tephritidae) from Miyako and Yaeyama Island by the male annihilation method. pp. 220-231. *In* Fruit Flies in the Tropics. S. Vijaysegraan and A.G. Ibrahim (eds.). Kuala Lumpur, Malaysia.

Permalloo, S., S.I. Seewooruthun, A. Joomaye, A.R. Soonnoo, B. Gungah, L. Unmole and R. Boodram. 1998. An area wide control of fruit flies in Mauritius. pp 203-210. *In* Proceedings of the Second Annual Meeting of Agricultural Scientists. J.A. Lalouette, D.Y. Bachraz, N. Sukurdeep & B.D. Seebaluck (eds.). *12-13 August, 1997.* Reduit, Mauritius: Food and Agricultural Research Council.

Seewooruthun, S.I., P. Sookar, S. Permalloo, A. Joomaye, M. Alleck, B. Gungah and A.R. Soonnoo. 1998. An attempt at the eradication of the Oriental fruit fly, Bactrocera dorsalis (Hendel) in Mauritius. pp. 211-218. *In* Proceedings of the Second Annual Meeting of Agricultural Scientists. J.A. Lalouette, D.Y. Bachraz, N. Sukurdeep and B.D. Seebaluck (eds.). 12-13 August, 1997. Reduit, Mauritius: Food and Agricultural Research Council.

Soonnoo, A.R., E.S.C. Smith, A. Joomaye, S. Permalloo and B. Gungah. 1995. A large scale fruit fly control programme in Mauritius. pp. 52-60. *In Problems and Management of Tropical Fruit Flies.* T.H. Chua and S.G. Khoo (eds.). Kai Wah Press, Kepong, Malaysia

Steiner, L.F., W.G. Hart, E.J. Harris, R.T. Cunningham, K. Ohinata and D.C. Kamakahi. 1970. Eradication of the Oriental fruit fly from the Mariana Islands by the methods of male annihilation and sterile insect release. J. Econ. Entomol. 63: 131-135.

White, I.M. and M.E. Elson-Harris. 1992. Fruit flies of economic significance: their identification and bionomics. CAB International in association with ACIAR: 165.

# Regional Programme for the Eradication of the Carambola Fruit Fly in South America

*Aldo Malavasi*[1]*, Alies van Sauers-Muller*[2]*, David Midgarden*[3]*, Victorine Kellman*[4]*, Dominique Didelot*[5]*, Phillippe Caplong*[6] *and Odilson Ribeiro*[7]

[1] IICA Office in Suriname, Letitia Vriesdelaan 11, Paramaribo, Suriname
[2] Ministry of Agriculture, Animal Husbandry and Fisheries, Letitia Vriesdelaan 10, Paramaribo, Suriname
[3] FAO, Letitia Vriesdelaan 11, Paramaribo, Suriname
[4] IICA Office in Guyana, 18 Brickdam, Stabroek, Georgetown, Guyana, Francais
[5] Ministère de l'Agriculture et la Pêche, Jardin Botanique, BP 5002, Cayenne, Guyana Francaise
[6] Federation Departementale de Groupements de Defense des Cultures de la Guyane, Cayenne, French Guyana
[7] Ministério da Agricultura, D.D.I.V. Brasília DF 70.235-070, Brazil

## INTRODUCTION

*Bactrocera carambolae* Drew & Hancock, the Carambola fruit fly (CFF), was probably introduced into Suriname from Indonesia in the 1960s or 1970s. The most likely mechanism of introduction was people arriving at Suriname from Indonesia by air, through Amsterdam. Any other method of transport would be too lengthy. Air travel was not commonly available to the general Surinamese population before the 1960s. About one-fifth of the Surinamese population is of Indonesian origin, and many strong ties remained between the countries. These ties are loosening with the increasing number of generations after immigration, which occurred in the late nineteenth and early twentieth century.

The first recorded *Bactrocera* found in Suriname was in 1975, when flies were reared from a market fruit and preserved unidentified in the Ministry of Agriculture's insect collection. *Bactrocera* were not recorded again until 1986, when infested fruits were brought to the attention of the Ministry by a homeowner. These specimens were sent to the United States for identification and were identified as *Dacus dorsalis*. *B. carambolae* was formally described in 1994 as a species belonging to the *B. dorsalis* complex (Drew and Hancock 1994). At that time, in 1986, little importance was given to the finding in the United States, perhaps because the identifier was unaware that Suriname is in South America rather than Asia. The international community would only become aware of the establishment of a *Dacus/Bactrocera* species in the Americas four years later. The population of flies in the Guyanas has now been identified as *B. carambolae*, and its establishment in South America is a threat to the production and marketing of fruits throughout the tropical and subtropical Americas and the Caribbean (Hancock 1989) (see Table 1).

It might be expected that the newly established *B. carambolae* would move rapidly into the tropical forests where there are many species of the native *Anastrepha* fruit flies and, presumably, many potential hosts. Contrary to this expectation, the process of geographic expansion of the CFF in South America was, and is, slow. Their host range remains restricted to a small number of introduced fruit species, which are found only in areas inhabited or cultivated by humans. There is no evidence that the fly is established in the forest, although male flies have been caught in Jackson traps baited with methyl eugenol as far as 60 km from known sites of infestation (the flies are highly mobile). Data from West Suriname and Amapá (Brazil) strongly suggest that populations of the fly can be eradicated by applications of the male annihilation technique (MAT) in villages and human settlements (but not in the forest) and that re-infestation does not occur in the forest. A similar situation occurs in Peninsular Malaysia (Vijaysegaran and Osman 1991).

Table 1. Timetable of dispersion.

| | |
|---|---|
| 1975 | First record in Paramaribo, Suriname, from market produce. |
| 1986 | Dispersal along the routes of transport east to French Guyana and west to the districts of Saramacca and Coronie. |
| 1989 | First detection attempted in French Guyana – about 200 km from Paramaribo. No flies were found as far away as Cayenne. |
| 1993 | CFF trapped in Orealla, Guyana, at the Surinamese border, 220 km from the origin. |
| March 1996 | First detection attempted in the Brazilian city of Oiapoque at the border with French Guyana and 500 km from Paramaribo. Flies were found the first day. |

If we consider 1975 to be approximately the establishment year in assessing the spread of CFF in South America (Figure 1), then the species took 21 years to expand east and west 500 km under control conditions. As with most fruit fly infestations, the most likely mechanism for introduction to a new area is through human activity (transporting infested fruits) although dispersal undoubtedly accounts for the spread to some areas. MAT was initiated in 1990 in West Suriname by one of us (ASM); this prevented the permanent establishment of CFF in that area and Guyana (van Sauers-Muller 1993). The movement of people and goods among the four countries is low, and only one road (interrupted by six ferries, one of which does not permit vehicles) links the cities from Georgetown (capital of Guyana) to Cayenne (capital of French Guyana). This will, however, change in the near future; a road from Cayenne to Oiapoque is under construction which will provide a link between heavily infested areas and the current pilot project in Oiapoque. A new ferry for transporting vehicles is also almost operational between Guyana and Suriname. These new routes of transport will create stronger links to uninfested areas and increase the probability that the CFF will become established. We expect that infestation will occur much more rapidly in countries, such as Brazil and Venezuela, which have a much more developed transport infrastructure.

## CFF HOSTS IN SOUTH AMERICA

The hosts of CFF are species that have been introduced to South America. The two primary hosts are carambola and java apple (*Syzygium samarangense* (Blume) Merr. & Perry. They are both exotic species. Java apple, although common in Suriname, is uncommon in French Guyana and extremely rare in Oiapoque and Guyana. The secondary hosts, mango, guava, sapodilla and West Indian cherry (*Malpighia punicifolia* L.) and the primary hosts, carambola, are found throughout the region but only in locations inhabited or disturbed by humans. In South America, the hosts of CFF remain low in number.

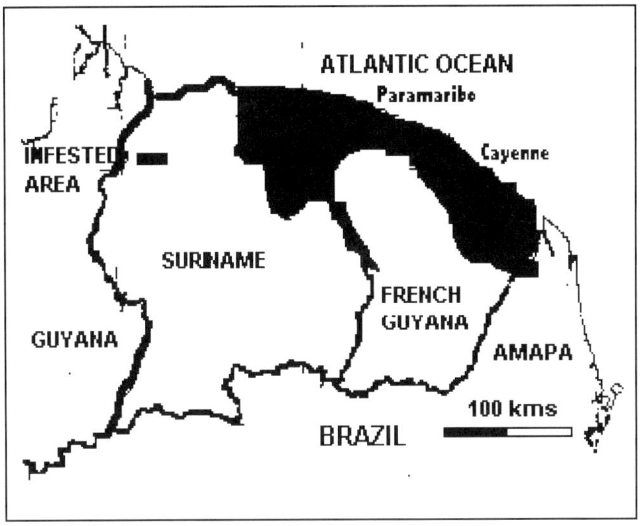


Figure 1. *Bactrocera carambolae* infested area in South America.

## INITIAL EFFORTS TO CONTROL CFF THROUGH AN AREA-WIDE PROGRAMME

A number of activities was carried out by the government of Suriname with the assistance of the IICA (Inter-American Institute for Cooperation on Agriculture), the Food and Agriculture Organization of the United Nations (FAO), the government of Brazil and the United States Department of Agriculture (USDA) to initiate a programme to control CFF and to prevent its dispersion to the rest of South America. However, difficulties of another nature emerged.

Isolated efforts were made by international organisations and governments in order to provide a reliable detection system, which delimitates the distribution of CFF in the countries. Efforts and funds were also made available to gather information on MAT under South American conditions. In the first instance, the pilot phase began in Suriname to provide technical information about the feasibility of the methodology. Brazil applied the methodology developed in Suriname, and worked on eradication since the first detection. After seven months, eradication was reached.

Only in 1996 with the seed money from the International Fund for Agricultural Development (IFAD), an area-wide programme was launched and implemented in August 1997 by IICA. All affected countries – Guyana, Suriname, France (French Guyana) and Brazil – are funding the programme through the Ministries of Agriculture and more resources are expected from donor countries such as the USA and the Netherlands.

The regional programme is coordinated from Paramaribo, where its headquarters is located, and keeps in close contact with the four national coordinators of each affected country.

## APPLIED METHODOLOGY

The methodology is based on MAT, which has been applied in different parts of the world for the *Bactrocera* species where the male fly is strongly attracted by methyl eugenol (Chiu and Chu 1991, Liu 1991). It consists of the distribution of fibre blocks soaked in a mixture of methyl eugenol and insecticide. The blocks are hung or placed in host trees at a rate that varies from 4-20 per hectare. At the moment, only ground dispersion has been applied and aerial release is planned for inaccessible areas. In addition, ME + malathion is sprayed with Panama pumps on poles and onto trees to attract and kill the males. In areas with a high larval or adult population, the application of a protein bait spray is planned.

The population is monitored with Jackson traps for males and McPhail traps for both sexes.

## PRESENT STATUS

### Guyana

Approximately 1,000 traps are used in Guyana to detect any introduction. Only occasional CFF findings are made at the border with Suriname, where a "hot spot" remains. The Ministry of Agriculture of Guyana is actively inspecting its border with Suriname where the largest traffic of products and people occurs.

### Suriname

A large portion of the country remains CFF-free. In the western part, only an isolated village remains infested. The latest trapping data show that the sparsely inhabited interior (the vast majority of the area) of Suriname is free of CFF. Less than 20% of the territory is infested at present.

### French Guyana

The coastal area is heavily infested. A total of 193 Jackson traps was installed in the city of Saint Georges and along the left margin of the Oiapoque River at the French side of the Brazilian border. The captures have shown an average of 5 (November 97) and 9 (February 1998) flies/trap/month with a high density in Saint Georges (20 flies/trap/week).

### Brazil

CFF is not found in the urban areas. Flies have been captured only in the traps placed at the right margin of the Oiapoque River, adjacent to the heavily infested city of Saint Georges. A detection system is implemented in the most important ports of entries from Suriname and French Guyana.

## PERSPECTIVES AND TIMETABLE

The methodology used by the programme has been effective and the timetable for the area-wide effort extends until the middle of 2000. High priority is the necessity to keep Guyana and Brazil free of CFF so as to prevent its spread into Venezuela and south Brazil.

Suriname can be CFF-free by mid-2000, depending upon the availability of funds. French Guyana is expected to be CFF-free by the end of 2000. All efforts are being made to reach total eradication in three years. Post-eradication activities are now planned to ensure that all regions are free and protected from new introductions.

## REFERENCES

Chiu, H.T. and Y.I. Chu. 1991. Male annihilation operation for the control of Oriental fruit fly in Taiwan, pp. 52-60. *In* Proceedings of the international symposium on the biology and control of fruit flies. K. Kawasaki, O. Iwahashi and K.Y. Kaneshiro (eds.). The University of Ryukyus, The Okinawa Precfectural Government, Okinawa, Japan.

Drew, R.A.I. and D.L. Hancock. 1994. The *Bactrocera dorsalis* complex of fruit flies (Diptera: Tephritidae: Dacinae) in Asia. Bull. Entomol. Res. Suppl. 2: 1-68.

Hancock, D.L. 1989. Identification of the *Dacus* species (Diptera: Tephritidae) present in Suriname. Food and Agriculture Organization of the UN, Tech. Rep. 1.

Liu, Y.C. 1991. Current status and control of the Oriental fruit fly, *Dacus dorsalis* Hendel of Taiwan, pp. 81-86. *In* Proceedings of the international symposium on the biology and control of fruit flies. K. Kawasaki, O. Iwahashi and K.Y. Kaneshiro (eds.). The University of Ryukyus, The Okinawa Precfectural Government, Okinawa, Japan.

van Sauers-Muller, A. 1993. Pilot eradication project for the Carambola fruit fly in Coronie, Suriname, pp. 439-442. *In* Fruit flies: biology and management. M. Aluja and P. Liedo (eds.). Spring-Verlag, New York, NY.

Vijaysegaran, S. and M.S. Osman. 1991. Fruit flies in Peninsular Malaysia: their economic importance and control strategies, pp. 105-115. *In* Proceedings of the international symposium on the biology and control of fruit flies. K. Kawasaki, O. Iwahashi and K.Y. Kaneshiro (eds.). The University of Ryukyus, The Okinawa Prefectural Government, Okinawa, Japan.

# Status Report on "The Integrated Fruit Fly Management Based on the Sterile Insect Technique in Guimaras Island, Philippines"

*S.A. Covacha[1], H.G. Bignayan[1], E.G. Gaitan[1], N.F. Zamora[1], R.P. Marañon[1], E.C. Manoto[2], G.B. Obra[2], S.S. Resilva[2] and M.R. Reyes[2]*

[1] BPI-NMRDC, Jordan, Guimaras, Philippines
[2] DOST, PNRI, Diliman, Quezon City, Philippines

## INTRODUCTION

Western Visayas has a large area planted with mangoes and is considered the major mango producing region of the country. As of 1992, about 10,000 hectares were devoted to the crop with a total production of 88,727 metric tons. The bulk of mango production comes from Guimaras Island with 54,944 bearing and 165,852 non-bearing trees. Major markets for Philippine mangoes are Japan, Hong Kong and Singapore. The country accounts for more than 90% of Japan's fresh mango imports. Exports to Japan also show an average increase of 20% yearly while those to Hong Kong have increased by 23%. However, expansion in the market of mangoes and other fruits is greatly restricted by the presence of *Bactrocera philippinensis* Drew and Hancock, a sibling species of the Oriental fruit flies, *Bactrocera dorsalis* (Hendel), in the country. The pests cause large economic losses to producers and are a major deterrent to the free movement of fresh fruits in the world market.

The control of *B. philippinensis* pests using insecticides cannot be relied upon because of problems like development of insect resistance, undesirable environmental contamination and resurgence of secondary pests. On the other hand, disinfestation treatments for fresh fruits are either expensive or not accepted by importing countries. Japan, for instance, accepts only vapour heat treated fruits from the Philippines (Merino et al. 1986). To facilitate the growth of the fruit industry, an effective area-wide eradication of fruit flies as achieved by Japan in its southern island is therefore needed. This involves the use of the male annihilation technique (MAT) and the sterile insect technique (SIT).

The probability of having similar success in the use of the male annihilation technique and the sterile insect technique in eradicating fruit flies from the island of Guimaras is not far from reality. Fulfilling requirement of an "isolated area", the geographical location of Guimaras is therefore a unique feature that will satisfy the implementation of these new techniques. In addition, the body of water that surrounds the island discourages easy invasion of fruit flies from neighbouring places and facilitates the imposition of quarantine in the area.

Prior to the implementation of MAT and SIT, there was felt a need to study the basic ecology of the subject species (identification of the host, fruit infestation record, population dynamics and fly movement). This information was gathered earlier in a cooperative experiment between the National Mango Research and Development Center, the Department of Agriculture (NMRDC-DA), the Philippine Nuclear Research Institute and the Department of Science and Technology (PNRI-DOST), in a project entitled "A Feasibility Study of Integrated Control of the Oriental Fruit Flies" (Manoto et al. 1998).

Furthermore, this study was conducted to establish Guimaras as an Island free of *B. philippinensis* through the introduction of an integrated pest management based on SIT and increase production and exportation of tropical fruits particularly mangoes.

## ECOLOGICAL STUDIES

### Host Phenology and Population Dynamics

A brief result of the basic ecological studies, which involved the determination of host fruits and the island-wide *B. philippinensis* seasonal population dynamics during a four year pre-treatment survey, is shown in Figures 1 and 2. From 1992 to 1996, the relative abundance of *B. philippinensis* in Guimaras had been affected by the availability of host fruits and the onset of the rainy season. The *B. philippinensis* population starts to build up in April when mangoes are most available and fruiting of other hosts commences. The peak population occurs in the months of June and July which coincide with the abundant fruiting of santol (*Sandoricum koetjape* (Burm. f.) Nakai) and sineguelas (*Spondias purpurea* L.) which are widely grown under backyard and natural vegetation conditions. The population of *B. philippinensis* fluctuates towards August to September with the lowest during the months of November to December.

### Estimates of *B. philippinensis* Population

Based on the results of the "mark, release, recapture" method of fruit fly density determination in the three agroecosystems (natural vegetation, commercial orchards and backyard), the total number of *B. philippinensis* in the island of Guimaras was estimated at 150,895,871. According to Dr. Koyama, J. (1995), IAEA Japanese Expert, population suppression by application of the male annihilation technique will be necessary to reduce the wild population prior to the release of sterile *B. philippinensis*

## CONTROL STRATEGIES IMPLEMENTED

### Male Annihilation

Beginning in February 1997, an island-wide application of MAT was done through the use of fibre board squares (FBS) (5x5x1.8 cm) impregnated with 10 - 12 ml of lure toxicant containing 90% methyl eugenol and 10% malathion. This was done to reduce the field population of male *B. philippinensis* prior to the implementation of SIT.

For ground application, FBS were tied with GI wire prior to being soaked with the lure toxicant. An allocated number of FBS per barangay (village) was placed in black vinyl bags and transported to target areas. The area covered for ground application included areas planted with various fruit trees, such as coconuts, as well as bamboo orchards, residential and school sites. Ground application was done by hanging FBS preferably on host trees, if present, at the density of 4 FBS per hectare.

Aerial application was done using a light Cessna plane from the Bureau of Plant Industry (BPI). Impregnated FBS were packed in screen bags containing 200 pieces per bag and transported to Iloilo Airport a day before the release schedule. This was done in forested areas at the rate of 2 FBS per hectare. A flight could be completed within 30 -

40 minutes, depending on the area covered with a flight velocity ranging from 40 - 70 kph at an altitude of 300 - 600 ft above the ground.

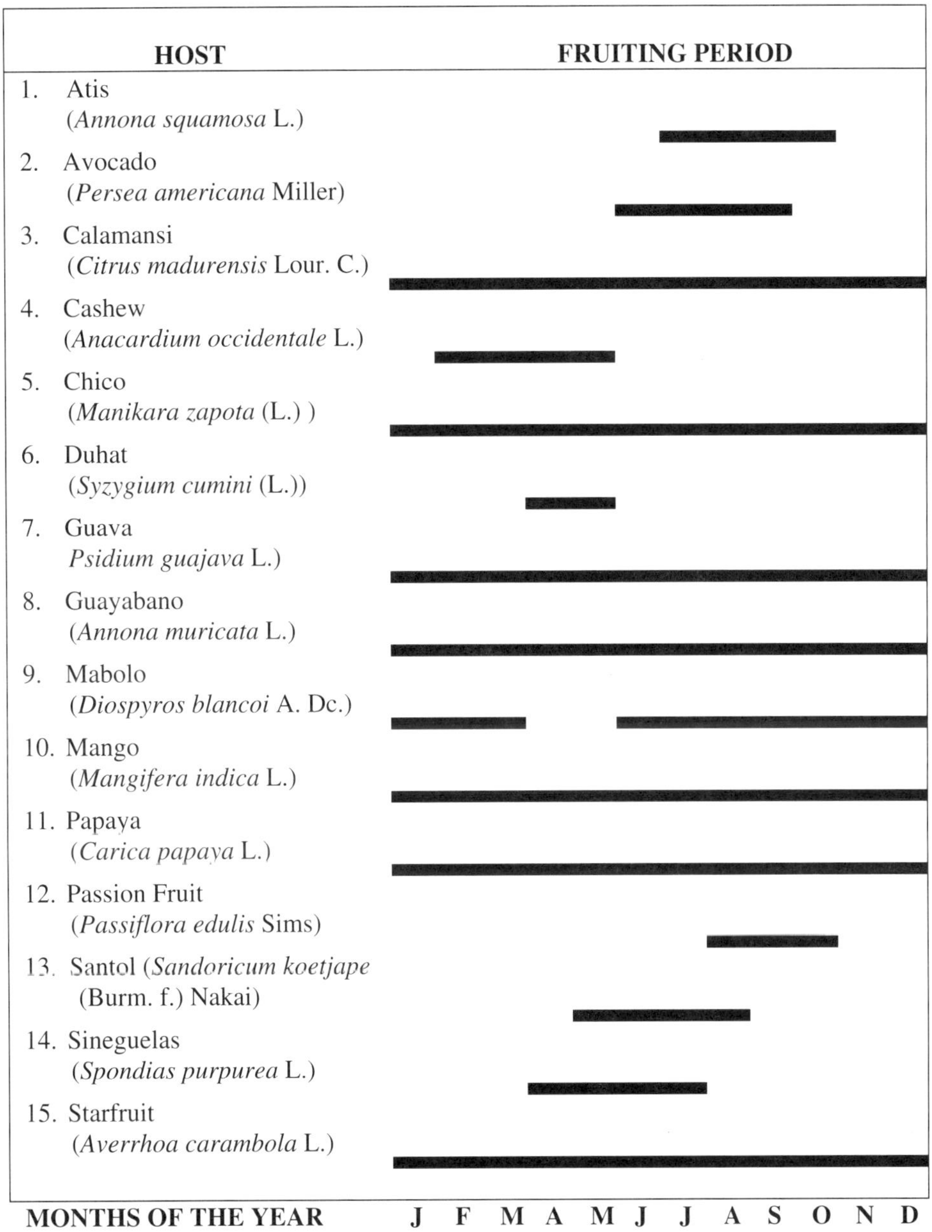


Figure 1. Fruiting period of different hosts of the *B. philippinensis*.

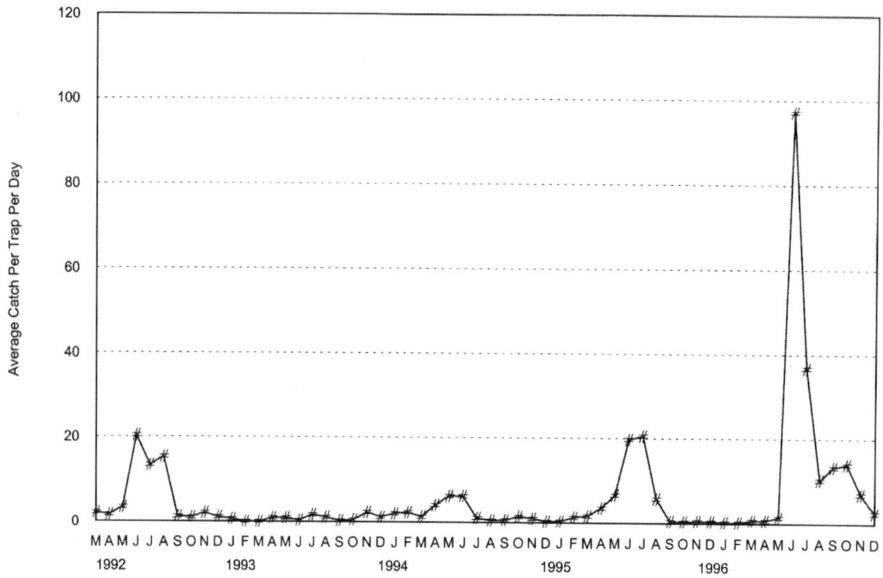


Figure 2. Seasonal occurrence of the *B. philippinensis* as indicated by the average number of flies captured.

### Estimates of *B. philippinensis* Population

Based on the results of the "mark, release, recapture" method of fruit fly density determination in the three agroecosystems (natural vegetation, commercial orchards and backyard), the total number of *B. philippinensis* in the island of Guimaras was estimated at 150,895,871. According to Dr. Koyama, J. (1995), IAEA Japanese Expert, population suppression by application of the male annihilation technique will be necessary to reduce the wild population prior to the release of sterile *B. philippinensis*

## CONTROL STRATEGIES IMPLEMENTED

### Male Annihilation

Beginning in February 1997, an island-wide application of MAT was done through the use of fibre board squares (FBS) (5x5x1.8 cm) impregnated with 10 - 12 ml of lure toxicant containing 90% methyl eugenol and 10% malathion. This was done to reduce the field population of male *B. philippinensis* prior to the implementation of SIT.

For ground application, FBS were tied with GI wire prior to being soaked with the lure toxicant. An allocated number of FBS per barangay (village) was placed in black vinyl bags and transported to target areas. The area covered for ground application included areas planted with various fruit trees, such as coconuts, as well as bamboo orchards, residential and school sites. Ground application was done by hanging FBS preferably on host trees, if present, at the density of 4 FBS per hectare.

Aerial application was done using a light Cessna plane from the Bureau of Plant Industry (BPI). Impregnated FBS were packed in screen bags containing 200 pieces per bag and transported to Iloilo Airport a day before the release schedule. This was done in forested areas at the rate of 2 FBS per hectare. A flight could be completed within 30 - 40 minutes, depending on the area covered with a flight velocity ranging from 40 - 70 kph at an altitude of 300 - 600 ft above the ground.

Ground and aerial applications were done at 50-day intervals from February to November 1997. A total of six applications of MAT was done with 525,534 pieces of lured FBS distributed in Guimaras Island (ground - 347,934 and aerial - 177,600). The application of MAT by both methods approximately covered 54% of the total land area of Guimaras. The rest of the area not covered are planted with crops which are non-target hosts of fruit flies (rice land, fish ponds, etc.)

The efficacy of the male annihilation in reducing the male *B. philippinensis* population was evaluated based on monthly catch per trap per day (CPTD) in 1996 when MAT was not applied and monthly CPTD in 1997 during the application of MAT. The monthly reduction in the male *B. philippinensis* population is shown in Figure 3. From March to June, a significant reduction in the population was observed. A slight fluctuation with a downward trend occurred in August. The highest percentage reduction was observed in October at 98.36%. Generally, the application of MAT at regular intervals significantly reduce the male fruit fly population in the island. These results further showed that planned SIT could be followed.

## The Sterile Insect Technique

### Transport of sterile *B. philippinensis* pupae

Sterile pupae mass reared, marked with fluorescent dye and irradiated at PNRI, were transported in carton boxes to Iloilo Airport via a Philippine Airlines (PAL) commercial flight. Upon arrival at Iloilo Airport, the pupal boxes were loaded again to an air-conditioned land transportation and then to a hired pumpboat going to the Buenavista wharf. Finally they were delivered to a *B. philippinensis* holding laboratory located near the Guimaras airstrip in McLain using a NMRDC vehicle. The transport of sterile pupae from PNRI to Iloilo Airport was done in two batches at one-day intervals for every release schedule.

### Bagging and holding of sterile pupae

About 100 g pupae were placed into each paper bag containing paper strips and sugar paper for the adult diet. The bag was then folded twice and stapled securely across the top. Bagged pupae were placed in the pupal racks arranged in a uniform distance for proper air circulation. The temperature of the holding room was maintained at 23°C to 27°C.

### Release

Aerial release was done using a BPI Cessna plane with a funnel installed at the rear left side opening of the aircraft as passage for bag droppings. About 230 - 250 pupal bags were loaded in each flight.

In areas where CPTD was more than one and at places with fruit infestation, ground release of sterile flies was conducted. This was done one day after adult emergence.

The use of SIT as complementary to MAT for the control of *B. philippinensis* in

Guimaras started in November 1997. As of April 1998, a total of 47 million sterile flies were released, both aerially and on the ground.

Effectiveness of sterile fly release was evaluated based on the percentage reduction of wild population as shown in Fig. 3 the continuous release of sterile flies successfully reduced wild population. The percentage reduction showed an upward trend from December to February. In March, when MAT was implemented in the previous year, the release of sterile flies complimented the control where further reduction of male fly population was achieved.

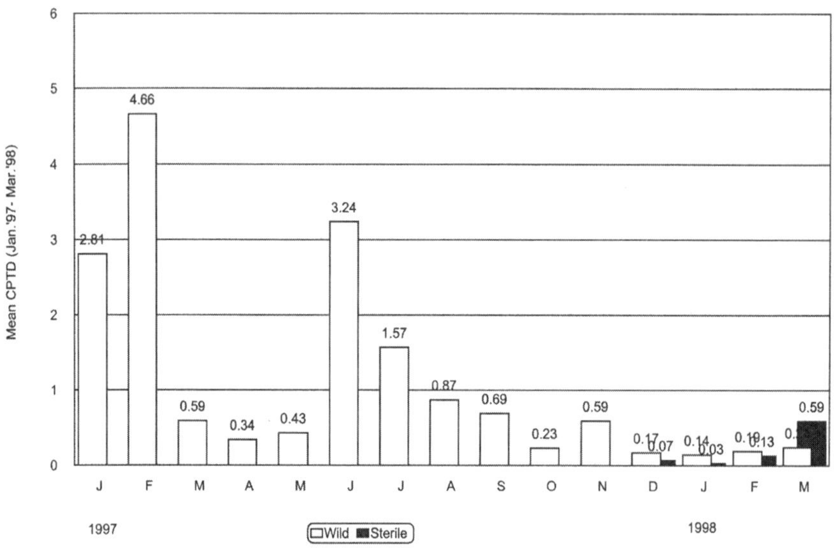


Figure 3. Monthly change in abundance (CPTD-catch/trap/day) of wild and sterile *B. philippinensis* caught in Guimaras.

## Quality control test

The percentage of emergence of sterile pupae was generally satisfactory while the percentage of flies that emerged under field conditions were below acceptable standards (60.0%) (Table 1). This could be attributed to several factors: 1) fluctuation of temperature in the holding room and during transport, 2) overstocking of bags inside the aircraft, and 3) delayed release of flies where supply inside pupal bags was not enough. Moreover, with these situations, more deformed flies were also observed.

## Monitoring Fruit Fly Population

The monthly monitoring of the wild fly population and sterile fly recaptures is presented in Figure 3. The population of wild flies recorded in October was suppressed to a low level after the fifth application of MAT with less than one male CPTD. A slight increase in the mean CPTD was recorded in November with the highest in Sibunag (1.61) where fruiting of native guava was abundant during this period. When release of sterile flies was started in the last week of November, wild fly catches monitored in December decreased in all areas. In the succeeding months, the mean CPTD still fell below one male/trap per day. This meant that a low population of wild flies was present in the field. Sterile fly recaptures obtained in December - February were comparatively lesser than wild fly catches as in some cases it was zero. In March, where more ground releases of sterile flies were done, a higher number of sterile to wild fly catches was monitored in the traps. Although the density of the sterile flies released in the field was low, the suppression of the wild fly population, after male annihilation followed by sterile fly releases, was achieved.

Table 1. Results of quality control test done under field and laboratory conditions (December 1997 - March 1998).

| Batch No. | Release Date Date | No. of Released Flies | % Emergence | | % Fliers | |
|---|---|---|---|---|---|---|
| | | | Field* | Lab | Field* | Lab |
| 1 | 29 Nov. 1997 | 2,613,800.00 | 91.72 | 93.70 | 89.68 | 89.66 |
| 2 | 14 Dec. 1997 | 4,168,345.00 | 99.86 | 88.70 | 98.69 | 84.00 |
| 3 | 16 Dec. 1997 | 6,338,541.00 | 94.34 | 95.00 | 87.67 | 92.33 |
| 4 | 27 Jan. 1998 | 4,764,448.00 | 84.64 | 88.48 | 29.62 | 84.33 |
| 5 | 9 Feb. 1998 | 8,046,558.00 | 82.80 | 76.50 | 54.80 | 64.20 |
| 6 | 23 Feb. 1998 | 3,253,828.00 | 84.70 | 95.20 | 25.10 | 81.50 |
| 7 | 9 Mar. 1998 | 7,140,293.00 | 91.90 | 91.80 | 42.90 | 73.00 |
| 8 | 20 Mar. – 4 Apr. 1998 | 10,797,238.00 | 87.40 | 89.10 | 60.90 | 68.70 |

* Taken from bag samples returned after flight and opened for one hour to allow adults to fly out from the bag. Adults flown out of the bags were considered absolute fliers.

### Fruit Collection/Inspcction

The degree of *B. philippinensis* infestation on the different host fruits inspected showed that starfruit (80.2) yielded the highest number of pupae per kilogram fruit samples followed by guava (74.2). Generally, infestation was high on fallen than on picked fruits.

The monthly fruit infestation in almost all areas decreased after the implementation of male annihilation followed by sterile fly release.

## CONCLUSION AND RECOMMENDATIONS

The implementation of control strategies in the island through the use of the sterile insect technique as complementary to the male annihilation technique can further reduce the wild male population.

Since the eradication of *B. philippinensis* is a long and continuous process, it is recommended that long term support, both political and financial, be extended for the successful implementation of the programme.

## REFERENCES

Koyama, J. 1995. End of mission report on feasibility study of integrated control of fruit flies. IAEA-RU-5189, 25 pp.

Manoto, E.C., S.S. Resilva, G.P. Bilog, H.G. Golez, S.A. Covacha, H.G. Bignayan. 1998. Feasibility of sterile insect release method based on integrated oriental fruit fly management programme in Guimaras Island. Philipp. Nucl. Journal 12: 7-18.

Merino, S.R., M.M. Eugenio, A.D. Ramos and S.P. Hernandez. 1986. VHT an alternate method of fruit fly disinfestation of Manila super mangoes. Plt. Indust. Bull. 1(9): 4.

# Suppression of Mediterranean Fruit Fly Populations over Mountainous Areas Through Aerial Phloxine B – Protein Bait Sprays: Regional Medfly Programme in Guatemala

*Grant T. McQuate and Steven L. Peck*

USDA-ARS, Tropical Fruit, Vegetable and Ornamental Crop Research Laboratory
Hilo, Hawaii, USA

## INTRODUCTION

The Mediterranean fruit fly, *Ceratitis capitata* (Wiedemann), was discovered in southern Mexico sometime in 1977 near Tapachula, Mexico. Farmers in Texas and other states of the United States became concerned that the Mediterranean fruit fly would spread northward through Mexico and into the US. In response to this threat to US agriculture, funds were appropriated by Congress to be used by the US Department of Agriculture (USDA) – Animal and Plant Health Inspection Service (APHIS) to eradicate the Mediterranean fruit flies from Mexico and establish a barrier zone in Guatemala to keep the Mediterranean fruit flies from spreading northward into Mexico. In Mexico and Guatemala, the organisation called "Mosca del Mediterráneo" (MOSCAMED) was created to support the programme. Aerially applied malathion bait sprays were used in the suppression programme beginning in Mexico in 1982 and beginning in Guatemala in 1985. Mexico has been free of the Mediterranean fruit fly since 1982, except for outbreaks in the southernmost state of Chiapas, adjacent to Guatemala. The spraying of malathion was banned by the government of Guatemala in early 1996 because of concern regarding possible adverse effects on honey bees. By this time, research had been started to evaluate the use of xanthene dyes as a potential alternative to malathion in protein bait sprays for the suppression of the Mediterranean fruit fly (Liquido et al. 1995). Light-activated toxicity of xanthene dyes has been documented for more than two dozen insect species overall (Heitz 1997). Field trials of xanthene dyes, as a safer alternative to malathion in bait sprays targeting the Mediterranean fruit fly, were begun in Hawaii in 1994 and in Guatemala in 1996 and proved to be promising. By the end of 1996, xanthene dyes were registered as a substitute for malathion to suppress/eradicate Mediterranean fruit flies in the barrier zone. In January, 1997, MOSCAMED-Guatemala began a spray programme with xanthene dyes as the toxicant in a protein bait spray. A second spray cycle was begun in December 1997, targeted to start before the annual Mediterranean fruit fly population increase. The sprays were continued until early February 1998, after which a weekly release of sterile Mediterranean fruit flies was started and was planned to continue through mid-June, 1998.

Here, we describe the 1997-1998 large area phloxine B – protein bait spray programme in southwestern Guatemala. We present the catch results of traps monitored throughout the weekly spray programme, as well as trap catch results from the initial ten weeks of sterile fly release which followed the spray programme. Additionally, we discuss issues which may affect the success of large area suppression programmes which target Mediterranean fruit fly or other tephritid fruit fly species using xanthene dye – protein bait sprays.


**Area-Wide Control of Fruit Flies and Other Insect Pests,**
ed. K.H. Tan. Penerbit Universiti Sains Malaysia, Penang, 2000.

## MATERIALS AND METHODS

### Site Location

The spray trial was located in mountainous coffee growing areas of southwestern Guatemala (Figure 1). The elevation in the spray area ranged from about 400 m to 1300 m. Aerial sprays of the phloxine B – protein bait solution were applied to give complete coverage to 16,700 hectares and alternate swath coverage to an additional 3,800 hectares. An unsprayed 600 hectare control area was located adjacent to the complete coverage spray area (Figure 1).

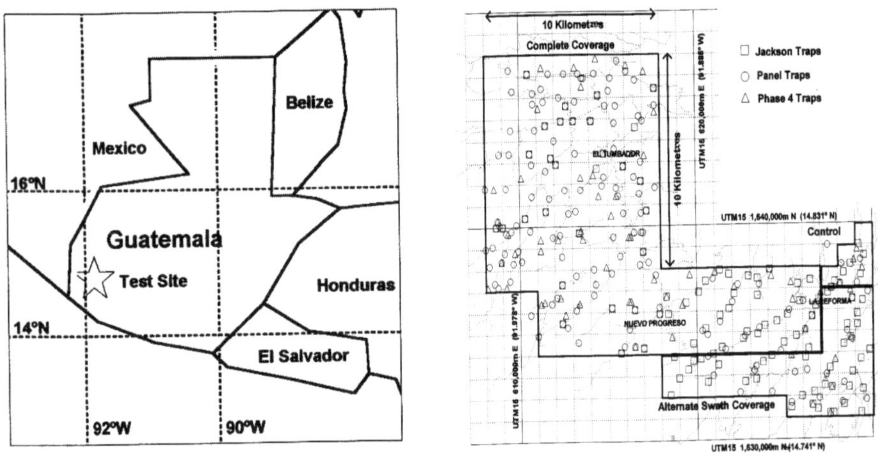


Figure 1. Location of site of aerial phloxine B – protein bait field trial for suppression of Mediterranean fruit fly and dimensions of complete coverage spray area, alternate swath spray area and control area.

### Formulation and Application Rate

The protein source used in the phloxine B – protein bait formulation was Mazoferm E802 (Corn Products, Argo, IL). The overall formulation consisted of 40% Mazoferm E802, 20% high fructose corn syrup (which contained about 60% fructose), 0.5% phloxine B, and 39.5% water. The phloxine B – protein bait was applied at a rate of 3.35 litres/ha in the complete coverage spray area and 1.68 litres/ha in the alternate swath spray area. For the complete coverage spray area, this provided about 1570 g Mazoferm E802, 1550 g water, 785 g high fructose corn syrup and 19.6 g of phloxine B per hectare.

### Assessing Effectiveness and Persistence of Bait Spray

Each week from the 4$^{th}$ to the 8$^{th}$ week of spraying in the complete coverage spray area, coffee leaves were retrieved shortly after the day's spray was completed. Leaves containing at least fifty 2 mm diameter spray droplets were placed in a plexiglass cubicle (30 x 30 x 30 cm) holding 25 male and 25 female (3-4 day old)

Mediterranean fruit flies. All cubicles had a supplemental water source and half of them had a supplemental food source. Comparable cages were set up using the same number of leaves taken from unsprayed coffee plants. Cages were set outside below a 50% shade cloth to mimic shade provided by coffee foliage. Fly mortality was recorded in these cages periodically over a 24-hour period. Each week, this test was replicated with leaves taken 48 hours and 96 hours after a spray.

## Application of Bait Spray

Phloxine B – protein bait solution was sprayed weekly by fixed-wing aircraft. Each plane had six wing-mounted spray nozzles set to provide coverage over a 50 m wide spray swath. Accuracy of application was facilitated through the use of a Global Positioning System (GPS) satellite navigation guidance and recording system.

### *Complete Coverage Area*

In the complete coverage spray area (16,700 hectares), eight weekly sprays were completed from 15 December 1997 to 6 February 1998.

### *Alternate Swath Area*

In the alternate swath spray area (3,800 hectares), six weekly sprays were completed from 5 January to 13 February 1998.

When "hot spots" were detected through weekly trap recoveries in the spray areas, additional localised spraying was performed using helicopters.

## Release of Sterile Flies

In the complete coverage spray area, 1,500 sterile male Mediterranean fruit flies of the temperature sensitive lethal (TSL) strain were released weekly per hectare prior to the first spray and subsequent to the last spray. In the alternate swath spray area, 3,000 normal strain sterile flies (male and female) were released weekly per hectare before, during, and after the spray programme. Both releases were scheduled to continue until 15 June 1998.

## Monitoring Mediterranean Fruit Fly Population

Three types of traps were placed throughout the spray and control areas: Jackson traps, yellow panel traps and phase 4 traps. These trap types are illustrated and described in Figure 2. Figure 1 shows the location of each trap type throughout the complete coverage spray area, the alternate swath spray area and the control area from a typical trap week. These traps were serviced on a weekly basis starting several weeks before the initial spray. Trap monitoring was scheduled to continue beyond the completion of sterile fly release.

## Monitoring Larval Infestation of Coffee Cherries

Coffee cherries were harvested regularly from the complete coverage spray area, the alternate swath spray area and the control area before, during, and after the weekly sprays to assess infestation levels by Mediterranean fruit flies.

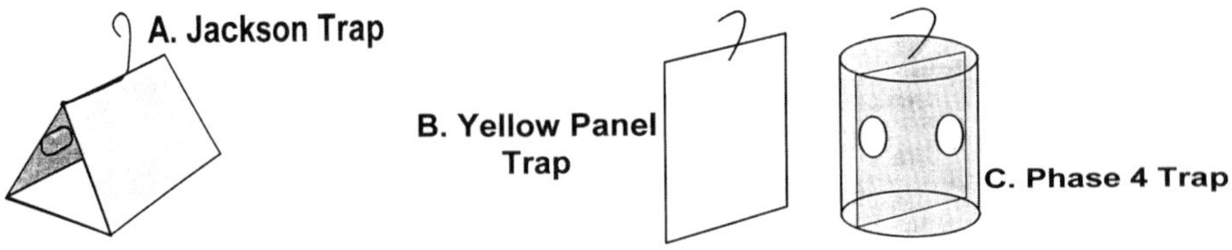


Figure 2. Trap types used to monitor Mediterranean fruit fly population density: A) *Jackson Traps* baited with trimedlure (a male attractant), B) *Yellow Panel Traps* baited with trimedlure, and C) *Phase 4 Traps* baited with ammonium acetate, 1,4-diaminobutane (putrescine) and trimethylamine, a synthetic food bait attractant which attracts both males and females.

## RESULTS

### Effectivenesss of Bait Spray

Percentage mortalities of Mediterranean fruit flies, by treatment, for the field cage tests, averaged over five weeks of tests, are presented in Figure 3. Mortality after 24 hours in cages receiving freshly sprayed leaves and no supplemental food source averaged 86.5%, with weekly averages ranging from 66.5% to 98%. In cages where a supplemental food source was added, average mortality after 24 hours was only reduced to 83%. In the control cages, average 24 hour mortality was 7.2% and 20.1% with and without added food source, respectively. This suggests that the bait spray should have been effective in the field even considering the presence of alternative food sources.

### Persistence of Bait Spray

Average 24 hour mortality of Mediterranean fruit flies in cages without a supplemental food source dropped to 62.6% and 61.2% with leaves collected 48 hours and 96 hours, respectively, after an aerial spray (Figure 3b,c). This showed that the baits maintained significant effectiveness at least four days after a spray application, in the absence of heavy rains (as was the case throughout most of this spray programme).

### Population Monitoring: Complete Coverage Spray Area

#### *Spray Period – Adults*

By week 2 in 1998, the wild Mediterranean fruit fly population had begun to increase considerably in traps in the control area (see Figure 4a). This was especially apparent in yellow panel trap catches which increased, on an average, from 1.75 to 150.5 flies per trap from week 1 to week 6. However, in the complete coverage spray area, average yellow panel trap catch decreased from 0.683 to 0.052 flies per trap in this same time interval (Figure 4b). By week 6, the week after the last spray, only three flies were trapped on yellow panel traps throughout the complete coverage spray area. The average catch of sterile Mediterranean fruit flies on yellow panel traps in the complete coverage spray area dropped by the third week of spraying and remained less than 5.0 during weeks 1-5, the remaining weeks of spraying, as sterile releases were stopped over the complete coverage spray area during the weeks of spraying (Figure 5b).

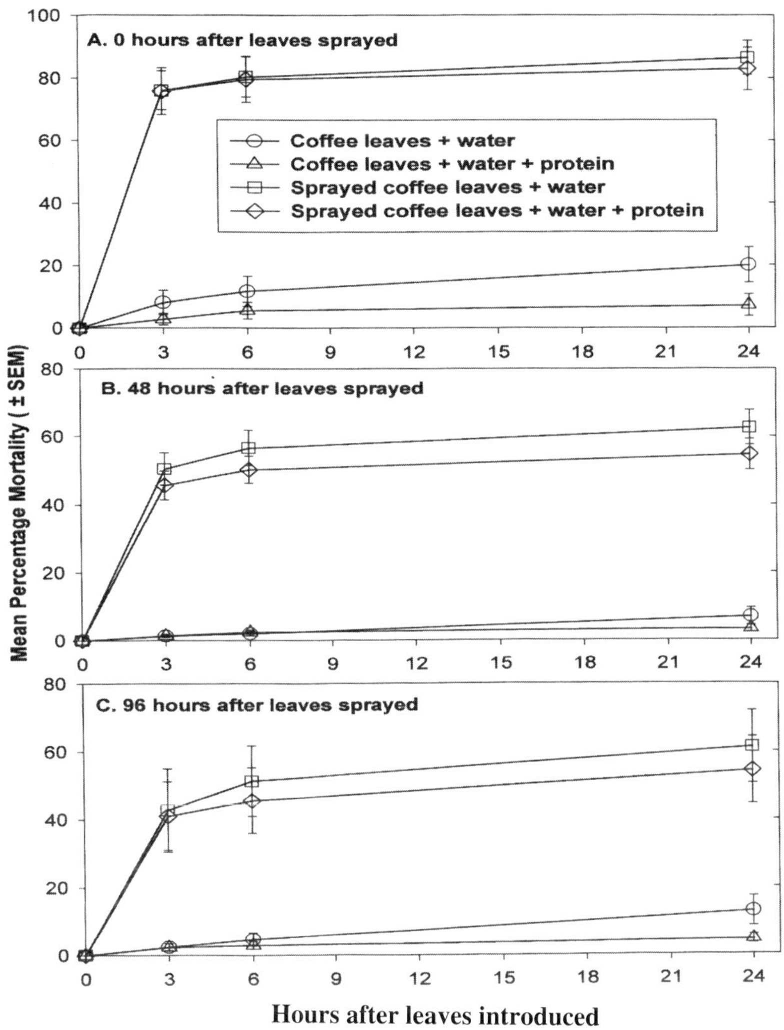


Figure 3. Mean percentage mortality (± SEM) of Mediterranean fruit flies in field cages up to 24 hours after introduction of leaves removed both from coffee plants sprayed by a phloxine B – protein bait spray and from unsprayed coffee plants. Leaves were presented in cages both with and without a supplemental protein source. Leaves from sprayed plants were taken: A) immediately after the aerial spray, B) 48 hours after the aerial spray, or C) 96 hours after the aerial spray.

### *Post-Spray, Sterile Release Period – Adults*

The wild fly catch increased during the post-spray period throughout the complete coverage spray area, although it was still quite low relative to the fly catch in the control area (Figure 4a, b). The average wild fly catch on yellow panel traps in the complete coverage spray area reached a maximum of 5.66 flies in week 11, decreasing

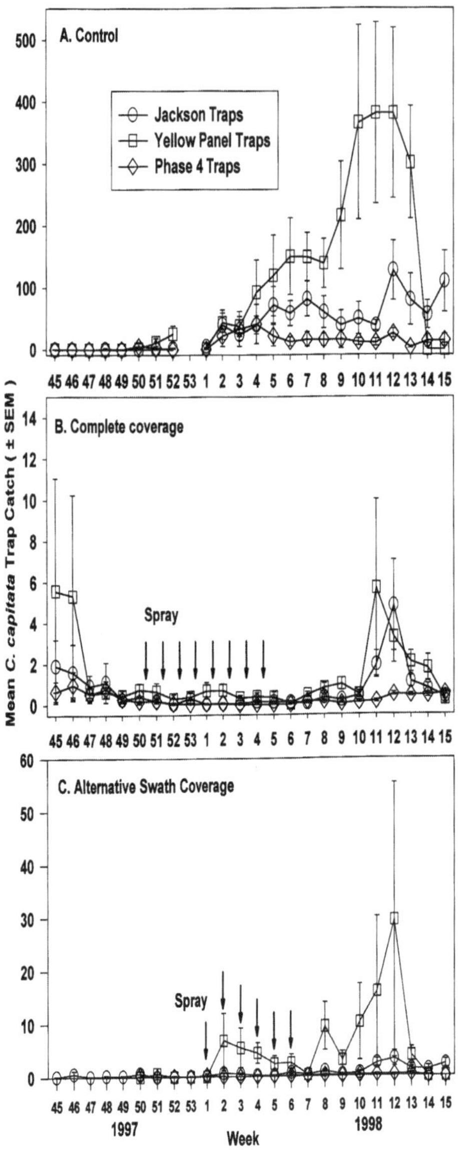


Figure 4. Mean weekly catch (± SEM) of wild Mediterranean fruit flies in Jackson traps, yellow panel traps, and phase 4 traps in: A) the (unsprayed) control area, B) the complete coverage spray area, and C) the alternate swath spray area.

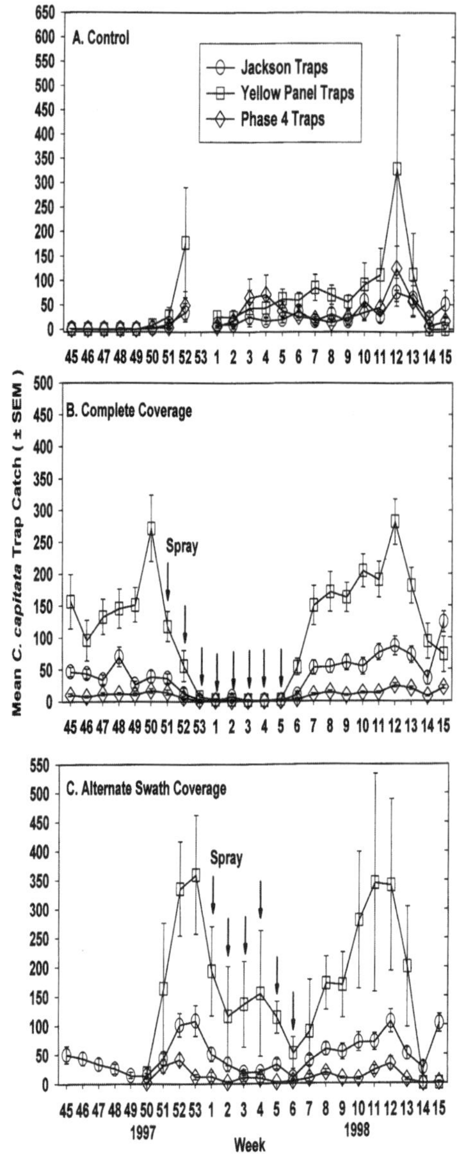


Figure 5. Mean weekly catch (± SEM) of sterile Mediterranean fruit flies in Jackson traps, yellow panel traps, and phase 4 traps in: A) the (unsprayed) control area, B) the complete coverage spray area, and C) the alternate swath spray area.

to 0.24 flies by Week 15 (Figure 4b). In week 11, the average fly catch on yellow panel traps in the control area was 381 and declined thereafter, apparently reflecting the time of natural population decline (Figure 4a). The average catch of sterile Mediterranean fruit flies on yellow panel traps in the complete coverage spray area returned to over 100 flies by week 7, the second week following the cessation of spraying (Figure 5b).

*Larval Infestation*

Larval infestation of coffee cherries collected from the complete coverage spray area averaged less than 1.0 larva per kg of coffee cherries from the beginning of the weekly sprays through week 6, the 1st week after the sprays ended. By week 7, however, it had begun to increase considerably (Figure 6b). By comparison, larval infestation from the beginning of the weekly sprays through week 6 was consistently greater than 22 larvae per kg of coffee cherries collected from the control area, reached 178 larvae per kg in week 6, and increased further in subsequent weeks (Figure 6a).

## Population Monitoring: Alternate Swath Spray Area

*Spray Period – Adults*

The Mediterranean fruit fly population began to increase in the alternate swath spray area at about the same time as in the control area, but the magnitude of increase was considerably less. During the course of the sprays, the average yellow panel trap catch decreased from 6.75 in week 2 to 0.25 in week 7, the week after the last spray (Figure 4c). In the control area, the average yellow panel trap catch increased from 44.0 to 149.75 during the same period (Figure 4a).

*Post-Spray, Sterile Release Period – Adults*

The wild fly catch increased throughout the post-spray period in the alternate swath spray area (Figure 4c), although it was still quite low relative to the fly catch in the control area (Figure 4a). Sterile fly release had not been stopped during spray weeks in this area, so sterile fly trap catches continued to be about the same (generally greater than 100 flies per yellow panel trap), following the completion of the spray cycle (Figure 5c).

*Larval Infestation*

The number of larvae recovered per kg of coffee cherry collected from the alternate swath spray area during the weeks of spraying (Figure 6c), was reduced compared to the level of infestation found in the control area (Figure 6a), but the reduction was much greater in coffee cherries collected from the complete coverage spray area (Figure 6b).

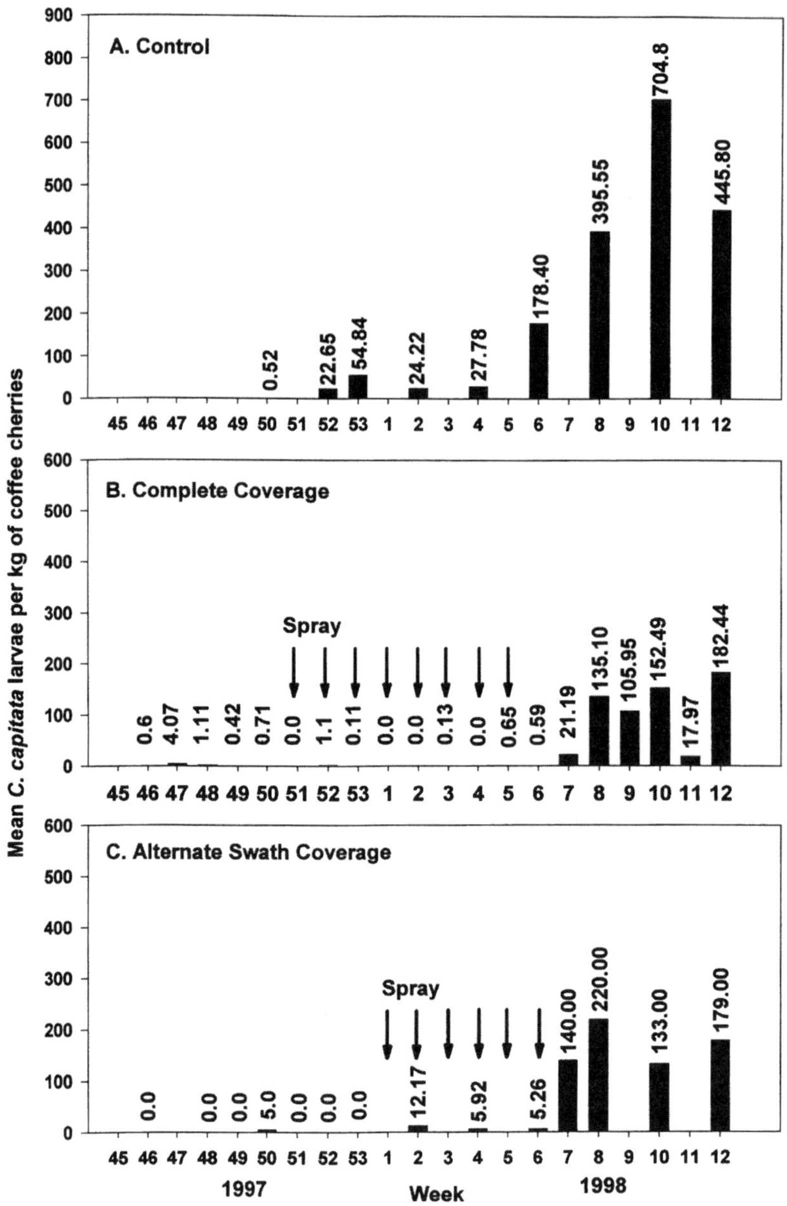


Figure 6. Mean number of Mediterranean fruit fly larvae per kg of coffee cherries harvested weekly from: A) the control area, B) the complete coverage spray area, and C) the alternate swath spray area.

## DISCUSSION AND CONCLUSION

### Effectiveness of Suppression

The results presented here show that phloxine B – protein bait sprays provide good suppression of Mediterranean fruit flies in mountainous coffee growing areas, especially during dry weather conditions such as prevailed throughout the present study. Applying the spray in alternate swaths provided some suppression of Mediterranean fruit fly populations, but much less than that provided by a complete coverage spray. The complete coverage spray programme, combined with the release of sterile flies, was, however, not able to achieve eradication of Mediterranean fruit flies in the area of coverage.

### Potential Factors Affecting Effectiveness of Suppression

#### *Weather*

Weather conditions were very favourable for this study as there was little problem with rain following the sprays. Rain soon after a spray can severely impact the effectiveness of the spray programme. As seen in this study, under dry conditions, the dye-bait solution can remain effective for several days after the spray.

#### *Time of Spray (Time of Day)*

In the spray area, updrafts which occur later in the day can adversely affect the uniform distribution of the dye-bait solution. Consequently, MOSCAMED aimed to spray only during morning hours and began to reverse the order of spraying in successive weeks, so that the same area was not always the last to be sprayed.

#### *Time of Spray (Seasonality)*

The spray programme was started prior to the anticipated annual Mediterranean fruit fly population increase. This timing is important to maximise spray programme effectiveness.

#### *Duration of Spray*

Good suppression was shown in the course of the spray programme, but population increase was observed after the cessation of the spray despite resuming the release of sterile flies. It would have been desirable to extend the spray programme until after the natural seasonal population reduction was observed in the control area. We recognise that the additional programme cost could be problematical.

#### *Delivery Problems*

- *Areas Where Shade Trees are Well Developed*
  Shade trees are routinely planted throughout the coffee growing areas. In places where these trees are well developed, the dye-bait solution may not readily reach the coffee plants and thus, may not be located by the Mediterranean fruit flies. It was not clear from this spray trial whether this

was actually a problem in certain sections of the spray area. This potential problem needs further attention.

- *Localised Wind Patterns*
  The complex topography involved here raises questions of whether localised wind patterns kept the dye-bait solution from reaching certain areas. Micro-climatic wind patterns could tend to keep the dye-bait solution from reaching certain areas within the intended coverage area.

### *Low Light Problems*

It is possible that where there is sufficiently reduced light conditions on steep north-facing slopes, mortality rates could be significantly reduced, especially considering that these slopes will also have less dye-bait solution applied per unit area relative to flat areas. Further trappings in these areas are needed to investigate the significance of this possible limitation.

### Recommendations for Future Spray Programmes

The results here suggest that it is necessary to prolong the duration of weekly sprays in areas where Mediterranean fruit fly populations are established until after fly populations in the control area reach their natural seasonal decline. Phloxine B bait sprays appear to have some limitations on their effectiveness. Many of these need further study. We recommend that conditions which could adversely impact the effectiveness of a spray programme be identified as soon as possible and that measures be included in the spray programme to assess the effect of these potential problems. For example, in this programme, additional trapping might have been used in areas where coverage was suspect (e.g., due to such factors as slope, wind pattern, and overstory) to determine if flies were escaping treatment in these areas.

## ACKNOWLEDGEMENTS

This paper would not have been possible without the help of all the staff of MOSCAMED-Guatemala; Danny Gates, Felipe Jeronimo, Pedro Rendon, and John Stewart of USDA-APHIS and Paul Barr of USDA-ARS.

## REFERENCES

Heitz, J.R. 1997. Pesticidal applications of halogenated xanthene dyes. Phytoparasitica 25:89-92.

Liquido, N.J., G.T. McQuate and R.T. Cunningham. 1995. Light-activated toxicity of phloxine B and uranine to Mediterranean fruit fly, *Ceratitis capitata* (Wiedemann) (Diptera: Tephritidae), adults. pp. 82-107. *In* Light-activated pesticides. J.R.Heitz and K.R. Downum (eds.). American Chemical Society Symposium Series 616, American Chemical Society, Washington, D.C.

# PART III.

## APPLICATION OF AREA-WIDE CONCEPT TO FRUIT FLIES

## B) MANAGEMENT

# Novel Insecticide Strategies Such as Phototoxic Dyes in Adult Fruit Fly Control and Suppression Programmes

*Daniel S. Moreno and Robert L. Mangan*

USDA, ARS, Kika de la Garza Subtropical Agricultural Research Center, Crop Quality and Fruit Insects Research Unit, 2301 South International Boulevard, Weslaco, TX 78596, USA

## INTRODUCTION

The problems of public acceptance, ecological impact, and integration with pest management programmes associated with use of broad spectrum insecticides in bait sprays for fruit flies are being addressed in our laboratory by our development of more precisely targeted bait systems which use insecticides which are less toxic to non-target organisms. Historically, bait and insecticide sprays to control fruit flies have been used since the beginning of the 20th century. Initially, inorganic insecticides were recommended. After the Second World War, chlorinated hydrocarbon insecticides replaced inorganic ones only to be replaced by the organic ones that are used at present. Back and Pemberton (1918) stated that baits used for fruit fly control were first recommended by Mally in South Africa for the control of the Mediterranean fruit fly, *Ceratitis capitata* (Wiedemann), in 1908-1909 and by Berlese in Italy for the control of the olive fruit fly, *Bactrocera oleae* (Gmelin). The methods were improved by Lounsboury in South Africa in 1912 for the control of *C. capitata* and by Newman during 1913-1914 in Australia for the control of the Queensland fruit fly, *Bactrocera tryoni* (Froggatt). In 1910, Marsh used low-volume insecticide applications against the melon fly, *Bactrocera cucurbitae* (Coquillett), in Hawaii. Thereafter, other investigators adopted the low-volume approach to kill fruit flies. Whenever baits were used, they added carbohydrates and fermenting substances such as sugars, molasses, syrups, or fruit juices. In the 1930s, McPhail (1937), while working with attractants, found that sugar-yeast solutions attracted flies, and, in 1939 found that protein lures were attractive to *Anastrepha* species, especially to the guava fruit fly, *A. striata* Schiner (Baker et al. 1944). It was not until 1952, however, when Steiner demonstrated the use of hydrolysed proteins and partially hydrolysed yeast in combination with organophosphate insecticides to control fruit flies, that researchers began to test the idea of attracticides. Protein-hydrolysate baits were first used in Hawaii for the control of the Oriental fruit fly, *Bactrocera dorsalis* (Hendel). The spray mixture contained protein hydrolysate, sugar, and parathion (Steiner 1952). In 1956, malathion (a fumigant, contact, stomach poison) was combined with protein bait to eradicate the *C. capitata* from Florida (Steiner et al. 1961). A mixture of technical grade malathion ($\geq$ 95%) and NuLure (Protein Insecticide Bait 7 or PIB-7) at a ratio of 1:4 (malathion:NuLure) was used to eradicate *C. capitata* from Brownsville, Texas (Stephenson and McClung 1966). Lopez et al. (1969) also used a malathion-bait combination at the same ratio to control the Mexican fruit fly *Anastrepha ludens* (Loew), applied at discrete locations; however, they did not elucidate the rationale for their 1:4 ratio. Later, Harris et al. (1971) showed that effectiveness of malathion to kill fruit flies was directly proportional to the ratio of malathion to NuLure; a ratio of 1:4 was more effective than other ratios tested. Thus, flies responding to the attracticide only needed fume exposure, contact, taste, or ingest mixture, whereupon they died in a short time. This may be the reason why the adequacy of the bait in the malathion-bait sprays was not questioned. The main reason for the wide use of malathion is its low mammalian toxicity as well as its low price. However, ultralow-volume (1 litre/ha) sprays contain concentrations of technical malathion near 20% a.i.

## NOVEL CHEMICAL APPROACHES

### Plant Growth Regulators

Plant resistance is a well-documented fact and plants can resist pest infestations as long as these are not severe. Resistance can be physical, nutritional, or chemical. With respect to fruit flies, these generally do not develop in green fruit. The reason may be that green fruit does not attract gravid fruit flies but if they arrive and oviposit, eggs or neonate larvae die due to natural chemical defenses. Without exception, all known commercial citrus fruit are not attacked by fruit flies until their physiological maturity – before the fruit begins to change colour. When fruit is completely ripe, it becomes very susceptible to fruit fly attack (Greany et al. 1983). However, chapote amarillo, *Sargentia gregii* Llave et Lex. (Baker et al. 1944), and white zapote, *Casimeroa edulis* Llave et Lex. (in the same family as citrus), are attacked by *A. ludens* from the time fruit sets. Larvae develop inside the seed of chapote amarillo if fruit is attacked when tender (Plummer 1941). With the discovery by Coggins (1973) that giberellic acid delays maturity of the rind in citrus fruits (maintains rind green for an extended period of time and inhibits pathogens from penetrating), Greany et al. (1987) proposed that the impact of giberellic acid could be used to inhibit oviposition by the Caribbean fruit fly, *Anastrepha suspensa* (Loew) on treated fruit. By that time the practice of applying giberellic acid to protect fruit against pathogens in orchards in California and Florida was well established. Greany et al. (1987) found that grapefruit and oranges were not infested with *A. suspensa* until fruit were over-ripe. They found the principle factors for resistance were thickness of flavedo (below the epidermis and contains most of oil cells), high concentrations of oxygenated terpenoids in oils and absolute quantity of oils in the rind; also, that *A. suspensa* oviposits principally in the flavedo and frequently found flies ovipositing in oil cells. Then, by spraying fruit on the tree with giberellic acid *A. suspensa* should not be able to oviposit. Greany et al. (1987) found that spraying fruit before it coloured, the rind maintained its firmness and in laboratory studies was less infested than control fruit. However, the model *(A. suspensa)* was not adequate to establish a principle because *A. suspensa* does not oviposit in fruit until it is over-ripe. Nguyen et al. (1992) confirmed this by sampling orchard fruit. They sampled 136,875 fruit in one zone and 201,000 in another and found that fruit infestation in both zones exceeded probit 9 ($\leq$ 0.0032%), these were 0.0022% and 0.0014%, respectively. The orchards received only general owner management. As Aluja (1996) points out, two facts can randomise results, orchards have to be perfectly managed and other flies can oviposit in batches and increase the number of "inadequate" fruit. *A. ludens* normally oviposits in batches between the albedo and the flavedo and *C. capitata* not only oviposits in batches but takes advantage of punctures made by other flies to oviposit (Rossler and Greany 1990). We have to be careful in adopting new strategies; though sophisticated, the effectiveness of a treatment can change from region to region due to factors that change tree physiology such as variety, weather profile, soil conditions and orchard management.

## Stomach Poisons

The use of stomach poisons against insects is also well established; what has changed over time is availability of toxicants and society's attitude towards them. Metcalf et al. (1962) point to various stomach poisons among the inorganic and organic insecticides. Here, we refer to compounds that are strictly stomach poisons and must be ingested to cause damage; consequently, they must be mixed in a bait before an insect ingests it. Borax has been used to kill larvae of the house fly, *Musca domestica* L., in manure and garbage and to prevent development of mosquitos in water. Boric acid is used to kill cockroaches. Ostensibly the mode of action of these two compounds is that boron crystals tear peritrophic membranes of the midgut causing dehydration. Chambers et al. (1987) proposed that boric acid or borax could be used to kill fruit flies instead of malathion. They found that 10% boric acid or borax caused similar mortality to malathion, except for its delayed action. They also found that flies fed 30% borax-NuLure only once caused more than 90% kill after 7 d, though flies were fed a normal diet. Also, surviving females lost sexual interest after 2 d. thus indicating that it is unnecessary to kill flies rapidly because they will not be able to reproduce even if they live an extended period of time.

Enkerlin et al. (1993) conducted laboratory and field tests using a mixture of boric acid-borax (3:1) on *A. ludens*. The mixture of these two compounds was based on the report by Kirk et al. (1979) which stated that in a concentrated solution, a mixture of boric acid-borax is more soluble in water. They found that by feeding the mixture in hydrolysed yeast and water to flies, this caused 90% kill in 22 h; they attributed this rapid kill to the boric acid-borax mixture as compared to the results obtained by Chambers et al. (1987). However, the results are not comparable because different proteins and concentrations were used. Results in field tests were similar among boric acid-borax mixture (8% a.i.) and malathion (0.57% a.i.). No phytotoxicity was noted. Although promising, boric acid and borax are poisonous and would affect any organism that ingest them and their long stability could cause long-term impact on the environment.

## Insect Growth Regulators

The use of insect growth regulators (IGRs) during susceptible periods in the development of insects can cause abnormal changes in morphology or physiology and cause death of treated insects. Unfortunately, the majority of IGRs have not been tested on fruit flies because immatures are not available targets, these being the most susceptible stages. Therefore, the only times that flies can be targeted is when they exit fruit to enter soil and these emerge as adults. The first instance is too short and for this reason, chemical control has targeted adults.

IGRs used to control insect infestations fall into two categories, those that mimic juvenile hormone and those that interfere with chitin synthesis. Several reports indicate that diflubenzuron, penfluron, triflumuron, and substituted benzylphenyls can sterilise dipterans (Knapp and Herald 1983, Broce and Gonzaga 1987). However, the effect of these types of compounds had not been studied until Lawrence (1983) tested diflubenzuron on *A. suspensa*.

Albrecht and Sherman (1987) used avermectin $B_1$ (avermectin $B_{1a}$ and avermectin $B_{1b}$) extracted from *Streptomyces avermectilis* on *C. capitata*, *B. dorsalis*, and *B. cucurbitae*. The compound has toxic properties but at the same time inhibits reproduction in a number of insect species. They made acetonic topical applications of avermectin on flies and caused great kill with high concentrations in 2-5 d. At low concentrations, kill occurred after 5 d and stabilised at 21 d after treatment. The $LD_{50}$/g for females was 0.021µg (15 d), 0.042 µg (15 d), 0.29 µg (10 d) for *B. dorsalis*, *B. cucurbitae* and *C. capitata*, respectively. According to Albrecht and Sherman (1987) *C. capitata* is more susceptible to avermectin $B_1$ than malathion, DDT, or heptachlor. They also found that fecundity and fertility were reduced and were irreversible. Avermectin $B_1$ did not affect mating and treated males mated to females did not affect fecundity. Hennessey and King (1996) tested abamectin (= avermectin $B_1$) on *A. suspensa*. They mixed 25% yeast hydrolysate with 75% sugar and fed it to flies with the desired amounts of abamectin mixed in. They killed 95% of females and males with 12.2 and 19.6 ppm, respectively, in 48 h. No field trials were conducted but the idea was interesting because abamectin is registered against the citrus rust mite, *Phyllocoptruta oleivora* Ashmead, in Florida (McCoy et al. 1994). King and Hennessey (1996), using the same methodology, also tested the efficacy of spinosad on *A. suspensa* and killed 99% of adult females and males with 9.4 and 5.8 ppm, respectively.

Cyromazine is an IGR that inhibits development and produces body larval distention, larviform pupae, and causes partial emergence of adults (Pochon and Casida 1983, Awad and Mulla 1984), and is effective against a great number of dipterans (Hart et al. 1982). Given this information Martínez and Moreno (1991) tested cyromazine on *A ludens*. Cyromazine was provided as a mixture in hydrolysed yeast-sugar (1:4) or in NuLure. Fecundity was greatly reduced in treated females, independent of male treatment. A single meal with 0.5% cyromazine in NuLure was sufficient to greatly reduce oviposition. In a separate study, Moreno et al. (1994) found that cyromazine caused great mortality at concentrations > 5%; surviving females (treated when < 4 d-old) found it difficult to oviposit on Parafilm sachets or grapefruit rinds. Later, a correlation was found among oviposition and hardness of substrate. Females fed 5% cyromazine oviposited only 5.5% that of control females on grapefruit rind with a hardness of 20.2 Newtons; there was no oviposition on navel or Valencia oranges nor in mango or peach, or plums with registered hardness of 22.4, 21.3, 18.3, 15.6, and 10.8 Newtons, respectively. This indicated that hardness per se is only part of the key in the interference. It was observed that part of the problem was physiological, as females wobble from side to side while trying to insert their ovipositors into substrates. If they succeeded, they were unable to inject eggs into the substrate. It was also found that when fructose was added to NuLure, lower concentrations of cyromazine were more effective. Thus, indicating that flies were consuming more food. Laboratory findings showed that cyromazine killed flies, reduced fecundity, and reduced larval development of $F_1$. In field cage studies, at a concentration of 1%, though there were few eggs laid, larvae died and no pupae were formed. In orchard spray tests, at a concentration of 5% cyromazine, no larvae were found in fruit sampled from trees and these were significantly reduced in fruit samples from ground as compared to a control. Results from treatments of 1% cyromazine and 5% malathion could not be separated from the control. As cyromazine was apparently more effective in the laboratory when fructose was added to NuLure, it indicates that the formulation can be improved and as positive results were obtained in small plots (three trees), better results can be expected by improving formulations. Díaz et al. (1996) found young (starting at 2 d old) fruit flies of West Indian, *A. obliqua* (Macquart), *A. serpentina* (Wiedemann), or *A. ludens* fed continuously on concentrations of 0.01-0.1% cyromazine, their fertility was affected. *A. ludens* fed 1% cyromazine-NuLure laid eggs in mangos but numbers of larvae were greatly reduced compared to a control. This indicates that a better understanding of cyromazine coupled with an improved bait formulation could produce improved results in the field.

Stark et al. (1992) tested cyromazine against three fruit fly species and their parasitoids in Hawaii and compared results to effects of diazinon. They mixed cyromazine with batches of clean sand and then introduced 3rd stage un- or parasitised larvae. They found that cyromazine did not inhibit pupation but diazinon did. At their $LC_{50}$, *B dorsalis* and *B cucurbitae* were more susceptible to cyromazine than diazinon. *C. capitata* was susceptible to both compounds. Parasitoid eclosion from larvae in diazinon-treated sand was affected in relation to concentration of diazinon, whereas, the proportion of parasitoids eclosing from larvae in cyromazine-treated sand was similar to eclosion in a control. Fecundity and fertility of *B. cucurbitae* surviving 0.5 ppm cyromazine were seriously affected. Diazinon affected $F_1$ production of *Diachasmimorpha tryoni* (Cameron) but this did not occur with cyromazine. This information indicates that cyromazine could work as well or better than diazinon but we do not know how well cyromazine can work in complex soils nor do we know if cyromazine would be present at the time when larvae exit fruit to bury themselves. Cyromazine is stable in soil from 3-6 months and is $\geq$ 10x less toxic than diazinon but its efficacy in soils is questionable. Stark et al. (1990) also studied the effect of azadiractin on same fruit flies in sand using same methodology as before; they found that azadiractin completely inhibited emergence of *C. capitata* and *B. dorsalis*, at concentrations of 14 mg/litre and of *B. cucurbitae* at 10 mg/litre. Interestingly, *B. cucurbitae* was more susceptible than other flies in both studies but, unfortunately, azadiractin is less stable than cyromazine and suffers from the same management problems.

## Phototoxic Dyes

A number of insect species exhibit photooxidative reactions when exposed to certain dyes (Heitz and Downum 1987, 1995). Some dyes, such as erythrosine B and phloxine B are food or pharmaceutical and cosmetic additives which have light-induced toxicity to certain species of insects. Several economic dipterans such as *M. domestica* and *M. autumnalis* (De Geer) have shown susceptibility to photoactive dyes (Fondren and Heitz 1978, Fondren et al. 1978, Burg et al. 1989, Pimprikar et al. 1980). Krasnoff et al. (1994) showed that erythrosin B had a light-induced toxicity to apple maggot, *Rhagoletis pomonella* Walsh. Erythrosin B has an oral $LD_{50}$ in rats of $\geq$ 6700 mg/kg as compared to malathion of $\geq$ 1375 mg/kg indicating a much greater safety margin. Phototoxic dyes could be classified as stomach poisons; yet, they are not poisons per se. They do not participate in the toxicological process, rather, they facilitate singlet oxygen which is highly oxidative. This means that dye has to be delivered in an acceptable manner to the target species and must be ingested. Moreno et al. (1994) encountered problems with food consumption with *A. ludens* in their cyromazine studies. This problem was largely overcome by adding 20% fructose to a NuLure formulation. They hypothesised that addition of fructose increased phagostimulatory properties of NuLure-cyromazine mixture. Therefore, the bait formulation, as the vehicle for the toxicant should: attract flies, be acceptable to the flies, be phagostimulant, and have nutritional characteristics for rapid physiological acceptance and absorption of the toxicant. Also, formulations should fall within conventional practices of handling similar products, be safe to mammals, and be relatively inexpensive.

## Formulation

As a consequence of a test in which 0.5% phloxine B in NuLure failed to reduce field populations of *A. ludens* as compared to a NuLure check and malathion-bait spray, Moreno and Mangan (1995) addressed the bait question and succeeded in developing a multicomponent bait that attracted flies, found the bait acceptable and phagostimulatory as compared to NuLure (attested by consumption studies), and is potentially physiologically acceptable as all the components are used by the food industry. They proposed a formulation consisting of 40 to 70% Mazoferm 802 as the attractive protein, 20% invert sugar as the phagostimulant, 1% SM9 as a penetrating adjuvant, 1% soybean oil as a stabiliser, 0.6% acetic acid for added attractiveness and preservative and optionally 0.4 to 1% Xanthan gum as a thickening agent. The photoactive dye phloxine B was added to the bait and it was found to be phototoxic to *A. ludens*. In addition, Moreno and Mangan (unpublished data) found 34 phototoxic dyes in laboratory tests on *A. ludens* and 11 of them are as, or more, toxic than phloxine B. The most phototoxic compound was methyl eosin; it was 10,000 times more toxic than phloxine B.

Also, Liquido et al. (1995) studied the effect of phloxine B with uranine (1:1) or phloxine B alone with various proteins alone or with sugars on *C. capitata* and concluded that 100% mortality of females was achieved when high dye concentrations ($\geq$ 1% dye) were mixed in the more phagostimulatory 20% autolysed ICN yeast hydrolysate or 20% Mazoferm, with or without 20% fructose. The dye did not perform as well in 10% molasses or 1% NuLure. In addition, they found that by combining phloxine B at concentration of 3.6 and 7.1% with methyl eugenol (1,2-dimethoxy-4-[2-propenyl]benzene), a potent attractant for males of *B. dorsalis*, they could kill this fly in less than 2 h.

## Adjuvants

In the search to increase the effectiveness of phloxine B, we addressed the question of membrane penetration. It is known that biological membranes are composed of lipid bilayers, integral proteins, and peripheral protein molecules held together by non-covalent interactions; these are highly selective permeability barriers. If interfacial tension can be broken, penetration of the membrane by hydrophilic molecules and ions can be increased. We tested many adjuvants but the most effective with phloxine B were Kinetic and polysorbate 60 (Figure 1). The adjuvants greatly increased the efficacy of phloxine B; $LC_{50}$ without any adjuvant was 1445 mg/litre, with SM-9 240 mg/litre, and with polysorbate 60 50 mg/litre (unpublished data). Field-cage tests verified laboratory results and showed that the addition of an adjuvant greatly increased the efficacy of phloxine B (Mangan and Moreno 1995). Also, Heitz, Mangan and Moreno (1998) showed that 0.5% phloxine B killed a mean of 80% *A. ludens* by using BaitStation in field cages; phloxine B killed flies faster with adjuvants than without them. In addition field tests with BaitStations showed that phloxine B killed flies at about the same rate as those seen in field cages.

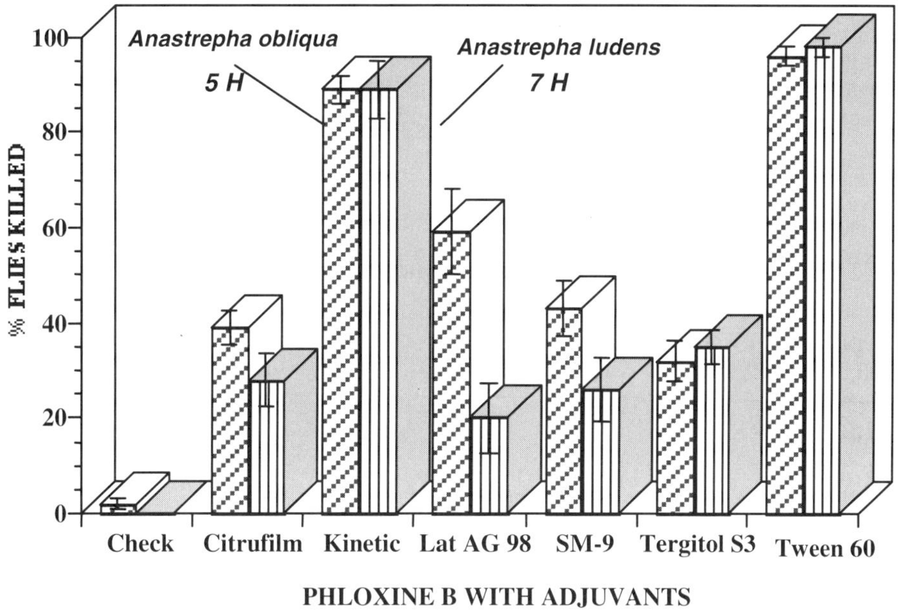


Figure 1. Mortality over time of West Indian fruit fly and Mexican fruit fly after feeding on 240 mg/litre of phloxine B in a Mazoferm 802 bait containing 1% adjuvants Citrufilm, Kinetic, Latron AG 98, SM-9, Tergitol S3, or Tween 60; flies were exposed to artificial light and received ≤ 300 µmol $s^{-1}$ $m^{-2}$ at the testing surface supplied by four 215 watt, 1500 ma ultra-high output fluorescent Vita-Lite bulbs (Duro-Test Corp.).

## Field Tests

From previous studies an improved dye-bait (0.5% phloxine B-70% Mazoferm 802 plus additives) formulation (unpublished data) was aerially released on *A. ludens*, *A. obliqua*, and *C. capitata* in Ataulfo mango orchards in Tapachula, Chiapas, Mexico, and found the efficacy of phloxine B as effective as that of malathion. However, dyes mixed with Captor 300 (product very much like NuLure) did not reduce populations below those of the check. The dye-bait was applied with backpack sprayers on feral *A. ludens* and *A. obliqua* in mixed plantings of mango orchards in Pololcingo, Guerrero, Mexico and found the dye-bait as effective as malathion-bait sprays in reducing fly populations. The dye-bait was applied with a high pressure (16.2 kg/sq. cm)) orchard sprayer as spot sprays (ca. 36 ml per tree, spot ca. 1 square metre) on *A. ludens* in orange orchards in Hualahuises, Nuevo Leon, Mexico and the dye-bait was as effective as malathion-bait sprays (Figure 2). The dye-bait was applied via fixed-wing aircraft on *A. ludens* in grapefruit orchards in Mission, Texas and it was found that the dye-bait was as effective as malathion-bait sprays. The dye-bait was applied via helicopter on *A. suspensa* in orange orchards in Arcadia, Florida and found to be as effective as malathion-bait sprays. Sterile flies were released in all studies, except those in Guerrero. In addition, application of the dye-bait via ground with high pressure equipment on feral *C. capitata* in citrus orchards in Morocco showed that the dye-bait controlled fly populations as well as a mixture of malathion-dimethoate-bait sprays. Field work conducted in Guatemala by McQuate and Peck (chapter within this book) with feral populations of *C. capitata* indicates that 1% phloxine B-uranine (1:1) or 0.5% phloxine B alone mixed with 40% Mazoferm and 20% fructose can reduce fly populations. However, there were no comparisons to the standard malathion-bait.

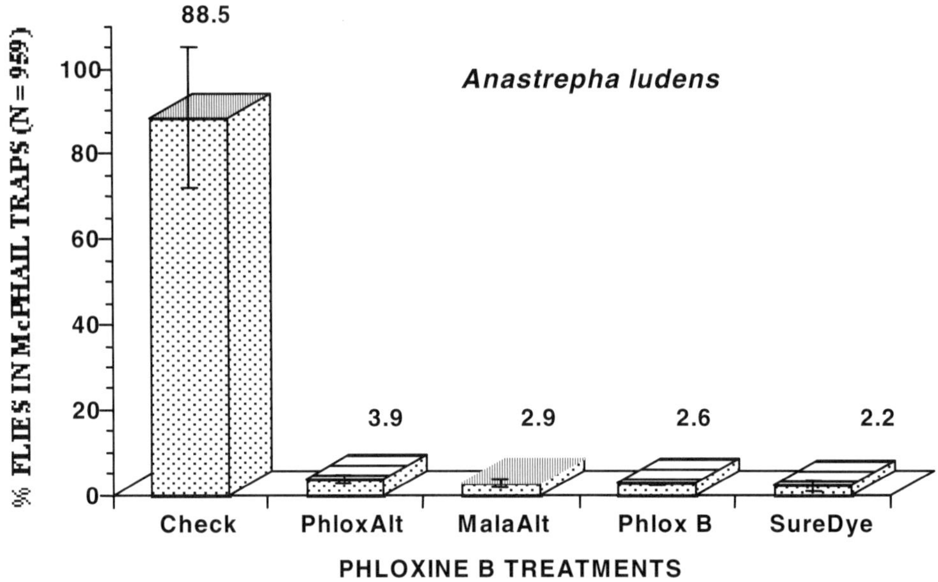


Figure 2. Effect of 0.5% phloxine B as alternate (Alt) and uniform treatments and SureDye uniform and malathion-NuLure alternate treatments, compared to a check Mazoferm 802 bait on released, sterile Mexican fruit flies in orange orchards, Hualahuises, Nuevo Leon, Mexico, 4 November 1997.

## Novel Chemistries

Novel insecticide chemistries, other than organophosphate or carbamate, are under evaluation with *A. ludens*, *A. obliqua* and *A. serpentina*, such as the abamectins, nicotinoids, pyrrols and others. Relatively recently a novel, naturally derived compound called spinosad has been found to be toxic against a large number of insect species. Spinosad is in a novel class of macrocyclic lactones produced by the soil actinomycete *Saccharopolyspora spinosa*. Spinosad is a contact and stomach poison and is particularly effective against Lepidoptera and Diptera. Spinosad has very low mammalian toxicity and is considered safe for many non-target organisms. Spinosad was tested against adult *C. capitata* at a concentration of 10 mg/litre and found that all flies died within 48 h (Adan et al. 1996). Spinosad kills slowly compared to malathion. Before initiating research with *A. ludens* we had to readdress the bait question because the one used in the field greatly reduced the efficacy of methyl eosin and because *R. pomonella* did not recognised such a bait. Once we improved the bait, we determined the $LC_{50}$ of phloxine B abamectin, emamectin, spinosad, imidacloprid, thiamethoxam and MK 243 on *A. ludens*. The data indicate that an estimated $LC_{50}$ = 0.159 mg/litre, was obtained for spinosad 72 h after feeding flies and this is possible by improving the bait developed for phloxine B studies (unpublished data). Because spinosad is a naturally-derived compound, it has very low mammalian toxicity, and great efficacy ($LC_{50}$ < 10 mg/litre) demonstrated in the laboratory against three different species of fruit flies, it warrants further investigations in the field.

## CONCLUSIONS

The information presented here indicates that novel chemical approaches can be used judiciously to control fruit flies. However, each approach should be selected according to agroecological needs, environmental impact, and societal acceptance of selected actions. The day has arrived when the indiscriminate use of insecticides has to cease and the few insecticides that we have left should be managed in a way that insect resistance is not induced.

From what we know about plant growth regulators, giberellic acid is the most effective compound for citrus fruit but 2-4-D could also be used alone or in combination with giberellic acid. However, we cannot say that giberellic acid would work in the same way as in citrus with mango or guava fruit because of their morphological and physiological ripening characteristics; certainly, the rind where all or most of the impact of giberellic acid occurs in citrus is different from the other two fruits. A few stomach poisons and IGRs appear promising, under certain conditions, to control fruit fly populations, as long as an appropriate bait is available for their use. The same can be said for phototoxic dyes, though most of the advances that we have today with baits were stimulated by the need to induce flies to eat a dye-laden bait, especially in situations where the toxicant has no contact activity. The efficacy of the toxicant is totally dependent on its ingestion and this can only happen if the toxicant has the qualities that were discussed previously. Baits are still under development and under the best of conditions we can hope to have a generic bait for most fruit flies. Dyes have a great potential from the ideological point of view because dyes are not poisonous per se. Dyes facilitate the evolution of singlet oxygen, when impacted by the sun's energy, which forms free radicals and these in turn damage cell membranes, organelles, enzymes, and cause oxidation of lipoidal tissues. When this happens, cell intra- and intercommunication ceases and physiological processes are curtailed, causing death of the affected insect. What we normally observe is that when a dye-fed fly exposes itself to the sun's energy, its activity increases, becomes agitated and disoriented, tries to regurgitate and defecate, extends its extremities forwards and backwards and with its front and back legs, attaches itself to leaf tissues or screens, and lastly appears to become paralysed and lays outstretched on its side and dies. Many dyes are food or pharmaceutical and cosmetic additives and properly formulated, do not attract beneficial insects such as honey bees, are rapidly degraded by the sun's energy, and do not persist long in the environment. Unfortunately, none of these strategies can be used alone to suppress fruit flies to the point of creating fruit fly-free zones or eradicate fly infestations. Every strategy has to be used in conjunction with other agronomic strategies in the proper integrated management of a crop, including requisites for targeted markets. Only a conscientious effort in the proper management of crops and pests can maintain an acceptable equilibrium in the ecosystem in which we live.

## REFERENCES


Adan, A., P. Del Estal, F. Budia, M. Gonzalez and E. Vinueala. 1996. Laboratory evaluation of the novel naturally derived compound spinosad against *Ceratitis capitata*. Pestic. Sci. 48: 261-268.

Albrecht, C.P. and M. Sherman. 1987. Lethal and sublethal effects of Avermectin $B_1$ on three fruit fly species (Diptera: Tephritidae). J. Econ. Entomol. 80: 344-347.

Aluja, M. 1996. Future trends in fruit fly management. pp. 309-320. *In* Fruit Fly Pests: A World Assessment of their Biology and Management. B.A. Mcpherson & G.J. Steck (eds). St. Lucie Press, Delray Beach, Florida.

Awad, T.I. and M.S. Mulla. 1984. Morphogenetic and histopathological effects induced by the insect growth regulator cyromazine in *Musca domestica* (Diptera: Muscidae) J. Med. Entomol. 21: 419-426.

Back, E.A. and C.E. Pemberton. 1918. The Mediterranean fruit fly in Hawaii. USDA Bull. 538, pp. 118.

Baker, A.C., W.E. Stone, C.C. Plummer and M. McPhail. 1944. A review of studies on the Mexican fruit fly and related Mexican species, USDA Misc. Publ. 521, pp. 155.

Broce, A.B. and V.G. Gonzaga. 1987. Effect of substituted benzylphenols and triflumuron on the reproduction of the face fly (Diptera: Muscidae) J. Econ. Entomol. 80: 37-43.

Burg, J.G., J.D. Webb, F.W. Knapp and A.H. Cantor. 1989. Field and laboratory efficacy studies of erythrosin B for *Muca domestica* (Diptera: Muscidae) and *Drosophila robusta* (Diptera: Drosophilidae) control. J. Econ. Entomol. 82: 171-174.

Chambers, D.L., H. Lorraine and R.V. Dowell. 1987. Boron as an alternative toxicant in protein bait for Mediterranean fruit flies. *In* Fruit Flies of Economic Importance. R. Cavalloro (ed.). A. A. Balkema, Rotterdam.

Coggins, C.W., Jr. 1973. Use of growth regulators to delay maturity and prolong shelf life of citrus. Acta Horticulturae, 34: 469-472.

Dethier, V.G. 1976. The Hungry Fly. Harvard University Press, Harvard. pp. 489.

Diaz, F., J. Toledo, W. Enkerlin and J. Hernandez. 1996. Cyromazine: Effects on three species of *Anastrepha*. pp. 333-337. *In* Fruit Fly Pests: A World Assessment of their Biology and Management. B.A. Mcpherson & G.J. Steck (eds.). St. Lucie Press, Delray Beach, Florida

Enkerlin, W, J. Reyes and R. Villalobos. 1993. Use of a mixture of boric acid, borax, hydrolyzed protein, and water to control *Anastrepha* fruit flies. *In* Fruit Flies: Biology and Management. M. Aluja & P. Liedo (eds.). Springer-Verlag

Fondren, J.E., Jr. and J.R. Heitz. 1978. Xanthene-induced toxicity in the adult face fly, *Musca autumnalis*. Environ. Entomol. 7: 843-846.

Fondren, J.E., Jr., B.R. Norment and J.R. Heitz. 1978. Dye-sensitised photoxidation in the house fly, *Musca domestica*. Environ. Entomol. 7: 205-208.

Greany, P.D., R.E. McDonald, P.E. Shaw, W.J. Schroeder, D.F. Howard, T.T. Hatton, P.L. Davis and G.K. Rasmussen. 1987. Use of gibberellic acid to reduce grapefruit susceptibility to attack by the Caribbean fruit fly, *Anastrepha suspensa* (Diptera: Tephritidae). Trop. Sci. 27: 261-270.

Greany, P.D., S.C. Styer, P.L. Davis, P.E. Shaw and D.L. Chambers. 1983. Biochemical resistance of citrus to fruit flies: Demonstration and elucidation of resistance to the Caribbean fruit fly, *Anastrepha suspensa*. Entomol. Exp. Appl. 34: 40-50.

Harris, E.J., D.L. Chambers, L.F. Steiner, D.C. Kamakahi and M. Komura. 1971. Mortality of tephritids attracted to guava foliage treated with either malathion or naled plus protein-hydrolyste bait. J. Econ Entomol. 64: 1213-1216.

Hart, R.J., W.A. Cavey, K.J. Ryan, M.B. Strong, B. Moore, P.L. Thomas, J.C. Boray and M. von Orelli. 1982. CGT-72662 – a new blowfly insecticide. Aust. Vet. J. 59: 104-109.

Heitz, J.R. and K.R. Downum, eds. 1987. Light activated pesticides. ACS Symposium Series 339, American Chemical Society, Washington, D.C.

Heitz, J.R. and K. Downum, eds. 1995. Light Activated Pest Control. ACS Symposium Series 616; American Chemical Society: Washington, D.C.

Heitz, J.R., R.L. Mangan and D.S. Moreno. 1998. Phototoxic insecticidal composition and method for controlling insect populations. U.S. Patent 5,798,112. Filed Sept.26,1996 issued 25 August, 1998.

Hennessey, M.K and J.B. King. 1996. Abamectin bait for Caribbean fruit fly (Diptera: Tephritidae). J. Econ. Entomol. 89: 987-989.

King, J.R and M.K. Hennessey. 1996. Spinosad bait for the Caribbean fruit fly (Diptera: Tephritidae. Fla. Entomol. 79: 526-531.

Kirk, E.R., F.O. Othner and A. Standen. 1979. Enciclopedia Tecnica Quimica. Mexico UTEHA, Tomo III.

Knapp, F. W. and F. Herald. 1983. Mortality of eggs and larvae of the face fly (Diptera :Muscidae) after exposure of adults to surface treated with BAY SIR 8514 and penfluron. J. Econ. Entomol. 76: 1350-1352.

Krasnoff, S.B., A.J. Sawyer, M. Chapple, S. Chock and W.H. Reissig. 1994. Light-activated toxicity of erythrosin B to the apple maggot (Diptera: Tephritidae) and reevaluation of analytical methods. Environ. Entomol. 23: 738-743.

Lawrence, P.O. 1983. Age-specific fecundity and offspring survivorship in the Caribbean fruit fly, *Anastrepha suspensa*, after diflubenzuron treatment. Insect Sci. Appl. 4: 285-290.

Liquido, N.J., G.T. McQuate and R.T. Cunningham. 1995. Light-activated toxicity of phloxine B and fluorescein in methyleugenol to Oriental fruit fly, *Bactrocera dorsalis* (Hendel) (Diptera: Tephritidae), males. *In* Light Activated Pest Control. J.R Heitz. & K. Downum (eds.). ACS Symposium Series 616; American Chemical Society: Washington, D.C.

Lopez-D, F., D.L. Chambers, M. Sanchez-R. and H. Kamasaki. 1969. Control of the Mexican fruit fly by bait sprays concentrated at discrete locations. J. Econ. Entomol. 62: 1255-1257.

Mangan, R.L. and D.S. Moreno. 1995. Development of phloxine B and uranine bait for control of Mexican fruit fly. *In* Light Activated Pest Control. J.R Heitz. & K. Downum (eds.). ACS Symposium Series 616; American Chemical Society: Washington, D.C.

Marsh, H.O. 1910. Report of the Assistant Entomologist, Board Commissioners Agr. Forest. Hawaii, pp. 152-159.

Martinez, A.J. and D.S. Moreno. 1991. Effect of cyromazine on the oviposition of Mexican fruit fly (Diptera: Tephritidae) in the laboratory. J. Econ. Entomol. 84: 1540-1543.

McCoy, C.W., C.C. Childers, P.A. Stansly, R.C. Bullock and J.L. Knapp. 1994. Citrus rust mite. *In* Florida Citrus Pest Management Guide. J.L. Knapp (ed.). University of Florida, Gainesville.

McPhail, M. 1937. Relation of time of day, temperature and evaporation to attractiveness of fermenting sugar solution to Mexican fruit fly. J. Econ. Entomol. 30: 793-799.

Metcalf, C.L., W.P. Flint and R.L. Metcalf. 1962. Destructive and Useful Insects. McGraw-Hill, New York.

Moreno, D.S. and R.L. Mangan. 1995. Response of the Mexican fruit fly (Diptera: Tephritidae) to two hydrolysed proteins and incorporation of phloxine B to kill adults, Supplement. *In* Light Activated Pest Control. J.R Heitz. & K. Downum (eds.). ACS Symposium Series 616; American Chemical Society: Washington, D.C.

Moreno, D.S., A.J. Martinez and M. Sanchez Rivello. 1994. Cyromazine effects on the reproduction of *Anastrepha ludens* (Diptera: Tephritidae) in the laboratory and in the field. J. Econ. Entomol. 87: 202-211.

Nguyen, R., C. Poucher and F.R. Brazzel. 1992. Seasonal occurrence of *Anastrepha suspensa* (Diptera: Tephritidae) in Indian River County Florida, 1984-1987. J. Econ. Entomol. 85: 813-820.

Pimprikar, G.D., J.E. Fondren, Jr. and J.R. Heitz. 1980. Small- and large-scale field tests of erythrosin B for house fly control in caged layer chicken houses. Environ. Entomol. 9: 53-58.

Plummer, C.C. 1941. The yellow chapote, a native host of the Mexican fruit fly. U.S. Dept. Agric. Tech. Bul. 775, pp. 12.

Pochon, J.-M. and J.E. Casida. 1983. Cyromazine-sensitive stages of house fly development: influence of penetration, metabolism and persistence on potency. Entomol. Exp. Appl. 34: 251-256.

Rossler, Y. and P.D. Greany. 1990. Enhancement of citrus resistance to the Mediterranean fruit fly. Entomol. Exp. Appl. 54:89-96.

Stark J.D., R.I Vargas, R.H. Messing and M. Purcell. 1992. Effects of cyromazine and diazinon on three economically important Hawaiian tephritid fruit flies (Diptera: Tephritidae) and their endoparasitoids (Hymenoptera: Braconidae). J. Econ. Entomol. 85: 1687-1694.

Stark, J.D., R.I. Vargas and R.K. Thalman. 1990. Azadirachtin: Effects on metamorphosis, longevity, and reproduction of three tephritid fruit fly species (Diptera: Tephritidae). J. Econ. Entomol. 83: 2168-2174.

Steiner, L.F. 1952. Fruit fly control in Hawaii with poison-bait sprays containing protein hydrolysates. J. Econ. Entomol. 45: 838-843.

Steiner, L.F., G.G. Rohwer, E.L. Ayers and L.D. Christenson. 1961. The role of attractants in the recent Mediterranean fruit fly eradication programme in Florida. J. Econ. Entomol. 54: 30-35.

Stephenson, B.C. and B.B. McClung. 1966. Mediterranean fruit fly eradication in the Lower Rio Grande Valley. Bull. Entomol. Soc. Am. 12: 374.

# Madeira-Med, a Sterile Insect Technique Programme for Control of the Mediterranean Fruit Fly in Madeira, Portugal

*R. Pereira, A. Barbosa, N. Silva, J. Caldeira, L. Dantas and J. Pacheco*

Programa Madeira-Med, Direcção Regional de Agricultura, Madeira, Portugal.

## INTRODUCTION

The islands of Madeira are located 980 km west-southwest from mainland Portugal and have a population of approximately 255,000. The islands are volcanic with very little level land suitable for large agricultural production. Approximately 47% of the land area is above 700 m. Thus the area likely to require Medfly control is about half of the islands.

Agricultural production is on small scale, frequently part-time and mostly terraced because of the volcanic nature of the land. Grapes for wine and bananas are the predominant fruit crops. Neither are primary Medfly hosts.

Citrus and tropical fruits are not produced in large quantities and are generally not of high quality. This is, to a large extent, because intensive Mediterranean fruit fly (Medfly), *Ceratitis capitata* (Wied.), attack has prevented the establishment of citrus and tropical fruit production. Medflies are present the year round on land below 300 m, resulting in the necessity of continuous control measures, usually insecticide bait sprays. Current annual losses from the Medflies in Madeira are estimated at US$3 million.

In 1992, the agricultural officials of Madeira applied for an European Union (EU) grant to eliminate the Medfly from Madeira using the sterile insect technique (SIT). After extensive discussions, the project was changed from eradication to control and approved in late 1993 with EU support of about 8 million ECU over a 7-year period. Subsequently, the Madeira officials applied for, and received, a technical assistance project from the International Atomic Energy Agency (IAEA). The IAEA support is primarily for training and consultant services.

Sterile female Medflies puncture fruits when they try to lay eggs. These punctures, called "sterile stings", result in a reduced crop value. For this reason, the Madeira-Med programme will use only sterile male Medflies in its SIT programme. This not only eliminates the sterile sting problem but also increases the efficacy of the sterile males from two to three fold. Further, there is a reduction in other programme costs when only sterile males are used.

The shift from eradication to control using SIT is significant as it represents the first planned use of SIT for Medfly control. Eradication has been the objective of all other Medfly SIT programmes.

The objective of this technique is the Medfly control in Madeira and Porto Santo Islands, to levels below the economic threshold.

Madeira-Med has operational units for field activities, Medfly mass production, quality control, fly handling and releases, public relations and administration.



**Area-Wide Control of Fruit Flies and Other Insect Pests,**
ed. K.H. Tan. Penerbit Universiti Sains Malaysia, Penang, 2000.

## FIELD ACTIVITIES

Ecological data on the Medfly population distribution, numbers, host preference and overwintering are being collected (Pereira et al. 1996). Surveys of Medfly adults and larvae, host species and phenology, temperature and Medfly history will all be used to guide the control programme, specifically the sterile fly releases.

### Climatic Conditions

The two populated islands of Madeira are located at 32° N in latitude and 17° W in longitude. Porto Santo is small (about 50km$^2$) with topographic and temperature conditions favourable to the Medfly. However, poor soil and low precipitation (380 mm a year) do not permit an abundance of host fruits. The maximum elevation is 517 m.

The major island (Madeira) is 740 km$^2$, has variable climatic conditions depending on altitude and/or whether it is north or south of the island. The north coast is cooler than the south (sea level in the north corresponds to 300 m elevation in the south). The maximum elevation is 1900 m.

Conditions in Funchal (the capital city) and areas in low altitudes are very favourable for the breeding of the Medfly throughout the year. Eight generations per year of *C. capitata* are produced in the Funchal area (Vieira 1952).

### Adult Survey

A study was started in 1994 to determine relative Medfly populations at different altitudes. Two replicate sites, each at four elevations (0 - 200 m; 200 – 400 m; 400 – 600 m and 600 – 800 m) both on the north and south coasts, as well as at Porto Santo, were included in the study. Each replicate consisted of 20 Jackson traps, 10 baited with trimedlure and 10 with protein hydrolysate (Pereira and Carvalho 1994, 1996). The traps were placed in the most prevalent hosts at the selected locations and remained in these hosts throughout the study. The 340 Jackson traps were serviced bi-weekly. The protein bait was replaced during each visit and the trimedlure every two months.

A total of 226,393 Medflies were captured in 1995, 189,249 in 1996 and 275,061 in 1997.

Fly captures decreased as elevation increased, mainly because of cooler temperatures. In general, Medfly populations are higher in the south than in the north because of the warmer weather and a wider variety of hosts in the south.

In the winter months, the populations are high up to 400 m above sea level in the south and up to 200 m above sea level in the north. In March and April, the pest exists only below 400 m in the north and 600 m in the south. From the beginning of May, the Medfly can be found at all altitudes.

In Porto Santo, the population remains low but constant during the first seven months of the year. Increases in August are because of ripening figs, the most important host on Porto Santo.

### Larval Survey

Larval surveys were started in late 1995 to evaluate the level of infestation in different hosts at different altitudes. The fruits were held in the laboratory for the larvae to mature and leave the fruits. Larvae were recorded as number of larvae per fruit and number of larvae per kg of fruit.

In 1996 2,412 fruit samples, totalling about 660 kg, were collected and yielded 30,702 larvae.

In 1997, (until July) 2,088 fruit samples weighing about 726 kg, yielded 78,349 larvae.

The Medfly was reported to attack more than 40 species of fruit in the Madeira Islands (Pereira et al., in press, Vieira 1952).

## MASS PRODUCTION

To implement the Madeira-Med programme, it was necessary to build a factory to produce 50 millions insects per week. This estimated production level was based on other Medfly SIT programmes and knowledge of Medfly biology, host numbers and damage in Madeira.

The objective of the mass production unit is to produce high quality sterile male Medflies for the Madeira-Med Medfly control programme. This involves mass Medfly production, process-control of the rearing ingredients and sterilisation of the Medflies that will be released.

This Medfly factory is the only one in Europe and the first designed to produce a temperature sensitive lethal *("tsl")* strain of *C. capitata*. A *tsl* strain allows the production of males only.

The IAEA has developed Medfly strains (genetic sexing strains) which permit the release of sterile males only (Hendrichs et al. 1995). Data to date demonstrate that the release of males only increases the efficacy of SIT from two- to four-fold. Thus the number of sterile males required when using a genetic sexing strain is much less than when a bisexual strain is used. Aerial release costs are reduced with the genetic sexing strain. If a genetic sexing strain based on killing females in the egg or at the neonate larval stage is used, rearing costs are reduced (Kerremans and Franz 1994).

The Madeira-Med rearing factory is self-contained with a biologically secure entrance and exit to prevent the escape of fertile Medflies. It is anticipated the factory will operate seven days a week (Barbosa et al., in press).

After they are mass produced, Medfly male pupae are sterilised with 145 Gy (14.5 krad) in a Cobalt 60 source.

Production in May 1998 is about 18 million/week. The plan is to increase production to about 40 million in the fourth quarter of 1998. The requirements for the Madeira-Med programme are shown in Table 1. Sterile males not used in Madeira will be made available to other programmes on a non-profit basis.

## QUALITY CONTROL

Quality control is exercised to ensure that high quality sterile male Medflies are released. This includes quality of the Medfly strain, diet ingredients, rearing conditions, sterile flies at the factory door and sterile flies released.

Field cage tests to measure the compatibility of the mass produced genetic sexing strains with wild Madeira Medflies have been conducted. The strains are compatible.

Competitiveness of the factory strain against the wild Madeira strain is evaluated in field cage tests.

The standard quality control tests of the factory produced insects are based on the results of an IAEA sponsored meeting held in May 1997. This meeting provided guidelines for standard evaluation of insect quality based on laboratory and field cage tests. The tests used in Madeira are: weight of the pupae; sex ratio; percentage of adult emergence; percentage of adults that can fly; longevity without food and water and competitiveness of sterile males with wild population in field cages.

## FLY HANDLING AND RELEASES

After the mature pupae are sterilised with gamma radiation, they are transferred to the fly handling facility for packaging, emergence and preparation for dispersal. A chilled fly dispersal system developed by the US Department of Agriculture is used. The sterile male Medflies are dispersed by air on pre-determined flight patterns, altitudes and frequencies. Sterile flies will be dispersed twice per week over all parts of the islands receiving SIT treatments.

Table 1 and Figure 1 show treatment areas, release densities and weekly total male releases per week.

Table 1. Area and release density of sterile males per hectare per week in the autonomous region of Madeira.

| Coast and area of releases | Area ($km^2$) | Release density (Nº males/ $km^2$) | Nº of sterile males to release (x $10^3$) |
|---|---|---|---|
| 0-400 m, North and South Coast | 175 | 150,000 | 26,250,000 |
| 400-600 m, South Coast | 79 | 50,000 | 3,950,000 |
| Porto Santo | 33 | 50,000 | 1,650,000 |
| Total | 287 | - | 31,850,000 |

Preliminary aerial dispersal started in May 1998 with a twin engine contract aircraft. Full coverage of Madeira and Porto Santo is planned for the fourth quarter of 1998.

The Madeira-Med programme will use aerial dispersal because it is more efficient, more effective and cheaper than ground release. Ground release will be used in a few locations on Madeira where aerial release is not advisable.

## PUBLIC RELATIONS

The Medfly has been present in Madeira for a long time. Rural residents and producers of fruit that are Medfly hosts are well aware that the Medfly is a serious pest. However, the majority of the Madeira population are urban and are not aware of the damage caused by the Medfly.

The objective of public relations activities for Madeira-Med activities is to ensure public support for the project. These activities must thus sensitise the Madeira people to the Medfly problem and encourage their direct assistance in the project. Video, radio, print and personal contacts are currently used in the public relations campaign. It is planned to use television in the near future.

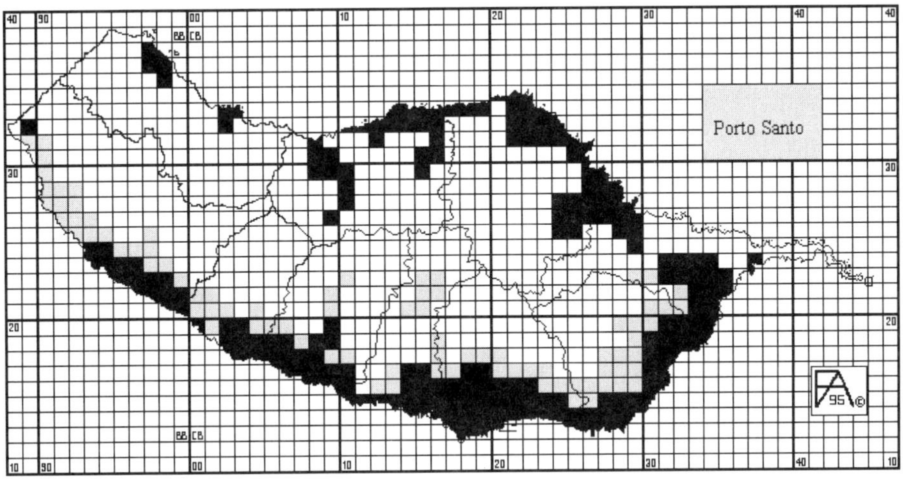


Low density (50,000 flies/ $km^2$/week) - 112 $km^2$, 6 600 000 (including Porto Santo 33 $km^2$ ).

High density (150,000 flies/ $km^2$/week)-175 $km^2$, 26 250 000.

Figure 1. Planned releases (aerial and ground) on Madeira and Porto Santo Islands at maximum activity.

The main contacts are through school children, agricultural fairs and meetings, other rural meetings, various expositions, and limited group tours through the Madeira-Med facilities. Included in these presentations are demonstrations of trapping and fruit sampling and a presentation of a video and slides describing the Madeira-Med programme. Posters and pamphlets are used extensively.

## CONCLUSION

The overall strategy of Medfly control for Madeira is to apply sterile males in areas of the islands where Medfly overwintering occurs and the pest is thus active the year round. If correctly conducted, the Medfly will not move into areas where temperatures are too cold for year round reproduction. It is estimated that about half of the islands will require treatment and of this, about half will require year round treatment. Confirmation of these assumptions must await collection of additional ecological data and sterility data from initial releases of sterile male Medflies.

Of particular importance in Madeira is the Medfly population in urban areas. Madeira is densely populated. Most people live in individual houses with Medfly hosts in their gardens. This poses problems with regard to reducing the Medfly population sufficiently low so that sterile males can be used economically for Medfly control. Special provisions must thus be made to reduce Medfly populations in urban areas. This will require strong support from the urban human population.

## ACKNOWLEDGEMENTS

MADEIRA-MED is financially supported by the government of Madeira, the European Union through POSEIMA (Programa de Opções Especificas para Fazer Face ao Afastamento e Insularidade da Madeira e Açores), REGIS II and the IAEA (International Atomic Energy Agency).

## REFERENCES

Barbosa, A., J.P. Carvalho and A. Brazão. 1995. The Madeira Medfly factory. 1st International Symposium on Biological Control in European Islands. 23-29 Set 1995 Ponta Delgada - Açores.

Hendrichs, J., G. Franz and P. Rendon. 1995. Increased effectiveness and applicability of the sterile insect technique through male-only releases for control of Mediterranean fruit flies during fruit seasons. J. Appl. Entomol. 119: 371-377.

Kerremans, P. and G. Franz. 1994. Cytogenetic analysis of chromosome 5 from the Mediterranean fruit fly, *Ceratitis capitata*. Chromosoma, 103: 142-146.

Pereira, R. and J.P. Carvalho. 1994. Utilização em Portugal de sistemas de armadilhas na luta contra Ceratitis capitata (Wied.) em citrinos. Bulletin OILB, 17 (6): 139-148.

Pereira, R. and J.P. Carvalho. 1996. Trap utilisation on study of Mediterranean fruit fly populations at citrus groves in Portugal. pp 135-140. *In* Fruit fly pests - a world assessment of their biology and management B.A. McPheron and G.J. Steck (eds.). St. Lucie Press, Delray Beach, Florida.

Pereira, R., A.M. Barbosa, A. Brazão, J.P. Carvalho, and D. Lindquist. 1996. Area-wide control of the Mediterranean fruit fly on Madeira with sterile insect technique. Proc. Int. Soc. Citriculture 568-572.

Pereira, R., A. Barbosa, A. Brazão, N. Silva, J. Pacheco, A. Rodrigues, J. Caldeira, L. Dantas, J.P. Carvalho and D. Lindquist. 1997. Madeira-Med, SIT programme against Medfly in Madeira, Portugal. II Internacional Open Meeting on mouches des fruits d'importance economique. Lisboa, Sep.

Vieira, R. 1952. A mosca da Fruta (*Ceratitis capitata*, Wied.) na Ilha da Madeira. Grémio dos Exportadores de Frutas da Ilha da Madeira.

# Regional Approach to the Management of Fruit Flies in the Pacific Island Countries and Territories

*Allan Allwood*

Project on Regional Management of Fruit Flies in the Pacific, Secretariat of the Pacific Community, Private Mail, Suva, Fiji

Of the 4,500 species of fruit flies (family Tephritidae) world-wide, over 350 species occur in the Pacific region. Of these, at least 25 species are regarded as being of major economic importance to fruit and vegetable production and to international trade within the region.

Recognition of the economic importance of fruit flies to horticultural production and trade increased markedly in the 1980s due to the imposition of restrictions on the use of ethylene dibromide (EDB) fumigation by trading partners. This treatment was the mainstay of quarantine treatments for fresh fruits and vegetables susceptible to fruit fly infestations and destined for markets in Australia, New Zealand, USA, Japan and Canada. Small, but economically significant, markets for fresh fruits and vegetables in the Pacific rim countries disappeared because alternative quarantine treatments for EDB fumigation were not available. Countries, such as Cook Islands, Fiji, Tonga and Samoa, looked for modern technologies to overcome these constraints to export. As well as quarantine treatment technologies, procedures new to the Pacific Island countries, such as quality assurance systems and quarantine pathways, had to be included into the production and marketing chains. Quarantine surveillance, particularly for exotic fruit flies, became a prerequisite for trade in fresh fruits and vegetables.

The emphasis on fruit flies also regionally increased because of the increasing number of incursions of exotic fruit flies into the region over the past 10-12 years. Outbreaks of exotic fruit flies in the Solomon Islands (1984-85), Nauru (1984-85), Northern Australia (1995 and 1998), New Zealand (1996), French Polynesia (1995-96), and Palau (1995-96) demonstrated the vulnerability of the Pacific Island countries and territories (PICTs) to these incursions. To address the increased threat of introduction of exotic fruit flies through increased tourism and regional travellers, a regional approach to the management of fruit flies was adopted.

This paper gives a brief insight into the region encompassing the PICTs. It also summarises the reasons for involvement in fruit fly research activities in the region, the agencies involved in the funding and research, the approaches adopted to overcome the constraints to production and trade and the challenges in the future.

## THE PACIFIC ISLAND REGION

The Pacific Island region includes the countries and territories from Palau in the west (130° east longitude) to Pitcairn Island in the east (125° west longitude), from the Northern Mariana Islands in the north (20° north latitude) to New Caledonia in the south (25° south latitude). It encompasses 22 countries and territories. The region straddles the Equator and also the International Date Line which runs between Samoa and Tonga.

The PICTs are separated culturally into Micronesia, Polynesia and Melanesia (Table 1). Although the people in these countries are different culturally, there are many similarities and common attitudes. For example, the custom of taking fruits and vegetables to friends and relatives as gifts in other islands or countries is common throughout the Pacific and poses a serious threat to quarantine security.

Table 1. Pacific island countries and territories listed in cultural groupings.

| Micronesia | Polynesia | Melanesia |
|---|---|---|
| Federated States of Micronesia | American Samoa | Fiji |
| Guam | Cook Islands | New Caledonia |
| Kiribati | French Polynesia | Papua New Guinea |
| Marshall Islands | Niue | Solomon Islands |
| Nauru | Pitcairn | Vanuatu |
| Northern Mariana Islands | Samoa | |
| Palau | Tokelau | |
| | Tonga | |
| | Tuvalu | |
| | Wallis and Futuna | |

The PICTs vary in geographic size, populations and population densities (Table 2). Several of the small islands are highly populated, e.g. Nauru. The economies are agriculturally-based, with subsistence production to guarantee food security a high priority. Many governments have developed policies to increase food production, especially production of fresh fruits and vegetables, to overcome nutritional health disorders, such as diabetes, coronary diseases and vitamin A deficiency.

Table 2. Land areas, populations and population densities of selected Pacific Island countries and territories.

| Country | Land area ($km^2$) | Population | Population density (People per $km^2$) |
|---|---|---|---|
| Papua New Guinea | 462,243 | 3,951,500 | 9 |
| Fiji | 18,272 | 777,700 | 43 |
| French Polynesia | 3,521 | 218,000 | 62 |
| Tonga | 747 | 98,000 | 150 |
| Nauru | 21 | 10,600 | 505 |

Vegetation on each island group varies markedly with the larger islands in the western areas, such as Papua New Guinea (PNG) and Solomon Islands, having greater plant diversity than the smaller islands in the east. The larger islands have large areas of tropical and sub-tropical rainforests at various altitudes. These differences in the plant diversity are reflected in the numbers of fruit fly species present. For example, PNG and Solomon Islands, with very rich and diverse flora, have over 180 and 48 species of fruit flies respectively. In contrast, Fiji in the central Pacific and Cook Islands in the eastern Pacific have five and two species of fruit flies, respectively.

All countries are subject to natural disasters such as cyclones and floods. Droughts caused by the effects of El Nino resulted in major food and water shortages, especially in small atoll countries, during 1997 and 1998. Some of the low atoll countries are in danger from rising ocean levels.

## HISTORY OF FRUIT FLY ACTIVITIES AND AGENCIES AND PICTs INVOLVED SINCE 1990

The PICTs recognised that they had neither the technical capacity nor the human and financial resources to resolve the problem caused by the presence of economically important fruit fly species and the quarantine restrictions imposed by trading partners. Consequently, the governments of the PICTs sought assistance from the Secretariat of the Pacific Community (formerly South Pacific Commission) (SPC), and the Food and Agriculture Organisation of the United Nations (FAO) to assist with developing a project and to initiate the research. Funding was obtained initially from the Australian Agency for International Development (AusAID) and the United Nations Development Programme (UNDP) from 1990-91. The Australian Centre for International Agricultural Research (ACIAR) and the Crawford Fund for International Agricultural Research (CFIAR) contributed to the funding, research and training from 1991-92. The New Zealand government and the United States Department of Agriculture-Agricultural Research Service (USDA-ARS) provided technical support and/or funding from 1991 and 1992, respectively. Expertise has been provided through FAO, USDA-ARS, the Queensland Department of Primary Industries (QDPI), Griffith University, the New Zealand government and SPC.

One of the strengths of the fruit fly projects in the Pacific region has been the cooperative effort between the donors identified above, the agencies that provided technical assistance and the national staff within each country or territory. The cooperation has resulted in minimal duplication of effort and resources and effective inputs by a wide range of experts covering quarantine surveillance, identification of species, biology/ecology, field control, quarantine treatment technology, quality assurance, quarantine preparedness, emergency response planning, eradication techniques and training.

The countries and territories involved in the fruit fly activities under the auspices of the FAO/AusAID/UNDP/SPC Project on Regional Management of Fruit Flies in the Pacific (RMFFP) are:

| | |
|---|---|
| 1990 - 1993 | Cook Islands, Fiji, Tonga, Samoa. |
| 1994 - 1997 | Cook Islands, Fiji, Tonga, Samoa, Federated States of Micronesia (FSM), Solomon Islands, Vanuatu. |
| 1997 - 2000 | All 22 PICTs.<br>[Major effort in PNG by ACIAR/Griffith University (1998 - 2002) and RMFFP (1997 - 2000)] |

## MAJOR ACHIEVEMENTS OF THE RMFFP AND OTHER PROJECTS

The objectives of the RMFFP and other projects in the Pacific are to improve the knowledge of fruit flies in the Pacific region, to develop environmentally sound field control systems for fruit flies, to establish quarantine surveillance systems nationally and regionally, to improve emergency response preparedness nationally and regionally, to formulate quarantine treatments to replace EDB fumigation and to upgrade technical skills of national staff.

### Improved Knowledge on Fruit Flies in the Pacific Region

Permanent trapping stations using methyl eugenol and cuelure are located in most countries, set up initially to assist with fruit fly fauna surveys. Once the fauna is determined, the trapping intensity and array are revised and traps relocated into high risk areas as an early warning system for exotic fruit flies. The early warning systems or quarantine surveillance systems focus on tourist resorts, diplomatic missions, urban areas, refuse dumps, airports, yacht and commercial harbours and agricultural production areas.

Detailed information on the species of fruit flies present in each country is now available (Drew and Romig 1996, Purea et al. 1996, Heimoana et al. 1996, Tora Vueti et al. 1996, Leblanc 1996, Amice and Sales 1996, Allwood et al. 1996, Vagalo et al. 1996, Tenakanai 1996). The trapping programme also provides data on seasonal abundance. Table 3 contains the numbers of species present in each country, with the major economic species listed.

Since 1990, over 25,000 samples of commercial/edible and wild fruits and vegetables, amounting to over 14 tonnes, have been collected and held in laboratories in seven project countries. Data on host ranges have proved valuable in demonstrating the pest status of fruit fly species during negotiations with trading partners. Specific studies on the losses caused by fruit flies have also been undertaken in Cook Islands, Fiji, FSM, Papua New Guinea, Tonga, Solomon Islands and Vanuatu (Allwood and Leblanc 1996, Hamacek 1996). Depending on the country and the fruit fly species, losses to guavas range from 40% to 99%; to mangoes from 20% to 27%; to breadfruit from 35% to 71%; to capsicum (excluding chillies) in Tonga from 97% to 100%; and to chillies in Tonga from 95% to 97%.

Data on the parasitoid fauna of fruit flies in project countries have been determined from the host surveys. *Fopius* (*Biosteres*) *arisanus* (Sonan), the egg parasite, is the most common species. Natural parasitism levels are generally low (about 10%), but increase towards the end of major fruiting periods of fruits such as Pacific almond *(Terminalia catappa* L.) and guavas. Levels of parasitism up to 60% have been recorded in Cook Islands (Kassim, personsal communication). Waterhouse (1993) indicated that parasitoids were established in nine PICTs. However, it seems that many of the parasitoid releases made in the 1950s were not successful or *F. arisanus* has become dominant.

## FIELD CONTROL SYSTEMS

Field control systems for fruit flies have been transferred into the Pacific region. These were modified from systems used in Malaysia and Australia. Farmers and

Table 3. Numbers of species of fruit flies (Tephritidae: Dacinae) in the Pacific region and a list of major economic species in each country.

| Country | Number of species | Major economic species |
|---|---|---|
| American Samoa | 4 | *Bactrocera kirki (Froggatt), B. xanthodes (Broun)* |
| Cook Islands | 2 | *B. melanotus* (Coquillett), *B. xanthodes* |
| FSM | 1 | *B. frauenfeldi* (Schiner) |
| Fiji | 5 | *B. passiflorae* (Froggatt), *B. xanthodes* |
| French Polynesia | 4 | *B. dorsalis* (Hendel), *B. kirki, B. tryoni* (Froggatt) |
| Guam | 2 | *B. cucurbitae* (Coquillett) |
| Kiribati | 1 | *B. frauenfeldi* |
| Nauru | 4 | *B. cucurbitae, B. dorsalis, B. frauenfeldi, B. xanthodes* |
| New Caledonia | 13 | *B. curvipennis* (Froggatt), *B. psidii* (Froggatt), *B. tryoni, B. umbrosa* (Fabricius) |
| Niue | 4 | *B. kirki, B. passiflorae, B. xanthodes* |
| Northern Mariana Islands | 2 | *B. cucurbitae* |
| Palau | 4 | *B. dorsalis, B. frauenfeldi, B. umbrosa* |
| Papua New Guinea | 180 | *B. atrisetosa* (Perkins), *B. bryoniae* (Tyron), *B. cucurbitae, B. decipiens* (Drew), *B. frauenfeldi, B. moluccensis* Perkins, *B. musae* (Tyron), *B. neohumeralis* (Hardy), *B. papayae* Drew & Hancock, *B. strigifinis* (Walker), *B. trivialis* (Drew), *B. tryoni, B. umbrosa* |
| Pitcairn | 1 | *B. tryoni* |
| Samoa | 7 | *B. kirki, B. xanthodes* |
| Solomon Islands | 48 | *B. cucurbitae, B. frauenfeldi, B. umbrosa, Dacus solomonensis Malloch* |
| Tokelau | 1 | *B. passiflorae* |
| Tonga | 5 | *B. facialis* (Coquillett), *B. kirki, B. xanthodes* |
| Tuvalu | 1 | *B. passiflorae* |
| Vanuatu | 13 | *B. trilineola* Drew, *B. umbrosa* |
| Wallis & Futuna | 5 | *B. kirki, B. passiflorae, B. xanthodes* |

villagers have been encouraged to use a combination of the following methods of control.

## Sound Crop Hygiene

In island situations, destruction of windfall, over-ripe and damaged fruits is strongly encouraged as part of the quality assurance scheme that has become a prerequisite for export. Farms are not registered for export or may have their export registration suspended if crop residues are not adequately destroyed. The aim is to have clean cultivation for fruits and vegetables destined for export.

## Early Harvesting

Some fruits are not susceptible to fruit fly infestation at the green mature or colour-break stages. In most countries of the world, bananas may be harvested at the green mature stage and exported without further quarantine treatment for fruit flies. The exception to this is in areas where the banana fruit fly *Bactrocera musae* Tryon and the Asian papaya fruit fly *B. papayae* Drew and Hancock occur, e.g., in Papua New Guinea. In the Pacific, papayas are harvested at colour-break to reduce the chance of infestation by fruit flies and then subjected to a forced hot air treatment.

## Physical Protection

The technique of bagging fruits in paper bags made from double layers of newspapers or from special rice-paper bags is being strongly encouraged in all PICTs as an inexpensive, effective method of control against fruit flies. Emphasis is being placed on the use of this method at the village or subsistence level of production. In the Pacific, this method of control has the added advantage of reducing damage caused by the fruit sucking moth, *Othreis spp.*

## Protein Bait Sprays

The inclusion of the application of protein bait sprays into the quality assurance schemes for export gives a guarantee that levels of damage in export fruits and vegetables will be very low. Extensive testing of Mauri's Pinnacle Protein Insect Lure (420 g of protein per litre; Low Salt) in Cook Islands, Fiji, FSM, Samoa, Solomon Islands, Tonga and Vanuatu proved that this control method is very effective, environmentally sound and inexpensive. Table 4 contains summaries of some of the results that clearly show its effectiveness.

In an attempt to reduce the costs of the protein source which has to be imported from Australia, waste yeast from the Royal Brewery in Tonga was modified by heating and adding papain enzyme and potassium sorbate to increase shelf-life (Lloyd and Drew 1996). Field trials on several formulations resulted in the commercial release of Royal Tongalure in March 1998. Results using this protein source are also shown in Table 4. Other breweries in the Pacific such as Fiji, Samoa, Solomon Islands and Vanuatu are very interested in converting waste yeast into a usable protein bait for fruit fly control.

Table 4. Summary of results of protein bait spray testing against fruit flies (Tephritidae: Dacinae) in Fiji, Tonga and Vanuatu.

| Country | Crop | Damage Levels (%) | |
|---|---|---|---|
| | | Untreated | Treated |
| Fiji | Guava[1] | 4.16 | 5.1 |
| | Mango[1] | 26.9 | 1.0 |
| Tonga | *Capsicum[1] | 97 - 100 | < 10 |
| | *Capsicum[2] | 100 | < 5 |
| | Chilli[1] | 95 - 97 | < 7 |
| Vanuatu | Guava[1] | 83 - 95 | < 10 |

1 Protein source used was Mauri's Pinnacle Protein Insect Lure

2 Protein source used was modified brewery waste yeast (Tongalure) from Royal Brewery in Tonga

* Excluding chilli

### Conservation of Parasitoids

The adoption of an integrated field control system based on sound crop hygiene, harvesting at an appropriate stage of maturity, physical protection, application of protein bait sprays as a spot spray to fruit trees and as a band to every third row in row crops such as capsicum, facilitates the conservation of parasitoids.

In the Pacific, inundative release of parasitoids is not carried out as a normal practice, except in countries where parasitoids do not occur. For example, *Psyttalia fletcheri* Coquillett was introduced into Solomon Islands from Hawaii to assess its effect on the control of the melon fly *B. cucurbitae.* Similarly, *F. arisanus* and *Diacasmimorpha longicaudata* Ashmead were introduced and released in Pohnpei and Kosrae (FSM), respectively, for the control of the mango fruit fly *B. frauenfeldi*.

## QUARANTINE TREATMENTS

### Early Harvesting of Fruits

As in other countries, harvesting bananas at the green mature stage is a common practice for the export of bananas, e.g., Samoa. Providing banana fruit fly or Asian papaya fruit fly is not present, both bananas and plantains may be exported to New Zealand, so long as the fruits arrive at the green stage. Early harvesting of other fruits is used to avoid high infestations by fruit flies as a component of the quality assurance schemes in place in Tonga, Cook Islands and Fiji. For example, it is a requirement under the bilateral quarantine agreement with New Zealand that papayas be harvested at no later than the colour-break stage to reduce the chance of eggs or larvae being in export fruits.

## Non-host Status

The RMFFP and the New Zealand Ministry of Agriculture and Forestry (MAF) Regulatory Authority developed the MAF Regulatory Standard 155.02.02: Specification for Development of Host Status as a Treatment. This Standard was designed to allow a combination of laboratory and field cage tests to be performed to determine whether a particular variety of fruit or vegetable at a specified stage of maturity is a host or non-host to various fruit fly species within a country. Since the Standard was legalised, collectively PICTs such as Cook Islands, Fiji, FSM, Tonga, Samoa and Vanuatu have tested 36 varieties of fruits and vegetables. As a result, two varieties of chillies, "Red Fire" and "Hot Rod" are being exported to New Zealand under the non-host status. Limes and Yapese lemons (*Citrus hystrix* DC.) in FSM and squash (pumpkin) in Fiji and Vanuatu are non-hosts to the fruit fly species that occur in each country. Other commodities that are likely to be non-hosts in several PICTs include cucurbits such as zucchini, bitter gourd, wax gourd, bottle gourd, "Ripley Queen", "Viamama" and "Smooth Cayenne" pineapples and limes. The non-host status of commodities remains valid provided the fruit fly fauna in countries where tests were carried out is unchanged. For this reason, quarantine surveillance systems have to be operated continuously.

## Forced Hot Air Treatments

Research and technology on forced hot air as a quarantine treatment to replace EDB fumigation was transferred from the USDA-ARS Fresh Fruit, Vegetable and Ornamental Laboratory, Hilo, to Fiji, Cook Islands and Tonga. Research to determine the most heat tolerant species in each country was conducted by national staff. Concurrently, forced hot air units were established in Cook Islands, Fiji and recently, in Tonga. A treatment, based on raising the fruit core temperature to 47.2°C and holding it at that temperature for 20 minutes, has been accepted by New Zealand. Using this quarantine treatment, papayas, mangoes and eggplants are being exported to New Zealand from Fiji. Research data from confirmatory tests on breadfruit in the commercial forced hot air unit have been submitted to New Zealand for technical assessment. Data from confirmatory tests on papayas, mangoes and eggplants have also been submitted to Australian authorities for assessment. Papayas and mangoes are being exported from Cook Islands to New Zealand, using the same forced hot air treatment.

The Horticulture and Food Research Institute of New Zealand Ltd. (Hort + Research) assisted Cook Islands, New Caledonia and Samoa with research into the heat tolerance of immature stages of fruit fly species in each country. The New Caledonia government will install a commercial forced hot air unit in June-July 1998. The government of Samoa will receive assistance from RMFFP and SPC to assess the feasibility of setting up a forced hot air facility there.

As the volume of data on the heat tolerance of immature stages of fruit flies increases throughout the region, the prospects for adopting generic treatments across fruit fly species and commodities look very promising. Negotiations between the New Zealand MAF Regulatory Authority and the RMFFP resulted in the acceptance of the concept of a generic forced hot air treatment. The acceptance of the treatment based on raising the fruit core temperature to 47.2°C and holding it at that temperature for 20 minutes is based on comparisons between the heat tolerance data for the most heat tolerant stages in each country. Provided the most heat tolerant stage of species in a country is less heat tolerant than a prescribed standard, the above mentioned forced hot treatment may be used.

This research and technology have allowed small PICTs to compete in niche markets where there are competitive advantages. The loss of EDB fumigation and the non-acceptance of chemical dipping or spray treatments, in fact, resulted in the PICTs adopting new technologies and improving, markedly, the quality of fresh commodities destined for export markets.

## REGIONAL QUARANTINE SURVEILLANCE AND PREPAREDNESS

### Quarantine Surveillance

The RMFFP and national governments have established quarantine surveillance as early warning systems in fourteen of the 22 PICTs. Early warning systems in the other eight PICTs will be established by August 1998. Spread over some 86 islands in the fourteen countries are 1,040 trapping stations. Each station comprises a pair of modified Steiner traps, one baited with methyl eugenol and one with cuelure. Most traps are serviced every two weeks.

### Emergency Responses and Eradication Programme

The RMFFP provides assistance in the documentation of the quarantine surveillance systems, emergency response plans and eradication procedures for all PICTs. Two eradication programmes are currently being run in the Pacific. The Oriental fruit fly, *B. dorsalis* , was recorded on Moorea and Tahiti in French Polynesia in 1996. The eradication programme is based on the distribution of blocks of coconut husk soaked in 3:1 methyl eugenol and malathion, by ground at 250-400 blocks per km $^2$ and at the equivalent of 400 blocks per km$^2$ by helicopter. The male annihilation effort was combined with targeted protein bait spraying of hot spots. The programme commenced in January 1997 and, by November, no Oriental fruit flies were recorded on Tahiti. The programme cost about US$1.6 million.

The second eradication programme will commence in Nauru in August 1998. Nauru has four species of fruit flies, namely, the Oriental fruit fly, melon fly, mango fruit fly *B. frauenfeldi* and *B. xanthodes*. The Oriental fruit fly and melon fly have been introduced during the past 11-12 years. The aim is to eradicate the Oriental fruit fly and *B. xanthodes,* using male annihilation with methyl eugenol/insecticide soaked compressed fibre-board blocks distributed over the 21 km$^2$ island. Eradication of the melon fly and mango fruit fly may be undertaken later in 1998. A combination of destruction of hosts for the melon fly only, male annihilation using cuelure and broad scale protein bait spraying from the ground will be adopted.

The programme will be used as a regional training exercise for plant protection and quarantine staff from as many of the 22 PICTs as possible. National staff will spend time in Nauru, working as part of the technical teams actually running the eradication activities. While in Nauru, each national staff member will develop an emergency response plan and eradication procedures, relevant to their respective countries.

## SUMMARY

The RMFFP and other projects have made considerable progress in understanding and managing fruit flies in the Pacific region. Countries and territories are in a much stronger technical position to negotiate quarantine protocols and trade now than they were in the late 1980s and early 1990s. Trading partners also have greater confidence in the quality of research that is being carried out and the data that are generated. Governments in the Pacific are genuinely committed to overcoming problems related to the presence of fruit flies and the private sector sees the enormous advantages of maintaining quarantine surveillance and strongly promoting research on fruit flies. This sound base will ensure a degree of sustainability, which may be strengthened by fostering further linkages with fruit fly research programmes in Southeast Asia, Hawaii, Australia, mainland USA and New Zealand.

## REFERENCES

Allwood, A.J. and L. Leblanc. 1996. Mango fruit fly *(Bactrocera frauenfeldi):* why so many in Federated States of Micronesia. pp. 125-130. *In* ACIAR Proceedings on Regional Symposium: Management of Fruit Flies in the Pacific. A.J. Allwood and R.A.I. Drew (eds.). Brown, Prior & Anderson, Melbourne.

Allwood, A.J., T. Tumukon, D. Tau and A. Kassim. 1996. Fruit fly fauna in Vanuatu. pp. 77-80. Ibid.

Amice, R. and F. Sales. 1996. Fruit fly fauna in New Caledonia. pp. 68-76. Ibid.

Drew, R.A.I. and M. Romig. 1996. Overview - Tephritidae in Pacific and South-East Asia, pp. 46-53. Ibid.

Hamacek, E. 1996. Host records of fruit flies in the South Pacific. pp. 102-104. Ibid.

Heimoana, V., F. Tunupopo, E. Toleafoa and C. Fakanaiki. 1996. Fruit fly fauna in Tonga, Western Samoa, American Samoa and Niue. pp. 57-59. Ibid.

Leblanc, L. 1996. Fruit fly fauna in Federated States of Micronesia, Guam, Palau, Kiribati, Northern Marianas and Marshall Islands. pp. 64-67. Ibid.

Lloyd, A. and R.A.I. Drew. 1996. Modification and testing of brewery waste yeast as a protein source for fruit fly baits. pp. 192-198. Ibid.

Purea, M., R. Putoa and E. Munro. 1996. Fauna of fruit flies in Cook Islands and French Polynesia. pp. 54-56. Ibid.

Tenakanai, D. 1996. Fruit fly fauna in Papua New Guinea. pp. 87-94. Ibid.

Tora Vueti, E., A.J. Allwood, L. Leweniqila, L. Ralulu, A. Balawakula, A. Malau, F. Sales and K. Peleti. 1996. Fruit Fly Fauna in Fiji, Tuvala, Wallis and Futuna, Tokelau and Nauru. pp. 60-63. Ibid.

Vagalo, M., R. Hollingsworth and F. Tsatsia. 1996. Fruit fly fauna in Solomon Islands. pp. 81-86. Ibid.

Waterhouse, D.F. 1993. Pest fruit flies in Oceanic Pacific, pp. 4-47. Ibid.

# Feasibility of Eradicating *Ceratitis* spp. Fruit Flies from the Western Cape of South Africa by the Sterile Insect Technique

*Brian N. Barnes and David K. Eyles*

ARC-Fruit, Vine and Wine Research Institute (INFRUITEC Center), Stellenbosch, South Africa

## INTRODUCTION

Fruit production (deciduous fruit, table and wine grapes, and citrus) is a major export-based industry in the Western Cape, with more than 200,000 ha under cultivation. The gross value of these fruits (excluding wine) exceeds US$400 million per annum. Deciduous fruit and table grapes make up the major portion of the industry, with approximately 110,000 ha under production.

The Western Cape is host to two species of fruit flies, the Mediterranean fruit fly, *Ceratitis capitata* (Wiedemann), and the Natal fruit fly, *C. rosa* Karsch. One or both of these species attack at least 19 species of fruits in this area. Both species have very similar life cycles and habits, and can cause enormous crop losses especially to fruits, but also to some vegetables. Both commercial and resource-limited farmers are affected by fruit flies.

Control of fruit flies is currently based on ground applications of insecticides, either as full-cover foliar sprays or low-volume bait sprays. Control costs and crop losses for deciduous fruit and table grapes alone are estimated at US$4 million annually.

South Africa is the only southern hemisphere deciduous fruit-exporting country that is not fruit fly-free or is not currently engaged in a project to eradicate fruit flies. Unless similar steps are taken, this situation is likely to threaten the competitiveness of the Western Cape's industry.

The Joint FAO/IAEA Division of Nuclear Techniques in Food and Agriculture of the United Nations recently approved funding for the INFRUITEC Centre of the ARC-Fruit, Vine & Wine Research Institute in Stellenbosch to undertake a study to determine the feasibility of eradicating fruit flies from the Western Cape by the sterile insect technique (SIT).

Most fruits in the Western Cape are produced in valley systems, and many valleys are isolated from one another to a greater or lesser extent. The Western Cape is itself well isolated from other areas with fruit fly hosts: by the ocean on the eastern, southern and western side, and by an extensive arid and semi-desert region on much of the northern side. The Western Cape as a whole, and a number of its individual fruit production valleys, are considered to be good candidates for the use of SIT to eradicate fruit flies (J. Hendrichs, G. Tween, J. Mumford, personal communication).

As part of the feasibility study, a pilot SIT project has been initiated in the Hex River Valley (HRV), South Africa's principal table grape production area. The Hex River valley comprises about 4,000 ha farmed by about 140 growers, with an average farm size of 18 ha. Most of the area is under table grapes, with a very small but increasing area under soft citrus. It is one of the more isolated of the production areas, has only one through road, and one small village. A cost-benefit study has shown it to be economically the most favourable of all areas in the Western Cape for an SIT programme, based on a comparison of current control costs versus estimated SIT costs.

This article describes progress with the pilot project to date, and factors likely to influence the success of SIT in this and other areas. Main areas of activity in the HRV have included the establishment of an extensive monitoring system, the implementation of a coordinated fruit fly baiting programme, and trial shipments of sterile *C. capitata* pupae.

## MATERIAL AND METHODS

### Monitoring System

With the aid of geographical positioning system instruments, 300 Jackson traps baited with trimedlure were deployed throughout the valley, at a density of 1 trap per 25 ha. This trap density was doubled in the village. Traps were placed in all areas and on as many different fruit fly host plants as possible. From March to September 1997, trap readings were made by Infruitec. Thereafter, traps were serviced by three permanent monitors appointed by the Hex River valley growers.

All traps were checked weekly, and lures and bottoms replaced every 6 wk (bottoms sooner if warranted). The data were stored in a computer database at Infruitec according to trap number, date of trap reading, number of flies per trap, fruit fly species, and the mean number of flies per week. Trap data were additionally sorted according to the host plant where the trap was located, and the mean number of flies per host plant per week. When data were sorted into host plant categories, only data from those traps which were not close to different host plants were used.

Once the limited supply of Jackson traps was exhausted, locally-made versions of the trap were supplied by an agrochemical company.

### Baiting Programme

The Hex River valley growers normally apply two to three organophosphate cover sprays and from seven to ten low volume alternate-row bait sprays per season against fruit flies. This is done on an uncoordinated basis, each grower deciding his own programme. All sprays are ground applied.

Growers have proved to be somewhat lax about baiting for fruit flies. This is possibly reflected in the very high fruit fly counts at the start of the SIT pilot project (Figure 1). For the SIT, project growers were therefore advised to have the baiting coordinated throughout the Hex River valley by a private contractor. This advice was not heeded due to the costs involved, and growers continued to apply their own bait sprays. However, they coordinated applications, which took place on the same day every week. The perimeters of all vineyards and every 5$^{th}$ vineyard row were treated. They were urged not to use mistblowers, but equipment that would deliver the bait in large droplets.

Shortly before harvest, when visible and toxic residues on the grapes would be a problem, hand-lances attached to spray pumps under low pressure were used for the vineyard applications, with large nozzle holes and the whirlplates removed. Every 20 to 25 m, streams of bait were directed into the air so that the droplets would fall onto the back side of the trellised vines. They also used this equipment for weekly treatments of all fruit fly host plants in their and their labourers' backyards.

Weekly applications of bait commenced in April 1997. The initial composition of the bait mixture was as registered by the regulatory authorities for use against fruit flies, and comprised 250 ml of a 42.5% solution of protein attractant (Hym-lure®) plus 175 ml mercaptothion in 100 litres of water. In January 1998, when it became clear that the registered bait mixture was insufficiently attractive to fruit flies, the amount of Hym-lure was increased to 2 litres per 100 litres water.

A contractor was, however, appointed to treat fruit fly hosts in the village. A hand-operated backpack sprayer and hand-lance were used, with the whirlplate removed and with a large nozzle hole. About 50 ml of bait mixture (as described above) was applied weekly in a single squirt to all fruit fly hosts and suspected hosts, as well as to other trees lining the streets.

### Trial Shipments of Sterile *C. capitata*

There is no *C. capitata* mass rearing facility in South Africa. For this reason, initial plans for the Hex River valley pilot project, therefore, made allowance for the weekly importation of sterile pupae from a foreign source until a local facility is in operation.

Three trial shipments of sterile *C. capitata* pupae were sent from the El Pino fruit fly facility in Guatemala to Infruitec in Stellenbosch. The purpose was to test the route and *in transit* conditions for future shipments, to become familiar with the permit and handling procedures at the port of entry (Cape Town International Airport), and conduct control tests. Two million *C. capitata* were shipped under low temperature and hypoxia on each occasion, 1 million each of the temperature sensitive lethal *(tsl)* strain and the bisexual strain. As the preferred carrier was KLM Airlines, the shortest route to Cape Town was via Amsterdam.

On arrival in Stellenbosch, the following aspects were noted or evaluated: duration in transit, percentage emergence of adults, flight ability of adults (using standard 10 cm high IAEA flight ability tubes), stress test (% survival after 48 h without food – *tsl* only), sterility test (% hatch of eggs from sterile male x fertile South African female – *tsl* only).

## RESULTS AND DISCUSSION

### Monitoring

More than 99.9% of all fruit flies recorded in the Hex River valley were *C. capitata*. Only 14 *C. rosa* specimens were recorded, and only at two trap sites, on guava and *Ficus* sp. (fig). Although it was very clear that the vast majority of trapped flies were *C. capitata*, the very many flies stuck to the trap insert and covered in gum on many occasions made it virtually impossible to be certain that no specimens of *C. rosa* went unrecorded. The size of the *C. rosa* population in the valley may therefore be a slight underestimation.

The fact that the fruit fly population in the Hex River valley comprises essentially only *C. capitata*, favours the use of sterile flies imported from a foreign source until such time as a local rearing facility comes into production.

Graphs of the fruit fly population fluctuation in the village and from specific host plants from March 1997 to April 1998 are shown in Figure 1. Very high *C. capitata* populations were present in the Hex River valley at the start of monitoring, especially in the village and on pome fruit. Fruit fly activity was very seasonal and correlated with ripening of hosts. Highest populations occurred from February to May (autumn). Little activity occurred during spring and summer. The village (comprising mixed fruits) was the greatest source of fruit flies, followed by guava, pome and stone fruit in farm gardens. Lowest populations occurred in citrus orchards and table grape vineyards.

Alarming increases in populations on *Ficus* spp. and guava occurred in the autumn of 1998, possibly as a result of inadequate baiting procedures. However, the decline in populations in the village are encouraging (see "Baiting programme"). Table grapes appear to be a poor host for fruit flies. The main focus of future bait treatments should therefore be non-vine hosts. The implications of the presence of very low numbers of *C. rosa* on the SIT programme are uncertain, and the situation should be carefully monitored.

Various problems were encountered with the Jackson traps and/or the locally-made version. Traps bodies often blew off the wire hanger in the wind. The plastic lure cage often slid to one end of the hanger. The local version soon started to collapse after exposure to rain, and had too much gum covering too little of the surface of the inserts. Growers are considering replacing all traps with a commercially available dry, bucket-type trap baited with trimedlure.

## Baiting

It became clear that the bait sprays applied by the growers were ineffective for two main reasons:

1) the application technique used did not deliver the bait in the required large droplets, and
2) the registered bait mixture was insufficiently attractive to the fruit flies. There was also some doubt as to whether all growers did in fact adhere to the weekly baiting schedule.

Despite warnings to the contrary, growers applied most bait by mistblowers, but with whirlplates removed and with large nozzle holes. Even so, droplet size was very small, and unattractive to, and ineffective against, fruit flies. In a crude test, pear leaves sprayed with bait in this manner and placed in a cage with *C. capitata* resulted in only 25% mortality after 24 h.

Applications to hosts in farm gardens also appeared to have been ineffective. Investigations suggested that incorrect equipment, application and bait concentration were to blame.

It is doubtful that 2 litres Hym-lure per 100 litres water is sufficiently attractive to fruit flies. This might be borne out by the fact that some fruit fly populations started to increase to higher-than-expected numbers in autumn 1998 (Figure 1). One encouraging trend was the substantial decrease in populations in the village (Figure 1). This might be testimony to the advisability of contracting out bait spray operations to a single operator.

The route via Amsterdam is considered to be too long in duration and too risky for the SIT programme. An alternative route via Miami, USA, is being investigated.

## Trial Shipments of Sterile Pupae

Table 1 summarises the shipment details and conditions of sterile *C. capitata* on arrival from Guatemala. In general, adult fruit flies of reasonable quality were obtained. However, the duration that the pupae were under hypoxia was longer than desired, as evidenced by the poor results of the flight ability and stress tests on Shipment 1 (70.5 h in transit). Shipment 3 was a failure due to the cold chain being broken somewhere en route because of a flight cancellation.

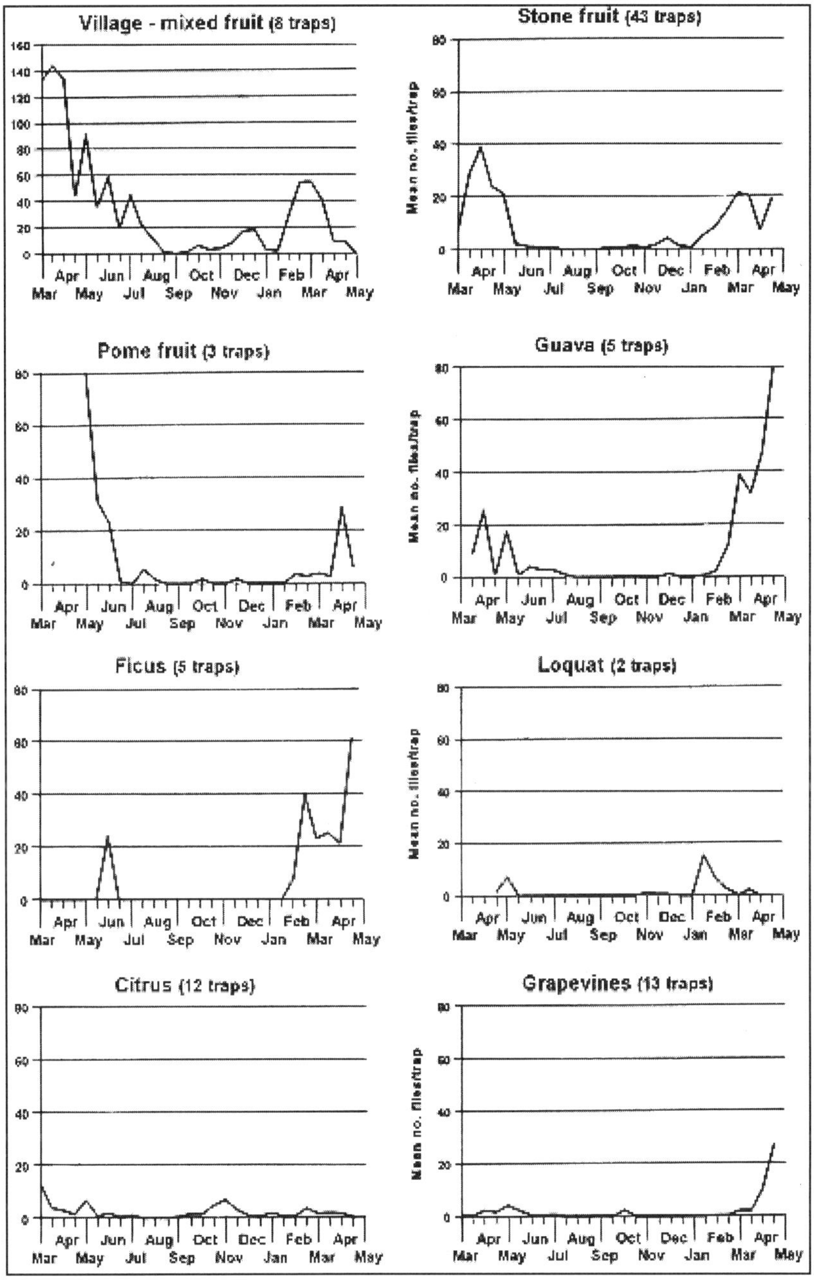


Figure 1. Seasonal population fluctuations of *Ceratitis capitata* on different hosts (March 1997 to April 1998) in the Hex River valley (vertical scale = mean number of flies/trap).

Table 1. Shipping duration and condition of sterile *C. capitata* on arrival in Stellenbosch after shipment from Guatemala.

| Shipment no. | Hours in transit | Emergence (%) | | Flight ability (% fliers) | | Stress test (% survival) | Sterility (% hatch) |
|---|---|---|---|---|---|---|---|
| | | tsl | bisex | tsl | bisex | Tsl | tsl |
| 1 | 70.5 | 83.6 | 82.2 | 59.0 | 34.4 | 19.0 | 0.63[a] |
| 2 | 46.5 | 87.4 | 92.2 | 70.4 | 67.8 | 59.9 | 0 |
| 3[b] | 53.5 | 55.0 | 83.4 | 38.2 | 57.8 | 48.6 | 0 |

[a] 1 egg out of 163. Suspected contamination with a fertile mated female from the South African colony.
[b] The cold chain was broken and flies were emerging on arrival.

Releases of sterile *C. capitata* in the Hex River valley have been postponed due, inter alia, to the unexpected unavailability of sterile pupae from overseas sources. Releases are now expected to start in September 1999, directly after winter.

## Factors Influencing the Establishment of SIT in South Africa

### *Positive Factors*

The relative isolation of the Western Cape, and of many of its fruit production areas, favour the establishment of SIT. In addition, the fruit industries of the Western Cape are highly organised with an excellent infrastructure.

Potentially great export market benefits will accrue to the fruit industries if one or more fruit fly-free areas can be established. The benefit of being fruit fly-free would not only be to commercial growers. A host of informal, resource-limited growers would also benefit from the absence of fruit flies.

The Western Cape fruit industries are well served by research organisations. Good training facilities are also available at universities in the area.

The interest, expertise and funding provided by the Joint Division of the FAO/IAEA is of enormous benefit to the SIT initiative currently underway in South Africa.

In the Hex River valley, positive factors include its isolation; the presence of effectively a single fruit fly species and a single crop; the high income of the growers; the cost-effectiveness of SIT versus current control methods, and the willingness of the growers to fund the running costs of an SIT operation. The growers have raised US$240,000 through an export carton levy for the first year of the pilot project, and funding by this mechanism will continue. An SIT coordinator and three monitors have been appointed in the Hex River valley. The Hex River valley growers are negotiating for the purchase of a suitable aircraft for aerial releases of *C. capitata*.

Although there is currently no local *C. capitata* rearing facility in South Africa, a building at INFRUITEC is being converted into a mass rearing facility for *C. capitata*. With an expected production of 10 million flies per week, it should be able to provide the Hex River valley with the required 6 million sterile flies per week, thereby saving highly expensive shipping costs for imported flies.

*Negative Factors*

To date, there has been effectively no input by the government, either local or national. Without this input, it will be impossible to carry out an SIT programme over the whole Western Cape. Growers in the different production areas will have to fund the running costs themselves, as in the Hex River valley.

Although SIT would be cost-effective over the Western Cape as a whole, taken production area for production area, SIT would be less expensive than current control methods in only two or three areas.

In many areas, a number of different fruit crops is grown. The presence of wine grapes in many areas also impacts negatively on SIT as wine grape growers are not concerned by fruit fly infestation.

The absence of a local rearing facility for *C. capitata* makes SIT extremely expensive due to the necessity to import sterile flies.

The presence of *C. rosa* in most areas besides the Hex River valley complicates SIT. *C. rosa* is not mass reared anywhere in the world.

The deciduous fruit industry has recently undergone deregulation. The resultant disruption in activities has negatively impacted on the SIT initiative. Deregulation has also increased the number of export organisations from one to about 140.

There is a certain lack of understanding amongst many growers about the urgency to have one or more fruit fly-free areas in the industry.

The following factors are vital to the success of the Hex River valley pilot project: the appointment of a full-time SIT coordinator; well-trained and dedicated monitors; a solution to the problems with fruit fly traps; a more effective baiting programme; a suitably equipped and staffed operations centre; continued funding for the project; and addressing the issue of informal street vendors bringing fruits into the village from outside the Hex River valley.

## ACKNOWLEDGEMENTS

J. Hendrichs, G. Tween, J. Mumford, J. Worley, D. Gates, T. Billak, J. Rudig and J-P. Cayol for encouragement and sound advice during IAEA consultancies and on other occasions; Muriel Knipe and N. du Plessis of INFRUITEC for technical assistance during monitoring; the Joint FAO/IAEA Division of Nuclear Techniques in Food and Agriculture of the United Nations, for continued financial support (Project SAF/5/002) and interest.

# Monitoring Guidelines Improve Control of Walnut Husk Fly in California

***Susan B. Opp[1], Katherine M. Reynolds[2], Carolyn Pickel[3] and William Olson[4]***

[1] Department of Biological Sciences, California State University, Hayward, CA, USA
[2] USDA/ARS, Western Regional Research Center, Albany, CA, USA
[3] Sutter/Yuba County Cooperative Extension, University of California, Yuba City, CA, USA
[4] Butte County Cooperative Extension, University of California, Oroville, CA, USA

## INTRODUCTION

The walnut husk fly (WHF), *Rhagoletis completa* Cresson, is a key pest of walnuts (*Juglans* spp.) in California, where over 95% of the US and approximately two-thirds of the world's commercial walnuts are produced. The primary hosts of this monophagous fruit fly are *J. regia* L. (commercially grown English walnut), *J. californica* S. Wats. var. *hindsii* (northern California black walnut), *J. californica* var. *californica* (southern California black walnut) and *J. nigra* Thunb. (eastern black walnut). Some cultivars of the English walnut are more susceptible than others; the most heavily infested varieties of English walnut include Eureka, Franquette, Hartley, Mayette and Payne.

Neither English walnuts nor the walnut husk fly are native to California. So-called "English" walnuts are sometimes more appropriately called "Persian" walnuts, in reference to Persia, the origin of *J. regia*. English walnuts were first planted in southern California in the 1860s. In contrast, the native range of WHF is the mid- and south-central United States where it attacks *J. nigra* (Boyce 1934). The fly was likely to have been introduced into southern California in the mid-1920s by tourists travelling from Kansas, New Mexico, Texas or Oklahoma.

WHF was first documented in California in 1926 in the San Bernardino County when maggots were found in the husks of English walnuts (Boyce 1929). The fly gradually spread throughout walnut growing regions of California. In 1928, only three or four orchards in the San Bernardino County were known to be infested. By 1932, the fly was also found in the Los Angeles and Orange Counties (Boyce 1933), and by 1954, it was found in Ventura, Riverside, and the San Diego Counties, in addition to the northern California county of Sonoma (Anonymous 1966). The spread of the fly in northern California was rapid. By 1958, WHF was found in San Joaquin County; in 1963, the fly was in Amador, Lake, Solano, Tulare and Yolo Counties; in 1964, it was found in Fresno, Mendocino, Monterey and Sutter Counties. By 1965, populations of WHF were found in Kings, San Francisco, Marin, Colusa, Glenn and Butte Counties, essentially having invaded every county in California by the end of that year (Anonymous 1966).

The history of WHF in California is an interesting case of invasion biology because a smooth outward spread from the point of initial introduction did not occur. Instead, some regions of northern California, such as Sonoma, Amador, and Lake Counties, were colonised prior to more central walnut growing regions, such as Fresno County. Armed with this historical information and based on allozyme work which suggested WHF in their native range were more genetically similar to each other than to flies in California, Berlocher (1984) suggested that WHF was introduced into California just once. The subsequent spread of the fly throughout the state was likely to have taken place by secondary introductions facilitated by human movements of infested walnuts from one or more of the established California populations (Berlocher 1984).

## Natural History

The walnut husk fly is univoltine, like other members of the temperate genus *Rhagoletis*. Female WHF oviposit in the husk (exocarp) of the developing walnut, usually in the upper third of the husk near the stem end. Eggs are deposited in batches of about fifteen below the surface of the husk, where they develop for four to seven days before hatching. The larvae feed on the husk for three to five weeks after which they exit the husk and burrow into the soil to overwinter as pupae. In most areas of California, adults eclose from late June through September, although flies may emerge as early as mid-May in some regions (Opp, unpublished). Up to 25% of pupae may remain in the soil for two years, and a few may not emerge for three or four years (Boyce 1934).

Females usually emerge first and will predominate for the first couple of weeks, although males typically predominate during the remainder of the season (Boyce 1934, Riedl and Hoying 1980). In the field, the average walnut husk fly may live 30 to 40 days as an adult (Boyce 1934). Six to eight days after emergence, adult WHF are capable of mating, although females typically have a ten to 20 day preoviposition period (Boyce 1934). During this period, females must have an extrinsic source of nitrogen (protein) for egg development (Tsiropoulos 1978).

Mating occurs on or near the host plant. Typically, males wait on walnuts to mate with females searching for oviposition sites. WHF is a polygamous species, with both males and females mating repeatedly. In addition, male WHF exhibit both contact and non-contact post-copulatory mate guarding by remaining near or in contact with a female following mating but prior to female oviposition (Opp et al. 1996). This guarding serves to ensure paternity because up to 90% of a female's offspring will be sired by her last mate (Telang 1995).

## Monitoring and Control

WHF is a key, indirect pest of walnuts. Economic damage to walnuts results from WHF larvae feeding on the husk, causing the husk to become soft and black, staining the developing nutshell. Larvae do not feed on the nut kernel, but in severe, early season infestations, kernels may become shriveled, darkened and mouldy (Boyce 1934, Hislop et al. 1981).

Various methods have been used to control WHF since its introduction into California over 70 years ago. Bait (Staley's No. 7) plus insecticide sprays were first reported used by Barnes and Ortega (1959) to control WHF. Over the years, various systemic and pyrethroid insecticides have been tested for the control of adult and larvae within husks (Nickel and Wong 1966, Hislop et al. 1981). In general, insecticides combined with baits have proved to be the best adulticides and are recommended for control (Flint 1987).

Early attempts to monitor WHF emergence and adult flight activity used liquid food-based attractants such as open pan traps containing casein-sodium hydroxide or glycine (Boyce and Bartlett 1941). Barnes and Osborn (1958) developed a dry sticky trap consisting of a sticky food carton baited with powdered ammonium carbonate. This trap was modelled after one developed by Hodson (1948) and later used by Frick (1952) for other species of fruit flies. This became the trap of choice because of its ease of use, despite the fact that it was not as effective as the glycine pan trap. Later, the Pherocon AM trap (a fluorescent-yellow rectangular panel trap coated with adhesive and baited with hydrolysed protein and ammonium acetate), developed for monitoring *R. pomonella* (Prokopy, 1975), was found to be a better monitoring tool for WHF than the food carton trap (Riedl and Hoying 1980).

Despite modifications and improvements to WHF monitoring traps over the years, control decisions based on trap captures were difficult to make because adult flies could be captured 40-70 days prior to the onset of damage (Riedl and Hoying 1980). In the 1980s, Riedl and associates tested a variety of modifications to WHF traps including a green sphere. Yellow rectangles were generally found to be preferred over green spheres, although ammonium carbonate could be added to green spheres to boost trap captures, but with a loss of selectivity for WHF (Riedl and Hislop 1985). Later studies showed that mature, fully-fed WHF were more quickly attracted to ammonia-baited traps than were young or protein-deprived flies (Reynolds et al. 1996). This further reinforced the need for refined monitoring tools that would allow growers to treat before egg laying damage had occurred.

Hamersky (1996) and Light and Jang (1996) tested the relative attractiveness of walnut volatiles (sesquiterpenes and monoterpenes) to WHF. Using combinations of electroantennograms and field tests, they found that one sesquiterpene, Beta-caryophyllene, was very attractive to male walnut husk flies. Although male WHF were attracted to caryophyllene-baited traps at the approximate time when female egg laying began, these traps provided inadequate lead time for growers to spray to control damage.

Our goals, in a series of studies conducted in commercial and unmanaged English walnut orchards, were to develop improved monitoring guidelines for WHF that would allow growers to make control decisions on an orchard by orchard basis. In addition, we wanted to disseminate these guidelines in a user-friendly format that would provide farm advisors and growers with information that could be implemented in a timely fashion. Finally, we planned to evaluate both the impact of our updated monitoring guidelines and the format of information dissemination on WHF control in commercial walnut orchards in a number of different walnut growing regions of California.

## MATERIALS AND METHODS

### Field Research

Field bioassays were conducted in commercial walnut orchards in the Sacramento Valley of California during the summers of 1995 and 1996. In both years, Pherocon® AM-NB (no bait) yellow panel traps were baited with Trece® Supercharger lures (ammonium carbonate). Traps were hung, two per tree, from each of three blocks in each of three orchards per year. One of the traps in each tree was hung high in the tree canopy (ca. 5m from the ground), while the second trap per tree was hung low in the canopy (ca. 2m from the ground). The orchards were located in Vina (Tehama County), Gridley (Butte County) and Esparto (Yolo County) in 1995. In 1996, the orchards were located in Vina (Tehama County), Gridley (Butte County) and Yuba City (Sutter County).

Male and female WHF were removed from traps twice a week. Females were dissected to determine if mature eggs were present in their ovaries. Ammonia lures were replaced every three weeks, and traps were rotated to new trees within each block once a week. In 1995, 25 walnuts were randomly collected in each of the three blocks per orchard twice a week and were dissected to determine when egg laying began (the onset of damage) and the extent of walnut infestation. In 1996, 50 walnuts were collected per block twice a week and were dissected as in the previous year.

### Implementation

With the results from these field studies, we developed a video containing information on biology, proper placement of monitoring traps, use of trap information for treatment timing, and the most appropriate control methods for WHF. In 1997, the video was shown at cooperative extension meetings and in conjunction with field meetings where techniques discussed in the video were demonstrated. At the cooperative extension meetings, participants were asked to fill out a questionnaire and quiz about WHF both before and after viewing the video. The quiz addressed the following: the preferred trap type for WHF, where traps should be placed in the orchard and within individual trees, where to look for WHF "hot spots" in orchards, how often traps should be monitored, the best treatment timing tool, and how many traps should be placed per acre. In addition, we asked whether or not growers had WHF problems, how they rated the video, what sections of the video they found least and most helpful, and what information in the video they thought might help with monitoring and control of WHF. Finally, regional demonstration plots were established in 15 orchards state-wide to assess the validity of the monitoring and control guidelines promoted in the video.

## RESULTS

### Field Research

In 1995, trap height in the canopy was shown to significantly affect the capture of WHF in both the Gridley and Esparto orchards, but not in the Vina orchard (Table 1). Although the effect of height was not significant in Vina, high traps captured greater numbers of both sexes of flies than low traps, similar to the effect observed in Esparto. However, in Gridley, high traps only captured more flies than low traps early in the season, but after the first treatment, low traps captured more WHF than high traps.

The date of detection of the first females with mature eggs in their ovaries varied among orchards (25 July for Vina, 21 August for Esparto, and 1 August for Gridley). In Vina, only one female with eggs was captured on that early date; three more gravid females were not captured until 4 August. The date of the detection of the first eggs in walnuts also varied among orchards (15 August for Vina, 1 September for Esparto, and 11 August for Gridley). Therefore, the time between the detection of two or more gravid females on traps and the detection of eggs in walnuts varied little (11 days in Vina, 10 days in Esparto, and 10 days for Gridley), with a minimum elapsed time of 10 days. All three orchards suffered less than 6% damage due to WHF as determined by visual inspection at harvest time.

Table 1. Effects of trap height on walnut husk fly captures for each of three commercial walnut orchards in 1995 (ANOVA on log (X+1) transformed counts of flies per trap per day).

a) Vina

| | | MALES | | FEMALES | |
|---|---|---|---|---|---|
| Source | df | F | **P** | F | *P* |
| Height | 1 | 1.41 | 0.24 | 2.86 | 0.09 |
| Total | 527 | | | | |

b) Gridley

| | | MALES | | FEMALES | |
|---|---|---|---|---|---|
| Source | df | F | **P** | F | *P* |
| Height | 1 | 23.54 | 0.00 | 10.32 | 0.001 |
| Total | 527 | | | | |

c) Esparto

| | | MALES | | FEMALES | |
|---|---|---|---|---|---|
| Source | df | F | **P** | F | *P* |
| Height | 1 | 81.76 | 0.00 | 108.47 | 0.00 |
| Total | 431 | | | | |

In 1996, the traps placed high in the canopy once again captured significantly greater numbers of WHF than traps placed low in the canopy (Table 2). Because we were working in commercial orchards (as in 1995), and we had information from the previous year, we recommended that our growers spray within 10 days of the first discovery of gravid females on traps, but we continued to monitor for gravid females and eggs in walnuts.

As in the previous year, the date of detection of gravid females on traps varied among orchards (29 July in Vina, 26 July in Yuba City, and 5 August in Gridley). A single gravid female was detected in Vina at first, and more gravid females were not captured until 5 August, seven days later. In Vina, the grower sprayed within 10 days after the detection of two or more gravid females, and we did not find eggs in walnuts until 4 ½ weeks later. In Gridley, the grower sprayed within the 10-day period, as recommended, and we did not detect the first WHF eggs in walnuts until late in the season, near harvest time. Unfortunately, the Yuba City grower could not spray until more than two weeks after the detection of gravid females. Eleven days after the first detection of gravid females in traps, eggs were found in walnuts in the Yuba City orchard. A second treatment for WHF was necessary in this orchard less than a month later, as indicated by a second presence of gravid females on traps. In 1996, the three test orchards suffered less than 2% damage due to WHF as determined by visual inspection at harvest time.

Table 2. Effects of trap height on walnut husk fly captures for each of three commercial walnut orchards in 1996 (ANOVA on log (X+1) transformed counts of flies per trap per day)

a) Vina

| | | MALES | | FEMALES | |
|---|---|---|---|---|---|
| Source | df | F | **P** | F | *P* |
| Height | 1 | 26.90 | 0.00 | 21.94 | 0.00 |
| Total | 239 | | | | |

b) Gridley

| | | MALES | | FEMALES | |
|---|---|---|---|---|---|
| Source | df | F | **P** | F | *P* |
| Height | 1 | 16.82 | 0.00 | 5.11 | 0.02 |
| Total | 319 | | | | |

c) Yuba City

| | | MALES | | FEMALES | |
|---|---|---|---|---|---|
| Source | df | F | **P** | F | *P* |
| Height | 1 | 12.82 | 0.00 | 13.48 | 0.00 |
| Total | 447 | | | | |

## Implementation

Based on our studies, we modified existing recommendations for WHF monitoring and control. In the video, we suggested that commercial orchards have a minimum of 2 Pherocon® AM-NB traps baited with ammonium carbonate Supercharger® lures per 10 acres of walnuts. These traps should be placed in WHF "hot spots" defined as areas with: 1) historical WHF damage, 2) dense shade and large trees, 3) moisture or nearby creeks, and/or 4) nearby black walnuts. Within trees, traps should be placed in the upper one-third of the tree canopy on the north side of trees. Traps should be hung by 15 May in coastal areas and by 1 July in valley locations, and should be monitored twice a week, with three times a week preferred, until the first treatment is applied. Thereafter, once a week monitoring should suffice. The best tool for timing treatments is the detection of mature eggs in female flies. Females can be identified by their light brown femurs, while males have black to dark brown femurs. Female WHF should be removed from a trap, placed on a card or dark-coloured paper, and gently dissected with a probe or knife blade to determine if mature eggs are present, i.e., if females are gravid. If eggs are present, they will appear as 5-15 white objects, less than 1 mm in length and shaped like tiny rice grains. After detection of gravid females, treatment (consisting of bait plus insecticide applied with large droplet size sprays to every other row of trees) should occur within one week to ten days. Monitoring should continue following treatment because fly emergence continues over several weeks and populations may build up to damaging levels a second or third time, requiring additional treatment.

The 22 minute WHF video, illustrated with field photos and video footage (available through the DANR Communication Services at the University of California, Davis), was shown in 1997 at nine cooperative extension meetings. In addition, seven summer field meetings were held to give hands-on demonstrations of the techniques shown in the video. Each cooperative extension office with walnut responsibilities in California was given a complimentary copy of the video to loan or rent to growers or other interested parties. In total, over 1,150 viewers, representing over 50,000 acres of walnuts saw the video. In addition, 72 copies of the video were sold by the DANR Communication Services to chemical company representatives, pest management companies, PCAs, etc.

Of the viewers, slightly over half indicated they had a WHF problem in their orchards (state-wide, WHF affects approximately half of commercial orchards). The "before and after" video quiz was completed by 324 viewers. In the approximately 30 minutes that elapsed between the "before and after" quizzes, scores improved by over 80%, indicating a large increase in general knowledge about WHF. In general, viewers thought the information in the video was very helpful, and 90% thought information in a video format was more useful than a slide presentation on the same topic. The new information on sexing and determining the egg load in flies was considered some of the most useful, but over one-third of the participants felt all the information was useful. The introductory and damage information sections were considered the least useful, but fewer than 10% of the people cited either of these sections as being the least useful.

Following the 1997 season, 82 of the video participants were reached to determine how useful the video information had been for WHF control. All (100%) felt the video was useful, but only 62% reported attempting to sex flies. Nevertheless, 65% reported improvements in spray timing, with reduced numbers of WHF treatments and less damage in 1997 than in 1996.

## DISCUSSION

The walnut husk fly continues to be a difficult pest for many walnut growers in California largely because of problems in predicting the timing of emergence and damage on either a regional or orchard by orchard basis. Most likely, this unpredictability arises because both the insect and its host (English walnut) are not native to California. While the long-term climate of California is predictable, year to year variations in climate may be considerable. For example, in 1997, WHF appeared in orchards several weeks earlier than normal due to an unusually mild and warm spring. For some growers, this resulted in greater numbers of insecticide applications for WHF than usual. The information provided in the video allowed growers to be prepared for such unusually early events and to avoid unnecessary early season damage. But in 1998, largely due to the lingering effects of the 1997-98 El Niño, WHF appeared in some orchards up to three weeks later than usual, although overall effects of this late emergence are unknown at this time. Again, the use of the information provided by the video allowed growers to avoid unnecessary early season treatments for WHF. In general, growers found the information presented in video format to be more appealing and useful than the same information presented in written or slide presentation format. In addition, cooperative extension farm advisors who participated in the video and demonstration presentations felt they learned about WHF through their involvement. Finally, in 1997, the video was awarded the State Communications Award from the National Association of County Agricultural Agents.

## ACKNOWLEDGEMENTS

We appreciate the expert video work and photography of Mike Poe, DANR Communication Services Video Director, UC, Davis, and support from the University of California State-wide IPM grant 95BU015 and Smith-Lever IPM grant 97SL005.

## REFERENCES

Anonymous. 1966. The fly is back stronger than ever! *In* Diamond Walnut News, pp. 16-17.

Barnes, M.M. and J.C. Ortega. 1959. Experiments with protein hydrolysate bait sprays for control of the walnut husk fly. J. Econ. Entomol. 52: 279-285.

Barnes, M.M. and H.T. Osborn. 1958. Attractants for the walnut husk fly. J. Econ. Entomol. 51: 686-689.

Berlocher, S.H. 1984. Genetic changes coinciding with the colonisation of California by the walnut husk fly, *Rhagoletis completa*. Evolution 38: 906-918.

Boyce, A.M. 1929. The walnut husk fly (*Rhagoletis juglandis* Cresson). J. Econ. Entomol. 22: 861-866.

Boyce, A.M. 1933. Control of the walnut husk fly, *Rhagoletis completa* Cress. J. Econ. Entomol. 26: 819-825.

Boyce, A.M. 1934. Bionomics of the walnut husk fly, *Rhagoletis completa*. Hilgardia 8: 363-579.

Boyce, A.M. and B.R. Bartlett. 1941. Lures for the walnut husk fly. J. Econ. Entomol. 34: 318.

Flint, M.L. (ed.) 1987. Integrated pest Management for Walnuts. 2nd ed. University of California, Publ. 3270, 96 pp.

Frick, K.E. 1952. Determining emergence of the cherry fruit fly with ammonium carbonate bait traps. J. Econ. Entomol. 45: 262-263.

Hamersky, W. 1996. Electrophysiological and behavioral responses of walnut husk flies to walnut volatiles. M.S. Thesis, California State University, Hayward.

Hislop, R.G., H. Riedl and J. Joos. 1981. Control of the walnut husk fly with pyrethroids and bait. Calif. Agric. 35: 23-26.

Hodson, A.C. 1948. Further studies of lures attractive to the apple maggot. J. Econ. Entomol. 41: 61-66.

Light, D.M. and E.B. Jang. 1996. Plant volatiles evoke and modulate tephritid behavior, pp. 123-133. *In* Fruit Fly Pests, A World Assessment of their Biology and Management. B.A. McPheron and G.J. Steck (eds.). St. Lucie Press, FL.

Nickel, J.L. and T.Y. Wong. 1966. Control of the walnut husk fly, *Rhagoletis completa* Cresson, with systemic insecticides. J. Econ. Entomol. 59: 1079-1082.

Opp, S.B., S. Spisak, A. Telang and S. Hammond. 1996. Comparative mating systems of two *Rhagoletis* species: the adaptive significance of mate guarding, pp. 43-49. *In* Fruit Fly Pests, A World Assessment of their Biology and Management. B.A. McPheron and G.J. Steck (eds.). St. Lucie Press, FL.

Prokopy, R.J. 1975. Selective new trap for *Rhagoletis cingulata* and *R. pomonella* flies. Environ. Entomol. 4: 420-424.

Reynolds, K.M., S.B. Opp, M. Moen and K. Denham. 1996. Mark-recapture studies of walnut husk flies attracted to food-based lures, pp. 141-146. *In* Fruit Fly Pests, A World Assessment of their Biology and Management. B.A. McPheron and G.J. Steck (eds.). St. Lucie Press, FL.

Riedl, H. and R. Hislop. 1985. Visual attraction of the walnut husk fly (Diptera: Tephritidae) to colored rectangles and spheres. Environ. Entomol. 14: 810-814.

Riedl, H. and S.A. Hoying. 1980. Seasonal patterns of emergence, flight activity and oviposition of the walnut husk fly in Northern California. Environ. Entomol. 9: 567-571.

Telang, A. 1995. Sperm precedence in the walnut husk fly, *Rhagoletis completa* Cresson. M.S. Thesis, California State University, Hayward.

Tsiropoulos, G.J. 1978. Holidic diets and nutritional requirements for survival and reproduction of the walnut husk fly. J. Insect Physiol. 24: 239-242.

# Occurrence and Managing of Fruit Flies in Cuba

*Luis L. Vázquez*[1], *Isabel Pérez*[2], *Aurelio Navarro*[2] *and Juan C. Casin*[2]

[1] Instituto de Investigaciones de Sanidad Vegetal (INISAV), Calle 110# 514 e/5ta B y 5ta F. Playa. Ciudad de La Habana, Cuba
[2] Centro Nacional de Sanidad Vegetal (CNSV), Ciudad de La Habana, Cuba

## INTRODUCTION

Fruit flies (Diptera: Tephritidae) are considered to be pests of potential importance because they damage fruits or limit their commercialisation. They therefore constitute one of the principal problems for the fruit industry world-wide.

Of particular interest are the species of the genus *Anastrepha* Schiner widely distributed in the neotropical region where, according to Norrbom and Chung (1988), more than 180 species are known. In Cuba, five members of this genus are reported in addition to *Toxotrypana curvicauda* Gerstaeker and up to now, no other species are known (Alayo and Garcia, 1983).

Although *Ceratitis capitata* Wiedeman or pest fruit flies from the genus *Anastrepha* and *Bactrocera* do not exist in the country, a defence programme has been operating for many years which includes among other measures, fruit sampling and trapping. The present work was undertaken in order to study the status of these insects in Cuba.

## MATERIALS AND METHODS

During the periods of 1979-1980, 1989-1991 and 1995-1996, evaluations were made in different areas within the country, planted with fruit trees, in order to estimate the incidence of Tephritidae and some aspects of its behaviour.

Simultaneously, as a part of the defence programme against fruit flies, periods were established for fruit sampling and for trapping flies on a national scale (National Plant Health Centre 1994). This programme guaranteed a high level of sampling and trapping, the latter with Jackson, Yellow Panel, Rebell and McPhail traps, and with attractants specific to exotic species of quarantine interest. This system was implemented by specialists of the Territorial Plant Protection Stations and the External Quarantine Border Post, with the support of fruit farmers and the supervision and technical assistance from the provincial directorates of Plant Health and the Citrus and Fruit Trees Institute, all of them connected with the Ministry of Agriculture.

The detections of fruit flies as a result of the implementation of this programme are reported as well as the summarised characteristics of the programme and the level of trapping and sampling conducted in 1997.

## RESULTS AND DISCUSSION

### Species of Fruit Flies and their Host Plants

In Cuba, five species of fruit flies of the genus *Anastrepha* have been reported, namely *A. insulae* Stone, *A. interrupta* Stone, *A. obliqua* Macquart, *A. ocresia* Walker and *A. suspensa* Loew (Alayo and Garcia 1983). *A. obliqua* and *A. suspensa* are more common and can be found readily (Brunner et al. 1975, Vázquez et al. 1980, Plá 1983). Considering the high level of representation of the genus in the neotropical region (Norrbom and Chung 1988), it is evident that the number of species in the island must be high and all of them should be further studied.

Another member of this family is *Toxotrypana curvicauda* Gerstaecker that is peculiar to papaya (Brunner et al. 1975). However, where the genus *Anastrepha* is concerned, we do not know whether any host of the species *A. insulae* and *A. interrupta* exists in the country.

*A. obliqua* has been previously detected in guava *(Psidium guajava* L.), orange *(Citrus sinensis* Osbech*)*, mango *(Mangifera indica* L.*)*, ciruela *(Spondias cirouella* L.*)* and rose-apple *(Syzygium jambos* (L.) Alston) (Bruner et al. 1975, Alayo and Garcia 1983). The coffee tree *(Coffee arabica* L.*)*, the peach *(Prunus persica* (L.) Stokes) and the coloured mammee *(Calocarpum sapota* (Jacq) Merr.) (Table 1) are the new plant hosts found in the country.

*A. suspensa* has also been found in guava, orange, cainito *(Chrysophyllum cainito* L.) and peach (Bruner et al. 1976, Alayo and Garcia 1983). The coffee tree, the jobo *(Spondias mombin* L.) and the ciruela are new plant hosts in Cuba (Table 1). Moreover, among these plants, the coffee tree had not been reported as a host by Weems (1965) or by Norrbom and Chung (1988). This is therefore a new report for this plant.

*A. ocresia* is peculiar to the sapodilla *(Manikara zapota* (Jacq) Gilly) and it has been reported in this zapotacea by various authors (Bruner et al. 1975, Alayo and Garcia 1983).

### Behaviour

A particular analysis of the seasonal occurrence of *A. obliqua* and *A. suspensa* in guava reveals that the higher detections (larvae in fruits + adults in traps) during the months of May and September (Figure 1) coincided with the principal period of fruiting of this plant and other fruit trees such as the mango, the coffee tree and wild fruits, all of them represented in the diversity of our principal agroecosystems.

It is interesting that these species appear generally in low level populations both in trapping and fruit samplings. We think this is due to various factors, among which are the following:

- An efficient natural regulation by the parasitoid *Bracanastrepha anastrephae* (Viereck) (Hymenoptera: Braconidae).
- The predatory activity on larvae developed within the soil by a diverse and abundant formicofauna (Hymenoptera: Formicidae).
- The high mortality at the pupal stage due to soil texture and humidity impeding a deeper penetration, thus favouring the action of high temperatures and general predators.
- The harvesting systems and the phytosanitary defence programme with preventive approaches.

Table 1. Principal species of fruit flies in Cuba and their host plants.

| **Family** | Species of hosts plants | ***A. obliqua*** | ***A. ocresia*** | ***A. suspensa*** | ***T. curvicauda*** |
|---|---|---|---|---|---|
| **Anacardiaceae** | *Mangifera indica* L. | * | | | |
| | *Spondias purpurea* L. | * | | ** | |
| | *Spondias mombin* L. | * | | ** | |
| **Myrtaceae** | *Psidium guajava* L. | * | | * | |
| | *Eugenia axillaris* (Sw) Willd | * | | | |
| | *Eugenia uniflora* L. | * | | | |
| | *Syzygium jambos* (L.) Alston | * | | | |
| | *Psidium cattleianum* Sabine | | | | |
| **Rutaceae** | *Citrus sinensis* (L.) Osbeck | * | | * | |
| **Rosaceae** | *Prunus persica* (L.) Stokes | ** | | * | |
| | *Prunus occidentalis* Sw | | | * | |
| | *Eriobotrya japonica* Lindl. | * | | | |
| **Sapotaceae** | *Chrysophyllum cainito* L. | | | * | |
| | *Manikara zapotilla* (Jacq) Gilly | | * | | |
| | *Calocarpum sapota* (Jacq) Merr. | ** | | | |
| **Caricaceae** | *Carica papaya* L. | | | | * |
| **Rubiaceae** | *Coffea arabica* L. | ** | | ** | |
| **Lauraceae** | *Persea americana* Mill | | | | |
| **Flacurtiaceae** | *Casearia hirsuta* Sw. | | | * | |

* Reported by other authors for Cuba, namely Alayo and Garcia (1983), Bruner et al. (1975), Norrbom and Chung (1988), Vázquez et al. (1980), Rodríguez and Hernández (1996).

** New hosts in the country.

## Incidence in Cultivated Fruit Trees

Until now, *A. obliqua* and *A. suspensa* are the species which appear in cultivated fruit trees, both widely distributed.

The guava is the fruit tree with the largest number of detections (Figure 2) and a higher incidence of *A. obliqua*. These results do not coincide with those obtained by Vázquez et al. (1980) and Plá (1983) who found *A. suspensa* predominating on this plant.

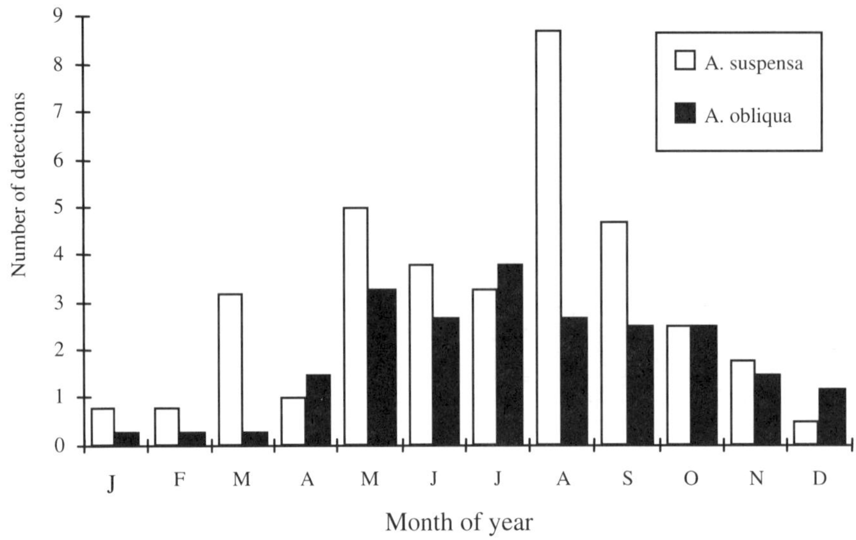


Figure 1. Seasonal occurrence of *Anastrepha* in guava.

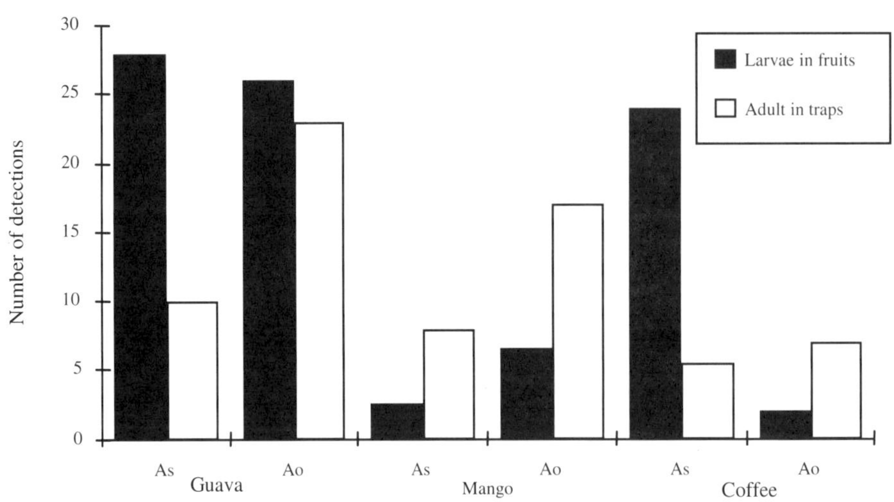


Figure 2. Comparative analysis of the number of yearly detections of *Anastrepha suspensa* (As) and *A. obliqua* (Ao) in different fruit trees.

By order of incidence, the mango where *A. obliqua* predominates comes second. These results are similar to previous reports stating that it was the most severely attacked fruit tree in the country by this species (Martinez 1963, Brunner et al. 1975). The third plant to host *Anastrepha* populations in Cuba is the coffee tree, which shows the presence of both species, although *A. suspensa* predominates.

Both *Anastrepha* species have been more recently monitored, with special programmes for official free area declarations. It is not detected in citrus fruits (Figure 2) and this collaborates with the results obtained by Vázquez et al. (1980) who, in a study carried out all over the country, did not find larvae in these fruits either.

The detection of adults in traps is greater in guava and mango tree groves, the average number of adults per trap being generally lower than one. This shows that the population levels are relatively low and principally not detected in traditional varieties. We think this is due to various factors: the grossness of the rind peel, the late fructification (in different time of the traditional varieties) and the crop in premature state.

## Programme of Phytosanitary Defence

There is in Cuba a tradition in the prevention of infestations by fruit flies, to reduce the risk of the danger of the possible introduction of *C. capitata,* a species present in other nearby countries.

In fact, the existing programmes are characterised by a strong component of detection by means of trapping and fruit inspection both in cultivated areas and in air and sea border posts. There are also strict legal regulations on the entrance of fresh fruits into the national territory (Plant Health National Center 1994).

The current programme is applied over approximately 263,579.43 hectares of citrus and other fruit trees and is stronger in the areas intended for fresh fruit export, mainly of citruses, where special monitoring systems are enforced to ensure infestation-free crops.

Basically, the Cuban programme covers two areas: the species existing in the country and the exotic ones for the purpose of quarantine (Figure 3).

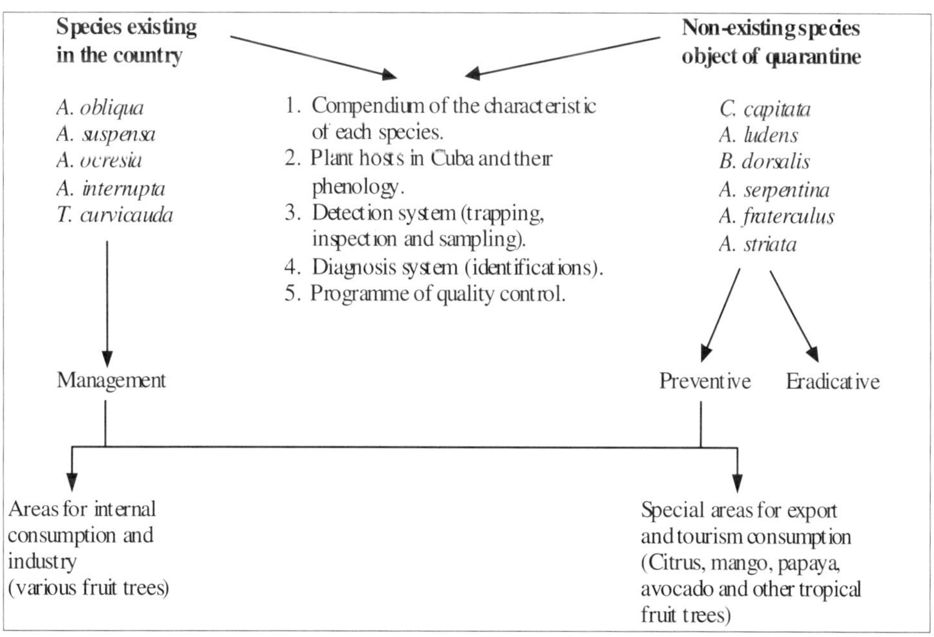


Figure 3. National fruit fly detection and management programme.

Both areas are interrelated in five instruments or procedures, mainly the system of detection and the system of diagnosis which jointly contribute towards the national alertness for these pests.

For the exotic species, the programme is essentially preventive and enacts a large number of actions of an eradicative character in case some of these species are detected.

The species of fruit flies existing in the country are managed using preventive tactics mostly of an agroecological character (Table 2), since regular applications of insecticides are inconceivable.

Table 2. Principal practices in the management of fruit flies of the genus *Anastrepha.*

| Preventive | Curative |
|---|---|
| - Trapping, fruit inspection and sampling.<br>- No new plantations of host fruit trees closer than 35 km to plantations where fresh fruits are grown for export.<br>- Good agrotechnical care of the plantations (pruning, weeding, fertilisation).<br>- Continuous gathering of fruits ending with total harvest.<br>- Preventive divulgation and training of technicians and farmers.<br>- Practice of the integrated pest management programme (IPM) in citrus.<br>- Avoidance of unnecessary application of insecticides in order to promote the activity of the bioregulators. | - Intensification of trapping, fruit inspection and sampling.<br>- Determination of the magnitude and intensity of the affected area.<br>- Total removal of fruits (from plant and fallen fruits).<br>- If there were some important foci, application of insecticide (Malathion + hydrolysed protein) only in affected areas.<br>- Regulation of fruit destination from affected areas.<br>- Treatments with Diazinon in the soils, under the affected trees. |

The efficiency of this programme is made evident by the fact that there was no detection of larvae in fresh fruits exported to other countries (oranges, grapefruit, mango, avocado, papaya, etc.) nor were there foci or affected areas in the country, indicating that insects have been managed with the preventive strategies.

The system of trapping, inspection and fruit sampling has proved to be a valuable source of national information for decision making, since it ensures the monitoring of these pests. A general idea of the implementation of this system is that in 1997, 5,325 traps were placed and serviced and 586,134.54 hectares were inspected where fruit samples were also taken.

## REFERENCES

Alayo, P. and I. Garcia. 1983. Lista anotada de los dipteros de Cuba. Edit. Clientifico Técnica (La Habana). pp. 83-84.

Bruner, S.C., L.C. Scaramuzza and A.R. Otero. 1975. Catálogo de los insectos que atacan a las plantas económicas de Cuba. 2da. edición revisada y aumentada. Instituto de Zoologia (La Habana). 288 pp.

Centro Nacional de Sanidad Vegetal. 1978. Metodologia para insepctores de Cuarentena Vegetal, DGSV, MINAGRI (La Habana). 258 pp.

Centro Nacional de Sanidad Vegetal. 1994. Programma de detección y manejo de moscas de la fruta. Ministerio de la Agricultura (La Habana). 59 pp.

Norrbom, A.L. and K. Chung. 1988. A list of the reported host plants of the species of *Anastrepha* (Diptera: Tephritidae). APHIS 81-52, USDA. 114 pp.

Plá, D. 1983. Distribution de *Anastrepha suspensa* (Diptera: Tephritidae) en las provincias de La Habana y Ciudad de La Habana en el cultivo de la guayaba *(Psidium guajava).* Cienc. Técn. Agric., P. de Plantas 6 (2): 83-88.

Rodríguez, Delys y Ana M. Hernández. 1996. Notas sobre *Anastrepha* Shiner. (Diptera: Tephritidae) en Cuba. I. Taller sobre moscas de la fruta (Ciudad Habana).

Vázquez, L., L. Larrinaga, I. Rodríguez, X. Fernández and R. Jiménez. 1980. La mosca del Caribe, *Anastrepha suspensa* (Loew) (Diptera: Tephritidae) en Cuba: VII Seminario Cientifico CENIC. Resúmenes (La Habana). pp. 18.

Weems, H.V. 1965. *Anastrapha suspensa* (Loew) (Diptera: Tephritidae). Entomology Circular No. 38. Fla. Dept. Agric., Div. of Plant Industry. 4 pp.

# An Overview of Quarantine for Fruit Flies

*E.R. Frampton*

Ministry of Agriculture and Forestry, Gerald Street, Lincoln, New Zealand

## INTRODUCTION

What is meant by "quarantine for fruit flies"? The Collins dictionary describes "quarantine" as *a period of isolation or detention, especially of persons or animals arriving from abroad, to prevent the spread of disease*. In providing an overview of quarantine for fruit flies, a broader definition needs to be applied, that is, the combination of activities required to maintain the fruit fly status of a particular geographical area – perhaps better referred to as a "quarantine system". Familiarity with New Zealand's quarantine system for fruit flies (Diptera: Tephritidae) provides a useful basis for subsequent comparison with other countries' systems where some fruit fly species may be present.

But, why have "quarantine for fruit flies"? The multivoltine life history of many species. combined with a relatively long-lived adult stage and highly fecund females, results in a high potential for rapid population increase (Bateman 1979, Fletcher 1987). These factors and the close association of fruit flies with harvested fruit or vegetables explain the high quarantine profile of these insects. However, there is no international requirement for a country to have a quarantine system and unless there are natural quarantine barriers (e.g., mountain range, oceans, deserts) that can be utilised, effective quarantine by an individual country may be an impossible task. The implementation of a successful quarantine system is very expensive and therefore, it would be expected that any benefits attained outweigh the costs (Ivess 1998). Ivess (1998) listed the following benefits from the implementation of an effective quarantine system:

- minimising production costs (including post harvest treatments),
- maintaining competitive advantages for market access due to the ongoing freedom from particular pests of quarantine significance,
- an environment free from many pests harmful to plant health, and
- the maintenance of ecosystems.

## NEW ZEALAND'S FRUIT FLY QUARANTINE SYSTEM

New Zealand is an island country and as such, enjoys the benefits of effective, natural quarantine barriers. It is also one of the few horticultural exporting countries that remains free from economically significant species of fruit flies. However, New Zealand's fruit fly free status is not maintained without significant effort. As a trading country that imports a wide range of fruit fly host material from a number of countries from all parts of the world, there is a continual threat to its fruit fly free status. In order that trade can continue without jeopardising New Zealand's status, an integrated approach (the quarantine system), ranging from offshore treatment to exotic pest response, has been developed.

Phytosanitary measures applied to imported fruit fly host material include:

- the application of a pre-export treatment which has been developed in accordance with the New Zealand Ministry of Agriculture and Forestry (MAF) standard, and
- an official bilateral quarantine arrangement, between MAF and the supply country's national plant protection organisation, which includes a description of each approved pre-export treatment.

Under New Zealand legislation, the Biosecurity Act 1993, any plant/plant product considered for importation (referred to as risk goods in the legislation) that could carry unwanted organism(s) (quarantine pest(s)) must be covered by an import health standard. An import health standard specifies the requirements to be met for the effective management of risks associated with the importation of risk goods, and includes those requirements that must be met before the commodity may be imported. Because of the different plant health status of supply countries, import health standards and hence phytosanitary measures are prepared on a country:commodity basis. Consequently, import health standards for fruit fly host material necessarily specify (any) approved pre-export treatments. The development by supply countries of such pre-export treatments is guided by reference to MAF Regulatory Authority standards, formerly referred to as NASS standards, such as:

- MAF Regulatory Authority Standard 155.02.02: Specification for Determination of Fruit Fly Host Status as a Treatment (16 September 1994),
- MAF Regulatory Authority Standard 155.02.03: Specification for the Determination of Fruit Fly Disinfestation Treatment Efficacy (22 November 1994),
- NASS Standard 158.03.06: Specification for Fruit Fly Area Freedom Monitoring Within a Country Where Harmful Species of Fruit Fly Exist (6 May 1992), and
- NASS Standard 158.03.07: Specification for Fruit Fly Area Freedom Monitoring in a Geographically Isolated Area (23 July 1992).

While New Zealand is one of the few countries to go as far as commiting to writing its requirements for the development of pre-export treatments, the New Zealand Ministry of Agriculture and Forestry also recognises the principle of equivalence and is prepared to consider alternative submissions by a supplier. Furthermore, MAF's interpretation of the term "treatment" is somewhat wider than that (i.e., treatment = officially authorised procedure for the killing, removal or rendering infertile of pests) used in the FAO Glossary of Phytosanitary Terms.

Being an island country, New Zealand is able to implement effective phytosanitary control systems. Border inspection is used to check compliance with New Zealand's import health standards for all possible pest entry pathways (i.e., passengers, vessels, mail and commercial cargo consignments). Before imported produce can be released for general (uncontrolled) distribution within New Zealand, it must first be given biosecurity clearance by an inspector under the Biosecurity Act 1993. Biosecurity clearance can only be given if an inspector is satisfied that all the requirements of the import health standard have been met.

Surveillance and exotic pest response are also components of the integrated approach and in the case of fruit flies, provide a means of early detection and eradication of these pests. New Zealand's fruit fly surveillance system is based on a national trapping network comprising some 7386 traps distributed in 45 "high risk" locations, including populated areas serving as centres for tourism and/or trade, areas of significant horticulture and areas assessed as being climatically conducive to the establishment of fruit flies. The trap used is a non-sticky, pot type lure trap, containing one of three synthetic male attractants. Cuelure, methyl eugenol or trimedlure-baited traps are respectively placed 400 metres, 1200 metres and 400 metres apart in the "high risk" locations, giving 3345 cuelure, 696 methyl eugenol and 3345 trimedlure-baited traps in total. These traps are inspected every 14 days.

If a fruit fly is detected in New Zealand in one of the traps, a response is initiated within 24 hours of the fruit fly's identification. Three response zones are defined: zone A, zone B and zone C. Zone A has a minimum radius of 200 m around the fruit fly find, zone B has a minimum radius of 1.5 kilometres around the find while zone C has a radius defined at the time of the response. There are three possible levels of response depending on the circumstances of the find. Response level 1 essentially comprises a delimiting survey. Subsequent finds can result in raising the response to level 2 or 3. An eradication programme is based on a pre-defined management strategy. A level 3 response would result in the implementation of wide area control methods such as the male annihilation technique or sterile insect technique, as appropriate. Specific techniques adopted to contain or eradicate a fruit fly outbreak would inevitably depend on the circumstances of the incursion.

In recent years, different fruit fly species have been detected: in 1990, *Bactrocera passiflorae* (Froggatt) was detected in the Auckland suburb of Otahuhu; in 1995 in Whangarei, *Bactrocera tryoni* (Froggatt) [Queensland fruit fly] was detected; and in March 1996, *Bactrocera tryoni* and *Bactrocera papayae* Drew and Hancock (referred by some in Southeast Asia as the papaya fruit fly) were detected on the same day in nearby suburbs in Auckland. On each occasion, the initiation of a level 1 response confirmed that these species had not been established.

However, the detection of two Mediterranean fruit flies (*Ceratitis capitata* (Wied.)) on 2 May 1996 rapidly led to the implementation of a level 2 response. As a result of the captures, an A zone of 200 metres radius around the finds and a B zone of 1.5 kilometres around the finds were defined. On 4 May 1996, additional traps were placed in both the A and B zones. In the A zone, trimedlure traps were placed in fruiting host trees, with at least one trap on each of the 78 properties comprising the A zone. In addition, 47 bait traps (primarily to detect the presence of female flies) were placed in fruiting host trees in the A zone. On 12 May, a further 21 trimedlure traps were placed in an area of waste ground located within the A zone, bringing the total number of trimedlure traps in the A zone to 106. In the B zone, traps were placed in fruiting host trees (where possible) at a density of between 40-55 traps/km$^2$, giving a total of 231 lure traps in the B zone. Fruit monitoring, involving the regular collections of ripe fruit from specified hosts from all A zone properties and examination for larvae, was also initiated.

On 5 May 1996, 14 more Mediterranean fruit flies (five males in five trimedlure traps, three males and six females in two bait traps) were collected from seven properties neighbouring the property where the original two males were caught (i.e., within the A zone). As a result of these additional trap captures, spot spraying of protein bait mixed with maldison insecticide was initiated in both the A and B zones. This involved the application of a minimum of one hundred 100 ml bait spots per hectare, with all host trees with fruit treated. On 6 May 1996, 13 more Mediterranean fruit flies (11 males in four trimedlure traps; one male and one female in a bait trap) were collected from four properties, all within 200 m of the original trap captures (i.e., within the A zone) on 2 May 1996. As a result of fruit monitoring, a larval infestation was located in a feijoa, *Feijoa sellowiana* Berg, located in one of the properties immediately adjacent to the property on which the original trap captures were made. Cover and ground sprays were applied to any trees with infested fruit. Subsequent finds of larvae-infested fruits brought the total number of larvae detected to 85, extracted from twelve fruits (eight feijoa, two tangelos - *Citrus reticulata* Blanc. x *C. paradisi* Mad. and two grapefruits - *Citrus paradisi* Mad) collected from just five A zone properties. There have been no larval finds since 23 May 1996.

The number of Mediterranean fruit flies trapped totalled 41 (31 males and 10 females); all fruit fly finds were from within the A zone. The last adult trap capture was on 15 May 1996.

## FRUIT FLY QUARANTINE SYSTEMS OF OTHER COUNTRIES

Many other fruit exporting countries have fruit fly quarantine systems. There are, of course, similarities to New Zealand's system, particularly in the surveillance component of the system, although the trapping grid configurations necessarily differ. However, with the exception of Chile (in recent years), those countries' systems need to take account of the presence of endemic (in some circumstances, introduced) fruit fly populations – of one, possibly more species, in a particular part of the country, if not the whole country. Where particular species of fruit flies are well established in parts of a country, more often than not, the quarantine system – especially the surveillance component if the maintenance of fruit fly free area(s) can be utilised – is designed to meet importing countries'/states' phytosanitary requirements, as well as early detection of any incursion. Examples of fruit fly free areas maintained for such purposes include the states of Chihuahua and Sonora in Mexico, the Murrumbidgee Irrigation Area of New South Wales in Australia, California, Florida (citrus production areas) and Texas (Rio Grande Valley) in the United States, and Hebei Province (ya-pear production areas) in the People's Republic of China. In some countries, certain fruit fly species have become the subject of large-scale, long term eradication programmes – the Mediterranean fruit fly programme(s) in Mexico and Central America, and the melon fly [*Bactrocera cucurbitae* ( Coquillett)] eradication programme in the Okinawa Islands of Japan (Kakinohana et al. 1993) constitute well publicised successes of the application of the sterile insect technique.

Early detection facilitates eradication, and such successes necessitate the implementation of comprehensive quarantine systems (incorporating various components) in these countries to ensure that the benefits of eradication are not squandered. Regardless of whether eradication has been achieved or whether there are just a few endemic fruit fly species present, countries must remain alert to the possible incursion of other species. Pacific Island countries are now very well aware of these risks – the establishment of *B. tryoni* in New Caledonia and French Polynesia some years ago served as a warning but more recently, the spread of the melon fly through the Solomon Islands and the arrival of *Bactrocera dorsalis* (Hendel) [the Oriental fruit fly] in French Polynesia highlight the increasing risks. French Polynesia had just one native fruit fly species, *Bactrocera kirki* (Froggatt) (Purea et al. 1997). Some Pacific Island countries have now implemented (e.g., New Caledonia), or are in the process of implementing, ongoing trapping regimes (e.g., Fiji, Vanuatu), the impetus provided by the Regional Fruit Fly Project. However, with over 4,500 known species in the family Tephritidae, more often than not, living with endemic fruit flies is the only option.

Integrated management within individual orchards, or "area control" methods which seek to suppress breeding populations over extensive areas and keep them at levels below economic pest status (Bateman 1989) may be the more appropriate strategy than eradication in many geographic or socio-economic situations.

## REFERENCES


Bateman, M.A. 1979. Dispersal and species interaction as factors in the establishment and success of tropical fruit flies in new areas. *Proc. Ecol. Soc. Australia 10:* 106-112.

Bateman, M.A. 1989. Preface. *In* Fruit Flies. Their Biology, Natural Enemies and Control. A.S.Robinson and G. Hooper (eds.).Elsevier, Amsterdam.

Fletcher, B.S. 1987. The biology of the dacine fruit flies. *Annu. Rev. Ent. 32:* 115-144.

Ivess, R. 1998. Protecting New Zealand's Agricultural Security: The Ministry of Agriculture and Forestry's Plant Biosecurity for New Zealand. AIC Biosecurity Conference, 30 April-1 May 1998.

Kakinohana, H., H. Kuba, M. Yamagishi, T. Kohama, K. Kinjyo, A. Tanahara, Y. Sokei and S. Kirihara. 1993. The eradication of the melon fly from the Okinawa Islands, Japan: II. Current Control Program. *In* Fruit Flies: Biology and Management. M. Aluja and P. Liedo (eds.). Springer Verlag New York, Inc. 492 pp.

Purea, M., R. Putoa and E. Munro, E. 1997. Fauna of fruit flies in the Cook Islands and French Polynesia. pp. 54-56. *In* Management of Fruit Flies in the Pacific. A regional symposium. A.J. Allwood and R.A.I. Drew (eds.). Nadi, Fiji 28-31 October 1996. ACIAR Proceedings No. 76. 267 pp.

# PART III.

## APPLICATION OF AREA-WIDE CONCEPT TO FRUIT FLIES

## C) BIOTECHNOLOGY, GENETICS AND MOLECULAR BIOLOGY OF FRUIT FLIES

# Population Genetics and Cryptic Species

*Bruce A. McPheron*

Pennsylvania State University, University Park, Pennsylvania, USA

## INTRODUCTION

Does the definition of a species matter for pest management purposes? Taxonomists provide us with tools – usually morphological characters – to identify a group of organisms that we call a species. The implication of this identification is that all of the individuals that fit the provided description are members of the species in question. The taxonomists have considered the range of variation among individuals in defining the species, but this variation is often forgotten when we take the concept of species to the level of management. Just as there is morphological variation among individuals, there is also variation in practically any character we might imagine, which has implications for the short and long term success of our management tactics. The rich literature on insecticide resistance should be a constant reminder of the fact that the pressure on pest survival and reproduction applied by our management approaches frequently leads to evolutionary changes within the pest species.

The degree of variation within a particular species is a defining characteristic of that species. This level of variability may have very important implications for successful management, so it is very important to measure variation and, whenever possible, the genetic basis of that variation, in a target species. Population genetic approaches can provide evidence of genetic structure (or lack thereof) among populations of a species. These types of data can be used to discuss the movement of pest populations on a local or global scale. In other cases, we may have a complex of species that share some, but not all, characteristics. Species complexes that share morphological characters (i.e., cannot be easily distinguished) but not biological characters are referred to as sibling or cryptic species.

## POPULATION GENETIC APPROACHES

An example of the use of pest population structure to track dispersal comes from the Mediterranean fruit fly, *Ceratitis capitata* (Wied.). The Mediterranean fruit fly has a sub-Saharan African origin, but it is now found widely throughout tropical and subtropical regions of the world (Headrick and Goeden 1996), where it is a constant threat to the production and distribution of a variety of agricultural products. Area-wide control programmes, either suppression or eradication, are ongoing in many locations. Successful eradication creates a need to distinguish the source of any re-infestations – are they novel introductions from outside the programme area (and, if so, what is the source), or do subsequent outbreaks represent persistence of the pre-existing population (Carey 1991)? This is an ideal situation for application of genetic methods, but successful application requires a substantial amount of background research. First, samples representing the range of variation within the pest must be assembled. Second, methods to discriminate population structure must be developed. Various genetic approaches can be complementary in this effort, because different genetic markers are subjected to different molecular evolutionary histories and may prove to be informative at different levels for unravelling the colonisation pattern of the insect (see Avise (1994) for numerous examples of the power of different classes of genetic markers). A combination of approaches often provides the greatest power of resolution. Third, methods of analysing the genetic data obtained must be developed or refined. Many questions in pest management, and this is particularly true of pest tephritids, involve evaluating probabilities associated with rare individual interceptions. Small sample size thus complicates the application of traditional statistical approaches.

Our research group has invested a significant effort in the study of the Mediterranean fruit fly mitochondrial DNA (Sheppard et al. 1992, McPheron et al. 1994, Gasparich et al. 1995, 1997). Mitochondrial DNA (mtDNA), which is maternally inherited in a haploid fashion without recombination, has been used widely in studies of colonisation (Avise 1994 and references therein). Our surveys began with the collection of material suitable for genetic analysis from nearly 100 populations world-wide (an effort that continues, often with the help of collaborators to whom we are very grateful). Initial genetic analysis revealed two mitochondrial variants (Sheppard et al. 1992), while subsequent analyses have expanded to 10 patterns useful as genetic markers (Gasparich et al. 1995, 1997). We are continuing to look for additional variation and have hope from recent studies that report other mtDNA polymorphisms (Silva 1996, Kourti 1997, Reyes and Ochando 1998).

The results of our study of the Mediterranean fruit fly mtDNA (Gasparich et al. 1997) demonstrate the highest level of variation within populations from sub-Saharan Africa (eight of the 10 mtDNA patterns), with lower levels of intraregional variation elsewhere in the species' range. Different locations around the world are more or less distinct; for example, in the New World, Central American populations have a pattern different from that which has been found in South America. That in the Andean countries is different from that of Brazil and Venezuela, with Argentina displaying some similarities to both Brazil and Peru. This variation has permitted us to examine several issues of interest to managers of area-wide control programmes. However, while the mtDNA variation reveals some geographic patterns among Mediterranean fruit fly populations, it lacks sufficient variability to answer many questions about single vs. multiple introductions into certain locations of quarantine interest.

Several research groups have been working to develop additional genetic markers for this species based upon nuclear DNA variation. Isozymes have not proved sufficiently variable to provide the resolution necessary to address colonisation issues on a global scale (Huettel et al. 1980, Malacrida et al. 1992, Baruffi et al. 1995). DNA-based analyses of nuclear genetic variation have used random amplified polymorphic DNA (RAPD), intron sequences, and allelic variation at microsatellite DNA loci. Baruffi et al. (1995) and Haymer et al. (1997) have published results from RAPDs. These studies reveal variation within and among populations, but, in large part because these analyses are restricted only to distance-based algorithms, RAPD analysis has not been widely adopted as a diagnostic tool for this species. Villablanca et al. (1998) described variation in intron sequences from four loci in Mediterranean fruit fly. Sequence data can be analysed by phylogenetic methods, leading to the possibility of applying phylogeography (phylogeny across geographically-defined populations, Avise (1994) and references therein) to the problem of Mediterranean fruit fly colonisation patterns. It may be possible to develop alternative data collection approaches, such as conformational polymorphism or restriction site polymorphism, for many of the alleles identified by sequencing, which would facilitate this type of analysis. Finally, our group has developed microsatellite loci for population analysis (Gasparich et al., unpublished; Meixner et al., unpublished). Microsatellite loci generally display numerous alleles, a fact that results in more potential markers to assort in different populations. Many alleles are widely distributed among samples from different geographic areas, while others are seen only more rarely among populations (Gasparich et al., unpublished; Douglas Prasher, USDA-APHIS, reports a similar finding based upon unpublished data; Prasher, personal communication). Given that a relatively high mutation rate is suggested as the mechanism whereby new microsatellite alleles are generated, we are less certain about the homology of apparently identical alleles from different places with microsatellite variation than we are with other types of variants, such as intron or mtDNA sequences. Thus, the greatest utility for microsatellites may be for analysis of populations in relatively restricted space and time dimensions, for example, the repeated outbreaks of Mediterranean fruit flies in California during the 1990s (Meixner et al., unpublished).

The end result is that genetic methods can inform us about the colonisation history of the Mediterranean fruit fly on a global scale. However, the most thorough answer will require multi-locus genotyping – using all of the available markers – and comparison with a consistent, comprehensive reference database. For example, it will be less useful if we continue to use one set of populations for mtDNA comparisons, but another, only partially overlapping set for intron sequences, than it would be to use the same world-wide data base for both sets of genetic markers. Furthermore, the reference database needs to be constantly updated. If we erect as our null hypothesis the idea that new populations are constantly becoming established in previously uninfested regions, we cannot assume that no new invasions are occurring in our potential source populations. Finally, we must continue to seek out new statistical approaches, especially statistical methods for dealing with very small sample sizes, to analyse our genetic data (see Roderick and Villablanca 1996). It is only with this combined programme that we will be able to provide useful information to project managers.

This approach could be taken with practically any species in which there is sufficient interest to invest in the development of appropriate markers. Berlocher (in press) summarises efforts to date to investigate population genetic structure across the family Tephritidae, both for pests and non-pests. The most intensive studies have been of *Rhagoletis pomonella* Walsh, the apple maggot fly, which is of interest both for its pest status and its role in evolutionary questions (see below and Feder 1998). Most of the genetic studies to date have used isozyme methods, which have been both inexpensive and powerful. DNA-based investigations are just beginning for most tephritid species.

## CRYPTIC SPECIES IN THE TEPHRITIDAE

The evolution of unique biological traits – ecologies or behaviours – with little or no accompanying morphological change leads to the existence of cryptic or sibling species complexes. This is a somewhat difficult, but very important, subject in tephritid biology, due to both the questions it raises for taxonomy and research and the implications it holds for management decisions. Diehl and Bush (1984) defined a variety of evolutionary levels, from variation due to non-genetic factors up through the level of species, that might lead to the recognition of different entities within a taxon previously recognised as a single species. A major problem here is that the boundaries between these different levels of organisation are often vague, and, frequently, the necessary studies have not been done to determine the genetic control of the observed variation and the amount of gene flow among the different variants in the complex. Berlocher (in press) has proposed a key for evaluating where a population might lie on the continuum between a single species with a genotype permitting plasticity of response to factors in the environment all the way through independently evolving species. Berlocher's suggestions clearly highlight the necessity of understanding the population structure discussed in the previous section of this paper.

Cryptic species complexes appear to exist in a variety of tephritid groups. Among the major pest taxa, the *Bactrocera dorsalis* Hendel complex is now a classic example of a group in which morphological change has been very slight despite significant biogeographic and host-associated evolutionary changes (Drew and Hancock 1994). This complex now consists of over 50 species, which vary in their degree of morphological distinctness and may actually be considered a complex of several cryptic species complexes. Other examples of cryptic species complexes include *Anastrepha fraterculus* Wied. (Steck 1991, Selivon 1996), *Chaetorellia jaceae* Robineau-Desvoidy (White and Marquardt 1989), *Euaresta* spp. (Berlocher 1984b) and *Aciurina* spp. (Dodson and George 1986).

Perhaps the most extensively studied cryptic species complexes occur within the genus *Rhagoletis*. Here we find a number of taxa where morphological characters do not cleanly diagnose species boundaries, but examination of ecological characters, most notably the choice of hosts for oviposition and larval development, define apparently independent gene pools. Examples include the *R. tabellaria* Fitch (Bush 1966, Berlocher 1984a), *R. cingulata* Loew (Bush 1966) and the *R. nova* Schiner (Frias 1992) species groups. The premier example of this problem is the *R. pomonella* species complex in North America, which has been studied intensively from morphological, genetic, behavioural, and ecological perspectives.

The *R. pomonella* species complex consists of four named species (Bush 1966), but a variety of other populations appear to be evolutionarily independent lineages and likely deserve species status (e.g., Berlocher et al. 1993, Payne and Berlocher 1995, Smith and Bush 1997) (Table 1). Morphological variation overlaps extensively for many pairs of taxa within this complex, although a few characters can be used with confidence (Bush 1966, Westcott 1982, Jenkins 1996). The primary basis for separating these taxa into putative species is their partitioning of host plants (Table 1). The division of taxa on this parameter is also reinforced by electrophoretic data, timing of life cycle and some studies of natural hybridisation (references in Berlocher et al. 1993, Berlocher, in press). The apparent evolution of species by adaptation to host plant in this species complex has led to proposals that the evolution of host use can provide the initial steps in speciation (reviewed by Bush 1992).

## IMPLICATIONS FOR PEST MANAGEMENT

The genetic structure of tephritid populations is clearly of interest to biologists interested in basic evolutionary questions, from the evolution of host use, to behaviour, morphology and physiology, to mechanisms of speciation. But, are there also implications for scientists who focus on pest management? The answer, of course, is yes. If we imagine a situation where a target pest species is actually a complex of species, there are several levels where this knowledge is critical.

Table 1. The *Rhagoletis pomonella* cryptic species complex, including four described taxa and five potential species-level taxa, showing the host genera and families used as larval hosts.

| *Rhagoletis* species | Host Genus | Host Family |
|---|---|---|
| *pomonella* Walsh | *Crataegus, Malus* | Rosaceae |
| *mendax* Curran | *Vaccinium, Gaylussacia* | Ericaceae |
| *zephyria* Snow | *Symphoricarpos* | Caprifoliaceae |
| *cornivora* Bush | *Cornus* | Cornaceae |
| "flowering dogwood fly" | *Cornus* (*florida*) | Cornaceae |
| "sparkleberry fly" | *Vaccinium* (*arboreum*) | Ericaceae |
| "Mexican apple maggot" | *Crataegus* (*mexicana*) | Rosaceae |
| "mayhaw fly" | *Crataegus* (*opaca* and related spp.) | Rosaceae |
| "plum fly" | *Prunus* | Rosaceae |

First, consider diagnosis. If specimens collected in surveys are a mix of species, a pest manager would require some way to determine the abundance of the actual pest in this mix. If this complex of species actually displays a variety of host associations, then the actual pest may represent only a fraction of the trap catch (important for establishing whether the pest is at a management threshold) or may be absent at certain times of year or in certain locations. Second, there are implications for the choice of management approaches. If a perceived pest population is actually a species complex, then those same issues of geographic and temporal heterogeneity apply. Management should be targeted at the actual pest, where and when it is found, which is dependent upon understanding those variables. Discussions of large-scale tactics, such as area-wide management using sterile insect techniques, are also affected by species complexes. If a species is to be mass produced for release programmes, it should be the proper species – a gene pool that is reproductively compatible with the target species. Moreover, with the global emphasis on biodiversity, there are likely to be demands that such management approaches also consider the possible negative effects on non-target species. This could include possible negative effects of sterile insect release on non-pest species in the species complex with which the factory produced sterile insects might interbreed.

Population structure, variation within what is truly a single species, also has implications for pest management plans. If a species exists in highly subdivided populations, then care must be taken to design control tactics, whether they be chemical, genetic, or biological, that reach all of these subpopulations. The potential for gene flow among populations of a pest must be considered with regards to the potential for evolution of resistance to the control method. Understanding of the natural population structure of a pest, for example, whether the pest is adapted to exist in small, highly fragmented populations, could be very important in devising ways to monitor populations after suppression or eradication campaigns. Population structure, often estimated by genetic variance, is also useful in estimating the risk of invasion by a particular pest or, as described in a previous section, evaluating the relative likelihood of particular pathways of dispersal.

Considerable effort has been placed on studies of the patterns of genetic variation in certain pest tephritids. This has involved a very productive partnership among scientists interested in basic genetic and ecological phenomena and scientists charged with pest management. Much of the funding for the basic aspects of fruit fly population structure has, in fact, come from action agencies. This long-term insight into the value of basic research for applied questions has led to a much better understanding of the pests we seek to manage and should continue to be encouraged.

## ACKNOWLEDGEMENTS

I thank the California Department of Food and Agriculture, the California Citrus Research Board, the Pennsylvania Department of Agriculture, the US Department of Agriculture National Research Initiative, the Florida Department of Agriculture and Consumer Services, the International Atomic Energy Agency, and the Mexican National Campaign against Fruit Flies for financial support of fruit fly research in my laboratory.

## REFERENCES

Avise, J.C. 1994. Molecular markers, natural history and evolution. Chapman and Hall, New York.

Baruffi, L., G. Damiani, C.R. Guglielmino, C. Mandi, A.R. Malacrida and G. Gasperi. 1995. Polymorphism within and between populations of *Ceratitis capitata*: comparison between RAPD and multi-locus enzyme electrophoresis data. Heredity 74: 425-437.

Berlocher, S.H. 1984a. A new North American species of *Rhagoletis* (Diptera: Tephritidae), with records of host plants of *Cornus*-infesting *Rhagoletis*. J. Kans. Entomol. Soc. 57: 237-242.

Berlocher, S.H. 1984b. Electrophoretic evidence for the validity of *Euaresta stelligera* Coquillett (Diptera: Tephritidae). J. Kans. Entomol. Soc. 57: 354-357.

Berlocher, S.H. (In press) Genetic population structure in Tephritidae. *In* Fruit flies (Tephritidae): phylogeny and evolution of behaviour. M. Aluja and A. L. Norrbom (eds.). CRC Press.

Berlocher, S.H., B.A. McPheron, J.L. Feder and G.L. Bush. 1993. Genetic differentiation at allozyme loci in the *Rhagoletis pomonella* (Diptera: Tephritidae) species complex. Ann. Entomol. Soc. Amer. 86: 716-727.

Bush, G.L. 1966. The taxonomy, cytology, and evolution of the genus *Rhagoletis* in North America (Diptera, Tephritidae). Bull. Mus. Comp. Zool. 134: 431-562.

Bush, G.L. 1992. Host race formation and sympatric speciation in *Rhagoletis* fruit flies (Diptera: Tephritidae). Psyche 99: 335-357.

Carey, J.R. 1991. Establishment of the Mediterranean fruit fly in California. Science (Wash. D. C.) 253: 1369-1373.

Diehl, S.R. and G.L. Bush. 1984. An evolutionary and applied perspective of insect biotypes. Annu. Rev. Entomol. 29: 471-504.

Dodson, G. and S.B. George. 1986. Examination of two morphs of gall-forming *Aciurina* (Diptera: Tephritidae): ecological and genetic evidence for species. Biol. J. Linn. Soc. 29: 63-79.

Drew, R.A.I. and D.L. Hancock. 1994. The *Bactrocera dorsalis* complex. Bull. Entomol. Res. Suppl. 2: 1-68.

Feder, J.L. 1998. The apple maggot fly, *Rhagoletis pomonella*. Flies in the face of conventional wisdom about speciation? pp. 130-144. *In* Endless forms: species and speciation. D. J. Howard and S. H. Berlocher (eds.). Oxford Univ. Press, New York.

Frias, L.D. 1992. Genetica, ecologia y evolucion de las especies chilenas del genero *Rhagoletis* (Diptera: Tephritidae). Acta Entomol. Chil. 17: 211-223.

Gasparich, G.E., W.S. Sheppard, H.-Y. Han, B.A. McPheron and G.J. Steck. 1995. Analysis of mitochondrial DNA and development of PCR-based diagnostic markers for Mediterranean fruit fly (*Ceratitis capitata*) populations. Insect Molec. Biol. 4: 61-67.

Gasparich, G.E., J.G. Silva, H.-Y. Han, B.A. McPheron, G.J. Steck and W.S. Sheppard. 1997. Population genetic structure of Mediterranean fruit fly (Diptera: Tephritidae) and implications for worldwide colonisation patterns. Ann. Entomol. Soc. Am. 90: 790-797.

Haymer, D.S., M. He and D.O. McInnis. 1997. Genetic marker analysis of spatial and temporal relationships among existing populations and new infestations of the Mediterranean fruit fly (*Ceratitis capitata*). Heredity 79: 302-309.

Headrick, D.H. and R.D. Goeden. 1996. Commentary: issues concerning the eradication or establishment and biological control of the Mediterranean fruit fly, *Ceratitis capitata* (Wiedemann) (Diptera: Tephritidae), in California. Biol. Control 6: 412-421.

Huettel, M.D., P.A. Fuerst, M. Maruyama and R. Chakraborty. 1980. Genetic effects of multiple population bottlenecks in the Mediterranean fruit fly (*Ceratitis capitata*). Genetics 94: s47-s48 (abstract).

Jenkins, J. 1996. Systematic studies of *Rhagoletis* and related genera (Diptera: Tephritidae). Ph. D. dissertation, Michigan State University, East Lansing, Michigan, USA.

Kourti, A. 1997. Comparison of mtDNA variants among Mediterranean and New World introductions of the Mediterranean fruit fly *Ceratitis capitata* (Wied.). Biochem. Genet. 35: 363-370.

Malacrida, A.R., C.R. Guglielmino, G. Gasperi, L. Baruffi and R. Milani. 1992. Spatial and temporal differentiation in colonising populations of *Ceratitis capitata*. Heredity 69: 101-111.

McPheron, B.A., G.E. Gasparich, H.-Y. Han, G.J. Steck and W.S. Sheppard. 1994. Mitochondrial DNA restriction map for the Mediterranean fruit fly, *Ceratitis capitata*. Biochem. Genet. 32: 25-33.

Payne, J.A. and S.H. Berlocher. 1995. Phenological and electrophoretic evidence for a new blueberry-infesting species in the *Rhagoletis pomonella* sibling species complex. Entomol. Exp. Appl. 75: 183-187.

Reyes, A. and M.D. Ochando. 1998. Use of molecular markers for detecting the geographical origin of *Ceratitis capitata* (Diptera: Tephritidae) populations. Ann. Entomol. Soc. Am. 91: 222-227.

Roderick, G.K. and F.X. Villablanca. 1996. Genetic and statistical analysis of colonisation, pp. 281-290. *In* Economic fruit flies: a world assessment of their biology and management. B.A. McPheron and G.J. Steck (eds.), St. Lucie Press, Delray Beach, Florida.

Selivon, D. 1996. Estudo sobre a diferenciação populacional em *Anastrepha fraterculus* (Wiedemann) (Diptera: Tephritidae). Ph. D. dissertation, Universidade de São Paulo, Brazil.

Sheppard, W.S., G.J. Steck and B.A. McPheron. 1992. Geographic populations of the medfly may be differentiated by mitochondrial DNA variation. Experientia 48: 1010-1013.

Silva, J.G. 1996. Demografia e estrutura genética de populações de *Ceratitis capitata* (Wiedemann) (Diptera, Tephritidae) e os processos de invasão e colonização no Brasil. Ph. D. dissertation, Universidade de São Paulo, Brazil.

Smith, J.J. and G.L. Bush. 1997. Phylogeny of the genus *Rhagoletis* (Diptera: Tephritidae) inferred from DNA sequences of mitochondrial cytochrome oxidase II. Molec. Phylogenet. Evol. 7: 33-43.

Steck, G.J. 1991. Biochemical systematics and population genetic structure of *Anastrepha fraterculus* and related species (Diptera: Tephritidae). Ann. Entomol. Soc. Am. 84: 10-28.

Villablanca, F.X., G.K. Roderick and S.R. Palumbi. 1998. Invasion genetics of the Mediterranean fruit fly: variation in multiple nuclear introns. Molec. Ecol. 7: 547-560.

Westcott, R.L. 1982. Differentiating adults of apple maggot, *Rhagoletis pomonella* (Walsh), from snowberry maggot, *R. zephyria* Snow (Diptera: Tephritidae) in Oregon.

White, I.M. and K. Marquardt. 1989. A revision of the genus *Chaetorellia* Hendel (Diptera: Tephritidae) including a new species associated with spotted knapweed, *Centaurea maculosa* Lam. (Asteraceae). Bull Entomol. Res. 79: 453-487.

# Sex Determination in Medfly: A Molecular Approach

*G. Saccone[1,2], A. Pane[1], G. Testa[1], M. Santoro[1], G. De Martino[1], F. Di Paola[1], C. Louis[2] and L.C. Polito[1]*

[1] Dipartimento di Genetica, Biologia Generale e Molecolare, Via Mezzocannone 8, 80134, Università "Federico II" di Napoli, Naples, Italy
[2] Dept. Of Biology, University of Crete and Institute of Molecular Biology and Biotechnology, FORTH, Heraklion, Crete, Greece

## INTRODUCTION

With the aim of developing new strategies of control to limit the damages inflicted on fruit crops by *Ceratitis capitata* (Wied.) (Medfly), a biotechnological approach is undertaken whereby female viability would be impaired or male viability would be improved following the introduction of specific genes into the genome of *C. capitata.* Only males will then be mass produced and released in the infested areas after sterilisation (Louis et al. 1987). Such conditional lethal or "advantageous" genes could be expressed in transgenic flies either female-specifically or male-specifically by using *cis* regulative sequences obtained from previously isolated endogenous *Ceratitis* genes (Saccone et al. 1996, 1998). By using molecular strategies based on a subtractive technique, we have recently isolated male-specifically expressed genes in the Medfly. Furthermore, we present the current status of the research on the *Ceratitis dsx* gene, showing sex-specific alternative splicing as in *Drosophila,* and on the *tra-inaZ* strategy to induce in *Drosophila* flies female-specific conditional lethality.

## RESULTS AND DISCUSSION

### A subtractive technique to isolate sex-specifically expressed genes in Medfly

An innovative method (Diatchenko et al. 1996, Gurskaya et al. 1996) to selectively amplify differentially expressed genes has been used to identify male-specifically expressed genes of *Ceratitis.* PolyA$^+$ RNA has been prepared using gonads-depleted sexed adult flies as starting material. cDNAs have been synthesised from male (tester) and female (driver) RNA (2 μg each) and used in a single round of subtractive hybridisation involving two hybridisation steps. First, an access of driver cDNA is added to two equivalent populations of tester cDNAs previously ligated with two different adaptors. The samples are then heat-denaturated and allowed to anneal. In the second step, the two primary hybridisation samples are mixed together with heat-denaturing and the addition of new fresh denatured driver cDNA. In the primary hybridisation step, the cDNAs common to both the driver and tester populations will anneal, while the differentially expressed sequences will be enriched in the single strand fraction. In the second step, only the subtracted single strand tester cDNAs are able to reassociate to form new kinds of hybrids with different adaptors at their ends. In a primary PCR reaction, only the cDNAs with different adaptors at each end can be exponentially amplified by using suppression PCR in which templates with the same adaptors at both ends will form pan-like structures (Clontech PCR-Select Kit). By using this method, we analysed about 400 clones. Plasmid DNAs prepared from each clone were digested, run on agarose gel and transferred by "sandwich Southern blot" onto twin Nylon membranes. We obtained 8 plus 8 blots representing the 400 clones in duplicate, and we hybridised them independently with two complex radioactive probes generated by reverse transcription using 1 μg of polyA$^+$ RNA prepared respectively from female and male adult flies. Three clones that were hybridising differently with the two probes have been analysed by Northern blot and by DNA sequencing. The Northern blot performed with each of the 3 clones revealed a transcript of about 0.9 kb (Figure 1). The sequence analysis of the 3 male-specifically expressed clones and a BLAST Database search showed that they are derived from a known *Ceratitis* gene *(Ccmasppro)* that is male-specifically expressed (Mintzas et al. 1996, Thymianou et al. 1998). This result confirmed at least that the subtractive technique was successful. Using as probes on the 8 blots from the 3 clones representing almost the entire transcription unit of the *Ccmasppro* gene, we were able to exclude all the remaining subtracted clones derived from this locus and to concentrate on new candidates of male-specifically expressed genes. In this way we have identified a *Ceratitis* cDNA fragment that according to Northern blot analysis which seems to be more expressed in males compared to females (Figure 2) and, by preliminary sequence and computer analysis, seems not to be clearly related to any known gene. Extensive sequence analysis, RACE strategy and genomic screening are underway to characterise the gene and its regulative sequences responsible for the male-specific expression. These results suggest that the technique we have chosen can be useful in isolating new sex-specifically expressed genes of *Ceratitis.* Inverting the driver/tester strategy used previously, we plan to use it to isolate new female-specifically expressed genes. Since we are using somatic tissues as the source of RNA, the genes expressed in the germ-lines are likely to be excluded from the analysis.

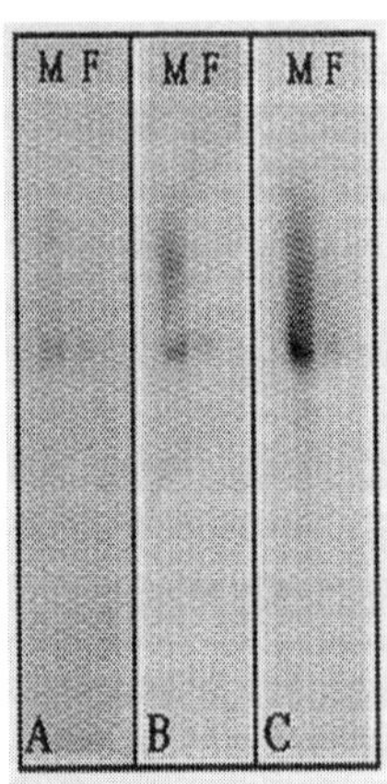


Figure 1. Northern blot analysis of polyA$^+$ RNA prepared from adult male and female *Ceratitis* flies showing a 0.9 Kb male-specific transcript identified by 3 different cDNA clones (A, B and C).

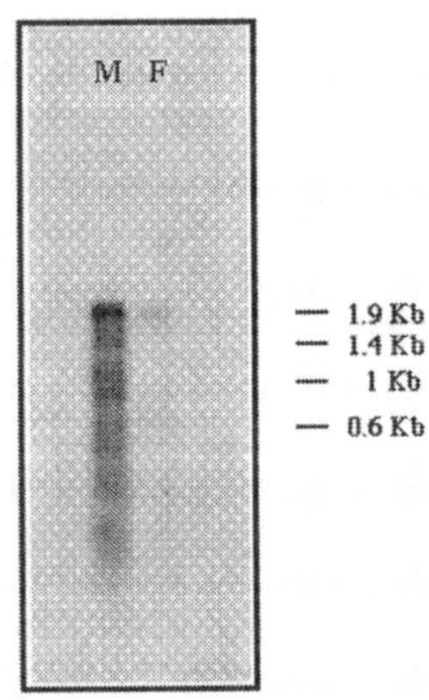


Figure 2. Northern blot analysis of polyA$^+$ RNA prepared from adult male and female *Ceratitis* flies showing the Ccmsp2 sex-specific transcriptional pattern.

## Further characterisation of the *Drosophila doublesex* homologue in *Ceratitis capitata*: a promise to drive female-specific expression of exogenous genes

We have previously isolated the *Ceratitis* homologue of the *Drosophila doublesex* sex-determining gene that shows high sequence conservation and a homologous sex-specific regulation by alternative splicing (Saccone et al. 1996). The sex-specific splicing of the *Drosophila dsx* is based on two features. First, a splicing enhancer containing 6 repeats of 13 nucleotides *(dsxRE),* in the 3' untranslated end of the female exon is required for activation of the female-specific 3' splice site by TRA (female-specific protein) and TRA-2. A sub-optimal splice site in the small intron is then required to prevent the use of this female-specific 3' splice site by the default splicing machinery in male cells (Burtis and Baker 1989). Interestingly, we found both of these features conserved in the *C. capitata dsx* homologue. In the female-specific exon, there are six well conserved copies of a 13 nucleotides repetitive element *(dsxRE)* and in the small intron, an unusual polypyrimidine tract interspersed with purines is present, which seems to be a poor acceptor site. These characteristic features seem to be also conserved in the *dsx* homologue recently isolated in *Bactrocera tryoni* Frogatt, another tephritid species (Shearman and Frommer, 1998). A genomic fragment containing the *Ccdsx* exon 3 and the female-specific exon has been inserted (under the control of *hsp-70)* in a *P* element vector to obtain *Drosophila* transgenic strains expressing the corresponding *Ceratitis* pre-mRNA. RT-PCR (reverse transcriptase-polymerase chain reaction) analysis on these strains could clarify whether the TRA and TRA-2 proteins, together with the default splicing machinery, will recognise and promote only the utilisation of the *Ceratitis dsx* female-specific exon in females only. This result could support the idea of an evolutionary conservation of the TRA and TRA-2 homologous proteins as regulators of the *Ccdsx* female-specific splicing in the Medfly. Furthermore, it could support the involvement of the *cis*-elements of dsx (found to be conserved in the sequence of the *Ccdsx* gene) in the sex-specific regulation. Similarly, a *Drosophila transformer* female-specific cDNA has also been inserted (under the control of an *hsp-70* promoter) in a *Minos* transformation vector, to obtain a *Ceratitis* transgenic strain producing the *Drosophila* TRA protein. An RT-PCR analysis of transgenic flies could clarify whether the ectopically expressed TRA is able to promote, in males, the female-specific splicing of the endogenous *Ccdsx* primary transcripts. In case the *Ccdsx* gene is also functionally conserved as a sex-determining gene, then the ectopic expression of DSXF (because of the presence of TRA) in males should cause an intersexual phenotype, as observed in some *Drosophila dsx* mutants and transgenic strains (Jursnich and Burtis 1993). These studies will lead us to understand the function and the regulation of the *Ceratitis dsx* gene and to more precisely define the *cis* and *trans* elements responsible for its sex-specific regulation. These regulative sequences could then be used to build up a transgene in which a conditional lethal function (such as the *inaZ* gene, see below) can be female-specifically expressed in female *Ceratitis* transgenic flies.

## A preliminary test in transgenic *Drosophila* strains for a strategy to female-specifically express exogenous genes

Our results so far suggest that a *transformer* homologue could be present in the *Ceratitis* genome as a female-specific splicing regulator of *Ccdsx,* therefore, cDNA library screenings are underway to isolate it. We are investigating both the possibilities of using *cis*-controlling elements of *Ccdsx* and a putatively conserved *Cctransformer* gene to drive, in transgenic Medflies, the female-specific expression of a conditional lethal gene. A potential candidate lethal gene is the ice nucleation gene *InaZ* from *Pseudomonas syringae* van Hall (Orser et al. 1985). As a preliminary step testing the functionality of an *InaZ* based sexing system, we are using regulatory sequences from the *Drosophila tra* gene to drive *InaZ* expression in transgenic *Drosophila* female larvae. This system will offer the possibility to verify the ability of INAZ to induce in females ice nucleation and cell lethality after cold shock at temperatures usually permissive for *Drosophila* larvae (between -5 and -8°C). We build up a chimeric construct bearing a *tra* genomic fragment fused to the procaryotic *inaZ* gene. In this way, the transgene pre-mRNA produced in *Drosophila* flies should be maturating either no-sex-specifically or female-specifically. Only the female-specific transcripts,though, will encode a full length chimeric TRA-INAZ protein. The construct was inserted in a pCaSper vector and 4 independent *Drosophila* transgenic lines were obtained. We started to analyse them molecularly: a RT-PCR experiment on total RNA extracted from one of the transgenic lines showed that the *tra-inaZ* pre-mRNA undergoes maturation in both sexes only in the no-sex-specific way, retaining the long stop codon-containing intron (Figure 3). The amplified products were analysed by gel electrophoresis, cloning and sequencing. It has been previously and independently shown by two groups of authors that all the sequences important for *tra* female-specific splicing are to be found within a genomic fragment of about 300 bp (Ernst 1991, O'Neil and Belote 1992). As we used in our chimeric transgene, a fragment containing these *cis* elements, our unexpected results suggest that the *inaZ* sequences may interfere with the female-specific activation of the cryptic female-specific splice site. Further studies are underway to ascertain if a higher quantity of SXL protein produced by an additional transgene will be sufficient to promote the female-specific splicing of the chimeric transgene, by competing with the repressive effect of the *inaZ* sequences.

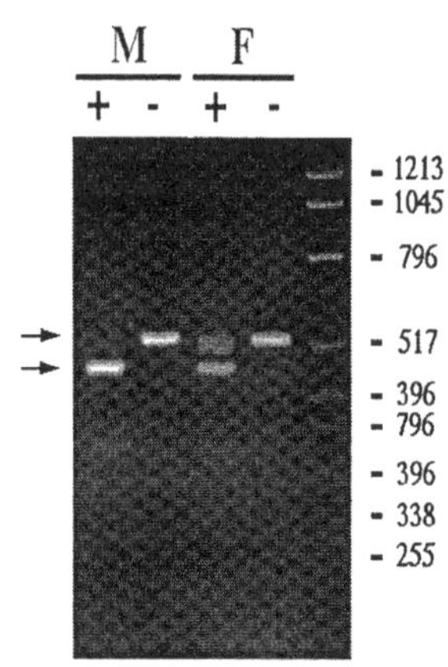


Figure 3. RT-PCR on total RNA extracted from a *Drosophila* transgenic strain carrying a *tra-inaZ* trangene and performed using a pair of primers specific for *tra* and *inaZ* sequences respectively. + and – indicate the presence/absence of Reverse Transcriptase in the reaction; the band present in the negative controls is derived from genomic DNA contamination. In males and in females, it is detected only as a major band of about 450 bp, although an additional 200 bp band was expected only in females as a product derived from a female-specifically spliced transcript.

## ACKNOWLEDGEMENTS

G. S. was supported in part by the Italian CNR (Centro Nazionale delle Ricerche; Comitato nazionale per le biotecnologie e la biologia molecolare). This work was supported by a grant P.O.P. 94/99 del. G. R. 6989 Regione Campania to C. P. and funds of the Hellenic Secretariat General for Research and Technology to C. L.

## REFERENCES

Burtis, K.C. and B.S. Baker. 1989. *Drosophila doublesex* gene controls somatic sexual differentiation by producing alternatively spliced mRNAs encoding related sex-specific polypeptides. Cell 56: 997-1010.

Diatchenko, K., Y.F.C. Lau, A.P. Campbell, A. Chenchik, F. Mogadam, B. Huang, S. Lukyanov, K. Lukyanov, N. Gurskaya, E.D. Sverdlov and P.D. Siebert. 1996. Suppression subtractive hybridisation: a method for generating differentially regulated or tissue-specific cDNA probes and libraries. Proc. Natl. Acad. Sci. USA 93: 6025-6030.

Ernst, U. 1991. Regulation of sexual differentiation in *Drosophila:* alternative splicing of the transformer primary transcript requires masking of the non-specific acceptor site in females. Inaugural PhD Dissertation, University of Zurich.

Gurskaya, N.G., L. Diatchenko, A. Chenchik, P.D. Siebert, G.L. Khaspekov, K.A. Lukyanov, L.L. Vagner, O.D. Ermolaeva, S.A. Lukyanov and E.D. Sverdlov. 1996. Equalising cDNA subtraction based on selective suppression of polymerase chain reaction: cloning of Jurkat cell transcripts induced by phytohemaglutinin and phorbol 12-myristae 13-acetate. Anal. Biochem. 240: 90.

Jursnich, V. A. and Burtis, K. C. 1993. A positive role in differentiation for the male Doublesex protein of Drosophilla. Dev. Biol. 155: 235-249.

Louis, C., C. Savakis and F.C. Kafatos. 1987. Possibilities for genetic engineering in insect of economic interest. pp. 47-57. *In* Fruitflies, Proc. II Intern. Symp., Economopoulos A.P. (Ed.). Elsevier, Amsterdam.

Loukeris, T.G., I. Livadaras, B. Arca, S. Zabalou and C. Savakis. 1995. Gene transfer into the Medfly, *Ceratitis capitata* with a *Drosophila hydei* transposable element. Science 270: 2002-2005.

Mintzas, A., K. Komitopoulou and A. Zacharopoulou. 1996. Structural analysis of heat shock and male-specific genes in the Medfly *Ceratitis capitata. In* Report: First Research Co-ordination Meeting on "Enhancement of the Sterile Insect Technique through Genetic Transformation using Nuclear Techniques", Joint IAEA/FAO, Vienna 31/9-4/10/96.

O'Neil, M.T. and J.M. Belote. 1992. Interspecific comparison of the *transformer* gene of *Drosophila* reveals an unsually high degree of evolutionary divergence. Genetics 131: 113-128.

Orser, C., B.J. Staskawicz, N.J. Panopoulos, D. Dahlbeck and S.E. Lindow. 1985. Cloning and expression of bacterial ice nucleation genes in *Escherichia coli.* J. Bacteriol. 164 (1): 359-366.

Saccone, G., I. Peluso, G. Testa, F. Di Paola, A. Pane and L.C. Polito. 1996. *Drosophila sex-lethal* and *doublesex* homologous genes in *Ceratitis capitata:* searching for sex-specific genes to develop a Medfly transgenic sexing strain. *In* Report: First Research Co-ordination Meeting on Enhancement of the Sterile Insect

Technique through Genetic Transformation using Nuclear Techniques, Joint IAEA/FAO, Vienna 31/9-4/10/96.

Saccone, G., I. Peluso, D. Artiaco, E. Giordano, D. Bopp and L.C. Polito. 1998. The *Ceratitis capitata* homologue of the *Drosophila* sex-determining gene *Sex-lethal* is structurally conserved but not sex-specifically regulated. Development 125: 1495-1500.

Shearman, D. C. and Frommer, M. 1998. The Bactrocera tryoni homologue of the *Drosophila melanogaster* sex-determination gene *doublesex*. Insect Mol. Biol. 7: 355-366.

Thymianou, S., Mavroidis, M., Kokolakis, G., Komitopoulou, K., Zacharopoulou, A. and Mintzas, A. C. 1998. Cloning and characterization of a cDNA and coding male-specific serum protein of the Mediterranean fruit fly, *Ceratitis capitata*, with sequence similarity to odourant-binding proteins. Insect Mol. Biol. 7(4): 345-53.

Zwiebel, L.J., G. Saccone, A. Zacharopoulou, N.J. Besansky, G. Favia, F.H. Collins, C. Louis and F.C. Kafatos. 1995. The *white* gene of *Ceratitis capitata:* a phenotypic marker for germline transformation. Science 270: 2005-8.

# A Population Analysis of the Queensland Fruit Fly *Bactrocera tryoni* Using Microsatellite Markers

*Hong Yu, Marianne Frommer, Merryl Robson and John Sved*

Fruit Fly Research Centre, School of Biological Sciences, A12, University of Sydney, NSW 2006, Australia

## INTRODUCTION

*Bactrocera tryoni* (Froggatt), the Queensland fruit fly or Q-fly, is the most economically important horticultural pest in Australia, infesting almost every commercial vegetable and fruit crop (Drew 1989). It is well established as a serious pest all along the east coast of Australia, as far south as the east Gippsland area of Victoria (Drew 1989). *B. tryoni* has the potential to spread across Australia to South Australia, Victoria and the tropical regions of the Northern Territory (Meats 1989) and flies classified as *B. tryoni* have been identified in the Northern Territory (Osborne et al. 1997). Winter breeding of *B. tryoni* is believed to occur only in the northern half of the range, although winged adults are usually sufficiently hardy to survive the southern winter without reproducing (Meats 1989). The number of generations per year is also a function of temperature, ranging from about eight in northern Queensland to about three in the Sydney region (Fletcher 1989).

In recent years, there has been an increase in the frequency of outbreaks in horticulturally important areas, inland in the southeast of the continent, where irrigation systems have been in use (Bateman 1991). Small-scale outbreaks occur in Adelaide (Maelzer 1990), and a more substantial outbreak was eradicated from Perth (Fisher 1996). These outbreaks mean the suspension of fruit fly free status with severe financial implications for the regions affected. To assist with the control of outbreaks within the fly-free zones and to facilitate area-wide management programmes in the endemic areas, it would be useful to have molecular genetic markers capable of identifying population structure.

Population analysis requires markers which are capable of easy and repeatable scoring and which are as polymorphic as possible. Microsatellites are now widely regarded as the most useful molecular markers available for genetic typing of individuals for kinship or larger-scale population studies (Bruford and Wayne 1993, Jarne and Lagoda 1996). Kinnear et al. (1998) have described the isolation of microsatellite markers for use in analysing the distribution of *B. tryoni* in Australia. Molecular screening led to the isolation of 16 markers, or loci, with a variety of simple and complex microsatellites. Based on a limited set of flies, these markers were all shown to be polymorphic, to varying degrees.

The analysis of population structure depends on the classification of large numbers of flies for alleles which differ by as few as one or two base pairs in length. Such an analysis is best carried out using an automated system capable of high throughput. Here we describe the first study of this type in tephritid fruit flies, in which a sub-set of the available microsatellite markers was chosen and used to give a revealing picture of the distribution of *B. tryoni* in Australia.

## MATERIALS AND METHODS

### The Isolation of Microsatellite Markers

Oligonucleotide probes were used to isolate DNA clones containing homology to known microsatellite sequences (Kinnear et al. 1998). Nineteen clones, when sequenced, were found to contain a mixture of simple and complex microsatellites. The majority of repeats were dinucleotides, with relatively short repeat lengths. Polymerase chain reaction (PCR) primers spanning the microsatellite regions were each designed to amplify a product of 100-200 base pairs in length. Sixteen loci yielded amplifiable products.

Based on the initial estimates of polymorphism of the markers (Table 1 of Kinnear et al. 1998) and the ease of scoring using the automated system (see below), we chose six markers for the current study. These markers are a mixture of simple and complex microsatellites. The nature of the microsatellite seems to be little related to its utility as a marker. To simplify the notation, we have re-designated the loci (Table 1).

Table 1. Microsatellite markers used in the present study, with their original clone designations.

| Old designation | New designation | Repeated sequence in clone |
|---|---|---|
| 11.7.1 | Bt10 | $(CA)_3T(CA)_2A(CA)_7$ |
| 11.2.5B | Bt11 | $(GT)_2AT(GT)_7$ |
| 12.8.1B | Bt14 | $(GTT)_6$ |
| Bt15 | Bt15 | $(TG)_8GTG$ |
| 3.3.5 | Bt17 | $(TG)_2CG(TG)_9$ |
| Bt32 | Bt32 | $(TG)_{11}$ |

Chromosomal localisation of the microsatellites (J.T. Zhao, personal communication) has revealed that Bt14 and Bt32 are loosely linked, and that Bt10 and Bt15 are reasonably tightly linked, within 1cM or so. We have confirmed significant levels of linkage disequilibrium between the latter pair in some, but not all, populations. As a first approximation, we have ignored the possibility of this association in the overall tests, in which chi-square values from different loci are summed to test for divergence between populations.

### Genomic DNA Preparations

For PCR screening of single flies, the head of each fly was macerated in a microfuge tube with a satay stick. DNA was extracted using a quick Chelex 100 (Bio-Rad) extraction method, based on the protocol of Walsh et al. (1991) and Moritz et al. (1992) and stored at 4°C. The remaining body and wings of each fly were individually stored in a -70°C freezer for future DNA preparation and/or checking of identification.

## Population Samples

Osborne et al. (1997) have described the trapping of wild adult male *B. tryoni* throughout eastern and central Australia using cuelure/malathion baits. Members of the CSIRO's Double Helix Club were invited to participate in a "National Fruit Fly Experiment" to collect fruit fly species attracted to cue-lure. More than 900 samples of *B. tryoni* flies from throughout the endemic region were collected in February - April 1994. Each summer since 1994, a selected number of participants from the 1994 sampling has agreed to replicate the experiment. These collections are considerably smaller in number, but span the endemic distribution.

We used flies sampled in 1994 for defining amplification conditions and for the characterisation of variability of the microsatellites. The 1995 and 1996 samples were then classified using these conditions, and the analysis reported below involves only those two years. The analysis will eventually be extended to 1994, 1997 and 1998.

The sampling sites used in the analysis are shown in Figure 1. The sites are numbered approximately from south to north for use in the analysis. It is to be noted that, although only a single sampling site is marked for Alice Springs (22), this represents the pooled flies from at least three samples for each year. In general, a maximum of 20 flies only was analysed from any one trap, and sites either comprise samples from three to four independent traps per year or are located relatively close to other sampling sites.

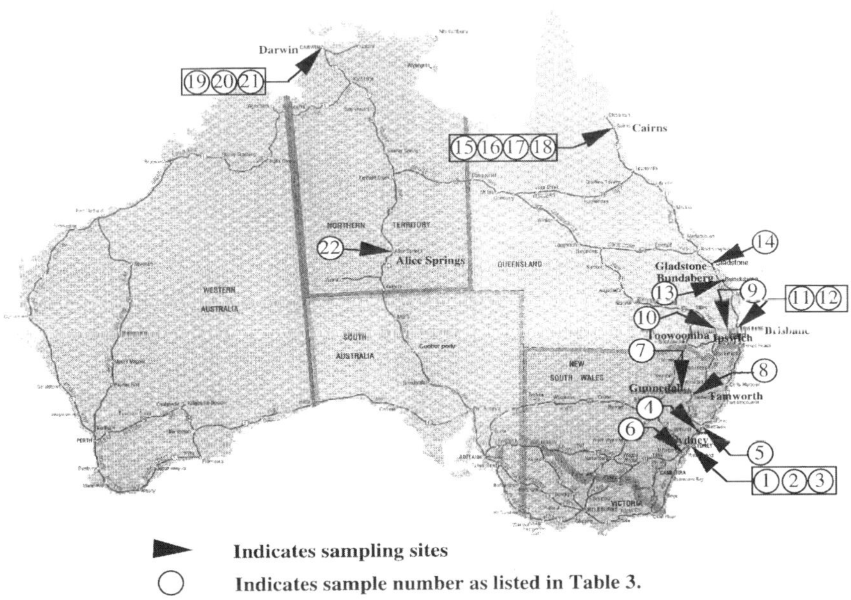


Figure 1. Map of Australia showing sampling sites.

## PCR Screening

Scoring of microsatellite alleles was carried out using a Corbett Automated DNA analyser. A method for rapid PCR with fluorescently-labelled primers and gel separation of PCR products was developed for this system. The equipment laser scans a very thin acrylamide gel for fluorescence and converts the image to a computer file, which can be displayed on a screen. The microsatellite alleles are scored and the information transferred directly to a data file. We have also developed a computer programme to combine data files for single acrylamide gels into an Excel spreadsheet entered as alleles of each microsatellite for each individual fly.

All PCR primers were synthesised by Bresatec Ltd. One primer from each pair was synthesised with a fluorescent label HEX (hexachloro-fluorescein) at the 5' end. The 10μl PCR reactions included 1x *Tth* plus buffer, 2mM $MgCl_2$, 0.2mM dNTPs, 0.15μM each primer, 0.165U *Tth* plus DNA polymerase and 3μl of a Chelex DNA supernatant.

PCR reactions were set up in 96-well plates and carried out in a HYBAID Thermal Cycler, under the following conditions: 1 cycle of 94°C for 3 min; 30 cycles of 94°C for 30 sec, 50°C for 45 sec, 72°C for 50 sec; 1 cycle of 72°C for 5 min.

2μl of PCR products were diluted into 4μl loading buffer (80% deionised formamide and approx. 50:1 bromphenol-blue), denatured at 95°C for 3 min, then placed in ice. The diluted samples (1.5μl) were run on an ultra-thin 6% polyacrylamide gel at 43°C, at 1400 volts for 1 hour, with size standards of GENESCAN-350 TAMRA standard (Applied Biosystems) and standard microsatellite alleles.

The laser-scanned gel image was analysed using ONE-Dscan (Scanalytics), one-dimensional electrophoresis analysis software and the data were transferred into the Excel spreadsheet for population analysis.

Accuracy of scoring in the samples is of primary importance. In particular, rare alleles can contribute disproportionately to the analysis, and their veracity needs to be confirmed. It is also important that different samples which are being compared should be scored under comparable circumstances. We found that, with experience, consistent results could be obtained, using the machine and analysis software.

## Methods of Analysis

Considerable literature has accumulated on methods of analysis for microsatellite markers (e.g., Feldman et al. 1997). The geographical complexity of the data set in our case, however, led us to undertake a more empirical analysis based on simple chi-square analyses. In particular, we found that the heterogeneity chi-square tests gave revealing results in the comparison of samples from different years and within and between regions. We have not attempted to estimate genetic distances or divergence times from the data.

We also used a chi-square method to test each microsatellite within each location for agreement with Hardy-Weinberg frequencies. The test was actually a test of the overall frequency of homozygotes and heterozygotes. We used the observed frequency at each locus, $p_i$, and calculated the expected frequency of homozygotes as $H = \sum_{i=1}^{k} p_i^2$, where k is the number of alleles at the locus, and the frequency of heterozygotes as 1 - H. We then calculated a one-degree-of-freedom chi-square for each microsatellite in each region. This test involves some approximation, although simulation has shown that the mean of the calculated chi-square statistic is close to its expectation of unity.

## RESULTS AND ANALYSIS

The primary data set consists of 427 flies classified from 1995 samples and 459 flies classified from 1996. However, there are some missing assignments, and approximately 2% and 3% respectively of loci from the two years were not assigned. Altogether 787 flies were classified for all six loci. We conducted the analyses on both the total data set and on the sub-set with complete classification, with very little difference in the outcome. The analyses reported below are on the sub-set with complete classification.

### Departures from Hardy-Weinberg Expectations

Each microsatellite within each location was tested for agreement with Hardy-Weinberg frequencies. The test computed and compared the overall frequency of homozygotes and heterozygotes in the samples. Given the mobility of fruit flies, any departures from random mating expectations are unlikely to be due to population structure. Similarly, since microsatellites are usually within non-gene regions, an effect of selection seems unlikely. However, the test is sensitive to two aspects of the data:

- Admixture of flies from a different population or species.
- Validity of the markers, in particular whether markers can be scored consistently and whether there are null alleles.

The sample contained flies from four samples from suburbs of Cairns. We noted significant excesses of homozygotes for four of the six markers (Table 2). Further investigation revealed that the Bt14 locus had unique alleles which were not present in any other population, and that the same flies had unusual alleles at other loci. We went back to the flies used in the analysis, and they were identified as *B. breviaculeus*, a non-pest species morphologically similar to *B. tryoni*. The analysis illustrates the use of microsatellites in species identification.

Table 2. Excess of homozygotes compared to expectation for four samples from the Cairns region.

| Sample number | Site | Microsatellite | | | | | |
|---|---|---|---|---|---|---|---|
| | | Bt10 | Bt11 | Bt14 | Bt15 | Bt17 | Bt32 |
| 15 | Edge Hill | * | * | * | | | |
| 16 | Edmonton | * | | * | | | |
| 17 | Freshwater | | | * | | | |
| 18 | Kanimbla | | | * | * | | |

* = 5% significance

With the removal of the aberrant flies from the sample, the significance of the tests for departure from Hardy-Weinberg equilibrium is shown in Table 3. All significant values were due to an excess over expectations of individuals scored as homozygotes. The data for Bt15 are noteworthy. The probable explanation is that a significant percentage of individuals scored as homozygotes are actually heterozygotes with one visible and one null allele.

Table 3. Chi-square values testing for agreement with Hardy-Weinberg expectations for the number of homozygotes in each sample.

| Sample number | Site | Microsatellite | | | | | |
|---|---|---|---|---|---|---|---|
| | | Bt10 | Bt11 | Bt14 | Bt15 | Bt17 | Bt32 |
| 1 | Asquith | 0.2 | 0.1 | 0.1 | 1.7 | 0.0 | 0.2 |
| 2 | Newtown | 0.3 | 3.3 | 0.2 | 0.0 | 0.7 | 0.4 |
| 3 | Leonay | 0.7 | 0.8 | 0.0 | 0.0 | 0.1 | 0.5 |
| 4 | Narara | 3.5 | 0.8 | 1.7 | 0.1 | 0.8 | 0.0 |
| 5 | Fassifern | 0.4 | 2.8 | 1.8 | 5.5 * | 0.8 | 1.3 |
| 6 | Ruse | 0.5 | 1.3 | 1.3 | 1.1 | 1.7 | 0.1 |
| 7 | Gunnedah | 1.3 | 2.3 | 0.5 | 1.7 | 0.0 | 1.0 |
| 8 | Tamworth | 0.7 | 2.4 | 0.1 | 1.7 | 0.8 | 0.2 |
| 9 | Ipswich | 0.0 | 0.9 | 0.4 | 0.1 | 0.5 | 1.0 |
| 10 | Balmoral | 1.3 | 1.1 | 2.1 | 1.1 | 0.0 | 0.1 |
| 11 | Sunnybank | 0.1 | 0.2 | 3.6 | 3.1 | 0.5 | 1.6 |
| 12 | Toowoomba | 0.0 | 0.1 | 0.4 | 2.4 | 0.1 | 0.0 |
| 13 | Bundaberg | 0.0 | 0.4 | 0.2 | 0.0 | 1.6 | 0.1 |
| 14 | Gladstone | 0.2 | 0.8 | 0.0 | 10.8 ** | 0.2 | 2.9 |
| 15 | Edge Hill | 1.3 | 0.4 | 0.3 | 2.0 | 0.3 | 0.0 |
| 16 | Edmonton | 0.0 | 0.3 | 0.4 | 1.0 | 0.0 | 0.8 |
| 17 | Freshwater | 0.2 | 0.5 | 0.8 | 1.2 | 0.3 | 0.9 |
| 18 | Kanimbla | 0.0 | 0.3 | 0.9 | 6.2 * | 1.9 | 0.0 |
| 19 | Howard Springs | 1.2 | 2.0 | 0.1 | 0.5 | 0.2 | 1.2 |
| 20 | Leanyer | 0.6 | 0.1 | 1.2 | 2.6 | 1.4 | 0.0 |
| 21 | Stuart Park | 0.2 | 0.0 | 0.0 | 0.0 | 0.0 | 1.3 |
| 22 | Alice Springs | 0.0 | 0.1 | 0.4 | 1.0 | 0.2 | 0.0 |
| TOTAL | | 12.3 | 20.9 | 16.4 | 43.7 ** | 12.2 | 13.5 |

* = 5% significance, ** = 1% significance

Despite the significance of the result for Bt15, the level of the null allele(s) is low – sufficiently low that it was not detected in the initial screen (with 50 flies), but required the large-scale screen to become evident. Fortunately, it is unlikely that the tests for differences within and between populations are significantly affected by such a low level of null alleles. However, in future studies, new primers should be tested, in order to minimise the occurrence of null alleles in Bt15.

## Heterogeneity between Years

Heterogeneity chi-square tests were carried out for the difference between 1995 and 1996 samples, for each microsatellite and each location separately. The chi-square values for the different microsatellites are essentially independent of each other, and can thus be pooled, along with their numbers of degrees of freedom. Table 4 gives the results for each of 21 locations summed over the six microsatellites.

Table 4. Chi-square values testing for heterogeneity between the 1995 and 1996 samples.

| Sample number | Site | Chi-square | d.f. | Sample size |
|---|---|---|---|---|
| 1 | Asquith | 26.4 | 29 | 240 |
| 2 | Newtown | 27.7 | 31 | 468 |
| 3 | Leonay | 21.0 | 28 | 240 |
| 4 | Narara | 41.4 | 35 | 408 |
| 5 | Fassifern | 30.4 | 33 | 420 |
| 6 | Ruse | 32.1 | 31 | 324 |
| 7 | Gunnedah | 39.2 | 37 | 408 |
| 8 | Tamworth | 50.6 | 40 | 456 |
| 9 | Ipswich | 43.7 | 41 | 420 |
| 10 | Toowoomba | 47.7 | 57 | 1380 |
| 11 | Balmoral | 53.6 | 44 | 444 |
| 12 | Sunnybank | 47.7 | 39 | 300 |
| 13 | Bundaberg | 49.2 | 39 | 436 |
| 14 | Gladstonc | 48.8 | 33 | 240 |
| 16 | Edmonton | 47.8 | 40 | 420 |
| 17 | Freshwater | 38.5 | 31 | 204 |
| 18 | Kanimbla | 43.9 | 39 | 324 |
| 19 | Howard Springs | 32.9 | 32 | 420 |
| 20 | Leanyer | 33.6 | 31 | 384 |
| 21 | Stuart Park | 40.8 | 31 | 336 |
| 22 | Alice springs | 21.7 | 21 | 1104 |
| TOTAL | | 818.7 | 742 | 9376 |

* Sample 15 of Figure 1, Edge Hill from the Cairns region, has been omitted because of the lack of a 1996 sample.

The data reveal a remarkable homogeneity between the two years of the sample. None of the samples is significant individually. The overall chi-square summed over all samples, is however, on the borderline of significance. Given the size of the overall sample, the power of this test is quite high, and indicates that, if there is any real change between years, it is quite small. The result needs to be extended for more years, but the picture given by the 1995 and 1996 samples is of stability of the populations.

**Heterogeneity Within and Between Regions**

We used a variety of tests to define regions. Nearby samples were tested for homogeneity in pairs and then in larger groupings. Each test was carried out for each of the six microsatellites, and the chi-square values were then combined over years for an overall test.

A sample of heterogeneity tests is shown in Table 5. There was generally homogeneity between samples from neighbouring areas. For example, the first six samples come from Sydney and its environs, and are homogeneous. Two samples from further inland in NSW, separated from Sydney by the Great Dividing Range, are also homogeneous. We tested separately for samples from southern Queensland, ranging from Brisbane to Gladstone, and for northern Queensland, around Cairns. The one exception to the homogeneity within a region came from the three samples from Darwin. A low, but nevertheless significant, level of heterogeneity was found here. We were unable to find any simple reason for this, e.g., admixture of flies of different strains/species. It might, however, be attributable to non-independence of flies from a single trap, e.g., if flies from a single brood contributed heavily to a particular trap.

Table 5. Significance tests within and between regions. Sample sites combined for the analysis are as shown in Figure 1.

| | | Chi-square | d.f. | Significance |
|---|---|---|---|---|
| WITHIN REGIONS | | | | |
| Sydney | 1, 2, 3, 4, 5, 6 | 249.9 | 250 | n.s. |
| Inland NSW | 7, 8 | 59.5 | 48 | n.s. |
| South Qld. | 9, 10, 11, 12, 13, 14 | 275.5 | 315 | n.s. |
| North Qld. | 15, 16, 17, 18 | 175.1 | 150 | n.s. |
| Queensland | 9 - 14, 15 – 18 | 73.6 | 68 | n.s. |
| Darwin | 19, 20, 21 | 128.0 | 86 | * |
| BETWEEN REGIONS | | | | |
| Sydney - inland NSW - South Qld. | | 343.4 | 146 | *** |
| Sydney - inland NSW | | 92.2 | 59 | ** |
| Inland NSW - South Qld. | | 144.1 | 71 | *** |
| Sydney - South Qld. | | 218.9 | 69 | *** |
| All 5 regions | | 1501.9 | 316 | *** |

* = 5% significance, ** = 1% significance, *** = 0.1% significance

Between regions, there was clear heterogeneity between the three southernmost regions (line 1 of the Between Region tests), and between any two of these (lines 2 - 4). We were, however, somewhat surprised by the homogeneity of flies in the Queensland sample. The distances between samples are greater in this case than the distances between heterogeneous samples from Sydney, inland NSW and southern Queensland. It may be significant, however, that the Queensland samples come from segments of the range where flies are most abundant and where breeding occurs throughout the year, whereas the more southerly samples represent marginal parts of the range. The results point to the existence of genetic differentiation in the marginal regions.

The final line of Table 5 shows the very high heterogeneity for the combined sample from the five regions. Table 6 shows that this heterogeneity holds for each of the six loci for each of the two years sampled.

Table 6. Tests of significance between the five regions conducted separately for each microsatellite and year.

| | 1995 | | | 1996 | | |
|---|---|---|---|---|---|---|
| Locus | Chi-square | d.f. | Significance | Chi-square | d.f. | Significance |
| Bt10 | 289.3 | 115 | *** | 445.6 | 140 | *** |
| Bt11 | 94.8 | 50 | *** | 254.6 | 75 | *** |
| Bt14 | 50.6 | 20 | *** | 252.9 | 30 | *** |
| Bt15 | 71.2 | 25 | *** | 71.5 | 25 | *** |
| Bt17 | 141.8 | 50 | *** | 148.7 | 55 | *** |
| Bt32 | 302.2 | 105 | *** | 379.3 | 110 | *** |

*** = 0.1% significance

As an illustration of the differences between regions, the summed data from 1995 and 1996 for microsatellite Bt32 are presented in Figure 2. The significant differences between regions are clearly reflected by the differences in patterns. The data appear to indicate a cline of frequencies along the east coast, although similar trends were not as evident for all of the other markers. The Darwin and Alice Springs populations are clearly very different from the others.

## Alice Springs

The data for this population are of particular interest in one respect. The data for Bt32 show that the rare alleles, summed as "rest" in the diagram, are completely missing from the Alice Springs population. A similar tendency was seen for the absence of the rarer alleles at each of the six loci. The homogeneity of the 1995 and 1996 season results shows that the same trend is found in each year.

As an empirical test of the presence of low frequency alleles, we summed, over all loci, the frequency of alleles whose total frequency was less than 5%. The numbers for the five regions were:

| Sydney | inland NSW | Queensland | Darwin | Alice Springs |
|---|---|---|---|---|
| 4.3% | 6.9% | 8.5% | 5.6% | 0.1% |

The data appear to be consistent with the notion that:

- The Alice Springs population was founded from a relatively small outbreak.
- The population remains isolated from the rest of the Australian population, at least from the east coast and Darwin populations.

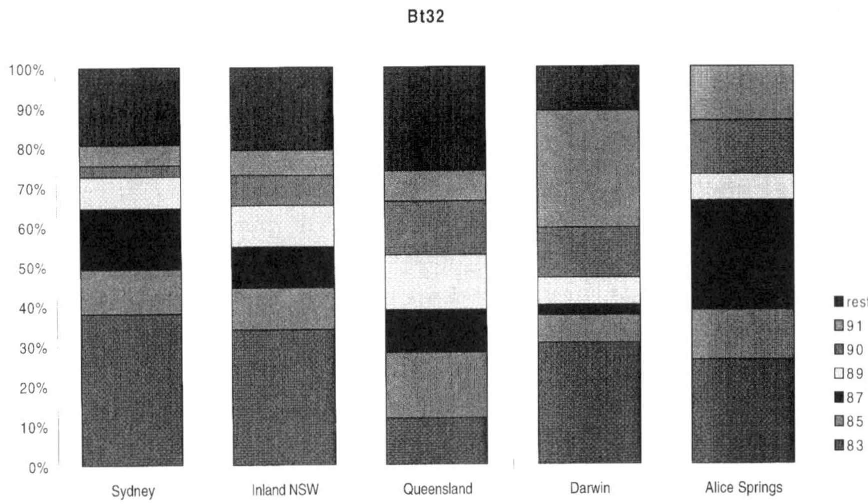


Figure 2. Frequencies of Bt32 alleles in five regions.

## DISCUSSION

Six polymorphic microsatellite markers which gave reproducible and easily scorable bands were used to analyse Australian endemic *B. tryoni* populations. It was found that:

- The endemic population varies significantly between regions with the greatest molecular differences between flies found in the Northern Territory and those found on the east coast.
- The east coast population can be divided into subpopulations and these subpopulations remained stable in 1995 and 1996.
- The Alice Springs population is most probably the result of a single introduction which has persisted.

This shows that microsatellites have sufficient variability to define population subdivisions, even in a continuous distribution of fruit flies, such as that along the east coast of Australia.

The DNA microsatellites could, therefore, be used to address specific problems of the horticultural industries, such as:

- In areas which are normally fly-free, but which suffer occasional outbreaks, the problem of establishing whether outbreaks in successive years are derived from an overwintering population or from separate introductions.
- In areas where flies are endemic, the problem of defining to what extent fruit flies move, or are transported, between different geographical areas. DNA microsatellite analysis would be used to map the population structure within endemic areas, to allow effective area-wide planning.
- In areas where the sterile insect technique (SIT) is used as a control measure, the problem of differentiating between wild and sterile factory flies.

We have made a preliminary investigation of the possibility of identifying the source population of a single fly drawn at random from one of the populations sampled in this study. The simulations indicate that six microsatellites are not sufficient to identify the population of origin with any degree of confidence, except possibly in the case of a fly from Alice Springs, where greater than 90% success is achieved. A set of about 30 microsatellites will be required. We are currently testing isolated outbreaks in a fly-free area to identify overwintering populations.

## ACKNOWLEDGEMENTS

We would like to thank Professor R.A.I. Drew for assistance with species identification and Dr J.T. Zhao for allowing us to use her data on the chromosomal location of microsatellites. Funding contributions were received from Woolworths Supermarkets and the Horticultural Research & Development Corporation.

## REFERENCES

Bateman, M.A. 1991. The impact of fruit flies on Australian horticulture. Horticultural Policy Council Report No. 3. Dept. of Primary Industries, Canberra.

Bruford, M.W. and R.K. Wayne. 1993. Microsatellites and their application to population genetic studies. Curr. Opin. Genet. Dev. 3: 939-943.

Drew, R.A.I. 1989. The tropical fruit flies (Diptera: Tephritidae: Dacinae) of the Australasian and Oceanic regions. Mem. Queensl. Mus. 26: 1-521.

Feldman, M.W., A. Bergman, D.D. Pollock and D.B. Goldstein. 1997. Microsatellite genetic distances with range constraints: analytic description and problems of estimation. Genetics 145: 207-216.

Jarne, P. and P.J.L. Lagoda. 1996. Microsatellites, from molecules to populations and back. Trends Ecol. Evol. 11: 424-429.

Fisher, K. 1996. Queensland fruit fly (*Bactrocera tryoni*): eradication from Western Australia. pp. 535-541. *In* "Fruit Fly Pests – A World Assessment of their Biology and Management, B.A. McPheron and G.J. Steck (eds.). St. Lucie Press, Florida.

Fletcher, B.S. 1989. Life history strategies of tephritid fruit flies. pp. 195-208. Vol. 3B. *In* Fruit flies: their biology, natural enemies and control. A.S.Robinson. and G.Hooper (Eds.). Elsevier, Amsterdam.

Kinnear, M.W., H.S. Bariana, J.A. Sved and M. Frommer. 1998. Polymorphic microsatellite markers for population analysis of a tephritid pest species, *Bactrocera tryoni*. J. Mol. Ecol. 7: 1489-1496.

Maelzer, D.A. 1990. Fruit fly outbreaks in Adelaide, S. A., from 1948-49 to 1986-87. I. Demarcation, frequency and temporal patterns of outbreaks. Aust. J. Zool. 38: 439-452.

Meats, A. 1989. Bioclimatic potential. pp 241-252. Vol. 3B. *In* Fruit flies: their biology, natural enemies and control. A.S.Robinson. and G.Hooper (Eds.). Elsevier, Amsterdam.

Moritz, C., C.J. Schneider and D. Wake. 1992. Evolutionary relationships within the *Ensatina enchscholtzii* complex confirm the ring species interpretation. Syst. Biol. 41: 273-291.

Osborne, R., A. Meats, M. Frommer, J.A. Sved, R.A.I. Drew and M.K. Robson. 1997. Australian distribution of 17 species of fruit flies (Diptera: Tephritidae) caught in cue lure traps in February 1994. Aust. J. Entomol. 36: 45-50.

Walsh, P.S., D.A. Metzger and R. Higuchi. 1991. Chelex 100 as a medium for simple extraction of DNA for PCR-based typing from forensic material. BioTechniques 10, 507-509.

# PART III.

## APPLICATION OF AREA-WIDE CONCEPT TO FRUIT FLIES

## D) GENETIC SEXING AND THE STERILE INSECT TECHNIQUE OF FRUIT FLIES

# The "Combi Fly Concept" Revisited: How Much Radiation is Required to Sterilise Males of a Genetic Sexing Strain?

*Gerald Franz*

Entomology Unit, FAO/IAEA Agriculture and Biotechnology Laboratory, Agency's Laboratories, A-2444 Seibersdorf, Austria

## INTRODUCTION

In the "Combi Fly Concept", as proposed by Steffens (1982, 1983), the radiation dose necessary to sterilise males can be reduced if the males carry a translocation between the male determining chromosome (Y) and an autosome. Such a chromosome rearrangement is required in genetic sexing strains (GSS) to link a selectable marker to sex. Due to the translocation, GSS males are already partially sterile and it might be possible to use a lower dose to sterilise them for release in SIT programmes. An added bonus to this approach is that this partial sterility is inherited by males. A reduction in radiation dose would be desirable because it would increase the quality of the insect.

## MATERIALS AND METHODS

### Rearing

Larvae were reared in a carrot-yeast medium, adults were fed on a sugar-yeast medium and pupation occurred in fine sand. Flies were kept under a 12 h light-dark cycle.

### Strains

EgyptII (EgII) is a wild type strain that has been reared in the laboratory since 1983. Two Y-autosome translocation strains were used: a single, T(Y;5)1-61, and a double translocation, T(Y;3;5)1-56. Their isolation and characterisation are described in Franz et al. (1994). The translocation strains were maintained by crossing, at each generation, females of a white eye white pupae (*we wp*) strain with translocation males with wild type phenotype.

### Irradiation and Crosses

Pupae were irradiated in air one day before adult emergence in a $^{60}Co$ source. For each experiment, 100 irradiated males were mass mated with 20 females from a *we wp* strain in 12 litre cages. Eggs were collected 7, 9 and 15 days after irradiation and hatch was assessed four days after egg deposition; survival was determined at the pupal and adult stages.

## RESULTS AND DISCUSSION

### Inherited Sterility

Inherited sterility caused by the translocation alone was measured by mating translocation males with females and determining egg hatch and adult survival. This sterility is caused by the segregation behaviour of translocations during meiosis where either alternate or adjacent-1 segregation occurs (Figure 1). Only the former (types 1 and 2 in Figure 1) leads to genetically balanced and, therefore, fully viable adults. Adjacent-1 segregation (types 3 and 4 in Figure 1), which occurs roughly at the same frequency as alternate segregation, leads to two types of genetically unbalanced karyotypes; one carries a deletion of part of the translocated autosome (type 3 in Figure 1) while the other carries a triplication (type 4 in Figure 1). Individuals carrying a deletion are most severely affected and, in most cases, they die during embryonic development and reduce egg hatch. Many simple translocations, like T(Y;5)1-61, show a reduction in egg hatch of approximately 25% which is close to the theoretical value assuming that all deletion carrying embryos die (Franz and Kerremans 1993). More complex translocations produce an even lower hatch, e.g., 50% for T(Y;3;5)1-56 (Table 1) and 34% for T(Y;A;A;A) (Franz and Kerremans 1993). In contrast to the deletion karyotype, the triplication karyotype (type 4 in Figure 1) is usually viable at least until the pupal stage. Consequently, egg hatch alone cannot be used to measure the full extent of the inherited sterility of translocation carrying males. The results in Table 1 clearly show this effect. In the wild type strain, virtually all embryos develop into adults while both translocation strains show severe lethality between the larval and adult stage. The resulting overall sterility levels, measured at the adult stage, are in good agreement with the theoretical expectations (i.e., 50% for a single translocation and 75% for double translocation; see Franz and Kerremans 1993).

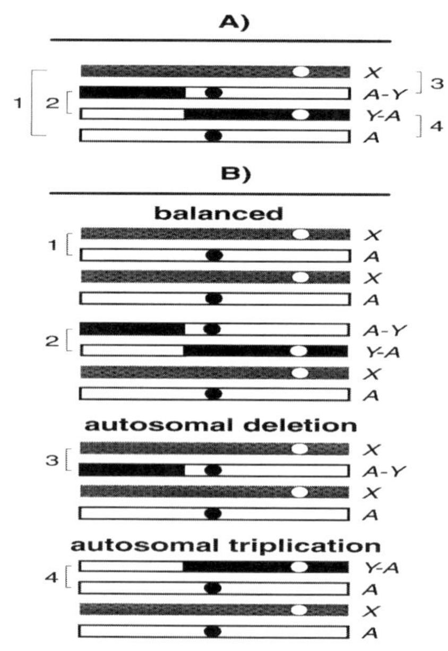


Figure 1. Karyotype of a Y-autosome translocation male (A) and its progeny (B).

For the "Combi Fly Concept", Steffens proposed that translocation strains involving two or ideally even three autosomes be used to achieve maximum levels of inherited sterility. However, even under small-scale laboratory rearing, such strains are very difficult to maintain. In case of a T(Y;A;A;A), only 12.5% of all eggs develop into adults. For mass rearing, this would require a tremendous colony size and does not appear to be practical. Furthermore, the complexity of the translocation has severe consequences for the stability of the strain and for the strain management. In the sheep blowfly, *Lucilia cuprina*, a T(Y;3;5) translocation was mass reared but spontaneous breakdown occurred rapidly (Foster et al. 1980). Based on the current practical experiences with the temperature sensitive lethal based GSS, developed for the Medfly, a simple T(Y;A) translocation is the most practical and economical solution, i.e., the maximum level of inherited sterility achievable is 50%.

Table 1. Effect of radiation on survival of larvae and adults of 3 Medfly strains.

| | wild type EgII | | | T(Y;5)1-61 | | | T(Y;3;5)1-56 | | |
|---|---|---|---|---|---|---|---|---|---|
| Dose (Gy) | eggs used | % eggs hatched | % adults recovered | eggs used | % eggs hatched | % adults recovered | eggs used | % eggs hatched | % adults recovered |
| 0 | 1,780 | 96.47 | 92.42 | 1,500 | 69.07 | 46.80 | 1,350 | 48.15 | 29.56 |
| 10 | 2,000 | 58.55 | 53.95 | 1,500 | 46.20 | 27.13 | 1,362 | 29.96 | 17.62 |
| 20 | 2,000 | 51.85 | 45.85 | 1,500 | 34.00 | 21.00 | 1,123 | 18.70 | 11.93 |
| 30 | 2,000 | 31.95 | 30.30 | 1,500 | 20.93 | 14.27 | 1,289 | 11.71 | 7.06 |
| 40 | 2,000 | 22.70 | 19.85 | 1,500 | 11.87 | 6.07 | 1,090 | 8.81 | 5.14 |
| 50 | 2,000 | 13.70 | 11.65 | 1,500 | 8.67 | 4.87 | 1,300 | 4.69 | 3.69 |
| 60 | 2,000 | 10.95 | 9.25 | 1,500 | 5.89 | 3.49 | 1,275 | 4.31 | 1.96 |
| 70 | 2,000 | 7.45 | 5.90 | 1,500 | 3.13 | 1.60 | 1,500 | 1.33 | 0.60 |
| 80 | 2,000 | 2.95 | 2.90 | 1,500 | 0.87 | 0.67 | 1,500 | 1.13 | 0.60 |
| 90 | 2,000 | 2.00 | 1.80 | 1,500 | 0.65 | 0.60 | 1,500 | 0.40 | 0.27 |
| 100 | 2,000 | 1.20 | 1.15 | 1,500 | 0.47 | 0.20 | 1,500 | 0.53 | 0.40 |

## COMBINED EFFECTS OF INHERITED AND RADIATION INDUCED STERILITY

Wild type and translocation males were irradiated with doses ranging from 10 to 100 Gy. The total sterility, radiation induced plus the genetic load from the translocation segregation, is shown in Figure 2. As described by Steffens (1982, 1983), at any given sterility level, the remaining fertility is reduced by 50% for a single translocation and by 75% for a double translocation. As a consequence, the sterility curves for the three types of males converge at higher doses/sterility levels. In other words, when sterility approaches 100%, the impact of the inherited sterility becomes negligible and the radiation doses required to reach such high levels of sterility are virtually the same for wild type and translocation carrying males. This is true irrespective of the level of inherited sterility, i.e., independent of the complexity of the translocation.

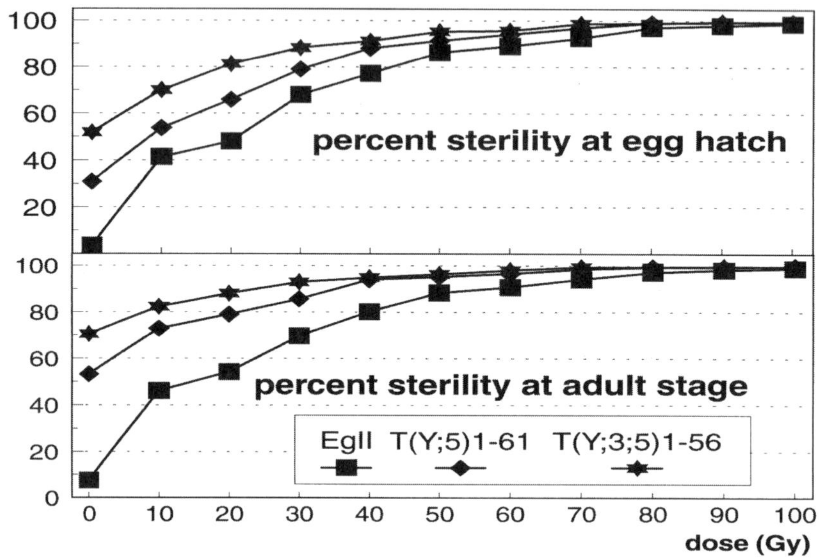


Figure 2. Sterility levels at different doses of gamma-irradiation for three Medfly strains.

**$F_1$ Sterility**

If complete sterility is required, this can only be achieved by irradiation treatment and the contribution of the inherited sterility is not utilised. As radiation has some negative impact on the quality of the released insect, Steffens (1983) proposed that 90% to 95% egg sterility can be achieved by a dose of ca. 60 Gy in case of a double translocation and 40 Gy in case of a triple translocation (wild type males require 85 Gy). This egg sterility does not show the full impact of the translocation induced sterility, i.e., the late lethality of the triplication type adjacent-1 offspring. If total sterility, measured at the adult stage, is taken into account, a level of 95% sterility is reached for a single translocation at a dose of 50 Gy, i.e., 25 Gy below the level required for wild type males and 50 Gy below the levels required for full sterility.

If males carrying a single translocation are released that are only 95% sterile as a consequence of the combined effects of translocation and radiation-induced sterility, the surviving 5% of $F_1$ offspring carry a significant genetic load, especially the $F_1$ males. They are at least 50% sterile as they have inherited the translocation from their fathers and, in addition, they will also carry new radiation-induced chromosome abnormalities (Table 2, scenario 2A).

Pericentric inversions are now being isolated to generate GSS with improved stability and into which genetic material can be easily introduced. Inversion heterozygotes eliminate recombinants resulting from recombination in the inverted segment. In females where the recombination frequency during meiosis is high, this leads to a significant level of sterility. In a GSS, with an inversion, the females would be homozygous and the males heterozygous for the inversion. If males which were 95% sterile were released and they carried an inversion in addition to the translocation, all male offspring would inherit the translocation and all the female offspring would be heterozygous for the inversion and have reduced fertility. Consequently, the $F_1$ offspring is at least 30% to 50% sterile (Table 2, scenario 2B) and the same is true for those flies in the following generations that receive these two types of chromosome rearrangements.

Steffens (1986) showed that irradiating wild type Medfly males with a sub-sterilising dose causes a substantial sterility in the $F_1$ generation. For example, males irradiated with 50 Gy produced partial sterility in up to 60% of the $F_1$ single pair families. This genetic load is in addition to that caused by the Y-autosome translocation and, if present, the inversion. All three effects taken together significantly reduce the overall reproductive potential of the released males despite the fact that they are not completely sterilised by the radiation.

Table 2. Comparison of releases with completely sterile GSS males versus partially sterilised GSS males.

| | P: | $F_1$: | |
|---|---|---|---|
| **SCENARIO 1**: irradiation to complete sterility | no offspring | | |
| **SCENARIO 2A**: rradiation to 95% sterility; T(Y;A) | 5% offspring ($F_1$) | a) males produce 50% sterile sperm due to T(Y;A)<br>b) additional sterility due to irradiation of parental male | a) females produce 100% fertile eggs<br>b) some sterility due to irradiation of parental male |
| **SCENARIO 2B**: rradiation to 95% sterility; T(Y;A) + inversion | 5% offspring ($F_1$) | a) males produce 50% sterile sperm due to T(Y;A)<br>b) additional sterility due to irradiation of parental male | a) females are heterozygous for the inversion and produce a reduced number of fertile eggs<br>b) additional sterility due to irradiation of parental male |

## CONCLUSIONS

Irradiation has a negative impact on the quality of the released insects (e.g., references in Steffens 1986). For example, Hooper and Katiyar (1971) showed that increasing the dose from 50 Gy to 110 Gy did not introduce more sterility into a population if competing un-irradiated males are present. The authors conclude that increased radiation-induced sterility is nullified by decreased sexual competitiveness. The "Combi Fly Concept" explores all sources of sterility that exist in a GSS, i.e., that caused by the translocation, the inversion and the $F_1$ sterility caused by the irradiation of the parental male. The data presented here show that even with a single translocation, the irradiation dose can be reduced by approximately 50%. The released males would be 95% sterile and the remaining fertility would lead to a significant genetic load in the $F_1$.

One of the reasons why SIT is expensive is the fact that very high overflooding ratios have to be maintained probably due to the reduced competitiveness of the irradiated insects. Consequently, it should be considered whether or not the same level of sterility can be achieved in the field if fewer but only partially sterilised males from a GSS are released. Furthermore, not all programmes are aimed at the eradication of the wild population. In control programmes, complete sterility in the field is not the goal and the wild population is only suppressed below a defined threshold. This level must be maintained by continuous releases. In principle, the "Combi Fly Concept", using GSS males with the components described above, would allow release ratios to be significantly reduced with the consequence that either programme costs can be reduced or larger areas can be treated.

Further research is required to reach the level of confidence that is required before such an important modification of the current procedures becomes acceptable for action programmes. This would include field cage tests as well as larger-scale field release tests to determine: 1) whether males, irradiated only with 50 Gy, are significantly more competitive, and 2) whether, indeed, fewer males can be released to reach the same sterility level in the field population.

## REFERENCES

Foster, G.G., R.H. Madern and A.T. Mills. 1980. Genetic instability in mass reared colonies of a sex-linked translocation strain of *Lucilia cuprina* (Wiedemann) (Diptera: Calliphoridae) during a field trial of genetic control. Theor. Appl. Genet. 58:169-175.

Franz, G., E. Gencheva and P.H. Kerremans. 1994. Improved stability of genetic sex-separation strains for the Mediterranean fruit fly, *Ceratitis capitata*. Genome 37:72-82.

Franz, G. and P.H. Kerremans. 1993. Radiation induced chromosome aberrations for the genetic analysis and manipulation of the Mediterranean fruit fly, *Ceratitis capitata*. pp. 187-194. *In* Proceedings of the International Symposium on Management of Insect Pests: Nuclear and Related Molecular and Genetic Techniques, 19-23 October 1992, Vienna, IAEA, Vienna. 669 pp.

Hooper, G.H.S. and K.P. Katiyar. 1971. Competitiveness of gamma-sterilised males of the Mediterranean fruit fly. J. Econ. Entomol. 64:1068-1071.

Steffens, R.J. 1982. The combi-fly, a new concept for genetic control of fruit flies. Naturwissenschaften 69:600-601.

Steffens, R.J. 1983. Combination of radiation- and translocation-induced sterility for genetic control of fruit flies. Entomol. Exp. Appl. 33:253-258.

Steffens, R.J. 1986. Autocidal control, pp. 139-154. *In* Fortschritte der Zoologie, Bd. 32. Biological Plant and Health Protection. G. Franz (ed.). Fischer Verlag, Stuttgart-New York.

# Comparison of Medfly Male-Only and Bisexual Releases in Large Scale Field Trials

*P. Rendón[1], D. McInnis[2], D. Lance[3] and J. Stewart[4]*

[1] USDA-APHIS-PPQ, Methods Development Station, c/o AmEmbassy – Guatemala City, APO AA 34024-3319, USA
[2] Tropical Fruit, Vegetable and Ornamental Crop Research Laboratory, USDA-ARS Honolulu, Hawaii 96823, USA
[3] USDA-APHIS-PPQ, P.O. Box 1064, Waimanalo, Hawaii 96795, USA
[4] Regional Medfly Program, Mexico-Guatemala-United States, USDA-APHIS-IS, GUATEMALA USDA-APHIS-IS, Guatemala

## INTRODUCTION

The sterile insect technique (SIT) has been used in several locations around the world for suppression or control of resident Mediterranean fruit fly (Medfly) *Ceratitis capitata* (Wied.) populations (Steiner et al. 1962, Rhode et al. 1971, Fisher et al. 1985). Successes in eliminating incipient infestations of this species have been reported (Cunningham et al. 1980, Wong et al. 1986).

In small-scale tests McInnis et al. (1994) showed that releases of only male Medflies has proved to be more efficient at inducing sterility into wild Medfly populations than the release of males and females together. When a mechanism for the elimination of Medfly females from SIT became available through the development of genetic sexing strains (GSS), it was suggested that economic advantages from the integration of male-only releases in control/eradication programmes were to be expected (Robinson et al. 1986). However, stability of strains under mass rearing conditions hampered their further use (Hooper et al. 1987). Improved GSS are now available (Franz et al. 1994) and allow for large-scale studies on the sterility induced into feral populations using GSS and different sex ratios of sterile Medflies.

Field studies, using released sterile males and females of bisexual strains, have varied in the area of their assessment from: 25$km^2$ (Fisher et al. 1985), 13 $km^2$ (Wong et al. 1986), 5 $km^2$ (Nitzan et al. 1993), 3.7 $km^2$ (Cirio et al. 1987) to 0.01-0.04 $km^2$ (McInnis et al. 1994). Release ratios (sterile:wild) have also varied from: 50:1 (Wong et al. 1986) to overflooding of 200:1 - 1,000:1 (Fisher et al. 1985, McInnis et al. 1994). In addition, sampling data used to validate population suppression by SIT have included adult trap catches, fruit infestation and percent egg hatch, with egg hatch being regarded as the most reliable index of success (Wong et al. 1986). Other studies have used the same indices but given them a different order of priority (Cirio et al. 1987, Nitzan et al. 1993, McInnis et al. 1994).

The current study's objective was to compare the induced sterility in the field when: 1) males and females of a standard strain (Petapa) are released together, and 2) only males from the VIENNA 4/Tol-94 GSS were released. During three consecutive years, induced sterility from releases of more than 1 million sterile Medfly/week of both VIENNA 4/Tol-94 and Petapa standard strain in different field sites (4 $km^2$) were studied.

**Area-Wide Control of Fruit Flies and Other Insect Pests,**
ed. K.H. Tan. Penerbit Universiti Sains Malaysia, Penang, 2000.

## MATERIALS AND METHODS

### Location of Testing Areas

The test area was in the Retalhuleu Department (southwestern Guatemala, 176 km from Guatemala City). Facilities for quality control tests and aerial releases were close to the test sites and an incipient pest population was present.

Two plots situated at about the same altitude (600 - 800 metres above sea level), having a size of 36 km² each (6 km x 6 km), were selected for the releases. The plots were 6 km apart. A core area 4 km² (2 km x 2 km, 700 - 800 m above sea level) in each plot was subjected to aerial releases of sterile Medfly. Each of the two core areas was subjected to several weeks of releases each of three years. One of the plots in a location known as Palajunoj (lat. 14°40'20", long. 91°37'23") was subjected during 1995 to releases of VIENNA 4/Tol-94. The second plot located in an area known as Asintal (lat. 14°40', long. 91°45') was subjected to releases of the standard Petapa strain (males and females). During the first two years, the areas for releases were reversed. Each plot had two buffer zones (B1-B2) surrounding the core area (C) (Figure 1). A third plot, in an area known as Mujulía (lat. 14°44'12", long. 91°47'28") located at 4.1 km from plot 2, served as control; from here, samples to correct sterility values were taken.

| | | | | | |
|---|---|---|---|---|---|
| (VIII)<br>B2 | B2 | B2 | (I)<br>B2 | B2 | (II)<br>B2 |
| B2 | B1(11) | B1(12) | B1(1) | B1(2) | B2 |
| B2 | B1(10) | a1 a2<br>Core<br>b1 b2 | a3 a4<br>Core<br>b3 b4 | B1(3) | B2 |
| (VII)<br>B2 | B1(9) | c1 c2<br>Core<br>d1 d2 | c3 c4<br>Core<br>d3 d4 | B1(4) | (III)<br>B2 |
| B2 | B1(8) | B1(7) | B1(6) | B1(5) | B2 |
| (VI)<br>B2 | B2 | B2 | (v)<br>B2 | B2 | (IV)<br>B2 |

Figure 1. Layout of each of the plots where either VIENNA 4/Tol-94 or standard strains were released.

## Trapping

Inside the core area, 16 Jackson (plus 16 McPhail traps during 1995) traps were placed, in order to determine: 1) the distribution of released insects, and 2) the ratio of sterile:fertile flies. Jackson and McPhail traps were paired for their location in the field.

Each of the four 1 km² blocks in the core area had 4 Jackson and 4 McPhail traps. Traps were placed at the four corners of each km², leaving a 250 m gap, perpendicular to the border. Each pair of traps was separated by a 500 m gap. Lines of traps (four in total) were identified (a-d) from northsouth and within each line, traps were numbered (1-4) from west to east (Figure 1). In addition to these traps, there were others in the buffer zones. Buffer zone B1 had one trap/km² located at the centre of the block; traps were numbered 1-12 clockwise. Buffer zone B2 had (8 traps; 0.4 traps/km²) located only at each of the ends of the row of blocks and in the central position; traps were numbered with Roman letters I-VIII clockwise. The outer buffer zones of the two plots were separated by a distance of 6 km in a straight line.

Jackson traps were relured with trimedlure plugs every four weeks. McPhail trap bait was refilled/or changed according to needs every week.

Trap location in the field was determined using a hand-held Geographical Positioning System (GPS) Magellan™ and geographical position (latitude/longitude).

In the core area (C), wind socks were placed in each of the corners (yellow colour) of the plot. An additional signal was placed in the centre of the four blocks (orange colour).

## Releases of Medfly

The release of sterile Medflies was performed by low-altitude aerial release from aircraft using paper bags. Each of the bags contained ca. 15,000 pupae. The bags contained crumpled paper to provide resting surfaces for emerging flies. As a food source, one square of paper (ca. 12 x 12 cm) was impregnated with a dried 10% sugar solution (Nadel et al. 1977). Bags containing pupae were closed by folding the tops and stapling them. Bags were then placed on wooden racks inside a temperature controlled room (24±1°C; 65±5% R.H.).

No light was provided during the 3-3.5 days prior to release. Release densities per plot per strain were planned to be 4,000 flying adults/ha/week for the Petapa (standard) and 2,000 for the VIENNA 4/Tol-94 strain. Actual released numbers were corrected by quality control data (pupal emergence and adult flight ability), as well as by the ratio of sterile:fertile flies observed in the field by trapping. Releases were carried out twice weekly (Wednesdays and Sundays) throughout each of the tests.

On the mornings of each release, 50% of the total number of bags required for the week were loaded into a temperature controlled vehicle and transported to the airstrip (10-15 min away). Bags were loaded into the rear of the airplane. Aircraft ferry time from the airstrip to plot was ca. 2 min. Releases were made by opening the bag with a blunt knife and pushing the bags through a tube placed at the rear of the airplane's fuselage. Differences in pressure ensured that the bag was sucked out so that it would land on the target plot. Precision was achieved by flying at optimum speed and altitude and by paying close attention to the wind socks on the ground for evidence of wind direction and velocity. The same procedure was carried out twice per week. The airplane flew at an altitude of 100-150 m. The cruising speed was 130-150 km/hr.

Flights were carried out one day per week from south to north and then the reverse (odd numbers of flight lanes first, even numbers second). On the second day of releases, flights were carried out from west to east. This procedure avoided possible bias in fly distribution. The distance between flight lanes was approximately 250 m.

### Sampling

Adult fly sampling was conducted once per week on the same day in both treatment locations (control area data were usually collected one day later). Samples for sterile:wild flies were taken from the traps. Trap inserts were removed from the Jackson traps and taken to the laboratory for analysis. Egg samples for the assessment of egg hatch in the field (to determine induced sterility) were taken from coffee berries. Ripe coffee berries were collected, taken to the laboratory, and held at temperatures of ca. 25-27°C and 60-70% R.H for 2-3 days. Berries were then dissected to find eggs and egg shells. Egg hatch was assessed to calculate induced sterility which was corrected for egg hatch found in the control area.

## RESULTS

### 1995 Evaluation

Results of initial monitoring of the wild population showed that levels in each plot were different (Figure 2-A). Palajunoj, the area where releases of VIENNA 4/Tol-94 were carried out, had a fly/trap/day (FTD) capture of 2.027 (week 15, 1995). Asintal, the area where the Petapa strain was released, had an FTD of 0.098 for the same week. Releases of sterile insects were carried out from weeks 15-27. Assessment of release rates showed that the actual release densities varied throughout the test in both locations (Figure 3-A). Sterile:wild ratios increased rapidly for the VIENNA 4/Tol-94 plot, while for the Petapa strain sterile:wild ratios increased slowly (Figure 4-A). Due to this result, release rates of the VIENNA 4/Tol-94 strain were decreased during the last weeks of the test (Figure 3-A). Sterile:wild ratios of VIENNA 4/Tol-94 from weeks 21-27 were higher than Petapa:wild by a factor greater than 2.5 (Figure 4-A).

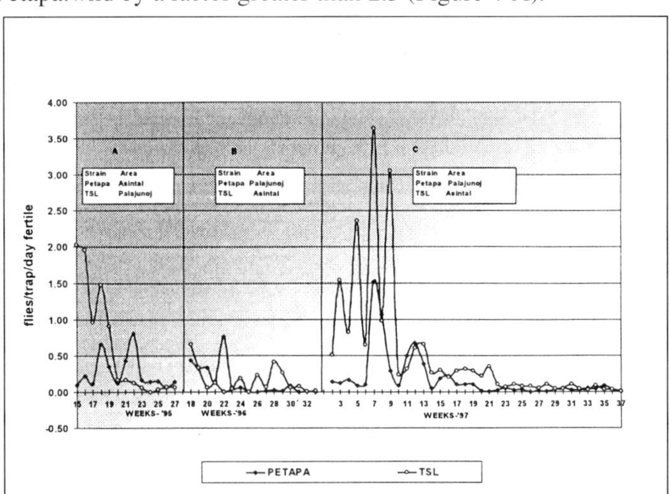


Figure 2. Wild fly population (flies /trap/day in areas where Petapa and TSL strains were released

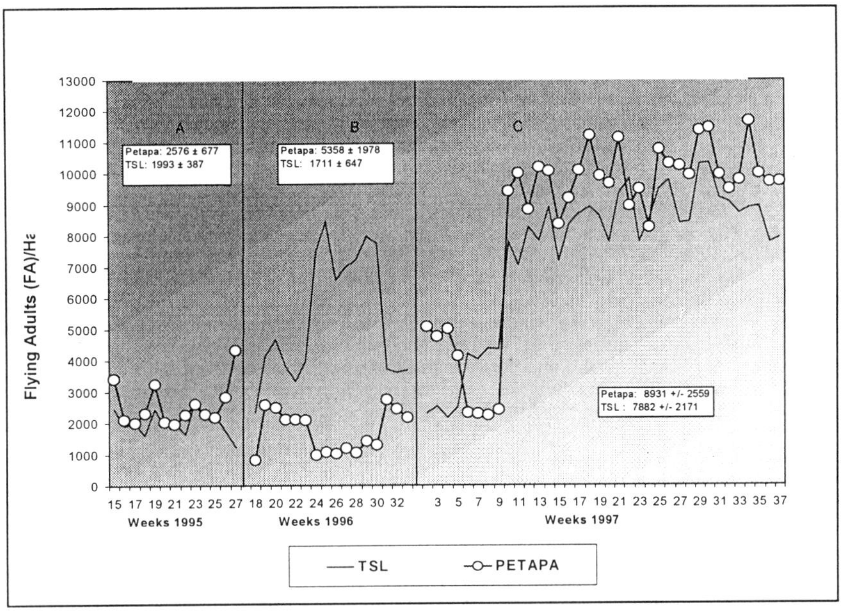


Figure 3. Released densities (FA/Ha), Petapa (males and females) and TSL (males-only) strains

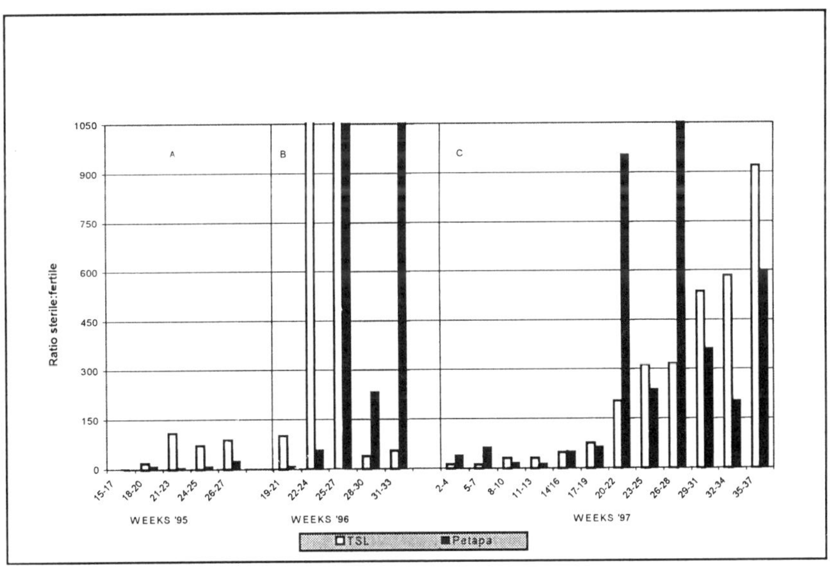


Figure 4. Ratio of sterile:wild males (based on FTD capture) during releases of Petapa and TSL strains

During the 1995 releases, the stability of the VIENNA 4/Tol-94 strain in terms of percentage of males in the releases, gradually decreased (Figure 5). In spite of this, the wild fly population levels (Figure 2-A) continued to decline from 2.027 FTD to 0.062 FTD in week 27.

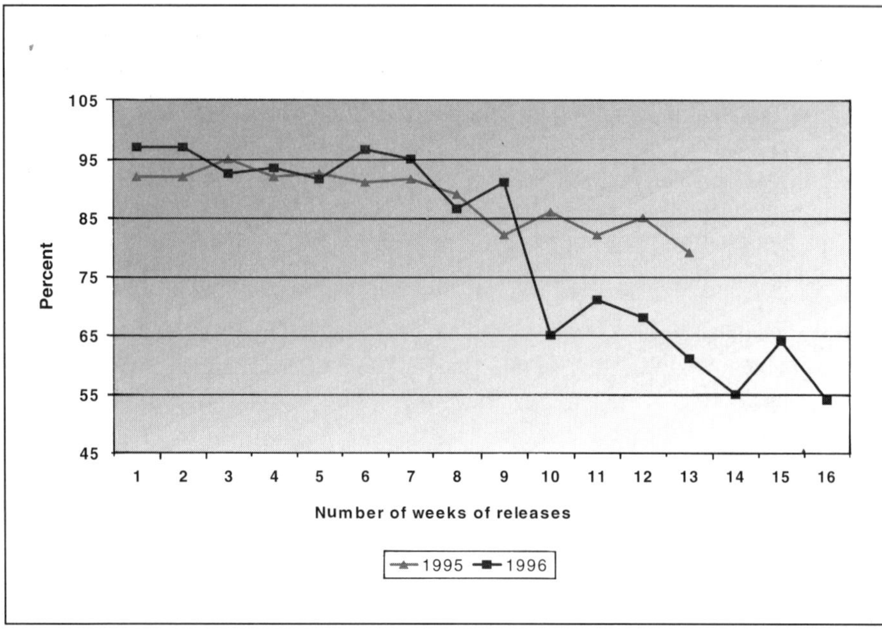


Figure 5. Percentage of TSL strain males in serial releases

Egg sampling results showed that greater egg sterility (3-4 times more) was found in the the VIENNA 4/Tol-94 plot, compared to either the control area or the standard Petapa strain plot (Figure 6).

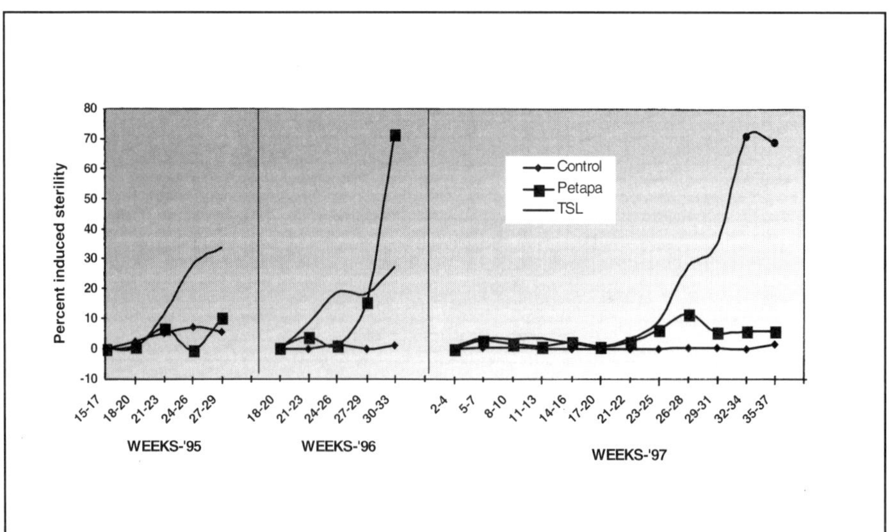


Figure 6. Comparison of induced sterility for TSL and Petapa strains.

## 1996 Evaluation

During 1996, the Petapa strain was released in Palajunoj and VIENNA 4/Tol-94 in the Asintal area. Initial population levels of wild (fertile) flies were similar (Figure 2-B) in both plots. Insect releases started during week 18. Release densities (flying adults/hectare) were higher for the Petapa strain than during 1995 (Figure 3-B). Mean released flies for the Petapa strain were 5,358 ±1978 while for VIENNA 4/Tol-94, the density remained similar to the previous year (1995=1993±387; 1996=1,711±647). In the Petapa plot, the resulting sterile:wild ratio was well above that for VIENNA 4/Tol-94 for the last 6 weeks of the test (Figure 4-B). The percentage of males in the VIENNA 4/Tol-94 releases decreased below the previous year, especially during the last 7 weeks of the evaluation (Figure 5). This fact, coupled with the higher sterile:wild ratios reached with the Petapa strain, produced similar induced sterility levels for weeks 27-29 in both plots (Figure 6).

## 1997 Evaluation

The Petapa strain was released in Palajunoj, VIENNA 4/Tol-94 was released in the Asintal area. Monitoring of the wild population levels showed that the VIENNA 4/Tol-94 plot had a higher infestation level than the Petapa plot (Figure 2-C). Sterile releases started during week 2. Released densities increased over previous years (Petapa= 8931±2559 and VIENNA 4/Tol-94= 7882±2171). High release densities produced an increase in sterile:wild ratios and by week 20, sterile:wild ratios had reached more than 100:1 for the remainder of the test (Figure 4-C).

Percentage males in the releases of VIENNA 4/Tol-94 remained greater than 99%, due to the introduction of the continuous filter rearing system (Fisher, 1996). Figure 6 shows that the VIENNA 4/Tol-94 plot reached high levels of induced sterility and is in direct relation to the increase in sterile:wild ratios (Figure 7).

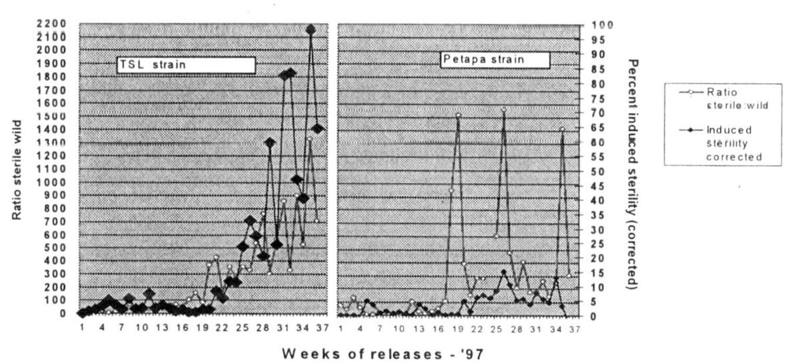


Figure 7. Comparison of sterile:wild ratio for each strain with corrected sterility.

## DISCUSSION

Large field evaluations of the bisexual (Petapa) and male-only (VIENNA 4/Tol-94) strains have been carried out during three consecutive years (1995-1997). With the exception of the second year, in which the bisexual strain achieved higher induced sterility levels during the last weeks of the test, results suggest that the male-only strain is several fold more effective under test conditions. In the second year, the male-only strain achieved higher levels of induced sterility during 12 out of 16 weeks, but during the last weeks of the evaluation, the bisexual strain achieved higher induced sterility than expected (based upon 1995 and 1997 results). This could be attributed to: 1) the higher sterile:wild ratio for the bisexual strain during the final 6 weeks, or 2) the lower percentage of males in the male-only strain, (during the final 7 weeks of the test). The latter raises the question as to the percentage of females that male-only releases could tolerate, before responding like releases of a bisexual strain.

During 1997, the male-only strain achieved higher induced sterility levels and sterile:wild ratios were in general higher during this test than in previous years for both strains. However, the bisexual strain did not exceed 15-20% induced sterility levels, in spite of sterile: wild ratios exceeding 500:1.

Additonal improvements from those already observed in favour of the male-only strain would be expected due to a reduction of the irradiation dose (Rendón 1996). During the current tests, the male-only strain was exposed to the same level of irradiation as the bisexual strain in order to compare responses on the same basis.

## REFERENCES

Cirio, U., M. Capparella, and A.P. Economopoulos. 1987. Control of Medfly *(Ceratitis capitata* Wied.) by releasing a mass reared genetic sexing strain. pp. 515-522. *In* Fruit Flies. Proceedings of the second international symposium. Colymbari, Crete, Greece. A.P. Economopoulos (ed.). Elsevier Science Publishers, Amsterdam.

Cunningham, R.T., W. Routhier, E.J. Harris, N. Cunningham, N. Tanaka, L. Johnston, W. Edwards, R. Rosander and J. Vettel. 1980. A case study: eradication of Medfly by sterile-male release. Citrograph 65: 63-69.

Fisher, K. T., A. R. Hill, and A. N. Sproul 1985. Eradication of *Ceratitis capitata* (Wiedemann) (Diptera:Tephritidae) in Carnarvon, Western Australia. J. Aust. Entomol. Soc. 24: 207-208.

Fisher, K. 1996. Review of development of the Medfly Toliman/Vienna 42 *TSL* strain in El Pino, Guatemala. 15-22 June, 1996. A report by FAO/IAEA Entomologist K. Fisher to Guatemala moscamed programme.

Franz, G., E. Gencheva, and Ph. Kerremans. 1994. Improved stability of genetic sex-separation strains for the Mediterranean fruit fly, *Ceratitis capitata*. Genome. 37: 72-82.

Hooper, G.H.S., A.S. Robinson and R.P. Marchand. 1987. Behaviour of a genetic sexing strain of Mediterranean fruit fly, *Ceratitis capitata*, during large scale rearing. pp. 349-362. *In* Fruit Flies. Proceedings of the second International Symposium. Colymbari, Crete, Greece. A.P. Economopoulos (ed.). Elsevier Science Publishers, Amsterdam, New York,

McInnis, D. O., S.Y.T. Tam, C. Grace and D. Miyashita. 1994. Population suppression and sterility rates induced by variable sex-ratio, sterile insect releases of *Ceratitis capitata* (Diptera: Tephritidae) in Hawaii. Ann. Entomol. Soc. Am. 87: 231 - 240.

Nadel, D.J., J. Monro, B.A. Peleg and H.C.F. Figidor. 1977. A method of releasing sterile Mediterranean fruit fly adults from aircraft. J. Econ. Entomol. 60: 899 - 902.

Nitzan, Y., Y. Roessler and A.P. Economopoulos. 1993. Mediterranean fruit fly: effective control of the Mediterranean fruit fly by genetic sexing male only sterile insect technique releases during 1989-1990. pp. 331-344. *In* Management of Insect Pests: Nuclear and Related Molecular and Genetic Techniques. IAEA-SM-327, Vienna, Austria.

Rendón, P.A. 1996. Development and evaluation of a temperature sensitive lethal (TSL) genetic sexing strain of the Mediterranean fruit fly *Ceratitis capitata* (Wied.) Unpublished Ph.D. thesis. Manchester University U.K. 267 pp.

Rhode, R.H., J. Simon, A. Perdomo, C. Gutierrez, F. Dowling and D.A. Lindquist. 1971. Application of the sterile-release technique in Mediterranean fruit fly suppression. J. Econ. Entomol. 64:708-713.

Robinson, A.S., U. Ciro, H.S. Hooper and M. Capparella. 1986. Field cage studies with a genetic sexing strain in the Mediterranean fruit fly, *Ceratitis capitata*. Entomol. Exp. Appl. 41: 231-235.

Steiner, L.F., W.C. Mitchell and A.H. Baumhover. 1962. Progress of fruit fly control by irradiation sterilisation in Hawaii and the Mariana Islands. J. Appl. Radiat. Isot. 13:427-434.

Wong, T.T.Y., R.M. Kobayashi and D.O. McInnis. 1986. Mediterranean fruit fly (Diptera:Tephritidae): Methods of assessing the effectiveness of sterile insect releases. J. Econ. Entomol. 79: 1501-1506.

# Dispersal and Survival of Sterile Male (TSL Strain) Mediterranean Fruit Flies

*S. Barbosa[1], A. Mexia[1] and R. Pereira[2]*

[1] Secção Autónoma de Protecção Integrada, Instituto Superior de Agronomia (UTL), Lisboa, Portugal
[2] Programa Madeira-Med, Direcção Regional de Agricultura, Madeira, Portugal

## INTRODUCTION

The Mediterranean fruit fly, *Ceratitis capitata* (Weid.), is a major pest of fruit in the Portuguese Autonomous Region of Madeira (Vieira 1952). The Medfly attacks more than 40 species of fruit (Vieira 1952, Pereira et al. 1996). It was reported as a pest primarily below 400 m but occurred up to 700 m on the south coast.

Madeira (32°N, 17°W) is located 980 km west-southeast of mainland Portugal. Its two principal islands (Porto Santo (50 km$^2$) and Madeira (740 km$^2$)) are populated by 255,000 people. Fruit and vegetable production is widespread on Madeira island but not on Porto Santo island because of poor soil and limited rainfall. The climate of Madeira is variable, depending upon altitude and location (northern/southern). On the whole, the climate is moderated by the effects of the surrounding sea.

The sterile insect technique (SIT) is a genetic method of insect control. Large numbers (frequently more than 50 million/week) of the target insect are reared in specially designed factories (Pereira et al. 1997). These insects are sterilised with gamma radiation and released by aircraft into the target area. Mating between the factory reared sterile males and fertile wild females produces no progeny. Thus, if sufficient sterile males are introduced into the target area on a continuous basis there is a very high probability that fertile wild females will mate with sterile males (Hendrichs et al. 1995). Under these conditions, the birth rate of the target species is greatly reduced and will rapidly reach zero if no fertile insects are brought into the target area.

Little work has been done with all male releases because it is only very recently that all male strains have been mass produced. It has been postulated that releasing only sterile male Medflies could result in better distribution and perhaps increased longevity of the released individuals.

To evaluate this hypothesis, tests of longevity and dispersion of sterile males were conducted. Ground releases were used as no aircraft was available for serial release.

## MATERIALS AND METHODS

The studies were conducted on the south coast of Madeira Island with the longevity study done in the Ribeira Brava area and the dispersal study in the Quebradas area (Figure 1). A temperature sensitive lethal (*tsl)* strain of Medflies (Vienna 60) that was being reared in the Madeira Medfly factory was used. This strain was developed by the IAEA in Vienna. The female eggs and neonate larvae were killed by moderate heat, permitting the production of males only.

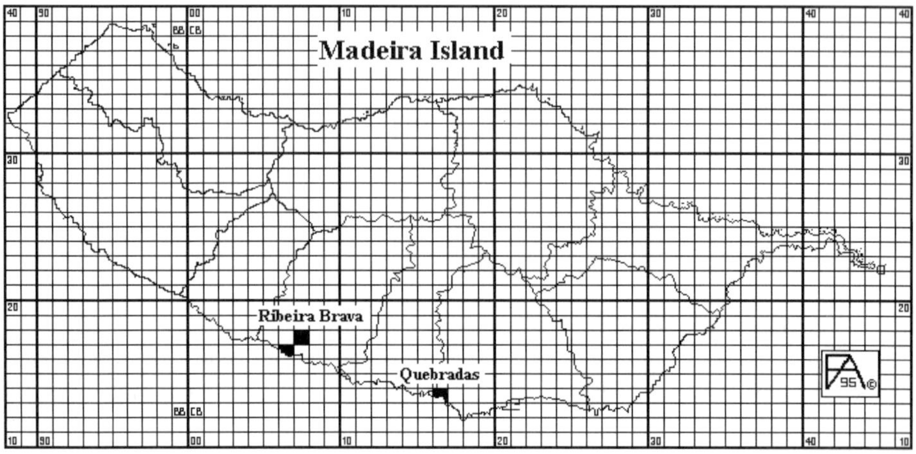


Figure 1. Location where the studies were done on Madeira Island (**longevity study**: Ribeira Brava; **dispersion study**: Quebradas).

## Longevity

The longevity study was conducted from 3-13 July 1997 in the Ribeira Brava area (Figure 1). Two-day old Day-Glo marked male flies were released from paper bags at ground level at a rate of 1,000 flies/hectare. Each bag contained the flies that had emerged from 6,000 pupae. Emergence was about 75%. Releases were made every 100 m in a grid pattern in the 3 $km^2$ test area. Thus a total of about 225,000 marked sterile male Medflies were released from 60 releases sites in the test area.

Twenty four hours after release, 10 trimedlure baited Jackson traps were randomly placed in the central part of the release area. The distance between each trap was a minimum of 100 m. The traps were installed at 17:00 and removed at 18:00; this is coincidental with the peak of Medfly daily activity at this time in the year. The same trapping procedure was followed for 9 days.

The captured flies were taken to the laboratory to determine if they were wild or marked released flies. This was done in the standard manner by observing the squashed fly heads under a microscope and UV light to determine if the flies were marked with Day-Glo dye.

To determine longevity of similar flies that were protected from some environmental pressures and prevented from leaving the test area, a "blank" test was run. The contents of one bag (flies that emerged from 6,000 pupae) were placed in each of 2 plexiglass cages (240 mm x 300 mm x 300 mm, standard quality control cages). Empty pupae, non-emerged and dead flies were removed.

The cages were placed in the shade under a citrus canopy. Water and food (sugar + hydrolysed protein, 3:1) were supplied. Dead flies were counted and removed from the cages daily for a 9-day period at the same time that the traps were serviced.

An additional test was conducted to compare longevity of flies fed on sugar only or sugar plus hydrolysed protein prior to release. Tests were conducted in plexiglass cages held in the same location as the "blank" test described above.

Sugar is the normal food provided to flies that are to be released in SIT programmes. Three day-old flies are most commonly released; thus the flies feed on the sugar for 48 to 72 hours.

Either sugar alone or a 3:1 mixture of hydrolysed protein and sugar was provided for flies during the holding period prior to release. Water was supplied. Paper bags containing 3,000 pupae were used. Two days after emergence, the flies were placed in the plexiglass cages, one bag/cage. Twelve cages were used, 6 with each treatment. Empty pupae, non-emerged and dead flies were removed.

The flies in six of the plexiglass cages were provided with water and 3:1 hydrolysed protein : sugar. The flies in the other 6 cages were given neither water nor food. The dead flies were removed daily for 10 days.

### Dispersion

The dispersion test was conducted 19-28 August 1997 in the Quebradas area (Figure 1). The flies were released from a single site in the centre of the test site. A total of 24,000 two day old Day-Glo marked sterile male Medflies were released. This is a common method for Medfly dispersion studies (Plant and Cunningham 1991). The same procedures regarding flies and trapping were used as described for the longevity tests.

Four trimedlure baited Jackson traps were placed in each of the 4 cardinal directions, one each at distances of 25 m, 50 m, 100 m and 200 m from the release site. The host material was plentiful, primarily tropical fruit, in the test site.

Traps were installed for a one-hour trapping period (17:00 to 18:00) each day, starting 24 hours after release of the Day-Glo marked sterile male Medflies. Determination of wild or released flies was as described for the longevity test.

## RESULTS AND DISCUSSION

### Longevity

Figure 2 shows the capture of sterile, wild and total male Medflies 10 days after releases (left graph) and the data from the "blank" test conducted in plexiglass cages (right graph). It is obvious that the apparent mortality of the released flies was much more rapid than the mortality of the caged flies. How much of this difference is a result of the released sterile flies leaving the test area is not known.

By the 4$^{th}$ day, the capture of sterile flies was 9% of the number captured on day 1. The numbers of wild flies captured was reasonably constant, indicating that weather and other factors did not influence trap catch significantly.

The data presented are similar to the results of a study conducted by Hendrichs et al. (1993) with the *tsl* strain Vienna-42. In both studies, captures on the 4$^{th}$ day after release declined to less than 10% of the day one capture rate.

These results clearly show that releases of sterile flies must be made twice weekly for a successful Medfly SIT programme.

The data from the test to evaluate the influence of feeding sterile Medflies on sugar only or on hydrolysed protein plus sugar prior to release are shown in Figure 3. The results indicate that feeding the hydrolysed protein had no effect on longevity, either on the flies without water and food or the flies given food and water after the initial 2-day feeding period in the paper bags.

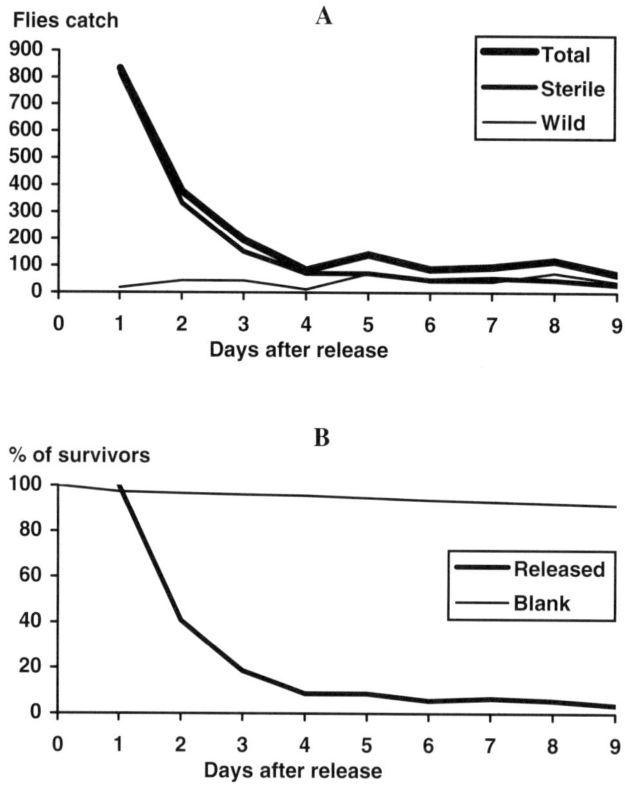


Figure 2. **A**: Total capture of male Medflies after release (in 10 Jackson trap with trimedlure);
**B**: Survival males of Medfly (the blank is in plexiglass cages).

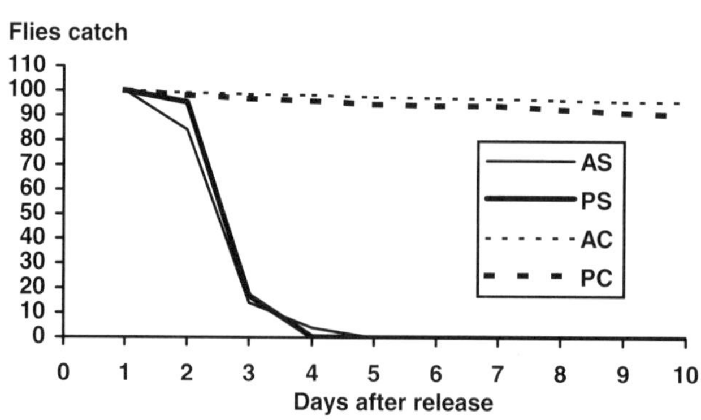


Figure 3. Survival males of Medfly in plexiglass cages (A-sugar before release; P-hydrolysate protein before the release; S-without food and water after release; C-with hydrolysate protein, sugar and water after releases).

**Dispersion**

Figure 4 shows the total and sterile fly captures for the dispersion test. The capture of sterile males remained remarkably constant from day 2 through to day 9, only dropping significantly on day 10. This apparent longevity of the sterile males (21% survival 4 days after release) is contradictory with the data in the longevity test. No explanation of this observation is obvious.

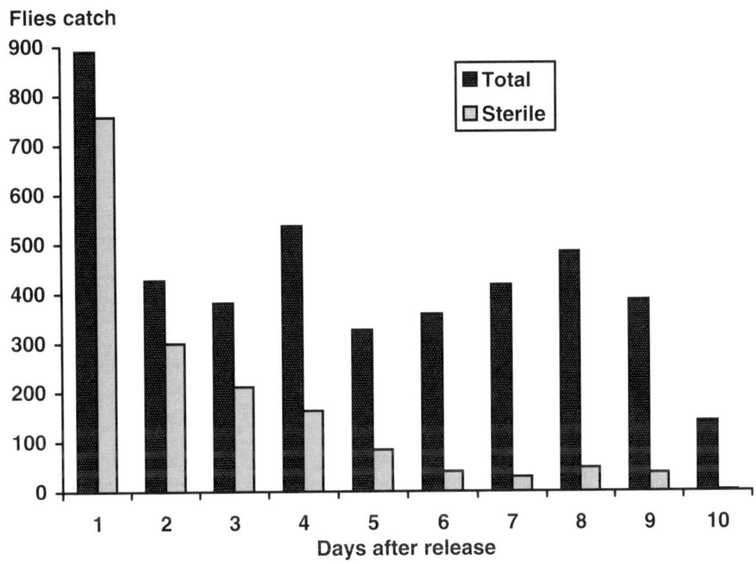


Figure 4. Total capture of male Medflies after release (in 16 Jackson trap with trimedlure).

Figure 5 shows dispersal data, both in distance and direction. The sterile male Medflies did not move very far after their release. Most of the flies were captured at 25 m and 50 m from the release point. Few flies were captured at 100 m and 200 m from the release point. These data strongly support the thesis that when Medflies (either wild or sterile) are in good host areas, they move very little. Thus, the dispersion of released sterile males for SIT programmes require that the dispersion be done primarily by ground release as the flies will then not disperse by moving away from the release point, at least when the release point is in a good host area.

The direction of the sterile fly movement was predominately south and west. There was no significant wind during the test period. The directional movement probably was related to the ripe mango grove in the south and west direction from the release point.

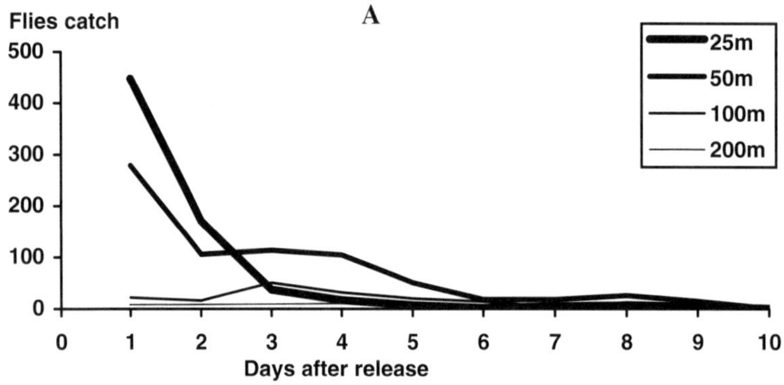


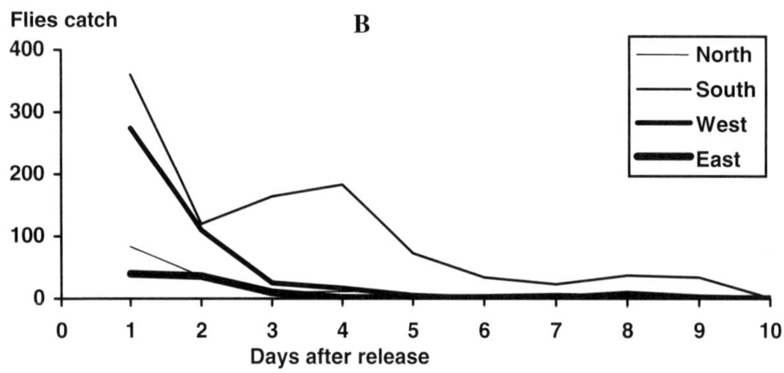


Figure 5. Total capture of male Medflies after release (in 16 Jackson trap with trimedlure); **A**: distance **B**: direction.

## CONCLUSION

- Sterile flies must be released 2 times each week in SIT programmes to achieve sufficient numbers of sterile flies in the treatment area to mate with the wild females.
- Feeding adult sterile Medflies during the first 2 days of their life with food that includes hydrolysed protein did not increase the survival of the flies.
- Sterile male Medflies do not disperse well when released in good host situations. Thus releases in SIT programmes must be very homogeneous over the treatment area.
- It would be useful if these tests could be repeated using aerial release.

## ACKNOWLEDGEMENTS

This work is supported by a PRODEP grant and the Madeira-Med Program.

## REFERENCES

Hendrichs, J., V. Wornoayporn, B.I. Katsoyannos and K Caggl. 1993. First field assessment of the dispersal and survival of mass reared sterile Mediterranean fruit fly males of an embryonal, temperature sensitive genetic sexing strain. *In* Management of insect pests: nuclear and related molecular and genetic techniques. IAEA-Vienna 453-462.

Hendrichs, J., G. Franz and P. Rendon. 1995. Increased effectiveness and applicability of the sterile insect technique through male-only releases for control of Mediterranean fruit flies during fruit seasons. J. Appl. Entomol. 119: 371-377.

Pereira, R., A.M. Barbosa, A. Brazão, J.P. Carvalho and D. Lindquist. 1996. Area-wide control of the Mediterranean fruit fly on Madeira with sterile insect technique. Proc. Int. Soc. Citriculture. 568-572.

Pereira, R., A. Barbosa, A. Brazão, N. Silva, J. Pacheco, A. Rodrigues, J. Caldeira, L. Dantas, J.P. Carvalho, and D. Lindquist 1997. Madeira-Med, SIT Program against Medfly in Madeira, Portugal. *II Internacional Open Meeting on mouches des fruits d'importance economique*. Lisboa, Sep. 1997.

Plant, R.E. and R.T. Cunningham. 1991. Analyses of the dispersal of sterile Mediterranean fruit flies (Diptera: Tephritidae) released from a point source. Environ. Entomol.. 20: 1493-1503.

Vieira, R. 1952. *A mosca da Fruta (Ceratitis capitata,* Wied.*) na Ilha da Madeira.* Grémio dos Exportadores de Frutas da Ilha da Madeira.

# Germline Transformation of the Mediterranean Fruit Fly, *Ceratitis capitata*

*Susan D. McCombs*

University of Hawaii at Manoa, Department of Entomology, 3050 Maile Way, Gilmore 310, Honolulu, HI 96822 USA

## INTRODUCTION

Gene transfer methodology for insects was first developed in *Drosophila melanogaster* Meigen using a transposon-mediated system based on the P element (Spradling and Rubin 1982, Rubin and Spradling 1982). In addition to the P element, three unrelated transposons have been used successfully in genetic transformation of *D. melanogaster*: *hobo* (Blackman et al. 1989), *Minos* (Loukeris et al. 1992), and *mariner* (Lidholm et al. 1993). Routine gene transfer in *Drosophila* created a great deal of optimism amongst researchers who sought to employ transgenic techniques in other arthropods. However, what followed were years of consistently disappointing results in other insect species. For example, the P element system was tried unsuccessfully in several species, but was eventually shown to be non-functional outside the genus *Drosophila* (O'Brochta and Handler 1988).

Ensuing research in non-drosophilids emphasised testing of other *Drosophila* systems and development of transposons isolated from other species. After nearly 15 years of intensive effort, the first successes have only recently been reported. Three *Drosophila*-derived transposon-based systems: *hobo* from *D. melanogaster*, *mariner* from *Drosophila mauritiana* Tsacas & David and *Minos* from *Drosophila hydei* Sturtevant have produced germline transformation in *Drosophila virilis* Sturtevant (Gomez and Handler 1997, Lozovskaya et al. 1996), *Aedes aegypti* (L). (Coates et al. 1998), and *Ceratitis capitata* (Wied.) (Loukeris et al. 1995), respectively. Germline transformation was accomplished with two transposon-based systems from non-drosophilids, *Hermes* from *Musca domestica* L. (Warren et al. 1994) and *piggyBac* from *Trichoplusia ni* Hübner (Cary et al. 1989) in *A. aegypti* (Jasinskiene et al. 1998) and *C. capitata* (Handler et al. 1998), respectively.

### Impact of Genetic Engineering on Insect Management

A primary goal of genetic studies on tephritid fruit flies, i.e., the Mediterranean fruit fly, has been the development of genetic sex sorting systems to remove females from mass reared flies produced for the sterile insect release method (SIRM). Sex sorting systems typically involve the linkage of a dominant selectable marker to the male-determining chromosome. Constructing such strains by traditional methods depends on generating translocations and is limited by two factors: 1) the availability of selectable markers, and 2) strain instability under mass rearing conditions. Use of molecular techniques may allow construction of sexing strains by integration of the selectable marker directly into the Y-chromosome, thereby avoiding the use of unstable translocations. Secondly, gene transfer technology should permit the use of cloned selectable markers from other species. This is especially important for species for which little genetic information is available.

Efficient methodology for introduction of exogenous DNA into the genome of *D. melanogaster* has resulted in great strides in understanding basic genetics and developmental processes in this species. Similarly, the ability to manipulate and insert genes of interest into tephritid fruit flies would provide important tools for studying biological and genetic properties. The hope is that this technology will lead to the development of highly efficient, species-specific control strategies for these pest insects. In addition to sexing systems, probable impacts of gene transfer technology on insect management programmes include: introducing stable and reliable marker systems to identify insects released in control programmes, simplifying sterilisation schemes for sterile insect release programmes and development and implementation of novel control methods, for example, autocidal biological control in which a conditional lethal gene is introduced and spreads in a population (Fryxell and Miller 1995). Genetic engineering would facilitate our understanding of chemical resistance mechanisms, sex determination and hormone action, all of which could be used to enhance current control techniques.

## METHODS

### Transposon-Based Gene Transfer Systems

Transposon-based gene transfer systems require two basic components: 1) a method for DNA delivery into the host genome, and 2) a marker for identification of individuals that have acquired this DNA. The major gene transfer systems are plasmid-based bipartite systems developed from Class II transposable elements i.e., *hobo*, P element, and *piggyBac*. These elements possess inherent mobility properties which allow them to efficiently integrate into host chromosomal DNA. One part of the gene transfer system, the vector, is constructed by removing or disrupting the transposase-encoding region of the transposable element. This region is replaced with a construct containing the selectable marker gene and the gene to be transferred. The second part, the helper plasmid, contains the transposase-encoding region of the transposon which is required for mobility of the vector. The transposase is typically put under heat-shock promoter control, and at least one of the terminal repeats is disabled. This disabled helper element mediates integration of the vector by producing the transposase, but cannot itself integrate into chromosomal DNA and so is lost from the organism. Because transposase is absent from the transformed strain, the stability of vector integrations is greatly enhanced.

### Microinjection Procedures

Successful incorporation of introduced DNA into the germ tissue requires that the vector and helper plasmids be co-injected into pre-blastoderm embryos prior to pole cell formation. During subsequent cell divisions the pole cells, progenitors of the germ tissue, envelop the plasmids and gene transfer occurs within their nuclei. The injection procedure is typically conducted at 22°C to delay embryonic development, thereby providing a one-hour window for completing injections. Eggs are collected from the host strain over a 15-minute period and held for an additional 15 minutes to allow the vitellin membrane and chorion to mature. Eggs are dechorionated in 30% Clorox bleach and rinsed with 0.02% Triton X-100 and distilled water. Following dechorionation, eggs are placed singly onto thin strips of double-stick tape held on 22 mm coverslips and allowed to desiccate to provide space for injected material. Eggs are covered with halocarbon oil and injected with the vector and helper plasmids in injection buffer (Handler et al. 1998) at 500 µg/ml and 150 or 300 µg/ml, respectively. Injected embryos, designated as the G0, are held under oxygen in a humidified chamber at 22°C until hatched larvae are removed to standard rearing media. G0 adults are individually backcrossed to the host strain and their progeny, the G1, are screened for marker gene expression.

## Transformant Selection

Detection of transformed individuals depends upon the introduction of a dominant-acting marker gene that can be easily identified by visual inspection or chemical selection. Dominant visible markers, such as the wild type eye colour genes *white* (Klemenz et al. 1987, Zweibel et al. 1995) and *rosy* (Spradling and Rubin 1982), allow detection of transformants having single gene insertions. These markers have been widely used because eye colour changes can be directly observed without additional manipulation. Wild type eye colour genes are, however, useful only when a mutant strain with a lesion at that locus is available. It was the isolation of white-eyed strains in the Mediterranean fruit fly (Saul and McCombs 1992, Rossler and Rosenthal 1992) that facilitated successful transformation in this species.

In most insect species there is a scarcity of genetic information, making dominant markers that do not require specific host strains essential for transformation detection. The dominant-acting green fluorescent protein (GFP) gene isolated from the jellyfish *Aequorea victoria* Murbach et Shearer (Prasher et al. 1992) has great potential for this application. The product of this gene, GFP, exhibits green fluorescence under long-wavelength ultraviolet light (Chalfie et al. 1994) in living organisms, e.g., the nematode *Caenorhabditis elegans* Murbach et Shearer and *D. melanogaster*. The wild type and several mutated variants of the GFP gene have been used as transformation and cellular markers in a wide variety of vertebrates and invertebrates. The high stability of the protein and the ease of detection in living tissue by simple examination under UV make GFP an excellent marker for the development of gene transfer systems in a variety of insects.

Chemical selection has the advantage of large-scale application, but typically requires addition of toxic materials to the larval media. The neomycin phosphotransferase gene (*neo*+) (Stellar and Pirrotta 1985) was tested extensively in non-drosophilids and found to be of limited use due to transient expression (McInnis et al. 1990). Antibiotic selection was most likely acting, not on the fruit fly host, but on its symbionts. The organophosphate dehydrogenase gene (*opd*), which confers resistance to organophosphate insecticides, has been used as a transformation marker in *D. melanogaster* (Benedict et al. 1994, 1995). Fear of introducing insecticide resistance into field populations of economically important species will likely limit the use of this marker. Two other chemical markers, the firefly luciferase gene and the bacterial B-galactosidase gene, have been used as markers in numerous species but their detection requires enzymatic reactions and subsequent screening. The destructive nature of detection renders them of little use as primary selection tools.

### Molecular Confirmation

G1 flies expressing the marker gene are designated as putative transformants and must be verified by molecular tests. Molecular confirmation of vector-mediated transformation is accomplished by: 1) Southern hybridisation to transformant genomic DNA using a probe derived from the vector, 2) sequencing of the integration sites by inverse polymerase chain reaction, and, if possible, and 3) *in situ* hybridisation to polytene chromosomes to identify the number and sites of vector integrations. Molecular confirmation is necessary to determine that a vector-mediated event occurred and that the selectable phenotype is due to the introduced marker gene, not contamination or reversion of the host strain mutation. Establishing that a transformation event is vector-mediated allows us to characterise the stability of the integration and gene expression.

## RESULTS

### *piggyBac*-mediated Transformation of the Mediterranean Fruit Fly

The *piggyBac* transposable element was originally discovered in the baculovirus genome following maintenance in *Trichoplusia ni* cell lines (Cary et al. 1989). It is a 2.5 kb short inverted repeat transposable element with 13 bp inverted terminal repeats and internal 19 bp inverted repeats. The 2.1 kb open reading frame codes for the transposase, a protein that facilitates the element's movement. Another interesting feature is that this element inserts at a specific target site, TTAA, duplicating this site in the process.

The *piggyBac* vector used in our work (Handler et al. 1998) incorporated the wild type white cDNA (*w*+) from the Mediterranean fruit fly (Zweibel et al. 1995) under control of the *D. melanogaster hsp*70 promoter. The helper plasmid contained the *piggyBac* transposase and the putative promoter region. Following coinjection of these plasmids into the white eye host strain, transformants were identified by visual screening of G1 offspring produced by backcrosses of injected individuals to the host strain. Six independent transformed lines, each with multiple G1 sibling transformants, were identified based on the presence of pigmentation in the compound eyes. The frequency of transformation was 3-5% in each of two independent transformation trials, a value which falls within the frequencies observed with the P element in *D. melanogaster* (Pirrotta 1988). Chromosomal integration of the *piggyBac* vector was confirmed in all lines by Southern hybridisation. Sequencing of the insertion sites in five transformed lines by inverse PCR (Ochman et al. 1993) indicated that the integrations were *piggyBac*-mediated (Handler et al. 1998). Each transformed line examined exhibited the TTAA target site duplication characteristic of *piggyBac* (Cary et al. 1989) and lacked plasmid or vector DNA in the insertion sites.

## CONCLUSIONS

Development of novel or enhanced biological control measures for agricultural pest insects by genetic engineering has great potential. Gene transfer technology should find wide and continued application for enhanced control measures directed at the

Mediterranean fruit fly and other tephritid species. Genetic markers may be introduced for stable and reliable identification of released flies. Simplified and less damaging male sterilisation schemes may be possible for the sterile insect release method applications. Construction of cold-resistant lines could lead to long-term storage of insects for release, thereby allowing smaller mass rearing facilities. As new genetic and molecular tools become available, these enhancements as well as novel control strategies are likely.

While recent success of gene transfer technology in the Mediterranean fruit fly provides promise, it must be used responsibly. The development of an efficient transformation system allows for rapid advances in genetic modification of this important agricultural pest, but with these advances comes the responsibility of making decisions concerning the safety of introducing this genetically-engineered plant pest into the environment. Precise definition of the inserted gene, genetic stability of transformed strains and horizontal transfer potential of the introduced DNA are fundamental characteristics of genetically-engineered organisms that must be considered in evaluation of environmental risk.

## REFERENCES

Benedict, M.Q., J.A. Scott and A.F. Cockburn. 1994. High-level expression of the bacterial *opd* gene in *Drosophila melanogaster*: improved inducible insecticide resistance. Insect Mol. Biol. 3: 247-252.

Benedict, M.Q., C.E. Salazar and F.H. Collins. 1995. A new dominant selectable marker for genetic transformation: *hsp*70-*opd*. Insect Biochem. Mol. Biol. 25: 1061-1065.

Blackman, R.K., M. Macy, D. Koehler, R. Grimaila and W.M. Gelbart. 1989. Identification of a fully-functional hobo transposable element and its use for germ-line transformation of *Drosophila*. EMBO J 8: 211-217.

Cary, L.C., M. Goebel, H.H. Corsaro, H.H. Wang, E. Rosen and M.J. Fraser. 1989. Transposon mutagenesis of baculoviruses: analysis of *Trichoplusia ni* transposon IFP2 insertion within the FP-locus of nuclear polyhedrosis viruses. Virology 172: 156-169.

Chalfie, M., Y. Tu, G. Euskirchen, W. Ward and D. C. Prasher. 1994. Green fluorescent protein as a marker for gene expression. Science 263: 802-805.

Coates, C.J., N. Jasinskiene, L. Miyashiro and A.A. James. 1998. Mariner transposition and transformation of the yellow fever mosquito, *Aedes aegypti*. Proc. Natl. Acad. Sci. USA 95: 3748-3751.

Fryxell, K.J. and T.A. Miller. 1995. Autocidal biological control: A general strategy for insect control based on genetic transformation with a highly conserved gene. J. Econ. Entomol. 88: 1221-1232.

Gomez, S.P. and A.M. Handler. 1997. A *Drosophila melanogaster hobo-white*$^+$ vector mediates low frequency gene transfer in *D. virilis* with full interspecific *white*$^+$ complementation. Insect Mol. Biol. 6: 165-171.

Handler, A.M., S.D. McCombs, M.J. Fraser and S.H. Saul. 1998. The lepidopteran transposon vector, *piggyBac*, mediates germ-line transformation in the Mediterranean fruit fly. Proc. Natl. Acad. Sci. USA. 95: 7520-7525.

Jasinskiene, N., C.J. Coates, M.Q. Benedict, A.J. Cornel, C.S. Rafferty, A.A. James and F.H. Collins. 1998. Stable transformation of the yellow fever mosquito, *Aedes aegypti,* with the Hermes element from the housefly. Proc. Natl. Acad. Sci. USA 95: 3743-3747.

Klemenz, R., U. Weber and W.J. Gehring. 1987. The white gene as a marker in a new P-element vector for gene transfer in *Drosophila*. Nucleic Acids. Res. 15: 3947-3959.

Lidholm, D.A., A.R. Lohe and D.L. Hartl. 1993. The transposable element *mariner* mediates germline transformation of *Drosophila melanogaster*. Genetics 134: 859

Loukeris, T.G., B. Arca, I. Livadaris, G. Dialektaki and C. Savakis. 1992. Introduction of the transposable element *Minos* into the germ line of *Drosophila melanogaster*. Proc. Natl. Acad. Sci. USA 92: 9485.

Loukeris, T.G., I. Livadaras, B. Arca, S. Zabalou and C. Savakis. 1995. Gene transfer into the medfly, *Ceratitis capitata*, with *Drosophila hydei* transposable element. Science 270: 2002-2005.

Lozovskaya, E.R., D.I. Nurminsky, D.L. Hartl and D.T. Sullivan. 1996. Germline transformation of *Drosophila virilis* mediated by the transposable element *hobo*. Genetics 142: 173-177.

McInnis, D.O. , D.S. Haymer, S.Y.T. Tam and S. Thanaphum. 1990. *Ceratitis capitata* (Diptera: Tephritidae): Transient expression of a heterologous gene for resistance to the antibiotic geneticin. Ann. Entomol. Soc. Am. 83: 982-986.

O'Brochta, D.A. and A.M. Handler. 1988. Mobility of P elements in drosophilids and nondrosophilids. Proc. Nat. Acad. Sci. USA 85: 6052-6056.

Ochman, H., F.J. Ayala and D.L. Hartl. 1993. Use of polymerase chain reaction to amplify segments outside boundaries of known sequences. Methods Enzymol. 218: 309-321.

Pirrotta, V. 1988. Vectors for P-mediated transformation in *Drosophila*. *In* Vectors: a survey of molecular cloning vectors and their uses. R.L. Rodriguez and D.T. Denhardt (eds.). Butterworths, Boston.

Prasher, D.C., V.K. Eckenrode, W. Ward., F.G. Prendergast and M.J. Cormier. 1992. Primary structure of the *Aequorea victoria* green-fluorescent protein. Gene 111: 229-233.

Rossler, Y. and H. Rosenthal. 1992. Genetics of the Mediterranean fruit fly (Diptera: Tephritidae): Morphological mutants on chromosome five. Ann. Entomol. Soc. Am. 85: 525-531.

Rubin, G.M. and A.C. Spradling. 1982. Genetic transformation of Drosophila using transposable element vectors. Science 218: 348-353.

Saul, S.H. and S.D. McCombs. 1992. Light eye color mutants as genetic markers for released populations of Hawaiian fruit flies (Diptera: Tephritidae). J. Econ. Entomol. 85: 1240-1245.

Spradling, A.C. and G.M. Rubin. 1982. Transposition of cloned P element into *Drosophila* germline chromosomes. Science 218: 341-347.

Stellar, H. and V. Pirrotta. 1985. A transposable element P vector that confers selectable G418 resistance to *Drosophila* larvae. EMBO J 4:167-171.

Warren, W.D., P.W. Atkinson and D.A. O'Brochta. 1994. The *Hermes* transposbale element from the house fly, *Musca domestica*, is a short inverted repeat-type element of the *hobo*, *Ac*, and *Tam3* (*h*AT) element family. Genet. Res. Camb. 64: 87-97.

Zweibel, L.J., G. Saccone, A. Zacharopoulou, N.J. Besansky, G. Favia, F.H. Collins, C. Louis and F.C. Kafatos. 1995. The *white* gene of *Ceratitis capitata*: a phenotypic marker for germline transformation. Science 270: 2005-2008.

# PART III.

# APPLICATION OF AREA-WIDE CONCEPT TO FRUIT FLIES

## E) REARING, QUALITY CONTROL AND NUTRITION OF FRUIT FLIES

# A Filter Rearing System for Mass Reared Genetic Sexing Strains of Mediterranean Fruit Fly (Diptera: Tephritidae)

*Kingsley Fisher*[1] *and Carlos Caceres*[2]

[1] FAO/IAEA Agriculture and Biotechnology Laboratory, Seibersdorf, A-2444 Austria
[2] USDA, El Pino Modular Mass Rearing Facility, Medfly Program, Guatemala

## INTRODUCTION

The Mediterranean fruit fly (Medfly), *Ceratitis capitata* (Wied.), is arguably the world's most widespread pest of fresh fruit production. With mounting controversy over using chemicals against insect pests, the sterile insect technique (SIT) has become increasingly more important as a successful technology in controlling or eradicating many insect pests. However, the wider adoption of SIT for Medflies has been hindered by damage to fruit from sterile female stings (Hendrichs et al. 1995). Moreover, the release of sterile females in SIT for Medflies is not efficacious (Hendrichs et al. 1995), a point validated in the field in Hawaii (McInnis et al. 1994) and Guatemala (Rendon, personal communication). Hendrichs et al. (1995) list many other advantages for releasing only male Medflies including improved economy, increased safety and improved field monitoring.

Genetic systems for the separation of sexes have been developed for Medflies (Franz and Kerremans 1994, Willhoeft et al. 1996) and they allow for large-scale releases of only males. Genetic sexing strains (GSS), as they are known, are based upon selectable characters linked to the male sex by using a Y-autosome translocation (Franz et al. 1996).

There are two types of GSS used in mass rearing. First, strains based upon a recessive mutation *(wp)* change the pupal colour from brown to white. In these strains, females emerge from white pupae and males from brown pupae. A machine is used to sort the pupae based upon colour. First described by Robinson and Van Heemert (1982), the most recent strain, SEIB 6-96 based upon the T(Y;5) 2-22 translocation, is relatively stable in small scale rearing (Franz et al. 1994). Second are the temperature sensitive lethal strains (*wp/tsl*) which carry a temperature sensitive lethal (*tsl*) mutation in addition to *wp*. In *tsl* strains, female embryos are killed by exposing eggs to a 34°C temperature during development (Franz et al. 1996). Male embryos are not temperature sensitive and survive the high temperature treatment. The inclusion of the *wp* marker enables the genetic integrity of the strain to be monitored. Current *tsl* strains in mass rearing include VIENNA 4/TOL-94, VIENNA 6-94 and VIENNA 7-97. These strains are based upon translocations T(Y;5) 1-61, T(Y;5) 2-22 and T(Y;5) 3-129 respectively (Kerremans and Franz 1995).

GSS have developed to a high level of sexing accuracy and stability that make possible their large-scale mass rearing. However, once the strain is transferred to large operational facilities producing several million or more insects per week, the sexing strain is subjected to a new set of constraints. Stability in operational mass rearing has been shown to erode over the long term e.g., in Guatemala, for the strain VIENNA 4/Tol-94 and in Argentina, for the strain SEIB 6-96 (Taret, personal communication), currently the two largest mass production facilities for GSS. This means a loss of the sex separation mechanism, but does not necessarily indicate a decrease in quality as such. Under the stress and selection pressures of mass rearing, recombinants gain a selective advantage, accelerating their accumulation. This accelerated buildup of recombinants in large-scale mass rearing is not well understood at this time, but is probably related to colony management practices.

In response to the instability in GSS mass rearing, a "filter rearing system" (FRS) was conceived and developed in the FAO/IAEA Agricultural and Biotechnology Laboratory in Seibersdorf to control the buildup of recombinants. The principle of the FRS is to maintain a small standby colony of GSS, with no recombinants, which can regularly refresh the mainstream of production with new material. The key factor is that no material from the mainstream is returned to the standby colony.

## MANAGING RECOMBINANTS IN MASS REARING

Recombinant individuals are identified as the "wrong sex" emerging from white or brown pupae, or in the case of *tsl* as females which survive the high temperature treatment (i.e., they have lost *tsl*). The accumulation of recombinants at a moderate scale of mass rearing has been monitored for three different GSS at Seibersdorf (Fig. 1).

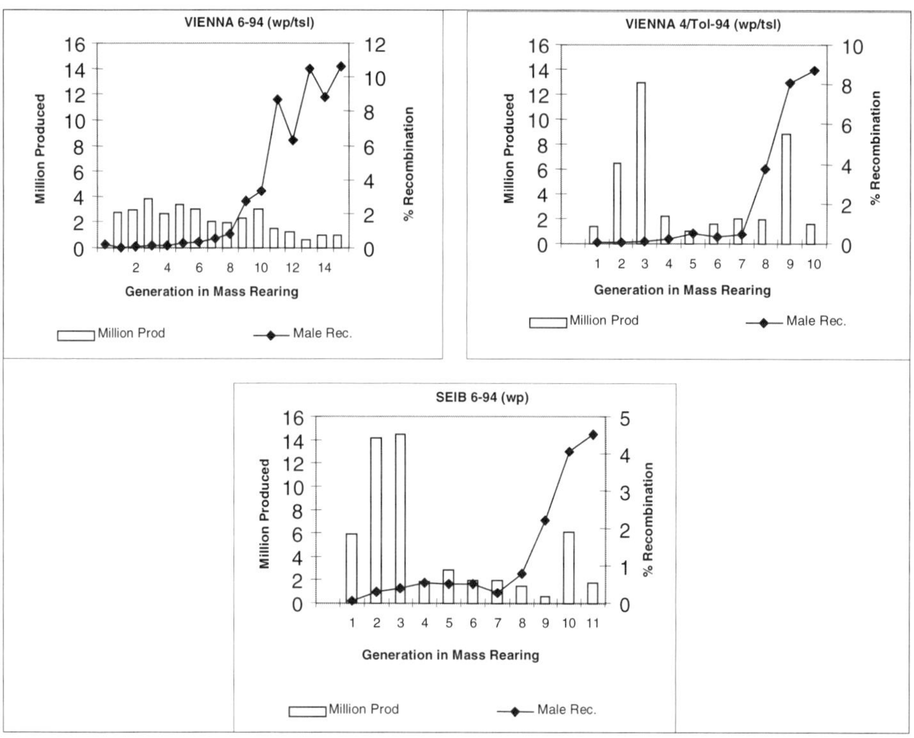


Figure 1. The buildup of recombinants in the FAO/IAEA Medfly facilities in Seibersdorf for 3 mass reared GSS (average productions: VIENNA 6-94 = 2.2 million pupae/wk, VIENNA 4/Tol-94 = 4.0 million pupae/wk, SEIB 6-95 = 4.8 million pupae/wk).

In order to be able to continue to use GSS effectively, a strategy for colony management is necessary to maintain stable male-only production. The management system must meet five criteria:

1. small enough to manage and remove recombinants before they breed,
2. large enough to amplify into the required operational colony,
3. able to match operational production continuity,
4. capable of providing a high degree of accuracy,
5. economically feasible.

Three basic management strategies were considered: dilution, replacement and filtering.

Diluting colonies with cleaned material does not meet management criteria 1, 2, 4 or 5. Dilution is done by continually removing recombinant individuals from a proportion of colony material, which is then added to the production colony. Based upon the rate of recombination observed in moderate scale mass rearing in Seibersdorf (Fig. 2), modelling suggests a colony would have to be diluted by 15% at each renewal, before the impact of recombinant accumulation could be slowed to tolerable levels. It means that in a colony of ca. 10 million flies (required to produce 100 million flying males/week), 1.5 million flies must be inspected and sorted weekly to ensure 15% dilution. It is conceivable in such large-scale operations that rates of recombination may require higher dilution than 15% because the strength of selective forces is difficult to estimate (this fails criterion 1 and has difficulty meeting criterion 2). At this level of sorting, human error increases and accuracy declines (fails criterion 4). The dilution strategy cannot reach levels of 99.5% purity as desired in some programmes (fails criterion 4). Simply diluting the effects of recombinant accumulation is expensive and requires a lot of resources (fails criterion 5).

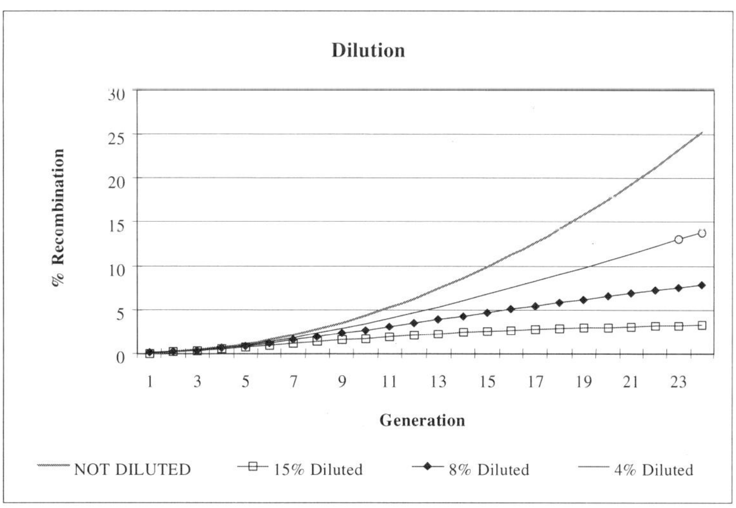


Figure 2. Results of a model which demonstrates the effect on recombination of diluting a production colony with 4, 8 and 15% recombinant-free material, each generation.

Replacing whole colonies at regular intervals is disruptive to ongoing production (does not meet management criteria 1, 3 or 5). It relies upon having a stock colony (which may not be on site) which is free of recombinants and the ability to raise a large new colony in a short time to replace the operational production colony, requiring additional resources or buying new colonies from a supplier. To avoid any down time during replacement, a facility would depend on having at least two large colony rooms, a rare commodity in facilities for economic and design reasons (fails criterion 5). The task of replacing colonies with recombinant-free colonies might be required 2-3 times annually and there remains the question of where to obtain such large colonies.

A filtering strategy, which allows only non-recombinant material to pass through to the main production colonies, meets all of the management criteria. Filtering can be effective in maintaining low recombinant levels in small manageable colonies for long periods of time, economically and without disrupting the normal production processes.

The principle of the FRS is to maintain a small standby line of GSS, with zero recombination, which can, at regular intervals, be amplified to refresh the mainstream of production with new genetic material. This principle, although developed for GSS, is also applicable to bisexual strain production (and other mass reared insects). The FRS allows for simple strain replacement, pre-adaptation to required parameters, low stress rearing, control over recombination in mass rearing, the selection of desirable traits, adaptation to new and existing technologies and the addition and testing of new genetic material at any time without affecting production.

## THE FILTER REARING SYSTEM (FRS)

The FRS consists of a small colony, physically cleaned of recombinants (i.e., recombinant-free) which is "bridged" to a large operational colony via a short series of colony amplifications, called an amplification bridge (Figure 3). Each component of the FRS is referred to as a "stream". Thus the recombinant-free colony is called a **clean stream**. The first stage of amplification is called the **initiation stream** and is not cleaned because of the sheer numbers involved. The second stage of amplification is called the **injection stream** which provides new replacement material to the colony stream, without interruption to operations. The **colony stream** provides material to a much larger colony called the **release stream** which is dedicated to providing all the flies for sterilisation and release and is maintained differently from the other streams.

The success of the FRS will not only be based upon its ability to control recombination in the mass rearing of GSS, but also on the opportunity to control other quality aspects of mass reared insects. The number of configurations of a filter rearing system is considerable, depending upon strain, resources, etc. and this in itself allows the system to be tailor-made to the needs of a facility. In a broad sense, however, there are two configurations of the FRS: continuous (a uni-directional rearing scheme) or periodic (a cyclic rearing scheme).

The continuous FRS operates its amplification bridge constantly (daily or weekly) to produce a release stream. The only cyclic colony is the clean stream, maintained recombinant-free. This is especially suited to strains where recombinants accumulate quickly or where there is a low tolerance in the field for sterile females.

A periodic FRS operates its amplification bridge when triggered by a monitored parameter such as recombination or longevity going beyond specified limits. In this configuration, a cyclic colony stream is added to maintain a large, low-stress colony for long periods (Figure 1 shows this could be 7-8 generations). A periodic FRS may also operate at routine intervals, for example, once every 4-6 months regardless of parameter values (providing, of course, that parameter values do not exceed acceptable limits in the mean time). Where a *wp* GSS is used and pupal colour sorting is the limitation on accuracy, a periodic FRS might be considered over a continuous FRS. The implementation of a FRS will require that the rearing processes within a facility be reconfigured. This is generally easier in large facilities than in small ones.

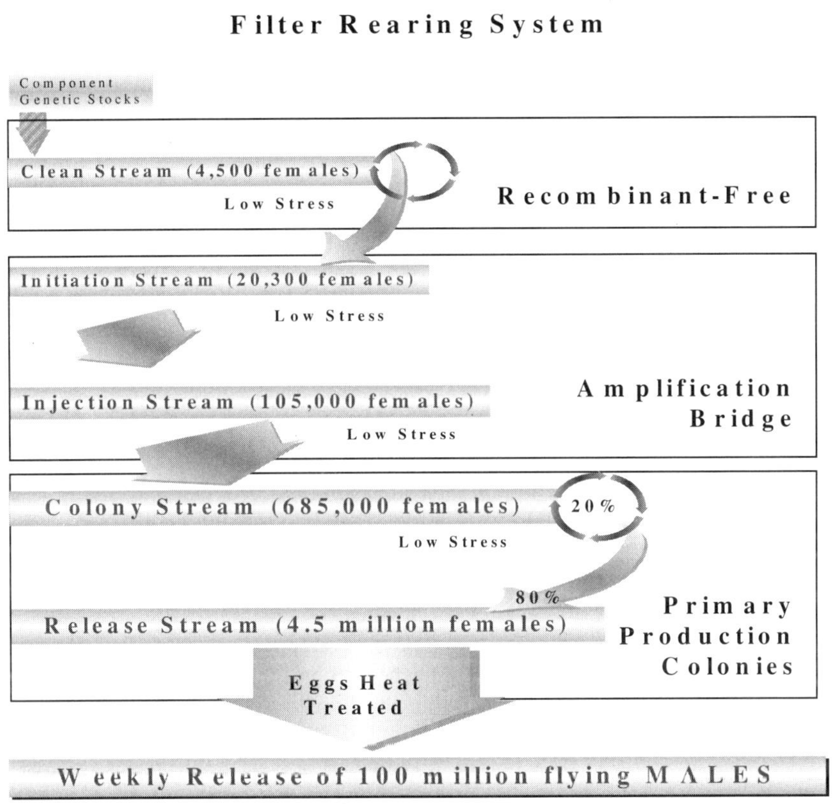


Figure 3. A diagrammatic representation of the filter rearing system (FRS) designed to control the accumulation of recombinants in Medfly GSS.

The FRS consists of five components or streams as shown in Figure 3, with the addition of the genetic stocks used to synthesise the GSS. Each of these components requires a different rearing strategy within a facility and components 4 and 5 require to be housed separately from each other and from components 1-3. Since the FRS was designed to cope with the obvious stability related to GSS, most of the following discussion will be confined to GSS.

The genetic stocks will not ordinarily be required in operational programme facilities, except where independence from an external supply of a GSS is necessary. Otherwise, it is probably more economical to buy a pre-constructed strain from a supplier like Seibersdorf, as required.

The elements of GSS (stocks) that are necessary to construct GSS, are maintained in separate small cultures. In the case of a *tsl*, three essential genetic stocks are required: *wp/tsl* stocks, translocation stocks and wild type stocks, often called the background of the GSS. The construction of a GSS takes many months but essentially needs to be done only once every 3-4 years. There are difficulties associated with the construction of GSS, especially *wp/tsl* GSS (Franz, personal communication) and specialised staff is required. Strain construction is probably not suitable for facilities involved in mass production for field programmes.

The FRS for operational programmes begins with the **clean stream**, which represents a small colony of the strain maintained under minimal stress conditions (for example, diurnal lighting, low densities, more males than females to maintain mating competitiveness (male:female ratios in cages of 1.2), optimal temperature). Care must be taken to maintain those adaptations, such as oviposition behaviour or appropriate mating behaviour, in order for large production lines to function efficiently and without changes in field capability. The clean stream is continuously maintained in discrete generations until required. It can be maintained at a very low level or size, depending upon the ultimate size of the production that is required. It must be "cleaned" every generation to maintain it free of recombinants. Cleaning involves the removal of either males from white pupae or females from brown pupae and must be accurate and reliable. In the case of *wp/tsl* GSS, it also includes regular temperature tests to verify the presence of the *tsl* mutation.

The key to the successful mass rearing of GSS is to accurately and economically manage a small colony which is big enough to be quickly amplified to replace large contaminated production colonies without interruption to field operations. The importance of the amplification bridge is its ability to transform a small manageable colony to a large production colony which matches the needs and continuity of large-scale field operations, whilst conserving genetic, behavioural and qualitative aspects.

Therefore, to meet these needs, a maximum of two amplifications are necessary (and this dictates the size of the clean stream) and begins with an **initiation stream** (Figure 3). The initiation stream simply amplifies the clean stream into a colony with consecutive days of production and 6-8 fold reproductive potential. For the periodic FRS, the initiation stream is only invoked when recombination needs to be reduced in the primary production colonies and on this basis, may only be required two times per year (Figure 1). All progeny from the initiation stream are placed in larger cages, which forms the injection stream. The initiation colony is not cleaned because of the large numbers involved, but the number of recombinants is expected to be very low, based on the natural recombination frequency (Franz et al. 1994). It provides the individuals needed for the second amplification or injection stream.

The **injection stream** forms a second amplification step between the small initiation stream and the colony stream (or release stream in the case of a continuous FRS), without compromising operations (Figure 3). The initiation stream is used to initiate the filter and the injection stream is used to inject new material into the primary production colonies. In the periodic FRS, the initiation stream is online for three consecutive weeks, which in turn provides three consecutive weeks of injection colony. This is necessary for the sequencing of the primary production colony replacement, since the filter is designed to replace recombined colonies over a short period, **not to dilute** them. Again, the injection colony is not cleaned because of the large fly numbers involved. It simply provides the stock needed to replace the colony (or release) stream with low recombinant stock. These streams (clean, initiation, injection and colony) are maintained under low stress conditions (e.g., low densities of flies in cages, male:females ratios of 1.2) to prevent selection advantages to recombinants.

In a periodic FRS, the **colony stream** is the centre of the FRS (a colony stream is not used in a continuous FRS). The colony stream runs collaterally with the release stream and interfaces with it. The colony stream produces stock for itself and for a much larger colony called the release stream. In the case of *tsl*, about 20% of eggs produced by the colony stream are recycled to maintain the colony stream. The remaining 80% of eggs produce individuals for the release stream. The colony stream is a middle-sized, low stress colony and is the only place that recombinants can accumulate because it is the only colony which cycles non-cleaned (unfiltered) progeny back to itself. Although the continuous FRS configuration does not use a colony stream, the gains (not having to manage a large cyclic colony) are to some extent offset by the need to have a much larger clean stream or many amplification stages (each of which could increase the frequency of recombinants). The continuous filter may be more simple but it is a less passive configuration compared with the periodic FRS. If the colony stream is kept under low stress and of a moderate size, then the accumulation of recombinants should be low for 7-8 generations (months).

The **release stream** produces the large number of eggs required for thermal treatment (*tsl*) or pupal colour separation (*wp*) to produce only male pupae (Figure 3). As a consequence, all of the release colony production goes in one direction to the field and is not recycled. In rare cases, a portion of this colony may be recycled to meet demands for sudden increases in production (an advantage of the periodic FRS) or to recover disastrous losses of other streams. The release stream can in fact be placed under stress because there is no chance for the selection of recombinants and there is no chance for the adaptation of flies to high stress because there is only one generation of production at any one time. For example, the release stream could be maintained under higher adult densities than in the other streams and a male:female ratio could decrease from 1.2 to 0.4.

The type of FRS chosen by a facility will depend heavily upon the resources dedicated to the task of running the clean and amplification streams. The reality in most facilities will be that staff resources are limited and that a periodic FRS will be the only option, but in any case, it is a good place to start.

An FRS for *wp* strains is more straight forward than for *tsl* strains, although the principles are the same. For example, there is no requirement for temperature tests to monitor the existence of the *tsl* mutation and the amplification bridge can be reduced to an injection stream, because of the higher reproductive potential of *wp* strains (unpublished data).

A continuous FRS is a high maintenance filter requiring rapid amplification from clean stream to release stream. Under this configuration, there will be continuous pressure on the clean stream to produce the required numbers for the injection stream each week, necessitating that a large clean stream be maintained by skilled staff. It becomes absolutely essential under these circumstances to make sure that the clean stream is maintained under minimal stress; therefore, many low density cages will be better than a few crowded cages. Cleaning must be *absolute* in this situation, since any recombinants will quickly be amplified in the injection stream.

At no stage can there be a backflow of material from the release stream to injection or clean streams. Under emergencies, backflow from injection to clean streams can be tolerated, as long as the material is cleaned of recombinants. This option is dependent upon what mutation is lost and is most appropriate for *wp* and the periodic FRS.

Using a FRS, the El Pino Modular Mass Rearing Facility in Guatemala has demonstrated low and controlled levels of recombination during the mass production of more than 120 million male pupae/wk from a *tsl* GSS (Caceres et al. 2000). Since installing an FRS, the ISCAMEN Medfly Production Facility in Mendoza (Argentina) has also demonstrated low and controlled levels of recombination during the mass production of 260 million pupae/wk of a *wp* GSS (Taret, personal communication). Facilities in Arica (Chile: Machuca, personal communication), Crete (Greece: Economopoulos, personal communication) and Madeira (Portugal: Pereira, personal communication) are in various stages of implementing an FRS for their respective GSS Medfly production.

## ACKNOWLEDGEMENTS

The willingness of international Medfly mass production organisations to work with this concept is acknowledged.

## REFERENCES

Caceres, C., K. Fisher and P. Rendon. 2000. Mass rearing of the Medfly temperature sensitive lethal genetic sexing strain in Guatemala. pp. 551-556. *In* Area-Wide Control of Fruit Flies and Other Insect Pests. K.H. Tan (ed.). Penerbit Universiti Sains Malaysia, Penang, Malaysia.

Franz, G. and P. Kerremans. 1994. Requirements and strategies for the development of genetic sex separation systems with special reference to the Mediterranean fruit fly *Ceratitis capitata*. pp. 113-122. *In* Fruit Flies and the Sterile Insect Technique. C. Calkins, W. Klassen and P. Liedo (eds.). CRC Press, Boca Raton, Florida 1994.

Franz, G., E. Gencheva and P. Kerremans. 1994. Improved stability of genetic sex-separation strains for the Mediterranean fruit fly, *Ceratitis capitata*. Genome. 37: 72-82.

Franz, G., P. Kerremans, P. Rendon and J. Hendrichs. 1996. Development and application of genetic sexing systems for the Mediterranean fruit fly based on a temperature sensitive lethal. pp. 185-191. *In* Fruit Fly Pests: a World Assessment of their Biology and Management. B. A. McPheron and G. J. Steck (eds.). St Lucie Press, Delray Beach, Florida,USA.

Hendrichs, J., G. Franz and P. Rendon. 1995. Increased effectiveness and applicability of the sterile insect technique through male-only releases for control of Mediterranean fruit flies during fruiting seasons. J. Appl. Entomol. 119: 371-377.

Kerremans, P. and G. Franz. 1995. Isolation and cytogenetic analyses of genetic sexing strains for the medfly, *Ceratitis capitata*. Theor. Appl. Genet. 91: 255-261.

McInnis, D.O., S. Tam, C. Grace and D. Miyashita. 1994. Population suppression and sterility rates induced by variable sex ratio, sterile insect releases of *Ceratitis capitata* (Diptera: Tephritidae) in Hawaii. Ann. Entomol. Soc. Am . 87: 231-240.

Robinson, A.S. and C. Van Heemert. 1982. *Ceratitis capitata* - a suitable case for genetic sexing. Genetica. 58: 229-237.

Willhoeft, U., G. Franz and D.O. McInnis. 1996. Towards the application of genetic sexing in Tephritid fruit fly SIT programmes. pp. 179-184. *In* Fruit Fly Pests: a World Assessment of their Biology and Management. B.A. McPheron and G.J. Steck (eds.). St. Lucie Press, Delray Beach, Florida, USA.

# Mass Rearing of the Medfly Temperature Sensitive Lethal Genetic Sexing Strain in Guatemala

*C. Cáceres[1], K. Fisher[2] and P. Rendón[3]*

[1] USDA, El Pino Modular Mass Rearing Facility, Medfly Program, Guatemala
[2] FAO/IAEA Agricultural and Biotechnology Laboratory, Seibersdorf, A-2444 Austria
[3] USDA-APHIS-PPQ, Methods Development Station, c/o AmEmbassy – Guatemala City, APO AA 34024-3319, USA

## INTRODUCTION

Field tests have demonstrated the increased efficiency of the sterile insect technique (SIT) for the Mediterranean fruit fly, *Ceratitis capitata* (Wied.), when only male Medflies are released (Robinson et al. 1986, Nitzan et al. 1993, McInnis et al. 1994, Rendón 1996). Genetic sexing strains (GSS) of Medflies, containing temperature sensitive lethal (*tsl*) and white pupae colour (*wp*) mutations (Franz et al. 1994) developed by FAO/IAEA, allow the separation of male flies from female flies. GSS technology has reached a stage where it is being used in large-scale operational programmes, such as the Moscamed Program in Guatemala. GSS based on the *wp/tsl* have the advantages of: 1) not requiring sophisticated equipment for sex separation, 2) a high accuracy of separation (> 99.5% males) is possible and, 3) separation is achieved during egg development, which excludes the unnecessary rearing of females (Franz et al. 1996).

It was shown by Franz et al. (1994) that *tsl* GSS are genetically stable for many generations under small-scale rearing conditions. However, under the large-scale rearing of operational programmes such as Moscamed (Hentze and Mata 1987), a gradual loss of the sex separation mechanism through recombination remains a problem, as has been demonstrated in Guatemala during 1994-1996. This in no way precludes the use of GSS technology, but it does mean that a management system must be used to control this gradual loss of stability; a strategy for colony management which maintains a stable and high level of accuracy of male-only production.

The El Pino facility, which mass produces sterile flies for the Guatemala Medflies SIT Program, has introduced a filter rearing system (FRS) (Fisher and Cáceres 2000), and has demonstrated in a Medfly *tsl* GSS known as VIENNA 4/Tol-94, that genetic stability can be maintained. We report the operation of the FRS and its impact upon genetic stability and male-only production. The concept of the FRS has the potential to improve the genetic quality of any mass reared insect, including bisexual strains of Medflies.

## MATERIALS AND METHODS

### The Facility

The El Pino facility, Guatemala, was conceived for the mass production of fruit



**Area-Wide Control of Fruit Flies and Other Insect Pests,**
ed. K.H. Tan. Penerbit Universiti Sains Malaysia, Penang, 2000.

flies (and their parasitoids). It was inaugurated in January, 1996 and has four production modules capable of producing ca. 1,600 million sterile Medflies pupae/week of a standard bisexual strain (equivalent to ca. 600 million sterile male flies), or 600 million sterile male pupae of a *tsl* GSS (equivalent to ca. 400 million sterile male flies).

## The Genetic Sexing Strain

The El Pino facility produces the VIENNA 4/Tol-94 GSS in 2 of the modules. The strain is based upon an outcross between a *wp/tsl* GSS (T(Y;5)1-61 translocation (Franz et al. 1994)) and a wild type with a genetic background from Tolimán, Guatemala (Franz et al. 1996, Rendón 1996).

## Mass Rearing

The *tsl* females are sensitive to high temperatures (32-35°C) in all developmental stages. Therefore, different rearing procedures are used for the production of colonies and of males only. Currently, both productions are maintained in separate modules.

Because of the temperature sensitivity of the *tsl* females, the oviposition room was maintained at 23±1°C and 65±5% r.h. Cages for egg production (0.9 $m^3$) each holding up to 75,500 females. Characteristically, VIENNA 4/Tol-94 females produce an average of 12±3 eggs/day during a 10-day oviposition period. Adults inside the cages were fed with a mixture (2.4 kg/cage) of 1 part hydrolysed protein (ICN®) to 3 parts cane sugar.

The larval diet, used for production, was a mixture of sugar cane bagasse (15%), corncob fractions (#30/80, Pulaski®: 3%), cane sugar (12%), whole wheat flour (2%), wheat germ (2%), vitamin mixture (Calier®: 0.1%), sodium benzoate (0.3%), hydrochloric acid (USP) (1.19%) and water (52%). In order to avoid overheating, diet trays are loaded with 4 kg for colony production, compared with 5 kg for male-only production (where excess heat was less problematic). Seeding densities in both cases averaged 14 viable eggs/gm of diet. As the *tsl* mutation has a pleiotrophic effect, the *tsl* female larvae were slower to development (11±2 days at 24ºC), than the heterozygous males (7±2 days at 24°C), and resulted in five larval collections to complete the collection of females for colony cages.

Male-only production was achieved by killing female embryos using a thermal treatment of 34ºC for 24 h, during egg development (Fisher 1998). Our studies have shown that improved male recovery can be obtained by reducing the thermal treatment time from 24 to 12 hrs at 34°C during the second day of embryogenesis (Cáceres, unpublished data). Egg-pupae development time for male-only production is ca. 7 days.

## The Filter Rearing System

The filter rearing system (FRS) was designed in the IAEA Laboratories in Seibersdorf (Fisher and Cáceres 1999), to maintain the genetic integrity of any GSS during mass production. It is based on the morphological *wp* selectable marker, where males are expected to emerge from brown pupae and females are expected to emerge from white pupae. Recombinant individuals are identified as males from white-pupae or females from brown-pupae and can be removed ("cleaned") to preserve GSS the integrity.

In order to assure the removal of recombinants, pupae are separated by colour 2-3 days before adult emergence. Individual pupae were kept in separate cells of plastic grids (100 x 1 $cm^2$ cells/grid). The grid is covered with a transparent plexiglass lid and the bottom of the grid is sealed with metallic fly-proof netting. Individual adults were allowed to emerge within the cells, and were examined under magnification to identify recombinants. Recombinant individuals were killed by a burst of gas, introduced via a fine nozzle activated by a foot-pedal (equipment by USDA-PPQ, Mission, Texas).

Since the large colonies of operational programmes cannot be cleaned for practical reasons, the FRS is based on the sequential amplification of a relatively small backup colony (or clean stream), which can be cleaned of recombinant individuals (Fisher and Cáceres 2000). The backup colony, cleaned each generation, must be large enough to produce sufficient offspring to provide for itself and to initiate the colony amplification sequence.

In order to guarantee the accurate production of only males (>99.5% males) during release operations, El Pino adopted a continuous FRS (for details see Fisher and Cáceres 2000). The operation of the continuous FRS in El Pino is shown in Figure 1.

The number of amplifications depended upon the size of the clean steam in relation to the size of the final colony (release stream) required to produce the desired number of males for release, and the rate at which recombinants accumulated at each amplification stage. In a continuous FRS, a cyclic colony stream (usually the stage prior to the release stream (Fisher and Cáceres 1999)) was not used. A large clean stream was preferred, in order to maintain a continuous system, with three generations between cleaned stocks and released flies.

## RESULTS

The mass rearing potential of *tsl* females during amplification stages is about 8-fold (i.e., 1 female fly will produce 8, F1 females). In El Pino, at the current production level of 150 million sterile male pupae/week, a clean stream of 30,000 *wp/tsl* females (i.e., females from white pupae) was required each week. A small amount of production from the clean stream (9.4%) was cleaned and returned to maintain the clean stream. The excess production from the clean stream was not cleaned and was used to initiate the amplification process. The initiation stream was maintained with 160,000 females weekly (Figure 1). The initiation stream provided 1.3 million females weekly to the injection stream, which in turn provided the 10.2 million females for the release stream. This size of release stream colony was able to produce sufficient eggs for the production of 150 million male pupae/week for sterilisation and release (Figure 1).

The level of recombination in each amplification stage of the FRS in El Pino was monitored (Table 1), and demonstrated that the management of recombination in GSS is possible, using the FRS. Since the introduction of the FRS late in 1996, El Pino has demonstrated the advantages of the FRS in maintaining low levels of recombination. During 14 months, an average of 78.2 million male pupae/week (Figure 2), with a purity of 99.8% (Figure 3) was produced.

A summary of key production parameters is shown in Table 2. Adult emergence and percentage flying males were depressed through the survival of adjacent-1 individuals to the pupal stage. Adjacent-1 individuals do not normally emerge (Franz et al. 1996). In general, these values conform to the standards set for the VIENNA 4/Tol-94 *tsl* GSS in the FAO/IAEA Medfly facility in Seibersdorf.

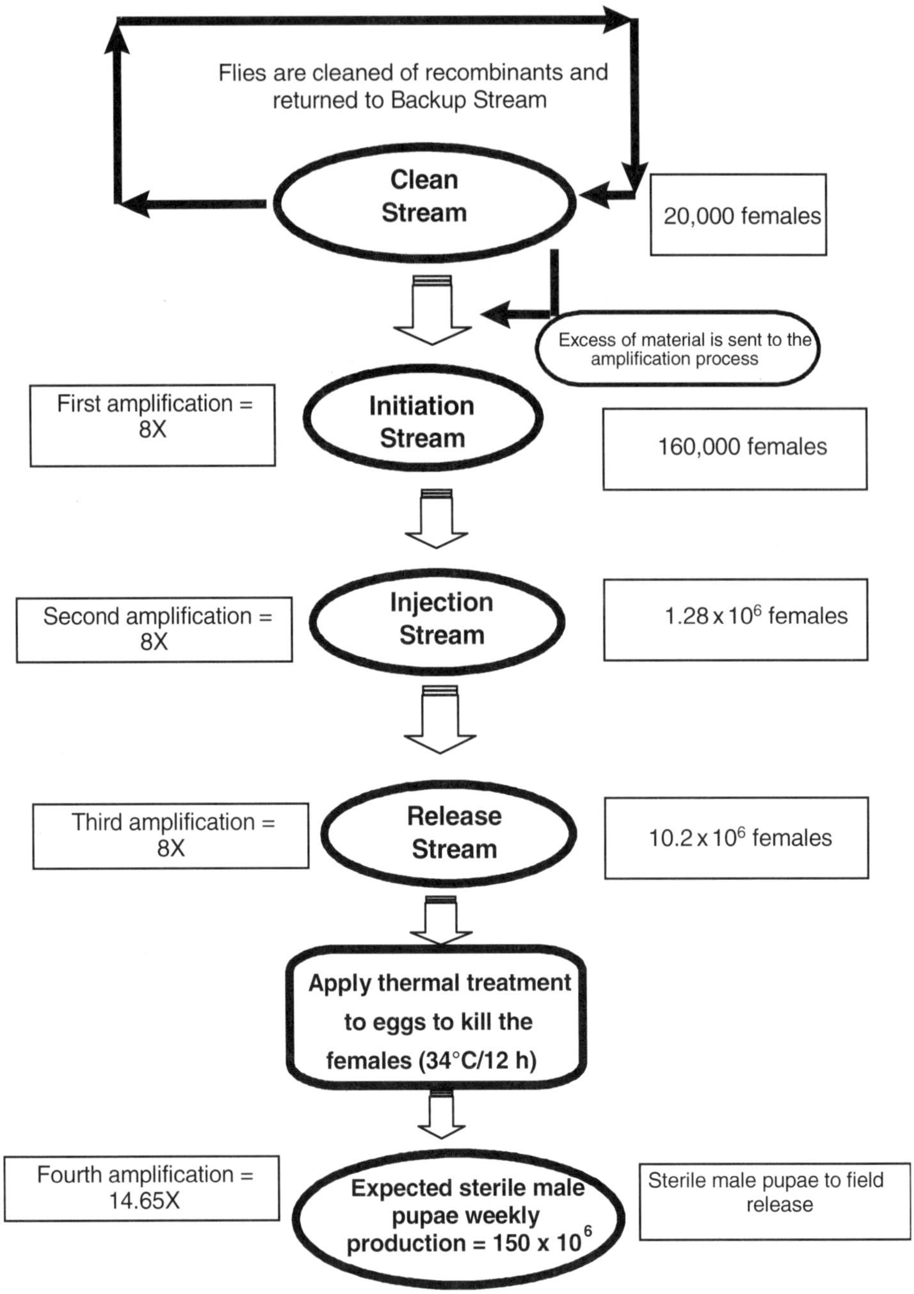


Figure 1. The continuous filter rearing system (FRS) for male-only production in Guatemala.

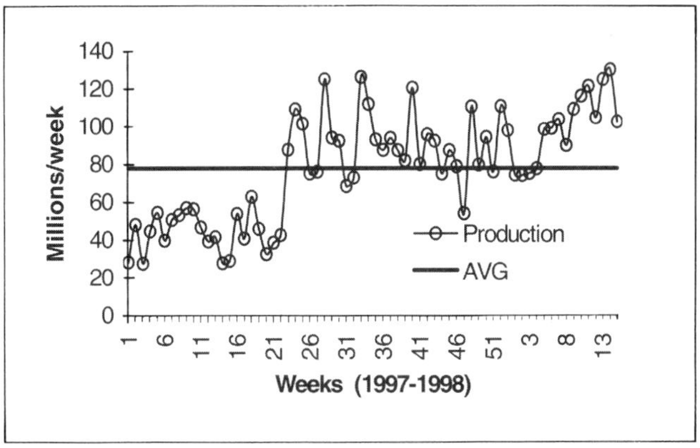


Figure 2. Weekly production of male pupae of VIENNA 4/Tol-94 in the El Pino facility.

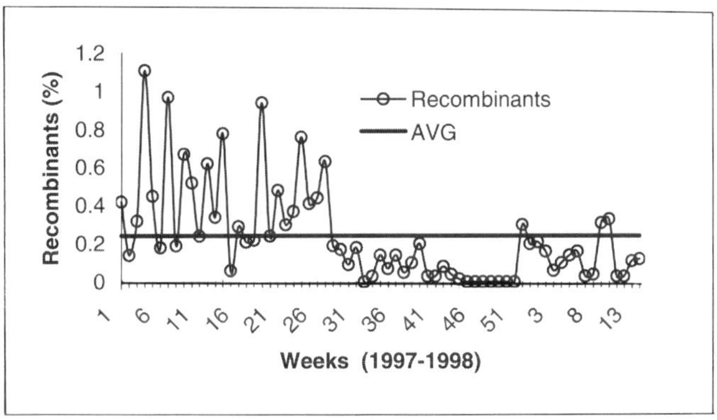


Figure 3. Recombination (females from brown pupae, which are not heat sensitive and survive to releases) in the male-only pupae production of VIENNA 4/Tol-94 in the El Pino facility.

Table 1. The level of recombination in the different levels of amplification of the FRS, after 14 months in mass rearing of VIENNA 4/Tol-94.

| Colony | % females from brown pupae | % males from white pupae |
|---|---|---|
| P generation | 0.001 ± 0.040 | 0.080 ± 0.060 |
| After cleaning | 0.000 ± 0.000 | 0.000 ± 0.000 |
| Initiation stream | 0.020 ± 0.014 | 0.240 ± 0.430 |
| Injection stream | 0.035 ± 0.080 | 0.570 ± 0.560 |
| Release colony | 0.090 ± 0.200 | 1.160 ± 0.590 |
| Male-only production | 0.220 ± 0.264 | 0.002 ± 0.008 |

Table 2. Summary of production and quality of the VIENNA 42/Tol-94 after the thermal treatment, at the El Pino Facility.

| Parameters | Male-only production |
|---|---|
| Average brown pupae/wk | 78.20 ± 27.86 |
| Efficiency eggs to pupae (%) | 21.00 ± 03.15 |
| Million of pupae per ton of diet | 11.55 ± 03.08 |
| Adult emergence (%) | 74.46 ± 06.50 |
| Flying sterile males (%) | 66.51 ± 07.04 |
| Pupal weight (mg) | 08.10 ± 00.33 |
| Sex ratio (males %) | 99.78 ± 01.29 |

## DISCUSSION

The parental material (P) used to reset the clean stream, always contained a few recombinant individuals which were removed to ensure the integrity of the clean stream with respect to *wp/tsl*. In general, the level of recombinants was low in all the stages of the continuous FRS in El Pino. Such low levels of recombination do not have any serious effects for the efficiency of GSS rearing (Rossler 1985, Hooper et al. 1987, Kafu et al. 1993b, Bush-Petersen et al. 1992). If it becomes necessary to increase production, two strategies can be used: 1) increase the size of the clean backup stream or 2) use (one more amplification stage prior to the release steam. According to our estimates, the level of recombination will not exceed more than 1%, if an additional amplification is added.

The normal consequence of recombination in GSS is to expect nearly equivalent frequencies of females from brown pupae and males from white pupae (Hooper et al. 1987). In all the amplification steps of the FRS in El Pino, more male recombinants (i.e., males from white puparia) were observed. This is likely to be caused by the loss of the translocation, which generates the occurrence of a free Y chromosome (G. Franz, personal communication). Individuals that contain a free Y chromosome are 100% fertile and they can accumulate rapidly, since they have a selective advantage in colony cages compared with the translocated males, which are partially sterile.

Since the *wp* and *tsl* mutations are located close to each other on the right arm of the chromosome 5 (59B and 59B-61D of the trichogen map, respectively (Kerremans and Franz 1994)), only a limited recombination can occur between the two mutations. When recombination does occur in this region, females from white pupae can lose the *tsl* gene (and therefore, the temperature sensitivity Franz et al. 1994). This might explain why a very small number of white pupae are recovered after the thermal treatment.

After 14 months of "stable" male-only production, the FRS demonstrated that it is an effective system to maintain the genetic integrity of the *tsl* GSS. The FRS did not increase the cost of rearing at the El Pino facility and represented 3% of the total costs of male-only production in VIENNA 4/Tol-94.The cost of production per million male pupae was not different between bisexual strain production and *tsl* strain production.

## ACKNOWLEDGEMENTS

The technical support of L. Andrade, C. Herrera, C. Romero, M. Perez, F. Pecorelli, F. Reyes, B. Ortega, and E. Portillo of the El Pino facility is greatly appreciated. The administrative support of J. Stewart and O. Recinos (USDA-APHIS, Guatemala) is also very much appreciated.

## REFERENCES

Busch-Petersen, E. and H. Baumgartner. 1992. Simulation of instability in genetic sexing strains: effects of wild type contamination in association with other biological parameters. Bull. Entomol. Res. 82: 183-189.

Fisher, K. and C. Cáceres. 2000. A filter rearing system for mass reared genetic sexing strains of Mediterranean fruit fly (Diptera: Tephritidae). pp. 551-558. *In* Area-Wide Control of Fruit Flies and Other Insect Pests. K.H. Tan (ed.). Penerbit Universiti Sains Malaysia, Penang, Malaysia.

Fisher, K. 1998. Genetic sexing strains of Mediterranean fruit fly (Diptera: Tephritidae): optimizing the high temperature treatment of mass reared temperature sensitive lethal . J. Econ. Entomol. 91: 1406-1414.

Franz, G., E. Gencheva and P. Kerremans. 1994. Improved stability of sex separation strains for the Mediterranean fruit fly *Ceratitis capitata*. Genome 37: 72-82.

Franz, G., P. Kerremans, P. Rendón, J. Hendrichs. 1996. Development and application of genetic sexing systems for the Mediterranean fruit fly based on a temperature sensitive lethal. pp. 185-191. *In:* Fruit Fly Pests. A world assessment of their biology and management. B.A. Mcpheron and G. Steck (eds.). St. Lucie Press, Delray Beach, Florida, USA.

Hentze, F. and R. Mata. 1987. Mediterranean fruit fly eradication programme in Guatemala. pp. 533-539. *In* Fruit Flies Proc. 2nd Int. Symp. Columbari, Crete, 1986. A.P. Economopoulos (ed.). Elsevier, Amsterdam.

Hooper, G.H.S., A.S. Robinson and P.R. Marchland. 1987. Behaviour of a genetic sexing strain of Mediterranean fruit cly, *Ceratitis capitata,* during large scale rearing. pp. 349-362. *In* Fruit Flies Proc. 2nd Int. Symp. Columbari, Crete, 1986. A.P. Economopoulos (ed.). Elsevier, Amsterdam.

Kafu, A. A., R.J. Wood, G.S. Mani and E. Busch-Petersen. 1993b. Spontaneous recombination between $wp^+$ and the translocation breakpoint in the T:Y($wp^+$)30C genetic sexing strain of *Ceratitis capitata* (Wied.). Heredity 71: 104-109.

Kerremans, Ph. and G. Franz. 1994. Cytogenetic analysis of chromosome 5 from the Mediterranean fruit fly, *Ceratitis capitata*. Chromosoma 103: 142-146.

McInnis, D.O., S.Y.T. Tam, C. Grace and D. Miyashita. 1994. Population suppression and sterility rates induced by variable sex-ratio, sterile insect releases of *Ceratitis capitata* (Diptera: Tephritidae) in Hawaii. Ann. Entomol. Soc. Am. 87: 231-240.

Nitzan, Y., Y. Rossler and A.P. Economopoulos. 1993. Mediterranean fruit fly: effective control by genetic sexing male only SIT releases during 1989-1990. pp. 331-344. *In* Management of Insect Pests: Nuclear and Related Molecular and Genetic Techniques. IAEA-SM 327, Vienna, Austria.

Rendón, P. 1996. Development and evaluation of a temperature sensitive lethal strain of the Mediterranean fruit fly *Ceratitis capitata* (Wied.). Unpublished Ph.D.Thesis. Manchester University, UK. 267 pp.

Robinson, A.S., U. Cirio, G.H.S. Hooper and M. Capparella 1986. Field cage studies with a genetic sexing strain in the Mediterranean fruit fly, *Ceratitis capitata*. Entomol. Exp. Appl. 41: 231-235.

Rossler, Y. 1985. Effects of recombination in males of the Mediterranean fruit fly (Diptera: Tephritidae) on the integrity of "genetic sexing" strains produced for sterile-insect releases. Ann. Entomol. Soc. Am. 78: 265-270.

# Mass Rearing of the Melon Fly in Okinawa, Japan - Special Reference to Quality Control

*Masaaki Yamagishii[1] and Hiroyuki Kakinohana[2]*

[1] Okinawa Prefectural Fruit Fly Eradication Project Office, Maji 123, Naha 902-0061 Japan
[2] Okinawa Prefectural Agricultural Expt. Station ,Naha, Okinawa 903-0814, Japan

## INTRODUCTION

The melon fly, *Bactrocera cucurbitae* (Coquillett), had been completely eradicated from Okinawa, Japan in 1993 (Yamagishi et al. 1993, Kakinohana 1994, Kuba et al. 1996). Following the expansion of target areas during the eradication campaign, the number of flies produced was increased from 5 million to 280 million per week. In the process of the eradication project, the mass reared strains had been replaced three times with new strains (Table 1).

Table 1. The three mass reared strains of the melon fly produced in Okinawa.

| Items | No. 1 Strain | No. 2 Strain | | No. 3 Strain |
|---|---|---|---|---|
| Areas of collection | Ishigaki Is. & Miyako Is. | Okinawa Is. | Ishigaki Is. | Okinawa Is. |
| Year of collection | 1973 | 1978 | 1979 | 1985 |
| Host fruits | ? | bitter gourd | squash & bitter gourd | bitter gourd |
| No. of founder populations | ? | 14,500 | 3,000 | 19,281 |
| Releasing period | 1975 – 1979 | 1979 - 1986 | | 1986 - |
| Released areas | Kume Is. | Kume Is. & Miyako Isles | | Okinawa Isles. & Yaeyama Isles |
| Location of facilities | Ishigaki | Ishigaki & Okinawa | | Okinawa |
| References | Nakamori et al., 1975 | Soemori and Nakamori, 1981 | | unpublished |

The aim of this paper is to show the changes in various traits of the third strain that were regularly monitored in the factory. First, unintentional and intentional artificial selections to which the strain was exposed are mentioned. Second, the changes in the monitored traits are shown, and finally, the relation between selection and the response to selection is discussed.

## ARTIFICIAL SELECTION

A procedure for mass rearing of the melon flies in Okinawa in 1988 - 1990, when the fly production was at the maximum level, is shown in Table 2. The third strain was reared under the artificial conditions indicated in Table 2, and inevitably exposed to selection different from that in nature. Possible differences in conditions include those in temperature, photoperiod, food, oviposition substrates and conditions during mating. Various genetic traits, therefore, should become different from those of the founder population in long-term mass rearing.

Table 2. Procedure for mass rearing of the melon flies in Okinawa in 1988 - 1990 when fly production was at a maximum level.

Adult rearing
- Rear adults during 6 weeks after emergence
- Rearing conditions: 26°C, 60% RH
- Scale: 62,500 pupae per box, 210 boxes per room x 2 rooms = 420 boxes
- Replace 60 boxes containing the oldest adults with new ones once a week

Egg collection
- Collect eggs laid by the 2nd - 6th week old adults once a week per room

Egg seeding
- Seed eggs 4 times per week
- Rearing scale: 88,800 eggs per tray containing 6 litres diets, 960 trays per room, 4 rooms per week

Larval rearing
- Rear larvae during 7 days after egg seeding
- Rearing conditions: 20 - 28°C, 80% RH

Collecting larvae
- Collect larvae on the 5th - the 7th day after egg seeding
- Spray water into diets on the last day to stimulate larval jumping

Setting larvae into vermiculite
- Rearing scale: 50,000 larvae per tray containing 10 litres of vermiculite, 5,400 trays per week

Pupal rearing
- Sift pupae from vermiculite on the 6th day after setting
- Rearing conditions: Before sifting, 25°C, 60% RH;<br>After sifting, 20 - 27°C, 60% RH

Irradiation
- Irradiate pupae at a dose of 70 Gy on the 3rd day before emergence

Transportation
- Transport pupae to the release centres immediately after irradiation

Marking
- Mark pupae with fluorescent dye

Release
- Put pupae into emergence boxes (40,000 pupae per box) and allow them to emerge
- Lower temperature below 5°C on the 2nd day after emergence, take out adults, put them into a release apparatus and release them by helicopter (4 million adults per flight)

Along with these unintentional selections, the third strain was subjected to an intentional selection for the timing of oviposition and the developmental rate of larvae (Figure 1). The selection depended on sampling of individuals to be used as parents. The adults of this strain were reared for six weeks after emergence, and eggs were collected from the 2 - 6 week old adults. Until the 34th generation, eggs laid by younger adults were selected for the offspring generation to hasten the reproductive activity of adults. From the 34th generation, the selection method was changed because of the reduction of adult longevity, that is, 1) no selection and 2) selection to reverse direction. In the former method, eggs for the next generation were collected in proportion to the volume of eggs laid by each age group of adults (no. 3-1 strain). In the latter method, eggs for the offspring generation were collected from only older adults to extend adult longevity (no. 3-2 strain). On the other hand, larvae of the strain were collected on the 5th, 6th, and 7th day after egg seeding. Undeveloped larvae in the larval diets on the last day of their collection were consequently eliminated from the population. In addition, quickly developing larvae were selected for the next generation until the 34th generation to shorten the larval period. The selection method was then changed; larvae were sampled proportionally to the number of larvae collected each day.

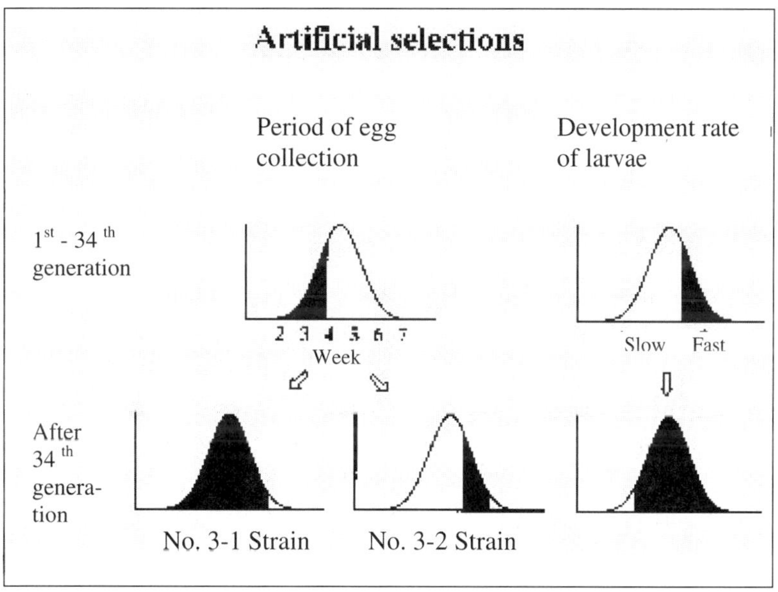


Figure 1. Artificial selections for mass rearing of the third strain. Eggs and larvae indicated by solid part of distribution were selected as the next generation.

## CHANGES IN TRAITS OF THE THIRD STRAIN

The traits of the third mass reared strain monitored for quality control are summarised in Table 3. Their genetic foundation is changed by unintentional and/or intentional selection. Excluding the accidental deterioration of the traits caused solely by environmental reasons, there were two types of traits. One was likely to be stable through mass rearing while the other was progressively changed, suggesting that it has a genetic basis.

## STABLE TRAITS

### Hatchability

The percentage of hatched eggs was consistently stable around 90% (90.24 ± 19.03 % = mean ± SD). Half of the eggs were kept in water below 5 °C and seeded the following day. The hatchability of these eggs had slightly deteriorated.

### Remains in Larval Diets

After larval collection, the number of larvae remaining in larval diets and that of pupae that accidentally pupated in the diets were counted to quantify the intensity of selection on the developmental period of larvae. Remaining larvae were either undeveloped or developed individuals that remained in the diets by accident. The percentage of those remaining was around 4 % (4.27 ± 2.84 % ). The ratio of larvae to pupae was roughly 3 : 1. The elimination of the slowly developed phenotype from the population acted as truncation selection on the developmental period of larvae, and resulted in a lesser mean value.

Table 3. Items for quality monitoring of the melon flies in Okinawa.

<u>Adults for egg collection</u>
- Number of generations
- Number of eggs laid by the 2nd - 6th week old adults

<u>Eggs</u>
- Hatchability

<u>Larvae</u>
- Number of eggs seeded (Calculated by the equation: 12,000 eggs per 1 ml x 7.4 ml per tray x number of trays)
- Temperature in larval diets (an indicator of larval development)
- Developmental rate
- Percentage of remains in diets

<u>Pupae</u>
- Weight and size of pupae
- Number of pupae produced
- Recovery rate (a ratio of number of pupae produced to number of eggs seeded)
- Production efficiency per diets
- Pupation period

<u>Adults produced</u>
- Emergence rate and percentage of flies that have flight ability (before and after irradiation, after marking and after transportation)
- Number of flies released
- Survival rate of non-irradiated flies for 10 weeks after emergence (every 5 weeks)
- Survival rate of irradiated flies during 5 weeks after emergence (once a month)
- Fertility of irradiated files (once a month)

### Pupal Weight

The values were consistently stable, and the faster the larvae emerged from the larval diets, the heavier the weight of the pupae (14.32 ± 0.62 mg on the 5th day, 13.72 ± 0.67 mg on the 6th day, and 12.45 ± 0.63 mg on the 7th day). The reason for the decrease of the pupal weight is not clear, but may be related to the deterioration of the diets.

### Recovery Rate of Pupae

The values fluctuated around 65% (64.72 ± 15.71 %). The utilisation efficiency of the larval diets was about 13 % by weight (volume 128.2 ± 12.0 g; number 9,663 ± 947 per one litre diet).

### Emergence Rate of Adults

The values fluctuated around 90% irrespective of the date of larval collection (90.9 ± 5.0 % on the 5th day, 92.1 ± 4.0 % on the 6th day, and 90.4 ± 3.8 % on the 7th day).

## PROGRESSIVELY CHANGED TRAITS

### Fecundity

The total fecundity per adult rearing box was gradually increased to about 150 ml as the generations advanced. The ratio of the volume of eggs laid by the younger adults compared to that laid by older ones became higher with the progress of the generations. However, the performance was slightly different for different strains. Strain no. 3-1 laid significantly more eggs than strain no. 3-2 only at the 2nd week while it laid fewer eggs from the 3rd to 6th weeks. This tendency appears to have continued until the present day.

### Volumc of Dicts Consumed by Adults

With the increase of fecundity, the volume of diets consumed by adults was gradually increased as the generations alternated. In comparison to the volume of sugar and autolised yeast consumed by the 3rd generation, those in the 92nd generation consumed about twice more sugar and three times more autolised yeast.

### Developmental Period of Larvae

As a result of the truncated selection which eliminated undeveloped larvae and selected faster developed ones, the developmental period of larvae became shorter – from 7.37 ± 0.72 days at the 2nd-5th generations to 5.91 ± 0.69 days at the 31st-35th generations. The values were then relatively stable irrespective of the strains.

### Survival Rate of Adults

The survival rate of adults was examined once a week for 10 weeks after emergence. The value at the 10th week after emergence became lower, from about 70 %

to 10 % as time passed (Figure 2). Between the two strains, the longevity of the no. 3-2 strain was slightly higher than that of the no. 3-1 strain. Females also exhibited greater longevity than males. Another observation to be mentioned is a change in the pattern of the survival curve with different generations. In earlier generations, the survival rate showed no drastic fall for 10 weeks, but as time passed, the degree of decline appeared obvious after 5 weeks.

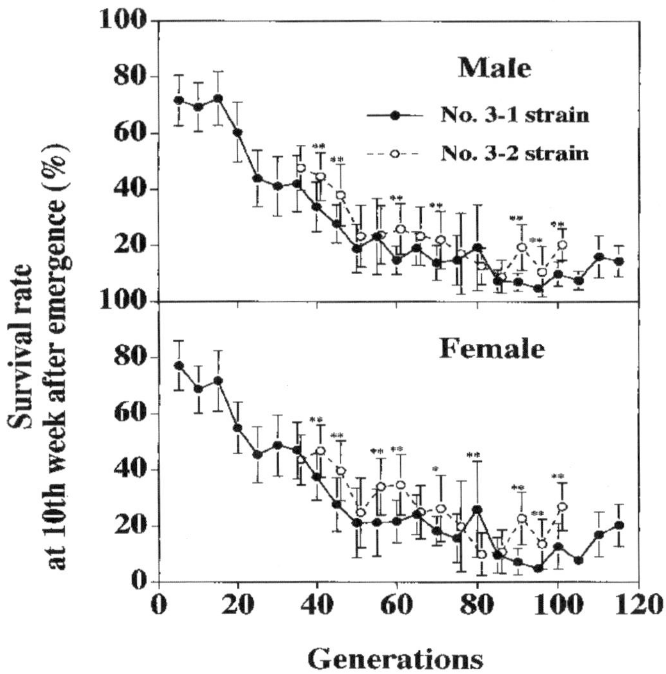


Figure 2. Survival rate at the 10th week after emergence (*; *P<0.05*, **; *P < 0.01*, Mann-Whitney *U-test*).

## DISCUSSIONS

The aim of quality monitoring of the mass reared flies in the factory is: 1) to produce a large number of flies of high quality as efficiently as the production of industrial products, 2) to clarify genetic changes in the traits that may ultimately be responsible for mating competitiveness in the fields, and 3) to identify the kind of rearing conditions or selection that causes genetic changes. Unexpected change in genetic traits may still occur due to unintentional selection or a correlated response to selection.

As mentioned in Table 2, the third mass reared strain was exposed to artificial conditions and the traits of early oviposition and fast development of larvae were intentionally selected. The changes to early oviposition and fast development of larvae were, of course, observed as the result of direct response to the intentional selections. However, an increase in early fecundity, total fecundity, the diet ingestion by adults and the shorter longevity of adults occurred unexpectedly. These changes seem to carry a genetic basis because of progressiveness as generations alternated. Recently, Miyatake (1997) showed genetic trade-off between early fecundity and longevity, using an artificial two-way selection on the timing of egg collection, suggesting that it was caused by antagonistic pleiotropy.

In the strain selected to reverse direction (no. 3-2 strain), the longevity was slightly extended but declined in the same manner as that of the non-selected strain (no. 3-1 strain) (Figure 2). The performance of propagation was, on the whole, similar to that of the non-selected strain, although the egg laying activity was slightly delayed. The process of collecting eggs in the third strain may essentially include selection to increase fecundity. Shorter longevity may also be caused by pleiotropy of the genes that control fecundity. The traits of large appetites in adults should genetically be related to the traits of fecundity.

Miyatake (1997) also indicated that the strain selected for early fecundity had a shorter larval period. The developmental rate of larvae of the third strain might, therefore, be exposed to both intentional selections. The larval period of both strains after the change in the selection method seems to be fixed and not too different from each other, although undeveloped larvae were still excluded from the population. It seems possible that the developmental rate of larvae in the strains had already reached the selection limit.

Using an artificial two-way selection, the correlated responses among traits in the melon fly have been reviewed by Miyatake (1996). It was recently reported that the developmental rate and circadian period were controlled by a gene that has a pleiotropic effect (Shimizu et al. 1997). Such genetical research is essential to improve methods in the mass rearing of insects.

## ACKNOWLEDGEMENTS

We wish to thank Drs. Teruya Tadashi and Koyama Juro for reading the draft of the manuscript.

## REFERENCES

Kakinohana, H. 1994. The melon fly eradication programme in Japan. pp. 223-236. *In* Fruit Flies and the Sterile Insect Technique C.O. Calkins, W. Klasssen and P. Liedo (eds.). CRC Press, Inc., Boca Raton, Florida, USA.

Kuba, H., T. Kohama, H. Kakinohana, M. Yamagishi, K. Kinjo, Y. Sokei, T. Nakasone and Y. Nakamoto. 1996. pp. 543-550. *In* Fruit Fly Pests: A World Assessment of Their Biology and Management. B. A. McPheron and G. J. Steck (eds.). St. Lucie Press, Delray Beach, FL, USA.

Miyatake, T. 1996. Artificial selection experiments in the melon fly: the status quo and problems. pp. 437-443. *In* Fruit Fly Pests: A World Assessment of Their Biology and Management. B. A. McPheron and G. J. Steck (eds.). St. Lucie Press, Delray Beach, FL, USA.

Miyatake, T. 1997. Genetic trade-off between early fecundity and longevity in *Bactrocera cucurbitae* (Diptera: Tephritidae). Heredity 78: 93-100.

Nakamori, H., H. Kakinohana and H. Soemori. 1975. Mass rearing of the melon fly, *Dacus cucurbitae* Coquillett. I. Effect of rearing density on the yield and quantity of flies. Okinawa Nogyo 13: 27-32. (In Japanese with an English summary).

Shimizu, T., T. Miyatake, Y. Watari and T. Arai. 1997. A gene pleiotropically controlling developmental and circadian periods in the melon fly, *Bactrocera cucurbitae* (Diptera: Tephritidae). Heredity 79: 600-605.

Soemori, H. and H. Nakamori. 1981. Production of successive generations of a new strain of the melon fly, *Dacus cucurbitae* Coquillett (Diptera: Tephritidae) and reproductive characteristics in mass rearing. Jap. J. Appl. Entomol. Zool. 25: 229-35. (In Japanese with an English summary).

Yamagishi, M., H. Kakinohana, H. Kuba, T. Kohama, Y. Nakamoto, Y. Sokei and K. Kinjo. 1993. Eradication of the melon fly from Okinawa, Japan, by means of the sterile insect technique, pp. 49-60. *In* Management of Insect Pest: Nuclear and Related Molecular and Genetic Techniques. IAEA, Vienna, Austria.

# Composition of Mediterranean Fruit Fly Third Instar Larvae (Diptera: Tephritidae) and Diet: Nutrient Balance Studies on Amino Acids, Minerals and Nutrient Composition in Fresh and Spent Mass Rearing Diets

*Harvey T. Chan Jr.*[1], *Harry Ako*[2], *Ruth Y. Niino-Duponte*[2], *James R. Carpenter*[2] and *Eric B. Jang*[1]

[1] Tropical Fruit, Vegetable and Ornamental Crops Research Laboratory, USDA-ARS, P.O. Box 4459, Hilo, Hawaii, 96720, USA

[2] University of Hawaii at Manoa, Department of Entomology, 3050 Maile Way, Gilmore 310, Honolulu, HI 96822 USA

## INTRODUCTION

Mass production of the Mediterranean fruit fly (Medfly) larvae, *Ceratitis capitata* (Wiedemann), requires a rearing diet (Tanaka et al. 1969 1970) of which the nutrient requirements and digestibility have not been established. Setbacks in rearing productivity from the expected 100% yield to as low as 3% yield may occasionally be directly attributed to insecticide contamination or a variety of possible cause(s) (Kobayashi, 1993). These causes include inadequate nutrition, poor diet formulation, overcrowding of either microorganisms or *Drosophila*, or to the inherent processes of oxidative or microbial deterioration of nutrients.

The purpose of this study was to establish the nutritional status of the Mediterranean fruit fly diet through a material balance study for changes in proximate composition (i.e., moisture, protein, fat, ash, carbohydrates), amino acids, minerals between fresh and spent diets, and in the fruit fly larvae themselves.

## MATERIAL AND METHODS

### Insects and Diet Material

*Ceratitis capitata* eggs were collected as described by Tanaka et al. (1969) over a 1 to 3 h period from adult stock colonies that were maintained on sugar, water and hydrolysed yeast. A 4.0 ml aliquot of eggs (ca. 150,000 eggs) were distributed over 5 kg of diet spread out on a tray (74.3 x 38.6 x 1.27 cm). The basic diet used in this study was developed by Tanaka et al. (1969). The major components (by weight) were wheat mill-run (26.0%), sugar (12%), dried torula yeast (3.6%), sodium benzoate (0.1%), methyl p-hydroxybenzoate (0.1%), and water (58%). Diet samples used in this study were taken from actual production runs at the USDA-APHIS Medfly rearing facility in Waimanalo, Hawaii, in 1993.

### Sampling for Chemical Analysis

Fresh and spent fruit fly diet were sampled at random from four batches at the USDA-APHIS Waimanalo Medfly Rearing Facility, Waimanalo, HI, and frozen for chemical analysis. Immediately after mixing the diet, a 1 kg sample of the fresh diet was frozen at -10°C. Fresh diet seeded with insect eggs were incubated at 27°C and 65% RH for 8 days; the eggs hatched within 48 h and developed into larvae. Additional water was sprayed onto the diets during larval development as needed to compensate for evaporative losses. Mature third instar larvae naturally egress 6 to 8 days after egg hatch from the diet. Larvae were collected daily (on the $6^{th}$, $7^{th}$ and $8^{th}$ day) weighed, and frozen at -10°C. After 8 days, approximately 1 kg of spent diet was collected and frozen at -10°C. Frozen samples of larvae, fresh and spent diets were freeze dried. The freeze dried samples were then stored at -10°C for further analysis.

### Nutrient Analysis of Media and Larvae

Media samples were dried in a forced air oven at 50°C to a constant weight, then ground through a Wiley mill to pass through a 1 mm mesh stainless steel screen. The freeze dried larvae were ground with mortar and pestle for homogeneity. Samples were analysed for dry matter, crude protein, ash, ether extract (AOAC 1980), neutral detergent fibre (NDF) (Robertson and Van Soest 1981), acid detergent fibre, permanganate lignin and cellulose (Goering and Van Soest 1970). Mineral analysis was done by inductively-coupled plasma atomic emissions spectroscopy (ICP-AES). Carbohydrate was determined by difference.

Amino acid analyses were conducted on dried samples which had been hydrolysed in 6 M HCl under an oxygen-free atmosphere obtained by alternately evacuating and flushing the sample with nitrogen. The internal standard norleucine was added to the sample and used to correct for incomplete recovery of amino acids (Boyer 1993). Amino acids were quantified on an HPLC equipped with an ion exchange column, eluted with a exponential gradient of sodium citrate buffer (0.02N, pH 3.15 initial to 1.0N, pH 7.4 final), and analysed with a ninhydrin post-column reactor. Proline and tryptophan concentrations were not measured.

## RESULTS AND DISCUSSION

Fresh diets at 5,777 g/diet tray produced 1094 g of third instar larvae which "popped out of the diet" and left a residual of 3,966 g of spent diet (Table 1).

The initial analysis showed that 717 g (12.4%) weight loss of diet occurred during the 8 days of larval development; however, this is not the "true" weight loss as the diet was "watered" as needed during larval development. The amount of water added was based upon the experienced judgement of the rearing technician. The true weight loss was calculated using the ash content as an internal standard. Ash is the carbon-free residue which remains after the foodstuff is ignited at 550-570°C in a muffle furnace and consists of the oxidised minerals. Since these minerals and their organic derivatives are known to be non-volatile under ambient temperatures, the ash content can be used as an immutable internal standard. The material balance, calculated using ash as the internal standard, was thus calculated as follows: The ash content (Table 2.) for fresh diet is 1.54%; a kg fresh diet produces 190 g of larvae (2.18% ash) or, since the net loss of ash per tray of diet is due only to the egressing larvae, the amount of ash in the spent diet is the difference between 15.4 g ash/kg fresh diet minus 4.14 g ash/190 g larvae which results in 11.3 g ash in spent diet. Hence, the true weight of the spent diet can be calculated (Equation 1) using the ash content in the spent diet of 2.13% (Table 1):

Table 1. Larval yield (n = 6).

| | Fresh diet wt. (g) | Spent diet wt. (g) | Corrected spent diet wt. (g) | Larval yield (g) | Larvae kg/kg fresh diet |
|---|---|---|---|---|---|
| Mean | 5776 | 3967 | 3050 | 1094 | 0.190 |
| SE | 189 | 236 | | 52 | 0.009 |

**Equation 1**:

$$15.4 \text{ g ash/kg fresh diet} = 4.14 \text{ g ash/190 g larvae} + 0.0213 \text{ (X g ash spent diet)}$$

$$\therefore 0.213 \text{ (X)} = 11.3 \text{ g}, \quad \text{X=528 g of spent diet}$$

Hence the material balance equation for the Medfly diet is as follows:

**Equation 2:**

$$1 \text{ kg fresh diet} = 190 \text{ g larvae} + 528 \text{ g spent diet} + 282 \text{ g metabolic/evaporative loss}$$

The results of the proximate analysis and amino acid analysis of Medfly larvae, fresh and spent diets are adjusted according to the material balance and listed in Tables 2 and 4 respectively.

Table 2. Proximate analysis of Medfly diet and Medfly larvae (n = 4).

| Proximate analysis | | | | | | |
|---|---|---|---|---|---|---|
| | Wet larvae | | Wet diet | | | |
| | | | Fresh | | Spent[a] | |
| | Mean | SE | Mean | SE | Mean | SE |
| % Water | 78.32 | 0.71 | 57.85 | 0.35 | 59.88 | 2.38 |
| % Protein | 9.52 | 0.05 | 6.39 | 0.06 | 7.19 | 0.40 |
| % Fat | 6.84 | 0.05 | 1.21 | 0.16 | 1.52 | 0.18 |
| % Ash | 2.18 | 0.01 | 1.54 | 0.05 | 2.13 | 0.15 |
| % Carbohydrates | 3.14 | 0.71 | 33.01 | 0.44 | 29.29 | 1.72 |

[a] Spent diet values were adjusted based on material balance equation

## Protein Analysis

Crude protein, as determined by the Kjeldahl method, measures total N which also includes non-protein nitrogen compounds produced by acid digestion and reduction to $NH_3$ (Jacobs 1958) as well as nitrogenous insect waste products, such as urea, thus accounting for the higher reported protein content by this method. The material balance for crude protein is as follows: 63.9 g protein/kg fresh diet-18.08 g protein/190 g larvae = 45.8 g calculated protein in spent diet; however, proximate analysis of spent diet showed 38.0 g protein/528 g spent diet (Table 3). Crude protein utilisation by fruit fly is 28% and total crude protein utilisation by insect and microorganisms is 45%.

Based on amino acid analysis, the calculated protein in the spent diet was 19.5 g, which is lower than the 38 g of protein determined by the Kjeldahl method (Table 4). Protein analysis via amino acid analysis showed: 51.34 g amino acids/kg fresh diet, 18.8 g amino acids/190g larvae, and 19.49 g amino acids/528 g spent diet. Based upon amino acid analysis protein utilisation in the diet by insects was 42.5% and utilisation by insects and microorganisms was 62%.

Table 3. Crude protein composition of fresh and spent Medfly diets and Medfly larvae.

| | | Mean | SE |
|---|---|---|---|
| Fresh diet (wet) | % protein | 6.39 | 0.06 |
| | protein (g/kg fresh diet) | 63.86 | 0.60 |
| Spent diet | % protein | 7.19 | 0.40 |
| | protein (g/528g spent diet) | 37.98 | 2.11 |
| Larvae on diet | % protein wet larvae | 9.52 | 0.048 |
| | protein (g/190g larvae) | 18.08 | 0.091 |

## Amino Acid Analysis

Glutamic acid was the most predominant amino acid in Medfly larvae and in fresh and spent diets (Table 4). Although glutamic acid losses amount to 78% during conversion to spent diet, the amino acid does not appear to be limiting as 5.3 g remained within the spent diet and larval uptake was only 11.9% glutamate. The limiting amino acids appeared to be those which were either not measureable or in negligible quantities in the fresh diet but were found to be in Medlfy insect larvae such as methionine, tyrosine and cystine. The uptake or utilisation of tyrosine by Medfly larvae was 177% which indicates that there was another source of tyrosine other than that supplied by the diet materials. The other sources of amino acids could be from microorganisms, such as *Acetobacter* or that Medflies are capable of biosynthesis of tyrosine from phenylalanine. Other amino acids which had high rates of uptake by the Medfly larvae and appeared to be partially limiting were lysine, valine, isoleucine, and histidine.

Spent fruit fly diet was higher in crude protein ($P<.01$), fat ($P<.05$), total ash ($P<.01$), P ($P<.05$), K ($P<.01$), Mn ($P<.01$) and valine ($P<.01$) than fresh diet (Tables 2, 4, 6). The higher percentage content of fat and protein in the spent diet may be due to the residual larvae that failed to egress and pupated in the diet. The higher proportion of ash and minerals in spent compared to fresh diet may be due to the metabolic conversion of carbonaceous material to $CO_2$ and water without any corresponding losses in mineral content due to their non-volatile state. Hence the proportion of minerals increases due to the metabolic conversion of organic materials $CO_2$ and water which are volatile weight losses.

Table 4. Amino acid content of fresh and spent Medfly diets and Medfly larvae listed relative to material balance of 1 kg fresh diet, 528 g spent diet, and 190 g Medfly larvae (n = 3).

| Amino acid | g fresh/kg | SE | g spent/ 528 | SE | g/190g larvae | SE | % Larval uptake |
|---|---|---|---|---|---|---|---|
| Aspartic | 3.79 | 0.78 | 1.72 | 0.08 | 1.86 | 0.15 | 49.0 |
| Threonine | 1.84 | 0.27 | 1.12 | 0.01 | 0.74 | 0.01 | 40.4 |
| Serine | 1.98 | 0.35 | 1.16 | 0.07 | 1.25 | 0.14 | 63.1 |
| Glutamic | 24.26 | 3.45 | 5.33 | 0.12 | 2.88 | 0.23 | 11.9 |
| Glycine | 2.47 | 0.19 | 1.33 | 0.06 | 1.37 | 0.11 | 55.5 |
| Alanine | 2.81 | 0.09 | 1.63 | 0.08 | 1.56 | 0.15 | 55.5 |
| Valine | 1.11 | 0.09 | 0.73 | 0.09 | 0.85 | 0.18 | 76.9 |
| Methionine | 0.00 | 0.00 | 0.21 | 0.02 | 0.89 | 0.05 | – |
| Isoleucine | 1.55 | 0.09 | 0.86 | 0.05 | 1.00 | 0.10 | 64.6 |
| Leucine | 3.32 | 0.96 | 1.51 | 0.04 | 1.22 | 0.07 | 36.8 |
| Tyrosine | 1.01 | 0.06 | 0.43 | 0.06 | 1.36 | 0.11 | 134.8 |
| Phenylalanine | 1.83 | 0.20 | 0.95 | 0.03 | 0.69 | 0.05 | 38.0 |
| Histidine | 0.97 | 0.06 | 0.58 | 0.04 | 0.58 | 0.07 | 60.1 |
| Lysine | 1.88 | 0.19 | 0.92 | 0.01 | 1.33 | 0.02 | 70.1 |
| Arginine | 2.51 | 0.09 | 1.01 | 0.04 | 1.19 | 0.08 | 47.2 |
| Cystine | | | | 0.05 | 0.44 | 0.10 | |
| Total | 51.34 | | 19.49 | | 18.79 | | 36.6 |

* calculations based on wet weight basis from the material balance Equation 2.

## Fat Analysis

Using the material balance equation (Equation 2), the fat content in the spent diet was calculated as follows: 12.1g fat/kg fresh diet – 12.99 g fat/190 g larvae = -0.89 g calculated to be remaining in the spent diet; however, 8.03 g of fat were found in 528 g spent diet which indicates a "surplus "of 8.9 g fat (Table 5). This so called "surplus" of fat is due to fatty acid synthesis during larval development and remains as part of the spent diet largely as larvae and pupae that had not egressed from the diet. Although this might indicate a deficiency of fat in the diet, a future study into individual fatty acid analysis to determine adequacy of essential fatty acids is warranted.

Table 5. Fat composition of 1 kg fresh and 528 g spent Medfly diets and 190 g Medfly larvae.

| | | Mean | SE |
|---|---|---|---|
| Fresh diet (wet) | % fat | 1.21 | 0.161 |
| | fat (g/kg fresh diet) | 12.12 | 1.613 |
| Spent diet | % fat | 1.515 | 0.177 |
| | fat (g/528g spent diet) | 7.998 | 0.932 |
| Larvae | % fat wet larvae | 6.839 | 0.049 |
| | fat (g/190g larvae) | 12.993 | 0.094 |

## Minerals

Mineral analyses of the Medfly larvae, fresh and spent diets are shown in Table 6. When the concentrations of the various minerals are expressed in percentages, greater amounts of the minerals are present in the spent diet compared to the fresh diet. When the percentage mineral values are converted to absolute values using the material balance equation (Equation 2.), the amount of minerals in the spent diets were less than in the fresh diets. Analysis of the mineral uptake shows that Medfly larvae acquired 49% of the available Mg and 38% of the available Ca and 27% of Zn from the diet (Table 7).

Table 6. Mineral composition of Medfly diet and Medfly larvae.

| Mineral composition | | | | | | |
|---|---|---|---|---|---|---|
| | Wet larvae | | Wet diet | | | |
| | | | Fresh | | Spent | |
| | Mean | SE | Mean | SE | Mean | SE |
| P % | 0.38 | 0.003 | 0.315 | 0.009 | 0.383 | 0.027 |
| K % | 0.28 | 0.003 | 0.341 | 0.010 | 0.458 | 0.027 |
| Ca % | 0.07 | 0.001 | 0.033 | 0.001 | 0.028 | 0.002 |
| Mg % | 0.31 | 0.003 | 0.120 | 0.003 | 0.123 | 0.008 |
| Na % | 0.03 | 0.001 | 0.040 | 0.002 | 0.052 | 0.004 |
| Mn mg/l | 29.20 | 1.328 | 42.361 | 1.642 | 59.184 | 4.669 |
| Fe mg/l | 55.03 | 0.813 | 46.576 | 4.198 | 53.166 | 6.671 |
| Cu mg/l | 2.36 | 0.108 | 5.480 | 0.311 | 5.618 | 0.318 |
| Zn mg/l | 48.83 | 0.655 | 34.352 | 3.954 | 37.015 | 2.317 |

Table 7. Mineral uptake by Medfly larvae.

| Mineral | g/190g larvae | g/kg fresh | g/528g Spent diet | % Larval uptake |
|---|---|---|---|---|
| P | 0.722 | 3.15 | 2.023 | 22.9 |
| K | 0.532 | 3.41 | 2.420 | 15.6 |
| Ca | 0.124 | 0.33 | 0.148 | 38.0 |
| Mg | 0.583 | 1.20 | 0.651 | 48.6 |
| Na | 0.058 | 0.40 | 0.275 | 14.4 |
| Mn | 0.0055 | 0.0424 | 0.031 | 13.1 |
| Fe | 0.0105 | 0.0466 | 0.028 | 22.4 |
| Cu | 0.0004 | 0.0055 | 0.003 | 8.2 |
| Zn | 0.0093 | 0.0344 | 0.019 | 27.0 |

## Carbohydrate Analysis

The fresh diet contained 330 g total carbohydrates of which 120 g was added sugar. Carbohydrate material balance analysis showed 330 g in fresh diet, less 5.97 g in larvae for a calculated net balance of 324 g in the spent diet. However, the actual net balance of total carbohydrates is 155 g/528 g spent diet (Tables 8, 9). The difference in carbohydrates between the actual and the calculated in 528 g of spent diet is 169 g which means that practically 51% of the carbohydrates was metabolised by the larvae and/or microorganisms and that the 155 g of non-utilised carbohydrates is 43.7% fibre (67.6g neutral detergent fibre). Fibre analysis classifies neutral detergent fibres as total fibre and acid detergent fibres as the cellulose/lignin fraction with the difference between neutral detergent and acid fibre comprising the hemicellulose and pectin fractions. The fresh diet contained 103g total fibre of which 37 g is the cellulose/lignin fraction and the remaining 66 g of fibre represent the hemicellulose/pectin fraction. The spent diet contained 67.6 g of total fibre of which 25.4 g are the cellulose and lignin components with the remaining 42.3 g comprising of hemicelluloses and pectins. A 34.7% loss in total fibre content from conversion of fresh to spent larval diets, with similar losses of 31.5% and 36.4% in the cellulose/lignon and hemicellulose/pectin fractions, indicates losses speculated to be due to microbial metabolism, as Medfly larvae are not known to metabolise cellulosic or hemicellosic materials.

## Effects of Microbial Activity on Diet Composition

In a previous publication (Jang and Chan 1993), acetic acid production in Medfly diet was determined to be 2.5% or 0.4 moles/kg spent diet. The amount of sugar added to diet is 120 g or 0.350 moles/kg fresh diet.

Microbial activity is hypothesised to be primarily responsible for the changes in carbohydrate concentration. Normally active acetic acid-producing bacteria in the diet produce about 25 g (0.417moles) of acetic acid/kg spent diet, or 0.218 moles/528 spent diet. The stoichiometric relationship of sucrose fermentation to acetic acid is that 1 mole of sucrose produces 4 moles of acetic acid. Therefore, 0.218 moles of acetic acid would need 0.055 moles (18.7 g) of sucrose as starting material. The onset of acetic acid production occurs on the second day after diet preparation and coincides with insect egg hatch (Jang and Chan 1993) and the production of acid with resulting decrease in pH has been shown to be essential to the survivability of Medfly larvae (Chan and Jang 1995). The rapid production of acetic acid within the two days infers a rapid fermentation of sucrose and concomitant depletion of carbohydrates. This raises the question of the quantity of carbohydrate needed in the diet for insect development. Sucrose is not needed to produce the acidic environment (pH<5.5) essential for larval development, as this can be artificially generated using hydrochloric or citric acid. The addition of hydrochloric or citric acid to lower the pH has the additional benefit of enhancing the antimicrobial activity of the benzoates.

Table 8. Carbohydrate composition of fresh and spent Medfly diets (wet basis).

| | % Fresh | SE | % Spent | SE | Fresh (g/kg) | Spent (g/528g) | Fresh minus Spent | % Used |
|---|---|---|---|---|---|---|---|---|
| Total CHO | 33.04 | 0.44 | 29.33 | 1.72 | 330.4 | 154.9 | 175.5 | 53.1 |
| AD (cellulose/lignin) | 3.71 | 0.17 | 4.80 | 0.30 | 37.1 | 25.3 | 11.8 | 31.7 |
| ND (total fibres) | 10.35 | 0.44 | 12.83 | 0.80 | 103.5 | 67.7 | 35.8 | 34.5 |
| ND-AD (hemicellulosel/pectin) | 6.64 | | 8.03 | | 66.4 | 42.4 | 24.0 | 36.1 |
| Non-fibre CHO (total CHO-ND) | 22.69 | | 16.50 | | 226.9 | 87.1 | 139.8 | 61.6 |

Table 9. Carbohydrate composition of Medfly larvae (wet basis).

| | % | SE | g/190g |
|---|---|---|---|
| Total CHO | 3.144 | 0.706 | 5.97 |
| AD (chitin) | 1.446 | 0.026 | 2.78 |
| ND (total fibres) | 2.395 | 0.088 | 4.55 |
| ND-AD (hemicellulose/pectin) | 0.949 | | 1.80 |
| Non-fibre CHO (total CHO-ND) | 0.749 | | 1.42 |

## Energy Balance

Analysis of the diets for energy source and expenditures can be attained by using energy balance equations and assigning energy values to the proximal analysis of fat and carbohydrate and amino acid analysis as protein in spent and fresh diets. Using the assigned energy values (kcal/gr) for carbohydrate (4.0), fat (9.0), and protein (4.0), the calculated values for 1 kg fresh diet are 1320 kcal for 330 g carbohydrate, 108.9 kcal for 12.1 g fat, 205.4 kcal for 51.3 g protein which totals to 1634 kcal. The calculated values for 528 g of spent diet are 619 kcal for 155 g of carbohydrate, 72.3 kcal for 8.03 g of fat, 78 kcal for 19.5 g of protein achieving a total of 769 kcal. From the compositional analysis of 190 g of Medfly larvae, the assigned energy values are 24 kcal for 5.97 g of carbohydrate, 117 kcal for 13.0 g of fat, and 75.2 kcal for 18.8 g of protein for a total of 216.2 kcal. The above information can be summarised in the following energy balance equation:

(fresh diet) 1634.3 kcal = (spent diet) 769.3 kcal + (larvae) 216.2 kcal + (X);

where X is the net metabolic expenditure, X= 649 kcal.

## Energy and Material Balance

Based on the material balance equation (Equation 2), there was 281.5 g weight loss during the 8-day larval development in the course of diet usage that was attributed to metabolic/evaporative causes. If the 649 kcal value from the energy balance is assigned to the 281.5 g metabolic weight loss, one can analyse the carbohydrate usage in the following terms: of the 330 g carbohydrate in the fresh diet, the sucrose component comprised 120 g which supplied 480 kcal. We can assume nearly complete metabolism of sugar as the spent diet showed that 67.7 g of the 155 g carbohydrate remaining was in the form of fibre with 87.1 g as non-fibre carbohydrate for 348 kcal while the fresh diet had 227g of non-fibre carbohydrate for 908 kcal. The difference between fresh and spent diet was 554 kcal of which 480 kcal was sugar which means 87% of the carbohydrate energy was supplied by sugar. Thus of the 628 kcal expended during the course of the diet utilisation, 76% of the energy can be attributed to sugar as the energy source.

## REFERENCES

AOAC. 1980. Official Methods of Analysis. ($13^{th}$ Ed.), Association of Official Analytical Chemists, Washington, D.C.

Boyer, R.F. 1993. Modern Experimental Biochemistry. Benjamin/Cummings, Redwood City, 555 pp.

Chan, H.T., Jr., J.D. Hansen and S. Y-T. Tam. 1990. Larval diets from different protein sources for Mediterranean fruit flies (Diptera: Tephritidae). J. Econ. Entomol. 83(5): 1954-1958.

Chan, H.T. and E.B. Jang. 1995. Diet pH effects on mass rearing of the Mediterranean fruit fly (Diptera: Tephritidae). J. Econ. Entomol. 88(3): 569-573.

Goering, H.K. and P. J. Van Soest. 1970. Forage fibre analysis (apparatus, reagents, procedures, and some applications). Agric. Handbook No. 379. ARS, USDA, Washington, D.C.

Jacobs, M.B. 1958. The chemical analysis of food and food products. D. van Nostrand Co., Inc. N.Y. 970 pp.

Jang, E.B. and H.T. Chan, Jr. 1993. Alleviation of acetic acid production during mass rearing of the Mediterranean fruit fly (Diptera: Tephritidae). J. Econ. Entomol. 86(2): 301-309.

Kobayashi, R. 1993. Evaluation and Analysis of Medfly Production at the Fruit Fly Rearing Facility in Hawaii (April 1990-December 1992). In house report, USDA-APHIS, PPQ Methods Development, Waimanalo Medfly Rearing Facility, Waimanalo, HI. 45 pp.

Robertson, J.B. and P.J. Van Soest. 1981. The detergent system of analysis and its application to human foods. *In* The analysis of dietary fibre in Food. W.P.T. James and O. Theander (eds.). pp. 123. Marcel Dekker, New York.

Tanaka, N., L., Steiner, K. Ohinata and R. Okamoto. 1969. Low cost larval rearing medium for mass production of Oriental and Mediterranean fruit flies. J. Econ. Entomol. 62: 967-968.

Tanaka, N., L., Steiner, K. Ohinata and R. Okamoto. 1970. Methods of mass rearing the Mediterranean fruit fly currently used by the USDA, pp. 19-23. *In* Sterile male technique for control of fruit flies. IAEA, Vienna, Austria.

# A Practical, Efficient and Low Cost Diet for Rearing the Mediterranean Fruit Fly Larvae

*A.G. Manoukas and E.N. Zografou*

Institute of Biology, NCSR "Demokritos", Agia Paraskevi 15130, Athens, Greece

## INTRODUCTION

The Mediterranean fruit fly *Ceratitis capitata* (Wied.) (Diptera: Tephritidae), has been artificially reared and used for the application of the sterile insect technique and other purposes, throughout the world. The larval diet used is rather expensive and it is mixed in the rearing facility. The most expensive ingredient used in this diet is yeast which is variable in composition and has a relatively short shelf life due mainly to its high nutritional value. This is particularly true for all countries like Greece which do not manufacture brewer's yeast. Also, it is widely known that the Mediterranean fruit fly larvae grow in a wide variety of fruits and artificial diets. These fruits and artificial diets, although very different in chemical/nutritional as well as physical/ecological parameters, are successfully tolerated and utilised by the larvae. These observations prompted the initiation of research into diets containing a variety of low cost ingredients widely used in the vertebrate feed industry and easily found in any country. To our knowledge, no one has tested complete diets produced by well-established feed manufacturers for larval rearing of this insect.

## MATERIALS AND METHODS

Table 1 presents the composition of the control and the experimental diet containing formula KA50 from VIOZOKAT S.A., GR. Formula KA50 is in pellet form and is made up of cereals and their by-products, soybean meal, alfalfa meal, sunflower meal, certain salts, amino acids and vitamins. The composition and approximate analysis of the control (C) and Formula KA50 diet are given in Table 1 and 2, respectively.

Formula KA50 was soaked in a portion of water containing sugar, sodium benzoate and citric acid for the required time. This was then placed in the mixer with the remaining amount of water and mixed well. Approximately 60g of each test medium were placed in plastic containers (replicates) with covers (KRIS-PAN Co, Greece), and three replicates each (treatment) were used. Twenty-five eggs per gram were used in all diets. Eggs were obtained from flies of the white pupa strain (genetic sexing strain GSWP-1-62 IAEA, Vienna) and maintained under standard conditions. The eggs were placed directly on the surface of the diet, except for 100 eggs/replicate, which were placed on a filter paper on the diet to record the hatchability of eggs.

The results were compared with those of a routinely used larval diet (Nade 1970). Statistical procedures were those of Steele and Torrie (1960).

Table 1. Composition of the control (C) and experimental diets KA50.

| Ingredients | C | KA50 |
|---|---|---|
| Water | 54.5 | 64.5 |
| Sugar | 17.5 | 9.1 |
| Yeast, Schwechat | 8.8 | - |
| Sodium benzoate | 0.5 | 0.5 |
| Citric acid | 0.6 | 0.6 |
| Wheat bran | 17.9 | - |
| Formula KA50 | - | 25.4 |

Table 2. Approximate analysis of diets.

| Component, % | Control[1] | KA50[2] |
|---|---|---|
| Moisture | 57.4 | 57.3 |
| Proteins (NX6.25) | 8.0 | 6.7 |
| Lipids | 0.9 | 0.8 |
| Ash | 1.7 | 3.5 |
| Nitrogen Free Extracts[3] | 32.0 | 31.7 |

[1] on the basis of the analysis of its ingredients
[2] on the analysis given by the manufacturer
[3] taken as a difference from 100

## RESULTS AND DISCUSSION

Table 3 presents the results of diet KA50 compared to the control. Mean numbers in the same column followed by different alphabets differ significantly at the probability level of 0.05.

It is clear that diet KA50 gave results equivalent to the control for hatchability, pupal weight and adult emergence and an 80% higher number of pupae/g diet compared to the control.

Table 3. Performance of the Mediterranean fruit fly larvae.

| Diet | Hatchability % | Pupae/g diet | Weight mg/pupa | Adults % on pupae |
|---|---|---|---|---|
| Control | 79.7 | 10.3[a] | 8.5 | 94.3 |
| KA50 | 82.7 | 18.0[b] | 8.1 | 95.2 |

Table 4 presents a comparison of performance and cost of diet KA50 and the control. The costs in ECU for a kg/diet and for a million pupae are about 0.2 and 11 respectively for KA50 compared to 0.9 and 90 for the control.

Table 4. Comparison of control and KA50 Diets for the Mediterranean fruit fly larvae.

| **Parameter recorded** | **Diets** | |
|---|---|---|
| | **Control** | **KA50** |
| Eggs/g medium | 25 | 25 |
| Eggs hatchability, % | 80 | 83 |
| Pupae/g diet | 10 | 18 |
| Pupal weight, mg | 8.5 | 8.1 |
| Adult emergence, % | 94 | 95 |
| ECU/kg medium | 0.9 | 0.2 |
| ECU/million pupae | 90 | 11 |

## CONCLUSION

Examination of the data suggests that KA50 could be a very low cost alternative in rearing Medflies for release or other purposes provided that insect quality is checked. All diets and ingredients should be monitored for certain quality parameters (Manoukas 1991). The data of approximate and other analyses also suggest that KA50 diet could be improved when more is known about the nutritional and other requirements of this insect.

## ACKNOWLEDGEMENTS

We would like to acknowledge the assistance of the European Union and the Greek Secretariat in research and development and financing this work as well as VIOZOKAT S.A. for the supply and quality control of KA50.

## REFERENCES

Manoukas, A.G. 1991. Dietary control in insects. The case of the olive fruit fly, *Dacus oleae* (GMEL.). pp. 174-182. *In* Quality Control of Mass Reared Anthropods. F. Bigler (ed). Workshop Proc. Int. Org. Biol. Control.

Nadel, D.J. 1970. Current mass-rearing techniques for the Mediterranean fruit fly pp. 13-19. *In* Sterile Male Technique for Control of Fruit Flies. Proceedings of a Panel Organised by the Joint FOA/IAEA Division of Atomic Energy in Food and Agriculture.

Steele, R.J. and Torrie, J.H. 1960. Principles and procedures of statistics. MacGraw Hill, London. 481 pp.

# PART III.

## APPLICATION OF AREA-WIDE CONCEPT TO FRUIT FLIES

---

### F) ECOLOGY-DEMOGRAPHY OF FRUIT FLIES

# Spring and Early Summer Phenology and Detection of *Ceratitis capitata* (Diptera: Tephritidae) in Northern Greece

*N.T. Papadopoulos*[1], *B.I. Katsoyannos*[1] *and J.R. Carey*[2]

[1] Laboratory of Applied Zoology and Parasitology, Department of Agriculture, University of Thessaloniki, Greece
[2] Department of Entomology, University of California, Davis, USA

## INTRODUCTION

The Mediterranean fruit fly (Medfly), *Ceratitis capitata* (Wiedemann) (Diptera: Tephritidae), is one of the most serious fruit pests world-wide, infesting more than 300 plant species (Liquido et al. 1991).

Many studies on population dynamics of *C. capitata* have been conducted in the tropics (Vargas et al. 1983, Nishida et al. 1985, Eskafi and Kolbe 1990, Harris et al. 1993) and in the Mediterranean area (Rivnay 1951, Benfatto et al. 1989, Campos et al. 1989, Fimiani 1989, Cayol and Causse 1992, Michelakis 1992, Israely et al. 1997, Katsoyannos et al. 1998a). However, there are no detailed studies on the seasonal occurrence and population dynamics of the fly in the most temperate parts of its distribution.

The population build up of the fly is mostly determined by host fruit abundance and availability, and by environmental conditions such as temperature and humidity. In northern Greece, which is within the northern limits of the fly's distribution, winter temperatures are unfavourable for *C. capitata* survival (Papadopoulos et al. 1996). In addition, there is a gap in host fruit availability from near December until the following May. However, *C. capitata* has developed a remarkable ability to survive in such climates (though suffering high mortality), predominantly as larvae within certain host fruits that become infested at the end of autumn and remain in the orchards until the following spring (Papadopoulos et al. 1996). The prolonged larval period, especially that of the 1st and the 2nd instars, due to low temperatures, enables the fly to survive long periods of unfavourable conditions (Papadopoulos et al. 1998). The few adults emerging in spring, may live as long as 3 months and can oviposit a high number of eggs in artificial oviposition substrates (Papadopoulos et al. 1996). The importance of some key factors – late spring and early summer maturing host fruits – for the re-establishment of the *C. capitata* population has been suggested (Cayol 1996, Israely et al. 1997, Katsoyannos et al. 1998a). However, there are no systematic and detailed studies on the occurrence and detection of *C. capitata* and on the importance of host fruits for breeding the first spring generation of Medfly, at least at the periphery of its distribution, like northern Greece.

In the area of Thessaloniki (northern Greece), using a regular non-dense trapping system of trimedlure baited Jackson traps (about 1 trap per ha), the first adults were detected during late August (unpublished data) from 1991 - 1995. Adults were being captured up to late November, but no captures were recorded from December through the following August. The adults emerging from the overwintering generation appeared in the area of Thessaloniki in April and May (Papadopoulos et al. 1996). The failure to detect adults from April to August was attributed to the low initial adult population of spring because of the high mortality (due to low temperatures) observed in the overwintering generation, and to the density of the trapping grid we used. Therefore, the need for a more effective trapping system for early detection of adults and even other methods for larval detection, such as fruit sampling, is obvious.

In the present study, we investigated the occurrence of the spring and early summer generation of Medfly in the area of Thessaloniki, and the possibility of their early detection using a dense grid of traps and systematic fruit sampling.

## MATERIALS AND METHODS

The study was carried out during 1993, 1994 and 1995 in Thessaloniki, northern Greece (40.3° northern latitude, 25.5° longitude), in an orchard of various host trees of *C. capitata.* In the experimental farm there were 3, 1.5, 1, 0.5 ha of apples (*Malus sylvestris* Mill.), pears (*Pyrus communis* L.), peaches (*Prunus persica* [L.] Batch), and apricots (*P. armeniaca* L.) respectively. There were also about 15 cherry (*P. avium* L.), 5 loquat (*Eriobotrya japonica* Thunb.), 10 fig (*Ficus carica* L.) and 5 oriental persimmon (*Diospyros kaki* L) trees. The climate of Thessaloniki is characterised by warm and dry summers and mild and wet winters, with temperatures often falling below freezing point. More detailed data concerning the experimental orchard and the climatic profile of the experimental area are given by Papadopoulos et al. (1996, 1998). No insecticide application against Medflies was applied in the experimental orchard up to August (when experiments were terminated).

Field infested Golden Delicious apples and apples of an unidentified local variety were collected during the autumn of 1993 and 1994, and placed outdoors in plastic containers on a layer of dry sand (3 cm thick). A total of 750 (Golden Delicious) and 600 (Golden Delicious, local variety) apples were collected in 1993 and 1994 respectively. Once a week, the sand was sieved and the recovered mature larvae and pupae were collected. Mature larvae and pupae were placed in plastic jars, covered with sand 3 cm thick and kept outdoors near the containers with the fruits.

The emergence of adults in spring, from those larvae and pupae held outdoors, was monitored. A sample of the adults (15 females and 15 males) that emerged in the period 15-30 April 1994, was held in pairs at an ambient temperature, in a cylindrical 1 litre capacity, transparent plastic cages provided with water and food (mixture of yeast hydrolysate, sugar and water). Survival and reproduction of these adults were monitored every 2 days. As oviposition substrate a petri dish lid (9 cm diameter) that bore seven 18 mm diameter ceresin wax domes was used in each cage (Prokopy and Boller 1970, Katsoyannos et al. 1986). Eggs were counted every 2 days and were then placed on moist blotting paper in capped petri dishes kept next to adults. Hatchability was determined 4-5 days later.

To detect the adult population in the experimental area, we installed, during the last week of March in 1994 and 1995, 30 Jackson traps baited with trimedlure (about 5 traps per ha). Traps were placed 1.5-2 m above the ground. Four, 5, 13, 3, 2, 2, and 1 traps were placed on apricot, peach, apple, fig, pear, plum (*Prunus domestica* L.) and oriental persimmon trees respectively in 1994, while in 1995, 4, 7, 11, 3, 2, 2, and 1 traps were placed on the same trees respectively. The number of males caught was recorded weekly.

Systematic fruit sampling, including all fruit species that had ripe or ripening fruits, was conducted in 1994, while in 1995 only apricots were sampled. Collected fruits were transferred to the laboratory (25°C), counted and placed in plastic containers

(30 by 40 cm base, 15 cm high) on a layer of dry sand (3 cm thick). Twice per week, the mature larvae and pupae that emerged from these fruits were collected.

## RESULTS

A total of 1,598 and 205 mature larvae emerged from 750 and 600 apples during the winter and spring of 1993-94, and 1994-95 respectively. In both periods, more than 80% of the larvae emerged from fruits up to the end of February (Figure 1), while the remaining ones emerged from March to the first days of May. In 1993-94, only 1.9% of the mature larvae obtained in winter yielded adults in spring, whereas almost 15% of those obtained in spring yielded adults. In 1994-95, no adults were obtained from pupae formed in winter, whereas 60% of the pupae obtained in spring yielded adults.

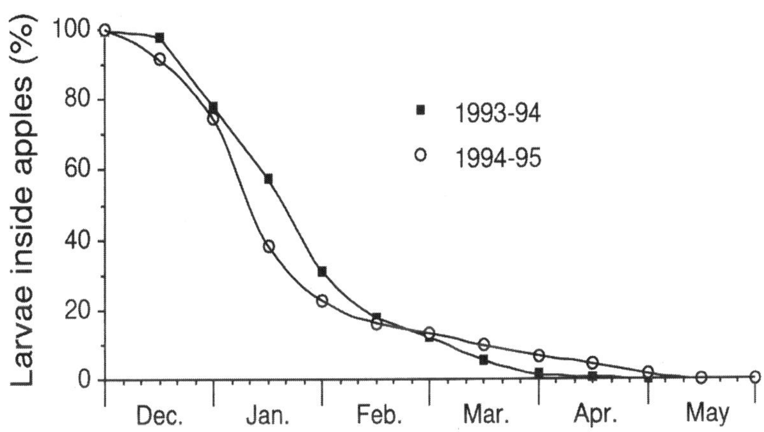


Figure 1. Emergence of mature larvae from field infested apples. (during the winter-spring period of 1993-94 and 1994-95. In both periods, apples were collected in November-December and were held outdoors in a rain-protected place)

In total, 59 and 12 adults were obtained in the spring 1994 and 1995 respectively. Adult emergence started at the end of March 1994 or in the first week of April 1995 (Figure 2). Almost 50% of these emerged in April and 30-50% in May in both years. The survival rate, under field conditions of a sample of 30 adults (15 males and 15 females) that emerge between 15-30 of April 1994, is given in Figure 3. These adults survived a maximum of 124 d (from May until September) while 50% of them lived almost 90 days. Five to 15 eggs per female per 2 days were oviposited by these flies in June and July and only a few (< 3 eggs per female per 2 days) in August. Obviously, the adults of the overwintering generation may survive throughout the summer and may infest ripe fruits in the beginning and middle of summer.

Loquats and cherries are the first fruits to ripen in the area of Thessaloniki in May. The ripening period of apricots begins in June and lasts until the first week of July. Ripe peaches appeared at the end of July. Of a total of 497 loquats and 550 cherries collected in 1994, no larvae were retrieved. The first pupae (5.3 pupae per 100 fruits) were retrieved from apricots collected from 16-23 June 1994 (Table 1). However, apricots collected earlier were not infested. In 1995, we conducted sampling only from apricot trees. The first infested sample (8.2 pupae per 100 fruits) was collected from 1-7 July. Consequently, apricots were first oviposited by adults of the overwintering generation and therefore were the most important host for the subsequent build up of the Medfly population in the summer.

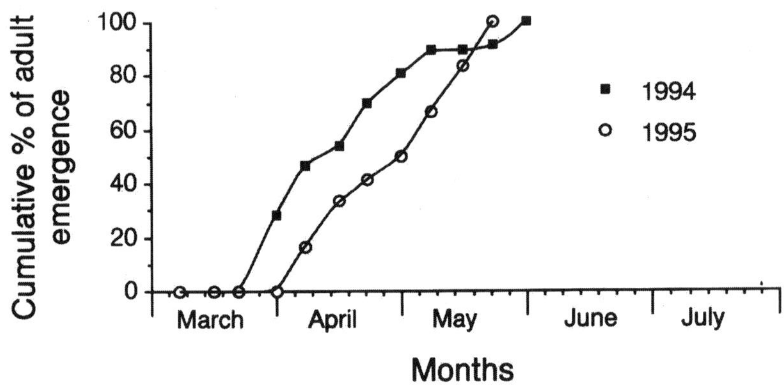


Figure 2. Adult emergence in spring 1994 and 1995, from pupae obtained during the winter-spring period from apples collected during November- December and maintained outdoors.

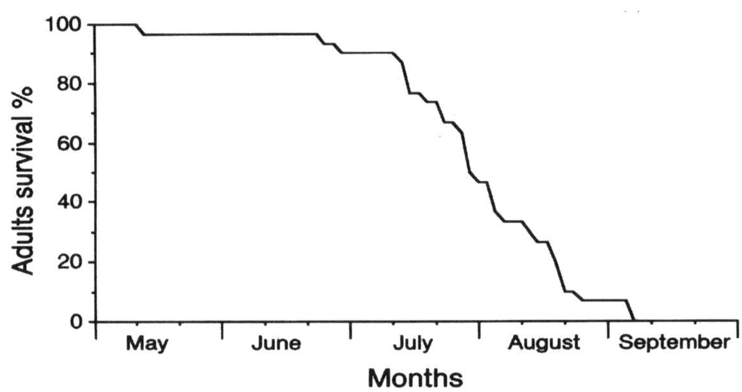


Figure 3. Outdoor survival of 30 adults (15 males + 15 females, emerged from 15 to 30 April 1994) of the overwintering generation.

The first adults were detected in the high density trapping grid (5 traps per ha) on 23 June and on 7 July by Jackson traps installed on apricot trees in 1994 and 1995 respectively (Figure 4). The first adults were detected on peaches 20 and 7 days later in 1994 and 1995 respectively. Traps placed on other tree species such as figs, pears, apples, etc. captured the first adults almost a month after the first adult detection. Later in the season, large numbers of adults were captured by traps placed on apple and other fruit trees (unpublished data).

Table 1. Infestation of apricots by *C. capitata* collected in the area of Thessaloniki, northern Greece, during June-July 1994 and 1995.

| Date of Collection | 1994 | | 1995 | |
|---|---|---|---|---|
| | Total no. fruits collected | Pupae per 100 fruits | Total no. fruits collected | Pupae per 100 fruits |
| 1 - 7 June | 104 | 0.0 | - | - |
| 8 - 15 June | 105 | 0.0 | 58 | 0.0 |
| 16 - 23 June | 95 | 5.3 | 46 | 0.0 |
| 24 - 30 June | 100 | 11.0 | 102 | 0.0 |
| 1 - 7 July | 100 | 33.0 | 85 | 8.2 |

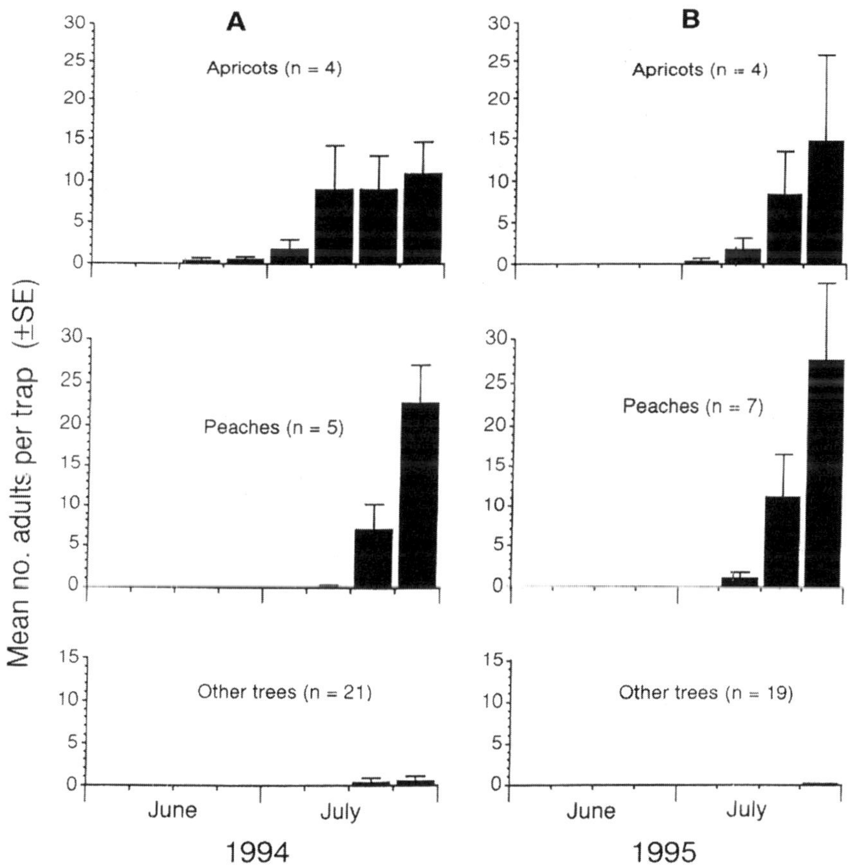


Figure 4. Captures of adults in trimedlure baited Jackson traps, placed on different host trees in June and July 1994 (A) and 1995 (B).

## DISCUSSION

In the area of Thessaloniki, the Medfly population during the first month of spring consisted of larvae inside apples (infested during the previous autumn), and pupae formed from mature larvae that emerge from fruits. These larvae and pupae yielded a small number of adults during April and May. The adults of the overwintering generation lived until September and oviposited a large number of eggs. However, the first adult detection using a regular trapping grid of trimedlure baited Jackson traps (about 1 trap per ha) occurred at the end of August (unpublished data). In this study, when more traps per ha (about 5) were used, we detected the first adults 1.5-2 months earlier than when a regular trapping grid was used. The first adult captures were recorded in both years in traps installed on apricot trees. Apricots were the first fruits infested, and the infestation (sampling of infested fruits) was detected almost at the same time as the first adult captures (end of June and beginning of July in 1994 and 1995 respectively). Since loquats and cherries, that ripen earlier (in May) than apricots, were not infested by the Medfly, we assume that apricots are the most important fruit species for breeding of the first overwintering generation. Adults that emerged from immatures bred on apricots may appear in the area of Thessaloniki in the last days of June and first days of July. Most probably, these adults, of the first summer generation, were those we detected first.

Using a dense grid of Jackson traps (about 5 per ha), we detected the first adults of the first summer generation, while the adults of the overwintering generation remained undetectable. Lack of detection of the adults of the overwintering generation is mostly attributed to the low effectiveness of Jackson traps as an adult detection tool under low population densities (Kaneshiro 1993, Carey 1991, Katsoyannos et al. 1998). Therefore, adopting a more powerful trapping system, such as that developed recently by Heath et al. (1998) and Katsoyannos et al. (1999), might be more effective in capturing the adults of the overwintering generation. Use of food-based attractants, such as ammonium acetate, putrescine and trimethylamine placed in McPhail type traps (Katsoyannos et al. 1998b) may increase the probabilities of capturing the first emerging adults. Our results show that, at least for Thessaloniki and other temperate areas, for adult detection programmes, traps should be placed on preferred early-ripening host trees such as apricots (Katsoyannos et al. 1998a). However, trap placement on trees whose fruits are overwintered may also be important for early detection (Israely et al. 1997).

Collection of large number of fruits in autumn and monitoring pupation and adult emergence in spring, can give an indication of the actual time of adult flight activity. If available in large numbers, collection in spring of fallen fruits infested the previous season might also give the same information, although due to high winter mortality, large numbers of such fruits (which may not be available) should be collected in order to obtain a few pupae. The initial fruit infestation, caused by the adults of the overwintering generation, could be also detected by visual inspection of the first infested fruits and also by collecting and incubating these fruits. However, visual inspection is extremely difficult, especially for apricots, since the oviposition sting is not visible, and because of the suspected very low initial infestation. Hence, this method is impractical. On the other hand, by adopting fruit sampling, the infestation from emerging pupae is verified with a delay (i.e., after fruit incubation) of several days.

Considering the above detection difficulties using fruit examination, we suggest that trapping is the easiest and most accurate method for Medfly detection under low population densities, provided that high numbers of highly efficient traps are placed on the most preferred host trees. However, fruit inspection and sampling might supplement trapping data. In such a case, preferred hosts, especially those exhibiting distinct symptoms of Medfly infestation, should be sampled.

## ACKNOWLEDGEMENTS

We thank N.A. Kouloussis (Aristotelian University of Thessaloniki) for reading the manuscript and making valuable suggestions. Research was supported by the Californian Citrus Research Board.

## REFERENCES

Benfatto, D., S. Longo and G. Siscaro. 1989. Four years' comparative analysis on population dynamics trend of *Ceratitis capitata* (Wied.) (Diptera, Tephritidae) on citrus, pp. 35-40. *In* Fruit flies of economic importance. R. Cavalloro (ed.). Proceedings CEC/IOBC International Symposium (Rome 7-10 April 1987). Balkema, Rotterdam.

Campos, M., P. Ramos and O.T. Jones. 1989. Monitoring population of *Ceratitis capitata* in the Granada province of Spain using three different trapping systems, pp. 401-404. *In* Fruit flies of economic importance. R. Cavalloro (ed.) Proceedings CEC/IOBC International Symposium (Rome 7-10 April 1987). Balkema, Rotterdam.

Carey, J.R. 1991. Establishment of the Mediterranean fruit fly in California. Science 253: 1369-1373.

Cayol, J.P. 1996. Box thorn, key early season host of the Mediterranean fruit fly. Int. J. Pest Managem. 42: 325-329.

Cayol, J.P. and R. Causse. 1993. Mediterranean fruit fly *Ceratitis capitata* Wiedemann (Dipt., Trypetidae) back in southern France. J. Appl. Entomol. 116: 94-100.

Eskafi, F.M. and M.E. Kolbe. 1990. Infestation patterns of commonly cultivated, edible fruit species by *Ceratitis capitata* and *Anastrepha* spp. (Diptera: Tephritidae) in Guatemala and their relationships to environmental factors. Environ. Entomol. 19: 1371-1380.

Fimiani, P. 1989. Mediterranean region, pp. 39-50. *In* Fruit flies: their biology, natural enemies and control. A.S. Robinson and G. Hooper (eds.). vol. 3A. Elsevier, Amsterdam.

Harris, E.J., R.I. Vargas and J.E. Gilmore. 1993. Seasonality in occurrence and distribution of Mediterranean fruit fly (Diptera: Tephritidae) in upland and lowland areas on Kauai, Hawaii. Environ. Entomol. 22: 404-410.

Heath. R.R. N.D. Epsky, B.D. Dueben, J. Rizzo and F. Jeronimo. 1998. Adding methyl-substituted ammonia derivates to a food-based synthetic attractant on capture of the Mediterranean and Mexican fruit flies (Diptera: Tephritidae). J. Econ. Entomol. 90: 1584-1589.

Israely, N., B. Yuval, U. Kitron and D. Nestel. 1997. Population fluctuations of adult Mediterranean fruit flies (Diptera: Tephritidae) in a Mediterranean heterogeneous agricultural region. Environ. Entomol. 26: 1263-1269.

Kaneshiro, K. 1993. Introduction, colonization and establishment of exotic insect populations: fruit flies in Hawaii and California. Am. Entomol. Spring 39: 23-29.

Katsoyannos, B.I., K. Panagiotidou and I. Kechagia. 1986. Effect of colour properties on the selection of oviposition site by *Ceratitis capitata*. Entomol. Exp. Appl. 42: 187-193.

Katsoyannos, B.I., R.R. Heath, N.T. Papadopoulos, N.D. Epsky and J. Hendrichs. 1999. Field evaluation of the Mediterranean fruit fly (Diptera: Tephritidae) female selective attractants for use in monitoring, mass trapping and sterile insect technique programmes. J. Econ. Entomol. 92: 583-589.

Katsoyannos, B.I., N.A. Kouloussis and J.R. Carey. 1998a. Seasonal and annual occurrence of Mediterranean fruit flies (Diptera: Tephritidae) on Chios island: differences between two neighbouring citrus orchards. Ann. Entomol. Soc. Am. 91: 43-51.

Katsoyannos, B.I., N.T. Papadopoulos, R.R. Heath, J. Hendrichs and N.A. Kouloussis. 1998b. Evaluation of synthetic food-based attractants for female Mediterranean fruit flies (Diptera: Tephritidae) in McPhail type traps. Fifth International Symposium on Fruit Flies of Economic Importance, Penang, Malaysia, 1-5 June 1998 (abstract).

Liquido, N.J., L.A. Shinoda and R.T. Cunningham. 1991. Host plants of the Mediterranean fruit fly (Diptera: Tephritidae): an annotated world review. Miscellaneous Publication 77. Entomol. Soc. Am. Lanham, MD.

Michelakis, S.E. 1992. Phenology of the Mediterranean fruit fly *Ceratitis capitata* Wiedemann in Crete. Israel J. Entomol. 25-26: 177-180.

Nishida, T., E. Harris, R.I. Vargas and T.T.Y. Wong. 1985. Distributional loci and host fruit utilisation patterns of the Mediterranean fruit fly, *Ceratitis capitata* (Diptera: Tephritidae), in Hawaii. Environ. Entomol. 14: 602-608.

Papadopoulos, N.T., J.R. Carey, B.I. Katsoyannos and N.A. Kouloussis. 1996. Overwintering of *Ceratitis capitata* (Diptera: Tephritidae) in northern Greece. Ann. Entomol. Soc. Am. 89: 526-534.

Papadopoulos, N.T., B.I. Katsoyannos and J.R. Carey. 1998. Temporal changes in the composition of the overwintering larval population of the Mediterranean fruit fly (Diptera: Tephritidae) in northern Greece. Ann. Entomol. Soc. Am. 91: 430-434.

Prokopy, R.J. and E.F. Boller. 1970. Artificial egging system for the European cherry fruit fly. J. Econ. Entomol. 63: 1413-1417.

Rivnay, E. 1951. The Mediterranean fruit fly in Israel. Bull. Entomol. Res. 41: 321-341.

Vargas, R.I., E.J. Harris and T. Nishida. 1983. Distribution and seasonal occurrence of *Ceratitis capitata* (Wiedemann) (Diptera: Tephritidae) on the island Kauai in the Hawaiian Island. Environ. Entomol. 12: 303-310.

# Speciation of *Bactrocera dorsalis* Complex Based on Aedeagal Length

*Osamu Iwahashi*

College of Agriculture, University of the Ryukyus, Nishihara, Okinawa 903-0213, Japan

## INTRODUCTION

A species complex of *Bactrocera dorsalis* (Hendel) in Southeast Asia is composed of 52 species (Drew and Hancock, 1994) and while some of these species are economically very important, distinguishing them based on morphological characters has been difficult (White and Elson-Harris 1992). Specifically, there is considerable difficulty in differentiating between males of two pairs of sympatric species, *B. philippinensis* Drew and Hancock /*B. occipitalis* (Bezzi) in the Philippines and *B. carambolae* Drew and Hancock /*B. papayae* Drew and Hancock in Indonesia. This may be, in part, because the evolutionary processes within this species complex are still very dynamic, and that natural hybridisation between sympatric species pairs might be occurring on a regular basis (He and Haymer 1997). Iwaizumi et al. (1997) developed a simple method to differentiate the two sets of sympatric species based on aedeagal lengths. However, these flies had been reared artificially under laboratory conditions and only a small number of specimens (n=5) was used. Consequently, they were not able to obtain a frequency distribution of the aedeagal length for each species. Iwahashi (1999a) measured a larger number of wild flies collected on Guimaras Is, Philippines, and found that flies with the aedeagal length of < 2.81 mm are *B. occipitalis*, while those of > 2.89 mm are *B. philippinensis*. Iwahashi (1999a,b) also showed that the measurement of the aedeagal length of fruit flies is a reliable characteristic for distinguishing between the 2 sympatric species pairs in the *B. dorsalis* complex. This being so, it may also be interesting to interpret phylogenetic relationships among *B. dorsalis* complex species based on the aedeagal length. Thus, aedeagal lengths of different populations of five *B. dorsalis* complex species are measured and their relationships discussed.

## MATERIALS AND METHODS

Wild flies were collected with methyl eugenol traps in Thailand, Vietnam, Indonesia, Philippines, Taiwan and Hawaii in 1997. These flies were kept in a paper bag or 75-99 % alcohol until measured. The specimens were soaked in boiling water and then dissected under a stereoscopic microscope. The aedeagus which is characteristically coiled was straightened on a microscope slide with a pair of forceps. The aedeagus and the wings of each male were covered with a cover slip so that they could be preserved as voucher specimens. The corresponding bodies of the flies were kept individually and labelled in a plastic tube with 95% or 99 % of alcohol. Measurements of the aedeagus were made using a digital micrometer (Digital Linear Gauge D-10S, Ozaki MFG, Tokyo).

## RESULTS

### Variation of Aedeagal length in Different Populations within a Species

Before examining the phylogenetic relationships among *B. dorsalis* complex species, the aedeagal lengths of different populations within a species were compared. There were no differences in the aedeagal length among 5 populations in Thailand and Vietnam except for the population from Chiang Mai, Thailand, where the flies with the longer type of aedeagus ( > 2.92 mm) lurked. This may be due to the small sampling size (n =12). On the other hand, the mean aedeagal lengths (±SD) of *B. dorsalis* in Hawaii (2.60 ± 0.08 mm) and Taiwan (2.61 ± 0.09 mm) were significantly shorter than those found in Thailand (2.71 ± 0.16 mm) ($P < 0.0001$, Bonferroni test).

Measurements of the aedeagal length of flies collected from Pakanbaru on the island of Sumatra, Indonesia, showed that males could be separated into 2 groups, i.e., those having the aedeagal length of < 2.60 mm and those of > 2.70 mm. The mean (± S.D.) of the shorter group was 2.45 ± 0.76 mm and the other 3.04 ± 0.13 mm, indicating that the former would belong to *B. carambolae* and the latter *B. papayae*. However, the criteria for distinguishing these two species based on the aedeagal length differed in other populations on the same island. Furthermore, the two groups overlapped each other and therefore, separating the two species was almost impossible in other populations. Fruit flies collected in Manila could be easily separated into two groups; again however, the distinguishing values (2.90 mm and 1.97 mm) were different from those of the island of Guimaras (2.81 mm and 2.90 mm). These results would appear to support the notion of He and Haymer (1997) that the evolution of the *B. dorsalis* complex is still continuing.

### Relationships among Five *B. dorsalis* Complex Species in Terms of the Aedeagal Length

When the aedeagal lengths of *B. dorsalis* in Chiang Mai, Thailand, and the two sympatric species of *B. carambolae* and *B. papayae* in Pakanbaru on the island of Sumatra, Indonesia, were compared, a clear relationship was observed. The mean (± S.D.) of the aedeagus in *B. dorsalis* was 2.71 ± 0.96 mm which is almost the same as the average (2.75 mm) of the means of *B. carambolae* (2.45 mm) and *B. papayae* (3.04 mm), suggesting that *B. carambolae* and *B. papayae* might be derived from *B. dorsalis*.

The second notable relationship is between *B. carambolae/B. papayae* in Indonesia and *B. occipitalis/B. philippinensis* in the Philippines. Flies in Menado, the northern end of Sulawesi Island, Indonesia, seemed to be separated into 2 groups: those with the aedeagal length of < 05) mm and that of the longer one was 2.87 (± 0.14). Using the flies collected on Guimaras Island, Philippines, it was judged that males with aedeagal length of < 2.81 mm are B. occipitalis, and those of > 2.89 mm are *B. philippinensis*. And the mean (± SD) were 2.69 (± 0.07) mm in *B. occipitalis* and 3.09 (± 0.09) mm in *B. philippinensis*. Again, the mean length of *B. papayae* (2.87 mm) was almost equal to 2.89 mm, the average of 2.69 mm and 3.09 mm, suggesting that *B. occipitalis* and *B. philippinensis* might be derived from *B. papayae*, if I consider the length of the aedeagus only.

The third notable relationship is between *B. dorsalis* in Taiwan and *B. occipitalis/B. philippinensis* in the Philippines. When flies from Manila and those from Taiwan were compared, the mean (± S.D.) of *B. dorsalis* was 2.61 ± 0.09 mm which was similar to that of *B. occipitalis* (2.66 ± 0.08 mm). However, it is unlikely that *B.*

*dorsalis* in Taiwan was derived from *B. occipitalis*, because morphological characters based on the descriptive score (White and Hancock 1997) of *B. dorsalis* were different from *B. occipitalis* but similar to *B. philippinensis*. Consequently, *B. dorsalis* in Taiwan could not have been derived from one of the two sympatric species in the Philippines but could have been derived from *B. dorsalis* of the Asian continent.

## DISCUSSION

During copulation, a male of *Bactrocera* species inserts his aedeagus through the tip of the aculeus into the vaginal duct. As the length of the male aedeagus is strongly correlated to the length of the female's terminalia in nine species of *Bactrocera* ($r$ = 0.97, $P < 0.05$, Iwaizumi et al. 1997), the male can achieve successful intromission and fertilisation of the female's gametes. There may be a genetic correlation between the ovipositor and penis lengths within a species. It is likely that a selective force works first on the ovipositor length of a species and, in turn, the length of the penis is selected because of the genetic correlation between the two characters. It may therefore be possible to discuss the evolution of the aedeagal length of a species in terms of the ovipositor length.

Several selective forces which can influence the ovipositor length may be possible. In 1997, a total of about 2,000 fruit fly pupae was reared from infested fruits from Subang on the island of Java and it was found that about 70 % of these flies were parasitised by *Fopius sp*. It was noted that females of this parasitoid had extraordinarily long ovipositors compared with their body size. It was nearly the same length as the ovipositor of the female fruit flies that emerged at the same time, suggesting that the parasitoid pressure would have played an important role in the evolution of the ovipositor length of the fruit flies.

A female which has a longer ovipositor can lay her eggs deeper within a fruit, which may decrease the possibility of her offspring being parasitised. However, such females, in turn, may take a longer time for oviposition, which may increase the possibility of them being attacked by a predator during oviposition. If these two selection pressures are strong enough, flies may have to be specialists to cope with one of them. Thus the possibility for a hybrid female, which has an intermediate length of ovipositor, to survive to reproduce the next generation may be low. Thus, assortative mating between females with longer ovipositors mating with males of longer penis lengths and females with shorter ovipositors with males of shorter penises may have evolved because of the lower fitness of intermediate forms.

Alternatively, the thickness of the skin of the host fruit through which a female inserts her ovipositor may have played a role in influencing the ovipositor length. If this hypothesis were valid, host specificity should have evolved as a result of the differential thicknesses of fruit skins. This might indeed be possible in some fruit fly species. However, mango and carambola fruits are infested by both species on Guimaras Island, suggesting that the thickness of the fruit skin may not be an important selective force in the differentiation of the ovipositor length in these two sympatric species.

## ACKNOWLEDGEMENTS

I thank K.Y. Kaneshiro for his critical reading of the manuscript. I also thank J. Koyama, S. Moriya and S. S. Tati-Subahar for helping collect specimens. This study was in part supported by the JSPS Cooperation Program under the Core University System and IAEA.

## REFERENCES

Drew, R.A.I. and D.L. Hancock. 1994. The *Bactrocera dorsalis* complex of fruit flies (Diptera: Tephritidae: Dacinae) in Asia. Bull. Entomol. Res. Suppl. Ser. 2 1-68.

He, M. and D.S. Haymer. 1997. Polymorphic intron sequences detected within and between populations of the Oriental fruit fly (Diptera: Tephritidae). Ann. Entmol. Soc. Am. 90: 825-831.

Iwahashi, O. 1999a. Distinguishing between the two sympatric species, *Bactrocera occipitalis* and *B. philippinensis* (Diptera: Tephritidae) based on aedeagus length. Ann. Entomol. Soc. Am. 92: 182-187.

Iwahashi, O. 1999b. Distinguishing between the two sympatric species, *Bactrocera carambolae* and *B. papayae* (Diptera: Tephritidae), based on aedeagal length. Ann. Entomol. Soc. Am. 92: 639-643.

Iwaizumi, R., M. Kaneda, and O. Iwahashi. 1997. Correlation of length of terminalia of males and females among nine species of *Bactrocera* (Diptera: Tephritidae) and differences among sympatric species of *B. dorsalis* complex. Ann. Entomol. Soc. Am. 90: 664-666.

Iwaizumi, R. , M. Kaneda, and O. Iwahashi. 1997. Correlation of length of terminalia of males and females among nine species of *Bactrocera* (Diptera: Tephilitidae) and differences among sympatric species of *B. dorsalis* complex. Ann. Entmol. Soc. Am. 90: 664-666.

White, I.M. and M.M. Elson-Harris. 1992. Fruit flies of economic significance. CAB, Wallingford, UK. 601 pp.

White, I.M. and D.L. Hancock. 1997. Indo-Australasian Dacini fruit flies, Computer Aided Biological Identification KEY, International. Institute of Entomolgy, London.

# Current Status and Perspectives for Management of *Anastrepha fraterculus* (Wied.) in Apple Orchards in Brazil

*Adalecio Kovaleski*[1], *Regina L. Sugayama*[2] *and Aldo Malavasi*[2]

[1] EMBRAPA/Centro Nacional de Pesquisa de Uva e Vinho. CP 177, CEP 95200-000, Vacaria, RS, Brazil
[2] Depto. Biologia, IBUSP. CP 11.461, CEP 05422-970, Sao Paulo, SP, Brazil

## CURRENT SCENARIO

Pomiculture is a recent activity in southern Brazil. The first apple orchards were installed in the early 1970s. Recently, the area grown with apples exceeded 30,000 ha, concentrated in the regions of Fraiburgo and Sao Joaquim (state of Santa Catarina) and Vacaria and Bom Jesus (state of Rio Grande do Sul). Part of the 600,000 tons that are harvested every year is exported to the USA and European countries.

Some exotic apple pests were unintentionally introduced, like the European red mite (*Panonychuls ulmi* Koch) and the Oriental fruit moth (*Grapholita molesta* Busck). Furthermore, some native species of insects became important pests, as in the South American apple leafroller (*Bonagota cranaodes* Meyrick) and the South American fruit fly (*Anastrepha fraterculus* (Wiedemann)).

The South American fruit fly is the best-studied pest of apples in Brazil regarding its biology and ecology. In this paper, we synthesise the information available and discuss the feasibility of adopting new control methods. Most experiments were conducted in Vacaria where *A. fraterculus* populations reach levels as high as 150 flies/trap day in some years.

Sixteen species of *Anastrepha* occur in the region of Vacaria and only *A. fraterculus* is considered economically important (Kovaleski et al. 1999). In addition to the typical morphology of *A. fraterculus*, the morphotype CSS (Selivon et al. 1996) was detected in McPhail traps and infested native fruits. The second most frequent species of *Anastrepha* is *A. dissimilis* Stone. It may be responsible for more than 20% of fruit flies in commercial apple orchards in some periods of the year (November-January) but does not attack apples (Kovaleski 1997).

Adult population fluctuation has been studied for the last four years using plastic McPhail traps containing grape juice at 25% (v/v) as attractant. It is more efficient than corn protein hydrolysate, vinegar, and sugarcane molasses (Kovaleski et al. 1995) and is widely used by apple growers as the standard attractant for fruit fly monitoring.

The highest population density is observed in mid-November and a second peak is observed in January-February in apple orchards and in forest areas surrounding these orchards (Kovaleski 1997). During this period, toxic bait sprays are applied on a weekly basis to prevent attack by ovipositing females.

Females show an intense oviposition activity since fruits are 15 mm in diameter, with preference for bigger fruits when they are available (Sugayama et al. 1997). However, at this stage, fruits are unsuitable as hosts and more than 99% mortality occurs in the egg or first instar (Sugayama 1995). During harvest, larval development occurs satisfactorily even though it is slower than in primary hosts. A cohort of adults reared in apples shows positive values of intrinsic rate of increase (r), demonstrating that the population could be maintained using apples as hosts (Sugayama et al. 1998c). Differences among cultivars were observed and effects are reflected as variations in larval development time and survivability (Branco 1998, Sugayama et al. 1998c). Fortunately, when apples are in the harvest stage, fruit fly population density is markedly lower than early in the season (Kovaleski 1997).

Four to six insecticide cover sprays further decrease the likelihood of development of *A. fraterculus* in apples. Even though these insecticides do not prevent females from ovipositing, they do prevent the development of immature stages inside host fruits and consequently the formation of galleries (Kovaleski 1995 and unpublished data). Since apples are highly unsuitable substrates during periods of highest fruit fly incidence and insecticides are able to inhibit immature development, it is concluded that there are no established populations in commercial orchards. Kovaleski et al. (1999) conducted two release-recapture experiments and demonstrated that *A. fraterculus* adults are able to move 600-1,000 m from areas of native forest into apple orchards.

In these forest areas, three Myrtaceae hosts may be found and the most infested ones are *Campomanesia xanthocarpa* Berg (locally known as "guabiroba") and *Feijoa sellowiana* Berg (false guava or pineapple guava). *Eugenia involucrata* DC (Rio Grande cherry) may also be highly infested but the number of fruit produced per tree and abundance of trees are lower than those of the two former hosts.

Guabirobas are susceptible to oviposition from mid-November to mid-December and some areas, locally known as "taperas" (abandoned domestic properties where several host trees may be found), produce a high number of heavily infested fruits. The climate during its fruiting season is mild and favours a faster development of immatures, as well as adult survival. Thus, a population increase following a fruiting period would be expected but the peak observed in February is markedly and consistently lower than that observed in spring (overwintering generation). According to Sugayama et al. (1998a), larval and pupal predation is intense in guabirobas (> 99%) which is a likely explanation for the reduction observed in the adult population density following a period of favourable abiotic conditions and high availability of host fruits.

According to Kovaleski et al. (1998), the abundance of false guavas in autumn (February-May) and winter severity determine the timing of the highest adult density during spring as well as the density itself. *A. fraterculus* may delay larval and pupal development to almost 6 months, a period completely lacking in host fruits with harsh climatic conditions for adult survival. The colder the winter, the later will adult emergence occur in spring and the higher the probability of adult survival until the next period of host availability in early December.

Parasitoids are not important mortality factors in feijoa and the mean parasitism is 1.6% (Kovaleski 1997). In guabirobas, even though overall parasitism may reach 29% in some years (Kovaleski 1997), more detailed studies showed that parasitism in the canopy is negligible early in the season, gradually increasing to 40% at the end of the fruiting season (Sugayama, unpublished data). Parasitism is greatly increased by eucoilids in guabiroba fruits that remain on the soil for seven days (Kovaleski et al.1995, Sugayama, unpublished data).

## PERSPECTIVES FOR MANAGEMENT

It is considered that the current practices guarantee a satisfactory control of *A. fraterculus* in apple orchards. Despite the inherent costs associated with chemical control (economical, ecological, and human safety), it is the only alternative we have to contain the damage by this aggressive pest. In the absence of insecticide sprays and depending on population density, losses may reach 100%.

We expect that the knowledge of pest biology and ecology may lead to a more judicious use of chemical insecticides, and propose that control should be undertaken in three different areas.

## In the Forest Areas, Before Adults Disperse into Commercial Orchards

### *Biological control*

Biological control by means of parasitoids should be encouraged during the fruiting season of guabirobas, when native species are absent and parasitism is close to 0%. *Diachasmimorpha longicaudata* (Ashmead) was able to increase parasitism on *A. fraterculus* dramatically within these fruits under laboratory conditions (Sugayama et al. 1998b) and could be considered as a good candidate species for inundative releases. Nevertheless, it is known that this species is poorly adapted to subtropical/temperate climates and that it shows a low efficiency in locating host patches (Sivinski, personnal communication) so that its establishment under conditions in Vacaria is unlikely. Another candidate species is *Aganaspis pelleranoi* Brèthes (Eucoilidae) that is already established in the region and shows intense activity at ground level. Sivinski (personal communication) also suggests the introduction of *Doryctobracon crawfordi* Viereck (Braconidae), since this species is better adapted to areas of high altitude and subtropical climate.

### *Bait sprays or bait station set up in host trees*

Conventional insecticides should not be used in areas of native forest, as they would have a severe impact on non-target organisms due to contact toxicity. The use of phototoxic dyes as insecticides combined with food attractants (Mangan and Moreno 1995) is considered a promising alternative to shut down the adult population emerging from *F. sellowiana* in spring. At this time, adults require proteinaceous food for sexual maturation and would be attracted by baits.

## In the Orchard Periphery, to Prevent Flies Dispersing into Inner Rows and Keeping Core Areas Free of Fruit Flies

### *Intensification of toxic bait sprays in orchard periphery rows*

Many growers already adopt this procedure.

### *Habitat manipulation and trap cropping*

Early cultivars could be used as trap crops, as *A. fraterculus* females prefer larger fruits when they are available. The advantage of using an apple cultivar as a trap crop instead of a highly suitable host species is that a high mortality in the immature stages is observed and application of insecticides could be reduced even in the trap crop. The main constraint to the implementation of this technique is operational, as it would require changes in orchard design which are sometimes not feasible.

### *Mass trapping*

Mass trapping in the orchard periphery would theoretically lead to a decrease in the population size that invades the orchard. This alternative will greatly depend on the development of efficient attractants.

## In the Orchard Periphery and Inner Rows

### *Identification of collected specimens of Anastrepha*

As previously mentioned, *A. dissimilis* is the second most frequent species captured in McPhail traps. This is a specialist species that exclusively attacks fruits of *Passiflora* spp. and is considered harmless to apples. An overestimation of population density occurs since on some occasions it may represent >20% of all fruit flies captured in commercial orchards. This practice is feasible as *A. fraterculus* and *A. dissimilis* show marked morphological differences.

### *Chemical control*

Determination of residual periods for registered insecticides - Applications could be reduced early in the season, as immature apples are highly unsuitable as substrates. Discovery of oviposition deterring substances is desirable.

## RESEARCH NEEDS

These are as follows:

- Development of a system to predict when population density is likely to increase.
- Biology and rearing of native species of parasitoids. This is strictly limited by the lack of mass rearing techniques for *A. fraterculus*.
- Formulation of phototoxic baits against *A. fraterculus*. The available formulation has negative impacts on the crop (phytotoxicity). Replacement of phytotoxic components or development of application methodology is desirable.
- Search for oviposition deterring substances. Even though the complete development of *A. fraterculus* is unlikely to occur in apples due to insecticides and natural fruit resistance, punctures lead to malformations and depreciation in the quality of the product.
- Search for efficient attractants. This is a key point in the implementation of control methods using phototoxic baits. It was recently observed that a combination of ammonium acetate, triethylamine and putrescine is not attractive to *A. fraterculus* under natural conditions. Butyl-hexanoathe was not attractive to *A. fraterculus* either.

## REFERENCES


Branco, E.S. 1998. Resistência de genótipos de macieira a mosca-das-frutas *Anastrepha fraterculus* (Diptera: Tephritidae). MSc Dissertation, University of Sao Paulo, Piracicaba, SP, Brazil, 119 pp.

Kovaleski, A. 1995. Avalição do dano causado pela mosca-das-frutas *Anastrepha fraterculus* Wied. (Diptera: Tephritidae) em maçãs com e sem tratamento químico. Resumos, 15o Congresso Brasileiro de Entomologia, Caxambu, MG, pp. 439.

Kovaleski, A. 1997. Processos adaptativos na colonização da maçã (*Malus domestica* L.) por *Anastrepha fraterculus* (Wied.) (Diptera: Tephritidae) na região de Vacaria, RS. PhD Thesis, University of Sao Paulo, SP, Brazil, 122 pp.

Kovaleski, A., L.G. Ribeiro, I. Nora and E. Humeres. 1995. Determinação da eficiência de atrativos alimentares na captura de moscas-das-frutas *Anastrepha fraterculus* (Wied., 1830) (Diptera: Tephritidae) em macieira no RS e SC. Resumos, 15o Congresso Brasileiro de Entomologia, Caxambu, MG, pp. 606.

Kovaleski, A., R.L. Sugayama and A. Malavasi. 1998. The colder, the better: overwintering in the South American fruit fly in Brazil. Fifth International Symposium on Fruit Flies of Economic Importance, Abstracts, p. 42.

Kovaleski, A., R.L. Sugayama and A. Malavasi. 1999. Movement of *Anastrepha fraterculus* from native breeding sites into apple orchards in Southern Brazil. Entomol. Exp. Appl. 91: 459-465

Kovaleski, A., K. Uramoto, R.L. Sugayama, N.A. Canal D. and A. Malavasi. 1999. A survey of *Anastrepha* Schiner (Diptera, Tephritidae) species in the apple growing area of the state of Rio Grande do Sul, Brazil. Revta. bras. Ent. 43(3/4):229-234.

Mangan, R.L. and D.S. Moreno. 1995. Development of phloxine-B and uranine bait for control of Mexican fruit fly, pp. 115-126. In Light-activated pest control. J.R. Heitz and K.R. Downun (eds.). ACS Symposium series 616; American Chemical Society, Washington, D.C.

Selivon, D., P. dos Santos, S.R. Matioli and J.S. Morgante. 1996. Morphological and genetic differences in sympatric populations of *Anastrepha fraterculus*. 2nd Meeting of the Working Group on Fruit Flies of the Western Hemisphere, Vi¤a del Mar, Chile, Proceedings, pp. 22-23.

Sugayama, R.L. 1995. Comportamento, demografia e ciclo de vida de *Anastrepha fraterculus* (Wied.) associada a três cultivares de maçã no sul do Brasil. MSc Dissertation, University of Sao Paulo, Sao Paulo, SP, Brazil, 97 pp.

Sugayama, R.L., E.S. Branco, A. Malavasi, A. Kovaleski and I. Nora. 1997. Oviposition behavior of *Anastrepha fraterculus* in apple and diel pattern of activities in an apple orchards in Brazil. Entomol. Exp. Appl. 83: 239-245.

Sugayama, R.L., A. Kovaleski and C. Francescato. 1998b. Would *Diachasmimorpha longicaudata* be able to parasitise *Anastrepha fraterculus* larvae inside its host fruits in southern Brazil? Fifth International Symposium on Fruit Flies of Economic Importance, Abstracts, p. 43.

Sugayama, R.L., A. Kovaleski, P. Liedo and A. Malavasi. 1998c. Colonisation of a new fruit crop by *Anastrepha fraterculus* in Brazil: a demographic analysis. Environ. Entomol. 27: 642-648.

Sugayama, R.L., A. Kovaleski and J. Sivinski. 1998a. Predation of larvae and pupae of *Anastrepha fraterculus* (Wied.) in Southern Brazil. Fifth International Symposium on Fruit Flies of Economic Importance, Abstracts, p. 44.

# Field Population Studies of the Oriental Fruit Fly *Bactrocera dorsalis* (Hendel) for the SIT Programme in Thailand

*P. Keawchoung[1], V. Limohpasmanee[1], R. Dokmaihom[2], A. Imyim[2] and S. Meecheepsom[2]*

[1] Atomic Energy for Peace of Thailand, 16 Vibhavadi – Rangsit Rd. Bangkok 10900, Thailand

[2] Plant Protection Service Division, Department of Agriculture Extension, Jatuchak, Bangkok 10900, Thailand

## INTRODUCTION

Pakchong district is a large area in the Nakornrajchasima province in Thailand (Figure 1) which produces many kinds of tropical fruits. As fruit flies are serious pests in fruit plantations in the area, the Department of Agriculture Extension has tried to control them by using the sterile insect technique (SIT) with complementary technology from the Office of Atomic Energy for Peace (OAEP). In order to obtain data required to plan the SIT programme to eradicate the fruit flies, subsequent field population studies were conducted.

## MATERIALS AND METHODS

### Traps

Steiner traps, each filled with 2ml methyl eugenol and insecticide (24:1 v/v), were used to obtain estimates of the fly population.

#### *Experimental Area*

Thirty-eight square km of three sub-districts in the Pakchong district.

### Flies

The Oriental fruit flies were reared at the mass rearing facility of the OAEP. A day before emergence, pupae were marked with fluorescent dye and irradiated at 90 Gy by Gamma beam-650 to sterilise the flies. They were reared in cardboard boxes after emergence with a mixture of 3 parts sugar and 1 part yeast hydrolysate for 10 days in the laboratory.

#### *Experimental Areas*

- Two square km of the Subtai sub-district in the Pakchong district.
- Eight square km of the Nongradon sub-district in the Pakchong district.

## Population Fluctuation

Seventy-six Steiner traps were set up in the orchards of 3 sub-districts in the Pakchong district. Each trap was set approximately 500 metres from each other. The flies were collected every week all year round. The lure was changed once a month throughout the period of the experiment. The flies collected from the traps were identified, counted and recorded.

## Population Estimation

Release centres and trapping points were set up in the experimental areas. Each release centre was in the middle of the trapping points in an experimental area. There were five traps and one release centre per square km. The marked flies were released at the release centre one week before setting up of the Steiner traps. The flies were collected every week for seven continuous weeks. The marked and unmarked flies were identified by the acetone technique. The data were calculated according to the Jackson Positive method, Hamada's method and Ito's method.

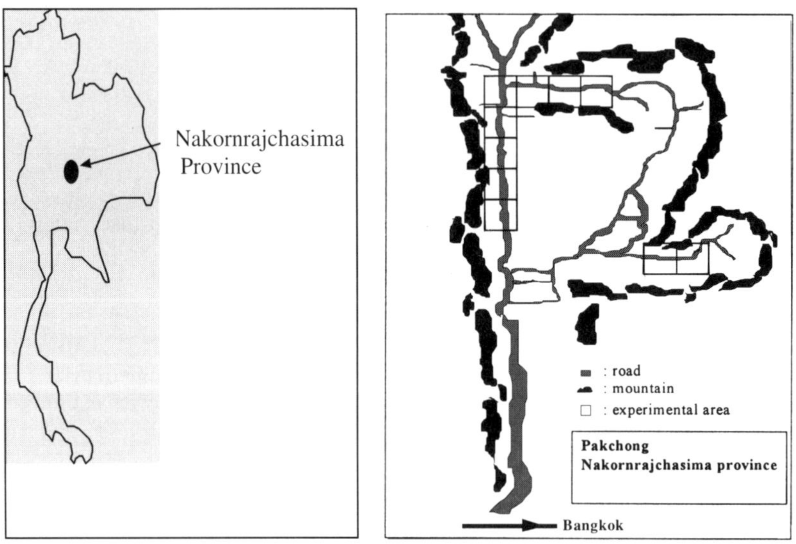


Figure 1. Map of Thailand and experimental areas.

## RESULTS

The population fluctuation studies showed the maximum and minimum number of the Oriental fruit fly trapped as 410 flies/trap/day on June and 4 flies/trap/day on January respectively (Figure 2). The number of the wild population of Oriental fruit flies was estimated by the "mark and recapture" method. Tables 1, 2 and 3 show details.

The sterile flies, irradiated at 90 Gy, could not compete with normal flies in a mating competition (Competitiveness = 0.77). The number of sterile flies, which should be released to eradicate the Oriental fruit fly, was 39,312 males/km$^2$/week.

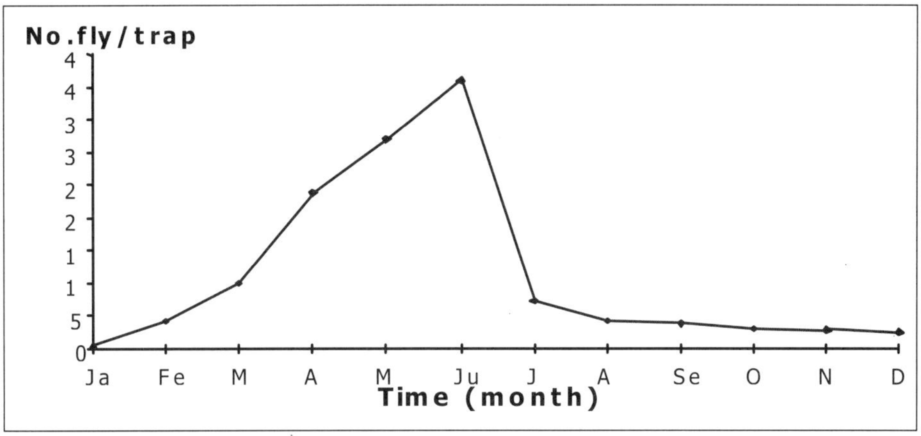


Figure 2. The population fluctuation of the Oriental fruit fly, *Bactrocera dorsalis* (Hendel) at the Pakchong district in 1992.

Table 1. The number of released and caught *Bactrocera dorsalis* (Hendel) in the Subtai sub-district.

| No. | No. released flies | Total caught flies | Marked flies | Unmarked flies |
|---|---|---|---|---|
| 1 | 3,673 | 557 | 103 | 454 |
| 2 | 3,570 | 349 | 105 | 244 |
| 3 | 3,465 | 284 | 54 | 230 |
| 4 | 3,411 | 171 | 11 | 160 |
| 5 | 3,400 | 299 | 31 | 268 |
| 6 | 3,369 | 128 | 5 | 123 |
| Total | | 1,788 | 309 | 1,479 |

Table 2. The number of released and caught *Bactrocera dorsalis* (Hendel) in the Nongradong sub-district.

| No. | No. released flies | Total caught flies | Marked flies | Unmarked flies |
|---|---|---|---|---|
| 1 | 7,075 | 1,662 | 632 | 1,030 |
| 2 | 6,443 | 581 | 82 | 499 |
| 3 | 6,361 | 517 | 60 | 457 |
| 4 | 6,301 | 278 | 23 | 255 |
| 5 | 6,278 | 795 | 73 | 722 |
| 6 | 6,205 | 483 | 43 | 440 |
| 7 | 6,162 | 1,226 | 72 | 1,154 |
| Total | | 5,542 | 985 | 4,557 |

Table 3. Number of wild *Bactrocera dorsalis* (Hendel) estimated by three methods.

| Method | Wild population | Recapture rate | Survival rate |
|---|---|---|---|
| **Subtai** | | | |
| Jackson | 5,620.1 | $8.413 \times 10^{-2}$ | 0.709 |
| Ito | 5,669.9 | $8.687 \times 10^{-2}$ | 0.721 |
| Hamada | 6,675.8 | $8.687 \times 10^{-2}$ | 0.686 |
| **Nongrdong** | | | |
| Jackson | 21,728.2 | 0.139 | 0.805 |
| Ito | 20,909.7 | 0.145 | 0.819 |
| Hamada | 16,112.0 | 0.145 | 0.819 |

The results of the Wanitch and Suchada experiment (unpublished data) on mating, sexual and sperm competitiveness of the sterile Oriental fruit fly (irradiated at 90 Gy) are as shown in Table 4.

Table 4. The number of mating of normal and irradiated males and competitiveness values in different ratios of normal and irradiated males.

| Female | Ratio of I/U | No. of copulation | No. of mated normal males | No. of mated irradiated males | Competitiveness |
|---|---|---|---|---|---|
| N | 0:30 | 9 | 9 | 0 | |
| N | 10:20 | 15 | 12 | 3 | 0.50 |
| N | 15:15 | 13 | 10 | 3 | 0.30 |
| N | 20:10 | 16 | 4 | 12 | 1.50 |
| N | 30:0 | 14 | 0 | 14 | |

## DISCUSSION

The population of the Oriental fruit fly was the lowest in January and increased from February to June because host plants such as sugar apple and mango were available during this period. After that, the population decreased drastically in July and maintained a low level until December.

In the population estimation experiment, the recapture rate, survival rate and population calculation methods were compared. The results showed that Ito's method and the Jackson Positive method were suitable to estimate populations of the Oriental fruit fly in the Nongradong sub-district because the recapture survival rates were not very low. The Hamada's method gave good results for the Oriental fruit fly population estimate in the Subtai sub-district because it avoided the effects of the low recapture survival rates by estimating the number of caught wild flies. However, to calculate the number of the released flies, the highest estimated population should be used.

## REFERENCE

Hamada, R. 1976. Density estimation by the modified Jackson's method. Appl. Entomol. Zool. II (3): 194-201.

Ito, Y. 1973. A method to estimation a minimum population density with a single recapture census. Res. Popul. Ecol. :159-168.

Jolly, G.M. 1965. Explicit estimates from capture-recapture data with both death and immigration-stochastic model. Biometrika 52, 1 and 2, 225-247.

Lewis, T. and L.R. Taylor. 1967. Estimating the size of populations. Introduction to experimental ecology. Academic Press Inc. New York, English. 123-124.

Poole, W.R. 1974. Mark-recapture methods. An introduction to quantitative ecology, McGraw-Hill, Inc. USA. 303-316.

Seber, G.A. 1962. The multi-sample single recapture census. Biometrika, 49, 3 and 4, 339-350.

Seber, G.A. 1965. A note on multiple-recapture census. Biometrika, 52, 1 and 2 249-259.

Yamamura, K., S. Wakamura and S. Kozai. 1992. A method for population estimation from a single release experiment. Appl. Entomol. Zool. 27(I): 9-17.

# PART III.

## APPLICATION OF AREA-WIDE CONCEPT TO FRUIT FLIES

## G) BIOCHEMISTRY AND PHYSIOLOGY OF FRUIT FLIES

# Physiological Control of Behaviour in Tephritid Fruit Flies

*Eric B. Jang*

USDA-ARS, Tropical Fruit, Vegetable and Ornamental Crop Research Laboratory, P.O. Box 4459, Hilo, Hawaii 96720, USA

## INTRODUCTION

Studies on the behaviour of tephritid fruit flies have historically focused on the interaction of external stimuli such as temperature, semiochemicals, seasonality, etc., or the interactions of flies between and among species for a number of observed behaviours such as mating, pheromone calling and oviposition. While descriptive behaviour represent much of what we know about these pest species, less is known about the underlying physiological mechanisms which function in priming or modulation of the observed behaviour. Central to our understanding of tephritid behaviour are the multiple and often complex internal factors which are involved, and the path/mechanisms by which external stimuli result in observed behaviour. Tephritid fruit fly physiology is a vastly understudied research area which may provide important information on how peripheral receptors receive information, the transduction and coding of information centrally and how behaviour is regulated biochemically. The integration of physiology disciplines to help explain behaviour is central to the goal of developing new technology which may be useful in fruit fly control.

In our laboratory, we have been studying the mechanisms of chemoreception and its link to behaviour in tephritids in such areas as olfaction, feeding, mating and oviposition. Our approach has been that tephritid behaviour can be largely influenced by their peripheral receptors which are responsible for receiving olfactory, gustatory, visual and tactile information inputs and their physiological state which controls internal modulation of behaviour. Thus, differences in behaviour between species might be explained on the basis of differences in their peripheral receptors, and the plasticity in which observed behaviour vary between the same species could very well be attributed to changes in their physiological state that are not readily apparent merely from visual observation.

The importance of the physiological state in behavioural studies has been recognised for years. In many papers, physiological states have been included as prerequisites to behavioural studies (i.e. ... "flies were 5-7 days old"..... or ... "only virgin flies were tested"... ) . However in many cases, prior knowledge of the physiological state is not well referenced in behavioural studies and as a result, subject to those variabilities which can result from changes in this state. These differences in the physiological state should be considered in the carrying out of behavioural bioassays.

## PERIPHERAL INPUTS

For the purpose of this paper, I will define peripheral inputs as environmental cues (inputs) and their associated morphological "receptors" present on the insect which "receive" these inputs. The nature of the receptor systems in tephritids has been looked at in some fruit flies (Dickens et al. 1988), but has not been thoroughly described or investigated for many other tephritid species. For example, behaviourally, tephritids can be classified within a continuum from monophagus to polyphagus, and from "highly chemoreceptive" to "mostly visual". However, what is known about the underlying receptor systems which accompany these differences in behaviour? Do highly chemoreceptive species have different systems from the non-chemoreceptive species for detecting semiochemicals? Do differences exist peripherally (the receptor) or more centrally (the brain)? Research to date have shown many similarities in receptor morphology (e.g., sensilla on the antennae of tephritids) but these studies have concentrated mostly on the more obvious organs such as the antennae. On the other "end", interesting variations appear in ovipositor morphology between tephritid species, but little is known about the physiology of these receptors. Labellar and tarsal "taste hairs" have also not been well studied in tephritids, either from the standpoint of morphology or physiology, which results often in use of generalities when these systems are discussed.

## INTERNAL MODULATORS

I consider all physiological processes which may influence behaviour as internal modulators. These modulators include factors such as age, mating status, nutritional history, as well as more specific molecules which may be involved either directly or indirectly in behaviour such as accessory gland fluids, sex peptides, hormones, etc. Internal modulators represent a vastly understudied area of research due in part to its secondary recognition as real modulators of behaviour. Over the last several years, we have seen increasing recognition of the influence of the physiological state on behaviour which has heightened our awareness of its importance to behaviour. For example, over the last five years, several papers have been written on the role of adult nutritional status and specific hunger on the responsiveness to lures and attractants. Additionally, biochemical studies, once the purview of molecular and cellular physiologists, are now beginning to provide explanations on observed behaviour for organismal biologists.

## PHYSIOLOGICAL CONTROL OF OLFACTORY BEHAVIOUR IN TEPHRITIDS

Semiochemicals serve important roles in the life history of many adult tephritids (Jang and Light 1996, Light and Jang 1996). Ecological studies which describe the life history of various tephritids are numerous. Specific behaviour such as feeding, pheromone calling, mating and oviposition are key factors determining the success of the species. Our research in Hawaii has focused on the factors which are responsible for olfactory-driven female behaviour. Female Mediterranean fruit flies (Medflies, *Ceratitis capitata* (Wied.)) are attracted to, and prefer, male pheromone over host fruit odours if the females have not yet mated. Electrophysiological studies of the peripheral antennal receptors did not reveal a significant difference in the response of virgin versus mated females to any of the major components of the Medfly male odour (Jang et al. 1989). Similarly, Medfly response to a variety of host fruit odours showed sensitivity to various ubiquitous fruit odours but selectivity to key "signature" volatiles were not apparent (Light et al. 1988). However, mating triggers profound physiological (and behavioural) changes resulting in a switch in preference of females to host fruit odours (Jang, 1995). While studying the multitude of factors which may influence such a change in behaviour, we looked at the effects of mating on subsequent olfactory behaviour. We found that in addition to the transfer of sperm from the male to the female as a result of mating, males also transfer fluid from the male accessory glands to the females. Females also receive physical information as a result of mating (stretch receptors). When females were injected (abdominally) with extracts of accessory gland fluids from males, they exhibited similar behaviour as their naturally mated counterparts, namely, the switch in behaviour from preference for male produced pheromones to host fruit odours. Females injected with saline only did not switch their behaviour and continued to be attracted to the pheromones over the host fruit odours. In addition to changes in the olfactory-stimulated switch in the attraction from pheromones to host fruit odours, females which were naturally mated or injected with the accessory gland fluid, laid significantly more eggs than non-mated or saline injected controls. Females did not exhibit the switch in behaviour immediately after mating, suggesting that secondary mechanisms may be involved in the regulation of the switch in behaviour. One proposed mechanism by which this change in behaviour may occur is through the action of hormones such as juvenile hormone (JH). Factor(s) from accessory gland fluids such as peptides may stimulate the release of JH from the corpora allata which in turn directly (or indirectly) stimulates changes in olfactory behaviour.

## EFFECTS OF ANTI-JH COMPOUNDS AND CHEMOSTERILANTS ON BEHAVIOUR

In Medfly, anti-juvenile hormone compounds such as precocene and the chemosterilant benzyl-1,3- benzodioxole have been shown to affect synthesis and release of JH from the corpora allata (Chang et al. 1994). This interference in JH production has been shown to affect sex attractancy of male Medflies (Chang and Hsu 1982) as well as ovarian development in females (Hsu et al. 1989). Preliminary studies with females treated with these compounds suggest that JH may also be involved in the modulation of olfactory behaviour.

Studies on the Oriental fruit fly (*Bactrocera dorsalis* (Hendel)) have confirmed a similar olfactory preference hierarchy of male pheromone to host fruit odour as a result of mating. Females treated with precocene a potent inhibitor of JH biosynthesis and then injected with the male accessory gland fluid did not exhibit the behaviour characteristics of mated flies. Similarly, the application of benzyl-1,3-benzodioxole (BBD), a chemosterilant for many tephritids, showed similar behaviour as precocene treated mated females.

## SEX AND THE STERILE MALE

If factors from male accessory glands are important in the control of female driven olfactory behaviour, what, if any, is the effect of irradiation on the ability of sterile males to change the behaviour of wild type females? To test this question, we set up a series of tests in which laboratory-reared normal, irradiated and wild type females were mated to conspecific males and observed for their response to pheromones or host fruit odour. These studies concluded that irradiated males were equally adept at altering female behaviour as non-irradiated males (Jang et al. 1998). Our studies have clearly shown a physiological role in modulation of olfactory behaviour in tephritids. Similar mechanisms are presumed to exist for other behaviours such as feeding, and oviposition behaviour. We suspect that oviposition behaviour is closely linked to ovarian development based on studies of chemosterilised and irradiated females.

## CONCLUSION

Chemoreception, transduction, age, mating status and nutritional state all play important roles in regulating behaviour. Hormonal activities and cellular homeostasis are further "downstream" in the regulatory process but no less important. In our example we have shown that mating and the transfer of accessory gland fluid had a direct impact on female Medfly olfactory behaviour which likely works through biochemical intermediates and possibly hormones. Improvements in our knowledge of these processes and their control will be the key to the development of control strategies which target behavioural processes against these pests.

## REFERENCES

Chang, F. and C.L. Hsu. 1982. Effects of precocene II on sex attractancy in the Mediterranean fruit fly, *Ceratitis capitata*. Ann. Entomol. Soc. Am. 75: 38-42.

Chang, F., E.B. Jang, C.L. Hsu, M. Ma and L. Jurd. 1994. Benzodioxole-1,3-benzodioxole derivatives and their effects on the reproductive physiology of insects. Arch. Insect Biochem. Physiol. 27:39-51.

Dickens, J.C., W.G. Hart, D.M. Light and E.B. Jang. 1988. Tephritid olfaction: Morphology of the antennae of four tropical species of economic importance. Ann. Entomol. Soc. Am. 81: 325-331.

Hsu, C.L., F. Chang, H.F. Mower, L.J. Groves and L. Jurd. 1989. Effect of orally administered 5-ethoxy-6-[-4-methoxyphenyl] methyl-1,3-benzodioxole on reproduction of the Mediterranean fruit fly. J. Econ. Entomol. 82: 1046-1050.

Jang, E.B. 1995. Effects of mating and accessory gland injections on olfactory-mediated behaviour in female Mediterranean fruit fly, *Ceratitis capitata*. J. Insect Physiol. 41: 705-710.

Jang, E.B and D.M. Light. 1996. Olfactory semiochemicals of tephritids pp. 73-90. *In* Economic Fruit Flies: A world assessment of their biology and management. B.A. McPheron and G. Steck (eds.) St. Lucie Press, Delray Beach, Florida, USA.

Jang, E.B., D.M. Light, R.A. Flath, J.T. Nagata and T.R. Mon. 1989. Electroantennogram responses of the Mediterranean fruit fly, *Ceratitis capitata* to identified volatile constituents from calling males. Entomol. Exp. Appl. 50:7-19.

Jang, E.B., D.O. McInnis, D.R. Lance and L.A. Carvalho 1998. Mating induced changes in olfactory-mediated behaviour of laboratory-reared normal, sterile and wild female Mediterranean fruit flies (Diptera:Tephritidae) mated to conspecific males. Ann. Entomol. Soc. Am. 90: 130-144.

Light, D.M. and E.B. Jang. 1996. Plant volatiles evoke and modulate tephritid behaviour. pp. 123-133. *In* Economic Fruit Flies: A world assessment of their biology and management. B.A. McPheron and G. Steck (eds). St. Lucie Press. Delray Beach, Florida, USA.

Light, D.M., E.B. Jang and J.C. Dickens. 1988. Electroantennogram responses of the Mediterranean fruit fly, *Ceratitis capitata* to a spectrum of plant volatiles. J. Chem. Ecol. 14:159-180.

# Kinetic Properties of the Two Alcohol Dehydrogenase (ADH) Isozymes of the Medfly *Ceratitis capitata* (Diptera: Tephritidae)

*C. Bonvicini, A.R. Malacrida and G. Gasperi*

Department of Animal Biology, Laboratory of Zoology, University of Pavia, Piazza Botta 9, I-27100 Pavia, Italy

## INTRODUCTION

Alcohol dehydrogenase (ADH; alcohol: $NAD^+$ oxidoreductase; EC 1.1.1.1) catalyses the reversible interconversion of a variety of alcohols and their corresponding aldehydes and ketones. Among insects, the ADH gene-enzyme system has been extensively studied in several species of *Drosophila* (Chambers 1988, Heinstra 1993, Ashburner 1998).

The best characterised ADH from a non-drosophilid insect is that of the Medfly, *Ceratitis capitata* (Wied.), based on data from molecular genetics (Malacrida et al. 1992, Gasperi et al. 1992, Brogna et al. 2000), biochemistry (Gasperi et al. 1994) and population genetics (Gasperi et al. 1992, Gomulski et al. 1998).

The primary interest in studying this enzymatic function in the Medfly was that the ADH system has been proposed, on the model of *Drosophila*, as a useful tool for genetic sexing strategies addressed to the biological control of this pest (Robinson et al. 1988). Moreover, molecular characterisation of *Adh* in a species like *C. capitata*, that diverged from the *Drosophilidae* more than 100 million years ago (Beverley and Wilson 1984), is of interest for studying the evolution of this protein in higher diptera.

The principal function of ADH in insect metabolism is to catabolise alcohols generated by microbial fermentation in larval and adult feeding sites; in *Drosophila melanogaster* Meigen, the presence of an active ADH is responsible for two different phenotypic traits, namely alcohol tolerance and alcohol utilisation (Van Delden 1982, David 1988). The ecological niche of *C. capitata* is different from that of *Drosophila* species, the first breeding on ripening fruits, the latter breeding on rotten plant material. Consequently, the physiological role of ADH may have diversified in these dipteran species.

## THE MEDFLY ADH SYSTEM

Initial studies revealed that ADH from Medfly differs from that of *D. melanogaster* (Figure 1). In fact, two well differentiated, genetically independent dimeric proteins, ADH-1 and ADH-2, are present in *C. capitata*. They are encoded by two tightly linked genes (0.49 cM) on the left arm of the 2nd chromosome (Malacrida et al. 1992), suggesting that the proteins are the product of gene duplication with divergence. The Medfly ADH proteins differ in several features, such as isoelectric point (pI), tissue localisation and development. The isoelectric point of ADH-1 (pI = 5.4) is different from that of ADH-2 (pI = 8.6), as well as from that of *D. melanogaster* ADH (pI = 7.6). The genes are not controlled coordinately during the development,

*Adh1* being expressed mainly in muscle and *Adh2* in the fat body, gut and ovary (Gasperi et al. 1992, Brogna et al. 2000).

The two peptides have the same molecular weight (ca. 27 kD) as the *D. melanogaster* ADH peptide (Gasperi et al. 1994). The Medfly ADH-1 is immunochemically similar to ADH-2; ADH-1-specific antibody cross-reacts with *Bactrocera oleae* extracts, while no reactivity occurs with *D. melanogaster* ADH (Gasperi et al. 1994). The lack of cross-reactivity between the *C. capitata* and *D. melanogaster* proteins suggests that the ADH isozymes of tephritids have diverged considerably during evolution from the drosophilid enzyme.

The two Medfly proteins are over 80% identical in primary sequences (Brogna et al. 2000), strongly supporting the hypothesis that they are the product of a gene duplication. The relatively high divergence between the two genes and the observation that most tephritid species have two or three isozymes (Matioli et al. 1992) would support the view that this duplication is a rather early event in the radiation of Tephritidae.

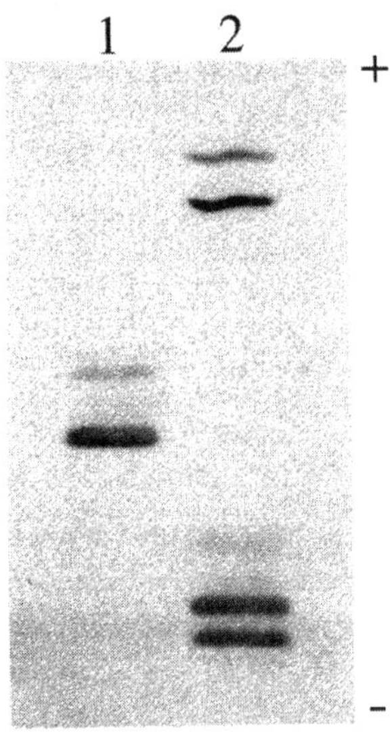


Figure 1. Isoelectrofocusing separation ( pH range 3-10) of ADH from *Drosophila melanogaster* (lane 1) and *Ceratitis capitata* (lane 2).
The isoelectric point (pI) values of ADH are 7.6 for *D. melanogaster* enzyme, and 5.4, 8.6 for ADH-1 and ADH-2 of *C. capitata*, respectively.

## *IN VITRO* DIFFERENCES BETWEEN PURIFIED PROTEINS (ADH-1 AND ADH-2)

The relatively high divergence between *Adh1* and *Adh2* genes is reflected in the kinetic and structural differences of their protein products. These differences were identified by *in vitro* experiments on purified protein preparations of ADH-1 and ADH-2.

## Differences in Catalytic Properties

*In vitro* differences were observed (Table 1) between ADH-1 and ADH-2 for the kinetic parameters which define the respective affinities (Km), turnover numbers (kcat), and specificity constants (kcat/Km) for the substrates (both primary and secondary alcohols), and the coenzyme ($NAD^+$).

ADH-2 is a more efficient catalyst than ADH-1 especially with respect to secondary alcohols. It reacts faster with propan-2ol compared to ethanol (kcat/Km = 1.5 ± 0.07 versus 0.5 ± 0.05). The kcat/Km of ADH-2 for propan-2ol is three times greater than that estimated for ADH-1 (kcat/Km = 1.5 ± 0.07 versus 0.4 ± 0.01). ADH-1 has a greater affinity for ethanol (Km = 3.9 ± 0.04) and exhibits a higher reaction rate at low concentration of this primary alcohol.

The catalytic differences between ADH-1 and ADH-2 are also evident in their interaction with the coenzyme $NAD^+$ for the production of NADH in the presence of ethanol. ADH-1 has much greater affinity for the coenzyme (Km = 19 ± 0.5 versus 77 ± 0.6), but the limiting factor is the turnover rate (kcat) which is four times lower than that assessed for ADH-2 (kcat = 0.76 ± 0.1 versus 3.4 ± 0.05).

Table 1. Kinetic parameters estimated for the two ADH isozymes of *Ceratitis capitata* and for two ADH allozymes of *Drosophila melanogaster*.
(The data are obtained from kinetic experiments at 25°C; and the ± values are standard deviations)

| | *Substrates* | | | | | | *Coenzyme* | | |
|---|---|---|---|---|---|---|---|---|---|
| | Ethanol | | | Propan 2-ol | | | $NAD^+$ | | |
| | $k_{cat}$ ($sec^{1}$) | $K_m$ (mM) | $k_{cat}/K_m$ ($sec^{-1.}$ $mM^{-1}$) | $k_{cat}$ ($sec^{-1}$) | $K_m$ (mM) | $k_{cat}/K_m$ ($sec^{-1.}$ $mM^{-1}$) | $K_{cat}$ ($sec^{-1}$) | $K_m$ (μM) | $k_{cat}/K_m$ ($sec^{-1.}$ $\mu M^{-1}$) |
| *C. capitata* ADH-1 | 3.1 ± 0.2 | 3.9 ± 0.4 | 0.8 ± 0.05 | 3.6 ± 0.9 | 8.6 ± 0.9 | 0.4 ± 0.01 | 0.76 ± 0.1 | 19 ± 0.5 | 0.04 ± 0.006 |
| *C. capitata* ADH-2 | 3.9 ± 0.3 | 7.1 ± 0.7 | 0.5 ± 0.05 | 4.1 ± 1.6 | 2.7 ± 0.9 | 1.5 ± 0.07 | 3.4 ± 0.05 | 77 ± 0.6 | 0.05 ± 0.004 |
| **D. melanogaster* Slow | 3.2 | 3.0 | 1.0 | 3.6 | 0.56 | 6 | | 135 | |
| **D. melanogaster* Fast | 4.9 | 4.0 | 1.2 | 10.4 | 0.83 | 12.5 | | 160 | |

* Data for *D. melanogaster* ADHs are derived from Heinstra (1993), Hovik *et al.* (1984) and Winberg and McKinley-McKee (1992).

**Differences in Thermodynamic Parameters**

The two isozymes differ also in their activation parameters of reaction such as the enthalpy (ΔH) and entropy (ΔS) of activation. Using ethanol (primary alcohol) as a substrate, the ΔH estimate is three times lower for ADH-1 in comparison with ADH-2 (ΔH = 5.9 ± 2 versus 18.9 ± 1.9), reducing the activation barrier to transfer the H from ethanol to the coenzyme $NAD^+$. ΔS is more negative for ADH-1 with respect to ADH-2 (ΔS = -216 ± 7.8 versus -169 ± 5.7), such that ethanol fits better in the substrate binding pocket of ADH-1.

**Differences in the Binding Interactions**

It is likely that the observed differences in reactivity between ADH-1 and ADH-2 depend on specific interactions between the proteins, substrates and cofactors. In fact, ADH-1 and ADH-2 differ in the mode of binding of substrates and coenzyme to the active sites, as assessed by the changes observed in the circular dichroism (c. d.) spectra of enzyme-substrate and enzyme-inhibitor complexes. The binding constants (K) of propan-2ol and ethanol to ADH-2 are higher than those evaluated for ADH-1 (Table 2), and the K value for the $NAD^+$/ADH-2 complex is four times greater than that of $NAD^+$/ADH-1.

Table 2. Apparent binding constants (K) of ADH1-donor and ADH2-donor complexes, evaluated by circular dichroism in the peptidic spectral region, at pH 7.5 in Tris-HCl buffer.

| Donors | K ($\mu M^{-1}$) | |
|---|---|---|
| | ADH-1 | ADH-2 |
| Ethanol | 0.015 | 0.037 |
| AMP + Ethanol | 0.034 | 0.033 |
| Propan 2-ol | 0.023 | 0.035 |
| AMP + Propan 2-ol | 0.023 | 0.020 |
| $NAD^+$ | 3.53 | 13.07 |
| AMP | 3.70 | 7.45 |
| NADH | 3.93 | 5.50 |

However, the cofactor $NAD^+$ differentially influences the binding of the alcoholic substrates with the two proteins. $NAD^+$ acts as a negative modulator for the propan-2ol/ADH-2 complex, as deduced by the binding behaviour of its analogous AMP (adenosine 5'-monophosphate), which reduces the K value from 0.035 to 0.020 for propan-2ol/ADH-2 complex. On the other hand, $NAD^+$ is a very effective cooperative modulator for the interaction of ethanol with ADH-1, because the presence of AMP doubles the K value (K = 0.015 to 0.034) for this complex. AMP does not influence the binding of the ADH-1/propan-2ol complex.

The greater versatility of ADH-1 for primary alcohols is also confirmed by the fact that another alcohol, such as methanol, does not compete with ethanol for the binding to the active site of ADH-1, while it does with propan-2ol.

## CONCLUSIONS

The divergence between the duplicated genes, *Adh1* and *Adh2*, has consequences with respect to the biochemical and structural properties of the two ADH isozymes of *C. capitata*. The *in vitro* differences in the kinetic, thermodynamic and structural properties observed between ADH-1 and ADH-2 clearly reflect a potential for a multifunctional action of the ADH system in the Medfly: ADH-2 acts as a generalist enzyme, while ADH-1 is a more specialised one, being more suitable for primary alcohols. These differences appear more significant when viewed in the context of the tissue localisation of these two isozymes, ADH-1 being in the larval and adult muscle, in contrast to ADH-2 which is mainly limited to the fat body and gut of both life stages. It seems justified to suggest differential functional roles for the two isozymes in the alcohol metabolism of the Medfly, with ADH-1 much more devoted to the ethanol utilisation for NADH production. The more generalist calalytic properties of ADH-2 seem to be more similar to the *Drosophila* ADH function, both in alcohol tolerance and alcohol utilisation (Heinstra 1993); the fat body and the gut are the main compartments of ADH activity in *Drosophila*.

## ACKNOWLEDGEMENTS

This research was supported by grants from the European Communities Commission (ECC) "Project IC18-CT96-0100", from the Italian Ministry of Agriculture, and from the Italian Ministry of the University & Scientific Research and Technology (M.U.R.S.T. "Funds ex-40%"). An Agreement with the International Atomic Energy Agency (Vienna, Austria) is supporting the Medfly Group at the Dept. of Animal Biology of the University of Pavia. This paper is part of the Ph.D. curriculum studies of C. Bonvicini.

## REFERENCES

Ashburner, M. 1998. Speculations on the subject of alcohol dehydrogenase and its properties in *Drosophila* and other flies. BioEssays 20: 949-954.

Beverley, S.M. and A.C. Wilson. 1984. Molecular evolution in *Drosophila* and the higher Diptera: II. A time scale for fly evolution. J. Mol. Evol. 21: 1-13.

Brogna S., Benos P.V., Gasperi G., and Savakis C. 2000. The *Drosophila* alcohol dehydrogenase gene may have evolved independently of the functionally homologous medfly, olive fly and flesh fly genes. (submitted).

Chambers, G.K. 1988. The *Drosophila* alcohol dehydrogenase gene-enzyme system. Adv. Genetics 25: 39-107.

David, J.R. 1988. Ethanol adaptation and alcohol dehydrogenase polymorphism in *Drosophila*: from phenotypic functions to genetic structures, pp. 163-172. *In* Population Genetics and Evolution. G. de Jong (ed.). Springer-Verlag, Berlin, Heidelberg.

Gasperi, G., L. Baruffi, A.R. Malacrida and A. Robinson. 1992. A biochemical genetic study of alcohol dehydrogenase isozymes of the Medfly, *Ceratitis capitata* Wied. Biochem. Genet. 30: 289-304.

Gasperi, G., D. Kafetzopoulos, A. Christodoulidou, V. Bouriotis and C. Savakis. 1994. Isolation and partial characterisation of two alcohol dehydrogenase isozymes from the Medfly, *Ceratitis capitata*. Insect Biochem. Mol. Biol. 24: 87-94.

Gomulski, L.M., K. Boutzis, S. Brogna, P.A. Morandi, F. Sebastiani, C. Torti, C. Bonvicini, C.R. Guglielmino, C. Savakis, G. Gasperi and A.R. Malacrida. 1998. Intron size polymorphism of the *Adh1* gene parallels the worldwide colonisation history of the Mediterranean fruit fly, *Ceratitis capitata*. Mol. Ecol. 7: 1729-1741.

Heinstra, P.W.H. 1993. Evolutionary genetics of the *Drosophila* alcohol dehydrogenase gene-enzyme system. Genetica 92: 1-22.

Hovik, R., J.O. Winberg and J.S. McKinley-McKee. 1984. *Drosophila melanogaster* alcohol dehydrogenase: substrate specificity of the *Adh*$^F$ alleloenzyme. Insect Biochem. 14: 345-351.

Malacrida, A.R., G. Gasperi, A. Zacharopoulou, C. Torti, E. Riva Francos and R. Milani.1992. Evidence for a genetic duplication involving alcohol dehydrogenase genes in *Ceratitis capitata*. Biochem. Genet. 30: 35-48.

Matioli, S.R., J.S. Morgante, V.N. Solferini and D.L. Frias. 1992. Evolutionary trends of alcohol dehydrogenase isozymes in some species of tephritid flies. Rev. Brasil. Genet. 15: 33-50.

Robinson, A.S., C. Savakis and C. Louis 1988. Status of molecular genetic studies in the Medfly *Ceratitis capitata*, in relation to genetic sexing. pp. 241-250. *In* Modern Insect Control: Nuclear Techniques and Biotechnology. International Atomic Energy Agency, Vienna, Austria.

Van Delden, W. 1982. The alcohol dehydrogenase polymorphism in *Drosophila melanogaster*: selection at an enzyme locus. Evol. Biol. 15: 187-222.

Winberg, J.O. and J.S. McKinley-McKee. 1992. Kinetic interpretations of active site topologies and residue exchanges in *Drosophila* alcohol dehydrogenases. Int. J. Biochem. 24: 169-181.

# Sex and Aggregation Pheromone Transport After Methyl Eugenol Consumption in Male *Bactrocera papayae*

*Alvin K.W. Hee and K.H. Tan*

School of Biological Sciences, Universiti Sains Malaysia, 11800 Penang, Malaysia

## INTRODUCTION

Amongst at least 52 sibling species complexes in the Oriental fruit fly, *Bactrocera dorsalis* Hendel (Diptera: Tephritidae), *B. papayae* (formerly Mal B) Drew and Hancock (Drew and Hancock 1994) is beginning to emerge as an economically important insect pest which poses a severe threat to the fruit cultivation in both subtropical and tropical countries. In Malaysia, *B. papayae* is one of the most damaging pests which infests many commercially grown fruits (Tan and Lee 1982).

Like the Oriental fruit fly and its sibling species complex, *B. carambolae* Drew and Hancock, *B. papayae* is also strongly attracted to, and compulsively feeds on, methyl eugenol (ME) (Tan 1993). Chemical analyses revealed that in *B. papayae* males, ME is converted to phenylpropanoids which are then selectively accumulated in the rectal gland. Of the three major volatile substances, 2-allyl-4,5-dimethoyphenol (allyl-DMP) was detected in higher quantities relative to the *trans*-coniferyl alcohol (4-(3-hydroxy-*E*-propenyl)-2-methoxyphenol) (CF) and *cis*-3,4-dimethoxycinnamyl alcohol (*cis*-DMC) (Nishida et al. 1988a, 1988b). Behavioural studies have also shown that allyl-DMP and CF function as male sex and aggregation pheromone in *B. papayae* (Tan and Nishida 1996, Hee and Tan 1998). Allyl-DMP was found to be the most attractive compound and *cis*-DMC the least attractive to the males (Tan 1996).

Consumption of ME enhances the mating competitiveness of males. This is demonstrated by the strong attraction of females to conspecific ME-fed males in wind tunnel experiments (Hee and Tan 1998). In male-male mating competition for virgin females, males that fed on ME performed significantly better (Shelly and Dewire 1994, Tan and Nishida 1996). Thus it appears that ME-fed males produced signals that were more attractive.

However, the characterisation and understanding of the functions of these phenylpropanoids have not been accompanied by studies of their physiological mode of transport in male flies. The current paper describes the sex and aggregation pheromone transport after methyl eugenol consumption in male *B. papayae*. Here, we investigate the presence of the pheromones and their precursor ME in the male digestive and circulatory system.

## MATERIALS AND METHODS

### Insects

Laboratory stocks of *B. papayae* were reared on an artificial medium using conditions as previously described by Hee and Tan (1998) at 27°C. Sexually mature virgin males (14-25 days after emergence) that responded maximally to ME (Tan et al.

1987) were used for experimentation. All experimentation on the males was performed in the morning during their peak period of response to ME (Tan 1985).

**Chemical Identification**

Methods used in identification of pheromones were slightly modified based on those used by Nishida et al. (1988a, 1988b). The GC-MS analyses were performed on a Hewlett Packard HP 589A mass spectrometer (70eV) coupled to a HP 5890 Series II gas chromatograph and connected to a GC column (30m x 0.25mm non-polar, fused silica capillary column coated with cross-linked-bonded methyl silicone Quadrex 007-1 0.25 μm thick) programmed from 80°C (approximately 1 min. holding) to 240°C at a rate of 10°C/min. Chemical identification was performed by comparison with the retention time, use of internal standards and mass spectra fragmentation pattern of authentic standards.

**Gut Tissue Dissection**

Sexually mature virgin males held on plastic netting by plasticine moulds were individually fed with ca. 0.1 μl of ME using a capillary micropipette to minimise contamination from ME present on the body. The gut tissues (excluding the crop and rectal gland) from 3 ME-fed males were dissected in Insect Clarke saline (6.5g NaCl, 1.4g KCl, 0.12g $CaCl_2$, 0.1g $NaHCO_3$, 0.01g $Na_2HPO_4$ in 1 litre of solution) six hours after initial feeding with ME. The gut tissues were carefully dissected from the decapitated males with a pair of fine forceps and extra precautions were taken to avoid puncturing the crop which may otherwise contaminate the tissues. The dissected tissues were then rinsed several times with saline and distilled water to remove any trace of contamination. A small strip of filter paper was used to absorb the excess saline and water. These tissues were then homogenised, using a small homogeniser which consists of a "V" glass tube and a small, fitting glass rod. 15 μl of redistilled ethanol was used as solvent. The homogenate was centrifuged using the MSE® benchtop centrifuge machine for 10 minutes at room temperature to ensure that a clear supernatant solution was obtained. The supernatant extract was then analysed using a gas chromatograph (GC) and gas chromatograph-mass spectrometer (GC-MS). Three replicates of the experiment were conducted. A similar procedure was repeated using ME-deprived males as a negative control.

**Haemolymph Extraction And Attractancy Study**

Haemolymph was obtained from a sexually mature virgin ME-fed male held on a green plastic netting supported by plasticine moulds. Using a sharp scapel, a small longitudinal slit was made along the centre of the meso and metathorax segments. A disposable glass capillary micropipette with a drawn-out tip attached to a micromanipulator (Narishige Co. Ltd, Japan) was used to collect the haemolymph. Between 10-13 μl of haemolymph from 15 ME-fed males was pooled into a 0.2 ml microcentrifuge tube (on ice) containing a few crystals of phenylthiocarbamide (1-phenyl-2-thiourea) (PTC) to inhibit tyrosinase activity. The collected haemolymph was centrifuged at 12,000g for 15 minutes at 4°C and the supernatant transferred to a "V" glass tube. 20 μl of redistilled ethanol was added to the extract. The mixture was again centrifuged using a MSE® benchtop centrifuge machine for 10 minutes at room temperature. The decanted supernatant was further concentrated under nitrogen flow to a final volume of ca. 5 μl which was used for GC and GC-MS analyses. The experiment was replicated in triplicates. In the control experiment, a similar procedure was repeated using ME-deprived males as a negative control and a different micropipette was used to prevent contamination from the haemolymph of ME-fed males.

In the attractancy study, sexually mature virgin males were exposed to 1 μl of ME-fed haemolymph containing trace amounts of PTC and ME-fed haemolymph alone respectively on a Whatman® No. 1 filter paper. ME-deprived male haemolymph was used as the control. The filter paper was then placed in a cage containing mature males. The attraction and behaviour of the flies were then examined.

**Parabiosis**

Parabiosis (back-to-back) between a ME-fed and ME-deprived male was accomplished in a laminar flow under aseptic conditions. The males were individually held in a ventral position onto a green plastic netting. The thoracic region of each male was exposed through a 2 $mm^2$ hole of the netting. Plasticine moulds were used to hold the wings of the flies to the netting. To prevent confusion, plasticine moulds of yellow colour for the ME-deprived male (Male A) and red colour for the male to be fed with ME (Male B) respectively were used. The males were then rapidly immersed in a 0.5% sodium hypochlorite solution before thorough rinsing with distilled water followed by drying on a tissue paper.

Melted paraffin wax was used to seal the mouth parts of Male A to prevent subsequent ingestion of ME and the regurgitated contents of the ME-fed male. A small incision was made on the dorsal mesothorax segment. Then, 2 μl of sterile Insect Clarke saline was immediately added to the wound to avoid dehydration.

Similarly, for Male B, this surgical procedure was repeated without the sealing of the mouth parts. Following this, a small piece of clean filter paper (ca. 1.5$mm^2$) was placed on the wound of Male A. Both males were then joined together in a back-to-back position. The pair was gently pushed together to ensure that no air bubbles were trapped and melted paraffin wax was used to seal the thorax joints. Subsequently, Male B was given ca. 0.1 μl of ME. The males were constantly sprayed with water to prevent dehydration. The rectal gland of Male A was dissected 6 hours after the initial feeding of Male B with ME. The procedure was performed for ten pairs of males of Type A and Type B to obtain ten rectal glands from Type B males. The preparation of rectal gland extracts for GC and GC-MS was conducted as previously described but 25 μl of ethanol was used as solvent. The experiment was replicated three times. For the control, a similar procedure was followed except that the males of Type B were not fed ME.

## RESULTS

In the ME-fed male *B. papayae*, chemical analyses of the dissected gut (excluding crop and rectal gland) extracts did not show any presence of ME or ME-metabolites. However, the ME-metabolites were detected in its haemolymph extract. The ME-metabolites were identified as the phenylpropanoids, allyl-DMP and CF by comparison with authentic standards. Identification of these phenylpropanoids was obtained from similar matches in the retention time, use of internal standards and by comparison of the mass spectra fragmentation pattern of : 1) the authentic synthetic standards (provided by R. Nishida, Kyoto University) (Figure 1), and 2) Wiley database library matches (HP-UX Chemstation).

When sexually mature virgin males of *B. papayae* were exposed to the haemolymph extract of conspecific males, they were strongly attracted to, and compulsively fed on, the ME-fed male haemolymph extract containing trace amounts of PTC. ME-fed male haemolymph attracted male flies when compared to that of ME-deprived males which did not (Figure 2).

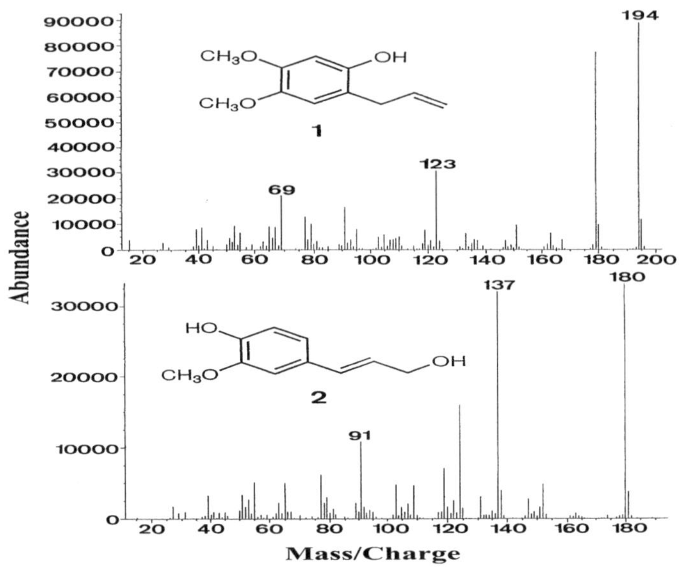


Figure 1. Mass spectra of the phenylpropanoids in the rectal glands of methyl eugenol-fed *B. papayae* males. Chromatogram done on a capillary non-polar Quadrex 007-1 column (30 m x 0.25 mm, programmed from 80°C to 240°C, 10°C/min). **1**: 2-ally-1-4,5-dimethoxyphenol, **2**: coniferyl alcohol.

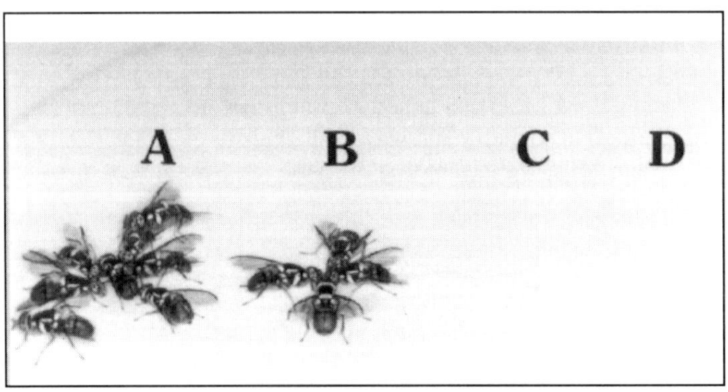


Figure 2. Attraction of sexually mature virgin males of *B. papayae* to the haemolymph extracts of conspecific males on a Whatman® No. 1 filter paper. **A** and **B**: *B. papayae* males feeding on the haemolymph extracts of ME-fed conspecific males with and without trace amounts of PTC respectively, **C**: Haemolymph extract of ME-deprived conspecific males with trace amounts of PTC, **D**: Haemolymph extract of ME-deprived conspecific males only.

In the parabiosis studies, phenylpropanoids were also detected in the rectal gland extracts of ME-deprived *B. papayae* males (Male A type) (Table 1). Using both GC and GC-MS analyses, these phenylpropanoid compounds were similarly identified as allyl-DMP and CF respectively.

Table 1. Amounts of ME-metabolites detected in A: haemolymph extracts of ME-fed males; and B: Rectal gland extracts of ME-deprived males used in parabiosis.

| Amounts of ME-metabolites present | | |
|---|---|---|
| Phenylpropanoids | A (μg/μl) | B (μg/gland) |
| 2-allyl-4,5-dimethoxyphenol | 0.02 ± 0.01 | 0.49 ± 0.08 |
| Coniferyl alcohol | 0.02 ± 0.02 | 0.33 ± 0.12 |

## DISCUSSION

Chemical analyses have shown that in the three species of the Oriental fruit fly complex, *B.dorsalis* and *B. papayae* males produced allyl-DMP, CF and *cis*-DMC after consuming ME. These were located in the rectal gland (Nishida et al. 1988a, 1988b, Tan and Nishida 1996). Besides producing endogenously large quantities of 6-oxononanol, *B. carambolae* also converts ME to CF (Tan and Nishida 1996). When *B. papayae* males were fed with excess amounts of ME, the intact ME was found in the body tissues but not in the rectal gland (Nishida et al. 1988b). Therefore, in investigating the transport of ME and its analogues, chemical analyses of ME-fed *B. papayae* males did not show the presence of these compounds in the gut extract excluding the crop and rectal gland. Hence, it is possible that the haemolymph may instead be involved in the transportation of ME and its analogues. The haemolymph of insects has been known to transport hormones, particularly juvenile hormone-binding proteins (JHBP) (Kanost et al. 1990). The detection of allyl-DMP and CF – which are characterised as male sex and aggregation pheromones in the haemolymph – suggests the involvement of certain types of haemolymph carrier proteins which are known as pheromone-binding proteins (PBP). The evidence for the transport of these male *B. papayae* pheromones by PBP is further shown from the parabiosis experiments where the rectal gland of ME-deprived males sequesters these phenylpropanoids.

*Cis*-DMC was detected in the ME-fed male *B. papayae* 1-3 days post feeding (Nishida et al. 1988b). However, this particular compound was not detected in our experiments. As the rectal phenylpropanoids are found in an increasing order of *cis*-DMC (trace) < CF < allyl-DMP in male *B. papayae*, the concentration of *cis*-DMC present might be too low to be detected.

Relatively larger quantities of allyl-DMP and CF were found in the rectal glands of ME-deprived males in parabiosis than in the haemolymph extract of ME-fed males. This phenomenon is due to the fact that the transportation of ME-metabolites in the haemolymph is constantly regulated and is limited by the amounts of PBP available. Therefore, these metabolites are selectively incorporated and gradually accumulate in the rectal gland.

It is apparent that the strong attraction of sexually mature virgin *B. papayae* males to both ME-fed male haemolymph containing trace amounts of PTC and ME-fed male haemolymph alone, demonstrates the non-inhibitory effects of the antioxidant to the haemolymph. We observed that the use of PTC is necessary to inhibit oxidation of the haemolymph which may otherwise occur rapidly. This will eventually result in the darkening and denaturation of the haemolymph.

From our experimental results, ME as the pheromone precursor was not detected in the gut (excluding the crop and rectal gland) and haemolymph of ME-fed males. ME was also not detected in males of Type A after parabiosis. While excess ME was found in the crop, the traced amounts of allyl-DMP were also detected in the crop extract after males had consumed ME (Hee and Tan, unpublished data). Thus, it is suggested that the crop is involved in the hydroxylation of ME and these ME-metabolites are transported by PBP in the haemolymph to the rectal gland. A study is currently being conducted the objective of which is to further investigate the male sex and aggregation pheromone transport system in male *B. papayae*.

## ACKNOWLEDGEMENTS

We wish to thank R. Nishida (Kyoto University) for providing us with chemicals, S.T. Tan and Y.C. Toong (Universiti Sains Malaysia) for their interest and discussion. We express our gratitude to C.C.H. Khoo, S.L. Wee and A.M. Charles for providing the laboratory cultures and C.L. Yee for technical assistance. This work is financially supported by the Malaysian Government R&D Grant MPKSN/IRPA 01-02-05-6004.

## REFERENCES

Drew, R.A.I. and D.L. Hancock. 1994. The *Bactrocera dorsalis* complex of fruit flies (Diptera: Tephritidae: Dacinae) in Asia. Bull. Entomol. Res. Suppl. No. 2. 1-68.

Hee, A.K.W. and K.H. Tan. 1998. Attraction of female and male *Bactrocera papayae* to conspecific males fed with methyl eugenol and attraction of females to male sex pheromone components. J. Chem. Ecol. 24: 753-764.

Kanost, M.R., J.K. Kawooya, J.H. Law, R.O. Ryan, M.C. Van Heusden and R. Ziegler. 1990. Insect haemolymph proteins. Adv. Insect Physiol. 22: 299-396.

Nishida, R., K.H. Tan, N.H. Lajis, A.M. Sukari, S. Takahashi and H. Fukami. 1988a. Accumulation of phenylpropanoids in the rectal gland of the Oriental fruit fly, *Dacus dorsalis*. Experientia. 44: 534-536.

Nishida, R., K.H. Tan and H. Fukami. 1988b. *cis*-3,4-Dimethoxycinnamyl alcohol from the rectal glands of male Oriental fruit fly, *Dacus dorsalis*. Chem. Exp. 3: 207-210.

Shelly, T.E. and A.M. Dewire. 1994. Chemically mediated mating success in male Oriental fruit flies (Diptera: Tephritidae). Ann. Entomol. Soc. Am. 87: 375-382.

Tan, K.H. 1985. Estimation of native populations of male *Dacus* spp. by Jolly's stochastic method using a new designed attractant trap in a village ecosystem. J. Plant Prot. Trop. 2: 87-95.

Tan, K.H. 1993. Ecohormones for the management of fruit fly pests - Understanding plant-fruit fly-predator interrelationships, pp. 495-503. *In* Proceedings, International Symposium on Management of Insect Pests: Nuclear and Related Molecular and Genetic Techniques, IAEA, Vienna, Austria.

Tan, K.H. 1996. Chemical ecology of fruit flies, *Bactrocera* species - An update, pp. 36- 46. *In* Problems and Management of Tropical Fruit Flies. T.H. Chua and S.G. Khoo (eds.). Kuala Lumpur, Malaysia.

Tan, K.H. and S.L. Lee. 1982. Species diversity and abundance of *Dacus* (Diptera: Tephritidae) in five ecosystems of Penang, West Malaysia. Bull. Entomol. Res. 72:709-716.

Tan, K.H. and R. Nishida. 1996. Sex pheromone and mating competition after methyl eugenol consumption in *Bactrocera dorsalis* complex, pp. 147-153. *In* Fruit fly pests: A world assessment of their biology and management. B.A. McPheron and G.J. Steck (eds.). St. Lucie Press, Delray Beach, Florida, USA.

Tan, K.H., L.G. Kirton and M. Serit. 1987. Age response of *Dacus dorsalis* Hendel to methyl eugenol in a) wind tunnel and b) trap set in a village, and its implication in population estimation, pp. 425-432. *In* Fruit Flies. A.P. Economopoulos (ed.). Proceedings Second International Symposium on Fruit Flies. G. Tsiveriotis Ltd.; Athens.

# PART III.

## APPLICATION OF AREA-WIDE CONCEPT TO FRUIT FLIES

### H) SEMIOCHEMICALS AND FIELD MONITORING OF FRUIT FLIES

# Roles of Semiochemicals in Mating Systems: A Comparison Between Oriental Fruit Fly and Medfly

*Ritsuo Nishida*[1], *Todd E. Shelly*[2], *Kenneth Y. Kaneshiro*[2] *and Keng-Hong Tan*[3]

[1] Laboratory of Chemical Ecology, Faculty of Agriculture, Kyoto University, Sakyo-ku, Kyoto 606-011, Japan
[2] Hawaiian Evolutionary Biology Program, University of Hawaii, USA
[3] School of Biological Sciences, Universiti Sains Malaysia, 11800 Penang, Malaysia

## INTRODUCTION

Males of tephritid fruit fly species show strong affinity to specific chemicals produced by plants. Amongst the economically important species in the Asian Pacific area, methyl eugenol acts as a potent attractant for males of the Oriental fruit fly, *Bactrocera dorsalis* (Hendel), and several other species within the dorsalis species complex (e.g., *B. papayae* Drew & Hancock, *B. carambolae* Drew & Hancock, etc.), cuelure [4-(4-acetoxyphenyl)-2-butanone] and the naturally occurring deacetyl derivative (raspberry ketone) act as specific attractants for flies such as the melon fly, *B. cucurbitae* (Coquillett) and the Queensland fruit fly, *B. tryoni* (Froggatt) (Metcalf 1990). These attractants have been successfully used as baits in mass trapping for monitoring populations during eradication programmes for these pests (Chambers 1977, Koyama et al. 1984). Likewise, trimedlure has been developed as a synthetic attractant for males of the Mediterranean fruit fly (Medfly), *Ceratitis capitata* (Wied.), while α-copaene has been known to be a naturally occurring attractant for the species. For most tephritids, however, the biological function of male attraction to these natural or artificial compounds remains unclear. Recent studies (Nishida et al. 1988 1997, Nishida and Fukami 1990, Tan 1993, Tan and Nishida 1996) have shown that males of *B. dorsalis* and related species ingest these compounds from natural sources, selectively incorporate them into the rectal glands, and use them to synthesise the sex pheromone and allomone. It appears that similar chemical compounds, when ingested, may provide pheromonal precursors in the melon fly as well (Nishida et al. 1993, Shelly and Villalobos 1995). In contrast, Medfly males do not feed on the source of chemical attractant. According to our observations, α-copaene strongly affected the courtship behaviour of the Medfly, which suggests that these natural compounds may possibly be involved in the formation of leks and the mating behaviour in natural populations. In this paper, we compare the roles of specific male attractants in the mating system of the Oriental fruit fly and the Medfly.

## ORIENTAL FRUIT FLY – ASSOCIATION WITH METHYL EUGENOL

Methyl eugenol (**1**) is a phenylpropanoid compound widely distributed in plants, and males in the *B. dorsalis* complex frequently congregate on "lure" plants such as *Cassia fistula* L., *Ocimum sanctum* L. and *Spathiphyllum cannaefolium* Schott. The males persistently lick the surface of the plant part that emit this compound (Kawano et al. 1968, Shah and Patel 1976, Lewis et al. 1988). Foraging behaviour can be observed in early daytime but such behaviour ceases just prior to sunset. Feeding on methyl eugenol, laboratory-reared males of *B. dorsalis* and *B. papayae* converted **1** into 2-allyl-4,5-dimethoxyphenol (**2**) and *trans*-coniferyl alcohol (**3**) and selectively stored the metabolites in the rectal glands in significant quantities (Nishida et al. 1988) (Figure 1). Thus the wild males must have consumed methyl eugenol from certain plant sources during foraging, and sequester the metabolites (**2** and **3**) in their body in varying quantities. The quantitative variation may be dependent upon the availability of methyl eugenol in their natural habitat and also the physiological age of the flies (Nishida et al. 1988). Males of *B. carambolae* was found to convert **1** selectively to **3** and stored it in the rectal gland together with an endogenous volatile, 6-oxo-1-nonanol (Tan and Nishida 1996). 1-(3,4-Dimethoxyphenyl)-2-propen-1-ol was identified as one of the components found in the rectal glands of *B. umbrosa* (Fabricius) males that fed on methyl eugenol (unpublished).

Rectal glands found in *Bactrocera* species have been suspected to be a reservoir of sex pheromones, and the males emit a "smoke" containing a series of compounds (fatty acids, trisodium phosphate, spiroketáls, alkylamides and other minor volatiles) at dusk, regardless of whether they have fed on natural or artificial sources of phenylpropanoid compounds (Ohinata et al. 1982, Perkins et al. 1992). The smoke production among the *Bactrocera* species is associated with the vigorous wing-fanning action of the male to atomise the sex pheromone emitted by the rectal gland (Kuba and Sokei 1988). After the incorporation of methyl eugenol, males of *B. dorsalis* and *B. papayae* released its metabolites in the air. In the experiment with *B. papayae,* compound **2** was recovered from the aeration extract only during sunset (17:00-19:00 pm) which coincided with the period when the males congregated and attracted females (Nishida and Fukami 1990).

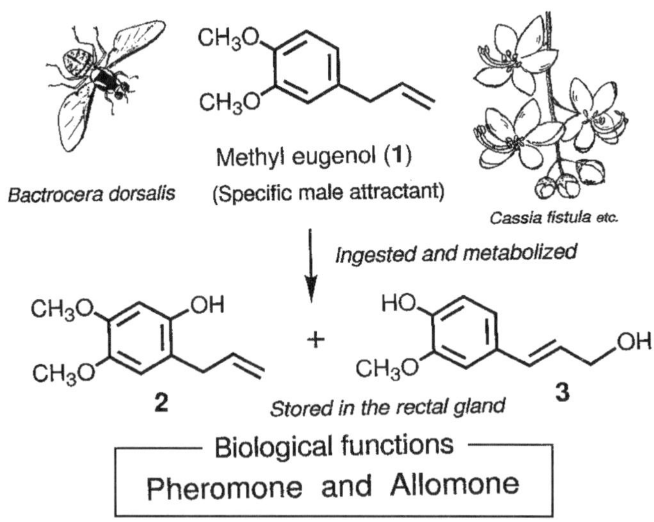


Figure 1. Acquisition of methyl eugenol and sequestration of its metabolites in the rectal glands by males of *Bactrocera dorsalis*.

Males of *B. dorsalis* that had fed on methyl eugenol were shown to have a mating advantage over the unfed males in the mating competition bioassay (Shelly and Dewire 1994, Tan and Nishida 1996). It was noted that the frequency of wing-fanning was significantly increased after ingestion of methyl eugenol, suggesting a physiological change in males directly affected by methyl eugenol feeding. On the other hand, the virgin females were strongly attracted to a synthetic mixture of **2** and **3**, indicating the pheromonal function of the rectal phenylpropanoids in the courtship sequence (Tan and Nishida 1996, Nishida et al. 1997). Females were stimulated to feed on the metabolites and occasionally extruded their ovipositor toward the chemical source. Such behavioural responses were seen exclusively during the period of sexual activity (i.e., at dusk between 17:00-20:00 pm) as shown in Figure 2. Similar behaviour was seen towards intact males scented with methyl eugenol metabolites, in that mating often took place after extrusion of the female's ovipositor. Thus, these compounds appeared to act as a short-range attractant and arrestant pheromone to elicit female acceptance of male courtship (Figure 1).

Enhancement of the mating success in *B. dorsalis* has also been shown by intact feeding on flowers of *C. fistula* and *Fagraea berteriana* A. Gray which produce methyl eugenol or derivatives of coniferyl alcohol (Nishida et al. 1997). Acquisition of phenylpropanoid compounds by *B. dorsalis* and related species appears to have evolved within the context of sexual selection, particularly female preference for males "scented" with chemicals derived from flowers. Female preference could reflect a runaway process; if the ability to acquire phenylpropanoid compounds has a heritable component, females mating with scented males may be increasing the likelihood that their male offspring will successfully locate and collect these compounds and thereby enjoy a mating advantage (Nishida et al. 1997).

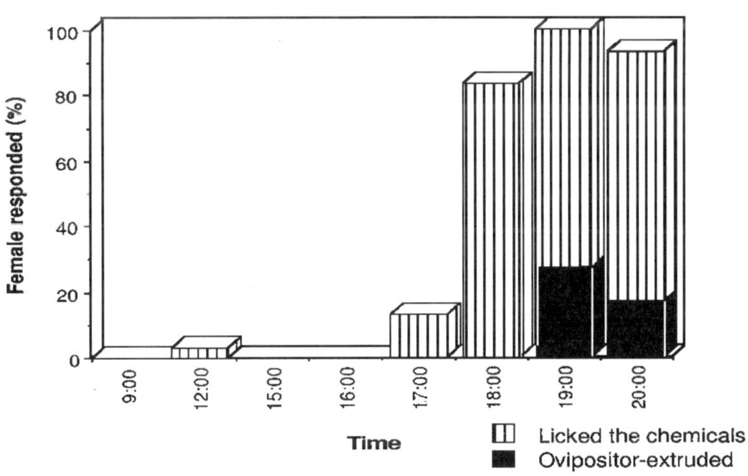


Figure 2. Female responses to synthetic rectal phenylpropanoids (2 + 3, 15 μg each/10mm$^2$ filter paper) at various time of a day *(Bactrocera dorsalis* virgin females, 21 day-old, N=30, 3 replicates).

## MEDITERRANEAN FRUIT FLY – ASSOCIATION WITH α-COPAENE

α-Copaene (Figure 3) is a sesquiterpene hydrocarbon found as a natural attractant for males of *Ceratitis capitata* initially from the essential oil of *Angelica archangelica* (Fornasiero et al. 1969). The compound is distributed in a wide diversity of plants, including several important hosts such as orange, guava, papaya, and mango (Teranishi et al. 1987, Warthen and McInnis 1989). Similar to the affinity of the Oriental fruit fly males to methyl eugenol, Medfly males are strongly attracted to α-copaene. However, in contrast to the behaviour of the Oriental fruit fly males which land and feed on the source of methyl eugenol, Medfly males approach the chemical source and perch motionless nearby without feeding.

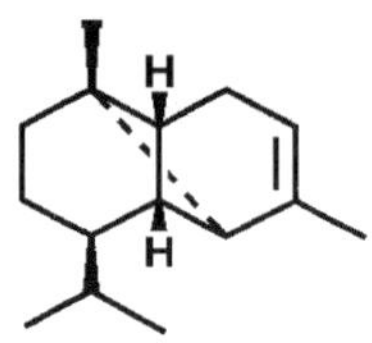


Figure 3. (+)-α-Copaene.

The mating system of *C. capitata* is based on what is referred to as lek behaviour which is characterised by an aggregation of male territories within the canopy of the tree to which females are attracted for mating (Prokopy and Hendrichs 1979). Each male defends his territory against rival intruders and everts his rectal epithelium from which the volatile pheromone is secreted (Arita and Kaneshiro 1986, Jang et al. 1994). Females enter the lek system and copulation occurs after a series of elaborated actions displayed by the males including wing-fanning and head oscillations. Although males use a variety of tree species, they appear to prefer *Citrus* spp. for lek formation (Hendrichs and Hendrichs 1990). Gas chromatographic analysis revealed the presence of small quantities of α-copaene in citrus leaves. It was also found that hydrocarbon fractions of citrus leaf extracts rich in α-copaene strongly attracted both sexes of *C. capitata* in a short-range bioassay. These preliminary observations prompted us to examine a possible involvement of α-copaene in their mating system.

Although (+)-α-copaene (predominantly (+)-form obtained from angelica seed oil) was attractive exclusively to males in outdoor cages where equal numbers of males and females were released, indoor observations showed that the compound provoked specific response from females as well. When virgin females were brought into the vicinity of α-copaene (few cm from its source), they are strongly attracted and arrested within the chemical atmosphere. The dose-response of males and females to α-copaene was separately examined by a short-range bioassay. Figure 4 clearly shows that not only males but also females responded to α-copaene, in which the response of females were relatively low compared to that of males. The behavioural patterns of males and females, however, were entirely different, in that females displayed "pseudo-male" calling posture (Arita and Kaneshiro 1983), raising the tip of their abdomen often accompanied by vigorous wing-fanning action and homosexual mounting of other females under exposure to α-copaene, as calling males would exhibit towards females. At a 1 μg dose, most of females responded in an agitated manner while displaying pseudo-male behaviour.

Males were quickly attracted to artificial plastic leaves treated with low concentrations of α-copaene (5 ng/$cm^2$) in an indoor cage. Although some occasional homosexual mountings were observed, most males perched motionless for more than 1 hour. Then they began to display calling behaviour by everting their rectal pheromone gland and defending the undersurface of individual leaves as their territories as might be observed during lek formation. When both sexes were released in a cage containing α-copaene-treated and untreated leaf models, matings took place exclusively on the treated leaves (unpublished). These results suggested that α-copaene serves as a chemical cue to facilitate orientation of the flies to the mating site.

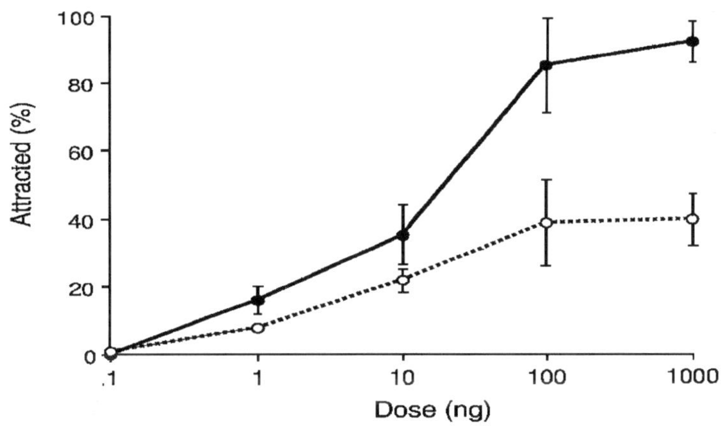


Figure 4. Attractant activity of (+)-α-copaene to virgin males (solid line) and females (broken line) of *Ceratitis capitata* at various doses (percent fly responded ± s.d., N = 40, 3 replicates).

α-Copaene is found either as (+)- or (-)-forms, or as a mixture of both enantiomers in plants. The (+)-form of α-copaene is known to be the potent attractant, whereas the (-)-enantiomer is inactive or much less active (Jacobson et al. 1987, Flath et al. 1994). It is necessary to determine the enantiomeric composition in the host plants (Teranishi et al. 1987, Takeoka et al. 1990) along with the actual release of the active enantiomer in the natural lek sites to support our speculation. The affinity of the Mediterranean fruit fly to α-copaene may have evolved partly because this brings them into close proximity with the host plants suitable for oviposition and thus to ensure the offspring quality. However, the mating system seems to involve much more complex components in relation to the male sex pheromone produced within the rectal epithelium (Jang et al. 1994). Further work is needed to enable better understanding of the adaptive nature of these compounds in the Medfly's biology in order that we may be able to develop more effective control technologies for this important agricultural pest.

## REFERENCES


Arita, L.H. and K.Y. Kaneshiro. 1983. Pseudomale courtship behaviour of the female Mediterranean fruit fly, *Ceratitis capitata* (Wiedemann). Proc. Hawaii. Entomol. Soc. 24: 205-210.

Arita, L.H. and K.Y. Kaneshiro. 1986. Structure and function of the rectal epithelium and anal glands during mating behaviour in the Mediterranean fruit fly male (*Ceratitis capitata*). Proc. Hawaii. Entomol. Soc. 26: 27-30.

Arita, L.H. and K.Y. Kaneshiro. 1989. Sexual selection and lek behaviour in the Mediterranean fruit fly, *Ceratitis capitata* (Diptera: Tephritidae). Pacific Sci. 43: 135-143.

Flath, R.A., R.T. Cunningham, T.R. Mon and J.O. John. 1994. Male lures for Mediterranean fruit fly (*Ceratitis capitata* Wied.): Structural analogs of α-copaene. J. Chem. Ecol. 20: 2595-2609.

Fornasiero, U., A. Guitto, G. Caporale, F. Baccichetti and L. Musajo. 1969. Identificazion della sostanza attrattiva per i maschi della *Ceratitis capitata*, contenunuta nell'olio essenziale dei semi di *Angelica archangelica*. Gazz. Chim. Ital. 99: 700-710.

Hendrichs, J. and M.A. Hendrichs. 1990. Mediterranean fruit fly (Diptera: Tephritidae) in nature: location and diet pattern of feeding and other activities on fruiting and nonfruiting hosts and nonhosts. Ann. Entomol. Soc. Am. 83: 632-641.

Jang, E.B., D.M. Light, R.G. Binder, R.A. Flath, and L.A. Carvalho. 1993. Attraction of female Mediterranean fruit flies to the five major components of male-produced pheromone in a laboratory flight tunnel. J. Chem. Ecol. 20: 9-20.

Kawano, Y., W.C. Mitchell and H. Matsumoto. 1968. Identification of the male Oriental fruit fly attractant in the golden shower blossom. J. Econ. Entomol. 61: 986-988.

Koyama, J., T. Teruya and K. Tanaka. 1984. Eradication of the Oriental fruit fly (Diptera: Tephritidae) from the Okinawa Islands by a male annihilation method. J. Econ. Entomol. 77: 468-472.

Kuba, H., and Y Sokei. 1988. The production of pheromone clouds by spraying in the melon fly, *Dacus cucurbitae* Coquillett (Diptera: Tephritidae) J. Ethol. 6: 105-110.

Lewis, J. A., J.C. Moor, M.T. Fletcher, R.A. Drew and W. Kitching. 1988. Volatile compounds from the flowers of *Spathiphyllum cannaefolium*. Phytochem. 27: 2755-1757.

Metcalf, R.L. 1990. Chemical ecology of *Dacinae* fruit flies (Diptera: Tephritidae). Ann. Entomol. Soc. Am. 83: 1017-1030.

Nishida, R. and H. Fukami. 1990. Sequestration of distasteful compounds by some pharmacophagous insects. J. Chem. Ecol. 16: 151-164.

Nishida, R., K.H. Tan, M. Serit, N.L. Lajis, A.M. Sukari, S. Takahashi and H. Fukami. 1988. Accumulation of phenylpropanoids in the rectal glands of males of the Oriental fruit fly, *Dacus dorsalis*. Experientia 44: 534-536.

Nishida, R., O. Iwahashi and K.H. Tan. 1993. Accumulation of *Dendrobium superbum* (Orchidaceae) fragrance in the rectal glands by males of the melon fly, *Dacus cucurbitae*. J. Chem. Ecol. 19: 713-722.

Nishida, R., T.E. Shelly and K.Y. Kaneshiro. 1997. Acquisition of female-attracting fragrance from a Hawaiian lei flower, *Fagraea berteriana*, by males of the Oriental fruit fly. J. Chem. Ecol. 23: 2275-2285.

Ohinata, K., M. Jacobson, R.M. Kobayashi, D.L. Chambers, M.S. Fujimoto and H. Higa. 1982. Oriental fruit fly and melon fly: biological and chemical studies of smoke produced by males. J. Environ. Sci. Health A17: 197-206.

Perkins, M.V,, M.T. Fletcher, W. Kitching, R.A. Drew and J.C. Moor. 1990. Chemical studies of rectal gland secretions of some species of *Bactrocera dorsalis* complex of fruit flies (Diptera: Tephritidae). J. Chem. Ecol. 16: 2475-2487.

Prokopy, R.J. and J. Hendrichs. 1979. Mating behaviour of *Ceratitis capitata* on a field-caged host tree. Ann. Entomol. Soc. Am. 72: 624-648.

Shah, A.H. and R.C. Patel. 1975. Role of tulsi plant (*Ocimum sanctum*) in control of mango fruit fly, *Dacus correctus* Bezzi (Tephritidae: Diptera). Curr. Sci. 45: 313-314.

Shelly, T.E. and E.M. Villalobos. 1995. Cue lure and the mating behaviour of male melon flies, *Bactrocera cucurbitae* (Diptera: Tephritidae). Fla. Entomol. 78: 473-482.

Shelly, T.E. and A.L.M. Dewire. 1994. Chemically mediated mating success in male Oriental fruit flies (Diptera: Tephritidae). Ann. Entomol. Soc. Am. 87: 375-382.

Takaoka, G., R.A. Flath, T.R. Mon, R.G. Buttery, R. Teranishi, M. Guntert, R. Lautamo and J. Szejtli. 1990. Further applications of permethylated β-cyclodextrin capillary gas chromatographic columns. J. High Resolution Chromatogr. 13: 202-206.

Tan, K.H. 1993. Ecohormones for the management of fruit fly pests – understanding plant-fruit fly-predator interrelationships. pp. 495-503. *In* Management of Insect Pests: Nuclear and Related Molecular and Genetic Techniques. IAEA-SM 327, Vienna, Austria.

Tan, K.H. and R. Nishida. 1996. Sex pheromone and mating competition after methyl eugenol consumption in the *Bactrocera dorsalis* complex. 147-153 *In* Fruit Fly Pests. B.A. MacPheron and G. J. Steck (eds.). St. Lucie Press, Delray Beach, Florida, USA.

Teranishi, R., R.G. Buttery, K.E. Matsumoto, D.J. Stern, R.T. Cunningham, and S. Gothilf. 1987. Recent developments in the chemical attractants for tephritid fruit flies. ACS Symp. Ser. 330: 431-438.

Warthen, J.D. and D.O. McInnis. 1989. Isolation and identification of male Medfly attractive components in *Litchi chinensis* stems and *Ficus spp.* stem exudates. J. Chem. Ecol. 15: 1931-1946.

# Pilot Experiment to Control Medfly, *Ceratitis capitata* (Wied.) (Diptera: Tephritidae) Using Mass Trapping Technique in a Cherimoyer *(Annona cherimola Mill*er) Orchard

*J.P. Ros[1], I. Escobar[2], F.J. Garcia Tapia[2] and G. Aranda[3]*

[1] Inst. Nal. Invest. Agrarias. José abascal 56, Madrid, Spain
[2] Caja Rural de Granada, Granada, Spain
[3] Del. Prov. Agric. Dep. Sanidad Vegetal, Malaga, Spain

## INTRODUCTION

Recently, as a result of assays coordinated by the International Atomic Energy Agency (and participated by Spain), it was decided that ammonium acetate, putrescine and trimethylamine be included in low release polyethylene bag dispensers (Biolure, Consep, Co) as the Mediterranean fruit fly (Medfly) females were greatly attracted by them. These synthetic substances are placed in traps at the frequency of one and a half months to two months. If Tephry traps are used, one DDVP wafer (containing Vapona or a similar substance) is enough to kill the flies that enter them. These attractants make it unnecessary to replenish the liquids in the Mcphail traps and remain effective throughout the entire fruit season.

The Caja Rural de Granada (a bank of farmers) encourages all techniques that increase crop profits for farmers. It is even more desirable if such crop profitability can be achieved without the application of insecticides which may result in the likely presence of toxic residues.

In the light of the results achieved by the attractants with regard to female Medflies, the Caja Rural de Granada, together with the National Institute of Agricultural Research Counselling (Agricultural Department), performed an experiment on mass trapping to confirm whether it is possible to protect a fruit plantation with the application of this biotechnical method.

Due to the great risk of this initial experiment, the farmers were free to use insecticides as often as they deemed necessary so that no damages due to any plague could be blamed on the experiment.

## MATERIALS AND METHODS

### Location

A Cherimoyer tree farm called "Cortijo San José", which was situated in Puntalón, a part of Motril (Granada, Spain), was selected for the experiment.

The farm has a size of 1.8 h with 289 trees (7.3 x 7.3 m) which are 17 years old. These trees are of the "Fino del Jete" variety.

### Insecticide Treatments

The insecticide treatments deployed were similar for the two sections the farm was divided into.

Bait sprays: solution in water of Buminal (0.8%) and malation (0.6%) was sprayed over approximately one metre of land.

Frequency: spraying was done on 19 August, 25 August, 2 September, 17 September, 30 September, 8 October, 22 October, 7 November and 22 November in 1997 (a total of nine treatments).

### Traps

The type of traps that was selected for this pilot experiment was one which had produced the best results: the Tephry traps. The baits chosen were the three synthetic components cited above (Biolure, Consep and Co) as well as the protein hydrolysate - nulure (9% dissolved in water) to fill the traps. To estimate the fly population at any one time, the trimedlure plug was used as an attractant for the male flies.

### Experimental Plots

The orchard was divided into two plots of approximately the same areas. One was treated with traps (mass trapping) while the other was used as a control. In each of the two plots, 75 traps that were baited with synthetic attractants – 74 with nulure and one with trimedlure – were placed, one per tree. In the control area, 5 traps were placed. Two of these were baited with nulure and one with trimedlure. Similarly, 5 traps were placed in the mass trapping area. The nulure traps were filled with liquid once a week and refilled once every 15 days. The synthetic attractants were replaced once every two months and the trimedlure once a month. All the traps were placed on the first day of August and removed in December, following the end of the harvest.

### Fly Population Evaluation

To evaluate the entire fly population during the duration of the experiment, the control traps located in both areas were examined every week.

### Fruit Harvest

The fruit harvest was carried out as scheduled. Harvesting commenced on 20 October and continued on 4, 11, 18 November and 1 December. The fruits were collected from the trees and placed in square boxes each with a holding capacity of 12 fruits.

### Fruit Sampling

Fruit sampling took place on the first day of harvesting to ascertain possible damage caused by the *Ceratitis* flies. The fruit pickers placed filled boxes at the bottom of each tree while the person responsible for the sampling (an expert technician) examined the three pieces of fruits from the three corners opposite him. If a fruit was found to have no apparent damage from the *Ceratitis* flies, it was replaced in the box. However, if it was found with an apparent ovipuncture, it was separated. At the end of the picking and sampling, a representative sample of separated fruits was kept at the environmental chambers in the laboratory where the fly punctures were accelerated. After seven days, an actual evaluation of the damage was made.

## RESULTS

The results were evaluated in two sections. The first represented the results of the fly population studies conducted during the experiment. Fly capture data in mass trapping and control areas using three types of bait are shown in Table 1. The second section shows the data resulting from the sampling made during the fruit picking and the subsequent treatment that was conducted to confirm the damages caused by the infestation (Table 2). Table 3 shows the percentage of damages and the number of punctures that occurred per 100 fruits.

Table 1. Number of flies (mean of two traps) captured in tephry traps baited with different attractants in the mass trapping and control areas.

| Date in 1997 | Trimedlure | | Nulure | | Synthetic Attractants | |
|---|---|---|---|---|---|---|
| | Mass trapping | Control | Mass trapping | Control | Mass trapping | Control |
| 11 Aug | 34 | 31 | 0 | 2.5 | 1 | 0 |
| 18 Aug | 23 | 16 | 0.5 | 5.5 | 4.5 | 7.5 |
| 19 Aug | Insecticide | Treatment | | | | |
| 25 Aug | 41 | 42 | 2 | 1.5 | 5 | 11.5 |
| 25 Aug | Insecticide | Treatment | | | | |
| 1 Sept | 31 | 36 | 3.5 | 11 | 4 | 15.5 |
| 2 Sept | Insecticide | Treatment | | | | |
| 9 Sept | 50 | 68 | 4 | 6.5 | 22.5 | 29 |
| 16 Sept | 41 | 45 | 0 | 2 | 7.5 | 15.5 |
| 17 Sept | Insecticide | Treatment | | | | |
| 22 Sept | 20 | 26 | 0 | 4.5 | 7 | 7 |
| 29 Sept | 46 | 30 | 1.5 | 0.5 | 3.5 | 6 |
| 30 Sept | Insecticide | Treatment | | | | |
| 6 Oct | 78 | 59 | 1.5 | 4.5 | 10 | 18 |
| 7 Oct | Insecticide | Treatment | | | | |
| 15 Oct | 117 | 127 | 7 | 3 | 8 | 25 |
| 20 Oct | 122 | 98 | 0.5 | 2.5 | 4 | 15.5 |
| 22 Oct | Insecticide | Treatment | | | | |
| 27 Oct | 237 | 175 | 1 | 10.5 | 9 | 35.5 |
| 4 Nov | 186 | 149 | 0 | 1.5 | 2 | 11 |
| 7 Nov | Insecticide | Treatment | | | | |
| 10 Nov | 138 | 75 | 0 | 0 | 10.5 | 6.5 |
| 17 Nov | 68 | 46 | 0 | 1 | 0 | 4.5 |
| 22 Nov | Insecticide | Treatment | | | | |
| 24 Nov | 35 | 29 | 0 | 0 | 2 | 3 |
| 2 Dec | 55 | 26 | 0 | 0.5 | 2.5 | 2 |

Table 2. Results of fruit sampling of damage in the field (apparent) and in the laboratory (real) caused by Medfly.

**MASS TRAPPING AREA**

| Date in 1997 | FIELD | | | LABORATORY | | | |
|---|---|---|---|---|---|---|---|
| | No. of Boxes/ Fruits Harvested | No. of Fruits Sampled | No. of Fruits with Apparent Damage | No. of Fruits Checked | No. of Punctures | No. of Fruits without Damage | No. of Fruits with Damage |
| 20 Oct | 166/1992 | 504 | 81 | **33** | **6** | 30 | **3** |
| 28 Oct | 151/1812 | 453 | 16 | **13** | **13** | 11 | **2** |
| 4 Nov | 169/2028 | 507 | 13 | **13** | **18** | 8 | **5** |
| 11 Nov | 157/1884 | 471 | 8 | **8** | **9** | 4 | **4** |
| 18 Nov | 172/2064 | 516 | 10 | **10** | **10** | 4 | **6** |
| 1 Dec | 194/2328 | 582 | 11 | **11** | **15** | 5 | **6** |

**CONTROL AREA**

| Date in 1997 | FIELD | | | LABORATORY | | | |
|---|---|---|---|---|---|---|---|
| | No. of Boxes/ Fruits Harvested | No. of Fruits Sampled | No. of Fruits with Apparent Damage | No. of Fruits Checked | No. of Punctures | No. of Fruits without Damage | No. of Fruits with Damage |
| 20 Oct | 194/2328 | 582 | 97 | | | | |
| 28 Oct | 162/1944 | 486 | 43 | **17** | **23** | 11 | **6** |
| 4 Nov | 76/912 | 457 | 33 | **20** | **32** | 7 | **13** |
| 11 Nov | 142/1704 | 426 | 32 | **25** | **57** | 9 | **14** |
| 18 Nov | 212/2544 | 636 | 37 | **23** | **42** | 10 | **13** |
| 1 Dec | 183/2196 | 549 | 11 | **11** | **21** | 5 | **6** |

Table 3. Percentage of damage and punctures/100 fruits in mass trapping and control areas.

| Date in 1997 | % damage Mass trapping area | % damage Control area | Punctures/ 100 fruits Mass trapping area | Punctures/ 100 fruits Control area |
|---|---|---|---|---|
| 20 Oct | **1.4** | | **2.9** | |
| 28 Oct | **0.5** | **3.1** | **3.5** | **2** |
| 4 Nov | **0.9** | **4.6** | **3.5** | **11.5** |
| 11 Nov | **0.8** | **4.2** | **1.9** | **17.2** |
| 18 Nov | **1.1** | **3.3** | **1.9** | **10.6** |
| 1 Dec | **1.0** | **1.0** | **2.5** | **3.8** |

Due to a misunderstanding, the first sample of damaged fruits from the control zone was not available for evaluation. This means that the figures for the overall infestation in the control area appear higher.

## CONCLUSION

In the plots where traps were more effective, puncture damages were five times less. The results of the experiment were considered positive as the damages were reduced to minimum levels (0.85 of the total crop). If bait sprays are not effective, only a mass trapping technique would be effective in the control of Medflies in fruit orchards. However, it is extremely difficult to evaluate if the insecticide treatments are effective and if so, the number of treatments that are necessary to control the fly population. It was also difficult to estimate if all female flies in the plantation could be captured in the absence of insecticide treatments. The question of the limits to the legal usage of insecticide treatments has also to be addressed.

## REFERENCES

Epsky, N., R. Heath, A. Guzman and W. Meyer. 1995. Visual cue and chemical cue interactions in a dry trap with food-based synthetic attractant for *C. capitata* and *Anastrepha ludens* (Diptera:Tephritidae). Environ. Entomol. 24: (6) 1387-1395.

Heath, R., N. Epsky, A. Guzman and B. Dueben. 1995. Development of a dry plastic insect trap with food-based synthetic attractant for the Mediterranean an Mexican fruit flies (Diptera:Tephritidae). J. Econ. Entomol. 88: (5). 1307-1315.

Ros, J.P., C. Garijo, L. Navarro and E. Castillo. 1996. Ensayos de campo con un nuevo atrayente de hembras de la mosca de la fruta *Ceratitis capitata* Wied. (Diptera: Tephritidae). Bol. San. Veg. Plagas no. 32. pp151-157. 1996 Ministerio de Agricultura España. .

Ros, J.P., E. Castillo, J. Crespo, Y. Latorre, P. Martín, M.A. Miranda, P. Moner and C. Sastre. 1997. Evaluación en campo de varios atrayentes sintéticos para la captura de hembras de la mosca mediterranea de la fruta.*Ceratitis capitata* Wied. Bol. San. Veg. Plagas. 23 (3): 393-402.

Ros, J.P., E. Wong, V. Castro and E. Castillo. 1997. La Trimetilamina: un efectivo potenciador de los atrayentes putrescina y acetato amónico para capturar las hembras de la mosca mediterranea de la fruta *Ceratitis capitata* Wied.(Diptera:Tephritidae). Bol. San. Veg. Plagas. 23 (4):515-521.

# PART III.

## APPLICATION OF AREA-WIDE CONCEPT TO FRUIT FLIES

## I) BEHAVIOUR OF FRUIT FLIES

# Behaviour and Chemical Ecology of *Bactrocera* Flies

*Keng-Hong, Tan*

School of Biological Sciences, Universiti Sains Malaysia, 11800 Penang, Malaysia

## INTRODUCTION

Many species of tephritid fruit flies have gained global status as pests of economic importance in fruit and vegetable cultivation. *Bactrocera* species are no exception. Males of most *Bactrocera* species are known to be attracted to either methyl eugenol (ME) or cuelure(CL)/raspberry ketone(RK) (Fletcher 1987, Metcalf 1987 and 1990). At the turn of the century, male fruit flies of both *B. diversa* (Coquillett) (formerly *Dacus diversus*) and *B. zonata* (Saunders) (formerly *Dacus zonatus*) were first observed to have a strong attraction to citronella oil (Howlett 1912). The chemical responsible for the attraction was discovered to be ME (Howlett 1915). Since that discovery, ME has been used successfully in monitoring and male annihilation programmes (Steiner et al. 1965), in estimating native population density and survival rates (Tan 1985, Tan and Jaal 1986, Tan and Serit 1994), and movements between ecosystems (Tan and Serit 1988). The unique characteristic of male *Bactrocera* flies is that not only are they strongly attracted to certain male attractants but they compulsively feed on them. This phenomenon was not fully understood (Fletcher 1987, Metcalf 1990, Metcalf and Metcalf 1992) until the early 1990s. Certain male attractants play a very important role in the behaviour and chemical ecology of *Bactrocera* flies, and aid in the understanding of the intricate interrelationships between plants, fruit flies and their predators (Tan 1993).

Every organism actively or passively secretes chemicals which act as a characteristic "body odour". This "body odour" affects behaviour of individuals, both intraspecies and interspecies, within a community and it is here referred to as ecomone (ecohormone) under a large group of semiochemicals (behaviour modifying chemicals). To understand the different roles of chemicals acting as a medium in communication between individuals and affecting behaviour of a receptive organism, a brief classification of semiochemicals is essential. Semiochemicals are divided into ecomone and para-ecomone, the former is released naturally into the environment, the latter is not. An ecomone with intraspecies activity is known as a pheromone. One with interspecies activity is generally grouped under allelochemicals. It is specifically known as: 1) an allomone when it benefits the releaser with detrimental effect on the receiver, 2) a kairomone when it benefits the receiver with detrimental effect on the releaser, 3) a synomone when it benefits both the releaser and receiver, or 4) an apneumone when released from dead or decaying material caused by microbial action. A para-ecomone may be either a constituent of an organism or a synthetic chemical not released naturally. It should be emphasised that a chemical can be an ecomone and a para-ecomone and, as an ecomone, may act as a pheromone as well as an allomone or a kairomone.

The study of an organism's ecomone in relation to the environment, interaction between individuals belonging to the same and/or different species, and how it affects behaviour constitutes the bulk of chemical ecology. Ecomones in applied entomology may be exploited as agents for 1) insect pest surveillance and monitoring, 2) trapping insect in population estimation or as an intervention technique such as the area-wide male annihilation technique, and 3) understanding and disrupting insect communication in a pest control or management programme. This paper presents an update of the behaviour within the context of chemical ecology of *Bactrocera* flies which is crucial in the understanding the flies' role in the complex communal interrelationships within Malaysian agro- and natural ecosystems as previously presented (Tan 1993).

## Para-ecomone

This is any synthetic compound or chemical that exists as a constituent of an organism and is not secreted naturally into the environment.

### *Male attractant/lure*

Synthetic male attractants (lures) have been used widely for ecological studies, population suppression and male annihilation of fruit flies. For most species of *Bactrocera* in the tropics, sexually mature males are attracted either to ME or CL/RK. These attractants are available commercially. Hitherto, only CL is not found to occur naturally. Initially, a male fruit fly shows a high level of excitement before taking flight when the appropriate attractant is perceived. On landing close to the chemical source, it actively searches for and then feeds on the chemical. A male *B. papayae* Drew and Hancock can consume 0.3-4 μl of neat ME (Tan and Nishida 1998). *B. papayae* and *B. dorsalis* (Hendel) males are more sensitive to ME than *B. carambolae* Drew and Hancock males. To date, in Southeast Asia, only males of *B. arecae* Hardy and Adachi are not attracted to any known male attractant (Chey and Tan 1985).

It was shown that male *B. papayae*, when exposed to ME, produces an antennal polypeptide which induces a series of behavioural patterns. The polypeptide induces feeding behaviour, such as probing and frequent extrusion of proboscis when injected into a male never exposed to ME. Injecting the antennal polypeptide, extracted from *B. papayae*, into a male *B. carambolae* induced the same behaviour (Lim, Tan and Tan, unpublished data).

### *Para-allomone*

Plants produce a diversity of compounds to deter phytophagous organisms. These plant constituents, not secreted naturally, act as a plant para-allomone by being a toxicant, deterrent, antifeedant, repellent, irritant and/or growth regulator. Plants containing ME play a role as a plant kairomone in fruit fly ecology (Metcalf 1990, Metcalf and Metcalf 1992). However, ME from plants does not act as a plant kairomone but as either a plant para-allomone or synomone (when secreted naturally discussed below). This is because ME or RK attracts male fruit flies in a manner which does not cause or directly do damage/harm to plants.

Essential oils from >200 plant species belonging to 32 families have been reported to contain ME in varying quantities (unpublished data from literature search). Several varieties of *Ocimum sanctum* (L.) Ridley and *O. basilicum* L. (Labiatae) contain different amounts of ME in their essential oils. Seven varieties of the former species store large quantities of ME and eugenol; these can be divided into eugenol and ME chemotypes (Nurdijati, Toong and Tan, unpublished data). A healthy *O. sanctum* plant is normally free from insect attack and does not attract fruit flies. When any part of the plant is damaged, it attracts male fruit flies. ME, found as a plant constituent, is believed to deter phytophagous insects because it is toxic to many insect species thereby acting as a repellent. Rat food pellets dipped in ME-acetone solution (1-10 mg/l) were avoided by the starved American cockroach, *Periplaneta americana* L. which fed on non-ME treated pellets (Tan, unpublished data). Therefore, ME in *O. sanctum* acts as a plant para-allomone to deter phytophagous insects.

## Ecomone

### *In plant-fruit fly relationship*

Plants, both host and non-host, form the focal point for all fruit fly activities. The common activities are host finding, oviposition, larval development, grooming, preening, foraging, pharmacophagy (consumption of non-essential biologically active chemicals), lekking, courtship, mating and mate competition. Most of these activities involve behavioural patterns that are mediated by ecomones, and have been reviewed (Fletcher 1987, Metcalf 1990, Metcalf and Metcalf 1992, Tan 1993). Different parts of a plant may release characteristic odours or fragrances that play an important role in influencing the behaviour of insects. In the plant-fruit fly relationship, it is the fruits and certain flowers that attract fruit flies through their natural fragrance acting as plant kairomone and synomone respectively.

### *Plant Synomone*

The dipteran flies, involved in pollination, are considered to be primitive pollinators because of their lapping or suctorial mouthparts. Of the many insect pollinators reported especially for orchids (Orchidaceae), no mention of the tephritid's involvement in pollination is made (Kevan and Baker 1983, Metcalf and Metcalf 1992). The tephritid flies in floral association are those that oviposit on flowers and their larvae feed on seeds of Compositae (Kevan and Baker 1983). There are many orchid species from several genus, excluding *Bulbophyllum*, that have eugenol and several contain ME in their floral scent (Kaiser 1993). A photograph showing a fruit fly on a *Bulbophyllum vinaceum* Ames and C. Schweinf. flower was published in the book dealing with *Bulbophyllum* (Vermeulen 1991). The observation that *Bactrocera* flies and males of pest and non-pest species, act as pollinators for orchid flowers was first reported in 1992 (Tan 1992).

The fruit fly orchid, *Bulbophyllum cheiri* Lindl., bears solitary flowers which contain eugenol, cis 2-allyl-4,5-dimethoxphenol (DMP) and a relatively large quantity of ME (Toong and Tan, unpublished data). The three phenyl-propanoids attract males of *B. carambolae, B. papayae* and *B. umbrosa* (Fabricius) (hereafter referred to as an ME-attracted species). Eugenol attracts flies but much less when compared with DMP or ME. In the forest, when a flower of *Bu. cheiri* blooms, it is covered with males of the ME-attracted species. The attracted males compulsively feed on the compounds secreted by the flower. Eventually, one of them will remove the pollinarium (a composite structure of pollinia, stipe and a sticky disc, the viscidium). The flower has a hinged seesaw lip that moves in response to a fly's weight, thus enabling the fly, after being tipped into the columnar cavity, to remove the pollinarium. Cross-pollination is ensured when the fly bearing the pollinarium visits and transfers the pollinia to the next flower (Tan 1993). In this case, the floral ME acts as a synomone because the fruit fly also benefits after ME-pharmacophagy (to be discussed under fruit fly-fruit fly interrelationship).

The flower of *Bulbophyllum patens* King, on the other hand, attracts one to several males from both the ME- and RK- [such as *B. albistrigata* (de Meijere), *B. caudatus* (Fabricius), *B. cucurbitae* (Coquillett) and *B. tau* (Walker)] attracted species. Initially, it was suspected that the flower contains both ME and RK. However, chemical analysis showed that a chemical - zingerone (Z), a ginger essence, with a structural resemblance to both ME and RK, was detected in the floral fragrance. Z was found to attract males of *B. cucurbitae*, and ME-attracted species in a wind tunnel. Thus for the first time, a chemical which is attractive to most species of *Bactrocera* flies was found (Tan and Nishida 2000). An attracted fly first lands on a lateral sepal and eventually moves to lap up the chemical on the floral lip which is hinged to the base of the column. The seesaw lip contains the highest content of Z. When the fly shifts its weight as it slowly laps along the seesaw lip, it is suddenly tipped into the columnar cavity. While retreating after being temporarily trapped between the column and lip, the fly removes the pollinarium from the column by touching the sticky viscidium or if it is bearing a pollinarium, the protruding pollinia are then deposited onto the stigma. Here Z is also acting as a synomone because, besides helping in pollination, it is consumed and used by the fly for pheromone production (Tan and Nishida 2000).

Flowers of *Bulbophyllum ecornutum* J. J. Sm. and *Bu. macranthum* Lindl. attract male flies of the RK-attracted species as pollinators. The floral scent of these two species contains RK. The pollination of these flowers by fruit flies is similar to that previously described for *Bu. patens*. However, the male melon fly feeds on the floral parts, especially the lip and lateral sepals which both secrete RK. The consumed RK is subsequently sequestered into the rectal gland, and acts as a component of the fly's aggregation and sex pheromone (Tan, unpublished data).

The four species of *Bulbophyllum* have non-resupinate flowers with a characteristic seesaw lip set at a higher plane than the floral column to facilitate pollination by fruit flies. Male fruit flies are the only visitors to these flowers. The different chemicals utilised as synomonal components to attract ME- or/and RK-sensitive *Bactrocera* species show that the four species of orchids have evolved to secure fruit flies as pollinators. Therefore, there is some form of co-evolution between fruit flies and the *Bulbophyllum* orchids, especially between *B. papayae* and the fruit fly orchid.

Several berries (*Ribes* and *Rubus* spp.) and plants contain RK but the chemical's role is unclear. The orchid flower of *Dendrobium superbum* Rchb. f. in Japan (Nishida et al. 1993) and three varieties of raspberry jam orchid *D. anosmum* Lindl. (synonymous with *D. superbum*) in Malaysia, also contain RK (Toong and Tan 1992). The synomonal role of RK in these orchids is still uncertain because we have yet to observe a male of RK-attracted species dislodge a pollinarium during feeding (confined to petals and sepals).

Varietal differences in synomonal components occur in flowers of the same species. The floral spike, bearing pollens, of an Australian fruit fly lily, *Spathiphyllum cannaefolium* Schott. (Araceae), emitted benzyl acetate and ME (Lewis et al. 1988). The variety found in Malaysia does not contain ME but instead high amounts of eugenol (Toong and Tan 1992). The floral fragrance generally attracts one or two male flies belonging to the ME-attracted species and the stingless bees, *Trigona* spp., to assist in pollination. Varietal differences in synomone also occur in the flowers of *Couroupita guianensis* Aubl. (Lecythidaceae). A variety in Hawaii was reported to contain ME

(Metcalf 1990). However, one grown in Penang contains eugenol that attracts male flies along with other insects to pollinate flowers (Toong and Tan 1992).

***Plant kairomone***

Female fruit flies locate host fruits through chemical and visual cues, to lay eggs. In a natural ecosystem, it is difficult or impossible for a fly to visualise a fruit from a distance or from above tree canopies. Each species of fruit, during ripening, emits its own scent that may act as a kairomone in attracting gravid female fruit flies to lay eggs in the fruit. Some fruits are more attractive to females of polyphagous species such as *B. carambolae, B. cucurbitae, B. dorsalis* and *B. papayae*. Fourteen different species of ripe non-punctured fruits were individually suspended below a canopy of a non-host plant for observation weekly over a period of 1-2 years. These experiments demonstrated that the majority of female *B. papayae* flies preferred to alight, search for a suitable site and attempt to oviposit, in decreasing order, on banana, *Musa sapientum* L., starfruit, *Averrhoa carambola* L., and papaya, *Carica papaya* L. (Tan 1993). A recent wind tunnel study using fruit juices, shows that mated and virgin females fly upwind to extracts of banana, starfruit, papaya and guava (*Psidium guajava* L.) (unpublished data). The behaviour of host finding and oviposition in *B. papayae* is similar to that described for the papaya fruit fly, *Toxotrypana curvicauda* Gerstaecker (Landolt and Reed 1990). Gravid female fruit flies are attracted via fruit scent from a distance without visual cue. The chemical nature of fruit kairomone is still unknown.

***Fruit fly-fruit fly interrelationship***

Fruit flies, as in many other insect species, communicate with each other via pheromones. Some insect species, for sexual selection and mating systems (Emlen and Oring 1977), also display lek, a communal display area, where males congregate to attract females for the purpose of mating. Lek formation and courtship involve the secretion and perception of pheromone.

***Aggregation pheromone***

The pharmacophagy (consumption of non-essential biological active chemicals) of attractants such as ME (in the case of ME-sensitive species) and CL or RK (in the RK-sensitive species), have not been fully understood for a long time. It has been shown that the attractant is either converted to other related chemical(s) or sequestered to act as aggregation pheromone to attract other males to form lek before courtship. In the case of *B. papayae*, the male either oxidises ME to DMP and trans-coniferyl alcohol (CF) that then act as aggregation pheromone (Hee and Tan 1998), or reduces Z to its alcohol that acts as a pheromone (Tan and Nishida 2000). In *B. carambolae*, consumed ME is converted to CF. While in the melon fly, RK is sequestered into the rectal gland after consumption to act as a aggregation pheromone (Nishida et al. 1990). The same process occurs for Z after pharmacophagy by the melon fly (Tan and Nishida 2000). In *B. tryoni* (Froggatt), RK – spontaneously converted from CL in the presence of moisture (Metcalf 1990) – is sequestered with small amounts converted to its alcohol which is also found in the rectal gland (Tan and Nishida 1995).

The aggregation pheromone is responsible in attracting other males to form lek, in which several males gather in a small space within a tree canopy, for several *Bactrocera* species such as *B. carambolae, B. cucurbitae, B. dorsalis* and *B. papayae*. It was suggested that lek formation, the display of males, is to attract females. However, in *Bactrocera* spp., the male secretes the aggregation pheromone to distract other males (which may be interested in the pharmacophagy of the attractant) from the attracted female, thereby increasing the mating success of the former. This was shown by the much higher mating success of ME-fed males in a male-male competition against ME-deprived males (Tan and Nishida 1996).

### *Sex pheromone*

The rectal gland of male fruit flies is known to be the source of the sex pheromone (Fletcher 1969). Chemical analyses of rectal glands for many species of *Bactrocera* have been conducted and reviewed (Fletcher and Kitching 1995). Generally, several volatile chemicals are found in a male rectal gland. It is important to show which chemicals are responsible for female attraction and submission, thus confirming the identity of the sex pheromone. This test is tedious and sometimes difficult, particularly when new compounds and their isomers need to be synthesised.

It is generally assumed that most *Bactrocera* species, when bred in the laboratory, are able to mate successfully without pharmacophagy of known male attractants. This assumption is true for *B. dorsalis. B. carambolae* and *B. papayae*. However, for *B. dorsalis* (Shelly and Dewire 1994, Tan and Nishida 1996) and *B. papayae* (Tan and Nishida 1996) males fed with ME competed significantly better than ME-deprived males for virgin females. Thus, the pharmacophagy of ME by male flies boosts the sex pheromone system to attract females. The mating behaviour of the Oriental fruit fly and the melon fly have been described (Kobayashi et al. 1978). The mating behaviour and mate selection have also been discussed by Nishida et al. in this proceedings.

The analysis of the rectal gland of wild males showed that many males of *B. papayae* contain DMP, CF and a small quantity of Z-3,4-dimethoxycinnamyl alcohol (DMC) (Nishida et al. 1988 a,b); and those of *B. carambolae* contain CF. In the laboratory, *B. dorsalis* (from Hawaii) males converted ME to the same phenyl propanoids as found in *B. papayae* (Tan and Nishida 1996). DMC and DMP showed weaker attractions than CF for *B. papayae* females in a wind tunnel (Khoo and Tan 1998, unpublished data). Mature females of *B. carambolae, B. dorsalis* and *B. papayae* belonging to the *B. dorsalis* complex, are attracted to CF at close range. Some attracted females even extend their ovipositors near the source of CF. They also extrude their ovipositors towards ME-fed males which have just begun copulation with other females. This behaviour, exhibited by the female, is a characteristic mating acceptance stance. Female *Bactrocera* flies, like females of other animals, choose their mates. When mounted by a male, the female extends her ovipositor during mating acceptance but kicks vigorously with her hind legs if she rejects the male. Copulation, when undisturbed, normally lasts from dusk to dawn although some mating pairs separate within five hours. Normally, the female kicks, with her hind legs, to dismount her mate to end copulation at the crack of dawn.

The same chemical components, acting as a sex pheromone, which are found in the rectal gland explains why *B. dorsalis* (from Hawaii) readily interbreeds with *B. papayae* in the laboratory especially after males have consumed ME. Hybrids of these two species are viable. $F_3$ offspring have been raised from the hybrids (unpublished data). In our laboratory, the *B. dorsalis* males were able to hybridise with *B. carambolae* females at a much lower frequency (ca. 16%) when compared with the frequency of mating with *B. papayae* (>70%). The reciprocal cross of the former was less frequent (ca. 5%). This is probably attributed to the fact that the *B. carambolae* male has difficulty inserting its relatively short aedeagus into the long ovipositor of the *B. dorsalis* female. The hybrids of *B. carambolae* and *B. dorsalis* were also viable. The ability of *B. dorsalis* to produce viable offspring when hybridised with both *B. carambolae* and *B. papayae* has great impact on fruit fly quarantine, management and SIT programmes.

Through the use of ME traps, native males, with intermediate morphology between *B. carambolae* and *B. papayae*, have been detected in increasing numbers from the field populations, since 1990 (unpublished data). The hybrids obtained in laboratory crosses are viable. Female *B. papayae* showed preference for conspecific males. On the other hand, *B. carambolae* females mated equally with *B. carambolae* and *B. papayae* males present in a large field cage (Wee and Tan 2000).

### *In fruit fly-predator relationship*

It is possible to further explain the strong attractancy of male fruit flies to ME. The endogenous production and secretion of an allomone after the pharmacophagy of ME demonstrate an anti-predation mechanism in the ME-attracted *Bactrocera* species. ME-fed males deter feeding by the Asian house gecko, *Hemidactylus frenatus* Dumeril and Bibron. Some geckos, after initial exposure to ME-fed males, would rather starve than to feed on the flies in a "no choice" experiment (Tan and Nishida 1992,1998). It was found that DMP>ME>CF, in a decreasing order of activity, act as a feeding deterrent against Japanese sparrows (Nishida and Fukami 1990). Thus CF, besides playing an important role as a sex pheromone, also acts as an allomone to deter vertebrate predators. It was shown that CF averts or causes repellence in the European starlings, *Sturnus vulgaris* L. The starlings also showed a strong aversion to DMC (Jakubas et al. 1992).

In another experiment using the king gecko, *Gekko monarchus* (Gray) Harrison, the gecko consumed less ME-fed males than ME-deprived males. In a "no choice" experiment over a period of two weeks, the dissection of all five geckos that fed on ME-males of *B. papayae*, revealed decolourised (anaemic) liver or/and creamish tumour(s) in the livers. In contrast, all geckos fed on ME-deprived males had dark red healthy livers (Wee and Tan, unpublished data).

For the RK-attracted species, RK may not play an additional role in anti-predation. On the other hand, the sequestered Z (a pungent compound) may deter predators and is currently being investigated. The anti-predation mechanism in the melon fly is achieved at sexual maturity which coincides with the endogenous synthesis of ethyl-hydroxybenzoate and 1,3-nonane-diol in the male rectal gland. The rectal compounds are detected after a week from eclosion, and significantly increase with age (Nishida et al. 1990). The nonane-diol produced endogenously in *B. cucurbitae* is a potent deterrent against gecko (Tan, 2000). It reduced the consumption of topically treated house fly, *Musca domestica* L. by the Asian house geckos (Tan 1992, 2000).

Sexually mature males of many *Bactrocera* species can spontaneously ejaculate their rectal contents, especially after the pharmacophagy of attractants, when under a stress such as via pinching or during carbon dioxide anaesthetisation (Tan and Nishida 1996). They have also been observed to release droplets of rectal contents as part of their escape response when disturbed. Such behaviour may startle or distract a potential predator and thereby provides the male flies an opportunity to escape predation.

## ACKNOWLEDGEMENTS

I am grateful to R. Nishida, S. T. Tan and Y.C. Toong for the many years of collaborative research and discussion; to A. Lamb for providing information and some orchid specimens; and to my research staff, post-graduate students for their assistance and interest in my research. Research facilities were provided by the Universiti Sains Malaysia and financial support was provided by the Malaysian Government R&D Grant MPKSN/IRPA 01-02-05-6004.

## REFERENCES

Chey, V.K. and K.H. Tan. 1985. *Dacus arecae* - (Diptera: Tephritidae) unattracted by cuelure and methyl eugenol. Malays. appl. Biol. 14: 55-56.

Emlen, S.T. and L.W. Oring. 1977. Ecology, sexual selection and the evolution of mating systems. Science 197: 215-223.

Fletcher, M.T. and W. Kitching 1995. Chemistry of fruit flies. Chem. Rev. 95: 789-828.

Fletcher, B.S. 1987. The biology of Dacinae fruit flies. Annu. Rev. Entomol. 32: 115-134.

Hee, A.K.W. and K.H. Tan. 1998. Attraction of female and male *Bactrocera papayae* to conspecific males fed with methyl eugenol and attraction of females to male sex pheromone components. J. Chem. Ecol. 24: 753-764.

Howlett, F.M. 1912. The effects of oil of citronella on two species of Dacus. Trans. R. Entomol. Soc. 60: 412-418.

Howlett, F.M. 1915. Chemical reactions of fruit flies. Bull. Entomol. Res. 6: 297-305.

Jakubas, W.J., P.S. Shah and J.R. Mason. 1992. Avian repellency of coniferyl and cinnamyl derivatives. Ecol. Applic. 2: 147-156.

Kaiser, R. 1993. The scent of orchids – olfactory and chemical investigations. Elsevier, Amsterdam, Lond., N.Y. 259 pp.

Kevan, P.G. and H.G. Baker. 1983. Insects as flower visitors and pollinators. Annu. Rev. Entomol. 28: 407-543.

Kobayashi, R.M., K. Ohinata, D.L. Chambers and M. Fujimoto. 1978. Sex pheromones of the Oriental fruit fly and the melon fly: mating behavior, bioassay method and attraction of females by live males and by suspected pheromone glands of males. Environ. Entomol. 7: 107-112.

Landolt, P.J. and H.C. Reed. 1990. Behavior of the Papaya fruit fly (Diptera:Tephritidae): Host finding and oviposition. Environ. Entomol. 19: 1305-1310.

Lewis, J.A., C.J. Moore, M.T. Fletcher, R.A.I. Drew and W. Kitching. 1988. Volatile compounds from the flowers of *Spathiphyllum cannaefolium*. Phytochem. 27: 2755-2757.

Metcalf, R.L. 1987. Plant volatiles as insect attractants. CRC Crit. Rev. Plant Sci. 5:251-301.

Metcalf, R.L. 1990. Chemical ecology of Dacinae fruit flies (Diptera:Tephritidae). Ann. Entomol. Soc. Am. 83:1017-1030.

Metcalf, R.L. and E.R. Metcalf. 1992. Fruit flies of the family Tephritidae. pp. 109-152. *In* Plant kairomones in insect ecology and control. R. L. Metcalf and E. R. Metcalf (eds.). Chapman and Hall, New York.

Nishida, R. and H. Fukami. 1990. Sequestration of distasteful compounds by some pharmacophagous insects. J. Chem. Ecol. 16: 151-164.

Nishida, R., O. Iwahashi and K.H. Tan. 1993. Accumulation of *Dendrobium superbum* (Orchidaceae) fragrance in the rectal glands by males of the melon fly, *Dacus cucurbitae*. J. Chem. Ecol. 19: 713-722.

Nishida, R., K.H. Tan and H. Fukami. 1988a. Cis-3,4-dimethoxycinnamyl alcohol from the rectal glands of male fruit fly, *Dacus dorsalis*. Chem. Express 3: 207-210.

Nishida, R., K.H. Tan, M. Serit, N.H. Lajis, S. Sukari, S. Takahashi and H. Fukami. 1988b. Accumulation of phenyl propanoids in the rectal glands of males of the oriental fruit fly, *Dacus dorsalis*. Experientia 44: 534-536.

Nishida , R., K.H. Tan, S. Takahashi and H. Fukami. 1990. Volatile components of male rectal glands of the melon fly, *Dacus cucurbitae* Coquillett (Diptera: Tehpritidae). Appl. Entomol. Zool. 25: 105-112.

Shelly, T.E. and A.L.M. Dewire. 1994. Chemically mediated mating success in male oriental fruit flies (Diptera:Tephritidae). Ann. Entomol. Soc. Am. 87: 375-382.

Steiner, L.F., W.C. Mitchell, E.J. Harris, T.T. Kozuma and M.S. Fujimoto. 1965. O.F.F. eradication by male annihilation. J. Econ. Ent. 58, 961-964.

Tan, K.H. 1985. Estimation of native populations of male *Dacus* spp. by Jolly's stochastic method using a new designed attractant trap in a village ecosystem. J. Pl. Prot. Tropics (2): 87-95.

Tan, K.H. 1992. Chemical ecology of Malaysian Dacinae fruit flies - Plant-fruit fly-predator interrelationships. 9th Annual Meeting of the International Society of Chemical Ecology, Section 5 - 1, July 6-9, 1992, Kyoto, Japan. pp. 35 (abstract).

Tan, K.H. 1993. Ecohormones for the management of fruit fly pests: understanding plant-fruit fly-predator interrelationships. pp. 495-503. *In* Management of Insect Pests: Nuclear and Related Molecular and Genetic Techniques. IAEA-SM 327, Vienna, Austria.

Tan, K.H. 2000. Sex pheromone components in the defense of the melon fly, *Bactrocera cucurbitae* against the Asian House gecko, *Hemidactylus frenatus*. J. Chem. Ecol. 26 (3): 697-704.

Tan, K.H. and R. Nishida. 1992. Fruit fly allomones against predation by lizard. Proc. Natl. IRPA Seminar, Kuala Lumpur, Malaysia 1: 237-238.

Tan, K.H. and R. Nishida. 1995. Incorporation of raspberry ketone in the rectal glands of males of the Queenland fruit fly, *Bactrocera tryoni* Froggatt (Diptera: Tephritidae). Appl. Entomol. Zool. 30: 494-497.

Tan, K.H. and R. Nishida. 1996. Sex pheromone and mating competition after methyl eugenol consumption in the *Bactrocera dorsalis* complex. pp 147-153. *In* Fruit fly pests - A world assessment of their biology and management. B. A. McPheron, and G. J. Steck (eds.). St. Lucie Press, Delray Beach, Florida, USA.

Tan, K.H. and R. Nishida. 1998. Ecological significance of male attractant in the defense and mating strategies of the fruit fly *Bactrocera papayae*. Entomol. Exp. Appl. 89: 155-158.

Tan, K.H. and R. Nishida. 2000. Mutual reproductive benefits between a wild orchid and fruit flies via a ginger essence. J. Chem. Ecol. 26(2): 533-546.

Tan, K.H. and M. Serit. 1988. Movements and population density comparisons of native male adult *Dacus dorsalis* and *Dacus umbrosa* (Diptera: Tephritidae) among three ecosystems. J. Pl. Prot. Tropics 5(1): 17-21.

Tan, K.H. and M. Serit. 1994. Adult population dynamics of *Bactrocera dorsalis* (Diptera:Tephritidae) in relation to host phenology and weather in two villages of Penang Island, Malaysia. Environ. Entomol. 23(2): 267-275.

Tan, K.H. and Z. Jaal. 1986. Comparison of male adult population densities of the Oriental and Artocarpus fruit flies, *Dacus* spp. (Diptera: Tephritidae), in two nearby villages in Penang, Malaysia. Res. Popul. Ecol. (28): 85-89.

Toong, Y. C. and K. H. Tan. 1992. Fruit fly attractants from local plants. Symposium on Tropical Fruit Flies, 18-20 May, 1992, Kuala Lumpur, Malaysia.

Vermeulen, J.J. 1991. Orchids of Borneo. Bentham-Moxon Trust, Toihaan Publishing Co. and The Sabah Society. vol. 2: 342 pp.

Wee S.L. and K.H. Tan. 2000. Interspecific mating of two siblings species of the *Bactrocera dorsalis* complex in a field cage. pp. 667-674. *In* Area-Wide Control of Fruit Flies and Other Insect Pests. K.H. Tan (ed.). Penerbit Universiti Sains Malaysia, Penang, Malaysia.

# World-wide Sexual Compatibility in Medfly, *Ceratitis capitata* (Wied.), and its Implications for SIT

*J.P. Cayol*

Entomology Unit, FAO/IAEA Agriculture and Biotechnology Laboratory, A-2444 Seibersdorf, Austria

## INTRODUCTION

The concept of the sterile insect technique (SIT), described by Knipling (1953), to control and/or eradicate insect pest populations has been applied to many Lepidoptera and Diptera species. Among Diptera species, the Mediterranean fruit fly, *Ceratitis capitata* (Wied.) (Diptera: Tephritidae), is often referred to as the most important agricultural pest in the world (Liquido et al. 1990) and it is also a major target of SIT action programmes world-wide.

The use of SIT requires that mass rearing facilities be developed to produce sterile insects for use in national programmes, with mass reared strains being established by colonising wild insects collected from the SIT target area. More recently, with the increasing demand for sterile Medflies and the limited number of production facilities, some rearing facilities began to export Medflies to other national/regional programmes. Eight facilities have now reached production levels which allow them to export sterile insects (Fisher and Cáceres 2000) on a regional or inter-regional basis. When this procedure is used, the flies released have to compete with wild flies of a different geographic origin.

The increasing use of Medfly genetic sexing strains (GSS) has also resulted in the same strain being used in different countries. To date, five rearing facilities in the world produce GSS (Fisher and Cáceres 2000). Since GSS are assembled from specific components, it is impossible to "colonise" them from each country where sterile GSS flies are needed. The GSS are often backcrossed with insects from the target population to increase the genetic variability (Franz et al. 1996), although in some cases this presents problems (Franz, personal communication). In practice, a single wild population is used as a basis for the synthesis of the GSS. Consequently, the same GSS based on the same wild genetic material may be used in various countries/continents and the question was raised concerning the sexual compatibility of these strains with wild Medfly populations in different countries.

In the present work, the sexual compatibility of wild populations originating from eight countries, representing five continents, was measured in pairwise comparisons under greenhouse and/or field cage conditions. In a second series of experiments, flies from four GSS were evaluated.

## MATERIALS AND METHODS

### Wild Material

Wild insects were collected as pupae from infested fruits in their country of origin. The location and the main host of these pupae are presented in Table 1. Pupae were shipped by express air mail to Seibersdorf, Austria, except for the field cage test run in Argentina (Table 1) where wild flies were tested locally (Cayol et al., 1999). On emergence, flies were sexed and kept in separate ventilated Plexiglass® cages (11 x 15.5 x 11 cm) until the test and provided with adult food (yeast and sugar) and water.

Table 1. Origin of wild flies tested.

| Country | Location | Main Host of Origin |
|---|---|---|
| Argentina | Patagonia | *Ficus carica* L.<br>*Prunus persica* L. |
| Australia | Perth | *Citrus sinensis* L. |
| Greece | Crete Island | *Citrus sinensis* L. |
| Guatemala | Antigua | *Coffea arabica* L. |
| Israel | Near Tel Aviv<br>Arava Valley | *Psidim guajava* L.<br>*Psidim guajava* L. |
| Kenya | Near Nairobi | *Coffea arabica* L. |
| Portugal | Madeira Island | *Psidim guajava* L. |
| France | Reunion Island | *Mimusops elengi* L. |

### Genetic Sexing Strains

GSS flies were obtained as pupae from the FAO/IAEA facility at Seibersdorf for greenhouse tests and from the KM8 facility in Mendoza (Argentina) for the field cage test run in Argentina. Four GSS were tested (Table 2). Emerged flies were kept under the same conditions as wild flies (see above).

Table 2. Origin of GSS flies tested.

| Strain | Selectable Marker | Facility |
|---|---|---|
| SEIB 6-96 | wp | Seibersdorf Mendoza, Argentina |
| VIENNA 4/Tol-94 | wp, tsl | Seibersdorf |
| VIENNA 7-97 | wp, tsl | Seibersdorf |
| AUSTRIA 6-97 | wp, tsl, yellow body | Seibersdorf |

### Test Cage

Flies were tested in a greenhouse located at the FAO/IAEA Agriculture and Biotechnology Laboratory (Seibersdorf, Austria). The greenhouse was temperature monitored (temperature ranging between 24 and 32°C). A cage made of netting material was placed inside the greenhouse. The cage contained six citrus trees (up to 1.8 metre in height) in a total volume of 15 $m^3$. In Argentina, flies were tested in field cages located in San Miguel de Tucuman (Tucuman Province). The cages were made of Saran® screen (20 x 20 mesh); they were cylindrical, with flat floor and ceiling, and measured 2.9 m in diameter and 2.0 m in height (Chambers et al. 1983). Each field cage contained one citrus tree. In both greenhouse and field cage tests, the cages were covered with a shading cloth to avoid the "greenhouse effect".

### Method

Strains (wild or GSS) were tested in pairwise comparisons (involving two wild populations or one wild population and one GSS at a time). Two days before being tested, healthy flies were selected and males and females, from alternatively one of the two populations were marked with a dot of water-based paint (Deka®) on the notum. On the day of the test, 30 flies of each sex and each strain were released into the cages at dawn. Males were released 30 minutes before females to give them time to establish their territory and start forming leks (Prokopy and Hendrichs 1979). The number and type of mating pairs were checked on a continuous census, and five minutes after initiation of mating, the pairs were collected and kept into vials (50 ml volume) to monitor mating duration. The mated flies were not replaced or released back into the cage after separation (Chambers et al. 1983). Tests lasted for 7-8 consecutive hours. At the end of the test, the mated females were dissected to check for the presence of sperms in the spermathecae. Tests were performed from March till December 1997.

## RESULTS

### Participation of Flies in Mating

This measures the suitability of the flies and the environment for mating. It represents the overall mating activity of the flies (Table 3).

If PM<0.2, the test is rejected (IAEA 1997). The mean PM values obtained in comparing wild populations for wild and GSS (PM=0.403 and 0.488 respectively, Table 4) confirmed that the test conditions were suitable for mating. In addition, it showed that a similar proportion of matings was achieved when only wild flies or when wild and GS flies were present in the cages.

### Comparison Among Wild Populations

Eight pairwise comparisons were run among wild Medfly populations (Figure 1), representing a total of 27 replicates. In all types of comparisons involving any of the wild populations tested, each of the four possible types of mating was encountered confirming that there was no absolute behavioural incompatibility between these populations. The sexual isolation among these wild populations was assessed using the

Isolation Index (ISI) (Cayol et al., 1999) as described in Table 3. The ISI ranges from -1 ("negative assortative mating", flies only mate with a "foreign" partner) to +1 ("positive assortative mating" or total sexual isolation, flies only mate with partner of the same origin), through an equilibrium at 0 (total sexual compatibility, no mating preferences). The overall mean value of ISI obtained for the eight wild populations (ISI=+0.234; Table 4) showed a slight tendency for homologous (male and female of the same origin) mating, but certainly no evidence of sexual isolation. The ISI value was relatively constant (ISI variance=0.088; Table 4) and showed that no major differences was found for the "mating preference" of wild populations.

Table 3. Indices used in the experiments (after Cayol et al. 1999, McInnis et al. 1996) ("ab": number of matings of "a" males with "b" females; in RSI: "L" for mass reared males and "W" for wild flies).

| To Measure | Index Formula | Range of Values | | |
|---|---|---|---|---|
| | | min | equilibrium | max |
| Participation in Mating | $PM = \frac{\text{No. of pairs collected}}{\text{No. of females released}}$ | 0 | +0.5 | +1 |
| Sexual Isolation | $ISI = \frac{(aa + bb) - (ab + ba)}{\text{Total No. of matings}}$ | -1 | 0 | +1 |
| Male Relative Performance | $MRPI = \frac{(aa + ab) - (bb + ba)}{\text{Total No. of matings}}$ | -1 | 0 | +1 |
| Female Relative Performance | $FRPI = \frac{(aa + ba) - (bb + ab)}{\text{Total No. of matings}}$ | -1 | 0 | +1 |
| Male Mating Competitiveness | $RSI = \frac{LW}{LW + WW}$ | 0 | +0.5 | +1 |

Table 4. Values of the indices obtained when comparing wild populations and wild and GSS strains (*: absolute value; variance in italics).

| Type of Comparison | Mean Values of the Indices | | | | |
|---|---|---|---|---|---|
| | PM | RSI | ISI | MRPI | FRPI |
| WILD vs. WILD<br>8 comparisons<br>27 replicates | 0.403<br>*0.010* | -<br>- | 0.234<br>*0.088* | 0.377 *<br>*0.050* | 0.293 *<br>*0.031* |
| WILD vs. GSS<br>8 comparisons<br>37 replicates | 0.488<br>*0.017* | 0.349<br>*0.051* | 0.225<br>*0.040* | -0.003<br>*0.099* | 0.285 *<br>*0.063* |

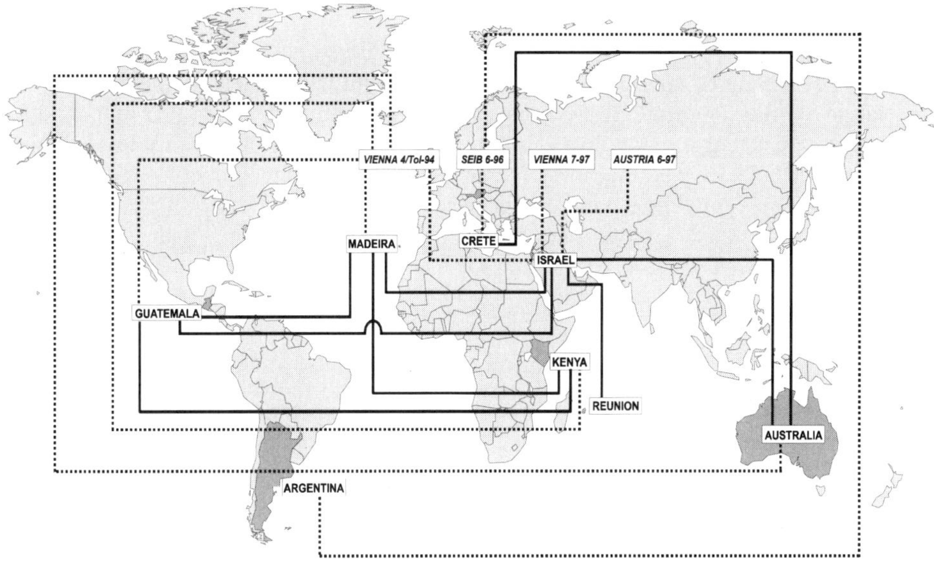


Figure 1. World map showing the type of pairwise comparisons assessed (Italicised names: GSS. Bold line: wild populations. Dotted line: wild population and GSS).

## Comparison Between Wild Populations and GSS

Nine pairwise comparisons were run between wild Medfly populations and GSS (Figure 1), representing a total of 37 replicates. As in the previous tests, in all comparisons between wild populations and GSS, each of the four possible types of mating was encountered confirming behavioural compatibility. The overall mean ISI value (ISI=+0.225; Table 4) showed a slight tendency for homologous mating, but no sexual isolation. As in the wild populations, the ISI value was stable among the strains tested (ISI variance=0.04; Table 4) and highlighted that no major difference was found among them. In addition, the ISI values obtained in the two types of tests (wild versus wild and wild versus GSS) were not significantly different (Tukey's HSD test; $P<0.05$) (Systat 1996). Consequently, wild populations are as behaviourally compatible with GSS as they are with other wild populations from various geographic origins. Two other indices which look at the relative mating performance of males (MRPI) and females (FRPI) of the two strains, regardless of their mating partners, were measured. These indices range between -1 (all matings achieved by one type of male (MRPI) or female (FRPI)) and +1 (all matings achieved by the other type of male (MRPI) or female (FRPI)) through an equilibrium at 0 (equal mating performance of males or females of the two strains) (Table 3). The overall MRPI value (MRPI=-0.003; Table 4) showed that GSS males were as efficient in mating as wild males but that GS females were more prone to mate than wild females (FRPI=0.285; Table 4).

### Mating Competitiveness of GSS Males

The mating competitiveness of GSS males with wild males was measured by the Relative Sterility Index (RSI) (McInnis et al. 1996) (Table 3). When RSI=0.5, wild and GS males are equally competitive. The overall RSI value (RSI=0.349; Table 4) obtained for the four GSS showed that mass reared GSS males achieved about one-third of the wild female matings. The mating competitiveness of GSS males was found satisfactory and relatively stable among the four GSS when competing with various wild populations (RSI variance=0.051; Table 4).

### Sperm Transfer

In these experiments, the transfer of sperm into the female spermathecae was considered the final confirmation of the success in mating. The proportion of matings which did not result in effective transfer of sperm (unsuccessful matings) was compared for each type of pair in both experiments. In wild/wild pairwise comparisons, the results showed no significant difference (Tukey's HSD tests, P<0.05) in the proportion of non-effective sperm transfer when males and females are from the same or from different geographic origins (Table 5).

In wild/GSS pairwise comparisons, the proportion of non-effective sperm transfer is not significantly different (±10%) from those obtained in wild/wild comparisons but slightly increased (>10%) when GSS females were involved (Table 5). The analysis also showed that significantly more homologous GSS/GSS than homologous wild/wild mating pairs resulted in non-effective sperm transfer (Table 5).

Table 5. Proportion of matings which did not result in effective sperm transmission in both series of experiments (numbers followed by the same letter do not differ significantly according to Tukey's HSD test P<0.05; variance in italics).

| | Proportion of Matings Without Effective Sperm Transfer | | | |
|---|---|---|---|---|
| Type of Comparison | Male and female of the same origin | Male and female of different origin | | |
| WILD vs. WILD<br>8 comparisons<br>27 replicates | 0.079 a<br>*0.016* | 0.090 ab<br>*0.025* | | |
| | Wild male wild female | GS male wild female | Wild male GS female | GS male GS female |
| WILD vs. GSS<br>8 comparisons<br>37 replicates | 0.087 ab<br>*0.019* | 0.099 ab<br>*0.036* | 0.154 ab<br>*0.023* | 0.0234 b<br>*0.063* |

## DISCUSSION

These data demonstrate that wild Medfly populations world-wide have not yet evolved specific sexual behaviours indicative of incipient pre-mating isolation mechanisms under local natural selection. In addition, it was shown that wild populations are as sexually compatible with GSS as they are with other wild populations. Heterologous matings between wild flies from different geographic origin or between wild and GSS flies did not result in a lower proportion of fertile matings (effective sperm transfer) than did homologous matings (flies from the same origin). This indicates that wild populations, as well as GSS, have not yet evolved any mating isolation mechanisms. The fact that GSS mass reared females are not as selective in mating as wild females (Cayol et al., 1999, Cayol, 2000) could explain the lower effective sperm transfer found when GSS females were involved; GSS females accept a mate more easily and would consequently not obviously reject a non-suitable (resulting in non-effective sperm transfer) male.

The relevance of mating behaviour studies to SIT has encouraged the IAEA to investigate this subject. A co-ordinated research programme was started in 1994 to look at the details of male courtship behaviour in wild populations from nine countries (IAEA 1994) from qualitative and quantitative standpoints, using slow motion video recording. To date, no significant differences have been found among the wild populations (IAEA 1997). However, it was shown that long term rearing could affect significantly the duration of the male courtship (Eberhard and Briceño 1996) and the male love songs (Briceño and Eberhard 1997).

Concerns about the sexual compatibility among Medflies from different origins represented somewhat of a threat to the movement of sterile flies from one country to another to support SIT programmes. These concerns become more pronounced when a GSS was proposed to be used in many different SIT programmes. The findings of the present experiments support the potential use of the same GSS anywhere in the world.

Gasparich et al. (1997) showed that the mitochondrial DNA of Medfly populations from 100 different origins was indeed variable and that it probably reflected the colonisation pattern of Medflies from its origin in Eastern Africa about 200 years ago. However, there was no evidence that substantial genetic differentiation had occurred. When a Medfly outbreak occurs, programme managers sometimes worry that the sterile flies released might not be from the same geographic origin and hence would not mate. A second concern is that the "foreign" flies might introduce new genetic material into the country. The fear is that "foreign" fertile flies would establish a new population with its own genetic and behavioural characteristics. However, the present work based on populations representative of five continents, clearly demonstrates that there are no population specific mating behaviour traits. This observation, together with the genetic data, suggests that the risk of introducing a more virulent form of Medfly into a specific country is remote.

In conclusion, strains to be used in SIT programmes in any country must be selected for the quality of the flies produced rather than the initial geographic origin of the strain.

## ACKNOWLEDGEMENTS

The author is grateful to the team leaders and staff who collected wild pupae in the different locations: Argentina, E. Rial (Programa Moscafrut Patagonia); Australia, R. Johnson (AWA); Crete, A. Economopoulos (University of Heraklion); Guatemala, P. Rendon (USDA); Israel, Y. Rössler (Citrus Marketing Board); Kenya, S. Lux (ICIPE); Madeira Island, R. Pereira (Madeira-Med Programme) and Reunion Island, S. Quilici (CIRAD-FLHOR).

## REFERENCES


Briceño, R.D. and B. Eberhard. 1997. Comparisons of the songs of successful and unsuccessful males of different genetic sexing strains of Medfly, *Ceratitis capitata* (Diptera: Tephritidae). 34 pp. *In* Medfly mating behaviour studies under field cage conditions: Report of the Third Research Co-ordination Meeting – Tel Aviv (Israel) 15-19 Sept. 1997. IAEA, Vienna.

Cayol, J.P. 2000. Changes in sexual behavior and some life traits of tephritid species caused by mass-rearing processes. pp. 843-860. *In* Fruit flies (Tephritidae): Phylogeny and evolution of behaviour. Aluja, M and A. Norrbom [eds.]. CRC Press, Boca Raton, 944pp.

Cayol, J.P., J. Vilardi, E. Rial and M.T. Vera. 1999. New indices and method to measure the sexual compatibility and mating performance of *Ceratitis capitata* (Diptera: Tephritidae) laboratory reared strains under field cage conditions. J. Econ. Entomol. 92(1): 140-145.

Chambers, D.L., C.O. Calkins, E.F., Boller, Y. Itô and R.T. Cunningham. 1983. Measuring, monitoring, and improving the quality of mass-reared Mediterranean fruit flies, *Ceratitis capitata* Wied. 2. Field tests for confirming and extending laboratory results. Z. Angew. Entomol. 95: 285-303.

Eberhard, B. and R.D. Briceño. 1996. Genetic assimilation of the effects of crowding on the duration of premounting courtship behaviour of male Medflies (Diptera: Tephritidae). 53pp. *In* Medfly mating behaviour studies under field cage conditions: Report of the Second Research Co-ordination Meeting - Tapachula (Mexico) 19-23 Feb. 1996. IAEA, Vienna.

Fisher, K. and C. Cáceres. 2000. A filter rearing system for mass-reared genetic sexing strains of the Mediterranean fruit fly (Diptera: Tephritidae). pp. 541-548. *In* Area-Wide Control of Fruit Flies and Other Insect Pests. K.H. Tan (ed.). Penerbit Universiti Sains Malaysia, Penang, Malaysia.

Franz, G., P. Kerremans, P. Rendon and J. Hendrichs. 1996. Development and application of genetic sexing systems for the Mediterranean fruit fly based on a temperature sensitive lethal. pp.185-191· *In* Fruit fly pests: a world assessment of their biology and management· McPheron, B.A. and G.J. Steck (eds.). St.Lucie Press, Delray Beach, Florida, USA.

Gasparich, G.E., J.C. Silva, H.Y. Han, B.A. McPheron, G.J. Steck and W.S. Sheppard. 1997. Population genetic structure of Mediterranean fruit fly (Diptera: Tephritidae) and implications for worldwide colonisation patterns. Ann. Entomol. Soc. Am. 90(6): 790-797.

[IAEA] International Atomic Energy Agency. 1994. Medfly mating behaviour studies under field cage conditions: Report of the First Research Co-ordination Meeting - Vienna (Austria) 4-7 Oct. 1994. IAEA, Vienna. 13 pp.

[IAEA] International Atomic Energy Agency. 1997. Medfly mating behaviour studies under field cage conditions: Report of the Third Research Co-ordination Meeting - Tel Aviv (Israel) 15-19 Sept. 1997. IAEA, Vienna. 36 pp.

Knipling, E.F. 1953. Possibilities of insect control or eradication through the use of sexually sterile males. J. Econ. Entomol. 48 (4): 459-462.

Liquido, N.J., R.T. Cunningham and S. Nakagawa. 1990. Hosts plants of the Mediterranean fruit fly (Diptera: Tephritidae) on the Island of Hawaii (1949-1985 survey). J. Econ. Entomol. 83: 1863-1868.

McInnis, D.O., D.R. Lance and C.G. Jackson. 1996. Behavioral resistance to the sterile insect technique by Mediterranean fruit fly (Diptera: Tephritidae) in Hawaii. Ann. Entomol. Soc. Am. 89(5): 739-744.
Prokopy, R.J. and J. Hendrichs. 1979. Mating behavior of *Ceratitis capitata* on a field caged host tree. Ann. Entomol. Soc. Am. 72: 642-648.
Systat Inc. 1996. Systat for Windows: statistics, Version 6 Edition. Evanston, Il: SYSTAT, Inc., 1996. 750 pp.

# Interspecific Mating of Two Sibling Species of the *Bactrocera dorsalis* Complex in a Field Cage

***Suk-Ling, Wee and Keng-Hong, Tan***

School of Biological Sciences, Universiti Sains Malaysia, 11800 Penang, Malaysia

## INTRODUCTION

The *Bactrocera dorsalis* (Hendel) species complex, comprises at least 52 species. Two closely related members of the complex, namely *B. papayae* Drew & Hancock and *B. carambolae* Drew & Hancock, were recently reclassified as new species (Drew and Hancock 1994). Under this taxonomic revision, *B. papayae* is now regarded as a distinct species from *B. carambolae* based on the differences of: 1) wing pattern of the costal band at apex $R^{4+5}$, 2) the presence of a dark spot on the fore femora and, 3) the pattern of the transverse black band on terga III-V. Chemical examination of the volatile components produced by the males of both species also revealed pronounced differences in the chemistry of their rectal gland secretions (Perkins et al. 1990).

In Malaysia, *B. papayae* has a wider distribution and a larger host range compared with *B. carambolae*. Starfruit (*Averrhoa carambola* L.) and various species of wax apple (*Syzygium spp.)* are the preferred hosts of *B. carambolae* whilst *B. papayae* attacks over 150 species but preferentially infests banana (*Musa spp.*), starfruit, mango (*Mangifera indica* L.), papaya (*Carica papaya* L.) and guava (*Psidium guajava* L.) in decreasing order (Tan 1993).

Data from field trapping studies using methyl eugenol (ME) in Penang Island, Malaysia, showed the presence of male flies with intermediate morphological characteristics between *B. papayae* and *B. carambolae* since 1992. Laboratory hybridisation showed that these two species are able to interbreed and produce viable offspring. The hybrids also possess a variety of intermediate characteristics between the two species (Unpublished data).

Therefore, the question arises as to whether *B. papayae* and *B. carambolae* should be categorised as different species, subspecies or even as different strains. Before a satisfactory conclusion can be achieved, there are some key issues that need to be addressed. Firstly, after ME consumption, males of *B. papayae* and *B. carambolae* produce a common sex pheromonal component, coniferyl alcohol, CF (Nishida et al. 1988a, Nishida et al. 1988b, Tan and Nishida 1996). Therefore, are females of *B. carambolae* and *B. papayae* attracted to heterospecific males in the absence of CF, a common sex pheromonal component? If so, do they mate interspecifically in a field cage, which simulates the natural environment, in the presence of conspecific males? Secondly, will the acquisition of ME enhance the interspecific mating competitiveness? Lastly, do the females show preference for their conspecific males or any males that have ingested ME in their diet and vice versa? The objective of the current investigation was to shed some light on these questions.

## MATERIALS AND METHODS

### Insect

Flies of *B. papayae* were cultured using an artificial larval medium whereas those of *B. carambolae* were cultured using fresh starfruit under ambient conditions in the insectary on a 12L: 12D cycle at 25-29°C and 83-90% relative humidity. Sexing of the males and females was performed 3-4 days after emergence. Virgin flies were maintained in separate cages and provided with a yeast-protein hydrolysate-sugar mixture and water. Mature virgin flies of 14-20 day old *B. papayae* (Tan et al. 1987) and 30-40 day old *B. carambolae* were used for experimentation.

### Chemical

Methyl eugenol (3,4-dimethoxyallyl benzene) (purity>98%) was supplied by International Pheromone System Limited, England.

### Interspecific Attraction in a Wind Tunnel

The wind tunnel was constructed as described by Jones and coworkers (1981), with slight modifications. The distance between upwind and downwind ends was 2 m and a transparent polythene tube (12-inch diam.) was used and changed regularly. Air speed was regulated at 14 -17 cm/s using voltage regulators. Studies were conducted between 17:00-20:00 h, 10 minutes each trial, alternated with 5 min for preparation. The room temperature was at 30-32°C with relative humidity of 89-93 %.

Research methods were as described previously by Hee and Tan (1998). Fifteen females of *B. papayae* were assayed for conspecific and heterospecific males. A positive response was recorded when flies performed zigzagging anemotaxis (as defined by Kennedy and Marsh, 1974) over at least 0.5 m to the source of attraction. The average light intensity (of up-, middle- and down-wind sections of the tunnel) was measured prior to each bioassay using a Lutron LX-101 lux meter. The experiment was replicated 6 times with different cohorts on different days. Similar procedure was employed to assay the attraction of *B. carambolae* females to *B. papayae* and conspecific males.

### Interspecific Mating Competition I

This test was conducted in a large outdoor cage (2.1 X 2.1 X 2.1 $m^3$). 140-150 males from each species were released to compete for the same number of virgin females of *B. papayae*. At 20:30-21:30 h, using a red light, mating pairs were individually collected in separate specimen tubes. The number of males *in copula* for each species was recorded. This procedure was repeated for virgin females of *B. carambolae* the following day. A total of three replicates were performed for females of each species.

### Interspecific Mating Competition II

This was conducted in the same manner as the previous test, except only *B. papayae* males were allowed to feed *ad libitum* for 30 min on ME which was dispensed on a piece of filter paper (Whatman® No.1), 24 hours prior to mating competition. The

ME-fed *B. papayae* males were then released into the outdoor cage together with ME-deprived males of *B. carambolae* to compete for the same number of *B. papayae* females. The number of males *in copula* from each species was recorded. Unless stated otherwise, each set of the test was performed in triplicate. A further set of experiments was performed using ME-fed *B. carambolae* males and ME-deprived *B. papayae*. The two sets of this test were also repeated by replacing *B. papayae* females with *B. carambolae* females.

### Statistical Tests

Comparisons of means for the data in interspecific mating competition I and II were performed using the Student's t-test. For the wind tunnel experiment, the data were transformed to a modified arcsine square root (Anscombe 1948) and subjected to analysis of variance and Duncan's multiple range tests (Sachs 1982).

## RESULTS

Our results showed that the females of *B. carambolae* and *B. papayae* responded positively to live conspecific as well as heterospecific males in a wind tunnel assay (Fig. 1 and Fig. 2). The attraction generally increased with the decrease in light intensity during the last two hours of photophase.

The attraction of *B. carambolae* females increased gradually as compared with *B. papayae* females with no obvious peak. There was no significant difference between the attraction to conspecific and heterospecific males. Approximately 50% attraction was observed from 19:15 to 19:30 h whereby males were observed to fan their wings and mount on other males. At 19:45 h, most *B. carambolae* males stopped courtship activity. Conspecific attraction was terminated earlier than in heterospecific attraction (Fig. 1).

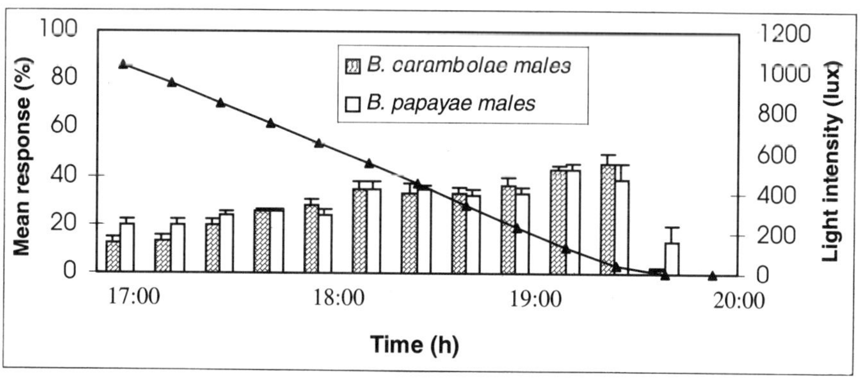


Figure 1. Mean percentage of *B. carambolae* female response to *B. papayae* and conspecific males in a wind tunnel (n=5; vertical bar = standard error).

In *B. papayae* female, the attraction increased and peaked at 19:30 h followed by 19:45 h in both conspecific and heterospecific males (Fig. 2). The peak attraction to conspecific males (above 85%) was significantly higher than in heterospecific males (*ca.* 65%). *B. papayae* males were involved in aggressive courtship behaviour by performing vigorous wing fanning and mounting on other males. At 20:00 h, approximately 10% of the females responded to conspecific males compared with heterospecific males (<3%). This was because a small percentage of *B. papayae* males were still engaged in wing fanning and mounting while *B. carambolae* males were observed to be sexually inactive (i.e. resting and preening) at 20:00 h (Fig. 2).

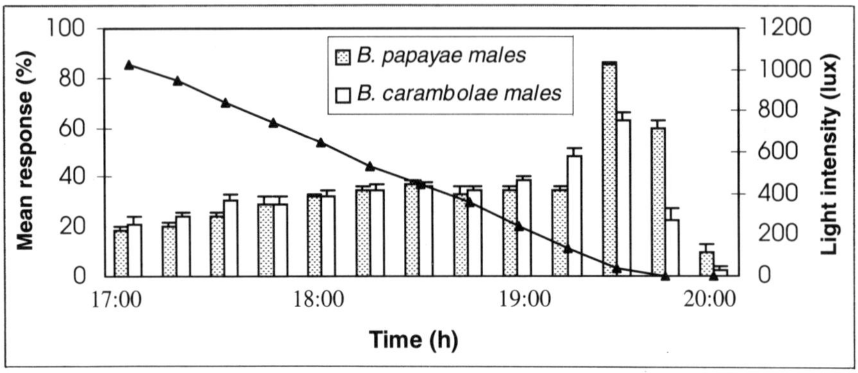


Figure 2. Mean percentage of *B. papayae* female response to *B. carambolae* and conspecific males in a wind tunnel (n=5; vertical bar = standard error).

Results demonstrate that females of *B. papayae* and *B. carambolae* mated readily without the acquisition of ME by the males in a field cage. However, *B. papayae* females showed a higher level of selectivity for conspecific males than *B. carambolae* females ($P<0.05$; Table 1). *B. carambolae* females were not highly selective and readily mated with any courting male within close proximity ($P>0.1$; Table 1).

Similar results were obtained for ME-fed males of either species in interspecific mating competition. Males of *B. papayae,* after feeding on ME, showed an increase in the total number of successful mating with *B. papayae* females when caged with ME-deprived males of *B. carambolae* (Fig. 2). However, *B. papayae* females still preferred to mate with conspecific males deprived of ME, in the presence of ME-fed *B. carambolae* males.($P<0.01$ in both cases; Table 2). Although the ratio of ME-fed *B. carambolae*: ME-deprived *B. papayae* males *in copula* increased from 2:1 to 3:1, there was no significant difference ($P>0.1$ in both cases; Table 2).

Table 1. Interspecific mating competition between *B. papayae* and *B. carambolae* in an outdoor cage (n = 3 each).

| Female species | *B. papayae* males (mean ± s.e.) | *B. carambolae* males (mean ± s.e.) | Comparison of means |
|---|---|---|---|
| *B. papayae* | 95.3 ± 9.5 | 8.0 ± 4.6 | $P<0.05$ |
| *B. carambolae* | 16.7 ± 5.2 | 16.0 ± 2.1 | $P>0.1$ |

Table 2. Interspecific mating competition between methyl eugenol (ME)-fed and ME-deprived males of *B. papayae* and *B. carambolae* (mean ± standard error).

| Female species | *B. papayae* ME-deprived males | *B. papayae* ME-fed males | *B. carambolae* ME-deprived males | *B. carambolae* ME-fed males | Comparison of means (n = 3) |
|---|---|---|---|---|---|
| *Bactrocera* | - | 106 ± 7.8 | 9.7 ± 4.3 | - | *P*<0.01 |
| *papayae* | 98.7 ± 5.4 | - | - | 4.7 ± 0.7 | *P*<0.01 |
| *Bactrocera* | - | 22.3 ± 5.9 | 41.3 ± 8.2 | - | *P*>0.1 |
| *carambolae* | 10.0 ± 2.9 | - | - | 38.0 ± 13.1 | *P*>0.1 |

## DISCUSSION

Coniferyl alcohol, a sex pheromonal component produced by conspecific males after pharmacophagy of ME, attracts *B. papayae* females at close range (Nishida et al. 1988a and b, Tan and Nishida 1996, Hee and Tan 1998). The same compound was also detected in the male rectal gland of *B. carambolae* after ME consumption (Tan and Nishida 1996). Tan (1997) has suggested that CF might be responsible for the interspecific attraction and copulation. This hypothesis was suggested after the detection of some male fruit flies with intermediate morphological characteristics between *B. papayae* and *B. carambolae* using ME-baited traps in Penang recently (unpublished data).

ME, present in some plant species, when ingested is converted to CF in *B. carambolae*, and to CF and 2-allyl-4, 5-dimethoxyphenol in *B. papayae* (Nishida et al. 1988a and b, Tan and Nishida 1996). CF being attractive to the females may be responsible for the interspecific mating. However, the wind tunnel study showed that females of both species were attracted to ME-deprived heterospecific males during dusk, just as they responded to conspecific calling males. Further field cage study demonstrated that both species mated readily without the consumption of ME, although the number of the successful interspecific mating differs from one species to another. This indicates that CF is not a critical factor in the interspecific crossing of these species.

*N*-3-methylbutyl acetamide, was detected in male rectal gland of both species (Nishida et al. 1988a, Tan and Nishida 1996). This amide is also found in the male rectal gland of *B. cucurbitae* (Coquillett) (Baker et al. 1981), *B. neohumeralis* (Hardy) (Bellas and Fletcher 1979), *B. tryoni* (Froggatt) (Tan and Nishida 1995) and *B. visenda* (Hardy) (Krohn et al. 1992). In addition, the compound was also found to increase quantitatively with sexual maturation in *B. tryoni* (Tan and Nishida 1995) and in *B. carambolae* (unpublished data).

*N*-3-methylbutyl acetamide was shown to elicit activation and increase flight activity in *B. cucurbitae* females in a wind tunnel (Baker et al.1982). It also attracted *B. papayae* females in a wind tunnel study (Hee and Tan 1998). Its potency of attraction in this case was greater than CF but less than 2-allyl-4, 5-dimethoxyphenol. Similar results were obtained in *B. carambolae* whereby 10-20% mature virgin females responded to the amide during dusk (unpublished data). Hence, *N*-3-methylbutyl acetamide, an endogenous compound blended into a multi-component pheromonal system, may also be responsible for the interspecific attraction.

Our work also demonstrated that *B. papayae* females are more species-specific than *B. carambolae* female. The former prefers to mate with conspecific males. The pharmacophagy of ME was found to increase the mating competitiveness in the males of *B. dorsalis* (Shelly and Dewire 1994, Tan and Nishida 1996) and *B. papayae* (Tan and Nishida 1996). It also elicited an early onset of mating behaviour in the *B. papayae* males (Hee and Tan 1998).

Although similar studies have not been carried out on *B. carambolae*, ME is presumed to function in the same manner as in the other sibling species. This is because males of *B. carambolae* are also strongly attracted to ME. Although *B. carambolae* males possess the advantage of having consumed ME, the total successful matings were not significantly higher than those of the ME-deprived males of *B. papayae*. This could be due to the recent findings that both sexes of *B. papayae* were found to perform better than *B. carambolae* in their sexual relationship in terms of first mating and subsequent mating over a 60-day period of observation (Wee and Tan, 2000). Flies of *B. papayae* performed their first mating within the first week of emergence whereas those of *B. carambolae* do so only two weeks after eclosion. Furthermore, *B. papayae* males were found to be more aggressive in their mating behaviour (Unpublished data) and performed more mating than *B. carambolae* males under the same conditions. Therefore, the efficacy of ME in enhancing the sexual ability of males may differ from one species to another.

Apart from the different aggressiveness in both the species, the terminalia length of the flies could be another factor that influenced the selectivity of *B. papayae*. Iwazumi et al. (1997) showed that aedeagus and aculeus of *B. papayae* were significantly longer than those of the *B. carambolae*. Therefore, in order to achieve successful intromission and fertilisation, the aedeagal length must be sufficiently long. A shorter aedeagus in the *B. carambolae* male might not enable it to transfer sperm successfully into the female of *B. papayae*. In order to ensure successful fertilisation, the female will select her mate. However, females of *B. carambolae* showed no preference for males of *B. papayae* or *B. carambolae*. This indicates that their receptivity is independent of the aedeagus length of the two species. This work has demonstrated that interspecific mating between *B. carambolae* and *B. papayae* is successful. The hybrids produced viable offspring at least up to $F_3$ (unpublished data). The fact that hybrids of both the species are able to produce viable offspring is of great concern. This suggests the possibility of the siblings and their hybrids' ability to penetrate into a broader geographical area, ensuring a wider host range with greater infestation and damage to the fruit industries in subtropical as well as tropical countries.

## ACKNOWLEDGEMENTS

We thank R. Nishida (Kyoto University) for providing the chemical, N-3-methylbutyl acetamide. This work is supported by the Malaysian R&D Grant MPKSN/IRPA 01-02-05-6004.

## REFERENCES


Anscombe, F.J. 1948. The transformation of Poisson, binomial and negative binomial data. Biometrika 35: 246-254.

Baker, R., R.H. Herbert and R.A. Lomer. 1982. Chemical components of the rectal gland secretions of male *Dacus cucurbitae*, the Melon Fly. Experientia 38: 232.

Bellas, T.E. and B.S. Fletcher. 1979. Identification of the major components in the secretion from the rectal pheromone glands of the Queensland fruit flies *Dacus neohumeralis*. J. Chem. Ecol. 5: 795.

Drew, R.A.I. and D.L. Hancock. 1994. The *Bactrocera dorsalis* complex of fruit flies (Diptera: Tephritidae: Dacinae) in Asia. Bull. Entomol. Res. Suppl. 2. 68pp.

Hee, A.K.H. and K.H. Tan. 1998. Attraction of female and male *Bactrocera papayae* to conspecific males fed with methyl eugenol and attraction of females to sex pheromone components. J. Chem. Ecol. 24: 753-764.

Iwazumi, R., M. Kaneda and O. Iwahashi. 1997. Correlation of length of terminalia of males and females among nine species of *Bactrocera* (Diptera: Tephritidae) and differences among sympatric species of *B. dorsalis* complex. Entomol. Soc. Am. 90: 664-666.

Jones, O.T., R.A. Lomer and P.E. Howse. 1981. Response of male Mediterranean fruit flies, *Ceratitis capitata* to trimedlure in a wind tunnel of novel design. Physiol. Entomol. 6: 175-181.

Kennedy, J.S. and D. Marsh. 1974. Pheromone-regulated anemotaxis in flying moths. Science 184: 999-1001.

Krohn, S., M.T. Fletcher, W. Kitching, C.J. Moore, R.A.I. Drew and W. Francke. 1992. Chemistry of fruit flies: glandular secretion of *Bactrocera (Polistomimetes )* *visenda* (Hardy). J. Chem. Ecol. 18: 2169-2176.

Nishida, R., K.H. Tan, N.H. Lajis, A.M. Sukari, S. Takahashi and H. Fukami. 1988a. Accumulation of phenylpropanoids in the rectal gland of the Oriental fruit fly, *Dacus dorsalis*. Experientia 44: 534-536.

Nishida, R., K.H. Tan and H. Fukami. 1988b. *Cis*-3,4-dimethoxycinnamyl alcohol from the rectal glands of male Oriental fruit fly, *Dacus dorsalis*. Chem. Exp. 3: 207-210.

Perkins, M.V., M.T. Fletcher, W. Kitching, R.A.I. Drew and C.J. Moore. 1990. Chemical studies of rectal gland secretions of some species of *Bactrocera dorsalis* complex of fruit flies (Diptera: Tephritidae). J. Chem. Ecol. 16: 2475-2488.

Sachs, L. 1982. Applied statistics: a handbook of techniques. (Trans. Z.Reynarowch). Spring-Verlag. New York. 706 pp.

Shelly, T.E. and A.M. Dewire. 1994. Chemically mediated mating success in male Oriental fruit flies (Diptera: Tephritidae). Ann. Entomol. Soc. Am. 87: 375-382.

Tan, K.H. 1993. Ecohormones for the management of fruit fly pests: understanding plant-fruit fly-predator interrelationships. pp. 495-503. *In* Management of Insect Pests: Nuclear and Related Molecular and Genetic Techniques. IAEA-SM 327, Vienna, Austria.

Tan, K.H., L.G. Kirton and M. Serit. 1987. Age response of *Dacus dorsalis* (Hendel) to methyl eugenol in (a) a wind tunnel and (b) traps set in a village, and its implication to population estimation. pp. 425-432. *In* Fruit Flies. A.P. Economopoulus (ed.). G. Tsiveriotis Ltd. Athens.

Tan, K.H. and R. Nishida. 1995. Incorporation of raspberry ketone in the rectal glands of males of the Queensland fruit fly, *Bactrocera tryoni* Froggat (Diptera: Tephritidae). Appl. Entomol. Zool. 30: 494-497.

Tan, K.H. and R. Nishida. 1996. Sex pheromone and mating competition after methyl eugenol consumption in the *Bactrocera dorsalis* complex. pp. 147-153. *In* Fruit Fly Pests: A World Assessment of Their Biology and Management. B.A. McPheron and G.J. Steck (eds.). St. Lucie Press, Delray, Florida, USA.

Wee, S.L. and K.H. Tan. 2000. Sexual maturity and intraspecific mating success of two sibling species of the *Bactrocera dorsalis* complex. Entomol. Exp. Appl. 94: 133-139.

# PART III.

## APPLICATION OF AREA-WIDE CONCEPT TO FRUIT FLIES

## J) BIOSYSTEMATICS AND BIODIVERSITY OF FRUIT FLIES

# Tephritid Taxonomy into the 21st Century – Research Opportunities and Applications

*R.A.I. Drew and M.C. Romig*

Australian School of Environmental Studies, Griffith University, Nathan Campus, Queensland 4111, Australia

## INTRODUCTION

We write with the firm conviction that taxonomic research forms the essential foundation for all other areas of investigation within the field of biology. This has been well demonstrated in the Tephritidae and is a position at which we have arrived through many years' experience in fruit fly systematic research covering taxonomy, behaviour, biology, ecology and pest control.

The importance of sound taxonomic research is highlighted at this time by the known presence of many sibling species complexes across the family. Within the Dacinae, for example, major pest species often occur within groups of closely related species, most of which are not pests. The *dorsalis* complex of Southeast Asia and the *musae* complex of Papua New Guinea are examples.

Tephritid taxonomy has a long history (over two centuries) and rich heritage with some 4,500 species having been described since the mid-1700s. This research has been carried out in major research centres around the world and particularly in Australia, Europe, Hawaii, mainland USA and South Africa.

In Mexico in February 1998, a significant meeting was held on the biology/behaviour and taxonomy of Tephritidae. Specialist researchers in this area presented valuable and interesting data on "Phylogeny and Evolution of Behaviour" in fruit flies. In summarising current knowledge on the taxonomy and biology of the Tephritidae, the meeting highlighted the outstanding achievements of taxonomy in its contributions to both basic research and pest management programmes over many decades of tephritid studies world-wide.

This presentation provides a link between the meetings in Mexico and Penang and enables us to present a summary of our current knowledge and genuine valuable applications of tephritid taxonomy to the overall fruit fly research and pest management effort. In doing this, this presentation also fits into the theme of this conference in Penang, "Fruit Flies – current global scenario".

## HISTORY OF FRUIT FLY TAXONOMY

Tephritid taxonomic research was pioneered by those forefathers of biology, Linnaeus and Fabricius. Since that time, some notable researchers have made major contributions to the discipline. The major workers and the periods within which they have worked are as follows (data from the US National Museum, Norrbom et al. 1998):

| Period | Major workers |
|---|---|
| 1758 – 1800 | Linnaeus, Fabricius |
| 1801 – 1850 | Meigen, Wiedemann, Macquart |
| 1851 – 1900 | Walker, Loew, Wulp |
| 1901 – 1950 | Bezzi, Hendel, Hering, Munro, Malloch |
| 1951 – 1998 | Munro, Hardy, Aczel, Foote, Drew |

It is interesting to group these taxonomic researchers by the region(s) in which they have concentrated their studies. These are set out below and show that more have investigated fauna in Europe and the Southeast Asian to Pacific areas than other parts of the world.

| Region | Taxonomic researchers (1758 – 1998) |
|---|---|
| Africa | Fabricius, Munro |
| Europe | Fabricius, Linnaeus, Meigen, Wiedemann, Macquart |
| Southeast Asia – Pacific | Walker, Bezzi, Hendel, Hering, Hardy, Malloch, Drew |
| North America – Central and South America | Loew, Wulp, Aczel, Foote |

As at 1998, there are not many active fruit fly taxonomists. The major contributors by region are as follows:

| Region | Major taxonomists (1998) |
|---|---|
| Africa | De Meyer, Freidberg |
| Europe | Merz, White, Korneyev |
| Southeast Asia – Pacific | Ho-Yeon Han, Drew, Hancock, White |
| North America – Central and South America | Norrbom, Carrol, Zucchi |

After two centuries of taxonomic research, one would expect that the world-wide fauna would be well known. This is the case for some biogeographic regions but not all. The status of our knowledge of the fauna is:

| Region | Level of known taxonomic research |
|---|---|
| Nearctic | Well known |
| Neotropical | Relatively unknown (except *Anastrepha*) |
| Palearctic | Well known |
| Afrotropical | Relatively unknown |
| Oriental | Reasonable knowledge |
| Australasian – Oceanic | Reasonable knowledge |

## CATEGORIES OF FRUIT FLY TAXONOMIC RESEARCH

At present, taxonomic research in the Tephritidae has been expanded to include more modern technology and to extend our knowledge into areas poorly known previously. For example, phylogenetic analyses and understanding of modes of speciation have been markedly enhanced by advances in technology in areas of genetics together with accompanying expansions in biological research efforts. The categories of taxonomic investigations that are being carried out world-wide, as at 1998, are as follows:

- Descriptions of species and classifications.
- Phylogenetic analyses based on morphology, DNA and allozymes.
- Modes of speciation including data on evolution of flora, host shifting and DNA.

## DNA RESEARCH RELATED TO TEPHRITID TAXONOMY

Currently, there are a number of laboratories around the world carrying out DNA research that has applications to taxonomic understanding of fruit flies. This work is summarised as follows:

### Australia

- *Curran – CSIRO, Canberra:* developing identification systems for specimens collected in quarantine surveys; work based on investigation of the internal transposable spacer (ITS) regions of the Ribosomal genes.
- *Frommer – University of Sydney:* studies on the *B. tryoni* species complex and testing the dacine generic and subgeneric classifications; work based on microsatellites, the mitochondrial cytochrome B gene and a nuclear eye colour gene.
- *Drew and Jing Ma – Griffith University, Brisbane:* studies designed to separate sibling species of *Bactrocera*; work based on microsatellites, 18S RNA, cytochrome B and D-loop sequences.
- *Yeates – University of Queensland:* phylogenetic studies.

**Italy**

- *Gasperi and Malacrida – University of Pavia:* allozyme and DNA studies related to generic and subgeneric classifications.

**New Zealand**

- *Armstrong – Lincoln University, Christchurch:* developing identification systems for specimens collected in quarantine surveys; work based on Restriction Fragment Length Polymorphisms in the ITS regions of the Ribosomal genes.

**Southeast Asia**

- Ho-Yeon Han – Yonsei University, Korea.

**USA**

- *McPheron – Pennsylvania State University:* studies on phylogenetic relationships in *Rhagoletis* and *Anastrepha*; work based on mitochondrial genes and intron sequences.
- *Smith – Michigan State University:* phylogenetic studies on *Rhagoletis.*
- *Haymer – University of Hawaii:* phylogenetic studies.
- *Roderick – University of Hawaii:* phylogenetic studies.
- *Smith and Srini – Kansas State University:* studies on *dorsalis* complex species.

It appears that considerably more work is needed in order to define the exact DNA sequences that will provide accurate separation of species. Also, there is a need to develop coordination of work and effective collaboration between the various laboratories.

## MAJOR TAXA BEING RESEARCHED WITHIN THE TEPHRITIDAE

At present, there is taxonomic research being undertaken on the genus *Anastrepha* and the subfamilies Tephritinae, Trypetinae, Ceratitinae and Dacinae. A summary of these groups, the work being carried out and future needs are as follows:

Genus *Anastrepha*

- Subtropical to tropical species.
- Hosts – fruit; endemic habitat is the neotropical rainforest ecosystem.
- This genus contains some major pests of commercial horticulture.
- Current research – investigations of phylogenetic relationships (Norrbom).
- Future needs – host and distribution data and larval taxonomy.

Subfamily Tephritinae

- Temperate climate species.
- Hosts primarily Asteraceae in open forest and grassland ecosystems.
- Some species are beneficial insects, being useful in biological control of weeds and flower pollination.
- Current research – descriptions of fauna in Europe (Korneyev, Merz) and the Middle East/Africa (Freidberg).

- Future needs – revisions of fauna in Africa, parts of Europe, Central and South America.

Subfamily Trypetinae
- Temperate to subtropical species.
- Hosts – fruit, stems of plants.
- Some species are major pests of commercial horticulture.
- Current research – Ho-Yeon Han in Korea covering Southeast Asian fauna.
- Future needs – revisions of fauna in most regions except North America and Australia and elucidation of sibling species in groups such as *Rhagoletis.*

Subfamily Ceratitinae
- Subtropical to Mediterranean species, best developed in Africa.
- Hosts – fruit.
- Some species are major pests of commercial horticulture.
- Current research – revision of genus *Ceratitis* in Africa (De Meyer).
- Future needs – continuing work on African fauna.

Subfamily Dacinae

- Subtropical to tropical species.
- Hosts – fruit; endemic habitat is the tropical/subtropical rainforest ecosystem.
- Many species are major pests of commercial horticulture.
- Current research – revision of the Southeast Asia and South Pacific fauna (Drew, Hancock).
- Future needs – completion of Southeast Asian revisions and complete revision of the African fauna.

In summary, we have an extensive knowledge of some of the major tephritid taxonomic groups, especially in Europe and North America, while the fauna of the other regions still requires considerable investigation. We currently have a few dedicated researchers involved but there is an ongoing need to maintain this research effort and, in some cases, expand it.

## FUTURE APPLICATIONS OF TEPHRITID TAXONOMIC RESEARCH

As we stated at the beginning, taxonomic research is the essential foundation for all other aspects of biological research. In tephritid programmes around the world, taxonomic research has proved vital to other research and pest management efforts over many decades. In the future, the need for accurate identification of species, and a clear understanding of their biology and behaviour, will remain unchanged and in some cases, be even more necessary. The areas where taxonomic research will be most useful are in preharvest field control, a range of quarantine programmes and enhancement of eradication strategies.

***In field control*** of fruit flies in horticultural crops, there is a definite movement away from cover sprays to protein bait sprays. While there is probably little chance that new, more attractive bait spray formulations will be developed, more effective field control can be achieved with existing baits by improvements in application technology. Such technological advances will depend on the accurate definition of species, often sibling species, and a thorough knowledge of their behaviour. For example, a sound knowledge of feeding behaviour including diurnal rhythms, feeding sites and dispersal patterns must all be considered when deciding upon the when, where and why of bait spray applications.

***In quarantine programmes,*** there is a significantly increasing demand to establish effective detection systems to prevent the establishment of new incursions of pest fruit fly species in countries. This is particularly so in North America, Southeast Asia and the Pacific region. Since 1995, there have been a number of outbreaks of major pest species in new areas. Species such as *Bactrocera papayae, B. dorsalis, B. cucurbitae* and *Ceratitis capitata* have been moving into new areas and have required expensive eradication programmes. A range of quarantine activities has been designed to prevent the spread and establishment of major exotic fruit fly pest species. These are summarised as follows:

- Monitoring for introduced exotic species – this requires a knowledge of endemic and exotic species present in the area under surveillance and efficient taxonomic identification services to service what are often large trapping programmes.
- Ongoing training for identification staff to ensure that skilled personnel are always available to deliver accurate and rapid species determinations.
- Development of market access technologies – successful establishment of international trade requires the acceptance of treatments that are usually species specific, e.g., chemical and physical treatments. Also, a knowledge of species distributions and their host records are a very important basis for host status testing and establishment of area freedom.

***In eradication programmes,*** there will always be an ongoing need to refine the various strategies in order to maximise efficacy of treatments and minimise environmental impacts. Sound taxonomic research which assists in studies such as comparative morphology of feeding, courtship and reproductive systems, will have vital applications in mass rearing, protein bait programmes and the sterile insect technique. Further, information on species biology and behaviour patterns will always be useful in the development of treatment application strategies.

## CONCLUSIONS

Taxonomic research and its applications have had, and always will have, a vital place in fruit fly programmes around the world, both for research and pest management. Quarantine programmes are, now and into the foreseeable future, dependent upon species identifications which usually have to be delivered quickly and accurately. A continuing supply of well trained staff will be vital to this work.

There are new and exciting prospects for scientists who are interested in genuine research effort. Our next generation of taxonomists will have to encompass an holistic approach where the morphology and biology of species are both used to define species populations. We remain excited as we see how the identification of species opens up an understanding of biology, and in turn how biological characteristics assist in defining species. We will need to bring together morphological analyses, biology and behaviour data, chemistry of pheromones and molecular biology (both allozyme and DNA studies) in order to accurately determine species, especially where sibling species are involved.

This approach was used in the revision of the *dorsalis* complex of Asia (Drew and Hancock 1994) and should be continued across the family. Only in this way will we establish genuine biological definitions of species.

The best way to demonstrate that the world needs high quality and genuine scientific research is to keep moving forward and producing it. In taxonomic research in the family Tephritidae, we must do this and keep producing the data that are so vitally needed in other research and pest management programmes.

## REFERENCES

Drew, R.A.I. and D.L. Hancock. 1994. The *Bactrocera dorsalis* complex of fruit flies (Diptera: Tephritidae: Dacinae) in Asia. Bull. Entomol. Res., Suppl. No. 2, 68 pp.

Norrbom, A. L., L. E. Carroll and A. Freidberg. 1998. Status of Knowledge. pp. 9-47. *In* Fruit fly expert identification system and systematic information database. Thompson, F. C. (ed.). Myia 9, ix + 524 pp.

# Taxonomy of *Rhagoletis* Population Associated with Wild Plums in Chile

*Daniel Frías and Andrés Alviña*

Instituto de Entomología, Universidad Metropolitana de Ciencias de la Educación, AV.J.P. Alessandri 774, Casilla 147, Santiago, Chile

## INTRODUCTION

In South America, there are about fifteen *Rhagoletis* species that live in association with wild and cultivated Solanaceae host plants (Foote 1981, Frías 1992). The principal information on taxonomy for these species is the morphology of adults. Thus, in the genus *Rhagoletis,* in general, there is little information about immature stages especially on first and second larva instars (Steck et al. 1990, Carrol and Wharton 1989, Steck and Wharton 1988, Persson 1963, White and Elson-Harris 1992, Hernandez–Ortíz 1992, 1993, Frías et al. 1993). Presently, in Chile, there are 4 species associated with Solanaceae host plants. *R. tomatis* Foote and *R. nova* (Schiner) are associated with cultivated Solanaceae *Lycopersicum esculentum* Miller or cultivated tomatoes and *Solanum muricatum* Aiton or sweet cucumber respectively. *R. conversa* Béthes has two *Solanum* host plants, *S. nigrum* L. and *S. tomatillo* (Remy) Phil. F.( Frías et al. 1984). The host for *R. penela* Foote is unknown. Moreover, in the last few years, a population on wild plums of the Myrobalan variety (Rosaceae) was detected (Gonzalez 1989). At present, there is no information about the origin and taxonomy of this population. In this work, we have studied the morphology of eggs, three instar larvae, pupae and adults of this population associated with wild plums as well as aspects of its geographical distribution in Chile.

## MATERIALS AND METHODS

Larvae were taken from fruits of wild plums. They were then killed by immersing in water and heated to boiling for about one minute. After cooling in water, they were stored in 70% ethanol (Steck and Wharton 1988). Their antenal sensory organ, mouthhook, preoral teeth, posterior and anterior spiracles and posterior tubercles were studied. Adults were collected on host plants and also from pupae in the laboratory. Their head setulae, scutum integumental pattern, wing pattern and ovipositor were studied in adults. These structures in the larvae and the wing in adults were dissected and mounted on slides. Preserved specimens were prepared for scanning electron microscopy. Wild plums fruits were collected from Regions IV to X of Chile (29° 54' S, 071° 15' W to 40° 58' S, 072° 52' W).

### Description of Immature Stages

#### Eggs

The eggs have short filaments similar to the sympatric *R. conversa* host race that live on *S. nigrum* near the host of wild plums (Figure 1).

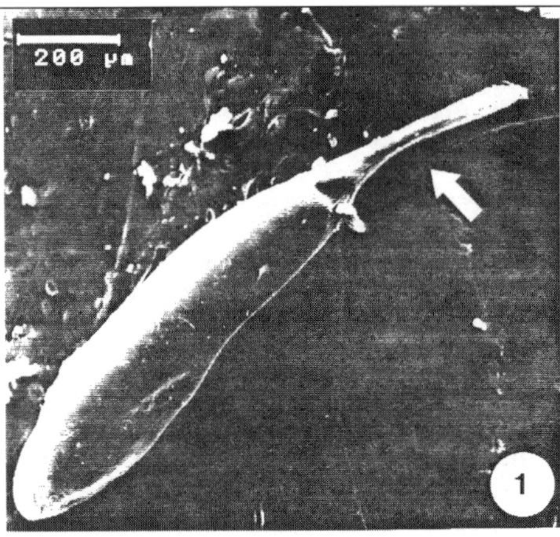


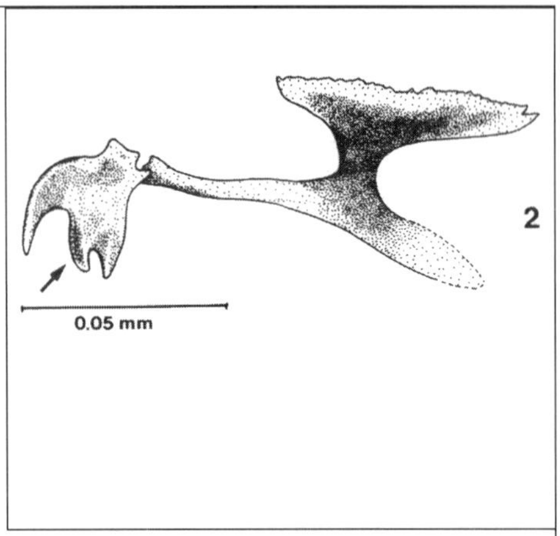


Figure 1. Eggs of *Rhagoletis* collected on fruits of wild plums.

Figure 2. Cephalopharyngeal skeleton of first instar larvae collected from fruits of wild plums.

**First instar larva**

The cephalopharyngeal skeleton is yellow. The mouthhook is big and the preapical tooth curved (Figure 2).

**Second instar larva**

The cephalopharyngeal skeleton is black with dental sclerite; the parastomal bar and dorsal and ventral cornu are curved and brown. The apical tooth is yellow and sharp. One small preapical tooth is yellow (Figure 3). The tubercles of the posterior region are shown in Figure 4.

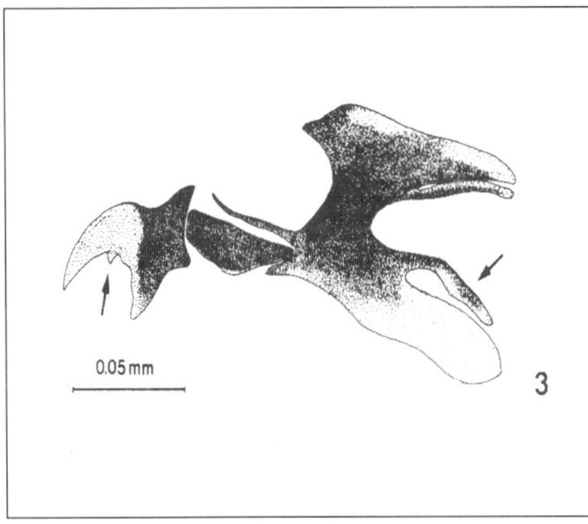


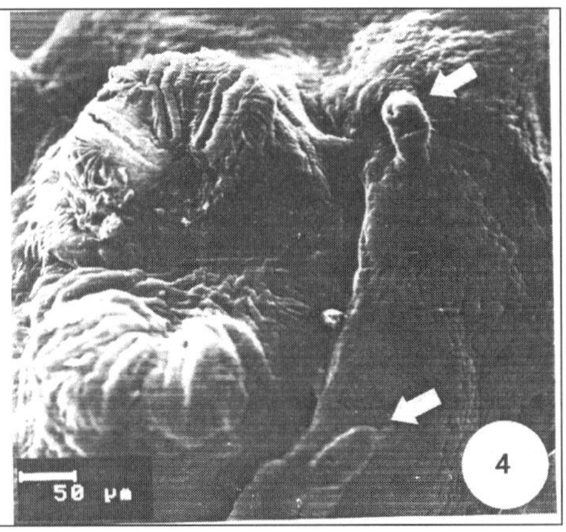


Figure 3. Cephalopharyngeal skeleton of second instar larvae collected from fruits of wild plums.

Figure 4. Tubercles of posterior region of second instar larvae.

### Third instar larva

The cephalopharyngeal skeleton is black, except for the dental sclerite; the dorsal and ventral cornu are yellow. The apical tooth is sharp and strongly curved. The hipopharyngeal sclerite with its basal projection is yellow. The ventral cornu is black and curved (Figure 5). There are four preoral teeth (Figure 6). The anterior spiracle is bilobed with 22 – 27 tubules (Figure 7). The posterior spiracles are as shown in Figures 8 and 9. The spiracular slits are about 5 times as long as the breadth. The dorsal and ventral bundles have 8 – 13 hairs while the lateral bundles have 7 – 10 hairs (Figure 10).

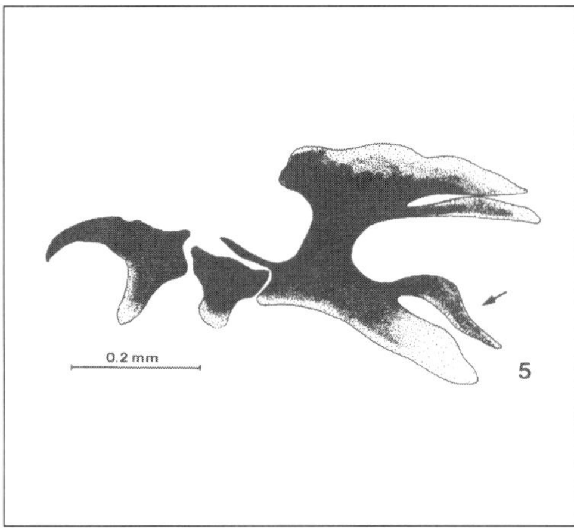


Figure 5. Cephalopharyngeal skeleton of third instar larvae collected from fruits of wild plums.

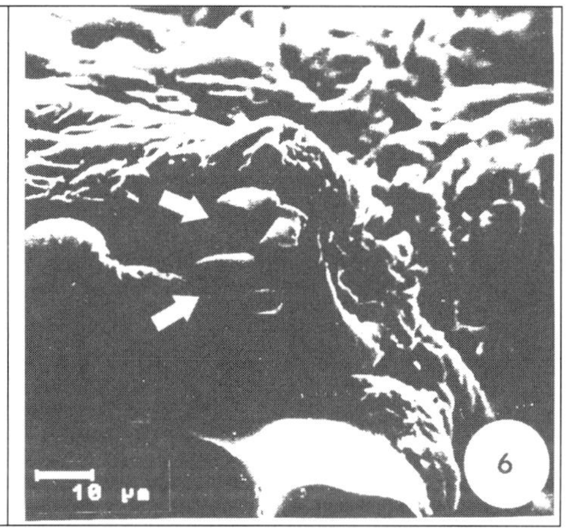


Figure 6. Preoral teeth of third instar Larvae.

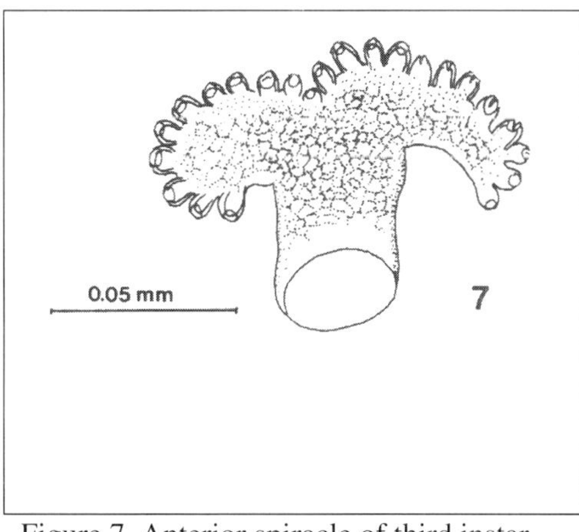


Figure 7. Anterior spiracle of third instar larvae.

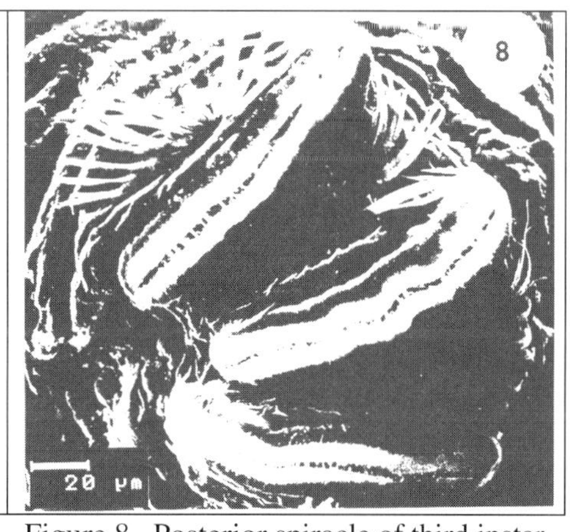


Figure 8. Posterior spiracle of third instar larvae: spiracular slits,dorsal, ventral and lateral bundless hairs.

Figure 9. Posterior spiracle of third instar larvae: spiracular slits.

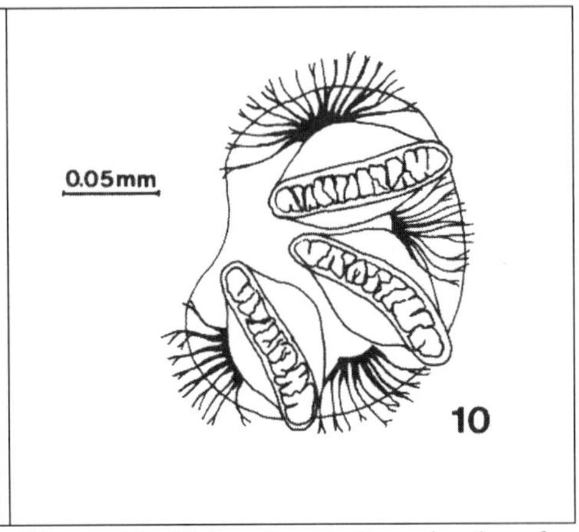


Figure 10. Hairs of posterior spiracles of third instar larvae.

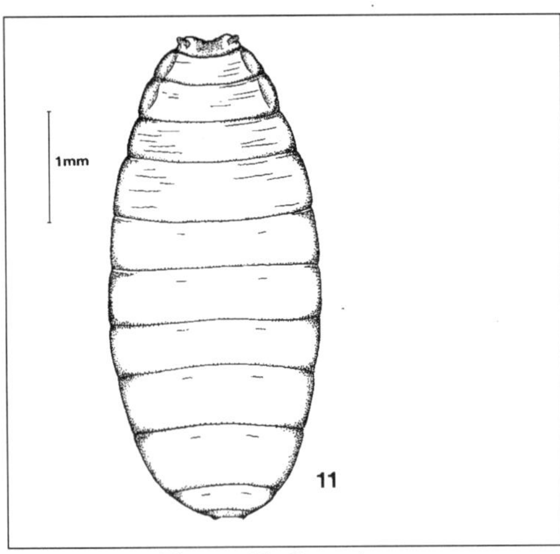


Figure 11. Puparium.

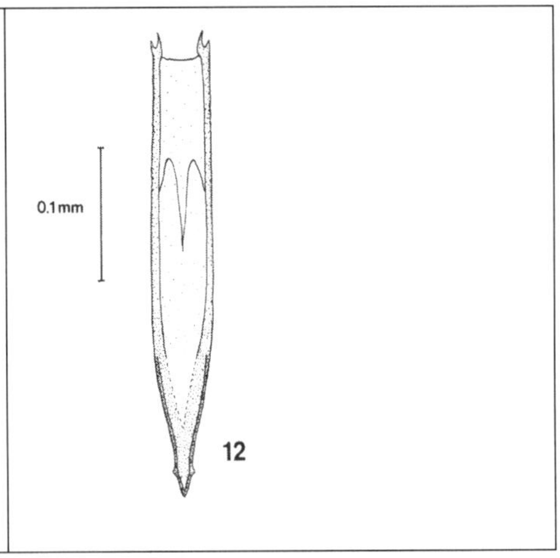


Figure 12. Ovipositor of female.

**Puparium**

This is dark brown (Figure 11).

**Ovipositor**

The ovipositor has minute lateral projections at the tip (Figure12).

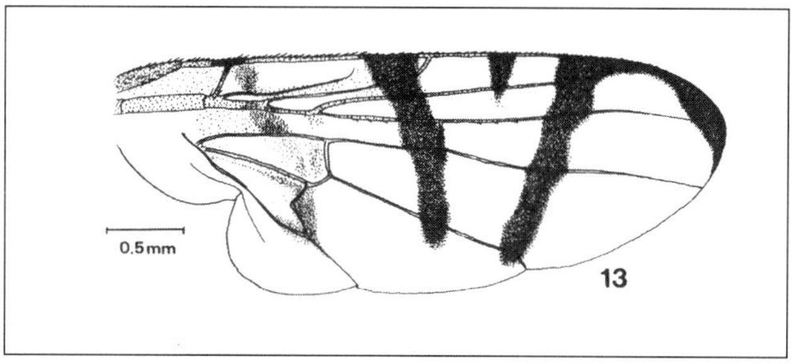


Figure 13. Wing pattern of male

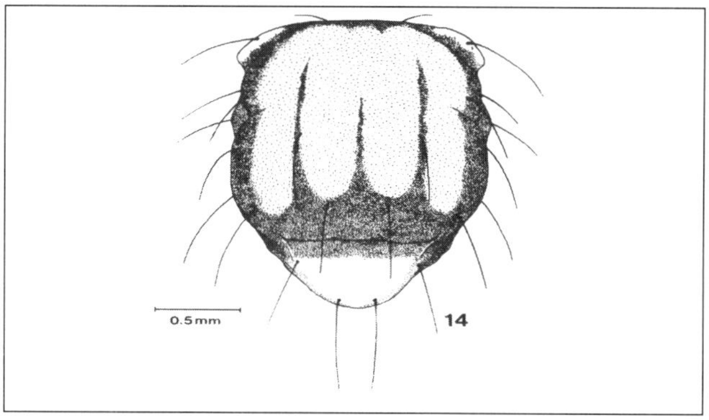


Figure 14. Thorax pattern of male

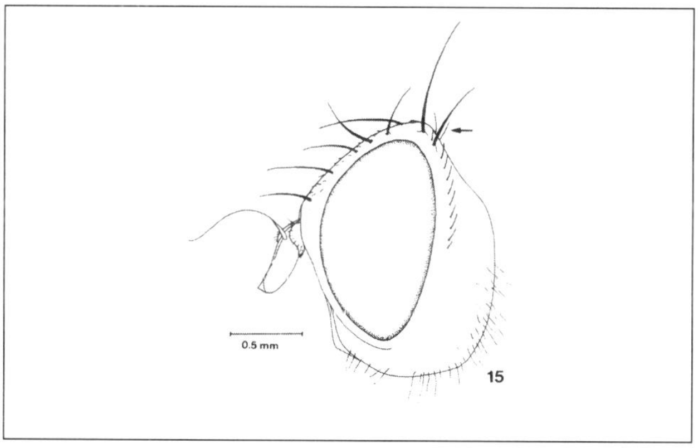


Figure 15. Lateral view of head of male

**Wing pattern**

The sub-apical band is connected to the anterior apical band. The accessory costal band is broad in the costal vein and narrower after the $R_{2+3}$ veins (Figure 13).

**Thorax pattern**

This black, scutal pattern consists of 4 longitudinal bars. The middle bars are united at the level of the transverse suture (Figure 14).

**Head in lateral view**

The post ocellars setulae are yellow. The third antennal segment is sharp apically. The ocellar triangle is black (Figure 15).

The map shows the localities where fruits of wild plums and flies were collected (29° 54' S, 071° 15'W to 40° 58' S, 072° 52' W) (Figure 16).

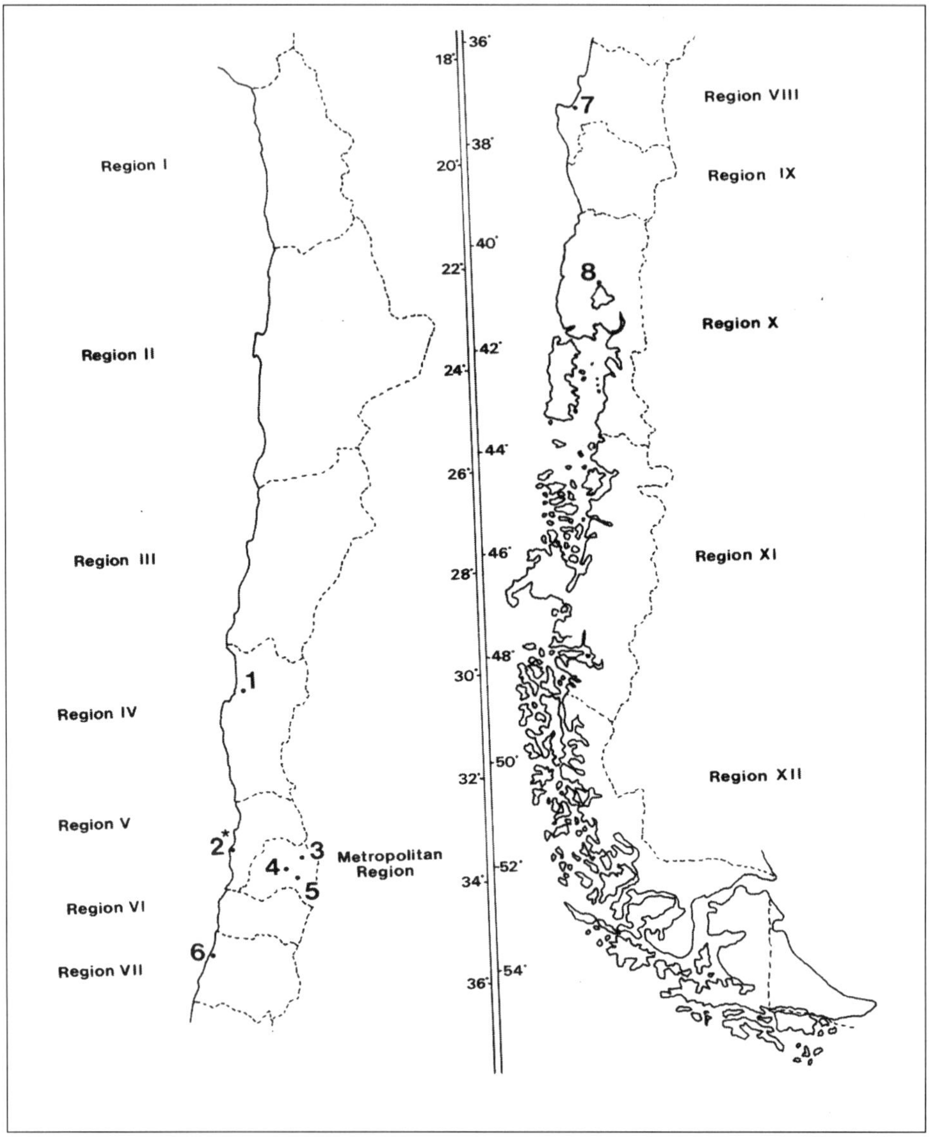


Figure 16. Map showing localities, dates and numbers of wild plum fruits of Myrobalan variety in different regions of Chile. In parentheses are indicated the data and number of fruit collected: 1) La Serena, Valle del Elqui, Region IV (18/01/97, N=50), 2) Valparaíso, Region V (27/02/97, N=70 and 30/01/98, N=41), 3) Farellones, Metropolitan Region (26/01/98, N=238), 4) Ñuñoa, Santiago, Metropolitan Region (27/01/98, N=150), 5) Las Vertientes, ajón del Maipo, Metropolitan Region (10/02/98, N=190), 6) Constitución, Region VII (01/02/98, N=78), 7) Arauco. Region VIII (02/02/, N=62) and 8) Puerto Octay, Region X (10/02/98, N=81).
* = Indicates the only locality where infected fruits with *Rhagoletis* were collected (33° 03' S, 071° 07' W).

## DISCUSSION

The wild plum population corresponds to the *nova* group. The adults are very similar to *R. conversa* that live on *Solanum nigrum* sympatrically. Nevertheless, at the immature stages, there are some differences in their morphology when compared with other Chilean *Rhagoletis* species in the following characteristics for the wild plum population. In the third instar larvae, the ventral cornu is black and curved; in contrast, this is straight in *R. conversa*, *R. nova* and *R. tomatis*. The anterior spiracles in third instar larvae is bilobed with 22 – 27 tubules, in *R. nova* with 15- 25, *R. tomatis*, 18 – 22, and *R. conversa,* 11-17 tubules (Frías 1986a, Frías and Martinez 1991a, Frías et al. 1993, Frías et al. 1996). With regard to the posterior spiracles, the wild plum population has dorsal and ventral bundles with 8 – 13 hairs. In *R. nova,* there are 6 – 13 and in *R. conversa,* 4 - 5 hairs. In the wild plums, the lateral bundles have 7 – 10 hairs, 6 – 9 hairs in *R. nova* and 4 – 5 hairs in *R. conversa.* In the third instar larvae of wild plums, we detected 4 preoral teeth, in *R. tomatis* 3, in *R. nova* 3 – 4 and in *R. conversa* 3 – 5 (Frías et al. 1993).

The geographical distribution of this population is localised only in Region V of Chile (33° 03' S, 071° 07' W) in sympatric conditions with *R. conversa* that live on *S. nigrum,* between Regions III to X (27° 22' S, 072° 20' W to 41° 28' S, 072° 56' W). Thus, as the geographical distribution of this plum population is included in the geographical distribution of their probable *R.. conversa* ancestral species, it is probable that this wild plum population corresponds to other host races or species with origins sympatrically similar to *R. nova* (Frías 1988, 1992) and *R. pomonella* host races (Bush 1969) in the framework of the hypothesis of Walsh (1864) and Smith (1966).

## REFERENCES


Bush, G.L. 1969. Sympatric host race formation and speciation in frugivorous flies of the genus *Rhagoletis* (Diptera: Tephritidae). Evolution 23 (2): 237-251.

Carroll, L.E. and R.A. Wharton. 1989. Morphology of the immature stages of *Anastrepha ludens* (Diptera: Tephritidae). Ann. Entomol. Soc. Am. 82: 201-214.

Frías, L.D. 1986a. Algunas consideraciones sobre la taxonomía de *Rhagoletis nova* (Schiner) (Diptera: Tephritidae). Rev. Chilena Entomol. 13: 59 - 73.

Frías, L.D. 1986b. Biología poblacional de *Rhagoletis nova* (Schiner) (Diptera: Tephritidae). Rev. Chilena Entomol. 13: 75 - 84.

Frías, L.D. 1988. Tiempos y Modos de especiación En: Manejo del tiempo en Biología y sus ingeniosas maneras de estudiarlo. I. Santibañez Editor. Fac. Med. U. de Chile. Copyright. Bibl. Nac.

Frías, L.D. 1992. Genética, ecología y evolución de las especies chilenas del género *Rhagoletis* (Díptera: Tephritidae). Acta Entomol. Chilena 17: 211-223.

Frías, L.D. and H. Martínez. 1991a. Estudio taxonómico en *Rhagoletis tomatis* Foote (Diptera:Tephritidae). Acta Entomol. Chilena. 16: 24 -254.

Frías, L.D., A. Malavasi and J.S. Morgante. 1984. Field observations of distributions and activities of *Rhagoletis conversa* (Diptera Tephritidae) on two hosts in nature. Ann. Entomol. Soc. Am. 77(5): 548 - 551.

Frías, L.D., I. Northland and J. Capetillo. 1991b. Aspectos de la Biología de *Rhagoletis tomatis* Foote (Diptera:Tephritidae) en poblaciones de la II Región de Chile. Acta Entomol. Chilena 16: 193-200.

Frías, LD., H. Martínez and A. Alviña. 1993. Morphological description of immature stages of *Rhagoletis tomatis* Foote (Díptera: Tephritidae). Acta Entomol. Chilena 18: 31-40.

Frías, L.D., B. Urzúa., C. Gonzalez., A. Henry and A. Alviña. 1996. Diferencias Morfológicas y Genéticas de las Especies de *Rhagoletis* de distribución chilena. Second Meeting of the Working Group on the Fruit Flies of the Western Hemisphere.I-26:42.Viña del Mar, Chile 3-8 Nov.

Foote, R.H. 1981. The genus *Rhagoletis* Loew South of the United States. USDA Tech. Bull. 1607, 77 pp.

González, R.H. 1989. Insectos y Acaros de Importancia Agrícola y Cuarentenaria en Chile. Editorial Ograma S.A. 310 pp.

Hernández-Ortíz, V.1992. El genero *Anastrepha* Schiner en México (Diptera: Tephritidae): Taxonomía, distribución y sus plantas huéspedes. Instituto de Ecología, Xalapa, Veracruz, México, Publ. N° 33 162 pp.

Hernández-Ortíz, V. 1993. Description of a new *Rhagoletis* from tropical Mexico (Diptera: Tephritidae). Proc. Entomol. Soc. Wash. 95: 418 - 424.

Persson, P.I. 1963. Studies on the biology and larval morphology of some Trypetidae (Diptera). Opusc. Entomol. 28: 33-69.

Smith, J.M. 1966. Sympatric speciation. Am. Nat. 100: 637-650.

Steck, G.J. and R.A. Wharton. 1988. Description of immature stages of *Anastrepha interrupta*, *A.limae* and *A.grandis* (Diptera: Tephritidae). Ann. Entomol. Soc. Am. 81(6): 994-1003.

Steck, G.J., L.E. Carroll, H. Celedonio-Hurtado and J. Guillen-Aguilar. 1990. Methods for identification of *Anastrepha* larvae (Diptera: Tephritidae), and key to 13 species. Proc. Entomol. Soc. Wash. 92 (2): 333-346.

Walsh, D.B. 1864. On phytophagic varieties and phytophagous species. Proc. Entomol. Soc. Phila. 3: 403-430.

White, I.M. and M. Elson-Harris. 1992. Fruit flies of economic significance. Their identification and bionomics. C.A.B., Aciar. Camberra 601 pp.

# A Checklist of the Species of *Anastrepha* with the Families of Their Host Plants and Hymenopteran Parasitoids in Brazil

*Roberto Antonio Zucchi*

Departamento de Entomologia, Escola Superior de Agricultura Luiz de Queiroz, Universidade de São Paulo, Brazil

## INTRODUCTION

Many surveys on fruit flies have already been carried out by several researchers in Brazil while others are still going on. Most of these surveys were conducted in areas where no studies had been previously done. With these surveys, new species and new records of species were found in Brazil. Also, in this decade, several surveys on fruit fly braconid parasitoids were conducted. These data have been summarised recently, because of the great interest in the biological control of fruit flies in Brazil. Research on eucoilid fruit fly parasitoids have been largely neglected. However, taxonomic studies are being conducted on eucoilids associated with frugivorous flies (Tephritidae and Lonchaeidae) in Brazil.

All the data on fruit flies, host plants and hymenopteran parasitoids are unfortunately scattered in the literature and frequently are only published as dissertations or congress abstracts which are not widely available. Even when submitted for publication, papers take a long time to come out in Brazil. Consequently, it is very difficult to get a list of the *Anastrepha* species in Brazil, or to determine which host plant species are associated with them. These data are of particular interest in the case of economically important species, especially those considered as quarantine pests. Therefore, such a list is very useful for regulatory entomologists and pest management programmes by listing the *Anastrepha* species in Brazil and their associated host plants and hymenopteran parasitoids.

The objective of this paper is to gather some available records of the *Anastrepha* species, their host plants and hymenopteran parasitoids (Braconidae and Eucoilidae) published in Brazil. Due to the space limitation of this paper, only families of the host plants of the *Anastrepha* species are presented. In fact, this work is part of a research which deals with the preparation of a database for the *Anastrepha* species in Brazil.

## MATERIALS AND METHODS

The records on the *Anastrepha* species herein presented are based on the list of host plants compiled by Norrbom and Kim (1988), and on papers (thesis, abstract of congress, short communications, full papers, etc.) published later on. Some records obtained by the author, not formally published, are also included. Only the primary records on host plants and parasitoids for the *Anastrepha* species in Brazil have been compiled. In order to list reliable data, only information based on reared specimens clearly indicated by original authors was taken into account. Data on braconid parasitoids are based on Leonel et al. (1995), which are updated here. Records on eucoilid parasitoids were compiled.

## RESULTS

### Fruit Flies

*Anastrepha* is the largest and most economically important genus of true fruit flies in the American tropics and subtropics (Norrbom and Kim 1988). Larvae of most *Anastrepha* species develop in the pulp of fleshy fruits, but larvae of a few species feed on the terminal buds and shoots of cassava or on the seeds.

Approximately 195 species are recognised in the genus *Anastrepha*. The most recent catalogue of the *Anastrepha* species and their host plants in Brazil was published as a dissertation (Zucchi 1978), in which 77 species were listed (two were synonymized later on). Only few results of this dissertation related to host plants were formally published. During the last 20 years, 17 *Anastrepha* species were found in Brazil. Hosts are known only for two of these species - *A. coronilli* and *A. turpiniae*. Ten species were recorded in the last three years (Table 1). Ninety two species of *Anastrepha* have been recorded in Brazil (Tables 2 and 4), including some species of agricultural importance such as the South American fruit fly *A. fraterculus* (Wied.), the West Indian fruit fly *A. obliqua* (Macquart) and the South American cucurbit fruit fly *A. grandis* (Macquart).

Table 1. Species of *Anastrepha* recorded in Brazil recently.

| | *Anastrepha* species | Reference |
|---|---|---|
| 1. | *aczeli* Blanchard, 1961 | Kovaleski et al.1996 |
| 2. | *alveatoides* Blanchard, 1961 | Uchoa et al. 1998 |
| 3. | *castanea* Norrbom, 1998 | Norrbom 1998 |
| 4. | *coronilli* Carrejo and González, 1993 | Ronchi-Teles et al. 1998 |
| 5. | *haywardi* Blanchard, 1961 | Uchoa et al. 1998 |
| 6. | *morvasi* Uramoto & Zucchi | Uramoto & Zucchi 1999 |
| 7. | *nigripalpis* Hendel, 1914 | Canal et al. 1996 |
| 8. | *tumida* Stone, 1942 | Uramoto et al. 1998 |
| 9. | *undosa* Stone, 1942 | Canal et al. 1996 |
| 10. | *zucchii* Norrbom, 1998 | Norrbom 1998 |

### Host Plants

For the 92 *Anastrepha* species in Brazil, hosts are known for 41 species (45%), and for 21 species (23%), only one host species is known. *A. fraterculus* is the most polyphagous species in Brazil with 58 species of host plants. *A. obliqua* is associated with 23 host species. Besides generalist species, e.g., *A. fraterculus* and *A. obliqua,* there are also specialist species, such as *A. grandis* which has been reared exclusively from Cucurbitaceae species, *A. pseudoparallela* mostly on *Passiflora* species (Passifloraceae*), A. pickeli, A. manihoti and A. montei* which breed in seeds of cassava plants (Euphorbiaceae) (Table 2).

Table 2. *Anastrepha* species for which hosts are known in Brazil.

| *Anastrepha* species | Family of host plant | No. species |
|---|---|---|
| 1. *amita* Zucchi | Verbenaceae | 1 |
| 2. *antunesi* Lima | Anacardiaceae; Myrtaceae; Rubiaceae | 3; 1; 1 |
| 3. *bahiensis* Lima | Myrtaceae; Moraceae | 3; 2 |
| 4. *barbiellinii* Lima | Cactaceae | 1 |
| 5. *barnesi* Aldrch | Sapotaceae | 1 |
| 6. *benjamin* Lima | Sapotaceae | 1 |
| 7. *bezzii* Lima | Sterculiaceae | 1 |
| 8. *bistrigata* Bezzi | Anacardiaceae; Myrtaceae; Sapotaceae | 2; 1; 1 |
| 9. *bondari* Lima | Sapotaceae; Myrtaceae; unknown families | 1; 1; 2 |
| 10. *consobrina* (Loew) | Passifloraceae | 1 |
| 11. *coronilli* Carrejo & González | Melastomataceae | 1 |
| 12. *curitis* Stone | Passifloraceae | 1 |
| 13. *dissimilis* Stone | Passifloraceae | 1 |
| 14. *distincta* Greene | Myrtaceae; Fabaceae; Anacardiaceae; Clusiaceae; Caesalpinaceae | 4; 3; 1; 1; 1 |
| 15. *duckei* Lima | Flacourtiaceae | 1 |
| 16. *fraterculus* (Wied.) | Myrtaceae; Rosaceae; Rutaceae; Sapotaceae; Fabaceae; Anacardiaceae;Annonaceae; Combretaceae; Ebenaceae; Hippocrateaceae; Malpighiaceae; Moraceae; Rubiaceae; Oxalidaceae; Passifloraceae | 17; 10; 10; 5; 2; 2; 2; 1; 1; 1; 1; 1; 1; 1; 1 |
| 17. *grandis* (Macquart) | Cucurbitaceae | 2 |
| 18. *greenei* Lima | Bombaceae | 2 |
| 19. *kuhlmanni* Lima | Passifloraceae | 1 |
| 20. *leptozona* Hendel | Sapotaceae; Icacinaceae; Myrtaceae; 3 unknown families | 6; 1; 1; 3 |
| 21. *lutzi* Lima | Passifloraceae | 1 |
| 22. *macrura* Hendel | Sapotaceae | 1 |
| 23. *manihoti* Lima | Euphorbiaceae | 1 |
| 24. *minensis* Lima | Myrtaceae | 1 |
| 25. *montei* Lima | Euphorbiaceae | 2 |

Continuation of Table 2.

| *Anastrepha* species | Family of host plant | No. species |
|---|---|---|
| 26. *obliqua* (Macquart) | Myrtaceae; Anacardiaceae; Combretaceae; Malpighiaceae; Oxalidaceae | 13; 7; 1; 1; 1 |
| 27. *parallela* (Wied.) | Sterculiaceae; unknown family | 1; 1 |
| 28. *phaeoptera* Lima | Moraceae | 1 |
| 29. *pickeli* Lima | Euphorbiaceae; Bombaceae | 1; 1 |
| 30. *pseudoparallela* (Loew) | Passifloraceae; Anacardiaceae; Myrtaceae | 3; 1; 1 |
| 31. *quararibae* Lima | Bombaceae | 2 |
| 32. *quiinae* Lima | Quiinaceae | 1 |
| 33. *rheediae* Stone | Clusiaceae | 1 |
| 34. *serpentina* (Wied.) | Sapotaceae; Myrtaceae; Clusiaceae; Hippocrateaceae | 5; 2; 1; 1 |
| 35. *sororcula* Zucchi | Myrtaceae; Anacardiaceae; Fabaceae; Rubiaceae | 9; 1; 1; 1 |
| 36. *striata* Schiner | Myrtaceae; Passifloraceae | 5; 1 |
| 37. *submunda* Lima | Annonaceae; Sapotaceae | 1; 1 |
| 38. *turpiniae* Stone | Anacardiaceae; Combretaceae; Fabaceae; Myrtaceae; Rosaceae; Rutaceae | 2; 1; 1 3; 1; 1 |
| 39. *xanthochaeta* Hendel | Passifloraceae | 1 |
| 40. *zenildae* Zucchi | Myrtaceae; Anacardiaceae; Combretaceae; Fabaceae; Melastomataceae; Rhamnaceae | 6; 1; 1; 1; 1; 1 |
| 41. *zernyi* Lima | Sapotaceae | 2 |

These records are results of several surveys on fruit flies which have been carried out in Brazil recently and on studies which are still going on. For example, approximately 100 records of fruit fly host plants have been found in this decade by several Brazilian researchers, e.g., Couturier et al. (1993), Salles (1995), Silva et al. (1996), Araujo et al. (1996), Canal (1997), Kovaleski (1997), Veloso (1997) and Uchôa (1999). Most of these most recent records were obtained from bearing fruit trees which occur in a typical Brazilian vegetation named "cerrado" (savanna), situated in the central part of Brazil (Veloso 1997; Uchôa 1999). For example, only three host plant species were known for *A. sororcula* and two for *A. zenildae.* However, based on studies carried out in "cerrado", nine new hosts for each one of these species were detected.

*Anastrepha* species in Brazil are associated with fruit bearing trees of 29 families. For 41 *Anastrepha* species associated with host plants, 37% of them feed on Myrtaceae and 24% on Sapotaceae. However, from 11 plant taxa (27%) just one unique species of *Anastrepha* is known (Table 3).

Table 3. Host plant families of *Anastrepha* species in Brazil.

| Host family | *Anastrepha* species |
|---|---|
| Anacardiaceae | *antunesi, bistrigata, distincta, fraterculus, obliqua, pseudoparallela, sororcula, turpiniae, zenildae* |
| Annonaceae | *fraterculus, submunda* |
| Bombaceae | *greenei, pickeli, quararibae* |
| Cactaceae | *barbiellinii* |
| Caesalpinaceae | *distincta* |
| Clusiaceae | *distincta, rheediae, serpentina* |
| Combretaceae | *fraterculus, obliqua, turpiniae, zenildae* |
| Cucurbitaceae | *grandis* |
| Ebenaceae | *fraterculus* |
| Euphorbiaceae | *manihoti, montei, pickeli* |
| Fabaceae | *distincta, fraterculus, sororcula, turpiniae, zenildae* |
| Flacourtiaceae | *duckei* |
| Hippocrateaceae | *fraterculus, serpentina* |
| Icacinaceae | *leptozona* |
| Malpighiaceae | *fraterculus, obliqua* |
| Melastomataceae | *coronilli, zenildae* |
| Moraceae | *bahiensis, fraterculus, phaeoptera* |
| Myrtaceae | *antunesi, bahiensis ,bistrigata, bondari, distincta, fraterculus, leptozona, minensis, obliqua, pseudoparallela, serpentina, sororcula, striata, turpiniae, zenildae* |
| Oxalidaceae | *fraterculus, obliqua* |
| Passifloraceae | *consobrina, curitis, dissimilis, fraterculus, kuhlmanni, lutzi, pseudoparallela, striata, xanthochaeta* |
| Quiinaceae | *quiinae* |
| Rhamnaceae | *zenildae* |
| Rosaceae | *fraterculus, turpiniae* |
| Rubiaceae | *antunesi, fraterculus, sororcula* |
| Rutaceae | *fraterculus, turpiniae* |
| Sapindaceae | *fraterculus* |
| Sapotaceae | *barnesi, benjamini, bistrigata, bondari, fraterculus, leptozona, macrura,serpentina, submunda, zernyi* |
| Sterculiaceae | *bezzii, parallela* |
| Verbenaceae | *amita* |

In spite of several studies carried out in the last decade in Brazil, there is still a need to sample fruit flies directly from their host plants. There are still many huge areas where no surveys on fruit flies have been conducted. For example, 30 *Anastrepha* species (Zucchi et al. 1996) are recorded in the Brazilian Amazon (3,867,886 km$^2$ and about 180 fruit tree species), but for one *species (A. duckei* Lima) only one unique host plant has been identified. In fact, for 51 *Anastrepha* species (55%), no host from Brazil is known (Table 4). Therefore, the host plants of many *Anastrepha* species in Brazil remain relatively unknown.

Table 4. *Anastrepha* species for which no host is known in Brazil.

| | |
|---|---|
| 1. | *aczeli* Blanchard |
| 2. | *alveata* Stone |
| 3. | *alveatoides* Blanchard |
| 4. | *amnis* Stone |
| 5. | *anomala* Stone |
| 6. | *atrigona* Hendel |
| 7. | *barrettoi* Zucchi |
| 8. | *belensis* Zucchi |
| 9. | *binodosa* Stone |
| 10. | *borgmeiri* Lima |
| 11. | *castanea* Norrbom |
| 12. | *caudata* Stone |
| 13. | *concava* Greene |
| 14. | *connexa* Lima |
| 15. | *costalimai* Autuori |
| 16. | *cruzi* Lima |
| 17. | *daciformis* Bezzi |
| 18. | *elegans* Blanchard |
| 19. | *ethalea* (Walker) |
| 20. | *fenestrata* Lima and Lutz |
| 21. | *fischeri* Lima |
| 22. | *flavipennis* Greene |
| 23. | *fumipennis* Lima |
| 24. | *furcata* Lima |
| 25. | *hamata* (Loew) |
| 26. | *hambletoni* Lima |

| | |
|---|---|
| 27. | *hastata* Stone |
| 28. | *haywardi* Blanchard |
| 29. | *integra* (Loew) |
| 30. | *longicauda* Lima |
| 31. | *luederwaldti* Lima |
| 32. | *matertela* Zucchi |
| 33. | *megacantha* Zucchi |
| 34. | *mixta* Zucchi |
| 35. | *morvasi* Uramoto & Zucchi |
| 36. | *nascimentoi* Zucchi |
| 37. | *nigripalpis* Hendel |
| 38. | *obscura* Aldrich |
| 39. | *perdita* Stone |
| 40. | *punctata* Hendel |
| 41. | *sagittifera* Zucchi |
| 42. | *shannoni* Stone |
| 43. | *similis* Greene |
| 44. | *simulans* Zucchi |
| 45. | *sinvali* Zucchi |
| 46. | *sodalis* Stone |
| 47. | *tenella* Zucchi |
| 48. | *townsendi* Greene |
| 49. | *tumida* Stone |
| 50. | *undosa* Stone |
| 51. | *zucchii* Norrbom |

## Braconid Parasitoids

The data related to fruit fly braconid parasitoids were summarised by Leonel et al. (1995). Seven braconid species have been associated with 15 species of *Anastrepha* in Brazil. The parasitoids belong to subfamily Opiinae, except for *Asobara anastrephae* which belongs to the subfamily Alysiinae. *Doryctobracon areolatus* is the most common braconid parasitoid of *Anastrepha* in Brazil, and it has been associated with 11 species (Table 5). *Opius sp.* represents either an undescribed species or a discrete colour morph of *Opius bellus.* Current knowledge of Brazilian tephritid parasitoids (e.g., field evaluation, distribution patterns, rearing techniques etc.) is still extremely inadequate.

## Eucoilid Parasitoids

Information is restricted to taxonomic studies or the relationship between the eucoilid and the fruit fly. Consequently, nothing is known about the role played by eucoilids in the biological control of tephritid pests in Brazil. Only two eucoilid species were associated to fruit-infesting tephritids in Brazil . However, as a result of surveys on eucoilid frugivorous fly parasitoids from several Brazilian states, other five species of eucoilids could be associated to *Anastrepha* species (Guimarães 1998) (Table 5).

Table 5. Hymenopteran parasitoids of *Anastrepha* species in Brazil.

| ***Anastrepha* species** | ***Braconid* species** | **Reference** |
|---|---|---|
| 1. *A. amita* Zucchi | *Doryctobracon areolatus*<br>*Doryctobracon brasiliensis* | Souza et al. 1996<br>Souza et al. 1996 |
| 2. *A. bahiensis* Lima | *Doryctobracon areolatus* | Canal et al. 1994 |
| 3. *A. bistrigata* Bezzi | *Dorctobracon areolatus* | Leonel et al. 1995 |
| 4. *A. distincta* Greene | *Opius* sp. | Canal et al. 1994 |
| 5. *A. fraterculus* Wied. | *Doryctobracon areolatus*<br>*Doryctobracon brasiliensis*<br>*Doryctobracon fluminensis*<br>*Opius bellus*<br>*Utetes anastrephae* | Lima 1937a<br>Lima 1937a<br>Lima 1938<br>Lima 1937b<br>Lima 1937a |
| 6. *A. leptozona* Hendel | *Doryctobracon areolatus*<br>*Opius* sp. | Canal et al. 1994<br>Canal et al. 1994 |
| 7. *A. manihoti* Lima | *Opius bellus* | Lima 1937a |
| 8. *A. montei* Lima | *Opius bellus* | Lima 1937a |
| 9. *A. obliqua* Macquart | *Asobara anastrephae*<br>*Doryctobracon areolatus*<br>*Opius bellus*<br>*Opius* sp.<br>*Utetes anastrephae* | Canal et al. 1994<br>Lima 1937b<br>Lima 1937a<br>Lima 1937a<br>Gonçalves 1938 |
| 10. *A. parallela* Wied. | *Doryctobracon fluminensis* | Lima 1948 |
| 11. *A. pseudoparallela* Loew | *Doryctobracon areolatus*<br>*Doryctobracon fluminensis* | Leonel 1995<br>Gonçalves 1938 |

Continuation of Table 5.

| ***Anastrepha* species** | ***Braconid* species** | **Reference** |
|---|---|---|
| 12. *A. serpentina* Wied. | *Doryctobracon areolatus*<br>*Doryctobracon brasiliensis*<br>*Opius bellus* | Lima 1937b<br>Gonçalves 1938<br>Lima 1938 |
| 13. *A. sororcula* Zucchi | *Doryctobracon areolatus*<br>*Doryctobracon brasiliensis*<br>*Utetes anastrephae* | Arrigoni 1984<br>Arrigoni 1984<br>Arrigoni 1984 |
| 14. *A. striata* Schiner | *Doryctobracon areolatus* | Canal et al. 1994 |
| 15. *A. turpiniae* Stone (as *A. fraterculus*) | *Doryctobracon areolatus* | Silva 1994 |
| | ***Eucoilid species*** | |
| 1. *Anastrepha* spp. | *Trybliographa* sp. | Guimarães 1998 |
| 2. *A. amita* | *Aganaspis pelleranoi*<br>*Dicerataspis flavipes*<br>*Lopheucoila anastrephae* | Guimarães 1998 |
| 3. *A. bahiensis* | *Aganaspis pelleranoi*<br>*Aganaspis nordlanderi* | Guimarães 1998 |
| 4. *A. fraterculus* | *Aganaspis pelleranoi* (as *Ganaspis carvalhoi*)<br>*Odontosema anastrephae* | Dettmer 1929<br>Borgmeier 1935 |
| 5. *A. pseudoparallela* | *Lopheucoila anastrephae* | Guimarães 1998 |
| 6. *A. serpentina* | *Aganaspis pelleranoi* (as *Ganaspis carvalhoi*) | Lima 1940 |
| 7. *A. striata* | *Aganaspis nordlanderi* | Guimarães 1998 |

## ACKNOWLEDGEMENTS

To Conselho Nacional de Desenvolvimento Científico e Tecnológico for a grant to carry out this research and to FAPESP (Fundação de Amparo à Pesquisa do Estado de São Paulo) for the financial support to attend the Fifth International Symposium on Fruit Flies of Economic Importance.

## REFERENCES

Araujo, E.L., R.A. Zucchi and N.A. Canal D. 1996. Caracterização e ocorrência de *Anastrepha zenildae* Zucchi (Diptera: Tephritidae) e seus parasitóides (Hymenoptera: Braconidae) numa nova planta hospedeira, no Rio Grande do Norte. An. Soc. Entomol. Brasil 25(1):147-150.

Arrigoni, E.B. 1984. Dinâmica populacional de moscas-das-frutas (Diptera, Tephritidae) em três regiões do Estado de São Paulo. Piracicaba. Ph. D. dissertation, ESALQ, 166 pp.

Borgmeier, T. 1935. Sobre alguns Cynipideos parasíticos e cecidogenos do Brasil (Hymenoptera: Cynipidae). Arch. Inst. Biol. Vegetal 2(1):97-124.

Canal, D., N.A. 1997. Levantamento, flutuação populacional e análise faunística das espécies de moscas-das-frutas (Dip., Tephritidae) em quatro municípios do norte do Estado de Minas Gerais. Ph. D. dissertation, ESALQ, 113 pp.

Canal D., N.A., R.A. Zucchi, N.M. Silva and F.L. Leonel Jr. 1994. Reconocimiento de las especies de parasitoides (Hym., Braconidae) de moscas de las frutas (Diptera: Tephritidae) en dos municipios de Estado de Amazonas, Brasil. Bol. Mus. Ent. Univ. Valle 2(1,2):1-17.

Canal D., N.A, R.A. Zucchi and C.D. Alvarenga. 1996. Nuevas especies y referencias de *Anastrepha* Schiner, 1868 (Diptera: Tephritidae) en el Brasil. *In* Meeting of the Working Group on Fruit Flies of the Western Hemisphere, 2nd, Viña del Mar, p. 30 (abstract).

Couturier, G., R.A. Zucchi, G. Saravia M. and N.M. Silva. 1993. New records of fruit flies of the genus *Anastrepha* Schiner, 1868 (Diptera: Tephritidae) and their host plants, in the Amazon region. Ann. Soc. Entomol. Fr.(N.S.), 29(2):223-224.

Dettmer, H. 1929. *Ganaspis carvalhoi* n.sp. (Hymenoptera – Cynipidae) um novo parasita da mosca das fructas (Anastrepha fratercula Wied.) Bol. Biol. 15(4-5).

Gonçalves, C.R. 1938. As môscas de frutas e seu combate. Publ. Serv. Def. San. Veg. 12: 43 pp.

Guimarães, J.A. 1998. Espécies de Eucoilinae (Hymenoptera: Figitidae) parasitóides de larvas frugívoras (Diptera: Tephritidae e Lonchaeidae) no Brasil. M. Sc. thesis, ESALQ, 86 pp.

Kovaleski, A. 1997. Processo adaptativos na colonização da maçã (Malus domestica L.) por *Anastrepha fraterculus* (Wied.) (Diptera: Tephritidae) na região de Vacaria, R.S. Ph. D. dissertation, Instituto de Biociências/USP, 122 pp.

Kovaleski, A., K. Uramoto, R.L. Sugayama, N.A. Canal D., and A. Malavasi. 1996. A survey of *Anastrepha* species present in apple growing area from Rio Grande do Sul, Brasil. *In* Meeting of the Working Group on Fruit Flies of the Western Hemisphere, 2nd, Viña del Mar, p. 32 (abstract).

Leonel Jr., F.L., R.A. Zucchi, and R.A. Wharton. 1995. Distribution and tephritid hosts (Diptera) of braconid parasitoids (Hymenoptera) in Brazil. Internat. J. Pest Managem., 41:208-213.

Lima, A. da Costa. 1937a. Novas moscas de frutas do genero *Anastrepha* (Dip., Tephritidae). O Campo, 8:34-38.

Lima, A. da Costa. 1937b. Novas moscas de frutas do genero *Anastrepha* (Dip., Trypetidae). O Campo, 8:60-64.

Lima, A. da Costa. 1938. Vespas parasitas de moscas de frutas (Hym., Braconidae). O Campo, 9:69-72.

Lima, A. da Costa. 1940. Alguns parasitos de moscas-das-frutas. An. Acad. Brasil. Scien., XII, p. 17.

Lima, A. da Costa. 1948. Entomofagos sul americanos (parasitos e predadores) de insetos nocivos a agricultura. Bol. Soc. Brasil. Agronomia 11:1-32.

Nascimento, A. S. 1990. Aspectos ecológicos e tratamento pós-colheita de moscas-das-frutas (Tephritidae) em manga, *Mangifera indica*. Ph. D. dissertation, Instituto de Biociências/USP. 97 pp.

Norrbom, A.L. 1998. A revision of the *Anastrepha daciformis* species group (Diptera: Tephritidae). Proc. Entomol. Soc. Wash. 100(1):160-192.

Norrbom, A.L. and K.C. Kim. 1988. A list of the reported host plants of the species of the *Anastrepha* (Diptera: Tephritidae). U. S. Dept. Agric. APHIS 81-52, 114 pp.

Ronchi-Teles, B., N.M. Silva, R.A. Zucchi. 1998. Constatação de *Anastrepha coronilli* (Dip., Tephritidae) na Amazônia Brasileira. *In* Congresso Brasileiro de Entomologia, 17th, Rio de Janeiro, RJ, p. 862 (abstract).

Salles, L.A.B. 1995. Bioecologia e controle da mosca-das-frutas sul-americana. EMBRAPA – CPACT, 58 pp.

Silva, N.M. 1994. Levantamento e análise faunística de moscas-das-frutas em quatro locais do Estado do Amazonas. Piracicaba. Ph.D. dissertation, ESALQ, 175 pp.

Silva, N.M., S. Silveira Neto, and R.A. Zucchi. 1996.The natural host plants of *Anastrepha* (Dip.: Tephritidae) in the State of Amazonas, Brazil. pp. 353-357. *In* Fruit Fly Pests. A world assessment of their biology and management. B.A. McPheron and G.J. Steck. (eds.). 616 pp., St. Lucie Press, Delray Beach, Florida, USA.

Souza Filho, M.F., A. Raga, N.A. Canal D. and R.A. Zucchi. 1996. Parasitóides (Hym., Braconidae) associados a *Anastrepha amita* (Dip., Tephritidae) em pombeiro *Cytharexylum myrianthum* (Verbenaceae). *In* Simpósio de Controle Biológico – Siconbiol, 5, Foz do Iguaçu, p. 287 (abstract).

Uchôa, F.M.A., I. Oliveira, R.M.S. Molina and R.A. Zucchi. 1998. Constatação de *Anastrepha alveatoides*, *A. castanea* e de *A. haywardi* (Diptera, Tephritidae) no Brasil. *In* Congresso Brasileiro de Entomologia, 17th, Rio de Janeiro, RJ, p. 72 (abstract).

Uchôa F., M.A. 1999. Biodiversidade de moscas frugívoras (Diptera, Tephritoidea), seus frutos hospedeiros e parasitóides (Hymenoptera) em áreas de cerrado do Estado de Mato Grosso do Sul. Ph. D. dissertation, ESALQ, 104 pp.

Uramoto, K., N.A. Canal and R.A. Zucchi. 1998. New records of *Anastrepha* species (Diptera: Tephritidae) in Brazil. *In* Fifth International Symposium on Fruit Flies of Economic Importance, 1-5 June 1998, Penang, Malaysia, p. 45 (abstract).

Uramoto, K. & R.A. Zucchi 1999. New species in the genus *Anastrepha* (Diptera: Tephritidae) from Brazil. Ann. Soc. Entomol. Brasil 28(1):85-89.

Veloso, V.R.S. 1997. Dinâmica populacional de *Anastrepha* spp. e *Ceratitis capitata* (Wied., 1824) (Diptera: Tephritidae) nos cerrados de Goiás. Ph.D. dissertation, Universidade Federal de Goiás, 115 pp.

Zucchi, R.A. 1978. Taxonomia das espécies de *Anastrepha* Schiner, 1868 (Diptera: Tephritidae) assinaladas no Brasil. Ph. D. dissertation, ESALQ, 105 pp.

Zucchi, R.A., N.M. Silva and S. Silveira Neto. 1996. *Anastrepha* species from the Brazilian Amazon: Distribution, Hosts, and Lectotype Designations. pp. 259-364. *In* Fruit Fly Pests - A world assessment of their biology and management. B.A. McPheron and G.J. Steck. (eds.). St. Lucie Press, Delray Beach, Florida, USA.

# Species Identification of Tephritids Across a Broad Taxonomic Range Using Ribosomal DNA

*Karen F. Armstrong and Charlotte M. Cameron*

Ecology and Entomology Group, Lincoln University, Canterbury, New Zealand

## INTRODUCTION

International trade and passenger travel are significant factors in the spread of economically important fruit fly species. The risk of accidental introduction via infested fruit is high, and in New Zealand the recent Medfly incursion in Auckland demonstrated the reality of this threat (Frampton, 2000). There are no economically important species of fruit fly established in New Zealand at present, but 31 are considered high risk in terms of their potential colonisation (refer to the Biosecurity (Notifiable Organisms) Amendment Order 1997). These are amongst a background of non-pest and low risk pest species that may also arrive in fruit from neighbouring countries or trading partners.

Quarantine officials closely monitor fruit fly host material at the New Zealand borders (Frampton, 2000). In terms of the action to be taken should an infestation be discovered, there is significant benefit from being able to accurately identify species from the immature life stages, or at least to distinguish the high and low risk groups (Armstrong et al. 1997a). The need for this quarantine application was also highlighted by White (1996) at the previous fruit fly symposium in Sand Keys, Florida, where he summarised the advances made in larval taxonomy over the last decade. Despite this, morphological keys such as those of Steck et al. (1990) and White and Elson Harris (1992), are still only available for about a third of ca. 250 pest species. For those species, even so, identification is not easy and only possible for good quality late instar larvae; there are no morphological characters for early instars or eggs.

Until recently in New Zealand, the identification of immature life stages depended entirely on rearing through to adults. This was time consuming and often unsuccessful (Armstrong et al. 1997b). A rapid molecular technique has since been described as a feasible alternative or supplementary quarantine tool (Armstrong et al. 1997a). The method is based on the polymerase chain reaction and restriction fragment length polymorphism (PCR-RFLP) analysis of ribosomal DNA (rDNA). Development of this technique has continued to "package" it for routine quarantine use. Here we present an overview of the improved method and summarise its capability as a diagnostic tool.

## MATERIALS AND METHODS

### Fruit Fly Specimens and DNA Extraction

The list of species included to date is given in Figure 2. Positively identified fruit flies were provided from appropriate localities around the world (see acknowledgments) and used to determine species-specific DNA markers. Voucher specimens are maintained at the Lincoln University's Entomology Museum. Genomic DNA was prepared from individual insects, as previously described, using standard proteinase K digestion and selective binding of the DNA to a solid matrix (Prep a gene, BioRad Laboratories) (Armstrong et al. 1997a). DNA from single insects was resuspended in 100 μl of tris-EDTA (TE) for adults or 2nd-4th instar larvae, and 20 μl for first instar larvae or eggs.

### PCR-RFLP Analysis

PCR amplification of the partial 18S plus complete Internal Transcribed Spacer (ITS)1 rDNA region was carried out in a GeneAmp PCR System 2400 (Perkin Elmer). The amplicon is equivalent to bases 1126 to 2748 on the *Drosophila melanogaster* Meigen (Diptera: Drosophilidae) rDNA map (Tautz et al. 1988). Reactions consisted of 0.6 μl of DNA (diluted 500-fold for eggs), 0.24 μM each primer, 0.15 mM each dNTP, 0.6U of Expand High Fidelity polymerase and 1x buffer (Boehringer Mannheim) in a total of 15 μl. The PCR primers used were NS15 (tephritid-specific, CAATTGGGTGTAGCTACTAC, Armstrong and Cameron, unpublished) and ITS6 (5.8S universal, AGCCGAGTGATCCACCGCT, Armstrong and Cameron, unpublished). The thermal profile was 2 min at 94°C followed by 31 cycles (35 for eggs) of 30 s at 94°C, 30 s at 60°C and 2 min at 68°C (no stepwise extension) and then a final extension of 5 min at 72°C. 100-200 ng (ca.1-2 μl) of the PCR product was used directly in 10 μl restriction digests using conditions recommended by the enzyme manufacturers. Products of the digestion were resolved on a 2% agarose (LE, FMC) gel containing ethidium bromide and visualised over UV.

## RESULTS

Figure 1 illustrates the type of restriction patterns that form the basis of the diagnostic procedure. In this example, the enzyme, *Rsa* I, shows how the difference in restriction patterns broadly reflects the genetic distance of the species being compared. For example, the distinctions between species of different genera or between the different *Bactrocera* subgenera are very clear; those between species within the *Bactrocera* subgenus are more subtle. Other restriction enzymes distinguish species that are not so easily differentiated here, e.g., *Ssp* I clearly separates *B. (Afrodacus) jarvisi* (Tryon) from *B. (Bactrocera) tryoni* (Froggatt) (data not shown). For diagnosis, several restriction enzymes are used to provide a haplotype of at least four unique or variable fragments (Armstrong et al. 1997a). The whole procedure takes less than 12 hours.

Thirty eight species have been analysed to date, including nine low risk species. Two other high risk species, *B. (B.) zonata* (Saunders) and *Ceratitis (Pterandrus) rosa* Karsch, have yet to be processed. A core of nine restriction enzymes, out of 21 screened, generates distinctive patterns for one or more species. Figure 2 summarises the number of restriction enzymes producing clear diagnostic patterns for any pairwise combination of species. This ranges from all nine distinguishing the most distantly related taxa, to none for the most closely related within the species groups. To date, 28 species have unique restriction pattern haplotypes. The remaining 10 fall into four groups of two or three species with identical haplotypes. These can often be distinguished by their geographic origin and/or the host fruit with which they were associated (see White and Elson Harris (1992) for host and distribution lists); nevertheless, they are all high risk species which would require the same quarantine action.

MW Cc Az Al Ao As Ag Af Bc Bx Bcb Bj MW Bt Bmu Bd Bm Btl Bk Bp Bl Bpd Bo Bu Ds Rc Cc MW

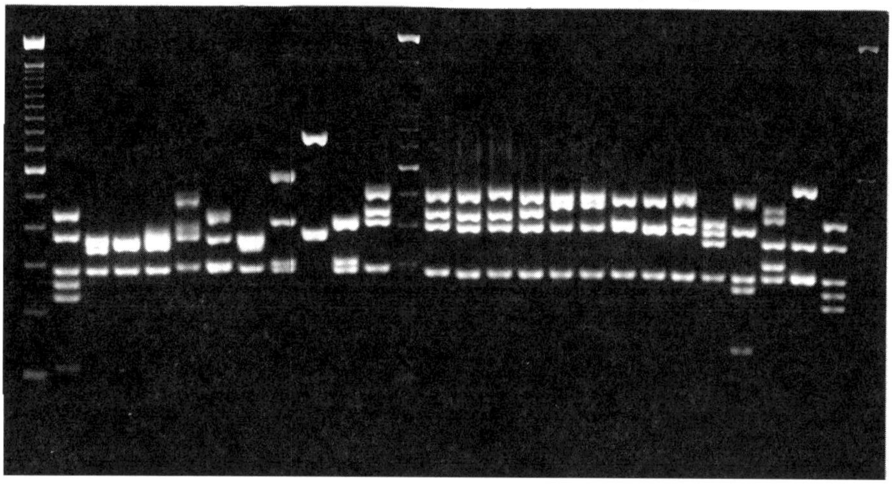

Figure 1. Diagnostic restriction markers. *Rsa* I RFLP of partial 18S plus ITS1 PCR product. Lanes 1, 13 and 28 are molecular weight markers (100 bp ladder, Gibco BRL). Cc, *Ceratitis (C.) capitata;* Az, *Anastrepha zenildae;* Al, *A. ludens;* Ao, *A. obliqua;* As, *A. serpentina;* Ag, *A. grandis;* Af, *A. fraterculus;* Bc, *Bactrocera (Au.) cucumis;* Bx, *B. (N.) xanthodes;* Bcb, *B. (Z.) cucurbitae;* Bj, *B. (Af.) jarvisi;* Bt, *B. (B.) tryoni;* Bmu, *B. (B.) musae;* Bd, *B. (B.) dorsalis;* Bm, *B. (B.) melanotus;* Btl, *B. (B.) trilineola;* Bk, *B. (B.) kirki;* Bp, *B.(B.) passiflorae;* Bl, *B.(B.) latifrons;* Bpd, *B. (B.) psidii;* Bo, *B. (B.) oleae;* Bu, *B.(B.) umbrosa;* Ds *Dacus (C.) solominensis;* Rc, *Rhagoletis completa.*

## DISCUSSION

### Protocol Development

This protocol differs from the previous (Armstrong et al. 1997a) in that only 1.5 kb of rDNA is amplified, rather than the 3.5 kb of the complete 18S+ITS1+5.8S+ITS2 region. Preliminary PCR-RFLP experiments (unpublished) have shown that the ITS1 provides the majority of diagnostic restriction sites for these taxa. Also, an 18S tag is required to be amplified with the ITS1 to detect the variable restriction sites at the ITS1 5' end. The length of the 18S tag (ca. 600 bp) was determined by the location of suitable tephritid-specific PCR priming sites. Specificity of the PCR, using a range of

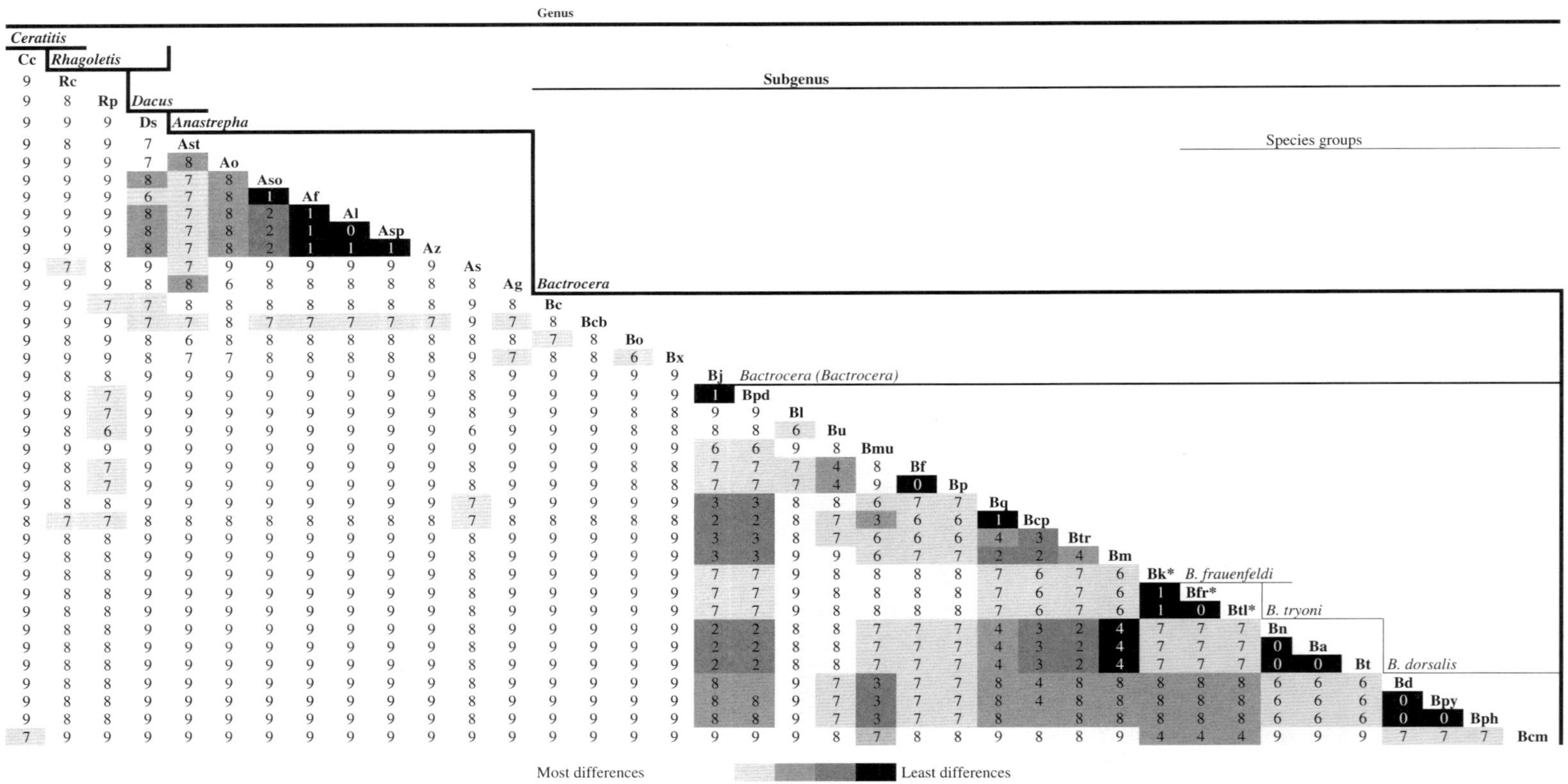


Figure 2. Number of restriction enzymes that distinguish each pairwise set of species. Cc, *Ceratitis (Ceratitis) capitata* (Wiedemann); Ds *Dacus (Callantra) solominensis* Malloch[†]; Rc, *Rhagoletis completa* Cresson[†]; Rp, *R. pomonella* (Walsh); Ast, *A. striata* Schiner; Ao, *A. obliqua* (Maquart); Aso, *A. sorocula* Zucchi[†]; Af, *A. fraterculus* (Wiedemann); Al, *A. ludens* (Loew); Asp, *A. suspensa* (Loew); Az, *Anastrepha zenildae* Zucchi[†]; As, *A. serpentina* (Wiedemann); Ag, *Anastrepha grandis* (Maquart)[†]; Bc, *Bactrocera (Austrodacus) cucumis* (French); Bcb, *B. (Zeugodacus) cucurbitae* (Coquillett); Bo, *B. (Daculus) oleae* (Gmlin)[†]; Bx, *B. (Notodacus) xanthodes* (Broun); Bj, *B. (Afrodacus) jarvisi* (Tryon); Bpd, *B. (B.) psidii* (Froggatt); Bl, *B.(Bactrocera) latifrons* (Hendel); Bu, *B.(B.) umbrosa* (Fabricius)[†]; Bmu, *B. (B.) musae* (Tryon)[†]; Bf, *B. (B.) facialis* (Coquillett); Bp, *B.(B.) passiflorae* (Froggatt); Bq, *B.(B.) quadriestosa (Bezzi)*[†]; Bcp, *B. (B.) curvipennis* (Froggatt); Btr, *B. (B.) trivialis* (Drew); Bm, *B. (B.) melanotus* (Coquillett); Bk, *B. (B.) kirki* (Froggatt); Bfr, *B. (B.) frauenfeldi* (Schiner); Btl, *B. (B.) trilineola* Drew; Bn, *B. (B.) neohumeralis* (Hardy); Ba, *B. (B.) aquilonis* (May); Bt, *B. (B.) tryoni* (Froggatt); Bd, *B.(B.) dorsalis* (Hendel); Bpy, *B. (B.) papayae* Drew and Hancock; Bph, *B. (B.) philippinensis* Drew and Hancock; Bcm, *B. (B.) carambolae* Drew and Hancock. [†] low risk species.

* The enzyme AccI distinguishing these species has not been tested with other species.

stringencies, has been demonstrated with DNA from fungi (Hyphomycetes), fruit (papaya) and other insects (Diptera, Coleoptera, Hemiptera and Lepidoptera) (unpublished data). Compared to the original technique, the PCR is more robust for use with DNA that is badly degraded or contaminated with fruit or fungi, and the restriction patterns are much simpler to interpret.

## Current Diagnostic Capability

The challenge in developing a single molecular test for species identification was to find a DNA region whose evolutionary rate appropriately reflected that of the taxa involved, i.e., to distinguish from between genera to within a species complex, but not between populations. Restriction sites within the ITS1 largely achieve this. The limit to its use appears to be at the level of species complexes where separation was generally poor. An exception to this was within the *B. (B.) dorsalis* complex where *B. (B.) carambolae* Drew and Hancock could be clearly separated. This indicates that resolution may be improved once the complete ITS1 sequence data are obtained for these species (see below). Certainly, other studies in insects using the PCR-RFLP analysis of the ITS have found diagnostic markers for cryptic species (Pfiefer et al. 1995, Beebe and Saul 1995, Severini et al. 1996). To our knowledge, however, there are no other cases published that utilise the ITS to distinguish so many species or over such a broad evolutionary range. Unexpectedly, clear separations have not yet been achieved between *B. (B.) passiflorae* (Froggatt) and *B. (B.) facialis* (Coquillett). These species are not considered closely related based on morphology (White and Elson-Harris 1992), but, in support of the restriction site data, ca. 600 bases of the ITS1 sequence are identical between them (unpublished, Genbank accession numbers AF121154 and AF121147 respectively). The remaining ca. 150 bases at the 5' end have yet to be sequenced. There is also very little sequence variation in the ITS2 of these two species, but a single restriction site has been shown to discriminate them (L. McKenzie and J. Curran, personal communication). Importantly, in the ITS1, there has been no evidence of diagnostic restriction site variation in populations from different geographic locations within any of the species (unpublished). An exception was *Anastrepha fraterculus* (Wiedemann), but these flies are now being proposed as comprising a species complex which is as yet poorly resolved morphologically (McPheron 2000).

## Reliability of the ITS1

For two species in this study, *B.(B.) aquilonis* (May) and *B. (B.) curvipennis* (Froggatt), intra-specific variation has been observed. This is manifested as either subtle differences in length or a "fuzzy" appearance of particular restriction fragments and is consistent with the presence of multiple sequence copies. Sequence data for the ITS of other insects, including the Medfly (Merrill and Campbell, unpublished, see Roderick 1996), have demonstrated that this can occur. The ITS is thought not to be free to evolve at a neutral rate due to the conserved regions necessary for transcription of the rRNA genes (Schlötterer et al. 1994). However, the intra-specific variation in the ITS of other insects has been associated with the length of simple sequence repeats. These are (ssr) thought to be located in DNA not critical for secondary structure (reviewed in Fenton et al. 1997). This could have implications for the reliability of the ITS as a diagnostic marker, but several factors, pertinent to this study, support its continued use. Primarily, for the large majority of fruit fly species here, there is no discernible intraspecific RFLP variation. Homogenisation through the PCR process and the use of restriction sites, i.e., small windows of sequence, possibly moderate any subtle sequence differences. The lack of sequence differentiation for some morphologically distinct species (see results above) also suggests that the ITS1 generally does not evolve very rapidly in tephritids. Prior knowledge of the variable fragments here, through the population analysis, precludes their use in diagnosis and using patterns from several enzymes safeguards against incorrect diagnoses. Finally, in contrast to point mutations that tend to occur between species (Fenton et al. 1997), changes in ssr length are unlikely to result in the addition or loss of a restriction site, at least for the recognition sequences of the enzymes used here. The only "populations" for which restriction *site* variation has been observed was for *A. fraterculus* (unpublished), but this would be consistent with its review as a species group (see above).

## CONCLUSION

The ITS1 is a versatile source of species molecular markers across a wide range of tephritid taxa. The PCR-RFLP procedure is practical as a diagnostic tool and is now used to identify all immature life stage tephritids intercepted at the New Zealand border. As with the use of any key, identification is made with caution in lieu of the hundreds of other species that have not yet been described. Nevertheless, in this particular application, as all high risk species are included, misdiagnosis would only be made for a low-risk species; thus quarantine decisions would not be compromised.

Molecular technology used in this way has the potential to produce a similar genetic database on fruit fly species to that established for world-wide Medfly populations by Steck et al. (1996) using mitochondrial DNA. Similar applications have been initiated by the Californian Department of Food and Agriculture to examine the *B. (B.) dorsalis* and *A. ludens* complexes and by the Australian quarantine and inspection service for a large number of native and exotic fruit flies. With the more widespread use of these techniques, the risk associated with different passenger routes or trade locations to be better assessed and a better evaluation of the kinetics affecting the global spread of fruit fly species may be possible.

## ACKNOWLEDGEMENTS

For the supply of fruit fly specimens, we are very grateful to Allan Allwood (Project on the Regional Management of Fruit Flies in the Pacific, Fiji), Suk-Ling Wee (Universiti Sains Malaysia), Stuart Smith (Northern Territories Department of Primary Industries, Australia), Harry Fay (Queensland Department of Primary Industries, Australia), Leslie McKenzie (CSIRO, Australia), Bruce McPheron (Pennsylvania State University, USA), Gary Steck (Florida Department of Food and Agriculture, USA), Eric Fisher (California Department of Food and Agriculture, USA), Aldo Malavasi and Roberto Zucchi (University de Sao Paulo, Brazil), and David Midgarden (FAO, Suriname). This project was funded by the NZ Ministry of Agriculture.

## REFERENCES

Armstrong, K.F., C.M. Cameron and E.R. Frampton. 1997a. Fruit fly (Diptera: Tephritidae) species identification: a rapid diagnostic technique for quarantine applicattion. Bull. Entomol. Res. 87: 111-118.

Armstrong, K.F., C.M. Cameron, E.R. Frampton and D.M. Suckling. 1997b. Aliens at the border and cadavers in the field: a molecular technique for species identification. Proc. 50th N.Z. Plant Protection Conf. Lincoln University: 316-321.

Beebe N.W. and A. Saul. 1995. Discrimination of all members of the *Anopheles punctulatus* complex by polymerase chain reaction-restriction fragment length polymorphism. Am. J. Trop. Med. Hyg. 53: 478-481.

Fenton B., G. Malloch and E. Moxey. 1997. Analysis of eriophyid mite rDNA internal transcribed spacer sequences reveals variable simple sequence repeats. Insect Mol. Biol. 6: 23-32.

Frampton, R. 2000. An overview of quarantine for fruit flies. pp. 475-479. *In* Area-wide Control of Fruit Flies and Other Insect Pests. K.H. Tan (ed.). Penerbit Universiti Sains Malaysia, Penang, Malaysia.

McPheron, B.A. 2000. Population Genetics and Cryptic Species. pp. 483-490. *In* Area-wide Control of Fruit Flies and Other Insect Pests. K.H. Tan (ed.). Penerbit Universiti Sains Malaysia, Penang, Malaysia.

Pfiefer, T.A., L.M. Humble, M. Ring and T.A. Grigliatti. 1995. Characterisation of gypsy moth populations and related species using a nuclear DNA marker. Can. Entomol. 127: 49-58.

Roderick, G.K. 1996. Population genetic studies of tephritid flies of economic importance. pp. 267-271. *In* Fruit fly pests: a world assessment of their biology and management. B.A. McPheron and G.J. Steck (eds.). St. Lucie Press, Delray Beach,Florida, USA.

Severini, C., F. Silvestrini, P. Mancini, G. La Rosa and M. Marinucci. 1996. Sequence and secondary structure of the rDNA second internal transcribed spacer in the sibling species *Culex pipiens* L. and *Cx. quinquefasciatus* Say (Diptera: Culicidae). Insect Mol. Biol. 5:181-186.

Schlötterer, C., M.T. Hauser, A. von Haeseler and D. Tautz. 1994. Comparative evolutionary analysis of rDNA ITS regions in *Drosophila*. Mol. Biol. Evol. 11: 513-522.

Steck, G.J., L.E. Carroll, H. Celedonio-Hurtado and J. Guillen-Aguilar. 1990. Methods for the identification of *Anastrepha* larvae (Diptera: Tephritidae), and key to 13 species. Proc. Entomol. Soc. Washington 92: 333-346.

Steck, G.J., G.E. Gasparich, H.-Y. Han, B.A. McPheron, and W.S. Sheppard. 1996. Distribution of mitochondrial DNA haplotypes among *Ceratitis capitata* populations worldwide. pp 291-296. *In* Fruit fly pests: a world assessment of their biology and management. B.A. McPheron and G.J. Steck (eds.). St. Lucie Press, Delray Beach, Florida, USA.

Tautz, D., J.M. Hancock, D.A. Webb, C. Tautz and G.A Dover. 1988. Complete sequences of the rRNA genes of *Drosophila melanogaster*. Mol. Biol. Evol. 5: 366-376.

White, I.M. 1996. Fruit fly taxonomy: recent advances and new approaches. pp 253-258. *In* Fruit fly pests: a world assessment of their biology and management. B.A. McPheron and G.J. Steck (eds.). St. Lucie Press, Delray Beach, Florida, USA.

White, I.M. and M.M. Elson-Harris. 1992. Fruit flies of economic significance: their identification and bionomics. CAB International, Wallingford. 601 pp.

# PART III.

## APPLICATION OF AREA-WIDE CONCEPT TO FRUIT FLIES

## K) NATURAL ENEMIES AND BIOCONTROL OF FRUIT FLIES

# Host Range and Reproductive Output of *Diachasmimorpha kraussii* (Hymenoptera: Braconidae), A Parasitoid of Tephritid Fruit Flies Newly Imported to Hawaii

*R.H. Messing*[1] *and M.M. Ramadan*[2]

[1] University of Hawaii, Department of Entomology, 7370 Kuamo'o Rd., Kapa'a, HI 96746, USA
[2] Hawaii Dept. of Agriculture, Plant Pest Control Branch, 1428 S. King St., Honolulu, HI 96814, USA

## INTRODUCTION

Four exotic tephritid fruit fly pests have colonised the Hawaiian islands over the past 100 years, where they have become major pests infesting hundreds of horticultural crops. The Mediterranean fruit fly, *Ceratitis capitata* (Wiedemann), Oriental fruit fly, *Bactrocera dorsalis* (Hendel), melon fly, *B. cucurbitae* (Coquillett), and Solanaceous fruit fly, *B. latifrons* (Hendel) are considered among the major obstacles to the development of a more robust agricultural economy in the state of Hawaii. Furthermore, the flies pose a continuous threat to agriculture in California and other areas in the southern United States, where it has been estimated that the establishment of the Medlfy alone would result in losses of over one billion dollars annually (Andrew et al. 1978).

Entomologists in Hawaii have conducted a number of classical biological control programmes against these tephritid pests over the years, resulting in the establishment of several parasitoid species and partial control of the flies in some crops (see reviews in Clausen et al. 1965, Wharton 1989). However, these programmes were conducted before the invasion of the state by the Solanaceous fruit fly; thus, there have been no biocontrol programmes targeted against this pest. Also, several entomologists have pointed out the potential of improved control over the other tephritid species in Hawaii by introducing new natural enemies (Gilstrap and Hart 1987, Messing 1995, Steck et al. 1986, Wharton 1989, Wong and Ramadan 1992).

We have therefore renewed efforts to import parasitoids from tropical and sub-tropical areas around the world to attack tephritid fruit flies in Hawaii. As part of this effort, we imported *Diachasmimorpha kraussii* Fullaway from Queensland, Australia, where it is an endemic parasitoid of *Bactrocera tryoni* (Froggatt) and several other endemic Australian tephritids. This paper reports the results of initial host range tests and studies on the reproductive output of *D. kraussii* in quarantine.

## MATERIALS AND METHODS

A cohort of 434 *D. kraussii* (74% female) from a laboratory colony in Queensland, Australia, was shipped by air freight to the quarantine facility of the Hawaii Department of Agriculture (HDOA) in Honolulu in April 1996. From this shipment, a rearing colony was established in quarantine using *C. capitata* as the rearing host. *C. capitata* larvae were provided by the USDA-ARS Tropical Fruit, Vegetable, and Ornamental Crops Research Laboratory in Honolulu. Parasitoid adults were maintained in 25 x 25 x 25 cm wooden screened cages, and given water and honey *ad libitum*.

Standard oviposition units (Wong and Ramadan 1992) were constructed of 9 cm diam. plastic petri dishes with tight-fitting lids covered with organdy, through which the wasps could readily oviposit. Tephritid larvae mixed in with their artificial, wheat-based rearing diets were packed into these units, to obtain larval densities ranging from ca. 200 (third instar) to 600 (first instar) *B. latifrons*, and 500 (third instar) to 900 (first instar) *C. capitata* larvae per unit. Oviposition units were placed beneath screened cutouts in the bottom of rearing cages containing ca. 200 male and 200 female *D. kraussii*. The age of parasitoid females used in these experiments ranged from 4-7 days. Parasitoids were allowed to oviposit freely into larvae in the units for 24 hours.

After exposure to the parasitoids, fly larvae were removed from the oviposition units and transferred to clean containers with extra larval medium, where they were held until the formation of fly puparia. After 7 days, puparia were sifted from the vermiculite and held in plastic containers with organdy covers. Adult flies emerged from unparasitised puparia after ca. 12 days; the remaining uneclosed puparia and all emerging parasitoids were counted and sexed.

## RESULTS AND DISCUSSION

For both *B. dorsalis* and *B. cucurbitae*, no viable offspring were produced by *D. kraussii* (Table 1). Dissections of parasitised puparia of these two species approximately 7 days after oviposition revealed that the eggs of *D. kraussii* were consistently encapsulated by the fly larvae, as evidenced by complete melanisation. No other parasite cadavers were found in uneclosed puparia. Both *B. dorsalis* and *B. cucurbitae* eclosed to yield morphologically normal adult flies.

Table 1. Percent parasitism by *D. kraussii* of four tephritid fruit fly species presented in laboratory oviposition units.

| Host | Instar | | | | |
|---|---|---|---|---|---|
| | First | Second | Early third | Mid-third | Late third |
| *Bactrocera dorsalis* | 0 | 0 | 0 | 0 | 0 |
| *B. cucurbitae* | 0 | 0 | 0 | 0 | 0 |
| *B. latifrons* | 8 | 41 | 34 | 58 | 12 |
| *Ceratitis capitata* | 10 | 48 | 64 | 28 | 21 |

*D. kraussii* successfully reproduced on all three instars of *C. capitata* and *B. latifrons* (Table 1). On Medflies, the largest number of parasitoid progeny was recovered from flies which were attacked as early third instar larvae (= 4-d old). Oviposition into older larvae resulted in fewer parasitoid offspring, but led to a greater percentage of uneclosed fly puparia, particularly when 6-d old larvae were attacked (Figure 1). The large number of uneclosed host puparia associated with oviposition into 5- and 6-d old late third instars is presumably due to the parasitoid killing the host but not having enough time to develop to maturity. Medfly larvae which were attacked in the early third instar also yielded the highest percentage of females for all larval age groups, and was the only group which avoided a preponderance of males (Figure 2).

In the Solanaceous fruit fly, recovery of parasitoid progeny was greatest when 8-d old middle third instar larvae were attacked. Oviposition into older third instar larvae again resulted in fewer parasitoid offspring and a greater percentage of uneclosed host puparia. The sex ratio in parasitoid progeny reared from the Solanaceous fruit fly was more female biased than in Medflies, presumably due to the larger size of *B. latifrons*. The highest percentage of female offspring in *B. latifrons* was obtained from 8-d old larvae (Figure 2).

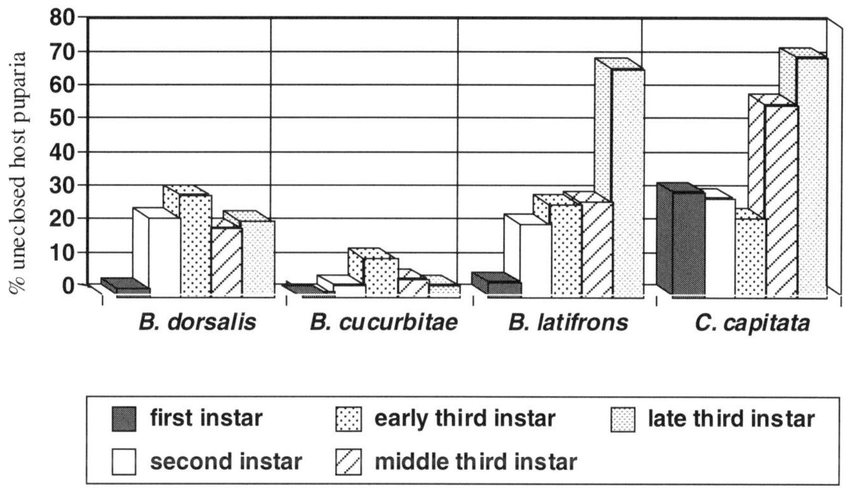


Figure 1. Effect of parasitism by *D. kraussii* on eclosion of 4 tephritid species in Hawaii.

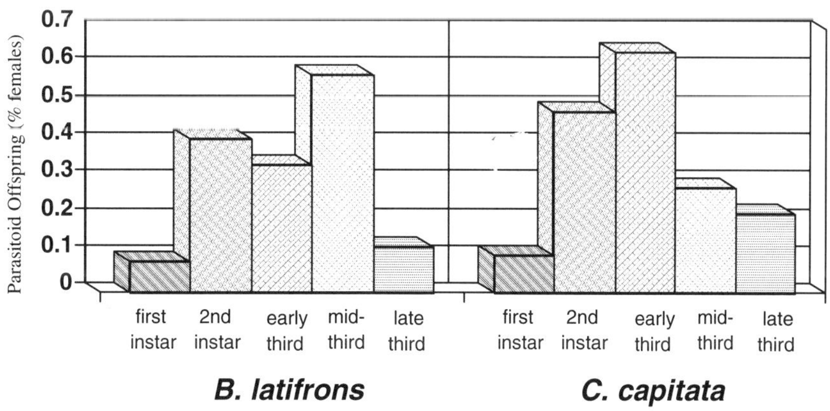


Figure 2. Sex ratio of *D. kraussii* offspring as a function of host larval age.

*D. kraussii* is a native of Queensland and the Northern Territory of Australia, where it parasitises not only *B. tryoni* but also *B. neohumeralis* (Hardy), *B. cacuminata* (Hering), *B. jarvisi* (Tryon), *B. kraussii* (Hardy), *B. halfordiae* (Tryon) and *B. melas* (Perkins & May) (Krisana 1994). It was previously introduced into Hawaii in 1949 for biological control of *B. dorsalis* (Clancy et al. 1952). Between 1950-1954, over 25,000 *D. kraussii* were reared in quarantine (using *C. capitata* as a factitious host) and released on all the major Hawaiian islands.

Chong (1952) reported that a single specimen of *D. kraussii* was recovered in the field from a melon fly from loquat on Maui Island. Assuming that the parasitoid was correctly associated with the host (which is sometimes questionable when several fly species emerge from the same fruit), this record may be explained by the fact that a parasitoid which successfully attacks the melon fly may sting a larva and break down the host's immune system, after which another parasitoid may successfully develop in the disabled host. This has been demonstrated in both the melon fly (Pemberton and Willard 1918) and in the Oriental fruit fly (Ramadan et al. 1994). For the single recovery reported by Chong (1952), we assume that the opiine parasitoid *Psyttalia fletcheri* Silvestri had previously stung a melon fly larva, disabling the immune system and allowing a later ovipositing *D. kraussii* to complete development.

Chong (1954) also reported that 10 *D. kraussii* adults were recovered from Jerusalem cherries (*Solanum pseudocapsicum* L.) at Waiki'i on the big island of Hawaii.This report did not include information on the host fly from which the parasitoids were reared, although the location and host plant suggest that the host insect was the Medfly. However, no further recoveries of *D. kraussii* were made in this location nor at any other location in Hawaii.

The tephritid biological control programme in Hawaii during the 1950s was aimed primarily at the Oriental fruit fly, although some of the imported parasitoids fortuitously attacked Medflies as well. When imported parasitoids, including *D. kraussii*, were released in the field, they were usually placed in locations where the primary target was abundant. Since the Oriental fruit fly predominates in lowland areas, and the Medfly (presumed to be displaced by Oriental fly: Bess 1953) tends to be more abundant at higher elevations, the failure of the establishment of *D. kraussii* in these early introductions may have been caused primarily by a lack of physiologically suitable hosts occurring in the immediate vicinity of most of the releases.

The distribution, diversity, and abundance of tephritid fruit flies in Hawaii have changed considerably since the large biological control programmes of the 1950s. The Solanaceous fruit fly was first reported in Hawaii in 1983, and has since been recovered on 15 host plants on all major islands. It was not available as a host during the previous importation of *D. kraussii*. Also, the abundance of Medflies, especially in lowland areas, has increased substantially with the expansion of coffee plantations on several islands. These changes make it more likely that *D. kraussii* can become established in Hawaii if released in a classical biological control programme, since solanum flies and Medflies now provide suitable host material in a wide range of habitats and elevations on all major islands.

The Medfly is currently attacked by several imported parasitoid species, including the opiine braconids *Fopius arisanus* (Sonan), *Diachasmimorpha tryoni* (Cameron), and, less frequently, *Biosteres vandenboschi* (Fullaway) and *Psytallia incisi* (Silvestri), as well as the gregarious eulophid *Tetrastichus giffardianus* Silvestri. We do not currently have enough information on climatic tolerances and host plant relations of these parasitoids to know if there are unexploited habitat niches which *D. kraussii* may more successfully occupy, nor to what extent, if any, competition between *D. kraussii* and the other wasp species will influence Medfly population dynamics. In previous competition studies, the egg parasitoid *F. arisanus* has been shown to be a superior intrinsic competitor over all other opiines in Hawaii, which are larval parasitoids (van den Bosch et al. 1951). Nevertheless, these larval parasitoids co-exist with *F. arisanus* in many habitats (Wong and Ramadan 1987), and even dominate in some high elevation sites (Ramadan, unpublished data). Although we did not simultaneously test Medfly larvae in the quarantine laboratory with both *D. kraussii* and the other most abundant larval opiines *(D. tryoni* and *D. longicaudata*), our data show a greater yield of *D. kraussii* offspring from standard ovipostiion units than either of the other two species (Wong et al. 1990, Wong and Ramadan 1992). Thus, in either classical (innoculative) or augmentative (Wong et al. 1992) release programmes, the newly imported parasitoid has the potential to contribute to fruit fly control in Hawaii.

## ACKNOWLEDGEMENTS

We thank M. Ross and G. Walter (University of Queensland) for providing *D. kraussii*, N. Mochizuki and J. Spencer (USDA-ARS) for providing larvae of *B. latifrons,* K. Teramoto (Hawaii Department of Agriculture) for use of quarantine facilities, N. Peabody (University of Hawaii) for assistance in rearing, and R. Wharton (Texas A & M University) for taxonomic assistance. Generic designations of opiine parasitoids are based on Wharton (1997).

## REFERENCES

Andrew, C.O., J.C. Cato and E.J. Prochaska. 1978. Potential economic impact of a fruit fly infestation on the U.S. citrus industry. Proc. Fla. State Hort. Soc. 36-46.

Bess, H.A. 1953. Status of *Ceratitis capitata* in Hawaii following the introduction of *Dacus dorsalis* and its parasites. Proc. Hawaii.. Entomol. Soc. 15: 221-234.

Chong, M. 1952. Notes and Exhibitions. Proc. Hawaii. Entomol. Soc. 14 (3): 361.

Chong, M. 1954. Notes and Exhibitions. Proc. Hawaii. Entomol. Soc. 15 (2): 270.

Clancy, D.W., P.E. Marucci and E. Dresner. 1952. Importation of natural enemies to control the Oriental fruit fly in Hawaii. J. Econ. Entomol. 45: 85-90.

Clausen, C.P., D.W. Clancy and Q.C. Chock. 1965. Biological control of the Oriental fruit fly *(Dacus dorsalis* Hendel) and other fruit files in Hawaii. USDA Tech. Bull. No. 1322: 1 102.

Gilstrap, F.E. and W.G. Hart. 1987. Biological control of the Mediterranean fruit fly in the United States and Central America. United States Department of Agriculture ARS-56: 1-64.

Krisana, R. 1994. The life cycle, mating behavior and sexual communication signals of *Diachamimorpha kraussii* Fullaway (Hymenoptera: Braconidae), a parasitoid of Dacine fruit flies (Diptera: Tephritidae). Doctoral Dissertation, University of Queensland, Brisbane.

Messing, R.H. 1995. Status and needs of biological control research for tephritid flies. pp. 365-367. *In* Fruit Flies Pests: A World Assessment of Their Biology and Management. B.A. McPheron & G.J. Steck, (eds.). St. Lucie Press, Delray Beach, Florida, USA.

Pemberton, C.E. and H.F. Willard. 1918. A contribution to the biology of fruit fly parasites in Hawaii. J. Agric. Res. 15: 419-466.

Ramadan M.M., T.T.Y. Wong and J.C. Herr. 1994. Is the Oriental fruit fly a natural host for the Opiine parasitoid *Diachamimorpha tryoni?* Environ. Entomol. 23: 761-769.

Steck, G.J., F.E. Gilstrap, R.A. Wharton and W.G. Hart. 1986. Braconid parasitoids of tephritidae infesting coffee and other fruits in west central Africa. Entomophaga 31: 59-67.

van den Bosch, R., H.A. Bess and F.H. Haramoto. 1951. Status of oriental fruit fly parasites in Hawaii. J. Econ. Entomol. 44: 753-759.

Wharton, R.A. 1989. Classical biological control of fruit infesting Tephritidae. Chapter 9.1. *In* Fruit Flies: Their Biology, Natural Enemies, and Control. A.S. Robinson and G. Hooper (eds.). World Crop Pests, Volume 3B, Elsevier, Amsterdam, The Netherlands.

Wong, T.T.Y. and M.M. Ramadan. 1987. Parasitisation of the Mediterannean and Oriental fruit flies (Diptera: Tephritidae) in the Kula area of Maui, Hawaii. J. Econ. Entomol. 80: 77-80.

Wong, T.T.Y. and M.M. Ramadan. 1992. Mass rearing biology of larval parasitoids of Tephritid flies in Hawaii. Chap. 23. *In* Advances in Insect Rearing for Research and Pest Management, T.E. Anderson and N.C. Leppla (eds.). Westview Press, Boulder, Colorado.

Wong, T.T.Y., M.M. Ramadan, D.O. McInnis and N. Mochizuki. 1990. Influence of cohort age and host age on oviposition activity and offspring sex ratio of *Biosteres tryoni*, a larval parasitoid of *Ceratitis capitata.* J. Econ. Entomol. 83: 779-783.

Wong, T.T.Y., M.M. Ramadan, J.C. Herr and D.O. McInnis. 1992. Suppression of Mediterranean fruit fly with concurrent parasitoid and sterile fly releases in Kula, Maui, Hawaii. J. Econ. Entomol. 85: 1671-1681.

# Biological Control of Fruit Flies (Diptera: Tephritidae) Through Parasitoid Augmentative Releases: Current Status

*Pablo Montoya[1] and Pablo Liedo[2]*

[1] Programa Moscamed, Comisión Nacional de Sanidad Agropecuaria, SAGAR, Apartado postal 368, 30700 Tapachula, Chiapas, Mexico
[2] El Colegio de la Frontera Sur, Apartado postal 36, 30700 Tapachula, Chiapas, Mexico

## INTRODUCTION

Fruit flies are among the main pests affecting the world fruit industry (Aluja 1993). Bait sprays have traditionally been used successfully to control them; however, the side effects on the environment and health hazards commonly associated with pesticides, have resulted in strong public opposition to the use of bait sprays. This is particularly so when sprays are applied in urban areas or in coffee plantations where, although Medflies are present, they do not pose a danger to crops. Alternative methods that are effective and environmental friendly to suppress fruit fly populations are highly desirable.

Biological control, the use of natural enemies to suppress pest populations, represents such an alternative. Some of the most successful cases of biological control are the control of *Icerya purchasi* Maskell (Homoptera: Margarodidae) by *Rodolia cardinalis* Mulsant (Coleoptera: Coccinellidae) in California (De Bach 1968, van den Bosch et al. 1982), and the control of *Aleurocanthus woglumi* Ashby (Homoptera: Aleyrodidae) mainly by *Encarsia (=Prospaltella) opulenta* Silv. (Hymenoptera: Aphelinidae) in Mexico (Jimenez 1961, 1971), both using the classical approach. However, this approach has been limited to certain conditions of environmental stability and biodiversity which are only found in a few ecosystems. Other factors, such as types of pests, the economic threshold and product quality requirements represent additional limitations.

The best option in many cases could be augmentative biological control, which could overcome some of the deficiencies of the classical approach (Sivinski 1996). According to Knipling (1992) and Barclay (1987), augmentative biological control can be considered as a formal alternative for suppressing pest populations and even for use in eradication programmes, after integration with the sterile insect technique (SIT). In this approach, mass production of natural enemies is required and this production has to be cost effective.

## AUGMENTATIVE BIOLOGICAL CONTROL

Greathead and Waage (1983) defined augmentative biological control as "the strategy in which natural enemies are mass reared for release at critical periods, aiming to suppress a pest population in a short period of time". In the past, some authors questioned the possibilities of using this method (Huffaker et al. 1977, Shumakov elements that support its use. According to this author, in order to effectively apply this

1977). Recently however, Knipling (1992) re-analysed this approach and provided new control method, the parasitoid releases have to be made before the pest population increases to economic threshold levels and conducted in an area-wide approach.

One of Knipling's (1992) main premises is that the parasitoids are able to identify hosts and complete their parasitising activity even at very low host density; otherwise this species would have been extinguished. Characteristics like a good search capacity mediated by the presence of specific kairomones and high host discrimination, give some basis to the premise that "at high parasitoid density, high parasitising rates could be obtained even at low host densities".

Another advantage of mass released parasitoids could be that they are not irradiated as in the case of sterile insects, in which irradiation is considered to be harmful. Parasitoids are also released into the field not to compete with their wild kin, but to complement them. The reproduction of the parasitoids under field conditions provides another advantage of this control method.

In the case of fruit flies, Knipling (1992) suggests that parasitoid inundative releases can be coupled with sterile fly releases in an eradication programme. For suppressing *Anastrepha suspensa* Loew using *Diachasmimorpha longicaudata* (Ashmead), Knipling's model suggests a parasitoid/fly ratio of 4:1, where 80,000 parasitoids should be released per square mile for a fly population density of 20,000/square mile, expecting to reach parasitism rates higher than 98% in the third generation.

Even though Knipling (1992) supports augmentative biological control against fruit flies, he also recognises that this method is considered just as another desirable ecological method completely untested in its approaches and expectations.

## FRUIT FLIES BIOLOGICAL CONTROL

The first biological control programme was established by the Australian government in 1902 (Wharton 1989). It started with the search for natural enemies against the Medfly *Ceratitis capitata* (Wied). This same fly was detected for the first time in Hawaii in 1910, but it was not until 1945 when the Oriental fruit fly *Bactrocera* (*Dacus*) *dorsalis* (Hendel) invaded the island, that the most important biological control programme against tephritids was established (Clausen et al. 1965). In this programme, the effectiveness of the parasitoid genus *Opius* (= *Biosteres* = *Diachasmimorpha*) was well established for controlling *Bactrocera* and *Ceratitis* fruit flies, where *D. longicaudata* (Ashmead), *B. vandenboschi* (Fullaway) and *Fopius (=Biosteres) arisanus* (Sonan) were the species most commonly recovered from 32 species released. (Bess et al. 1961).

That the biological control of *B. dorsalis* was considered a success could be attributed mainly to the action of *F. arisanus* (van den Bosch et al. 1982) even though expectations were not fully realised. Therefore, a mass rearing programme to release different species of fruit fly parasitoids was considered. Marucci and Clancy (1950) started the work of developing mass rearing technology for these parasitoids. Many developments were achieved during their 40-year period of research, reaching its current status of successful mass reared *D. longicaudata* in the Moscafrut facility in Metapa, Mexico. This facility produced up to 50 millions parasitoids per week (Cancino et al. 1996).

### Augmentative Releases

Wong et al. (1991) evaluated *D. tryoni* releases against *C. capitata,* recording important parasitism differences between the releasing zone (47.0%) and the control zone (14.2%). This was due to a very low adult emergence per fruit in the treated zone. In 1992, Wong et al. focused on evaluating suppression of the Medfly population by releasing parasitoids and sterile flies. Based on the differences of flies captured in traps, percentages of *C. capitata* egg hatching and parasitism percentages, it was concluded that simultaneous releases of parasitoids and sterile flies could be used in eradication programmes of tephritids, even though major research is needed to ensure its success under certain places and conditions.

Sivinski et al. (1996) and Burns et al. (1996), reported significant decreases in *A. suspensa* trap captures in the *D. longicaudata* releasing zones in Florida, when compared to historical data and to control zones. However, percentages of parasitism registered were low with the highest average at 8.83%.

In Latin America, Enkerlin et al. (1990) worked for about two years with releases of *D. longicaudata* and *D. tryoni* (Cameron) against *Anastrepha* spp. populations in Mazapa de Madero, Chiapas, Mexico, where a significant reduction in fruit infestation and high parasitism rates (> 90%) were obtained. Camacho (1989, 1994) emphasises the grower's participation in transferring the technology to reproduce *D. longicaudata* and *Pachycrepoideus vindenmiae* Rondani (Hymenoptera: Pteromalidae), which under an integrated pest management approach, have contributed to a reduction in Medfly infestations in Costa Rica.

Montoya et al. (2000) evaluated *D. longicaudata* augmentative releases against *Anastrepha* fruit fly populations in mango orchards (> 1600 ha). Parasitoids were released on a weekly basis at a density of ca. 1000 parasitoids/ha. Releases began at the onset of flowering and lasted 35 weeks until the end of the production cycle, with a control zone 7-8 km away. Samples of mangoes and other tropical fruits were collected from commercial orchards (growing Ataulfo mango cultivar) and backyard orchards where no management practices occurred. There were highly significant differences in the parasitism percentage between the release and the control zone. The trapping results indicated that *D. longicaudata* achieved a 2.7 fold suppression of *Anastrepha* spp. populations in backyard orchards, whereas in commercial orchards, such an effect was not observed, probably due to the very low densities of fruit flies.

These results indicate that suppression might be achieved under these ecological conditions and the parasitoid/fly ratio is to be established. It was also found that *Anastrepha obliqua* McQuart populations were suppressed more effectively than *Anastrepha ludens* Loew populations. This could be correlated to the type of host fruits used by each species.

## CONCLUSIONS

Even though most authors have considered that suppressing fruit fly populations as a product of parasitoid releases is of great potential, all the results have been discreet when considering this kind of biological control as part of an integrated and effective eradication programme. On the other hand, parasitoid releases against fruit flies have never been evaluated, as suggested by Knipling (1992), in large isolated areas or in island conditions, where the incidental movement or migration of fruit fly adults could be nullified or decreased considerably. According to Knipling, this is a fundamental concept and if total populations are not treated, one could expect misleading results.

In most empirical work, the parasitoid species evaluated have been *D. longicaudata* and *D. tryoni*, which are considered to have advantages such as adaptability to mass rearing conditions, as well as good performance in the field. However, they should not be considered as the only species available. The search for new parasitoid species against fruit flies continues (Messing 2000), together with the case of the egg parasitoid *F. arisanus*, which is considered to have great potential (Harris et al. 2000).

## REFERENCES


Aluja, M. 1993. Manejo Integrado de Moscas de la Fruta. 1ª ed. Edit. Trillas. México D.F. 251 pp.

Barclay, H.J. 1987. Models for pest control: complementary effects of periodic releases of sterile pest and parasitoids. Theor. Popul. Biol. 321: 76-89.

Bess, H.A., R. van den Bosch and F. Haramoto. 1961. Fruit fly parasites and their activities in Hawaii. Proc. Hawaii. Entomol. Soc. 17: 367-378.

Burns, R.E., , J.O. Díazand T.C. Holler. 1996. Inundative releases of the parasitoid *Diachasmimorpha longicaudata* for the control of the Caribbean fruit fly, *Anastrepha suspensa*. pp. 377-381. *In* Fruit Fly Pests. A World Assessment of their Biology and Management. B.A. McPheron and G.J. Steck (eds). St. Lucie Press. Delray Beach, Fla.

Camacho, H. 1994. The integrated use of the sterile flies and parasitoids in fruit fly control in Costa Rica. Abstracts of the "Fourth International Symposium on Fruit Flies of Economic Importance" Sand Key, Florida Xl-2 pp.

Camacho, H. 1989. El programa de Manejo Integrado de la mosca del Mediterráneo en Costa Rica. Seminario sobre la Técnica de los Insectos Estériles para el control o erradicación de la mosca de la fruta en América Latina. FAO IAEA/OEIA. Programas y Resúmenes. Guatemala C.A. 17-21 de abril de 1989. 37 pp.

Cancino, J., S. de la Torre, S. Ruiz, F. de M. Moreno, E. Hernández and M. Zenil. 1996. Establecimiento de la producción masiva de *Diachasmimorpha longicaudata* en Metapa de Domínguez, Chiapas, México, pp. 82-83. *In* Proceedings, 2nd Meeting of the Working Group on Fruit Flies of the Western Hemisphere, 3-8 November 1996. Viña del Mar, Chile.

Clausen, C.P., D.W. Clancy and Q.C. Chock. 1965. Biological control of the oriental fruit fly (*Dacus dorsalis* Hendel) and the other fruit flies in Hawaii. USDA. Technical Bulletin 1322. 103 pp.

De Bach, P. 1968. Control Biológico de las Plagas de Insectos y Malas Hierbas. CECSA. México. 948 pp.

Enkerlin, D., J., Cancino, J. Guillén and L. Martínez. 1990. Evaluaciones del efecto de las liberaciones de parasitoides sobre poblaciones de moscas de la fruta del género *Anastrepha* spp. (Diptera: Tephritidae) en Mazapa de Madero, Chiapas. XIII Reunión Nacional de Control Biológico. AMCB. (Datos sin publicar).

Greathead, D.J. and J.K. Waage. 1983. Opportunities for biological control of agricultural pests in developing countries. World Bank Technical Paper Number 11, Washington D.C. USA.

Harris, E., R.C. Bautista and J.S. Spencer. 2000. Utilisation of the egg-larval parasitoid *Biosteres arisanus* for augmentative biological control of tephritid flies. pp. 725-732. *In* Area-wide Control of Fruit Flies and Other Insect Pests. K.H. Tan (ed.). Penerbit Universiti Sains Malaysia, Penang, Malaysia.

Huffaker, C.B., R.L. Rabb and J.A. Logan. 1977. Some aspects of populations dynamics relative to augmentation of natural enemy action. pp. 3-38. *In* Biological Control by Augmentation of Natural Enemies. R.L. Ridgway and S.B. Vinson (eds.). Plenum Press. New York.

Jiménez, J.E. 1961. Situación actual de la mosca prieta de los cítricos. Fitófilo 16: 4-11.

Jiménez, J.E. 1971. Comportamiento de los enemigos naturales de la mosca prieta de los cítricos en la República Mexicana. Fitófilo 66: 2-6.

Knipling, E.F. 1992. Principles of insect parasitism analysed from new perspectives. Agriculture Handbook No. 693. ARS-USDA. Washington D.C. USA. 337 pp.

Marucci, P.E. and D.W. Clancy. 1950. The artificial culture of fruit flies and their parasites. Proc. Hawaiian Entomol. Soc. 14:163-166.

Messing, R.H. 2000. Newly imported parasitoid effective against the Malaysian fruit fly in Hawaii. pp. 713-718. *In* Area-wide Control of Fruit Flies and Other Insect Pests. K.H. Tan (ed.). Penerbit Universiti Sains Malaysia, Penang, Malaysia.

Montoya, P., P. Liedo, B. Benrey, J.F. Barrera and M. Aluja. 2000. Biological control of *Anastrepha* fruit flies (Diptera: Tephritidae) through augmentative releases of *Diachasmimorpha longicaudata* (Ashmead) (Hymenoptera: Braconidae). Biol. Control (in press).

Shumakov, E.M. 1977. Ecological principles associated with augmentation of natural enemies. pp. 39-70. *In* Biological control by augmentation of natural enemies. R.L. Ridgway and S.B. Vinson (eds). Plenum Press. New York.

Sivinski, J.M. 1996. The past and potential of biological control of fruit flies. pp. 365-375. *In* Fruit Fly Pests. A World Assessment of their Biology and Management. B.A. McPheron and G.J. Steck (eds.). St. Lucie Press. Delray Beach, Fla.

Sivinski, J.M., C.O. Calkins, R. Baranowsky, D. Harris, J. Brambila, J. Díaz, R.E. Burns, T. Holler and G. Dodson. 1996. Suppression of a Caribbean fruit fly (*Anastrepha suspensa* (Loew) Diptera: Tephritidae) population through augmented releases of the parasitoid *Diachasmimorpha longicaudata* (Ashmead) (Hymenoptera; Braconidae). Biol. Control 6: 177 185.

van den Bosch, R. Messenger, P.S. and A.P. Gutierrez. 1982. An introduction to biological control. Plenum Press. New York. 247 pp.

Wharton, R.A. 1989. Classical biological control of fruit-infesting Tephritidae. pp. 303-314 *In* Fruit Flies. Their biology, natural enemies and control. A.S. Robinson and G. Hooper (eds.). Vol. 3B. Elsevier, Oxford.

Wong, T.T.Y., M.M. Ramadan, D.O. McInnis, N. Mochizuki, J.I. Nishimoto and J.C. Herr. 1991. Augmentative releases of *Diachasmimorpha tryoni* (Hymenoptera: Braconidae) to suppress a Mediterranean fruit fly (Diptera: Tephritidae) population in Kula, Maui, Hawaii. Biol. Control 1:2-7.

Wong, T.T.Y., M.M. Ramadan, J.C. Herr and D.O. McInnis. 1992. Suppression of a Mediterranean fruit fly (Diptera: Tephritidae) population with concurrent parasitoid and sterile fly release in Kula, Maui, Hawaii, J. Econ. Entomol. 85: 1671-1681.

# Utilisation of the Egg-Larval Parasitoid, *Fopius (Biosteres) arisanus*, for Augmentative Biological Control of Tephritid Fruit Flies

*Ernest J. Harris, Renato C. Bautista and John P. Spencer*

USDA-ARS, Tropical Fruit, Vegetable and Ornamental Crop Research Laboratory, 2727 Woodlawn Drive Honolulu, Hawaii 96822, USA

## INTRODUCTION

In Hawaii, entomologists concerned about tephritid fruit fly control recognise and accept the fact that the introduction of tephritid fruit flies consisting of the melon fly *Bactrocera cucurbitae* (Coquillet), Mediterranean fruit fly, *C. capitata* (Wiedemann), Oriental fruit fly, *B. dorsalis* (Hendel) (Figure 1) and the Solanaceous fruit fly, *B. latifrons* (Hendel) required the introduction of many species of parasitoids into Hawaii (Clausen 1956) to reduce crop damage caused by tephritid fruit flies. The parasitoids established in the order of their succession were *Diachasimorpha longicaudata* (Ashmead), *Biosteres vandenboschi* (Fullaway), and *Fopius (Biosteres) arisanus* (Sonan). *F. arisanus* (Figure 2) was first discovered in Hawaii in 1949 in a guava fruit collection (van den Bosch and Haramoto 1951). In 1950, the rate of parasitism caused *F. arisanus* to increase and this insect spread and became the dominant and most widely distributed parasitoid in Hawaii (Haramoto and Bess 1970). Entomologists investigating fruit fly ecology in Hawaii recognised that the four species of tephritid fruit flies differ in their distribution, abundance and host utilisation patterns in different habitats. The rapid spread and distribution of *F. arisanus* in Hawaii indicated the reality that among the parasitoids, *F. arisanus* has the highest adaptation capabilities in the Hawaiian ecosystem comparable to that of *B. dorsalis* and *C. capitata,* the most persistent fruit fly species in Hawaii.

A strategy receiving high priority to improve biological control of tephritid fruit flies is foreign exploration to find new parasitoids for introduction into tephritid fruit fly invested areas including Guatemala and Hawaii. It is possible another species comparable to *F. arisanus* might be found. New introductions could increase the diversity of parasitoid species and result in the introduction of species more efficient for suppressing *B. latifrons* in Hawaii. The cost of parasitoid exploration is very expensive, US$100,000 or more. The countries supplying the insects could ask for payment for use of their natural resources. In some situations, there is strong opposition to parasitoid introductions from the lay and scientific communities due to the argument that introduced parasitoids could have adverse effects on beneficial insects which control weeds. When new introductions are successful, they must be evaluated in quarantine for months before they can be released into the field. There is no guarantee that new parasitoid species will find a niche in the tephritid fruit fly infested areas for competing well against the existing parasitoid species.

Figure 1. The Mediterranean fruit fly, *Ceratitis capitata*, on the left, the Oriental fruit fly, *Bactrocera dorsalis* on the right, and the melon fly, *B. cucurbitae*, on the bottom. *B. latifrons* is not shown.

Figure 2. The egg-larval parasitoid, *Fopius (Biosteres) arisanus.*

A strategy which we believe has been neglected is laboratory colonisation, mass rearing and augmentative release of *F. arisanus,* the parasitoid species which has proved to be most successful in the Hawaiian ecosystem. We envisioned that successful colonisation of *F. arisanus* could provide the insight and know-how for developing different strains of *F. arisanus* for different fruit fly species. This strategy would complement and support foreign explorations for tephritid fruit fly parasitoids. We describe how *F. arisanus* is colonised, explain the research developments leading to mass production and outline what we believe can be done to develop new strains of *F. arisanus.*

## RESEARCH METHODS

### Parasitoid Colony Development

A papaya trap (Figure 3) was developed (Harris and Okamoto 1983, Harris and Bautista 1994) to collect large numbers of live *F. arisanus* in the field to provide a constant supply of parasitoids for laboratory colonisation studies. To solve the problem of male biased progeny sex ratio, males and females were segregated at emergence and 10-15 days later, they were combined in cubicle cages (26 x 28 x 26 cm) at ratios of 1:4 (males:females) for mating. The mated females were provided *B. dorsalis* eggs in papaya fruit sections (8 x 3 x 1) containing 800-1000 eggs for oviposition (parastisation). The parasitised eggs in papaya sections were placed on an artificial diet for progeny production of flies and parasitoids. At the end of seven generations, the colony became laboratory adapted and males and females were combined at a 1:1 ratio (50 males:50 females) thereafter. The colony (referred to as the Harris strain) has been continuously reared in the laboratory for < 200 generations using a natural fruit substrate medium for oviposition (Harris and Okamoto 1991). The strain was reared using *B. dorsalis* as the hosts. This procedure is designed to maintain strain integrity to enhance field searching ability of the laboratory-colonised parasitoids.

Figure 3. The papaya fruit trap used to capture *F. arisanus* in the field.

### Laboratory Studies for Mass Production

We conducted four experiments to provide information to develop mass production technology for *F. arisanus* to evaluate augmentative biological control of tephritid fruit flies.

Experiment 1 compared the host suitability of *B. dorsalis, B. cucurbitae, C. capitata,* and *B. latifrons* for development of the egg parasitoid, *F. arisanus*. For the four fruit flies, eggs (3-4 h old) were inoculated separately into papaya fruit sections (8 x 3 x 1 cm) into 10 holes (4-5 mm deep) with the blunt end of a camel hair brush. A clutch of 110 eggs/hole was inoculated = 1,100 eggs/fruit section. The number of females that landed and oviposited in the fruit was counted at hourly intervals. Fruit sections were exposed separately to 50 males + 50 females of the four fruit fly species in cubical cages. An equal number of fruits was inoculated in another set of fruits and held in separate cages as controls. At the end of 24 h, fruit sections were retrieved and samples of 100 eggs were taken from each fruit and dissected to check for parasitism. Fruits with the remaining eggs were placed separately inside a 9 cm-diameter cup with a wheat diet for rearing fruit fly species. The cups were placed inside a 20-cm diameter plastic container containing beach sand. Two weeks later, the pupae were screened, counted, and the flies and parasitoids were recorded. This test was replicated 7 times.

Experiment 2 compared the effects of different fruit species (solo papaya, *Carica papaya* L., apple banana, *Musa sapientum* L., Haden mango, *Mangifera indica* L., false kamani, *Terminalia cattapa* L. and common lime, *Citrus aurantiifolia* (L.) Swing. on the parasitisation rate of *B. dorsalis*. The response of female parasitoids that landed and oviposited in the papaya substrate inoculated with *B. dorsalis* eggs was compared with apple banana, Haden mango, false kamani and lime. Except for false kamani, fruits were trimmed into sections (8 x 3 x 1 cm). Each fruit section was perforated with 10 holes = 30 eggs/hole = 300 eggs/fruit section. Inoculated fruits were arranged inside a screened cage (60 x 30 x 40 cm) in a completely randomised design. Cohorts of *F. arisanus* adults (500 males + 500 females) were aspirated into the cage and allowed to oviposit for 24 h. A total of 20 fruit samples/replication (4 fruits/species x 5 species) was assayed. The rate of parasitised *B. dorsalis* calculated from the number of emerged parasitiods divided by the total number of host pupae recovered was compared among fruit fly species. The fruits were processed as in the previous test. This test was replicated 5 times.

Experiment 3 determined if increasing the number of *B. dorsalis* host eggs to the female *F. arisanus* ratio would give increases in parasitoid progeny yields. Papaya fruit sections (8 x 3 x 1 cm) were inoculated with *B. dorsalis* eggs in increments of 250, 500, 1,000, 1,250, and 1,500 eggs to obtain egg female ratios of 5:1, 10:1, 20:1, 25:1, and 30:1, respectively. Eggs were distributed equally among 10 holes perforated in the fruit surface arranged in 2 rows of 5 holes each (25 eggs/hole = 250 eggs/cohort). Inoculated fruits were placed in separate cubical cages (26 x 28 x 26 cm) into which 50 males + 50 females were introduced. Females were allowed to oviposit into host eggs 24 h, after which fruit sections were retrieved and processed as described above. This test was replicated 10 times.

Experiment 4 compared the mating efficiency and percent parasitism of the laboratory (Harris) strain with the wild strain of *F. arisanus* collected in a papaya orchard. Papaya fruit sections (8 x 3 x 1 cm) were inoculated with *B. dorsalis* eggs and placed in 2 separate cages. One cage had 40 males + 40 females of the laboratory strain and the other had 40 males + 40 females of the wild strain. Cages of wild and laboratory *F. arisanus* were allowed to oviposit for 24 h, after which fruit sections were retrieved and processed as described above. This test was replicated 4 times.

## RESULTS

### Experiment 1

When a random sample of 100 eggs was dissected for parasitisation, the mean percent parasitisation was higher ($F = 6.4$; $df = 3, 24$; $P = 0.002$) in *B. dorsalis* (33.6 ± 3.9%) but not significantly different from *C. capitata* (28.7 ± 4.4%) or *B. latifrons* (21.2 ± 3.2%). Among the four fruit fly species, the percent egg parasitisation was lowest in *B. cucurbitae* but not significantly different from *C. capitata* or *B. latifrons*.

Among the four fruit fly species, the percent egg parasitisation was significantly lowest in *B. cucurbitae* (13.2 ± 3.6%) although not different from. *B. latifrons*. Significantly more ($F = 51.1$: $df = 3, 36$; $P < 0.0001$) parasitoid progeny (Figure 4) was produced in *B. dorsalis* than in the other fruit fly species.

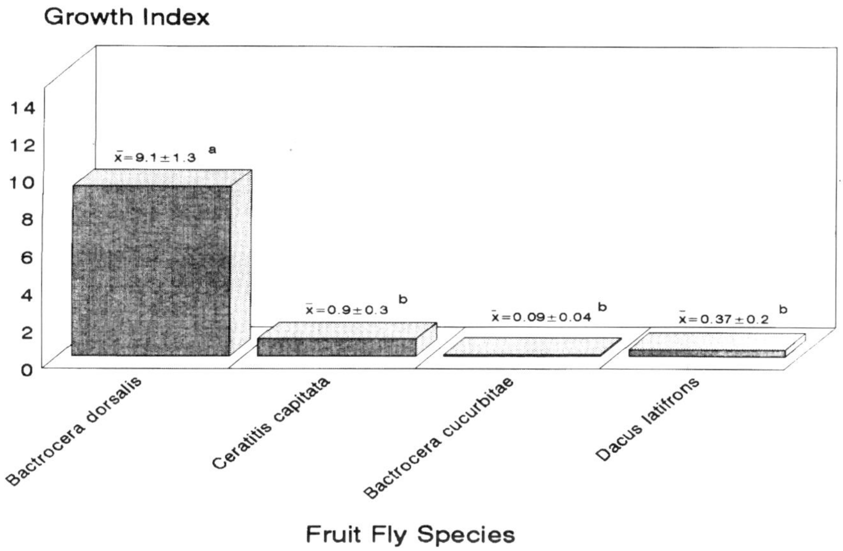


Figure 4. Mean progeny production, shown as growth index, of *F. arisanus* from *B. dorsalis*, *C. capitata*, *B. cucurbitae*, and *B. latifrons* hosts.

**Experiment 2**

The percent parasitism of *B. dorsalis* among fruit species were as follows (in descending order): banana, 70.6 ± 23%; false kamani 58.6 ± 7.9%; papaya 50.3 ± 6%; Haden mango 70.6 ± 6.2%; and common lime 12.1 ± 5.7%. The Chi-square test was not significant ($X^2 = 1.33$, $df = 3$, $P < 0.05$) which indicated that the host preference of *F. arisanus* varied with host fruit species.

**Experiment 3**

Increases in the density of host eggs that were exposed to gravid *F. arisanus* females corresponded with increases in parasitoid progeny yields (Figure 5). However, no significant differences in incremental yields beyond 20:1 host eggs to parasitoid female ratios were obtained.

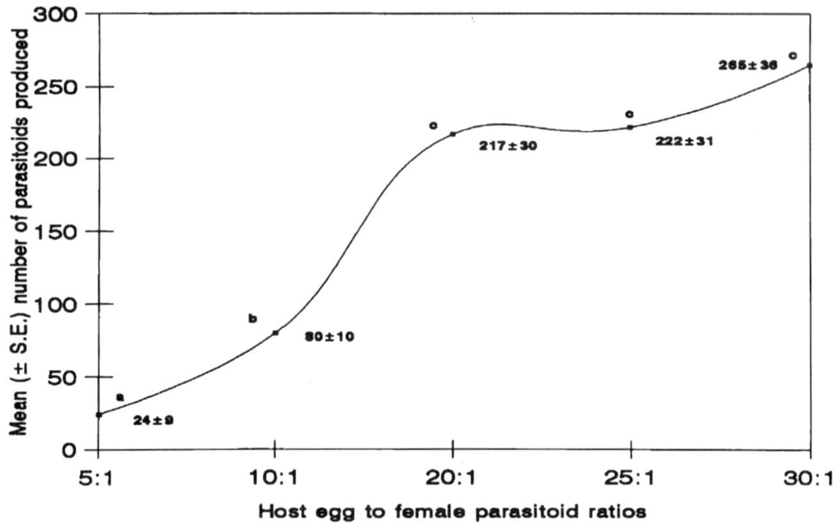


Figure 5. Mean number of *F. arisanus* produced at various host eggs to female parasitoid ratios.

## Experiment 4

Based on the percentage yield of female progeny produced (Figure 6), significantly more ($t_{0.05}$ = 6.3, P < 0.05) female progeny were produced by the Harris strain than by the wild strain. Based on dissections of 100 eggs in laboratory studies, the Harris strain of *F. arisanus* showed better parasitisation ability than the wild strain. The strains were equal in their ability to find fruit fly eggs in fruit sections.

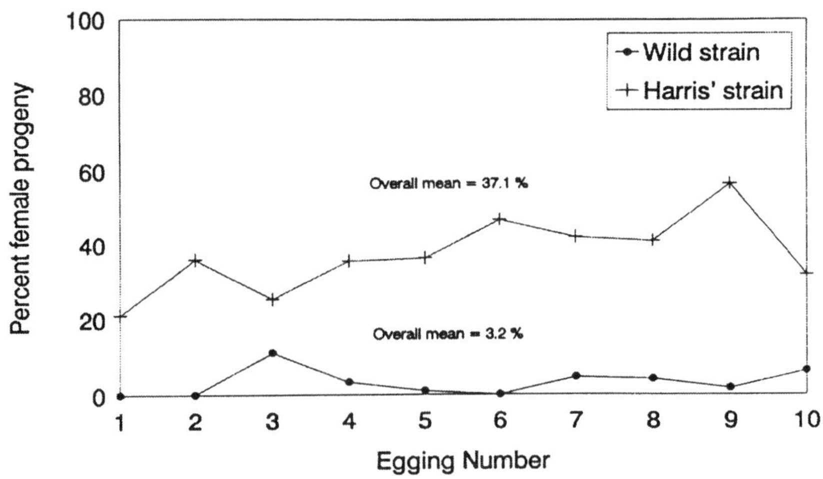


Figure 6. Mean percent female progeny produced by the laboratory and wild strain of *F. arisanus*.

## DISCUSSIONS

Initial efforts to rear *F. arisanus* in the laboratory involved collecting mated wild females from ripe papaya fruits in papaya orchards with a net and transporting the females to the laboratory for colonisation studies. Progeny production from mated wild females varied between collection sites and dates. From these results, it was apparent to us that we needed to develop a method of collecting females that was more consistent and independent of the field population. We developed and used the papaya fruit trap for consistency in collecting parasitoids in the field and rearing them in the laboratory to establish a *F. arisanus* colony. The collecting of wild females and progeny production from fruit traps proved to be a method for successful laboratory colonisation as well as a method to augment wild populations of *F. arisanus* by moving them from areas of high concentration to areas of low concentration in the field. Progeny production from wild females and from papaya fruit traps was male biased. We compensated by adjusting the sex ratio in cubicle cages to 2-4 times more females per cage than males. Using the fruit trap, a laboratory colony of *F. arisanus* was established from a field population collected in Poamoho, Hawaii. This colony was the source of the starter colony of *F. arisanus* now being successfully mass produced in Hawaii. Our experience colonising *F. arisanus* has shown that successful development of new *F. arisanus* strains on different species of tephritid fruit flies will depend on obtaining a consistent supply of *F. arisanus* to overcome the bottle neck in laboratory colonisation. In Hawaii, *F. arisanus* can find its fruit fly hosts among different species of host fruits in upland and lowland locations of agricultural, residential and urban areas and compete better than the larval parasitoids.

In Malaysia, Palacio et al. (1992) found in studies of interspecific competition among *B. persulcatus, F. arisanus,* and *D. longicaudata* parasitoids of *B. dorsalis* that *B. persulcatus* was dominant over *F. arisanus* and *B. longicaudata* in the field. The predominance of *F. arisanus* in Hawaii and *B. persulcatus* in Malaysia confirms that interspecific competition among parasitoids is little understood. Using the laboratory colony of *F. arisanus* available in Hawaii, we have a tool that we can use in Malaysia to study parasitoid interactions. There appears to be biological and behavioural differences in the Malaysian and Hawaiian *F. arisanus* strains which can be exploited for mutual benefit.

## CONCLUSIONS

The egg parasitoid, *F. arisanus,* in Hawaii is highly adaptable and is the dominant parasitiod of all species of tephritid fruit flies established in Hawaii throughout the range of habitats occupied by fruit flies. The diversity in fruit species selected for oviposition by the Hawaiian tephritids does not deter the host finding efficiency of *F. arisanus.* The conclusions which have emerged from our research with parasitoids is that *F. arisanus* is the species which offers the best potential for successful augmentative biological control of tephritid fruit flies alone or in combination with other compatible methods. The development of the Harris strain removed the most challenging problem we faced in successfully mass producing the egg parasitoid. Classical biological control methods were used to introduce the parasitoid species established in Hawaii and augmentative biological control offers the most promise for the future to make better use of the established parasitoids. Our experience has shown us that we must have in continuous development strains of *F. arisanus* reared by exposing fruit fly eggs in fruit sections to the parasitoids; this is to assure the production of laboratory reared parasitoids which can find fruit fly eggs in fruits in the field. The development of new *F. arisanus* parasitoid strains suggests an approach which may be used to rear exotic and native species of parasitoids for inundative and augmentative releases against tephritid fruit flies.

## REFERENCES


Bautista, R.C. and E.J. Harris. 1996. Effects of fruit substrates on the parasitisation of tephritid fruit flies by the parasitoid *Biosteres arisanus* (Sonan) (Hymenoptera: Braconidae) Environ. Entomol. 25: 470-475.

Clausen, C.P., D.W. Clancy and C.Q. Chock. 1965. Biological control of the Oriental fruit fly (*Dacus dorsalis)* Hendel and other fruit flies in Hawaii. U. S. Dept. Agric. Tech. Bull. No. 1322.

Haramoto, F.H. and H.A. Bess. 1970. Recent studies on the abundance of the Oriental fruit fly and Mediterranean fruit flies and the status of their parasites. Proc. Hawaii. Ent. Soc. 20: 551-566.

Harris, E.J. and R.C. Bautista. 1994. Fruit trap: a detection and collection tool for opiine parasitoids (Hym.: Braconidae) of the Oriental fruit fly, *Bactrocera dorsalis* (Dip.: Tephritidae) Entomophaga 39: 341-349.

Harris, E.J. and R.C. Bautista. 1996. Effects of fruit fly host, fruit species and fruit fly egg to female parasitoid ratio on laboratory rearing of *Biosteres arisanus*. Entomol. Exp. Appl. 79: 187-194.

Harris, E.J. and R.Y. Okamoto. 1983. Description and evaluation of a simple method for collection of the parasite *Biosteres oophilus* Hymen: Braconidae). Entomophaga 28: 241-243.

Harris, E.J. and R.Y. Okamoto. 1991. A method for rearing *Biosteres arisanus* (Sonan) (Hymenoptera: Braconidae) in the laboratory. J. Econ. Entomol. 84: 417-422.

Harris, E.J., R.Y. Okamoto, C.Y.L. Lee and T. Nishida, 1991. Suitability of *Dacus dorsalis* and *Ceratitis capitata* (Diptera: Tephritidae) as hosts of the Parasitoid *Biosteres arisanus* (Sonan) [Hymenoptera: Braconidae]. Entomophaga 36: 425-430.

Palacio, I.P., A.G. Ibrahim and R. Ibrahim. 1992. Interspecific competition among parasitoids of the Oriental fruit fly, *Bactrocera dorsalis* (Hendel). Phil. Entomol. 8: 1087-1097.

van den Bosch, R. and F.H. Haramoto. 1951. *Opius oophilus* Fullaway, an Egg-Larval Parasite of the Oriental fruit fly discovered in Hawaii. Proc. Hawaii. Ent. Soc. 14: 251-255.

# Comparisons of Demographic Parameters: Six Parasitoids (Hymenoptera: Braconidae) and their Fruit Fly (Diptera: Tephritidae) Hosts

*Roger I. Vargas*[1] *and Mohsen M. Ramadan*[2]

[1] USDA-ARS, Tropical Fruit,Vegetable and Ornamental Crop Research Laboratory, P.O. Box 4459, Hilo, HI 96720, USA
[2] Hawaii Department of Agriculture, 1428 King St., Honolulu, HI 96810, USA

## INTRODUCTION

Four economically important fruit flies have been introduced accidentally into the Hawaiian Islands. They are the melon fly, *Bactrocera cucurbitae* (Coquillett) (introduced in 1895), the Mediterranean fruit fly, *Ceratitis capitata* (Wiedemann) (in 1907), the Oriental fruit fly, *B. dorsalis* (Hendel) (in 1945) and the Solanaceous fruit fly, *B. latifrons* (Hendel) (in 1983). These fruit flies jeopardise development of a diversified tropical fruit and vegetable industry in Hawaii, cause exported fruits to undergo expensive quarantine treatment and provide a reservoir for introduction into mainland United States.

The establishment of fruit flies in Hawaii resulted in subsequent releases of numerous entomophagous insects. For example, Bess et al. (1961) listed a total of 32 natural enemies released between 1947 and 1952. Today, *Fopius* (=*Biosteres*) *arisanus* (Sonan), *Diachasmimorpha longicaudata* (Ashmead), *Biosteres vandenboschi* (Fullaway), *Psyttalia incisi* (Silvestri), *Diachasmimorpha tryoni* (Cameron) and *Psyttalia fletcheri* (Silvestri) are the most abundant species (Nishida 1953, Bess et al. 1961, Haramoto and Bess 1970, Stark et al. 1991, Vargas et al. 1991, Clausen et al. 1965). These species have played a major role in the reduction of fruit flies throughout the Hawaiian Islands. For example, as a result of parasitisation (60-79.1%) by *F. arisanus,* the average number of Oriental fruit fly larvae per guava *(Psidium guajava* L.) fruit declined from 8.5 in 1950 to 2.6 in 1955 (Clausen et al. 1965).

Demographic population analysis has diverse applications: analysing population stability and structure, estimating extinction probabilities, predicting life history evolution, predicting outbreaks in pest species and examining the dynamics of colonising or invading species (McPeek and Kalisz 1993). This study of the demography of Hawaiian fruit flies and their parasitoids is based on data from Vargas et al. (1984) and Vargas and Ramadan (unpublished data). This paper describes the comparative demography of *F. arisanus, B. tryoni, B. longicaudata, B. vandenboschi, P. incisi* and *P. fletcheri.*

## MATERIALS AND METHODS

Fruit flies and parasitoids were obtained from mass rearing stock maintained at the USDA Tropical Fruit and Vegetable Research Laboratory in Honolulu, Hawaii (Vargas 1989, Wong and Ramadan 1992). The following data were collected: 1) life-cycle survivorship, and 2) fecundity and fertility. Standard life table parameters

(Deevey 1947, Birch 1948, Ricklefs 1990) were calculated from daily records of mortality, fecundity and fertility of cohorts of *F. arisanus* (reared on *B. dorsalis*), *D. longicaudata* (reared on *B. dorsalis*), *B. vandenboschi* (reared on *B. dorsalis*), *P. incisi* (reared on *B. dorsalis*), *D. tryoni* (reared on *C. capitata*) and *P. fletcheri* (reared on *B. cucurbitae*) (Vargas et al. 1984, Vargas and Carey 1990, Carey et al. 1988). Definitions, parameter symbols and formulae are summarised in Tables 1 and 2. Calculations conform to Carey (1993).

Table 1. Definitions and formulae for various life table and demographic parameters (reproductive).

| Parameter | Definition | Formula |
|---|---|---|
| $x$ | Age interval in days | |
| $1_x$ | Proportion of females surviving to start of the age interval | |
| $m_x$ | Number of female eggs laid by average female at age $x$ | |
| $M_x$ | Average number of offspring produced by female at age $x$ | |
| Preoviposition period | Amount of time prior to eggs being laid | |
| Gross fecundity rate | Theoretical natality rate during lifetime or organism | $\sum_{x=\alpha}^{\beta} M_x$ |
| Net fecundity rate | Total number of fertile eggs laid by female during her lifetime | $\sum_{x=\alpha}^{\beta} 1_x M_x$ |
| Daily reproduction | Number of eggs produced per day in terms of entire female lifespan | $\sum_{x=\alpha}^{\beta} M_x/(\omega-\varepsilon)$ |
| Net reproductive rate ($R_o$) | Per generation contribution of newborn females to the next generation | $\sum_{x=\alpha}^{\beta} 1_x m_x$ |

Adopted from Carey (1993)

## RESULTS AND DISCUSSION

### Parasitoids

Survivorship values ($1_x$) for species were similar (45-50 d) except for the longer-lived *P. incisi* (133 d). Reproductive and population parameters for the six species of fruit flies are summarised in Table 3. Highest gross fecundity rates were obtained for *F. arisanus* (124.9 eggs/female) and lowest for *B. vandenboschi* (34.2 eggs/female). Mean generation times were similar for all species (27.2 to 30.3 d). Intrinsic rates of increase were highest for *D. longicaudata* (0.12) and *F. arisanus* (0.12) and lowest for *B. vandenboschi* (0.08). Highest net reproductive rates were obtained for *P. incisi* (29.4 females/gen.) and lowest for *B. vandenboschi* (10.1 females/gen.).

## Fruit Flies

Demographic parameters for *B. cucurbitae, C. capitata* and *B. dorsalis* reared on artificial diets are summarised in Vargas et al. (1984). *C. capitata* was the shortest lived species, while *B. cucurbitae* was the longest lived species. *C. capitata* with the shortest mean generation time (T = 31.5 d) and the second highest net reproductive rate ($R_0$ = 317.5 females/gen.) possessed the highest intrinsic rate of increase ($r_m$ = 0.18). *B. dorsalis* with the highest $R_0$ (418.5) but a longer generation time (T = 37.3 d) possessed the second highest $r_m$ (0.16). *B. cucurbitae* with a comparatively low $R_0$ (255.4 females/gen.) and a long generation time (T = 37.3 d) had the lowest $r_m$ (0.15).

Table 2. Definitions and formulae for various life table and demographic parameters (population).

| Parameter | Definition | Formula |
|---|---|---|
| Intrinsic rate *($r_m$)* | Rate of increase in a closed population | $1 = \sum_{x=\alpha}^{\beta} e^{-rx} l_x m_x$ |
| Finite rate of increase *($\lambda$)* | Factor by which a population increases in size from time t to time t+1 | $e^r$ |
| Intrinsic birth rate *(b)* | The per capita instantaneous rate of birth in the stable population | $1/ \sum_{x=1}^{\omega} e^{-rx} l_x$ |
| Intrinsic death rate *(d)* | The per capita instantaneous rate in the stable population | b-r |
| Mean generation time *(T)* | The time required for a newborn female to replace herself $R_o$ – fold | $(\log_e R_o)/r$ |
| Doubling time *(DT)* | The time for the population to increase two-fold | $(\log_e 2)/r$ |

Adopted from Carey (1993)

## CONCLUSION

The present study provides simple demographic models for six species of fruit fly parasitoids and their fruit fly hosts. Our study provides parameters to construct parasite/pest population growth models. Generally, parasite species were shorter-lived than their fruit fly host counterparts. Similarly, parasitoids had lower rates of increase than their fruit fly host. Models based on survivorship and fecundity, when validated, will be very useful for predicting population trends in the field and making pest management decisions with respect to the effects of parasites on pest populations. Furthermore, demographic parameters will be useful parameters for predicting field establishment and success of parasitoid species in classical and augmentative biological control programmes.

Table 3. Reproductive and population parameters for six parasitoid species: *Fopius arisanus (F. a.), Psyttalia fletcheri (P. f.), Psyttalia incisi (P. i.), Diachasmimorpha longicaudata (D. l.), Diachasmimorpha tryoni (D. t.),* and *Biosteres vandenboschi (B. v.).*

| | Species | | | | | |
|---|---|---|---|---|---|---|
| | *F. a.* | *P. f.* | *P. i.* | *D. l.* | *D. t.* | *B. v.* |
| **Reproductive Parameters** | | | | | | |
| Gross Fecundity Rate (eggs/female) | 124.9 | 75.3 | 108.8 | 97.9 | 54.0 | 34.2 |
| Net Fecundity (eggs/female) | 54.7 | 43.1 | 58.7 | 56.4 | 32.8 | 20.2 |
| Eggs per day (eggs/day) | 4.2 | 3.3 | 1.0 | 3.9 | 2.4 | 1.3 |
| Eggs per Insect-day (eggs/day) | 3.6 | 3.1 | 2.1 | 3.6 | 2.8 | 1.1 |
| Mean Age Gross Fecundity (days) | 31.5 | 29.5 | 38.4 | 28.2 | 28.2 | 31.3 |
| Mean Age Net Fecundity (days) | 28.6 | 28.8 | 34.8 | 27.5 | 28.0 | 30.9 |
| **Population Parameters** | | | | | | |
| Intrinsic Rate of Increase (l/t) | 0.12 | 0.11 | 0.10 | 0.12 | 0.10 | 0.08 |
| Finite Rate of Increase (per day) | 1.13 | 1.11 | 1.12 | 1.13 | 1.11 | 1.08 |
| Intrinsic Birth Rate (l/t) | 0.15 | 0.14 | 0.13 | 0.15 | 0.13 | 0.10 |
| Intrinsic Death Rate (l/t) | 0.03 | 0.03 | 0.03 | 0.03 | 0.03 | 0.02 |
| Net Reproductive Rate (per gen) | 27.35 | 21.54 | 29.36 | 28.19 | 16.41 | 10.12 |
| Mean Generation Time (days) | 27.33 | 28.34 | 33.38 | 27.21 | 27.75 | 30.27 |

## ACKNOWLEDGEMENT

We thank Dale Kanehisa (Tropical Fruit, Vegetable and Ornamental Crop Research Laboratory, USDA-ARS, Honolulu, HI) for her help in collecting and analysing data.

## REFERENCES

Birch, L.C. 1948. The intrinsic rate of natural increase of an insect population. J. Anim. Ecol. 17: 15-26.

Bess, H.A., R. van den Bosch and F.H. Haramoto. 1961. Fruit fly parasites and their activities in Hawaii. Proc. Hawaii. Entomol. Soc. 17: 367-378.

Carey, J.R. 1993. Applied demography for biologists with special emphasis on insects. Oxford University Press, N.Y.

Carey, J.R., T.T.Y. Wong and M. Ramadan. 1988. Demographic framework for parasitoid mass rearing case study of *Biosteres tryoni,* a larval parasitoid of tephritid fruit flies. Theor. Pop. Biol. 34: 279-296.

Clausen, C.P., D.W. Clancy and Q.C. Chock. 1965. Biological control of the Oriental fruit fly *(Dacus dorsalis* Hendel) and other fruit flies in Hawaii. U.S. Dep. Agric. Tech. Bull. 1322.

Deevey, E.S. Jr. 1947. Life tables for natural populations of animals. Q. Rev. Biol. 22: 283-314.

Haramoto, F.H. and H.A. Bess. 1970. Recent studies on the abundance of the Oriental and Mediterranean fruit flies and the status of their parasites. Proc. Hawaii. Entomol. Soc. 20: 551-566.

McPeek, M.A. and S. Kalisz. 1993. Population sampling and bootstrapping in complex designs: demographic analysis. 273-289. *In* Design and analysis of ecological experiments. S.M. Scheiner and J. Gurevitch (eds.). Chapman and Hall. N.Y.

Nishida, T. 1953. Ecological study of the melon fly *Dacus cucurbitae* Coq. in the Hawaiian Islands, Ph.D. dissertation, University of California, Berkeley.

Ricklefs, R.E. 1990. Ecology. 3rd edition. W.H. Freeman and Co., N.Y.

Stark, J.D., R.I. Vargas and R.K. Thalman. 1991. Diversity and abundance of parasitoids (Hymenoptera: Braconidae) of the Oriental Fruit fly (Diptera: Tephritidae) in guava orchards on Kauai, Hawaii. J. Econ. Entomol. 84: 1460-1467.

Vargas, R.I. 1989. Mass production of tephritid fruit flies. pp. 141-152. *In* Fruit flies, their biology, natural enemies, and control. *In* World crop pests. Vol. 3B. W. Robinson and G. Hooper (eds.). Elsevier Press. Amsterdam.

Vargas, R.I. and J.R. Carey. 1990. Comparative survival and demographic statistics for wild Oriental fruit fly, Mediterranean fruit fly, and melon fly (Diptera: Tephritidae) on papaya. J. Econ. Entomol. 83: 1344-1349.

Vargas, R.I., D. Miyashita and T. Nishida. 1984. Life history and demographic parameters of three laboratory-reared tephritids (Diptera: Tephritidae). Ann. Entomol. Soc. Am. 77: 651-656.

# Development and Reproductive Biology of the Egg-Pupal Parasite, *Fopius arisanus* in *Anastrepha suspensa*, a New Tephritid Host

*Pauline O. Lawrence*[1], *Ernest J. Harris*[2] *and Renato C. Bautista*[2]

[1] Department of Entomology and Nematology, University of Florida, Gainesville, FL. 32611-0620, USA

[2] USDA-ARS, Tropical Fruit, Vegetable and Ornamental Crop Research Laboratory, 2727 Woodlawn Drive, Honolulu, Hawaii 96822, USA

## INTRODUCTION

*Fopius (=Biosteres) arisanus* (Sonan) (formerly *Opius oophilus* Fullaway) (Hymenoptera: Braconidae) is a solitary egg parasite (parasitoid) that attacks tephritid fruit fly (Diptera: Tephritidae) eggs and first instars (Haramoto 1953, Clausen et al. 1965, Harris and Okamoto 1991). It completes its development within the host's larva and pupa and emerges from the latter as an adult and as such, is an egg-pupal endoparasite.

*F. arisanus* is known to attack at least seven tephritid fruit fly species (Wharton and Gilstrap 1983) and appears to be the only egg-pupal parasite of tephritids in the Western Hemisphere. It is considered to be the most successful of the parasites that attack the Oriental fruit fly, *Bactrocera dorsalis* (Hendel) and the Mediterranean fruit fly (Medfly), *Ceratitis capitata* (Weidmann) in Hawaii (Knipling 1995), resulting in 74-92% of total parasites recovered from both host species (Wong and Ramadan 1987). However, in Malaysia, Palacio et al. (1992) found that *F. arisanus* was outcompeted by the larval endoparasite, *Diachasmimorpha longicaudata* (Ashmead) (Hymenoptera: Braconidae) in multiple parasitised *B. dorsalis*.

While several larval parasites of tephritids had been cultured successfully in the laboratory (Ramadan 1991) and utilised in inundative release programmes, *F. arisanus* proved difficult to maintain in culture. In recent years, a laboratory strain of *F. arisanus* (termed the "Harris strain") has been established on *B. dorsalis* (Harris and Okamoto 1991). Efforts are currently in progress to mass rear this strain on the Medfly and other tephritid pests.

*F. arisanus* was first released into Florida from Hawaii in 1974-75 as a biological control agent against the Caribbean fruit fly (Caribfly) *Anastrepha suspensa* (Loew) (Diptera: Tephritidae) but this was unsuccessful (Baranowski et al. 1993). Interestingly, it was also introduced into Costa Rica from Hawaii and was subsequently reared from puparia of *Anastrepha* spp. (Wharton et al. 1981), indicating its establishment in that country.

Our research team has been evaluating the potential of *F. arisanus* as a biological control agent of the Caribfly in Florida. As a prelude to mass rearing and release attempts, we wished to determine whether the Caribfly was acceptable to, and/or suitable for, this parasite's development and reproductive success. To that end, we determined whether *F. arisanus* could successfully parasitise Caribfly eggs and develop without abnormalities. Further, we utilised the numbers of mature eggs in the ovaries as an indication of the parasite's biotic potential. Our results indicate that *F. arisanus* readily attacks Caribfly eggs and develops to the adult stage in synchrony with emerging fruit fly adults from unparasitised puparia. To date, up to six generations have been reared from Caribfly. Scanning electron microscopy of newly emerged *F. arisanus* indicates that *A. suspensa* supports normal development of the parasite. Taken together, our studies provide the first evidence that *F. arisanus* can be successfully cultured in the laboratory on a New World tephritid. This work lays the foundation for future studies on *F. arisanus* developmental physiology and on parasite productivity in the Caribfly.

## MATERIALS AND METHODS

### Rearing of Insects

Eggs of *A. suspensa* were supplied by the Division of Plant Industry (DPI), Florida Department of Agriculture, Fruit Fly Rearing Facility, Gainesville, Florida. The initial stock of *F. arisanus* adults was reared on *B. dorsalis* and provided by the Tropical Fruit and Vegetable Research Laboratory, Honolulu. *F. arisanus* adults were reared at 24-27°C and 50-70% RH under a 14:10 light:dark regime and provisioned with water and honey-agar. Parasite females were housed with males throughout and were presumed to be mated prior to oviposition. *A. suspensa* eggs (24-36-h old) at a rate of 25 per female, were inserted using a pipett, into 5 cm (ca. 2 in) grooves cut in the rind of guava (*Psidium guajava* L.), mango (*Mangifera indica* L.), peach (*Prunus persica* L.), papaya (*Carica papaya* L.) or loquat (*Eriobotrya japonicum* L.) which are hosts of Caribfly (Swanson and Baranowski 1972, Weems 1965, White and Elson-Harris 1992). The inoculated fruit slices were introduced into the cages with 10-day old *F. arisanus* females for 8 or 24 h. The parasitised eggs were then washed from the fruit and applied to filter paper or perforated cloth strips and incubated for 24 h under the conditions described above. The strips were then placed on the fruit fly larval diet of corn cob grits until hatching and larval development. The mature fruit fly larvae (3rd instars) were allowed to pupate in vermiculite moisted 50% (w/w) with 5% sodium benzoate or deionised water, as previously described (Lawrence et al. 1976). The moisture content was monitored throughout pupal development until adult emergence.

In order to determine the duration of wasp development, 500 females were housed with 500 males immediately after emergence and, at 10 days of age, were provisioned with 24-36-h old Caribfly eggs for 8 h as described above. The parasitised Caribfly eggs were then applied to larval diet and allowed to pupariate as described above. Samples of 20 eggs, larvae and puparia were dissected each day to determine percent parasitisation and the age and stage achieved by the parasite. From the first appearance of adult parasites, emergence was monitored daily and the sex and number of individuals noted. Three replicate experiments were performed and their means and standard errors calculated.

### Light Microscopy

Ovaries were removed from mated 6-day old, oviposition-naïve females by grasping the penultimate abdominal segment with a pair of fine forceps and gently pulling away from the abdomen. The ovaries were then excised from the ovipositor and adhering tissues and placed in a drop of Tris buffered saline [(TBS) 20 mM Tris, 500 mM NaCl, pH 7.4]. The intact ovary was drawn and measured under a dissecting microscope by means of a squared eyepiece micrometer calibrated with a stage micrometer. Ovaries and dissected ova were photographed under a Nikon Diaphot® inverted microscope.

### Scanning Electron Microscopy

To determine whether parasites reared on Caribfly larvae were malformed, particularly with respect to the aedeagus, ovipositor and locomotory appendages, 10 newly emerged males and females were washed in deionised water to remove debris and dried in a desiccator for 24 h. Whole insects were mounted on specimen stubs with double-sided sticky tape, then sputter-coated with gold-palladium alloy for about 20 sec and viewed under a Hitachi S570 scanning electron microscope at an accelerating voltage of 15 kV. Adults reared on *B. dorsalis* (the usual host of *F. arisanus*) were prepared in the same manner and served as controls.

### Biotic Potential of Female *F. arisanus*

The number of mature eggs in the ovaries (biotic potential) of *F. arisanus* females was determined by dissecting females of chronological ages that had had three 24 h oviposition experiences and those deprived of eggs (naïve). The first of the three oviposition experiences was initiated when females were 10 days old and continued on alternate days. Twenty four hours after the last oviposition experience, the ovaries from 20 females were dissected and the mature eggs in the egg chambers (vitellaria) counted. Dissections were performed daily until day 26 when <10% of the initial population remained. The experiment was replicated three times and the mean (± S.E.) number of mature eggs per female was calculated for each age.

## RESULTS

Our results indicate that *F. arisanus* females readily parasitise *A. suspensa* eggs (Figure 1). The overall mean development time over three generations of *F. arisanus* females from egg to adulthood in the Caribfly was 25.3 ± 1.2 days (Table 1). Male development was 23.75 ± 1.1 days (data not shown). Dissections of parasitised eggs,

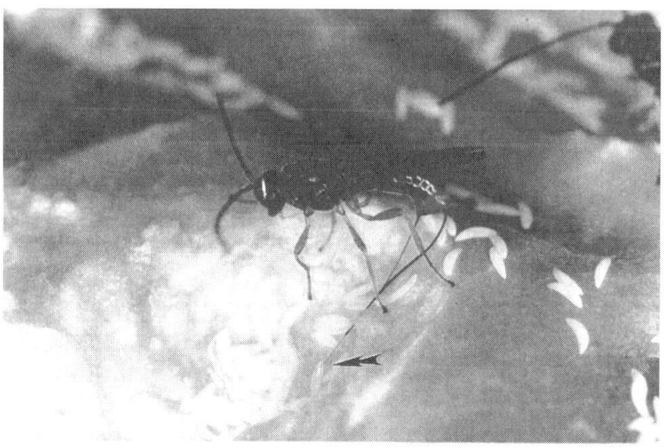

Figure 1. *Fopius arisanus* females ovipositing in 24-36-h old eggs (arrowheads) of the Caribbean fruit fly, *Anastrepha suspensa* on *Carica papaya.*

larvae and puparia revealed a mean (± S.E.) of 73 ± 8.7, 34.5 ± 8.5, and 21.9 ± 7.6% parasitisation, respectively. *F. arisanus* has four larval instars as evidenced by larval morphology and mouth hook measurements (Lawrence et al., in preparation). Up to 50% of *F. arisanus* larvae were encapsulated by host hemocytes. The encapsulated parasites were heavily melanised, thus preventing identification of their instars.

Table 1. Mean development time (days ± S.E.) of *Fopius arisanus* females reared through 3 generations in *Anastrepha suspensa**.

| Replicate | G-1 | G-2 | G-3 |
|---|---|---|---|
| 1 | 22.3 (0.8) | 23.4 (0.9) | 25.8 (1.0) |
| 2 | 23.7 (1.1) | 26.8 (0.6) | 24.0 (1.2) |
| 3 | 24.0 (1.3) | 23.3 (1.2) | 23.4 (0.8) |

* While up to 6 generations have been reared, the data represent the mean of three replicate experiments for three generations of females only. Males emerged an average of 1.5 ± 0.5 days earlier than females and are not included in these data.

Females have two polytrophic ovaries, each with two ovarioles. Each ovariole has a long distal germarium and a proximal enlarged egg chamber (vitellarium) containing mature ova (Figure 2). The two ovarioles appear to fuse at their bases, just

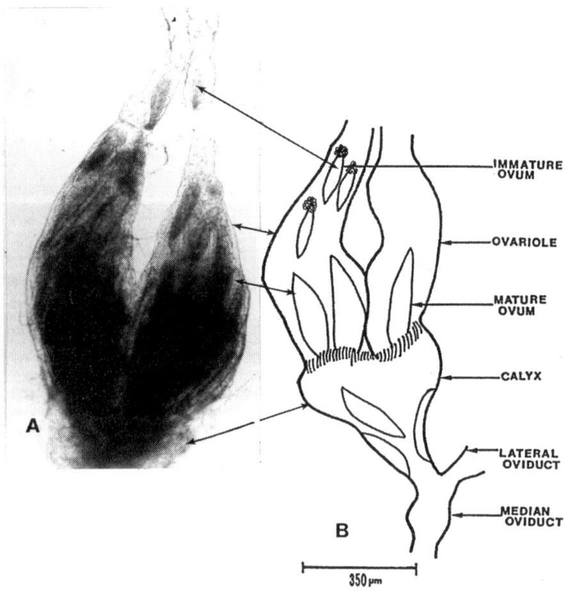


Figure 2. Light micrograph (A) and diagram (B) of the left ovary of 6-day-old *Fopius arisanus*, showing the two ovarioles, with immature ova within the germarium and mature ones in the egg chamber (vitellarium), lateral and median oviducts and the swollen, membranous calyx. Arrowed lines connect the parts in the light micrograph (A) with the corresponding labelled structures in the diagram (B).

below their respective egg chambers, where they form a swollen membranous calyx (Figure 2). The calyx is 0.35 - 0.47 mm in diameter (Figure 2) and under the dissecting microscope, has a tindal blue appearance. Immature oocytes, as expected, occur in the germarium with nutritive trophocytes attached to one end while the mature, chorionated eggs in the vitellarium lack trophocytes (Figure 3). Chorionated eggs pass through the lumen of the calyx into the lateral oviduct and then into the median oviduct (Figure 2) prior to entering the ovipositor.

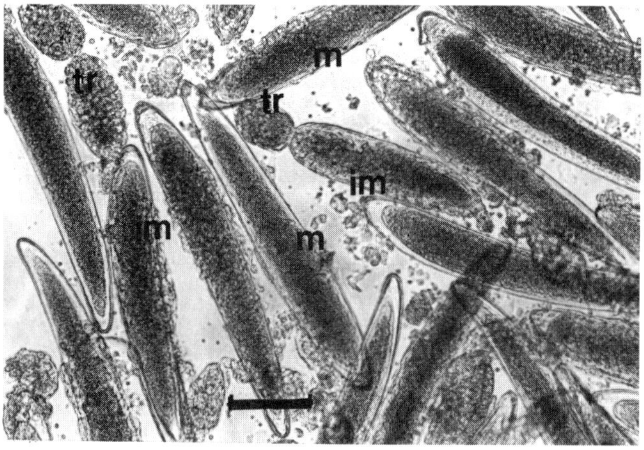


Figure 3. Eggs dissected from the ovary of a 6-day old *Fopius arisanus*. Mature, chorionated egg (m) and immature (im) eggs with attached trophocytes (tr). Bar = 50 μm.

The number of mature eggs in the ovaries of oviposition-naïve females increased between emergence and 6 days of age (figure not shown). Egg numbers continued to increase through day seven, declined gradually through day 18, then fell sharply thereafter through day 20 (Figure 4). Although egg numbers increased on day 21, they continued the overall downward trend through to 26 days of age when the experiment was terminated (Figure 4). For females with oviposition experience, the mean number of eggs in their ovaries 24 h post-oviposition (16 days old) was about 40% below that of their naïve counterparts (Figure 4). However on day 17, the number increased sharply and ultimately exceeded that of naïve females of corresponding ages (19-21 days old, Figure 4). Like naïve females, the experienced ones exhibited an overall decline in egg numbers from 23-26 days of age. Nevertheless, from 19-26 days of age, their biotic potential exceeded that of naïve females (Figure 4).

No differences were observed in the morphology or number and type of sensilla of the antennae, locomotory or external reproductive appendages of male and female *F. arisanus*, when compared with *B. dorsalis*-reared insects (figures not shown). Given the ovipositor's unique role of drilling through the chorion of the host's egg, its morphology is described here. The ovipositor (Figure 5) is 18.6 ± 0.9 (S.E.) μm wide and 2.1 ± 0.3 (S.E) mm long. It is comprised of a dorsal and two ventral processes. The former has a longitudinal groove 4.2 ± 0.05 μm wide along its ventral surface. Within the groove are several downward curved spines, arranged in ladder-like orientation along its length (Figure 5). The ventral processes consist of two elongate, adjacent sheaths of similar morphology, apposed to the long axis on the ventral surface of the dorsal process. Both

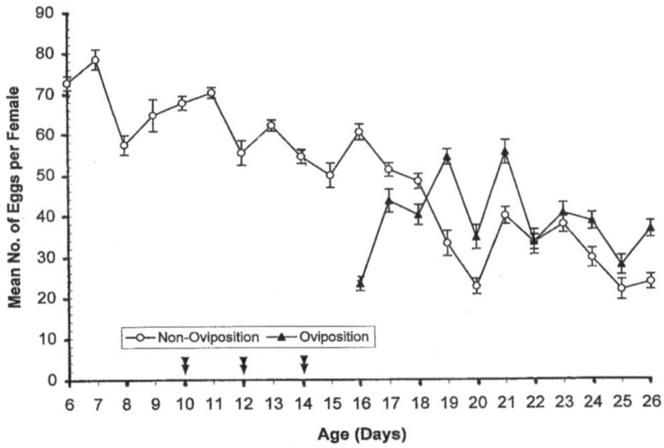


Figure 4. Mean number of mature (chorionated) eggs in the egg chambers (vitellaria) of 6-26-day old *Fopius arisanus* with (black triangles) and without (open circles) oviposition experience. Females with oviposition experience were provisioned with hosts as described in Materials and Methods. Bars represent standard errors of the means. Double arrowheads indicate 24 h oviposition experience at 10, 12 and 14 days of age.

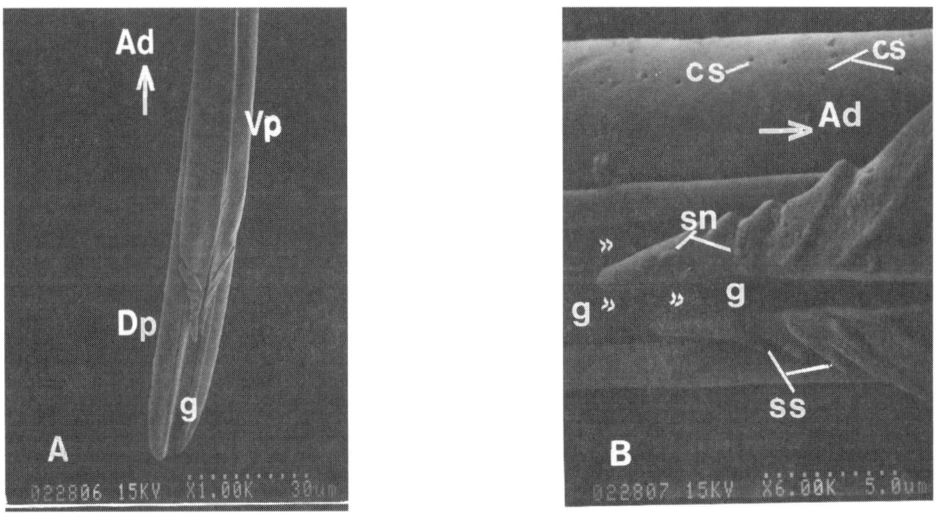


Figure 5. Scanning electron micrograph of the ventral side of the ovipositor of a newly emerged *Fopius arisanus*. Figure 5A shows the relationship between the dorsal (Dp) and ventral (Vp) processes. g = groove on ventral side of dorsal process. Range of dotted lines (lower right of micrograph) = 30 μm. Figure 5B. Higher magnification of the tips of the ventral processes, showing the teeth (ss) of the serrated edges and sensilla (sn), probably sensilla basiconica, at the base of each tooth. Within the groove (g) on the dorsal process are several downward directed spines (>>). Numerous coeloconic sensilla (cs) or pits occur along the outer ventral edges of the dorsal process. Ad = anterior of female wasp. Range of dotted lines (lower right of micrograph) = 5.0 μm.

ventral processes fit over the groove on the dorsal process. The distal tip of each ventral process resembles one-half of a serrated "arrowhead". Each "arrowhead" is comprised of five teeth, each with a sensillum at its base (Figure 5). All ten sensilla of both tips appear to be of the basiconic type (Figure 5).

## DISCUSSION

This paper provides the first evidence that *F. arisanus*, originally reared on *B. dorsalis*, can parasitise and develop in *A. suspensa*, a New World tephritid. When compared to published data, it appears that the duration of the parasite's life cycle is closely tied to that of the host. For example, *A. suspensa* completes its development in 23-25 days in the laboratory (Lawrence 1979) and in this study, *F. arisanus* adults emerged at the same time on average, as adult Caribflies from unparasitised puparia. *F. arisanus* reared on *B. dorsalis* developed in 16-20 days (Ibrahim et al. 1992) which is approximately the duration of the *B. dorsalis* development. The phenomenon of adult parasite synchronous emergence with the unparasitised host population has been reported in other tephritid-braconid host-parasite systems (Pemberton and Willard 1918, Lawrence et al. 1976, Lawrence 1982 and others). As reported previously for *F. arisanus* and other braconid parasites, the males begin to emerge about one day earlier than females (Pemberton and Willard 1918, Ramadan 1991).

The high percentage (up to 50%) of encapsulation of parasites by larval and pupal hosts accounts, in major part, for the differences between the percent parasitisation in host eggs (73 ± 8.7%) and larvae (34.5 ± 8.5) and puparia (21.9 ± 7.6%). Thus, these data suggest that *A. suspensa* is a semi-permissive host. Nevertheless, our ongoing studies indicate a decline in the rate of encapsulation with each successive *F. arisanus* generation.

Unlike most braconid larval-pupal parasites of the tephritids studied to date (Lawrence et al. 1978, Harris and Okamoto 1991, Ramadan 1991), *F. arisanus* has a distinct ovarian calyx (Figure 2). Other studies (Lawrence, unpublished) indicate that microbodies of unknown identity, along with proteins secreted from calyx cells, become associated with the egg and are introduced into the host at oviposition. However, the function of the calyx fluid is not known. In other braconid wasps, proteins and other substances from the ovarian calyx inhibit encapsulation of wasp eggs by the host (Webb and Luckhart 1996). The inability of *F. arisanus* to abrogate the host's cellular defences and the function of the calyx fluid deserve further investigation.

In Malaysia, *F. arisanus* females (0-20 days old) reared on *B. dorsalis* are reported to have a "reproductive capacity" of 75.0 ± 2.9 eggs per day over a lifespan of 16-20 days (Ibrahim et al. 1992). Ramadan (1991) and Ramadan et al. (1992, 1994), working on the same host-parasite system in Hawaii, also reported a mean of 72.8 eggs per female over a lifespan of up to 30 days (mean 18.6 days). The mean number of mature eggs in non-ovipositing females (6-26 days old) in this study was 50.3 ± 5.1 eggs per female per day over a 20-day period (Figure 4). Other studies in progress in our laboratory on 0-7-day old *F. arisanus* females reared on Caribfly and fed spun honey compared to honey-agar fed to adults in this study, indicate the presence of up to 90 mature eggs in 5-7-day old females (Lawrence, personal observation). The influence of adult food on the biotic potential of synovigenic insects like *F. arisanus*, which produce eggs after emergence, is well documented (Doutt 1964).

A comparison of mature eggs in the ovaries of experienced and naïve females suggests that oviposition stimulates egg production (Figure 4). This was evidenced by egg numbers that increased sharply within 48 h after oviposition, and ultimately exceeded those of their inexperienced counterparts (Figure 4). The phenomenon of egg replenishment is known among synovigenic females (Doutt 1964). Such females are also known to resorb their eggs in the absence of hosts (ovisorption, Doutt 1964) as is evidenced by the decline in egg numbers in non-ovipositing females (Figure 4).

The ovipositor of *F. arisanus* females is well adapted to drilling through the chorion of the host's egg. The serrated tips of the ventral processes apparently work in tandem as they move along the groove on the ventral side of the dorsal process during drilling (Figure 5). It is possible that the downward-curved rows of teeth in the groove of the dorsal process (Figure 4) serve as anchors or grips for the two processes as they enter the egg chorion. This could explain the rachet-like movement of the ovipositor when females are observed under the dissecting microscope during oviposition (Lawrence, personal observation). While a detailed ultrastructural analysis of the sensilla on the ovipositor has not been performed, it appears that they are of the basiconic type (Figure 5). Sensilla basiconica have been associated with contact chemoreception in insects (Romoser 1973). It is therefore likely that the presence of these sensilla at the tip of the drilling apparatus (ventral processes) facilitates the female's determination of the host's biochemical suitability for oviposition and development of the egg. Ramadan (1991) observed that an increase in the number of host eggs resulted in fewer instances of superparasitism (>1 parasite egg per host egg). If this is verified in subsequent studies, it would suggest that females are capable of host discrimination (Lawrence et al. 1978, van Lentren 1981), whereby they can detect substances within the host that indicate suboptimal conditions such as the pre-existence of another parasite. Hence, a female could avoid ovipositing her egg in that host to minimise competition known to occur in endoparasites (Lawrence et al. 1978, Lawrence 1988, van Lentren 1981).

In conclusion, this is the first report of *F. arisanus* acceptance of, and development in, *Anastrepha suspensa*. These results pave the way for additional basic studies which would facilitate evaluation of the efficacy of mass rearing this parasite on the Caribfly. This system can also serve as an excellent model for understanding the cellular defense responses to parasitism in insects.

## ACKNOWLEDGEMENTS

We thank Scott Erney, Tamara Landau and Mike Miller for insect rearing and Marco Toapanta and Bill Castner for photography. We also thank Steve Erwin and Denise Bonillo for the dissections and graph preparation and Bill Carpenter for assistance with the scanning electron microscope. Funding to P.O. Lawrence by the Florida Department of Citrus and the National Science Foundation is gratefully acknowledged.

## FOOTNOTES

Voucher specimens of *F. arisanus* have been deposited in the insect collection at the Florida Department of Agriculture, Doyle Conner Building, Gainesville, FL.

## REFERENCES


Baranowski, R., H. Glenn and J. Sivinski. 1993. Biological control of the Caribbean fruit fly (Diptera: Tephritidae). Fla. Entomol. 76: 246-251.

Clausen, C., D. Clancy and Q. Chock. 1965. Biological control of the Oriental fruit fly (*Dacus dorsalis* Hendel) and other fruit flies in Hawaii. Tech. Bull. USDA. 1322: 1-102.

Doutt, R.L. 1964. Biological characteristics of entomophagous adults. pp. 145-167. *In* Biological control of insect pests and weeds. P. DeBach (ed.). Reinhold Pub. Co., New York.

Haramoto, F. 1953. Observations on the biology of *Opius oophilus* Fullaway (Braconidae: Hymenoptera). Proc. Eighth Pac. Sci. Congr. 15: 1275-1289.

Harris, E. and R. Okamoto. 1991. A method for rearing *Biosteres arisanus* (Hymenoptera: Braconidae) in the laboratory. J. Econ. Entomol. 84: 417-422.

Ibrahim, A.G., I.P. Palacio and I. Rohani. 1992. The life-cycle of *Biosteres arisanus*, with reference to adult reproductive capacity on eggs of Oriental fruit-fly. Malays. Appl. Biol. 210: 63-69.

Knipling, E. 1995. Principles of Insect Parasitism Analysed from New Perspectives: Practical Implications for Regulating Insect Populations by Biological Means. USDA-ARS Agric Handbook No. 693, Washington, DC. 335 pp.

Lawrence, P., R. Baranowski and P. Greany. 1976. Effect of host age on development of *Biosteres (Opius) longicaudatus*, a parasitoid of the Caribbean fruit fly, *Anastrepha suspensa*. Fla. Entomol. 59: 33-39.

Lawrence, P., P. Greany, J. Nation and R. Baranowski. 1978. Oviposition behaviour of *Biosteres longicaudatus*, a parasite of the Caribbean fruit fly, *Anastrepha suspensa*. Ann. Entomol. Soc. Am. 71: 253-256.

Lawrence, P.O. 1979. Immature stages of the Caribbean fruit fly, *Anastrepha suspensa*. Fla. Entomol. 62: 214-219.

Lawrence, P.O. 1982. *Biosteres longicaudatus*: developmental dependence on host (*Anastrepha suspensa*) physiology. Exp. Parasitol. 53: 396-405.

Lawrence, P.O. 1988. Intraspecific competition between larvae of the parasitic wasp *Biosteres longicaudatus*: An investigation in vitro. Oecologia 74: 607-611.

Palacio, I., A. Ibrahim and R. Ibrahim. 1992. Interspecific competition among opiine parasitoids of the Oriental fruit fly *Bactrocera dorsalis* (Hendel). Phillipines Entomol. 8: 1087-1097.

Pemberton, C.E. and H.F. Willard. 1918. A contribution to the biology of fruit fly parasites in Hawaii. J. Agric. Res. 15: 419-465.

Ramadan, M. 1991. Reproductive strategies and mass production of opiinae (Hymenoptera: Braconidae) parasitoids of *Bactrocera dorsalis* (Hendel) (Diptera: Tephritidae). Ph.D. dissertation, University of Hawaii, Manoa.

Ramadan, M., T. Wong and J. Beardsley. 1992. Reproductive behaviour of *Biosteres arisanus* (Sonan) (Hymenoptera: Braconidae), an egg-larval parasitoid of the Oriental fruit fly. Biol. Control 2: 28-34.

Ramadan, M., T. Wong and D. McInnis. 1994. Reproductive biology of *Biosteres arisanus* (Sonan), an egg-larval parasitoid of the Oriental fruit fly. Biol. Control. 4: 93-100.

Romoser, W.S. 1973. The Science of Entomology. 2nd Ed. Macmillan Pub. Co., Inc., New York.

Swanson, R. and R. Baranowski. 1972. Host range and infestation by the Caribbean fruit fly *Anastrepha suspensa* (Diptera: Tephritidae) in South Florida. Proc. Fla. State Hort. Soc. 85: 271-274.

van Lentren, J.C. 1981. Host discrimination by parasitoids. pp. 153-179. *In* Semiochemicals: their role in pest control. D.A. Nordlund, R.L. Jones and W.J. Lewis (eds.). J. Wiley & Sons, Inc., New York.

Webb, B. and S. Luckhart. 1996. Factors mediating short-term and long-term immunosuppression in a parasitised insect. J. Insect Physiol. 42: 33-40.

Weems, H. 1965. *Anastrepha suspensa* (Diptera: Tephritidae). Fla. Dept. Agr. Div. Plant. Industry. Entomol. Circ. No. 38. pp. 1-4.

Wharton, R.A. and F.E. Gilstrap. 1983. Key to and status of opiine braconid (Hymenoptera) parasitoids used in biological control of *Ceratitis* and *Dacus* *s.l.* (Diptera: Tephritidae). Ann. Entomol. Soc. Am. 76: 721-742.

Wharton, R., F. Gilstrap, R. Rhode, M. Fischel-M and W. Hart. 1981. Hymenopterous egg-pupal and larval-pupal parasitoids of *Ceratitis capitata* and *Anastrepha* spp. (Dip.: Tephritidae) in Costa Rica. Entomophaga 26: 285-290.

White, I. and M. Elson-Harris. 1992. Fruit flies of economic importance: their identification and bionomics. CAB International and Austr. Centr. Internat. Agric. Res. Redwood Press, Ltd. Melksham, UK. 601 pp.

Wong, T.T. and M.M. Ramadan. 1987. Parasitisation of the Mediterranean and Oriental fruit flies (Diptera: Tephritidae) in the Kula area of Mauii, Hawaii. J. Econ. Entomol. 80: 77-80.

Wong, T.T.Y. and M.M. Ramadan. 1992. Mass rearing biology of larval parasitoids (Hymenoptera Braconidae: Opiniinae) of tephritid fruit flies (Diptera: Tephritidae) in Hawaii. *In* Advances in insect rearing for research and pest management. T.E. Anderson and N.C. Leppla (eds.). Westview, Boulder.

## FAO/IAEA INTERNATIONAL CONFERENCE ON AREA-WIDE CONTROL OF INSECT PESTS INTEGRATING THE STERILE INSECT AND RELATED NUCLEAR AND OTHER TECHNIQUES

### FINAL PROGRAMME

**Penang, Malaysia**
**28 May – 2 June 1998**

**Wednesday, 27 May 1998**

| | |
|---|---|
| 1400 – 2200 | REGISTRATION<br>Posters Set-up for Period May 28 - 31st |

**Thursday, 29 May 1998**

| | |
|---|---|
| 0900 | OPENING CEREMONY<br>Addresses:<br>Tan Sri Dr. Koh Tsu Koon, Chief Minister of Penang, Malaysia<br>Dr. Manase Peter Salema, Deputy Director, Joint FAO/IAEA Division of Nuclear Techniques in Food and Agriculture, Vienna, Austria |
| 0930 – 1000 | BREAK |

**SESSION A: Area-Wide Approach: Concepts and Economics**
*Moderator*: J. Hendrichs *Rapporteur*: A. Robinson

| | |
|---|---|
| 1000 – 1030 | **Lindquist**: Pest Management Strategies: Area-wide and Conventional |
| 1030 – 1100 | **Klassen**: Area-wide Approaches to Insect Pest Interventions: History and Lessons |
| 1100 – 1130 | **Mumford**: Economics of Area-wide Pest Control |
| 1130 – 1200 | **Griffin**: Trade Issues and Area-wide Pest Management |
| 1200 – 1230 | DISCCUSSION OF PRESENTATIONS |
| 1300 – 1400 | LUNCH |

**SESSION B: Tsetse and Screw-worm Programmes**
*Moderator*: E. Opiyo *Rapporteur*: U. Feldmann

| | |
|---|---|
| 1330 – 1350 | **Msangi** *et al.*: Success in Zanzibar: Eradication of Tsetse |
| 1350 – 1410 | **Assefa** & Feldmann: Integrating the Sterile Insect Technique to Eradicate Tsetse from the Southern Rift Valley in Ethiopia |
| 1410 – 1430 | Allosopp & **Phillemon-Motsu**: Current Tsetse Control Operations in Botswana and Prospects for the Future |
| 1430 – 1450 | **Kabayo** & Feldmann: Potential for Area-wide Control or Eradication of Tsetse in Africa |
| 1450 – 1520 | DISCCUSSION OF PRESENTATIONS |

| | |
|---|---|
| 1520 – 1540 | BREAK |
| 1540 – 1610 | VIEWING OF POSTERS OF SESSION B |
| | *Moderator*: J. Wyss *Rapporteur*: J.P. Kabayo |
| 1610 – 1640 | **Wyss**: Screw-worm Eradication in the Americas- Overview |
| 1640 – 1700 | **Grant** *et al.*: A Screw-worm Eradication Programme for Jamaica and Other Caribbean Nations |
| 1700 – 1720 | **Tweddle**: Insurance Against an Old World Screw-worm Fly Invasion |
| 1720 – 1740 | **Al Taweel** *et al.*: Incidence of Old World Screw-worm Fly, *Chrysomya bezziana* in Iraq |
| 1740 – 1810 | DISCCUSSION OF PRESENTATIONS AND POSTERS OF SESSION B |
| 1830 | COCKTAIL PARTY |

**Friday, 29 May 1998**

**SESSION C: Applications of the Area-Wide Concepts**
*Moderator*: W. Klassen *Rapporteur*: E. Opiyo

**SESSION C-1: Medical and Veterinary Pests**

| | |
|---|---|
| 0700 – 0800 | Breakfast in Poster Area |
| 0800 – 0820 | **Hougard** *et al.*: Vector Control Operations in the Onchocerciasis Control Programme in West Africa |
| 0820 – 0840 | **Mahon**: Genetic Control of the Australian Sheep Blow Fly, *Lucilia cuprina* |
| 0840 – 0900 | **Pegram** *et al.*: Progress in the Eradication of *Amblyomma variegatum* from the Caribbean |
| 0900 – 0920 | **Schofield**: Area-wide Control of Chagas Disease Vectors in Latin America |
| 0920 – 0940 | **Curtis** & Andreasen: Prospects for Large-Scale Mosquito Vector Control |
| 0940 – 1010 | DISCCUSSION OF PRESENTATIONS AND POSTERS OF SESSION C-1 |
| 1010 – 1040 | BREAK |

**SESSION C-2: Plant Pests**

*Moderator*: D. Lindquist *Rapporteur*: J.P. Cayol

| | |
|---|---|
| 1040 – 1110 | **Cunningham** & Grefenstette: Eradication of the Cotton Boll Weevil in the United States: A Successful Multi-regional Approach |

| | |
|---|---|
| 1110 – 1130 | **Chandler** *et al.*: Corn Rootworm Area-wide Management in the United States |
| 1130 – 1150 | **Lockwood**: Area-wide Pest Management of Locusts and Grasshoppers: The Striking Similarities of Problems and Solutions in Africa and the United States |
| 1150 – 1210 | **Loosjes**: The Sterile Insect Technique for Commercial Control of Onion Fly |
| 1210 – 1230 | **Calvitti** & Cirio: The Sterile Insect Technique in the Integrated Pest Management of White Fly Species in Greenhouses |
| 1230 – 1300 | DISCCUSSION OF PRESENTATIONS AND POSTERS OF SESSION C-2 |
| 1300 – 1400 | LUNCH |

**SESSION D: Lepidoptera and Augmentative Biological Control**

*Moderator*: Ocampo *Rapporteur*: A. Robinson

| | |
|---|---|
| 1400 – 1430 | **Carpenter**: Area-wide Integration of Lepidoptera $F_1$ Sterility and Augmentative Biological Control |
| 1430 – 1450 | **Anisimov**: Computer Simulation of Population Suppression by Irradicated and Mutant Lepidoptera Pests |
| 1450 – 1510 | **Walters** *et al.*: Pink Boll Worm Integrated Management Using Sterile Insects Under Field Trial Conditions |
| 1510 – 1530 | **Bloem**: SIT for Codling Moth Eradication in British Columbia, Canada |
| 1530 – 1600 | DISCCUSSION OF PRESENTATIONS |
| 1600 – 1630 | BREAK |

*Moderator*: R.T. Cunningham *Rapporteur*: P. Gomes

| | |
|---|---|
| 1630 – 1650 | **Calkins** *et al.*: Area-wide Population Suppression of Codling Moth |
| 1650 – 1710 | Greany & **Carpenter**: Uses of Nuclear Techniques in Biological Control |
| 1710 – 1740 | VIEWING OF POSTERS OF SESSION D |
| 1740 – 1810 | DISCCUSSION OF PRESENTATIONS AND POSTERS OF SESSION D |

**Saturday, 30 May 1998**

**SESSION E: Molecular Biology and Genetics in Relation to Area-wide Insect Control**

*Moderator*: D. O'Brochta *Rapporteur*: G. Franz

| | |
|---|---|
| 0700 – 0800 | Breakfast in Poster Area |
| 0800 – 0830 | **Marec**: Genetic Techniques in Insect Pest Control: An Overview |

| | |
|---|---|
| 0830 – 0900 | **Atkinson** & O'Brochta: Arriving at the Age of Pest Insect Transgenesis |
| 0900 – 0930 | **Lucchesi** & Pannuti: *Drosophila* as a Model for the Study of Sex Determination in Other Dipteran Species |
| 0930 – 1000 | **Crampton** *et al.*: Molecular Approaches to the Modification of Insect Pest Populations |
| 1000 – 1030 | DISCCUSSION OF PRESENTATIONS AND POSTERS OF SESSION E |
| 1030 – 1100 | BREAK |

**SESSION F: Supportive Technologies for Area-wide Control**

*Moderator*: R. Heath *Rapporteur*: P. Guerin

| | |
|---|---|
| 1100 – 1130 | **Jones** & Casagrande: The Use of Semiochemical Based Devices and Formulations in Area-wide Programmes - A Commercial Perspective |
| 1130 – 1150 | **Howse**: Environmentally Safe Pest Control Using Novel Bio-electrostatic Techniques: Initial Results and Prospects for Area-wide Usage |
| 1150 – 1210 | **Ujvary** *et al.*: Analogues of Insect Attractants as Promising Alternatives to Natural Semiochemicals |
| 1210 – 1230 | **Oloo** *et al.*: Integration of SIT and Autosterilisation/Lethality in Eradication of *Glossina fuscipes fuscipes* Newst. (Diptera: Glosssinidae) in Buvuma Islands in Lake Victoria, Uganda |
| 1230 – 1300 | DISCCUSSION OF PRESENTATIONS |
| 1300 – 1400 | LUNCH |

*Moderator*: C.O Calkins *Rapporteur*: K. Fisher

| | |
|---|---|
| 1400 – 1430 | **Leopold**: Cold Storage of Insects: Using Cryopreservation and Dormancy as an Aid to Mass Rearing |
| 1430 – 1500 | **Mahon** & Ahmad: Mass Rearing the Old World Screw-worm, *Chrysomya bezziana* |
| 1500 – 1520 | S. Bloem, **K. Bloem** & Calkins: Incorporation of Diapause into Codling Moth Mass Rearing: Production Advantages and Insect Quality Issues |
| 1520 – 1550 | BREAK |
| 1550 – 1610 | **Opiyo** *et al.*: New Systems for the Large-Scale Production of Male Tsetse Flies |
| 1610 – 1630 | **Mangan** & Moreno: Development and Use of Novel Insecticide Formulations such as Photoactive Dyes in Fruit Fly Adult Suppression Systems |
| 1630 – 1650 | **Slagell**: Awareness, Flight Guidance and Reliability: Impact of Technology upon Sterile Insect Technique |

| | |
|---|---|
| 1650 – 1720 | VIEWING OF POSTERS OF SESSION F |
| 1720 – 1800 | DISCCUSSION OF PRESENTATIONS AND POSTERS SESSION |

**Sunday, 31 May 1998**

| | |
|---|---|
| 1100 – 1400 | Removal of Posters for Peroid May 28 – 31st |
| 1400 – 2200 | Registration for the Fifth International Fruit Fly Symposium |
| | Posters Set-up for the Period June 1 – 3rd |

*Sessions of **1st June** and **2nd June, 1998** are shared sessions of *FAO/IAEA INTERNATIONAL CONFERENCE ON AREA-WIDE CONTROL OF INSECT PESTS INTEGRATING THE STERILE INSECT AND RELATED NUCLEAR AND OTHER TECHNIQUES* and *FIFTH INTERNATIONAL SYMPOSIUM ON FRUIT FLIES OF ECONOMIC IMPORTANCE.*

**FIFTH INTERNATIONAL SYMPOSIUM ON FRUIT FLIES OF ECONOMIC IMPORTANCE**

**FINAL PROGRAMME**

**Penang, Malaysia**
**1 - 5, June 1998**

**Sunday, 31 May 1998**

| | | |
|---|---|---|
| 1800 – 2200 | REGISTRATION | [Secretariat] |
| | Placement of Posters for Sessions 1- 6 | [Lobby & Piano Lounge] |

**Monday, 1 June 1998**

| | | |
|---|---|---|
| 0800 – 0900 | REGISTRATION | |
| 0900 | OPENING CEREMONY | [Bunga Cengkih Ballroom / Symposium Hall] |
| 0900 – 0910 | Chairman, Local Organising Committee | |
| 0910 - 0920 | Chairman, International Steering Committee | |
| 0920 - 0940 | Hon. Minister of Agriculture, Malaysia | |
| 0940 – 1000 | Tea/Coffee Break | |

**SESSION 1A: Area-Wide Fruit Fly Action Programmes**
*Moderator*: H. Kakinohana *Rapporteur*: P. Rendon

| | |
|---|---|
| 1000 – 1020 | Lobos & **Machuca**: Eradication of Medfly from Chile and Joint Programme in Southern Peru |
| 1020 – 1040 | De Longo, Colombo & **Gomez-Riera**: Use of Massive SIT for the Control of Medfly [*Ceratitis captitata* (Wied.)], Strain Seib 6-96, in Mendoza, Argentina |
| 1040 – 1100 | Villaseñor, Carillo, **Zavala**, Stewart, Lira & Reyes: Current Progress in the Medfly Programme Mexico-Guatemala |
| 1100 – 1120 | **Gomes**, Ravins & Bahdousheh: Area-Wide Control of Medfly in the Lower Jordan Rift Valley |
| 1120 – 1140 | **Dowell**, Siddiqui, Meyer & Spaugy: Mediterranean Fruit Fly Preventive Release Program in Southern California |
| 1140 – 1200 | **Reyes**, Santiago & Hernandez: Mexican Fruit Fly Eradication Program |
| 1200 – 1300 | VIEWING OF POSTERS |
| 1300 – 1400 | LUNCH |

**SESSION 1B: Area-Wide Fruit Fly Action Programmes**
*Moderator*: J. Koyama *Rapporteur*: K.H. Tan

| | |
|---|---|
| 1400 – 1420 | Broughton, Gleeson, **Hancock** & Osborne: Eradicating *Bactrocera papayae* Drew & Hancock from North Queensland |

| | |
|---|---|
| 1420 – 1440 | **Seewooruthun**, Permalloo, Gungah, Soonnoo & Alleck: Eradication of an Exotic Fruit Fly from Mauritius |
| 1440 – 1500 | **Malavasi**, Sauers-Muller, Midgarden, Kellman, Didelot, Caplong & Ribiero**:** Regional Programme for the Eradication of the Carambola Fruit Fly in South America |
| 1500 – 1520 | Manoto, **Covacha**, Bignayan, Gaitan, Zamora, Marañon, Obra, Resilva & Reyes**:** Status Report on Integrated Fruit Fly Management Based on the Sterile Insect Release Method in Guimaras Island |
| 1520 – 1600 | DISCUSSION |
| 1600 – 1630 | Tea/Coffee Break |

**SESSION 2: Area-Wide Approaches To Fruit Fly Management**

*Moderator*: A. Malavasi *Rapporteur*: T. Batkin

| | |
|---|---|
| 1630 – 1650 | **Pereira**, Barbosa, Silva, Caldeira, Dantas & Pacheco: Madeira-Med Programme, A Sterile Insect Technique Programme for Control of the Mediterranean Fruit Fly in Madeira, Portugal |
| 1650 – 1710 | **Allwood**: Regional Approach to the Management of Fruit Flies in the Pacific Island Countries and Territories |
| 1710 – 1730 | **Barnes** & Eyles: Feasibility of Eradicating *Ceratitis* spp. Fruit Flies from the Western Cape of South Africa by the Sterile Insect Technique |
| 1730 – 1750 | **Steck**: Use of Sterile Insect Technique in Mediterranean Fruit Fly Eradication and Caribbean Fruit Fly Control Programmes in Florida |
| 1750 – 1830 | DISCUSSION -- “Needs of Industry and Action Agencies” |

**Tuesday, 2 June 1998**

| | | |
|---|---|---|
| 0700 – 0800 | Breakfast | [Main Poster Area] |

**SESSION 3: Quarantine and Post-Harvest**

*Moderator*: S. Vijayasegaran *Rapporteur*: G. Hallman

| | |
|---|---|
| 0800 – 0830 | **Frampton**: An overview of Quanrantine for Fruit Flies |
| 0830 – 0900 | **Mangan** & Shellie: Integrated Use of Pre- and Post-Harvest Technology to Meet Quarantine Standards Against Fruit Flies |
| 0900 – 0930 | VIEWING OF POSTERS |
| 0930 – 1000 | Tea/Coffee Break |
| 1000 – 1030 | DISCUSSION of Oral Presentations and Posters |

**SESSION 4: Biotechnology, Genetics and Molecular Biology**
*Moderator*: D. Haymer *Rapportteur*: G. Steck

| | |
|---|---|
| 1030 – 1100 | **McPheron**: Population Genetics and Cryptic Species |
| 1100 – 1130 | Frommer, Lillemets, Robson and **Sved**: DNA Micro-satellites for Analysis of the Population Structure of Australian Native Fruit Flies |
| 1130 – 1200 | VIEWING OF POSTERS |
| 1200 – 1230 | DISCUSSION of Oral Presentations and Posters |
| 1230 – 1330 | LUNCH |

**SESSION 5: Genetic Sexing and the Sterile Insect Insect Technique**
*Moderator*: D. McInnis *Rapporteur*: A. Robinson

| | |
|---|---|
| 1330 – 1400 | **Franz** & Fisher: Modifications of Genetic Sexing Strains Based on the Experience Gained in the Mass Rearing of Such Strains |
| 1400 – 1430 | **Rendon**, McInnis, Lance & Stewart: Comparison of Medfly Male-only and Bisexual Releases in Large Scale Field Trials |
| 1430 – 1500 | VIEWING OF POSTERS |
| 1500 – 1530 | DISCUSSION of Oral Presentations and Posters |
| 1530 – 1600 | Tea/Coffee Break |

**SESSION 6: Rearing, Quality Control and Nutrition**
*Moderator*: J. Zavala *Rapporteur*: S. Opp

| | |
|---|---|
| 1600 – 1620 | **Fisher** & *Cáceres*: A Filter Rearing System for Mass Reared Medfly |
| 1620 – 1640 | ***Cáceres***, Fisher & Rendón: Mass Rearing of the Medfly Temperature Sensitive Lethal Genetic Sexing Strain |
| 1640 – 1700 | **Yamagishi** & Kakinohana: Mass Rearing of the Melon Fly in Okinawa, Japan-- Special Reference to Quality Control |
| 1700 – 1730 | VIEWING OF POSTERS |
| 1730 – 1800 | DISCUSSION of Oral Presentations and Posters |

**Wednesday, 3 June 1998**

| | |
|---|---|
| | Removal of Posters from Sessions 1-6, Placement of Posters for Sessions 7- 12 |
| | GROUP DISCUSSIONS |
| | ORGANISED TOURS |
| 2030 | **SYMPOSIUM DINNER** [Bunga Cengkih Ballroom] (Co-Sponsored by the State Government of Penang and Universiti Sains Malaysia) |

**Thursday, 4 June 1998**

| | | |
|---|---|---|
| 0700 – 0800 | Breakfast | [Main Poster Area] |

**SESSION 7: Ecology - Demography**

*Moderator*: R. Vargas *Rapporteur*: P. Liedo

| | |
|---|---|
| 0800 – 0830 | **Iwahashi**: Speciation of *Bactrocera dorsalis* Complex Based on Aedeagal Length |
| 0830 – 0900 | Papadopoulos, **Katsoyannos** & Carey: Spring and Early Summer Phenology of *Ceratitis capatata* (Diptera: Tephritidae) in Northern Greece |
| 0900 – 0930 | VIEWING OF POSTERS |
| 0930 – 1000 | Tea/Coffee Break |
| 1000 – 1030 | DISCUSSION of Oral Presentations and Posters |

**SESSION 8: Biochemistry and Physiology**

*Moderator*: A. Economopoulos *Rapporteur*: S.T. Tan

| | |
|---|---|
| 1030 – 1100 | Bonvicini, Monzani, Valentini, **Malacrida** & Gasperi: Kinetic Properties of the Two ADH Isozymes of the Medfly, *Ceratitis capitata* (Diptera: Tephritidae) |
| 1100 – 1130 | **Jang**: Physiological Control of Behaviour in Tephritid Fruit Flies |
| 1130 – 1200 | VIEWING OF POSTERS |
| 1200 – 1230 | DISCUSSION of Oral Presentations and Posters |
| 1230 – 1400 | LUNCH |

**SESSION 9: Semiochemicals and Field Monitoring**

*Moderator*: O. Jones *Rapporteur*: D. Frias

| | |
|---|---|
| 1400 – 1430 | **Nishida**, Shelly, Kaneshiro & Tan: Roles of Semiochemicals in Mating Systems: A Comparison between Oriental Fruit Fly and Medfly |
| 1430 – 1500 | **Heath** *et al.*: Evaluation of Female Mediterannean Fruit Fly (*Ceratitis capitata*) Synthetic Attractant Systems for Trapping and Sterilty Assessment in Thirteen Countries |
| 1500 – 1530 | VIEWING OF POSTERS |
| 1530 – 1600 | DISCUSSION of Oral Presentations and Posters |
| 1600 – 1630 | Tea/Coffee Break |

**SESSION 10: Behaviour**
*Moderator*: Y. Chu *Rapporteur*: P. Lawrence

1630 – 1700 **Tan**: Behaviour and Chemical Ecology of *Bactrocera* Flies
1630 – 1700 **Cayol**: Worldwide Sexual Compatibility in Medfly, *Ceratitis capitata* (Wied.) and its Implications for SIT

1730 – 1800 VIEWING OF POSTERS

1800 – 1830 DISCUSSION of Oral Presentations and Posters

**Friday, 5 June 1998**

0700 – 0800 Breakfast [Main Poster Area]

**SESSION 11: Biosystematics, Biodiversity and Evolution**
*Moderator*: S. Quilici *Rapporteur*: A. Allwood

0800 – 0830 **Drew** & Romig: Tephritid Taxonomy into the 21st Century– Research Opportunities and Applications

0830 – 0930 VIEWING OF POSTERS

0930 - 1000 DISCUSSION of Oral Presentation and Posters

1000 – 1030 Tea/Coffee Break

**SESSION 12: Natural Enemies, Biocontrol and Beneficial Fruit Flies**
*Moderator*: E. Harris *Rapporteur*: T. H. Chua

1030 – 1100 **Messing**: Newly Imported Parasitoid Effective Against the Malaysian Fruit Fly in Hawaii

1100 – 1130 VIEWING OF POSTERS

1130 – 1200 DISCUSSION of Oral Presentation and Posters

1200 – 1400 LUNCH

1400 – 1500 ROUND-UP DISCUSSIONS

1500 – 1630 CLOSING CEREMONY

# Taxonomic Index





Page number indicates first occurrence of the term in a chapter

Page number indicates first occurrence of the term in a chapter

Page number indicates first occurrence of the term in a chapter

Page number indicates first occurrence of the term in a chapter

# Author Index



Page number refers to the first page of a chapter in which the name appeared.

Page number refers to the first page of a chapter in which the name appeared.

Page number refers to the first page of a chapter in which the name appeared.

# Subject Index

Page number indicates first occurrence of the term in a chapter

Page number indicates first occurrence of the term in a chapter